ENGINEERING MECHANICS

DYNAMICS

THIRTEENTH EDITION

Library
North Highland College
Ormlie Road
THURSO
Caithness KW14 7EE

Withdrawn

SI Prefixes

Multiple	Exponential Form	Prefix	SI Symbol
1 000 000 000	10^9	giga	G
1 000 000	10^6	mega	M
1 000	10^3	kilo	k

Submultiple			
0.001	10^{-3}	milli	m
0.000 001	10^{-6}	micro	μ
0.000 000 001	10^{-9}	nano	n

Conversion Factors (FPS) to (SI)

Quantity	Unit of Measurement (FPS)	Equals	Unit of Measurement (SI)
Force	lb		4.448 N
Mass	slug		14.59 kg
Length	ft		0.3048 m

Conversion Factors (FPS)

1 ft = 12 in. (inches)

1 mi. (mile) = 5280 ft

1 kip (kilopound) = 1000 lb

1 ton = 2000 lb

ENGINEERING MECHANICS

DYNAMICS

THIRTEENTH EDITION

R. C. HIBBELER

Library
North Highland College
Ormlie Road
THURSO
Caithness KW14 7EE
Withdrawn

PEARSON

Upper Saddle River Boston Columbus San Francisco New York
Indianapolis London Toronto Sydney Singapore Tokyo Montreal Dubai
Madrid Hong Kong Mexico City Munich Paris Amsterdam Cape Town

Library of Congress Cataloging-in-Publication Data on File

Vice President and Editorial Director, ECS: Marcia Horton
Acquisitions Editor: Norrin Dias
Editorial Assistant: Sandra Rodriguez
Managing Editor: Scott Disanno
Production Editor: Rose Kernan
Art Director, Interior and Cover Designer: Kenny Beck
Art Editor: Gregory Dulles
Media Editor: Daniel Sandin
Operations Specialist: Lisa McDowell
Senior Marketing Manager: Tim Galligan
Marketing Assistant: Jon Bryant

About the Cover: Outdoor shot of electric train/© Nick M. Do / iStockphoto.com

© 2013 by R.C. Hibbeler
Published by Pearson Prentice Hall
Pearson Education, Inc.
Upper Saddle River, New Jersey 07458

All rights reserved. No part of this book may be reproduced or transmitted in any form or by any means, without permission in writing from the publisher.

Pearson Prentice Hall™ is a trademark of Pearson Education, Inc.

The author and publisher of this book have used their best efforts in preparing this book. These efforts include the development, research, and testing of the theories and programs to determine their effectiveness. The author and publisher shall not be liable in any event for incidental or consequential damages with, or arising out of, the furnishing, performance, or use of these programs.

Pearson Education Ltd., *London*
Pearson Education Australia Pty. Ltd., *Sydney*
Pearson Education Singapore, Pte. Ltd.
Pearson Education North Asia Ltd., *Hong Kong*
Pearson Education Canada, Inc., *Toronto*
Pearson Educación de Mexico, S.A. de C.V.
Pearson Education—Japan, *Tokyo*
Pearson Education Malaysia, Pte. Ltd.
Pearson Education, Inc., *Upper Saddle River, New Jersey*

Printed in the United States of America
10 9 8 7 6 5 4 3 2 1

ISBN-10:0-13-291127-2
ISBN-13:978-0-13-291127-6

To the Student

With the hope that this work will stimulate
an interest in Engineering Mechanics
and provide an acceptable guide to its understanding.

PREFACE

The main purpose of this book is to provide the student with a clear and thorough presentation of the theory and application of engineering mechanics. To achieve this objective, this work has been shaped by the comments and suggestions of hundreds of reviewers in the teaching profession, as well as many of the author's students.

New to this Edition

New Problems. There are approximately 35% or about 410 new problems in this edition. These new problems relate to applications in many different fields of engineering. Also, a significant increase in algebraic type problems has been added, so that a generalized solution can be obtained.

Additional Fundamental Problems. These problem sets serve as extended example problems since their solutions are given in the back of the book. Additional problems have been added, especially in the areas of frames and machines, and in friction.

Expanded Solutions. Some of the fundamental problems now have more detailed solutions, including some artwork, for better clarification. Also, some of the more difficult problems have additional hints along with its answer when given in the back of the book.

Updated Photos. The relevance of knowing the subject matter is reflected by the realistic applications depicted by the many photos placed throughout the book. In this edition 20 new or updated photos are included. These, along with all the others, are generally used to explain how the relevant principles of mechanics apply to real-world situations. In some sections they are incorporated into the example problems, or to show how to model then draw the free-body diagram of an actual object.

New & Revised Example Problems. Throughout the book examples have been altered or enhanced in an attempt to help clarify concepts for students. Where appropriate new examples have been added in order to emphasize important concepts that were needed.

New Conceptual Problems. The conceptual problems given at the end of many of the problem sets are intended to engage the students in thinking through a real-life situation as depicted in a photo. They can be assigned either as individual or team projects after the students have developed some expertise in the subject matter.

Hallmark Features

Besides the new features mentioned above, other outstanding features that define the contents of the text include the following.

Organization and Approach. Each chapter is organized into well-defined sections that contain an explanation of specific topics, illustrative example problems, and a set of homework problems. The topics within each section are placed into subgroups defined by boldface titles. The purpose of this is to present a structured method for introducing each new definition or concept and to make the book convenient for later reference and review.

Chapter Contents. Each chapter begins with an illustration demonstrating a broad-range application of the material within the chapter. A bulleted list of the chapter contents is provided to give a general overview of the material that will be covered.

Emphasis on Free-Body Diagrams. Drawing a free-body diagram is particularly important when solving problems, and for this reason this step is strongly emphasized throughout the book. In particular, special sections and examples are devoted to show how to draw free-body diagrams. Specific homework problems have also been added to develop this practice.

Procedures for Analysis. A general procedure for analyzing any mechanical problem is presented at the end of the first chapter. Then this procedure is customized to relate to specific types of problems that are covered throughout the book. This unique feature provides the student with a logical and orderly method to follow when applying the theory. The example problems are solved using this outlined method in order to clarify its numerical application. Realize, however, that once the relevant principles have been mastered and enough confidence and judgment have been obtained, the student can then develop his or her own procedures for solving problems.

Important Points. This feature provides a review or summary of the most important concepts in a section and highlights the most significant points that should be realized when applying the theory to solve problems.

Fundamental Problems. These problem sets are selectively located just after most of the example problems. They provide students with simple applications of the concepts, and therefore, the chance to develop their problem-solving skills before attempting to solve any of the standard problems that follow. In addition, they can be used for preparing for exams, and they can be used at a later time when preparing for the Fundamentals in Engineering Exam.

Conceptual Understanding. Through the use of photographs placed throughout the book, theory is applied in a simplified way in order to illustrate some of its more important conceptual features and instill the physical meaning of many of the terms used in the equations. These simplified applications increase interest in the subject matter and better prepare the student to understand the examples and solve problems.

Homework Problems. Apart from the Fundamental and Conceptual type problems mentioned previously, other types of problems contained in the book include the following:

- **Free-Body Diagram Problems.** Some sections of the book contain introductory problems that only require drawing the free-body diagram for the specific problems within a problem set. These assignments will impress upon the student the importance of mastering this skill as a requirement for a complete solution of any equilibrium problem.

- **General Analysis and Design Problems.** The majority of problems in the book depict realistic situations encountered in engineering practice. Some of these problems come from actual products used in industry. It is hoped that this realism will both stimulate the student's interest in engineering mechanics and provide a means for developing the skill to reduce any such problem from its physical description to a model or symbolic representation to which the principles of mechanics may be applied.

Throughout the book, there is an approximate balance of problems using either SI or FPS units. Furthermore, in any set, an attempt has been made to arrange the problems in order of increasing difficulty except for the end of chapter review problems, which are presented in random order.

- **Computer Problems.** An effort has been made to include some problems that may be solved using a numerical procedure executed on either a desktop computer or a programmable pocket calculator. The intent here is to broaden the student's capacity for using other forms of mathematical analysis without sacrificing the time needed to focus on the application of the principles of mechanics. Problems of this type, which either can or must be solved using numerical procedures, are identified by a "square" symbol (■) preceding the problem number.

The many homework problems in this edition, have been placed into two different categories. Problems that are simply indicated by a problem number have an answer and in some cases an additional numerical result given in the back of the book. An asterisk (*) before every fourth problem number indicates a problem without an answer.

Accuracy. As with the previous editions, apart from the author, the accuracy of the text and problem solutions has been thoroughly checked by four other parties: Scott Hendricks, Virginia Polytechnic Institute and State University; Karim Nohra, University of South Florida; Kurt Norlin, Laurel Tech Integrated Publishing Services; and finally Kai Beng, a practicing engineer, who in addition to accuracy review provided suggestions for problem development.

Contents

The book is divided into 11 chapters, in which the principles are first applied to simple, then to more complicated situations.

The kinematics of a particle is discussed in Chapter 12, followed by a discussion of particle kinetics in Chapter 13 (Equation of Motion), Chapter 14 (Work and Energy), and Chapter 15 (Impulse and Momentum). The concepts of particle dynamics contained in these four chapters are then summarized in a "review" section, and the student is given the chance to identify and solve a variety of problems. A similar sequence of presentation is given for the planar motion of a rigid body: Chapter 16 (Planar Kinematics), Chapter 17 (Equations of Motion), Chapter 18 (Work and Energy), and Chapter 19 (Impulse and Momentum), followed by a summary and review set of problems for these chapters.

If time permits, some of the material involving three-dimensional rigid-body motion may be included in the course. The kinematics and kinetics of this motion are discussed in Chapters 20 and 21, respectively. Chapter 22 (Vibrations) may be included if the student has the necessary mathematical background. Sections of the book that are considered to be beyond the scope of the basic dynamics course are indicated by a star (★) and may be omitted. Note that this material also provides a suitable reference for basic principles when it is discussed in more advanced courses. Finally, Appendix A provides a list of mathematical formulas needed to solve the problems in the book, Appendix B provides a brief review of vector analysis, and Appendix C reviews application of the chain rule.

Alternative Coverage. At the discretion of the instructor, it is possible to cover Chapters 12 through 19 in the following order with no loss in continuity: Chapters 12 and 16 (Kinematics), Chapters 13 and 17 (Equations of Motion), Chapter 14 and 18 (Work and Energy), and Chapters 15 and 19 (Impulse and Momentum).

Acknowledgments

The author has endeavored to write this book so that it will appeal to both the student and instructor. Through the years, many people have helped in its development, and I will always be grateful for their valued suggestions and comments. Specifically, I wish to thank all the individuals who have contributed their comments relative to preparing the thirteenth edition of this work, and in particular, James R. Morgan of Texas A & M University and Hongbing Lu of University of Texas at Dallas.

There are a few other people that I also feel deserve particular recognition. These include comments sent to me by D. Sullivan, H. Kuhlman and G. Benson. A long-time friend and associate, Kai Beng Yap, was of great help to me in preparing and checking problem solutions. A special note of thanks also goes to Kurt Norlin of Laurel Tech Integrated Publishing Services in this regard. During the production process I am thankful for the assistance of Rose Kernan, my production editor for many years, and to my wife, Conny, and daughter, Mary Ann, who have helped with the proofreading and typing needed to prepare the manuscript for publication.

Lastly, many thanks are extended to all my students and to members of the teaching profession who have freely taken the time to e-mail me their suggestions and comments. Since this list is too long to mention, it is hoped that those who have given help in this manner will accept this anonymous recognition.

I would greatly appreciate hearing from you if at any time you have any comments, suggestions, or problems related to any matters regarding this edition.

Russell Charles Hibbeler
hibbeler@bellsouth.net

your work...

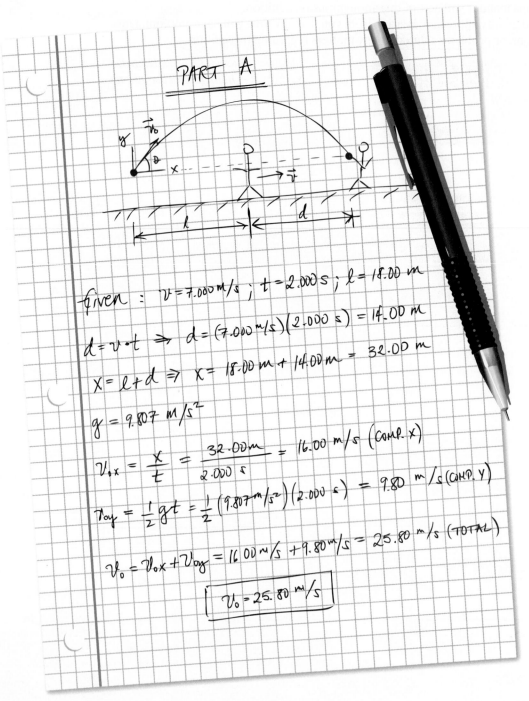

PART A

Given : $v = 7.000 \text{ m/s}$; $t = 2.000 \text{ s}$; $\ell = 18.00 \text{ m}$

$d = v \cdot t \Rightarrow d = (7.000 \text{ m/s})(2.000 \text{ s}) = 14.00 \text{ m}$

$x = \ell + d \Rightarrow x = 18.00 \text{ m} + 14.00 \text{ m} = 32.00 \text{ m}$

$g = 9.807 \text{ m/s}^2$

$v_{ox} = \dfrac{x}{t} = \dfrac{32.00 \text{ m}}{2.000 \text{ s}} = 16.00 \text{ m/s}$ (COMP. X)

$v_{oy} = \dfrac{1}{2} g t = \dfrac{1}{2} (9.807 \text{ m/s}^2)(2.000 \text{ s}) = 9.80 \text{ m/s}$ (COMP. Y)

$v_0 = v_{ox} + v_{oy} = 16.00 \text{ m/s} + 9.80 \text{ m/s} = 25.80 \text{ m/s}$ (TOTAL)

$$\boxed{v_0 = 25.80 \text{ m/s}}$$

your answer specific feedback

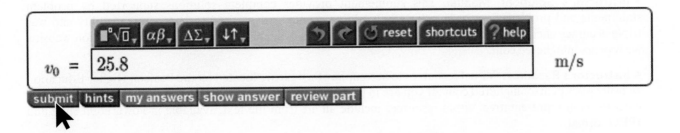

$v_0 = $ | 25.8 | m/s

submit   hints   my answers   show answer   review part

Try Again; 4 attempts remaining

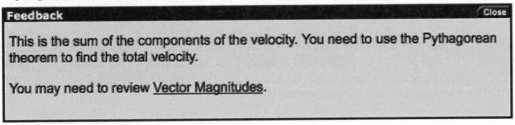

Feedback   Close

This is the sum of the components of the velocity. You need to use the Pythagorean theorem to find the total velocity.

You may need to review Vector Magnitudes.

www.MasteringEngineering.com

Resources for Instructors

- **MasteringEngineering.** This online Tutorial Homework program allows you to integrate dynamic homework with automatic grading and adaptive tutoring. MasteringEngineering allows you to easily track the performance of your entire class on an assignment-by-assignment basis, or the detailed work of an individual student.

- **Instructor's Solutions Manual.** This supplement provides complete solutions supported by problem statements and problem figures. The thirteenth edition manual was revised to improve readability and was triple accuracy checked. The Instructor's Solutions Manual is available on Pearson Higher Education website: www.pearsonhighered.com.

- **Instructor's Resource.** Visual resources to accompany the text are located on the Pearson Higher Education website: www.pearsonhighered.com . If you are in need of a login and password for this site, please contact your local Pearson representative. Visual resources include all art from the text, available in PowerPoint slide and JPEG format.

- **Video Solutions.** Developed by Professor Edward Berger, University of Virginia, video solutions are located on the Companion Website for the text and offer step-by-step solution walkthroughs of representative homework problems from each section of the text. Make efficient use of class time and office hours by showing students the complete and concise problem-solving approaches that they can access any time and view at their own pace. The videos are designed to be a flexible resource to be used however each instructor and student prefers. A valuable tutorial resource, the videos are also helpful for student self-evaluation as students can pause the videos to check their understanding and work alongside the video. Access the videos at www.pearsonhighered.com/hibbeler/ and follow the links for the *Engineering Mechanics: Dynamics*, Thirteenth Edition text.

Resources for Students

- **MasteringEngineering.** Tutorial homework problems emulate the instructor's office-hour environment, guiding students through engineering concepts with self-paced individualized coaching. These in-depth tutorial homework problems are designed to coach students with feedback specific to their errors and optional hints that break problems down into simpler steps.

- **Dynamics Study Pack.** This supplement contains chapter-by-chapter study materials, a Free-Body Diagram Workbook and access to the Companion Website where additional tutorial resources are located.

- **Companion Website.** The Companion Website, located at www.pearsonhighered.com/hibbeler/, includes opportunities for practice and review including:

  - **Video Solutions**—Complete, step-by-step solution walkthroughs of representative homework problems from each section. Videos offer fully worked solutions that show every step of representative homework problems—this helps students make vital connections between concepts.

- **Dynamics Practice Problems Workbook.** This workbook contains additional worked problems. The problems are partially solved and are designed to help guide students through difficult topics.

Ordering Options

The *Dynamics Study Pack* and MasteringEngineering resources are available as stand-alone items for student purchase and are also available packaged with the texts. The ISBN for each valuepack is as follows:

- *Engineering Mechanics: Dynamics* with Study Pack: ISBN: 0133028003
- *Engineering Mechanics: Dynamics* with MasteringEngineering Student Access Card: ISBN: 0133009564

Custom Solutions

Please contact your local Pearson Sales Representative for more details about custom options or visit

www.pearsonlearningsolutions.com, keyword: Hibbeler.

CONTENTS

12
Kinematics of a Particle 3

13
Kinetics of a Particle: Force and Acceleration 107

14
Kinetics of a Particle: Work and Energy 169

15
Kinetics of a Particle: Impulse and Momentum 221

Review

16
Planar Kinematics of a Rigid Body 311

17
Planar Kinetics of a Rigid Body: Force and Acceleration 395

18
Planar Kinetics of a Rigid Body: Work and Energy 455

19
Planar Kinetics of a Rigid Body: Impulse and Momentum 495

Review

20
Three-Dimensional Kinematics of a Rigid Body 549

21
Three-Dimensional Kinetics of a Rigid Body 579

22
Vibrations 631

Appendix

Fundamental Problems Partial Solutions and Answers 680

CREDITS

Chapter opening images are credited as follows:

Chapter 12, © Banol2007 | Dreamstime.com
Chapter 13, © Skvoor | Dreamstime.com
Chapter 14, © Alan Haynes.com / Alamy
Chapter 15, © Charles William Lupica / Alamy
Chapter 16, Faraways/Shutterstock.com
Chapter 17, © Toshi12 | Dreamstime.com
Chapter 18, © Oleksiy Maksymenko / Alamy
Chapter 19, iofoto/Shutterstock.com
Chapter 20, © Juice Images / Alamy
Chapter 21, © imagebroker / Alamy
Chapter 22, © Joe Belanger / Alamy
Cover © Nick M. Do / iStockphoto.com

Other images provided by the author.

ENGINEERING MECHANICS

DYNAMICS

THIRTEENTH EDITION

Chapter 12

Although each of these boats is rather large, from a distance their motion can be analyzed as if each were a particle.

Kinematics of a Particle

CHAPTER OBJECTIVES

- To introduce the concepts of position, displacement, velocity, and acceleration.

- To study particle motion along a straight line and represent this motion graphically.

- To investigate particle motion along a curved path using different coordinate systems.

- To present an analysis of dependent motion of two particles.

- To examine the principles of relative motion of two particles using translating axes.

12.1   Introduction

Mechanics is a branch of the physical sciences that is concerned with the state of rest or motion of bodies subjected to the action of forces. Engineering mechanics is divided into two areas of study, namely, statics and dynamics. *Statics* is concerned with the equilibrium of a body that is either at rest or moves with constant velocity. Here we will consider *dynamics*, which deals with the accelerated motion of a body. The subject of dynamics will be presented in two parts: *kinematics*, which treats only the geometric aspects of the motion, and *kinetics*, which is the analysis of the forces causing the motion. To develop these principles, the dynamics of a particle will be discussed first, followed by topics in rigid-body dynamics in two and then three dimensions.

12

Historically, the principles of dynamics developed when it was possible to make an accurate measurement of time. Galileo Galilei (1564–1642) was one of the first major contributors to this field. His work consisted of experiments using pendulums and falling bodies. The most significant contributions in dynamics, however, were made by Isaac Newton (1642–1727), who is noted for his formulation of the three fundamental laws of motion and the law of universal gravitational attraction. Shortly after these laws were postulated, important techniques for their application were developed by Euler, D'Alembert, Lagrange, and others.

There are many problems in engineering whose solutions require application of the principles of dynamics. Typically the structural design of any vehicle, such as an automobile or airplane, requires consideration of the motion to which it is subjected. This is also true for many mechanical devices, such as motors, pumps, movable tools, industrial manipulators, and machinery. Furthermore, predictions of the motions of artificial satellites, projectiles, and spacecraft are based on the theory of dynamics. With further advances in technology, there will be an even greater need for knowing how to apply the principles of this subject.

Problem Solving.
Dynamics is considered to be more involved than statics since both the forces applied to a body and its motion must be taken into account. Also, many applications require using calculus, rather than just algebra and trigonometry. In any case, the most effective way of learning the principles of dynamics is *to solve problems*. To be successful at this, it is necessary to present the work in a logical and orderly manner as suggested by the following sequence of steps:

1. Read the problem carefully and try to correlate the actual physical situation with the theory you have studied.
2. Draw any necessary diagrams and tabulate the problem data.
3. Establish a coordinate system and apply the relevant principles, generally in mathematical form.
4. Solve the necessary equations algebraically as far as practical; then, use a consistent set of units and complete the solution numerically. Report the answer with no more significant figures than the accuracy of the given data.
5. Study the answer using technical judgment and common sense to determine whether or not it seems reasonable.
6. Once the solution has been completed, review the problem. Try to think of other ways of obtaining the same solution.

In applying this general procedure, do the work as neatly as possible. Being neat generally stimulates clear and orderly thinking, and vice versa.

12.2   Rectilinear Kinematics: Continuous Motion

We will begin our study of dynamics by discussing the kinematics of a particle that moves along a rectilinear or straight-line path. Recall that a *particle* has a mass but negligible size and shape. Therefore we must limit application to those objects that have dimensions that are of no consequence in the analysis of the motion. In most problems, we will be interested in bodies of finite size, such as rockets, projectiles, or vehicles. Each of these objects can be considered as a particle, as long as the motion is characterized by the motion of its mass center and any rotation of the body is neglected.

Rectilinear Kinematics.   The kinematics of a particle is characterized by specifying, at any given instant, the particle's position, velocity, and acceleration.

Position.   The straight-line path of a particle will be defined using a single coordinate axis s, Fig. 12–1a. The origin O on the path is a fixed point, and from this point the *position coordinate s* is used to specify the location of the particle at any given instant. The magnitude of s is the distance from O to the particle, usually measured in meters (m) or feet (ft), and the sense of direction is defined by the algebraic sign on s. Although the choice is arbitrary, in this case s is positive since the coordinate axis is positive to the right of the origin. Likewise, it is negative if the particle is located to the left of O. Realize that position is a vector quantity since it has both magnitude and direction. Here, however, it is being represented by the algebraic scalar s since the direction always remains along the coordinate axis.

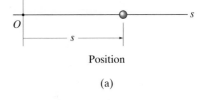

Position

(a)

Displacement.   The *displacement* of the particle is defined as the *change* in its *position*. For example, if the particle moves from one point to another, Fig. 12–1b, the displacement is

$$\Delta s = s' - s$$

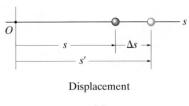

Displacement

(b)

Fig. 12–1

In this case Δs is *positive* since the particle's final position is to the *right* of its initial position, i.e., $s' > s$. Likewise, if the final position were to the *left* of its initial position, Δs would be *negative*.

   The displacement of a particle is also a *vector quantity*, and it should be distinguished from the distance the particle travels. Specifically, the *distance traveled* is a *positive scalar* that represents the total length of path over which the particle travels.

Velocity.   If the particle moves through a displacement Δs during the time interval Δt, the *average velocity* of the particle during this time interval is

$$v_{avg} = \frac{\Delta s}{\Delta t}$$

If we take smaller and smaller values of Δt, the magnitude of Δs becomes smaller and smaller. Consequently, the *instantaneous velocity* is a vector defined as $v = \lim_{\Delta t \to 0} (\Delta s / \Delta t)$, or

$$(\xrightarrow{+}) \qquad\qquad \boxed{v = \frac{ds}{dt}} \qquad\qquad (12\text{–}1)$$

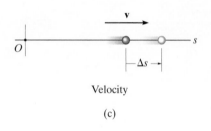

Velocity

(c)

Since Δt or dt is always positive, the sign used to define the *sense* of the velocity is the same as that of Δs or ds. For example, if the particle is moving to the *right*, Fig. 12–1c, the velocity is *positive;* whereas if it is moving to the *left*, the velocity is *negative*. (This is emphasized here by the arrow written at the left of Eq. 12–1.) The *magnitude* of the velocity is known as the *speed*, and it is generally expressed in units of m/s or ft/s.

Occasionally, the term "average speed" is used. The *average speed* is always a positive scalar and is defined as the total distance traveled by a particle, s_T, divided by the elapsed time Δt; i.e.,

$$(v_{sp})_{avg} = \frac{s_T}{\Delta t}$$

For example, the particle in Fig. 12–1d travels along the path of length s_T in time Δt, so its average speed is $(v_{sp})_{avg} = s_T / \Delta t$, but its average velocity is $v_{avg} = -\Delta s / \Delta t$.

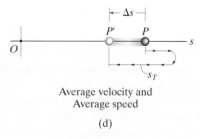

Average velocity and
Average speed

(d)

Fig. 12–1 (cont.)

Acceleration.   Provided the velocity of the particle is known at two points, the *average acceleration* of the particle during the time interval Δt is defined as

$$a_{\text{avg}} = \frac{\Delta v}{\Delta t}$$

Here Δv represents the difference in the velocity during the time interval Δt, i.e., $\Delta v = v' - v$, Fig. 12–1e.

The *instantaneous acceleration* at time t is a vector that is found by taking smaller and smaller values of Δt and corresponding smaller and smaller values of Δv, so that $a = \lim_{\Delta t \to 0}(\Delta v/\Delta t)$, or

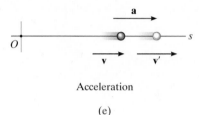

Acceleration

(e)

$(\xrightarrow{+})$

$$a = \frac{dv}{dt}$$ (12–2)

Substituting Eq. 12–1 into this result, we can also write

$(\xrightarrow{+})$

$$a = \frac{d^2s}{dt^2}$$

Both the average and instantaneous acceleration can be either positive or negative. In particular, when the particle is *slowing down*, or its speed is decreasing, the particle is said to be *decelerating*. In this case, v' in Fig. 12–1f is *less* than v, and so $\Delta v = v' - v$ will be negative. Consequently, a will also be negative, and therefore it will act to the *left*, in the *opposite sense* to v. Also, notice that if the particle is originally at rest, then it can have an acceleration if a moment later it has a velocity v'; and, if the *velocity* is *constant*, then the *acceleration is zero* since $\Delta v = v - v = 0$. Units commonly used to express the magnitude of acceleration are m/s^2 or ft/s^2.

Finally, an important differential relation involving the displacement, velocity, and acceleration along the path may be obtained by eliminating the time differential dt between Eqs. 12–1 and 12–2, which gives

Deceleration

(f)

Fig. 12–1 (cont.)

$(\xrightarrow{+})$

$$a\,ds = v\,dv$$ (12–3)

Although we have now produced three important kinematic equations, realize that the above equation is not independent of Eqs. 12–1 and 12–2.

12

Constant Acceleration, $a = a_c$. When the acceleration is constant, each of the three kinematic equations $a_c = dv/dt$, $v = ds/dt$, and $a_c\, ds = v\, dv$ can be integrated to obtain formulas that relate a_c, v, s, and t.

Velocity as a Function of Time.   Integrate $a_c = dv/dt$, assuming that initially $v = v_0$ when $t = 0$.

$$\int_{v_0}^{v} dv = \int_{0}^{t} a_c\, dt$$

($\xrightarrow{+}$)

$$\boxed{v = v_0 + a_c t}$$
Constant Acceleration

(12–4)

Position as a Function of Time.   Integrate $v = ds/dt = v_0 + a_c t$, assuming that initially $s = s_0$ when $t = 0$.

$$\int_{s_0}^{s} ds = \int_{0}^{t} (v_0 + a_c t)\, dt$$

($\xrightarrow{+}$)

$$\boxed{s = s_0 + v_0 t + \tfrac{1}{2} a_c t^2}$$
Constant Acceleration

(12–5)

Velocity as a Function of Position.   Either solve for t in Eq. 12–4 and substitute into Eq. 12–5, or integrate $v\, dv = a_c\, ds$, assuming that initially $v = v_0$ at $s = s_0$.

$$\int_{v_0}^{v} v\, dv = \int_{s_0}^{s} a_c\, ds$$

($\xrightarrow{+}$)

$$\boxed{v^2 = v_0^2 + 2a_c(s - s_0)}$$
Constant Acceleration

(12–6)

The algebraic signs of s_0, v_0, and a_c, used in the above three equations, are determined from the positive direction of the s axis as indicated by the arrow written at the left of each equation. Remember that these equations are useful *only when the acceleration is constant and when* $t = 0$, $s = s_0$, $v = v_0$. A typical example of constant accelerated motion occurs when a body falls freely toward the earth. If air resistance is neglected and the distance of fall is short, then the *downward* acceleration of the body when it is close to the earth is constant and approximately 9.81 m/s^2 or 32.2 ft/s^2. The proof of this is given in Example 13.2.

Important Points

- Dynamics is concerned with bodies that have accelerated motion.
- Kinematics is a study of the geometry of the motion.
- Kinetics is a study of the forces that cause the motion.
- Rectilinear kinematics refers to straight-line motion.
- Speed refers to the magnitude of velocity.
- Average speed is the total distance traveled divided by the total time. This is different from the average velocity, which is the displacement divided by the time.
- A particle that is slowing down is decelerating.
- A particle can have an acceleration and yet have zero velocity.
- The relationship $a\,ds = v\,dv$ is derived from $a = dv/dt$ and $v = ds/dt$, by eliminating dt.

During the time this rocket undergoes rectilinear motion, its altitude as a function of time can be measured and expressed as $s = s(t)$. Its velocity can then be found using $v = ds/dt$, and its acceleration can be determined from $a = dv/dt$.

Procedure for Analysis

Coordinate System.

- Establish a position coordinate s along the path and specify its *fixed origin* and positive direction.
- Since motion is along a straight line, the vector quantities position, velocity, and acceleration can be represented as algebraic scalars. For analytical work the sense of s, v, and a is then defined by their *algebraic signs*.
- The positive sense for each of these scalars can be indicated by an arrow shown alongside each kinematic equation as it is applied.

Kinematic Equations.

- If a relation is known between any *two* of the four variables a, v, s, and t, then a third variable can be obtained by using one of the kinematic equations, $a = dv/dt$, $v = ds/dt$ or $a\,ds = v\,dv$, since each equation relates all three variables.*
- Whenever integration is performed, it is important that the position and velocity be known at a given instant in order to evaluate either the constant of integration if an indefinite integral is used, or the limits of integration if a definite integral is used.
- Remember that Eqs. 12–4 through 12–6 have only limited use. These equations apply *only* when the *acceleration is constant* and the initial conditions are $s = s_0$ and $v = v_0$ when $t = 0$.

*Some standard differentiation and integration formulas are given in Appendix A.

12

EXAMPLE | **12.1**

The car on the left in the photo and in Fig. 12–2 moves in a straight line such that for a short time its velocity is defined by $v = (3t^2 + 2t)$ ft/s, where t is in seconds. Determine its position and acceleration when $t = 3$ s. When $t = 0$, $s = 0$.

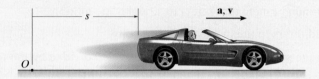

Fig. 12–2

SOLUTION

Coordinate System.   The position coordinate extends from the fixed origin O to the car, positive to the right.

Position.   Since $v = f(t)$, the car's position can be determined from $v = ds/dt$, since this equation relates v, s, and t. Noting that $s = 0$ when $t = 0$, we have*

($\xrightarrow{+}$)
$$v = \frac{ds}{dt} = (3t^2 + 2t)$$

$$\int_0^s ds = \int_0^t (3t^2 + 2t)dt$$

$$s \Big|_0^s = t^3 + t^2 \Big|_0^t$$

$$s = t^3 + t^2$$

When $t = 3$ s,

$$s = (3)^3 + (3)^2 = 36 \text{ ft} \qquad \qquad Ans.$$

Acceleration.   Since $v = f(t)$, the acceleration is determined from $a = dv/dt$, since this equation relates a, v, and t.

($\xrightarrow{+}$)
$$a = \frac{dv}{dt} = \frac{d}{dt}(3t^2 + 2t)$$

$$= 6t + 2$$

When $t = 3$ s,

$$a = 6(3) + 2 = 20 \text{ ft/s}^2 \rightarrow \qquad \qquad Ans.$$

NOTE: The formulas for constant acceleration *cannot* be used to solve this problem, because the acceleration is a function of time.

*The *same result* can be obtained by evaluating a constant of integration C rather than using definite limits on the integral. For example, integrating $ds = (3t^2 + 2t)dt$ yields $s = t^3 + t^2 + C$. Using the condition that at $t = 0$, $s = 0$, then $C = 0$.

EXAMPLE | 12.2

A small projectile is fired vertically *downward* into a fluid medium with an initial velocity of 60 m/s. Due to the drag resistance of the fluid the projectile experiences a deceleration of $a = (-0.4v^3)$ m/s^2, where v is in m/s. Determine the projectile's velocity and position 4 s after it is fired.

SOLUTION

Coordinate System.   Since the motion is downward, the position coordinate is positive downward, with origin located at O, Fig. 12–3.

Velocity.   Here $a = f(v)$ and so we must determine the velocity as a function of time using $a = dv/dt$, since this equation relates v, a, and t. (Why not use $v = v_0 + a_c t$?) Separating the variables and integrating, with $v_0 = 60$ m/s when $t = 0$, yields

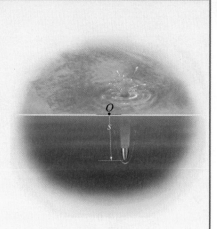

Fig. 12–3

$$(+\downarrow) \qquad\qquad a = \frac{dv}{dt} = -0.4v^3$$

$$\int_{60\text{ m/s}}^{v} \frac{dv}{-0.4v^3} = \int_0^t dt$$

$$\frac{1}{-0.4}\left(\frac{1}{-2}\right)\frac{1}{v^2}\Big|_{60}^{v} = t - 0$$

$$\frac{1}{0.8}\left[\frac{1}{v^2} - \frac{1}{(60)^2}\right] = t$$

$$v = \left\{\left[\frac{1}{(60)^2} + 0.8t\right]^{-1/2}\right\}\text{ m/s}$$

Here the positive root is taken, since the projectile will continue to move downward. When $t = 4$ s,

$$v = 0.559\text{ m/s}\!\downarrow \qquad\qquad Ans.$$

Position.   Knowing $v = f(t)$, we can obtain the projectile's position from $v = ds/dt$, since this equation relates s, v, and t. Using the initial condition $s = 0$, when $t = 0$, we have

$$(+\downarrow) \qquad\qquad v = \frac{ds}{dt} = \left[\frac{1}{(60)^2} + 0.8t\right]^{-1/2}$$

$$\int_0^s ds = \int_0^t \left[\frac{1}{(60)^2} + 0.8t\right]^{-1/2} dt$$

$$s = \frac{2}{0.8}\left[\frac{1}{(60)^2} + 0.8t\right]^{1/2}\Big|_0^t$$

$$s = \frac{1}{0.4}\left\{\left[\frac{1}{(60)^2} + 0.8t\right]^{1/2} - \frac{1}{60}\right\}\text{ m}$$

When $t = 4$ s,

$$s = 4.43\text{ m} \qquad\qquad Ans.$$

EXAMPLE | 12.3

During a test a rocket travels upward at 75 m/s, and when it is 40 m from the ground its engine fails. Determine the maximum height s_B reached by the rocket and its speed just before it hits the ground. While in motion the rocket is subjected to a constant downward acceleration of 9.81 m/s^2 due to gravity. Neglect the effect of air resistance.

SOLUTION

Coordinate System. The origin O for the position coordinate s is taken at ground level with positive upward, Fig. 12–4.

Maximum Height. Since the rocket is traveling *upward*, $v_A = +75$ m/s when $t = 0$. At the maximum height $s = s_B$ the velocity $v_B = 0$. For the entire motion, the acceleration is $a_c = -9.81$ m/s^2 (negative since it acts in the *opposite* sense to positive velocity or positive displacement). Since a_c is *constant* the rocket's position may be related to its velocity at the two points A and B on the path by using Eq. 12–6, namely,

$$(+\uparrow) \qquad v_B^2 = v_A^2 + 2a_c(s_B - s_A)$$

$$0 = (75 \text{ m/s})^2 + 2(-9.81 \text{ m/s}^2)(s_B - 40 \text{ m})$$

$$s_B = 327 \text{ m} \qquad\qquad Ans.$$

Velocity. To obtain the velocity of the rocket just before it hits the ground, we can apply Eq. 12–6 between points B and C, Fig. 12–4.

$$(+\uparrow) \qquad\qquad v_C^2 = v_B^2 + 2a_c(s_C - s_B)$$

$$= 0 + 2(-9.81 \text{ m/s}^2)(0 - 327 \text{ m})$$

$$v_C = -80.1 \text{ m/s} = 80.1 \text{ m/s} \downarrow \qquad Ans.$$

The negative root was chosen since the rocket is moving downward. Similarly, Eq. 12–6 may also be applied between points A and C, i.e.,

$$(+\uparrow) \qquad v_C^2 = v_A^2 + 2a_c(s_C - s_A)$$

$$= (75 \text{ m/s})^2 + 2(-9.81 \text{ m/s}^2)(0 - 40 \text{ m})$$

$$v_C = -80.1 \text{ m/s} = 80.1 \text{ m/s} \downarrow \qquad Ans.$$

NOTE: It should be realized that the rocket is subjected to a *deceleration* from A to B of 9.81 m/s^2, and then from B to C it is *accelerated* at this rate. Furthermore, even though the rocket momentarily comes to *rest* at B ($v_B = 0$) the acceleration at B is still 9.81 m/s^2 downward!

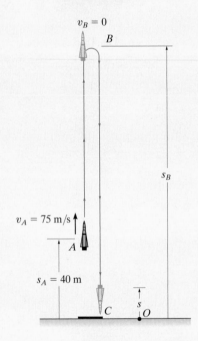

$v_B = 0$

B

s_B

$v_A = 75$ m/s

A

$s_A = 40$ m

C

s

O

Fig. 12–4

EXAMPLE | 12.4

A metallic particle is subjected to the influence of a magnetic field as it travels downward through a fluid that extends from plate A to plate B, Fig. 12–5. If the particle is released from rest at the midpoint C, $s = 100$ mm, and the acceleration is $a = (4s)$ m/s^2, where s is in meters, determine the velocity of the particle when it reaches plate B, $s = 200$ mm, and the time it takes to travel from C to B.

SOLUTION

Coordinate System. As shown in Fig. 12–5, s is positive downward, measured from plate A.

Velocity. Since $a = f(s)$, the velocity as a function of position can be obtained by using $v\, dv = a\, ds$. Realizing that $v = 0$ at $s = 0.1$ m, we have

$(+\downarrow)$

$$v\, dv = a\, ds$$

$$\int_0^v v\, dv = \int_{0.1\,\text{m}}^s 4s\, ds$$

$$\frac{1}{2}v^2 \Big|_0^v = \frac{4}{2}s^2 \Big|_{0.1\,\text{m}}^s$$

$$v = 2(s^2 - 0.01)^{1/2} \text{ m/s} \qquad (1)$$

At $s = 200$ mm $= 0.2$ m,

$$v_B = 0.346 \text{ m/s} = 346 \text{ mm/s} \downarrow \qquad \textit{Ans.}$$

The positive root is chosen since the particle is traveling downward, i.e., in the $+s$ direction.

Time. The time for the particle to travel from C to B can be obtained using $v = ds/dt$ and Eq. 1, where $s = 0.1$ m when $t = 0$. From Appendix A,

$(+\downarrow)$

$$ds = v\, dt$$

$$= 2(s^2 - 0.01)^{1/2}dt$$

$$\int_{0.1}^s \frac{ds}{(s^2 - 0.01)^{1/2}} = \int_0^t 2\, dt$$

$$\ln\left(\sqrt{s^2 - 0.01} + s\right)\Big|_{0.1}^s = 2t \Big|_0^t$$

$$\ln\left(\sqrt{s^2 - 0.01} + s\right) + 2.303 = 2t$$

At $s = 0.2$ m,

$$t = \frac{\ln\left(\sqrt{(0.2)^2 - 0.01} + 0.2\right) + 2.303}{2} = 0.658 \text{ s} \qquad \textit{Ans.}$$

NOTE: The formulas for constant acceleration cannot be used here because the acceleration changes with position, i.e., $a = 4s$.

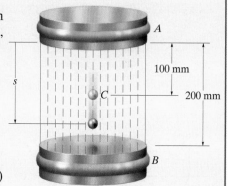

Fig. 12–5

12

EXAMPLE | 12.5

A particle moves along a horizontal path with a velocity of $v = (3t^2 - 6t)$ m/s, where t is the time in seconds. If it is initially located at the origin O, determine the distance traveled in 3.5 s, and the particle's average velocity and average speed during the time interval.

SOLUTION

Coordinate System. Here positive motion is to the right, measured from the origin O, Fig. 12–6*a*.

Distance Traveled. Since $v = f(t)$, the position as a function of time may be found by integrating $v = ds/dt$ with $t = 0$, $s = 0$.

$(\xrightarrow{+})$
$$ds = v \, dt$$
$$= (3t^2 - 6t) \, dt$$
$$\int_0^s ds = \int_0^t (3t^2 - 6t) \, dt$$
$$s = (t^3 - 3t^2) \text{ m} \qquad (1)$$

In order to determine the distance traveled in 3.5 s, it is necessary to investigate the path of motion. If we consider a graph of the velocity function, Fig. 12–6*b*, then it reveals that for $0 < t < 2$ s the velocity is *negative*, which means the particle is traveling to the *left*, and for $t > 2$ s the velocity is *positive*, and hence the particle is traveling to the *right*. Also, note that $v = 0$ at $t = 2$ s. The particle's position when $t = 0$, $t = 2$ s, and $t = 3.5$ s can be determined from Eq. 1. This yields

$$s|_{t=0} = 0 \quad s|_{t=2\,s} = -4.0 \text{ m} \quad s|_{t=3.5\,s} = 6.125 \text{ m}$$

The path is shown in Fig. 12–6*a*. Hence, the distance traveled in 3.5 s is

$$s_T = 4.0 + 4.0 + 6.125 = 14.125 \text{ m} = 14.1 \text{ m} \qquad Ans.$$

Velocity. The *displacement* from $t = 0$ to $t = 3.5$ s is

$$\Delta s = s|_{t=3.5\,s} - s|_{t=0} = 6.125 \text{ m} - 0 = 6.125 \text{ m}$$

and so the average velocity is

$$v_{avg} = \frac{\Delta s}{\Delta t} = \frac{6.125 \text{ m}}{3.5 \text{ s} - 0} = 1.75 \text{ m/s} \rightarrow \qquad Ans.$$

The average speed is defined in terms of the *distance traveled* s_T. This positive scalar is

$$(v_{sp})_{avg} = \frac{s_T}{\Delta t} = \frac{14.125 \text{ m}}{3.5 \text{ s} - 0} = 4.04 \text{ m/s} \qquad Ans.$$

NOTE: In this problem, the acceleration is $a = dv/dt = (6t - 6)$ m/s^2, which is not constant.

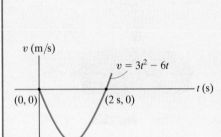

$s = -4.0$ m $\qquad s = 6.125$ m

O

$t = 2$ s $\qquad t = 0$ s $\qquad t = 3.5$ s

(a)

v (m/s)

$v = 3t^2 - 6t$

$(0,0)$ $\qquad (2\,s, 0)$ $\qquad t$ (s)

$(1\,s, -3$ m/s$)$

(b)

Fig. 12–6

It is highly suggested that you test yourself on the solutions to these examples, by covering them over and then trying to think about which equations of kinematics must be used and how they are applied in order to determine the unknowns. Then before solving any of the problems, try and solve some of the Fundamental Problems given below. The solutions and answers to all these problems are given in the back of the book. **Doing this throughout the book will help immensely in understanding how to apply the theory, and thereby develop your problem-solving skills.**

FUNDAMENTAL PROBLEMS

F12–1. Initially, the car travels along a straight road with a speed of 35 m/s. If the brakes are applied and the speed of the car is reduced to 10 m/s in 15 s, determine the constant deceleration of the car.

F12–1

F12–2. A ball is thrown vertically upward with a speed of 15 m/s. Determine the time of flight when it returns to its original position.

F12–2

F12–3. A particle travels along a straight line with a velocity of $v = (4t - 3t^2)$ m/s, where t is in seconds. Determine the position of the particle when $t = 4$ s. $s = 0$ when $t = 0$.

F12–4. A particle travels along a straight line with a speed $v = (0.5t^3 - 8t)$ m/s, where t is in seconds. Determine the acceleration of the particle when $t = 2$ s.

F12–5. The position of the particle is given by $s = (2t^2 - 8t + 6)$ m, where t is in seconds. Determine the time when the velocity of the particle is zero, and the total distance traveled by the particle when $t = 3$ s.

F12–5

F12–6. A particle travels along a straight line with an acceleration of $a = (10 - 0.2s)$ m/s^2, where s is measured in meters. Determine the velocity of the particle when $s = 10$ m if $v = 5$ m/s at $s = 0$.

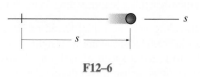

F12–6

F12–7. A particle moves along a straight line such that its acceleration is $a = (4t^2 - 2)$ m/s^2, where t is in seconds. When $t = 0$, the particle is located 2 m to the left of the origin, and when $t = 2$ s, it is 20 m to the left of the origin. Determine the position of the particle when $t = 4$ s.

F12–8. A particle travels along a straight line with a velocity of $v = (20 - 0.05s^2)$ m/s, where s is in meters. Determine the acceleration of the particle at $s = 15$ m.

12 **PROBLEMS**

12–1. A baseball is thrown downward from a 50-ft tower with an initial speed of 18 ft/s. Determine the speed at which it hits the ground and the time of travel.

12–2. When a train is traveling along a straight track at 2 m/s, it begins to accelerate at $a = (60 \, v^{-4}) \, \text{m/s}^2$, where v is in m/s. Determine its velocity v and the position 3 s after the acceleration.

Prob. 12–2

12–3. From approximately what floor of a building must a car be dropped from an at-rest position so that it reaches a speed of 80.7 ft/s (55 mi/h) when it hits the ground? Each floor is 12 ft higher than the one below it. (*Note:* You may want to remember this when traveling 55 mi/h.)

*12–4. Traveling with an initial speed of 70 km/h, a car accelerates at 6000 km/h² along a straight road. How long will it take to reach a speed of 120 km/h? Also, through what distance does the car travel during this time?

12–5. A bus starts from rest with a constant acceleration of 1 m/s². Determine the time required for it to attain a speed of 25 m/s and the distance traveled.

12–6. A stone A is dropped from rest down a well, and in 1 s another stone B is dropped from rest. Determine the distance between the stones another second later.

12–7. A bicyclist starts from rest and after traveling along a straight path a distance of 20 m reaches a speed of 30 km/h. Determine his acceleration if it is *constant*. Also, how long does it take to reach the speed of 30 km/h?

■*12–8. A particle moves along a straight line with an acceleration of $a = 5/(3s^{1/3} + s^{5/2}) \, \text{m/s}^2$, where s is in meters. Determine the particle's velocity when $s = 2$ m, if it starts from rest when $s = 1$ m. Use a numerical method to evaluate the integral.

12–9. If it takes 3 s for a ball to strike the ground when it is released from rest, determine the height in meters of the building from which it was released. Also, what is the velocity of the ball when it strikes the ground?

12–10. The position of a particle along a straight line is given by $s = (1.5t^3 - 13.5t^2 + 22.5t) \, \text{ft}$, where t is in seconds. Determine the position of the particle when $t = 6$ s and the total distance it travels during the 6-s time interval. *Hint*: Plot the path to determine the total distance traveled.

12–11. If a particle has an initial velocity of $v_0 = 12$ ft/s to the right, at $s_0 = 0$, determine its position when $t = 10$ s, if $a = 2 \, \text{ft/s}^2$ to the left.

*12–12. Determine the time required for a car to travel 1 km along a road if the car starts from rest, reaches a maximum speed at some intermediate point, and then stops at the end of the road. The car can accelerate at 1.5 m/s² and decelerate at 2 m/s².

12–13. Tests reveal that a normal driver takes about 0.75 s before he or she can *react* to a situation to avoid a collision. It takes about 3 s for a driver having 0.1% alcohol in his system to do the same. If such drivers are traveling on a straight road at 30 mph (44 ft/s) and their cars can decelerate at 2 ft/s², determine the shortest stopping distance d for each from the moment they see the pedestrians. *Moral*: If you must drink, please don't drive!

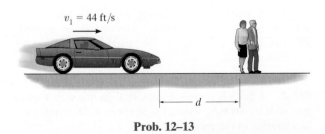

Prob. 12–13

12–14. A car is to be hoisted by elevator to the fourth floor of a parking garage, which is 48 ft above the ground. If the elevator can accelerate at 0.6 ft/s², decelerate at 0.3 ft/s², and reach a maximum speed of 8 ft/s, determine the shortest time to make the lift, starting from rest and ending at rest.

12–15. A train starts from rest at station A and accelerates at 0.5 m/s² for 60 s. Afterwards it travels with a constant velocity for 15 min. It then decelerates at 1 m/s² until it is brought to rest at station B. Determine the distance between the stations.

*12–16. A particle travels along a straight line such that in 2 s it moves from an initial position $s_A = +0.5$ m to a position $s_B = -1.5$ m. Then in another 4 s it moves from s_B to $s_C = +2.5$ m. Determine the particle's average velocity and average speed during the 6-s time interval.

12–17. The acceleration of a particle as it moves along a straight line is given by $a = (2t - 1)$ m/s², where t is in seconds. If $s = 1$ m and $v = 2$ m/s when $t = 0$, determine the particle's velocity and position when $t = 6$ s. Also, determine the total distance the particle travels during this time period.

12–18. A freight train travels at $v = 60(1 - e^{-t})$ ft/s, where t is the elapsed time in seconds. Determine the distance traveled in three seconds, and the acceleration at this time.

*12–20. The velocity of a particle traveling along a straight line is $v = (3t^2 - 6t)$ ft/s, where t is in seconds. If $s = 4$ ft when $t = 0$, determine the position of the particle when $t = 4$ s. What is the total distance traveled during the time interval $t = 0$ to $t = 4$ s? Also, what is the acceleration when $t = 2$ s?

12–21. If the effects of atmospheric resistance are accounted for, a freely falling body has an acceleration defined by the equation $a = 9.81[1 - v^2(10^{-4})]$ m/s², where v is in m/s and the positive direction is downward. If the body is released from rest at a *very high altitude*, determine (a) the velocity when $t = 5$ s, and (b) the body's terminal or maximum attainable velocity (as $t \rightarrow \infty$).

12–22. The position of a particle on a straight line is given by $s = (t^3 - 9t^2 + 15t)$ ft, where t is in seconds. Determine the position of the particle when $t = 6$ s and the total distance it travels during the 6-s time interval. *Hint*: Plot the path to determine the total distance traveled.

12–23. Two particles A and B start from rest at the origin $s = 0$ and move along a straight line such that $a_A = (6t - 3)$ ft/s² and $a_B = (12t^2 - 8)$ ft/s², where t is in seconds. Determine the distance between them when $t = 4$ s and the total distance each has traveled in $t = 4$ s.

*12–24. A particle is moving along a straight line such that its velocity is defined as $v = (-4s^2)$ m/s, where s is in meters. If $s = 2$ m when $t = 0$, determine the velocity and acceleration as functions of time.

12–25. A sphere is fired downwards into a medium with an initial speed of 27 m/s. If it experiences a deceleration of $a = (-6t)$ m/s², where t is in seconds, determine the distance traveled before it stops.

12–26. When two cars A and B are next to one another, they are traveling in the same direction with speeds v_A and v_B, respectively. If B maintains its constant speed, while A begins to decelerate at a_A, determine the distance d between the cars at the instant A stops.

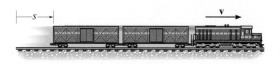

Prob. 12–18

12–19. A particle travels to the right along a straight line with a velocity $v = [5/(4 + s)]$ m/s, where s is in meters. Determine its position when $t = 6$ s if $s = 5$ m when $t = 0$.

A   B

d

Prob. 12–26

12–27. A particle is moving along a straight line such that when it is at the origin it has a velocity of 4 m/s. If it begins to decelerate at the rate of $a = (-1.5v^{1/2})$ m/s^2, where v is in m/s, determine the distance it travels before it stops.

*12–28. A particle travels to the right along a straight line with a velocity $v = [5/(4 + s)]$ m/s, where s is in meters. Determine its deceleration when $s = 2$ m.

12–29. A particle moves along a straight line with an acceleration $a = 2\,v^{1/2}$ m/s^2, where v is in m/s. If $s = 0$, $v = 4$ m/s when $t = 0$, determine the time for the particle to achieve a velocity of 20 m/s. Also, find the displacement of particle when $t = 2$ s.

12–30. As a train accelerates uniformly it passes successive kilometer marks while traveling at velocities of 2 m/s and then 10 m/s. Determine the train's velocity when it passes the next kilometer mark and the time it takes to travel the 2-km distance.

12–31. The acceleration of a particle along a straight line is defined by $a = (2t - 9)$ m/s^2, where t is in seconds. At $t = 0$, $s = 1$ m and $v = 10$ m/s. When $t = 9$ s, determine (a) the particle's position, (b) the total distance traveled, and (c) the velocity.

*12–32. The acceleration of a particle traveling along a straight line is $a = \frac{1}{4}\,s^{1/2}$ m/s^2, where s is in meters. If $v = 0$, $s = 1$ m when $t = 0$, determine the particle's velocity at $s = 2$ m.

12–33. At $t = 0$ bullet A is fired vertically with an initial (muzzle) velocity of 450 m/s. When $t = 3$ s, bullet B is fired upward with a muzzle velocity of 600 m/s. Determine the time t, after A is fired, as to when bullet B passes bullet A. At what altitude does this occur?

12–34. A boy throws a ball straight up from the top of a 12-m high tower. If the ball falls past him 0.75 s later, determine the velocity at which it was thrown, the velocity of the ball when it strikes the ground, and the time of flight.

12–35. When a particle falls through the air, its initial acceleration $a = g$ diminishes until it is zero, and thereafter it falls at a constant or terminal velocity v_f. If this variation of the acceleration can be expressed as $a = (g/v_f^2)(v_f^2 - v^2)$, determine the time needed for the velocity to become $v = v_f/2$. Initially the particle falls from rest.

*12–36. A particle is moving with a velocity of v_0 when $s = 0$ and $t = 0$. If it is subjected to a deceleration of $a = -kv^3$, where k is a constant, determine its velocity and position as functions of time.

12–37. As a body is projected to a high altitude above the earth's *surface*, the variation of the acceleration of gravity with respect to altitude y must be taken into account. Neglecting air resistance, this acceleration is determined from the formula $a = -g_0[R^2/(R + y)^2]$, where g_0 is the constant gravitational acceleration at sea level, R is the radius of the earth, and the positive direction is measured upward. If $g_0 = 9.81$ m/s^2 and $R = 6356$ km, determine the minimum initial velocity (escape velocity) at which a projectile should be shot vertically from the earth's surface so that it does not fall back to the earth. *Hint:* This requires that $v = 0$ as $y \to \infty$.

12–38. Accounting for the variation of gravitational acceleration a with respect to altitude y (see Prob. 12–37), derive an equation that relates the velocity of a freely falling particle to its altitude. Assume that the particle is released from rest at an altitude y_0 from the earth's surface. With what velocity does the particle strike the earth if it is released from rest at an altitude $y_0 = 500$ km? Use the numerical data in Prob. 12–37.

12.3   Rectilinear Kinematics: Erratic Motion

When a particle has erratic or changing motion then its position, velocity, and acceleration *cannot* be described by a single continuous mathematical function along the entire path. Instead, a series of functions will be required to specify the motion at different intervals. For this reason, it is convenient to represent the motion as a graph. If a graph of the motion that relates any two of the variables s, v, a, t can be drawn, then this graph can be used to construct subsequent graphs relating two other variables since the variables are related by the differential relationships $v = ds/dt$, $a = dv/dt$, or $a\, ds = v\, dv$. Several situations occur frequently.

The s–t, v–t, and a–t Graphs.   To construct the v–t graph given the s–t graph, Fig. 12–7a, the equation $v = ds/dt$ should be used, since it relates the variables s and t to v. This equation states that

$$\frac{ds}{dt} = v$$

$$\frac{\text{slope of}}{s\text{–}t \text{ graph}} = \text{velocity}$$

For example, by measuring the slope on the s–t graph when $t = t_1$, the velocity is v_1, which is plotted in Fig. 12–7b. The v–t graph can be constructed by plotting this and other values at each instant.

The a–t graph can be constructed from the v–t graph in a similar manner, Fig. 12–8, since

$$\frac{dv}{dt} = a$$

$$\frac{\text{slope of}}{v\text{–}t \text{ graph}} = \text{acceleration}$$

Examples of various measurements are shown in Fig. 12–8a and plotted in Fig. 12–8b.

If the s–t curve for each interval of motion can be expressed by a mathematical function $s = s(t)$, then the equation of the v–t graph for the same interval can be obtained by differentiating this function with respect to time since $v = ds/dt$. Likewise, the equation of the a–t graph for the same interval can be determined by differentiating $v = v(t)$ since $a = dv/dt$. Since differentiation reduces a polynomial of degree n to that of degree $n - 1$, then if the s–t graph is parabolic (a second-degree curve), the v–t graph will be a sloping line (a first-degree curve), and the a–t graph will be a constant or a horizontal line (a zero-degree curve).

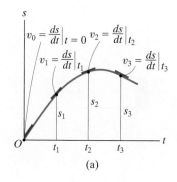

(a)

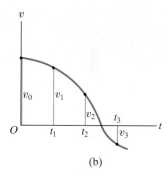

(b)

Fig. 12–7

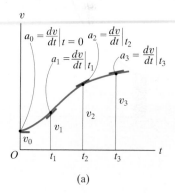

(a)

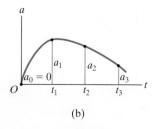

(b)

Fig. 12–8

12

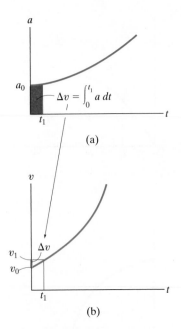

(a)

(b)

Fig. 12–9

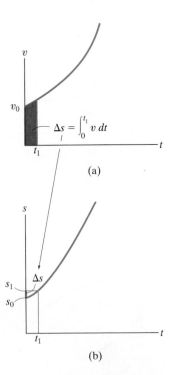

(a)

(b)

Fig. 12–10

If the a–t graph is given, Fig. 12–9a, the v–t graph may be constructed using $a = dv/dt$, written as

$$\Delta v = \int a\, dt$$

$$\text{change in velocity} = \text{area under } a\text{–}t \text{ graph}$$

Hence, to construct the v–t graph, we begin with the particle's initial velocity v_0 and then add to this small increments of area (Δv) determined from the a–t graph. In this manner successive points, $v_1 = v_0 + \Delta v$, etc., for the v–t graph are determined, Fig. 12–9b. Notice that an algebraic addition of the area increments of the a–t graph is necessary, since areas lying above the t axis correspond to an increase in v ("positive" area), whereas those lying below the axis indicate a decrease in v ("negative" area).

Similarly, if the v–t graph is given, Fig. 12–10a, it is possible to determine the s–t graph using $v = ds/dt$, written as

$$\Delta s = \int v\, dt$$

$$\text{displacement} = \text{area under } v\text{–}t \text{ graph}$$

In the same manner as stated above, we begin with the particle's initial position s_0 and add (algebraically) to this small area increments Δs determined from the v–t graph, Fig. 12–10b.

If segments of the a–t graph can be described by a series of equations, then each of these equations can be *integrated* to yield equations describing the corresponding segments of the v–t graph. In a similar manner, the s–t graph can be obtained by integrating the equations which describe the segments of the v–t graph. As a result, if the a–t graph is linear (a first-degree curve), integration will yield a v–t graph that is parabolic (a second-degree curve) and an s–t graph that is cubic (third-degree curve).

The v–s and a–s Graphs.   If the a–s graph can be constructed, then points on the v–s graph can be determined by using $v\,dv = a\,ds$. Integrating this equation between the limits $v = v_0$ at $s = s_0$ and $v = v_1$ at $s = s_1$, we have,

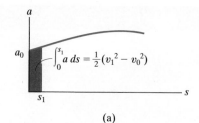

(a)

$$\tfrac{1}{2}(v_1^2 - v_0^2) = \int_{s_0}^{s_1} a\,ds$$

area under
a–s graph

Therefore, if the red area in Fig. 12–11a is determined, and the initial velocity v_0 at $s_0 = 0$ is known, then $v_1 = \left(2\int_0^{s_1} a\,ds + v_0^2\right)^{1/2}$, Fig. 12–11b. Successive points on the v–s graph can be constructed in this manner.

If the v–s graph is known, the acceleration a at any position s can be determined using $a\,ds = v\,dv$, written as

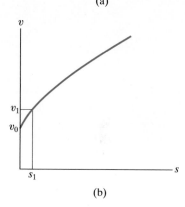

(b)

Fig. 12–11

$$a = v\left(\frac{dv}{ds}\right)$$

velocity times
acceleration = slope of
v–s graph

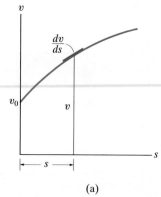

(a)

Thus, at any point (s, v) in Fig. 12–12a, the slope dv/ds of the v–s graph is measured. Then with v and dv/ds known, the value of a can be calculated, Fig. 12–12b.

The v–s graph can also be constructed from the a–s graph, or vice versa, by approximating the known graph in various intervals with mathematical functions, $v = f(s)$ or $a = g(s)$, and then using $a\,ds = v\,dv$ to obtain the other graph.

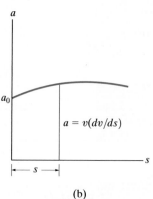

(b)

Fig. 12–12

EXAMPLE | 12.6

A bicycle moves along a straight road such that its position is described by the graph shown in Fig. 12–13a. Construct the v–t and a–t graphs for $0 \le t \le 30$ s.

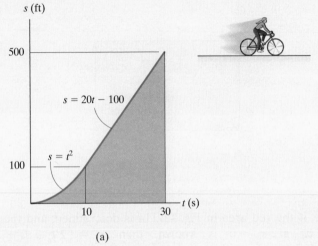

s (ft)

500

$s = 20t - 100$

100

$s = t^2$

10   30 t (s)

(a)

SOLUTION

v–t Graph. Since $v = ds/dt$, the v–t graph can be determined by differentiating the equations defining the s–t graph, Fig. 12–13a. We have

$$0 \le t < 10 \text{ s}; \qquad s = (t^2) \text{ ft} \qquad v = \frac{ds}{dt} = (2t) \text{ ft/s}$$

$$10 \text{ s} < t \le 30 \text{ s}; \qquad s = (20t - 100) \text{ ft} \qquad v = \frac{ds}{dt} = 20 \text{ ft/s}$$

The results are plotted in Fig. 12–13b. We can also obtain specific values of v by measuring the *slope* of the s–t graph at a given instant. For example, at $t = 20$ s, the slope of the s–t graph is determined from the straight line from 10 s to 30 s, i.e.,

$$t = 20 \text{ s}; \qquad v = \frac{\Delta s}{\Delta t} = \frac{500 \text{ ft} - 100 \text{ ft}}{30 \text{ s} - 10 \text{ s}} = 20 \text{ ft/s}$$

a–t Graph. Since $a = dv/dt$, the a–t graph can be determined by differentiating the equations defining the lines of the v–t graph. This yields

$$0 \le t < 10 \text{ s}; \qquad v = (2t) \text{ ft/s} \qquad a = \frac{dv}{dt} = 2 \text{ ft/s}^2$$

$$10 < t \le 30 \text{ s}; \qquad v = 20 \text{ ft/s} \qquad a = \frac{dv}{dt} = 0$$

The results are plotted in Fig. 12–13c.

NOTE: Show that $a = 2$ ft/s^2 when $t = 5$ s by measuring the slope of the v–t graph.

v (ft/s)

$v = 2t$

20

$v = 20$

10   30 t (s)

(b)

a (ft/s^2)

2

10   30 t (s)

(c)

Fig. 12–13

EXAMPLE 12.7

The car in Fig. 12–14a starts from rest and travels along a straight track such that it accelerates at 10 m/s² for 10 s, and then decelerates at 2 m/s². Draw the v–t and s–t graphs and determine the time t' needed to stop the car. How far has the car traveled?

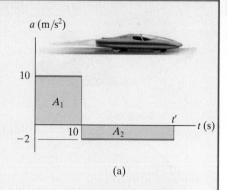

(a)

SOLUTION

v–t Graph. Since $dv = a\,dt$, the v–t graph is determined by integrating the straight-line segments of the a–t graph. Using the *initial condition* $v = 0$ when $t = 0$, we have

$$0 \le t < 10\text{ s}; \quad a = (10)\text{ m/s}^2; \quad \int_0^v dv = \int_0^t 10\,dt, \quad v = 10t$$

When $t = 10$ s, $v = 10(10) = 100$ m/s. Using this as the *initial condition* for the next time period, we have

$$10\text{ s} < t \le t'; a = (-2)\text{ m/s}^2; \int_{100\text{ m/s}}^v dv = \int_{10\text{ s}}^t -2\,dt, v = (-2t + 120)\text{ m/s}$$

When $t = t'$ we require $v = 0$. This yields, Fig. 12–14b,

$$t' = 60\text{ s} \qquad\qquad Ans.$$

A more direct solution for t' is possible by realizing that the area under the a–t graph is equal to the change in the car's velocity. We require $\Delta v = 0 = A_1 + A_2$, Fig. 12–14a. Thus

$$0 = 10\text{ m/s}^2(10\text{ s}) + (-2\text{ m/s}^2)(t' - 10\text{ s})$$

$$t' = 60\text{ s} \qquad\qquad Ans.$$

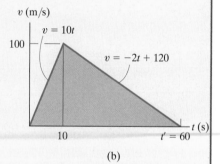

(b)

s–t Graph. Since $ds = v\,dt$, integrating the equations of the v–t graph yields the corresponding equations of the s–t graph. Using the *initial condition* $s = 0$ when $t = 0$, we have

$$0 \le t \le 10\text{ s}; \quad v = (10t)\text{ m/s}; \quad \int_0^s ds = \int_0^t 10t\,dt, \quad s = (5t^2)\text{ m}$$

When $t = 10$ s, $s = 5(10)^2 = 500$ m. Using this *initial condition*,

$$10\text{ s} \le t \le 60\text{ s}; v = (-2t + 120)\text{ m/s}; \int_{500\text{ m}}^s ds = \int_{10\text{ s}}^t (-2t + 120)\,dt$$

$$s - 500 = -t^2 + 120t - [-(10)^2 + 120(10)]$$

$$s = (-t^2 + 120t - 600)\text{ m}$$

When $t' = 60$ s, the position is

$$s = -(60)^2 + 120(60) - 600 = 3000\text{ m} \qquad Ans.$$

The s–t graph is shown in Fig. 12–14c.

NOTE: A direct solution for s is possible when $t' = 60$ s, since the *triangular area* under the v–t graph would yield the displacement $\Delta s = s - 0$ from $t = 0$ to $t' = 60$ s. Hence,

$$\Delta s = \tfrac{1}{2}(60\text{ s})(100\text{ m/s}) = 3000\text{ m} \qquad Ans.$$

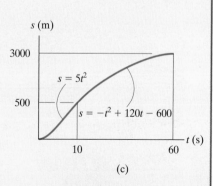

(c)

Fig. 12–14

12

12

EXAMPLE | 12.8

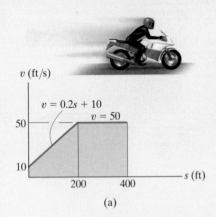

v (ft/s)

$v = 0.2s + 10$

$v = 50$

50

10

200   400

s (ft)

(a)

a (ft/s²)

$a = 0.04s + 2$

10

2

$a = 0$

200   400

s (ft)

(b)

Fig. 12–15

The v–s graph describing the motion of a motorcycle is shown in Fig. 12–15a. Construct the a–s graph of the motion and determine the time needed for the motorcycle to reach the position $s = 400$ ft.

SOLUTION

a–s Graph. Since the equations for segments of the v–s graph are given, the a–s graph can be determined using $a \, ds = v \, dv$.

$$0 \leq s < 200 \text{ ft}; \qquad v = (0.2s + 10) \text{ ft/s}$$

$$a = v\frac{dv}{ds} = (0.2s + 10)\frac{d}{ds}(0.2s + 10) = 0.04s + 2$$

$$200 \text{ ft} < s \leq 400 \text{ ft}; \qquad v = 50 \text{ ft/s}$$

$$a = v\frac{dv}{ds} = (50)\frac{d}{ds}(50) = 0$$

The results are plotted in Fig. 12–15b.

Time. The time can be obtained using the v–s graph and $v = ds/dt$, because this equation relates v, s, and t. For the first segment of motion, $s = 0$ when $t = 0$, so

$$0 \leq s < 200 \text{ ft}; \qquad v = (0.2s + 10) \text{ ft/s}; \qquad dt = \frac{ds}{v} = \frac{ds}{0.2s + 10}$$

$$\int_0^t dt = \int_0^s \frac{ds}{0.2s + 10}$$

$$t = (5\ln(0.2s + 10) - 5\ln 10) \text{ s}$$

At $s = 200$ ft, $t = 5\ln[0.2(200) + 10] - 5\ln 10 = 8.05$ s. Therefore, using these initial conditions for the second segment of motion,

$$200 \text{ ft} < s \leq 400 \text{ ft}; \qquad v = 50 \text{ ft/s}; \qquad dt = \frac{ds}{v} = \frac{ds}{50}$$

$$\int_{8.05 \text{ s}}^t dt = \int_{200 \text{ m}}^s \frac{ds}{50};$$

$$t - 8.05 = \frac{s}{50} - 4; \quad t = \left(\frac{s}{50} + 4.05\right) \text{ s}$$

Therefore, at $s = 400$ ft,

$$t = \frac{400}{50} + 4.05 = 12.0 \text{ s} \qquad\qquad Ans.$$

NOTE: The graphical results can be checked in part by calculating slopes. For example, at $s = 0$, $a = v(dv/ds) = 10(50 - 10)/200 = 2$ m/s². Also, the results can be checked in part by inspection. The v–s graph indicates the initial increase in velocity (acceleration) followed by constant velocity ($a = 0$).

FUNDAMENTAL PROBLEMS

F12–9. The particle travels along a straight track such that its position is described by the s–t graph. Construct the v–t graph for the same time interval.

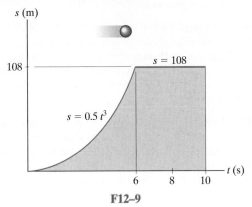

$s = 108$

108

$s = 0.5\,t^3$

$6 \quad 8 \quad 10$

t (s)

F12–9

F12–10. A van travels along a straight road with a velocity described by the graph. Construct the s–t and a–t graphs during the same period. Take $s = 0$ when $t = 0$.

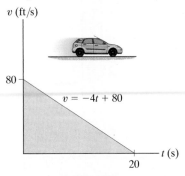

v (ft/s)

80

$v = -4t + 80$

20

t (s)

F12–10

F12–11. A bicycle travels along a straight road where its velocity is described by the v–s graph. Construct the a–s graph for the same time interval.

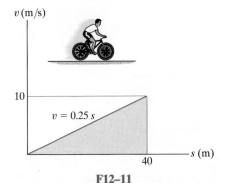

v (m/s)

10

$v = 0.25\,s$

40

s (m)

F12–11

F12–12. The sports car travels along a straight road such that its position is described by the graph. Construct the v–t and a–t graphs for the time interval $0 \le t \le 10$ s.

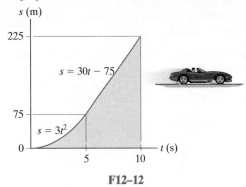

s (m)

225

$s = 30t - 75$

75

$s = 3t^2$

0

$5 \quad 10$

t (s)

F12–12

F12–13. The dragster starts from rest and has an acceleration described by the graph. Construct the v–t graph for the time interval $0 \le t \le t'$, where t' is the time for the car to come to rest.

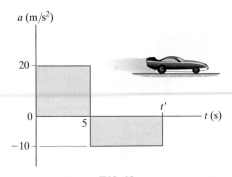

a (m/s²)

20

0

5

t'

t (s)

-10

F12–13

F12–14. The dragster starts from rest and has a velocity described by the graph. Construct the s–t graph during the time interval $0 \le t \le 15$ s. Also, determine the total distance traveled during this time interval.

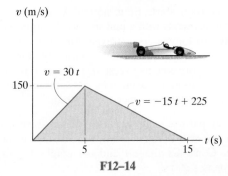

v (m/s)

$v = 30\,t$

150

$v = -15\,t + 225$

$5 \quad 15$

t (s)

F12–14

PROBLEMS

12–39. A freight train starts from rest and travels with a constant acceleration of 0.5 ft/s². After a time t' it maintains a constant speed so that when $t = 160$ s it has traveled 2000 ft. Determine the time t' and draw the v–t graph for the motion.

*12–40. A sports car travels along a straight road with an acceleration-deceleration described by the graph. If the car starts from rest, determine the distance s' the car travels until it stops. Construct the v–s graph for $0 \leq s \leq s'$.

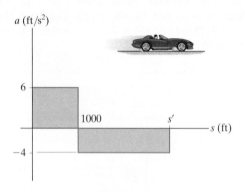

Prob. 12–40

12–41. A train starts from station A and for the first kilometer, it travels with a uniform acceleration. Then, for the next two kilometers, it travels with a uniform speed. Finally, the train decelerates uniformly for another kilometer before coming to rest at station B. If the time for the whole journey is six minutes, draw the v–t graph and determine the maximum speed of the train.

12–42. A particle starts from $s = 0$ and travels along a straight line with a velocity $v = (t^2 - 4t + 3)$ m/s, where t is in seconds. Construct the v–t and a–t graphs for the time interval $0 \leq t \leq 4$ s.

12–43. If the position of a particle is defined by $s = [2 \sin (\pi/5)t + 4]$ m, where t is in seconds, construct the s–t, v–t, and a–t graphs for $0 \leq t \leq 10$ s.

*12–44. An airplane starts from rest, travels 5000 ft down a runway, and after uniform acceleration, takes off with a speed of 162 mi/h. It then climbs in a straight line with a uniform acceleration of 3 ft/s² until it reaches a constant speed of 220 mi/h. Draw the s–t, v–t, and a–t graphs that describe the motion.

12–45. The elevator starts from rest at the first floor of the building. It can accelerate at 5 ft/s² and then decelerate at 2 ft/s². Determine the shortest time it takes to reach a floor 40 ft above the ground. The elevator starts from rest and then stops. Draw the a–t, v–t, and s–t graphs for the motion.

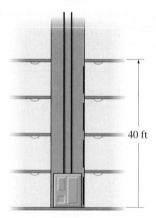

Prob. 12–45

12–46. The velocity of a car is plotted as shown. Determine the total distance the car moves until it stops ($t = 80$ s). Construct the a–t graph.

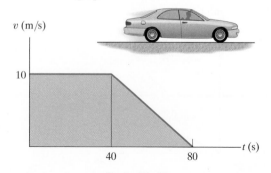

Prob. 12–46

12–47. The v–s graph for a go-cart traveling on a straight road is shown. Determine the acceleration of the go-cart at s = 50 m and s = 150 m. Draw the a–s graph.

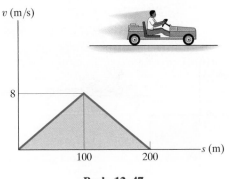

v (m/s)

8

100   200   s (m)

Prob. 12–47

12–50. The v–t graph of a car while traveling along a road is shown. Draw the s–t and a–t graphs for the motion.

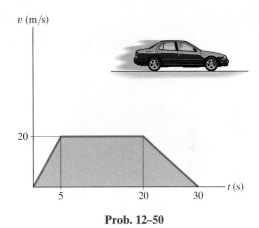

v (m/s)

20

5   20   30   t (s)

Prob. 12–50

*****12–48.** The v–t graph for a particle moving through an electric field from one plate to another has the shape shown in the figure. The acceleration and deceleration that occur are constant and both have a magnitude of 4 m/s². If the plates are spaced 200 mm apart, determine the maximum velocity v_{max} and the time t' for the particle to travel from one plate to the other. Also draw the s–t graph. When $t = t'/2$ the particle is at s = 100 mm.

12–49. The v–t graph for a particle moving through an electric field from one plate to another has the shape shown in the figure, where $t' = 0.2$ s and $v_{max} = 10$ m/s. Draw the s–t and a–t graphs for the particle. When $t = t'/2$ the particle is at s = 0.5 m.

12–51. The a–t graph of the bullet train is shown. If the train starts from rest, determine the elapsed time t' before it again comes to rest. What is the total distance traveled during this time interval? Construct the v–t and s–t graphs.

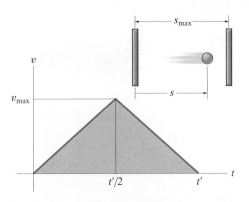

s_{max}

v

v_{max}

s

$t'/2$ t' t

Probs. 12–48/49

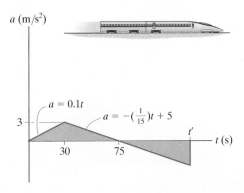

a (m/s²)

$a = 0.1t$

3

$a = -(\frac{1}{15})t + 5$

30   75 t' t (s)

Prob. 12–51

12

*12–52. The snowmobile moves along a straight course according to the v–t graph. Construct the s–t and a–t graphs for the same 50-s time interval. When $t = 0, s = 0$.

12–54. The dragster starts from rest and has an acceleration described by the graph. Determine the time t' for it to stop. Also, what is its maximum speed? Construct the v–t and s–t graphs for the time interval $0 \leq t \leq t'$.

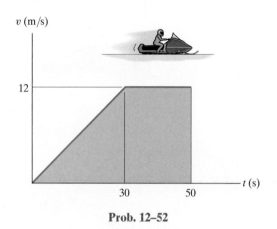

Prob. 12–52

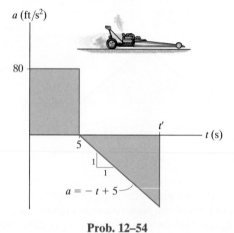

$a = -t + 5$

Prob. 12–54

12–53. A two-stage missile is fired vertically from rest with the acceleration shown. In 15 s the first stage A burns out and the second stage B ignites. Plot the v–t and s–t graphs which describe the two-stage motion of the missile for $0 \leq t \leq 20$ s.

12–55. A race car starting from rest travels along a straight road and for 10 s has the acceleration shown. Construct the v–t graph that describes the motion and find the distance traveled in 10 s.

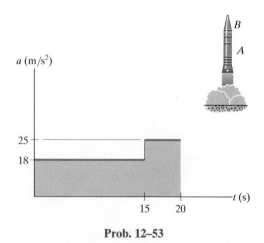

Prob. 12–53

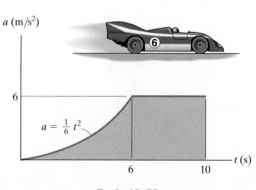

$a = \frac{1}{6} t^2$

Prob. 12–55

*12–56. The v–t graph for the motion of a car as if moves along a straight road is shown. Draw the a–t graph and determine the maximum acceleration during the 30-s time interval. The car starts from rest at $s = 0$.

12–57. The v–t graph for the motion of a car as it moves along a straight road is shown. Draw the s–t graph and determine the average speed and the distance traveled for the 30-s time interval. The car starts from rest at $s = 0$.

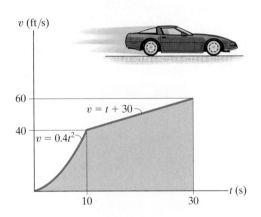

Probs. 12–56/57

12–58. The jet-powered boat starts from rest at $s = 0$ and travels along a straight line with the speed described by the graph. Construct the s–t and a–t graph for the time interval $0 \leq t \leq 50$ s.

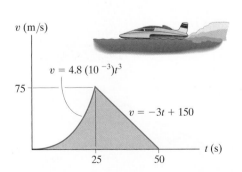

Prob. 12–58

12–59. An airplane lands on the straight runway, originally traveling at 110 ft/s when $s = 0$. If it is subjected to the decelerations shown, determine the time t' needed to stop the plane and construct the s–t graph for the motion.

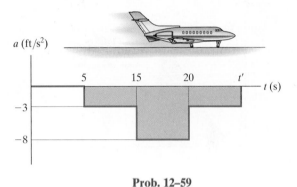

Prob. 12–59

*12–60. A car travels along a straight road with the speed shown by the v–t graph. Plot the a–t graph.

12–61. A car travels along a straight road with the speed shown by the v–t graph. Determine the total distance the car travels until it stops when $t = 48$ s. Also plot the s–t graph.

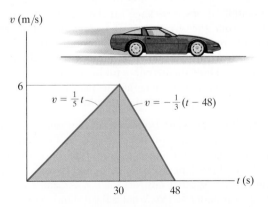

Probs. 12–60/61

12–62. A motorcyclist travels along a straight road with the velocity described by the graph. Construct the s–t and a–t graphs.

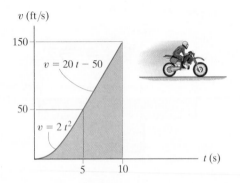

$v = 20\,t - 50$

$v = 2\,t^2$

Prob. 12–62

12–63. The speed of a train during the first minute has been recorded as follows:

t (s)	0	20	40	60
v (m/s)	0	16	21	24

Plot the v–t graph, approximating the curve as straight-line segments between the given points. Determine the total distance traveled.

*12–64.** A man riding upward in a freight elevator accidentally drops a package off the elevator when it is 100 ft from the ground. If the elevator maintains a constant upward speed of 4 ft/s, determine how high the elevator is from the ground the instant the package hits the ground. Draw the v–t curve for the package during the time it is in motion. Assume that the package was released with the same upward speed as the elevator.

12–65. Two cars start from rest side by side and travel along a straight road. Car A accelerates at 4 m/s² for 10 s and then maintains a constant speed. Car B accelerates at 5 m/s² until reaching a constant speed of 25 m/s and then maintains this speed. Construct the a–t, v–t, and s–t graphs for each car until t = 15 s. What is the distance between the two cars when t = 15 s?

12–66. A two-stage rocket is fired vertically from rest at s = 0 with an acceleration as shown. After 30 s the first stage A burns out and the second stage B ignites. Plot the v–t graph which describes the motion of the second stage for 0 ≤ t ≤ 60 s.

12–67. A two-stage rocket is fired vertically from rest at s = 0 with an acceleration as shown. After 30 s the first stage A burns out and the second stage B ignites. Plot the s–t graph which describes the motion of the second stage for 0 ≤ t ≤ 60 s.

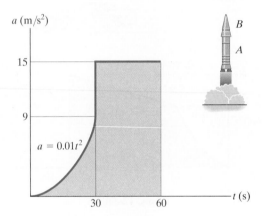

$a = 0.01t^2$

Probs. 12–66/67

*12–68.** The a–s graph for a jeep traveling along a straight road is given for the first 300 m of its motion. Construct the v–s graph. At s = 0, v = 0.

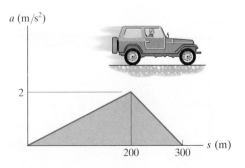

Prob. 12–68

12–69. The v–s graph for the car is given for the first 500 ft of its motion. Construct the a–s graph for $0 \leq s \leq 500$ ft. How long does it take to travel the 500-ft distance? The car starts at $s = 0$ when $t = 0$.

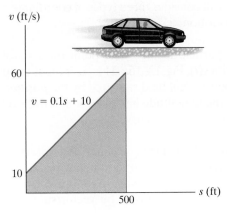

Prob. 12–69

12–71. The v–s graph of a cyclist traveling along a straight road is shown. Construct the a–s graph.

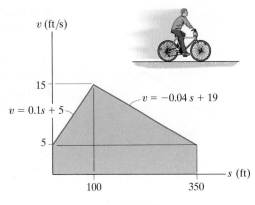

Prob. 12–71

12–70. The boat travels along a straight line with the speed described by the graph. Construct the s–t and a–s graphs. Also, determine the time required for the boat to travel a distance $s = 400$ m if $s = 0$ when $t = 0$.

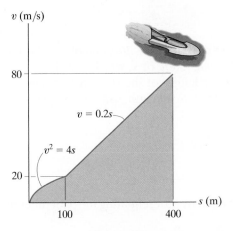

Prob. 12–70

*■12–72. The a–s graph for a boat moving along a straight path is given. If the boat starts at $s = 0$ when $v = 0$, determine its speed when it is at $s = 75$ ft, and 125 ft, respectively. Use a numerical method to evaluate v at $s = 125$ ft.

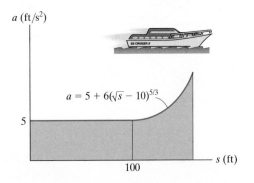

Prob. 12–72

12.4   General Curvilinear Motion

Curvilinear motion occurs when a particle moves along a curved path. Since this path is often described in three dimensions, vector analysis will be used to formulate the particle's position, velocity, and acceleration.* In this section the general aspects of curvilinear motion are discussed, and in subsequent sections we will consider three types of coordinate systems often used to analyze this motion.

Position.   Consider a particle located at a point on a space curve defined by the path function $s(t)$, Fig. 12–16a. The position of the particle, measured from a fixed point O, will be designated by the *position vector* $\mathbf{r} = \mathbf{r}(t)$. Notice that both the magnitude and direction of this vector will change as the particle moves along the curve.

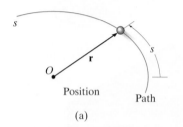

Position

(a)

Displacement.   Suppose that during a small time interval Δt the particle moves a distance Δs along the curve to a new position, defined by $\mathbf{r}' = \mathbf{r} + \Delta\mathbf{r}$, Fig. 12–16b. The *displacement* $\Delta\mathbf{r}$ represents the change in the particle's position and is determined by vector subtraction; i.e., $\Delta\mathbf{r} = \mathbf{r}' - \mathbf{r}$.

Velocity.   During the time Δt, the *average velocity* of the particle is

$$\mathbf{v}_{avg} = \frac{\Delta\mathbf{r}}{\Delta t}$$

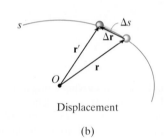

Displacement

(b)

The *instantaneous velocity* is determined from this equation by letting $\Delta t \to 0$, and consequently the direction of $\Delta\mathbf{r}$ *approaches* the *tangent* to the curve. Hence, $\mathbf{v} = \lim_{\Delta t \to 0}(\Delta\mathbf{r}/\Delta t)$ or

$$\mathbf{v} = \frac{d\mathbf{r}}{dt} \qquad (12\text{–}7)$$

Since $d\mathbf{r}$ will be tangent to the curve, the *direction* of $\mathbf{v}$ is also *tangent to the curve*, Fig. 12–16c. The *magnitude* of $\mathbf{v}$, which is called the *speed*, is obtained by realizing that the length of the straight line segment $\Delta\mathbf{r}$ in Fig. 12–16b approaches the arc length Δs as $\Delta t \to 0$, we have $v = \lim_{\Delta t \to 0}(\Delta r/\Delta t) = \lim_{\Delta t \to 0}(\Delta s/\Delta t)$, or

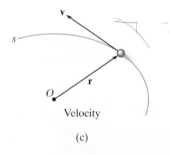

Velocity

(c)

Fig. 12–16

$$v = \frac{ds}{dt} \qquad (12\text{–}8)$$

Thus, the *speed* can be obtained by differentiating the path function s with respect to time.

*A summary of some of the important concepts of vector analysis is given in Appendix B.

Acceleration.

Acceleration.   If the particle has a velocity **v** at time t and a velocity $\mathbf{v}' = \mathbf{v} + \Delta\mathbf{v}$ at $t + \Delta t$, Fig. 12–16d, then the *average acceleration* of the particle during the time interval Δt is

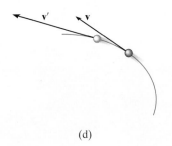

(d)

$$\mathbf{a}_{avg} = \frac{\Delta\mathbf{v}}{\Delta t}$$

where $\Delta\mathbf{v} = \mathbf{v}' - \mathbf{v}$. To study this time rate of change, the two velocity vectors in Fig. 12–16d are plotted in Fig. 12–16e such that their tails are located at the fixed point O' and their arrowheads touch points on a curve. This curve is called a *hodograph*, and when constructed, it describes the locus of points for the arrowhead of the velocity vector in the same manner as the *path s* describes the locus of points for the arrowhead of the position vector, Fig. 12–16a.

To obtain the *instantaneous acceleration*, let $\Delta t \rightarrow 0$ in the above equation. In the limit $\Delta\mathbf{v}$ will approach the *tangent to the hodograph*, and so $\mathbf{a} = \lim_{\Delta t \to 0}(\Delta\mathbf{v}/\Delta t)$, or

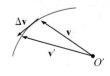

(e)

$$\mathbf{a} = \frac{d\mathbf{v}}{dt} \tag{12–9}$$

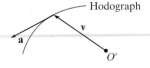

(f)

Substituting Eq. 12–7 into this result, we can also write

$$\mathbf{a} = \frac{d^2\mathbf{r}}{dt^2}$$

By definition of the derivative, **a** acts *tangent to the hodograph*, Fig. 12–16f, and, *in general it is not tangent to the path of motion*, Fig. 12–16g. To clarify this point, realize that $\Delta\mathbf{v}$ and consequently **a** must account for the change made in *both* the magnitude *and* direction of the velocity **v** as the particle moves from one point to the next along the path, Fig. 12–16d. However, in order for the particle to follow any curved path, the directional change always "swings" the velocity vector toward the "inside" or "concave side" of the path, and therefore **a** *cannot* remain tangent to the path. In summary, **v** is always tangent to the *path* and **a** is always tangent to the *hodograph*.

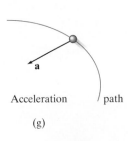

(g)

Fig. 12–16

12.5   Curvilinear Motion: Rectangular Components

Occasionally the motion of a particle can best be described along a path that can be expressed in terms of its x, y, z coordinates.

Position.   If the particle is at point (x, y, z) on the curved path s shown in Fig. 12–17a, then its location is defined by the *position vector*

$$\mathbf{r} = x\mathbf{i} + y\mathbf{j} + z\mathbf{k} \tag{12–10}$$

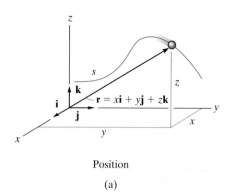

Position

(a)

When the particle moves, the x, y, z components of $\mathbf{r}$ will be functions of time; i.e., $x = x(t)$, $y = y(t)$, $z = z(t)$, so that $\mathbf{r} = \mathbf{r}(t)$.

At any instant the *magnitude* of $\mathbf{r}$ is defined from Eq. B–3 in Appendix B as

$$r = \sqrt{x^2 + y^2 + z^2}$$

And the *direction* of $\mathbf{r}$ is specified by the unit vector $\mathbf{u}_r = \mathbf{r}/r$.

Velocity.   The first time derivative of $\mathbf{r}$ yields the velocity of the particle. Hence,

$$\mathbf{v} = \frac{d\mathbf{r}}{dt} = \frac{d}{dt}(x\mathbf{i}) + \frac{d}{dt}(y\mathbf{j}) + \frac{d}{dt}(z\mathbf{k})$$

When taking this derivative, it is necessary to account for changes in *both* the magnitude and direction of each of the vector's components. For example, the derivative of the $\mathbf{i}$ component of $\mathbf{r}$ is

$$\frac{d}{dt}(x\mathbf{i}) = \frac{dx}{dt}\mathbf{i} + x\frac{d\mathbf{i}}{dt}$$

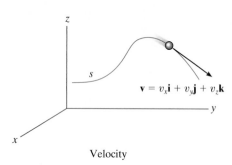

Velocity

(b)

Fig. 12–17

The second term on the right side is zero, provided the x, y, z reference frame is *fixed*, and therefore the *direction* (and the *magnitude*) of $\mathbf{i}$ does not change with time. Differentiation of the $\mathbf{j}$ and $\mathbf{k}$ components may be carried out in a similar manner, which yields the final result,

$$\mathbf{v} = \frac{d\mathbf{r}}{dt} = v_x\mathbf{i} + v_y\mathbf{j} + v_z\mathbf{k} \tag{12–11}$$

where

$$v_x = \dot{x} \quad v_y = \dot{y} \quad v_z = \dot{z} \tag{12–12}$$

The "dot" notation $\dot{x}, \dot{y}, \dot{z}$ represents the first time derivatives of $x = x(t)$, $y = y(t), z = z(t)$, respectively.

The velocity has a *magnitude* that is found from

$$v = \sqrt{v_x^2 + v_y^2 + v_z^2}$$

and a *direction* that is specified by the unit vector $\mathbf{u}_v = \mathbf{v}/v$. As discussed in Sec. 12–4, this direction is *always tangent to the path*, as shown in Fig. 12–17b.

Acceleration.
The acceleration of the particle is obtained by taking the first time derivative of Eq. 12–11 (or the second time derivative of Eq. 12–10). We have

$$\mathbf{a} = \frac{d\mathbf{v}}{dt} = a_x\mathbf{i} + a_y\mathbf{j} + a_z\mathbf{k} \qquad (12\text{–}13)$$

where

$$\begin{aligned} a_x &= \dot{v}_x = \ddot{x} \\ a_y &= \dot{v}_y = \ddot{y} \\ a_z &= \dot{v}_z = \ddot{z} \end{aligned} \qquad (12\text{–}14)$$

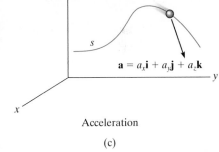

$\mathbf{a} = a_x\mathbf{i} + a_y\mathbf{j} + a_z\mathbf{k}$

Acceleration

(c)

Here a_x, a_y, a_z represent, respectively, the first time derivatives of $v_x = v_x(t), v_y = v_y(t), v_z = v_z(t)$, or the second time derivatives of the functions $x = x(t), y = y(t), z = z(t)$.

The acceleration has a *magnitude*

$$a = \sqrt{a_x^2 + a_y^2 + a_z^2}$$

and a *direction* specified by the unit vector $\mathbf{u}_a = \mathbf{a}/a$. Since $\mathbf{a}$ represents the time rate of *change* in both the magnitude and direction of the velocity, in general $\mathbf{a}$ will *not* be tangent to the path, Fig. 12–17c.

Important Points

- Curvilinear motion can cause changes in *both* the magnitude and direction of the position, velocity, and acceleration vectors.

- The velocity vector is always directed *tangent* to the path.

- In general, the acceleration vector is *not* tangent to the path, but rather, it is tangent to the hodograph.

- If the motion is described using rectangular coordinates, then the components along each of the axes do not change direction, only their magnitude and sense (algebraic sign) will change.

- By considering the component motions, the change in magnitude and direction of the particle's position and velocity are automatically taken into account.

Procedure for Analysis

Coordinate System.

- A rectangular coordinate system can be used to solve problems for which the motion can conveniently be expressed in terms of its *x, y, z* components.

Kinematic Quantities.

- Since *rectilinear motion* occurs along *each coordinate axis*, the motion along each axis is found using $v = ds/dt$ and $a = dv/dt$; or in cases where the motion is not expressed as a function of time, the equation $a\,ds = v\,dv$ can be used.

- In two dimensions, the equation of the path $y = f(x)$ can be used to relate the *x* and *y* components of velocity and acceleration by applying the chain rule of calculus. A review of this concept is given in Appendix C.

- Once the *x, y, z* components of **v** and **a** have been determined, the magnitudes of these vectors are found from the Pythagorean theorem, Eq. B-3, and their coordinate direction angles from the components of their unit vectors, Eqs. B-4 and B-5.

EXAMPLE | **12.9**

At any instant the horizontal position of the weather balloon in Fig. 12–18a is defined by $x = (8t)$ ft, where t is in seconds. If the equation of the path is $y = x^2/10$, determine the magnitude and direction of the velocity and the acceleration when $t = 2$ s.

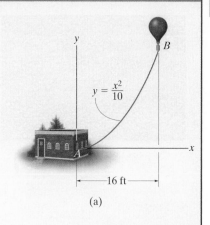

(a)

SOLUTION

Velocity. The velocity component in the x direction is

$$v_x = \dot{x} = \frac{d}{dt}(8t) = 8 \text{ ft/s} \rightarrow$$

To find the relationship between the velocity components we will use the chain rule of calculus. When $t = 2$ s, $x = 8(2) = 16$ ft, Fig. 12–18a, and so

$$v_y = \dot{y} = \frac{d}{dt}(x^2/10) = 2x\dot{x}/10 = 2(16)(8)/10 = 25.6 \text{ ft/s} \uparrow$$

When $t = 2$ s, the magnitude of velocity is therefore

$$v = \sqrt{(8 \text{ ft/s})^2 + (25.6 \text{ ft/s})^2} = 26.8 \text{ ft/s} \qquad Ans.$$

The direction is tangent to the path, Fig. 12–18b, where

$$\theta_v = \tan^{-1}\frac{v_y}{v_x} = \tan^{-1}\frac{25.6}{8} = 72.6° \qquad Ans.$$

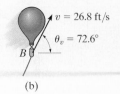

(b)

Acceleration. The relationship between the acceleration components is determined using the chain rule. (See Appendix C.) We have

$$a_x = \dot{v}_x = \frac{d}{dt}(8) = 0$$

$$a_y = \dot{v}_y = \frac{d}{dt}(2x\dot{x}/10) = 2(\dot{x})\dot{x}/10 + 2x(\ddot{x})/10$$

$$= 2(8)^2/10 + 2(16)(0)/10 = 12.8 \text{ ft/s}^2 \uparrow$$

Thus,

$$a = \sqrt{(0)^2 + (12.8)^2} = 12.8 \text{ ft/s}^2 \qquad Ans.$$

The direction of **a**, as shown in Fig. 12–18c, is

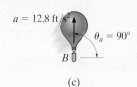

(c)

Fig. 12–18

$$\theta_a = \tan^{-1}\frac{12.8}{0} = 90° \qquad Ans.$$

NOTE: It is also possible to obtain v_y and a_y by first expressing $y = f(t) = (8t)^2/10 = 6.4t^2$ and then taking successive time derivatives.

EXAMPLE | 12.10

For a short time, the path of the plane in Fig. 12–19a is described by $y = (0.001x^2)$ m. If the plane is rising with a constant upward velocity of 10 m/s, determine the magnitudes of the velocity and acceleration of the plane when it reaches an altitude of $y = 100$ m.

SOLUTION
When $y = 100$ m, then $100 = 0.001x^2$ or $x = 316.2$ m. Also, due to constant velocity $v_y = 10$ m/s, so

$$y = v_y t; \qquad 100 \text{ m} = (10 \text{ m/s}) t \qquad t = 10 \text{ s}$$

Velocity. Using the chain rule (see Appendix C) to find the relationship between the velocity components, we have

$$y = 0.001x^2$$

$$v_y = \dot{y} = \frac{d}{dt}(0.001x^2) = (0.002x)\dot{x} = 0.002 x v_x \qquad (1)$$

Thus

$$10 \text{ m/s} = 0.002(316.2 \text{ m})(v_x)$$
$$v_x = 15.81 \text{ m/s}$$

The magnitude of the velocity is therefore

$$v = \sqrt{v_x^2 + v_y^2} = \sqrt{(15.81 \text{ m/s})^2 + (10 \text{ m/s})^2} = 18.7 \text{ m/s} \qquad \textit{Ans.}$$

Acceleration. Using the chain rule, the time derivative of Eq. (1) gives the relation between the acceleration components.

$$a_y = \dot{v}_y = (0.002\dot{x})\dot{x} + 0.002x(\ddot{x}) = 0.002(v_x^2 + x a_x)$$

When $x = 316.2$ m, $v_x = 15.81$ m/s, $\dot{v}_y = a_y = 0$,

$$0 = 0.002[(15.81 \text{ m/s})^2 + 316.2 \text{ m}(a_x)]$$
$$a_x = -0.791 \text{ m/s}^2$$

The magnitude of the plane's acceleration is therefore

$$a = \sqrt{a_x^2 + a_y^2} = \sqrt{(-0.791 \text{ m/s}^2)^2 + (0 \text{ m/s}^2)^2}$$
$$= 0.791 \text{ m/s}^2 \qquad \textit{Ans.}$$

These results are shown in Fig. 12–19b.

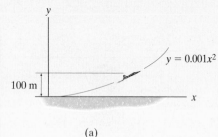

(a)

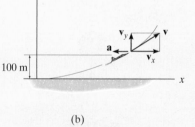

(b)

Fig. 12–19

12.6   Motion of a Projectile

The free-flight motion of a projectile is often studied in terms of its rectangular components. To illustrate the kinematic analysis, consider a projectile launched at point (x_0, y_0), with an initial velocity of $\mathbf{v}_0$, having components $(\mathbf{v}_0)_x$ and $(\mathbf{v}_0)_y$, Fig. 12–20. When air resistance is neglected, the only force acting on the projectile is its weight, which causes the projectile to have a *constant downward acceleration* of approximately $a_c = g = 9.81 \text{ m/s}^2$ or $g = 32.2 \text{ ft/s}^2$.*

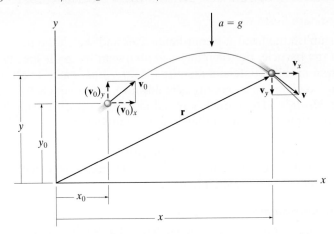

Fig. 12–20

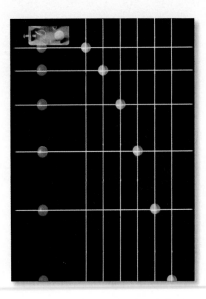

Each picture in this sequence is taken after the same time interval. The red ball falls from rest, whereas the yellow ball is given a horizontal velocity when released. Both balls accelerate downward at the same rate, and so they remain at the same elevation at any instant. This acceleration causes the difference in elevation between the balls to increase between successive photos. Also, note the horizontal distance between successive photos of the yellow ball is constant since the velocity in the horizontal direction remains constant.

Horizontal Motion.   Since $a_x = 0$, application of the constant acceleration equations, 12–4 to 12–6, yields

$(\xrightarrow{+})$ $v = v_0 + a_c t$ $v_x = (v_0)_x$

$(\xrightarrow{+})$ $x = x_0 + v_0 t + \frac{1}{2} a_c t^2;$ $x = x_0 + (v_0)_x t$

$(\xrightarrow{+})$ $v^2 = v_0^2 + 2a_c(x - x_0);$ $v_x = (v_0)_x$

The first and last equations indicate that *the horizontal component of velocity always remains constant during the motion.*

Vertical Motion.   Since the positive y axis is directed upward, then $a_y = -g$. Applying Eqs. 12–4 to 12–6, we get

$(+\uparrow)$ $v = v_0 + a_c t;$ $v_y = (v_0)_y - gt$

$(+\uparrow)$ $y = y_0 + v_0 t + \frac{1}{2} a_c t^2;$ $y = y_0 + (v_0)_y t - \frac{1}{2} g t^2$

$(+\uparrow)$ $v^2 = v_0^2 + 2a_c(y - y_0);$ $v_y^2 = (v_0)_y^2 - 2g(y - y_0)$

Recall that the last equation can be formulated on the basis of eliminating the time t from the first two equations, and therefore *only two of the above three equations are independent of one another.*

*This assumes that the earth's gravitational field does not vary with altitude.

12

To summarize, problems involving the motion of a projectile can have at most three unknowns since only three independent equations can be written; that is, *one* equation in the *horizontal direction* and *two* in the *vertical direction*. Once $\mathbf{v}_x$ and $\mathbf{v}_y$ are obtained, the resultant velocity $\mathbf{v}$, which is *always tangent* to the path, can be determined by the *vector sum* as shown in Fig. 12–20.

Procedure for Analysis

Coordinate System.

- Establish the fixed x, y coordinate axes and sketch the trajectory of the particle. Between any *two points* on the path specify the given problem data and identify the *three unknowns*. In all cases the acceleration of gravity acts downward and equals 9.81 m/s^2 or 32.2 ft/s^2. The particle's initial and final velocities should be represented in terms of their x and y components.

- Remember that positive and negative position, velocity, and acceleration components always act in accordance with their associated coordinate directions.

Kinematic Equations.

- Depending upon the known data and what is to be determined, a choice should be made as to which three of the following four equations should be applied between the two points on the path to obtain the most direct solution to the problem.

Horizontal Motion.

- The *velocity* in the horizontal or x direction is *constant*, i.e., $v_x = (v_0)_x$, and

$$x = x_0 + (v_0)_x t$$

Vertical Motion.

- In the vertical or y direction *only two* of the following three equations can be used for solution.

$$v_y = (v_0)_y + a_c t$$

$$y = y_0 + (v_0)_y t + \tfrac{1}{2} a_c t^2$$

$$v_y^2 = (v_0)_y^2 + 2a_c(y - y_0)$$

For example, if the particle's final velocity v_y is not needed, then the first and third of these equations will not be useful.

Gravel falling off the end of this conveyor belt follows a path that can be predicted using the equations of constant acceleration. In this way the location of the accumulated pile can be determined. Rectangular coordinates are used for the analysis since the acceleration is only in the vertical direction.

EXAMPLE | 12.11

A sack slides off the ramp, shown in Fig. 12–21, with a horizontal velocity of 12 m/s. If the height of the ramp is 6 m from the floor, determine the time needed for the sack to strike the floor and the range R where sacks begin to pile up.

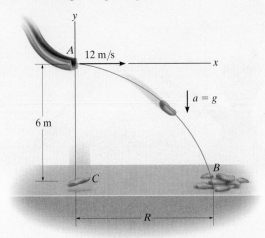

Fig. 12–21

SOLUTION

Coordinate System. The origin of coordinates is established at the beginning of the path, point A, Fig. 12–21. The initial velocity of a sack has components $(v_A)_x = 12$ m/s and $(v_A)_y = 0$. Also, between points A and B the acceleration is $a_y = -9.81$ m/s^2. Since $(v_B)_x = (v_A)_x = 12$ m/s, the three unknowns are $(v_B)_y$, R, and the time of flight t_{AB}. Here we do not need to determine $(v_B)_y$.

Vertical Motion. The vertical distance from A to B is known, and therefore we can obtain a direct solution for t_{AB} by using the equation

$(+\uparrow)$ $$y_B = y_A + (v_A)_y t_{AB} + \tfrac{1}{2} a_c t_{AB}^2$$

$$-6 \text{ m} = 0 + 0 + \tfrac{1}{2}(-9.81 \text{ m/s}^2) t_{AB}^2$$

$$t_{AB} = 1.11 \text{ s} \qquad\qquad Ans.$$

Horizontal Motion. Since t_{AB} has been calculated, R is determined as follows:

$(\xrightarrow{+})$ $$x_B = x_A + (v_A)_x t_{AB}$$

$$R = 0 + 12 \text{ m/s}(1.11 \text{ s})$$

$$R = 13.3 \text{ m} \qquad\qquad Ans.$$

NOTE: The calculation for t_{AB} also indicates that if a sack were released *from rest* at A, it would take the same amount of time to strike the floor at C, Fig. 12–21.

EXAMPLE 12.12

The chipping machine is designed to eject wood chips at $v_O = 25$ ft/s as shown in Fig. 12–22. If the tube is oriented at 30° from the horizontal, determine how high, h, the chips strike the pile if at this instant they land on the pile 20 ft from the tube.

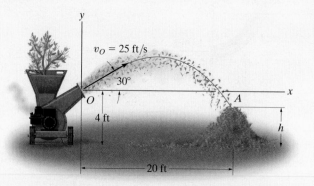

Fig. 12–22

SOLUTION

Coordinate System. When the motion is analyzed between points O and A, the three unknowns are the height h, time of flight t_{OA}, and vertical component of velocity $(v_A)_y$. [Note that $(v_A)_x = (v_O)_x$.] With the origin of coordinates at O, Fig. 12–22, the initial velocity of a chip has components of

$$(v_O)_x = (25 \cos 30°) \text{ ft/s} = 21.65 \text{ ft/s} \rightarrow$$

$$(v_O)_y = (25 \sin 30°) \text{ ft/s} = 12.5 \text{ ft/s} \uparrow$$

Also, $(v_A)_x = (v_O)_x = 21.65$ ft/s and $a_y = -32.2$ ft/s². Since we do not need to determine $(v_A)_y$, we have

Horizontal Motion.

$(\xrightarrow{+})$
$$x_A = x_O + (v_O)_x t_{OA}$$
$$20 \text{ ft} = 0 + (21.65 \text{ ft/s}) t_{OA}$$
$$t_{OA} = 0.9238 \text{ s}$$

Vertical Motion. Relating t_{OA} to the initial and final elevations of a chip, we have

$(+\uparrow)$
$$y_A = y_O + (v_O)_y t_{OA} + \tfrac{1}{2} a_c t_{OA}^2$$
$$(h - 4 \text{ ft}) = 0 + (12.5 \text{ ft/s})(0.9238 \text{ s}) + \tfrac{1}{2}(-32.2 \text{ ft/s}^2)(0.9238 \text{ s})^2$$
$$h = 1.81 \text{ ft}$$ *Ans.*

NOTE: We can determine $(v_A)_y$ by using $(v_A)_y = (v_O)_y + a_c t_{OA}$.

12

EXAMPLE 12.13

The track for this racing event was designed so that riders jump off the slope at 30°, from a height of 1 m. During a race it was observed that the rider shown in Fig. 12–23a remained in mid air for 1.5 s. Determine the speed at which he was traveling off the ramp, the horizontal distance he travels before striking the ground, and the maximum height he attains. Neglect the size of the bike and rider.

(a)

SOLUTION

Coordinate System. As shown in Fig. 12–23b, the origin of the coordinates is established at A. Between the end points of the path AB the three unknowns are the initial speed v_A, range R, and the vertical component of velocity $(v_B)_y$.

Vertical Motion. Since the time of flight and the vertical distance between the ends of the path are known, we can determine v_A.

$$(+\uparrow) \qquad y_B = y_A + (v_A)_y t_{AB} + \tfrac{1}{2} a_c t_{AB}^2$$
$$-1\ \text{m} = 0 + v_A \sin 30°(1.5\ \text{s}) + \tfrac{1}{2}(-9.81\ \text{m/s}^2)(1.5\ \text{s})^2$$
$$v_A = 13.38\ \text{m/s} = 13.4\ \text{m/s} \qquad\qquad Ans.$$

Horizontal Motion. The range R can now be determined.

$$(\xrightarrow{+}) \qquad x_B = x_A + (v_A)_x t_{AB}$$
$$R = 0 + 13.38 \cos 30°\ \text{m/s}(1.5\ \text{s})$$
$$= 17.4\ \text{m} \qquad\qquad Ans.$$

In order to find the maximum height h we will consider the path AC, Fig. 12–23b. Here the three unknowns are the time of flight t_{AC}, the horizontal distance from A to C, and the height h. At the maximum height $(v_C)_y = 0$, and since v_A is known, we can determine h *directly* without considering t_{AC} using the following equation.

$$(v_C)_y^2 = (v_A)_y^2 + 2a_c[y_C - y_A]$$
$$0^2 = (13.38 \sin 30°\ \text{m/s})^2 + 2(-9.81\ \text{m/s}^2)[(h - 1\ \text{m}) - 0]$$
$$h = 3.28\ \text{m} \qquad\qquad Ans.$$

NOTE: Show that the bike will strike the ground at B with a velocity having components of

$$(v_B)_x = 11.6\ \text{m/s} \rightarrow, \qquad (v_B)_y = 8.02\ \text{m/s}\downarrow$$

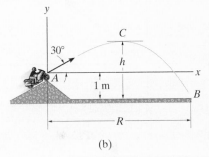

(b)

Fig. 12–23

FUNDAMENTAL PROBLEMS

F12–15. If the x and y components of a particle's velocity are $v_x = (32t)$ m/s and $v_y = 8$ m/s, determine the equation of the path $y = f(x)$. $x = 0$ and $y = 0$ when $t = 0$.

F12–16. A particle is traveling along the straight path. If its position along the x axis is $x = (8t)$ m, where t is in seconds, determine its speed when $t = 2$ s.

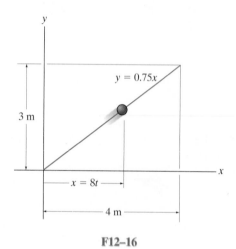

$y = 0.75x$

3 m

$x = 8t$

4 m

F12–16

F12–17. A particle is constrained to travel along the path. If $x = (4t^4)$ m, where t is in seconds, determine the magnitude of the particle's velocity and acceleration when $t = 0.5$ s.

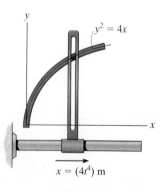

$y^2 = 4x$

$x = (4t^4)$ m

F12–17

F12–18. A particle travels along a straight-line path $y = 0.5x$. If the x component of the particle's velocity is $v_x = (2t^2)$ m/s, where t is in seconds, determine the magnitude of the particle's velocity and acceleration when $t = 4$ s.

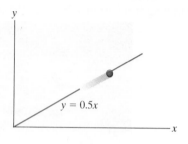

$y = 0.5x$

F12–18

F12–19. A particle is traveling along the parabolic path $y = 0.25x^2$. If $x = (2t^2)$ m, where t is in seconds, determine the magnitude of the particle's velocity and acceleration when $t = 2$ s.

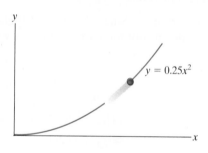

$y = 0.25x^2$

F12–19

F12–20. The box slides down the slope described by the equation $y = (0.05x^2)$ m, where x is in meters. If the box has x components of velocity and acceleration of $v_x = -3$ m/s and $a_x = -1.5$ m/s^2 at $x = 5$ m, determine the y components of the velocity and the acceleration of the box at this instant.

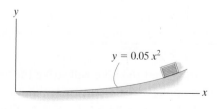

$y = 0.05 x^2$

F12–20

F12–21. The ball is kicked from point A with the initial velocity $v_A = 10$ m/s. Determine the maximum height h it reaches.

F12–22. The ball is kicked from point A with the initial velocity $v_A = 10$ m/s. Determine the range R, and the speed when the ball strikes the ground.

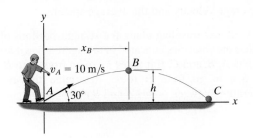

F12–21/22

F12–23. Determine the speed at which the basketball at A must be thrown at the angle of 30° so that it makes it to the basket at B.

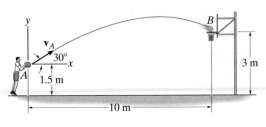

F12–23

F12–24. Water is sprayed at an angle of 90° from the slope at 20 m/s. Determine the range R.

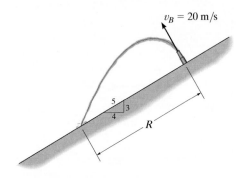

F12–24

F12–25. A ball is thrown from A. If it is required to clear the wall at B, determine the minimum magnitude of its initial velocity $\mathbf{v}_A$.

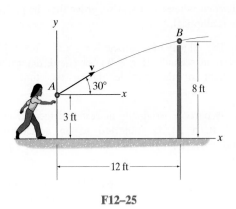

F12–25

F12–26. A projectile is fired with an initial velocity of $v_A = 150$ m/s off the roof of the building. Determine the range R where it strikes the ground at B.

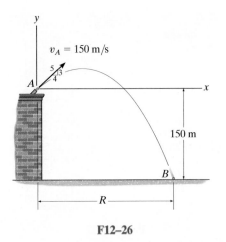

F12–26

PROBLEMS

12–73. The position of a particle is defined by $r = \{5(\cos 2t)\mathbf{i} + 4(\sin 2t)\mathbf{j}\}$ m, where t is in seconds and the arguments for the sine and cosine are given in radians. Determine the magnitudes of the velocity and acceleration of the particle when $t = 1$ s. Also, prove that the path of the particle is elliptical.

12–74. The velocity of a particle is $\mathbf{v} = \{3\mathbf{i} + (6 - 2t)\mathbf{j}\}$ m/s, where t is in seconds. If $\mathbf{r} = \mathbf{0}$ when $t = 0$, determine the displacement of the particle during the time interval $t = 1$ s to $t = 3$ s.

12–75. A particle, originally at rest and located at point (3 ft, 2 ft, 5 ft), is subjected to an acceleration of $\mathbf{a} = \{6t\mathbf{i} + 12t^2\mathbf{k}\}$ ft/s². Determine the particle's position (x, y, z) at $t = 1$ s.

*12–76. The velocity of a particle is given by $v = \{16t^2\mathbf{i} + 4t^3\mathbf{j} + (5t + 2)\mathbf{k}\}$ m/s, where t is in seconds. If the particle is at the origin when $t = 0$, determine the magnitude of the particle's acceleration when $t = 2$ s. Also, what is the x, y, z coordinate position of the particle at this instant?

12–77. The car travels from A to B, and then from B to C, as shown in the figure. Determine the magnitude of the displacement of the car and the distance traveled.

12–78. A car travels east 2 km for 5 minutes, then north 3 km for 8 minutes, and then west 4 km for 10 minutes. Determine the total distance traveled and the magnitude of displacement of the car. Also, what is the magnitude of the average velocity and the average speed?

12–79. A car traveling along the straight portions of the road has the velocities indicated in the figure when it arrives at points A, B, and C. If it takes 3 s to go from A to B, and then 5 s to go from B to C, determine the average acceleration between points A and B and between points A and C.

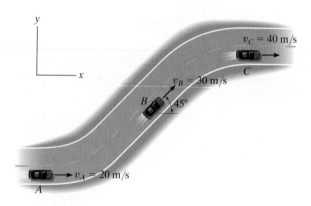

Prob. 12–79

*12–80. A particle travels along the curve from A to B in 2 s. It takes 4 s for it to go from B to C and then 3 s to go from C to D. Determine its average speed when it goes from A to D.

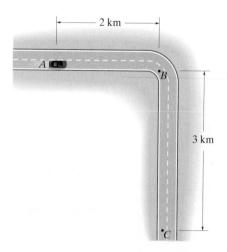

Prob. 12–77

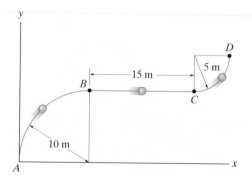

Prob. 12–80

12–81. The position of a crate sliding down a ramp is given by $x = (0.25t^3)$ m, $y = (1.5t^2)$ m, $z = (6 - 0.75t^{5/2})$ m, where t is in seconds. Determine the magnitude of the crate's velocity and acceleration when $t = 2$ s.

12–82. A rocket is fired from rest at $x = 0$ and travels along a parabolic trajectory described by $y^2 = [120(10^3)x]$ m. If the x component of acceleration is $a_x = (\frac{1}{4}t^2)$ m/s^2, where t is in seconds, determine the magnitude of the rocket's velocity and acceleration when $t = 10$ s.

12–83. The particle travels along the path defined by the parabola $y = 0.5x^2$. If the component of velocity along the x axis is $v_x = (5t)$ ft/s, where t is in seconds, determine the particle's distance from the origin O and the magnitude of its acceleration when $t = 1$ s. When $t = 0$, $x = 0$, $y = 0$.

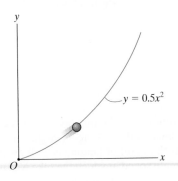

$y = 0.5x^2$

Prob. 12–83

*****12–84.** The motorcycle travels with constant speed v_0 along the path that, for a short distance, takes the form of a sine curve. Determine the x and y components of its velocity at any instant on the curve.

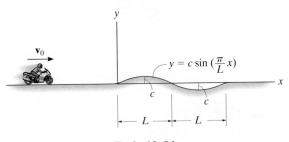

$y = c \sin \left(\frac{\pi}{L} x\right)$

Prob. 12–84

12–85. A particle travels along the curve from A to B in 1 s. If it takes 3 s for it to go from A to C, determine its *average velocity* when it goes from B to C.

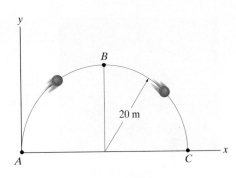

B

20 m

A C

Prob. 12–85

12–86. When a rocket reaches an altitude of 40 m it begins to travel along the parabolic path $(y - 40)^2 = 160x$, where the coordinates are measured in meters. If the component of velocity in the vertical direction is constant at $v_y = 180$ m/s, determine the magnitudes of the rocket's velocity and acceleration when it reaches an altitude of 80 m.

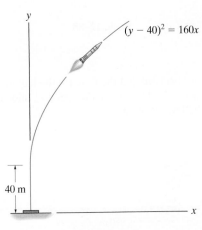

$(y - 40)^2 = 160x$

40 m

Prob. 12–86

12–87.   Pegs A and B are restricted to move in the elliptical slots due to the motion of the slotted link. If the link moves with a constant speed of 10 m/s, determine the magnitude of the velocity and acceleration of peg A when $x = 1$ m.

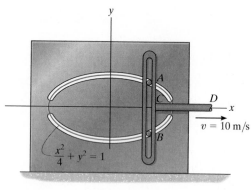

Prob. 12–87

12–90.   Determine the minimum initial velocity v_0 and the corresponding angle θ_0 at which the ball must be kicked in order for it to just cross over the 3-m high fence.

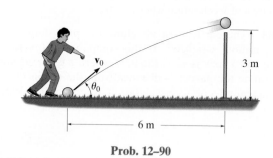

Prob. 12–90

*12–88.   The van travels over the hill described by $y = (-1.5(10^{-3})\, x^2 + 15)$ ft. If it has a constant speed of 75 ft/s, determine the x and y components of the van's velocity and acceleration when $x = 50$ ft.

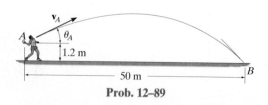

Prob. 12–88

12–91.   During a race the dirt bike was observed to leap up off the small hill at A at an angle of $60°$ with the horizontal. If the point of landing is 20 ft away, determine the approximate speed at which the bike was traveling just before it left the ground. Neglect the size of the bike for the calculation.

12–89.   It is observed that the time for the ball to strike the ground at B is 2.5 s. Determine the speed v_A and angle θ_A at which the ball was thrown.

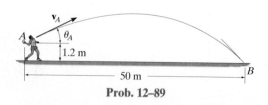

Prob. 12–89

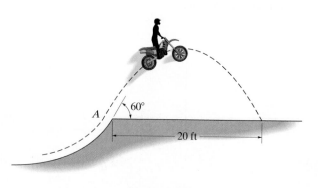

Prob. 12–91

*12–92.   The girl always throws the toys at an angle of 30° from point A as shown. Determine the time between throws so that both toys strike the edges of the pool B and C at the same instant. With what speed must she throw each toy?

12–95.   A projectile is given a velocity $\mathbf{v}_0$ at an angle ϕ above the horizontal. Determine the distance d to where it strikes the sloped ground. The acceleration due to gravity is g.

*12–96.   A projectile is given a velocity $\mathbf{v}_0$. Determine the angle ϕ at which it should be launched so that d is a maximum. The acceleration due to gravity is g.

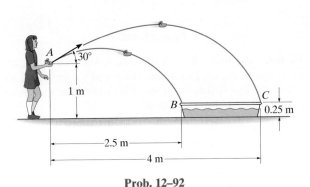

Prob. 12–92

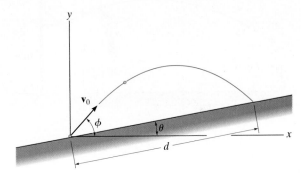

Probs. 12–95/96

12–97.   Determine the minimum height on the wall to which the firefighter can project water from the hose, so that the water strikes the wall horizontally. The speed of the water at the nozzle is $v_C = 48$ ft/s.

12–93.   The player kicks a football with an initial speed of $v_0 = 90$ ft/s. Determine the time the ball is in the air and the angle θ of the kick.

■12–98.   Determine the smallest angle θ, measured above the horizontal, that the hose should be directed so that the water stream strikes the bottom of the wall at B. The speed of the water at the nozzle is $v_C = 48$ ft/s.

12–94.   From a videotape, it was observed that a player kicked a football 126 ft during a measured time of 3.6 seconds. Determine the initial speed of the ball and the angle θ at which it was kicked.

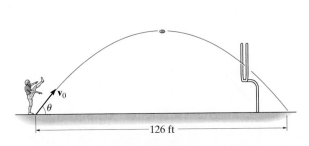

Probs. 12–93/94

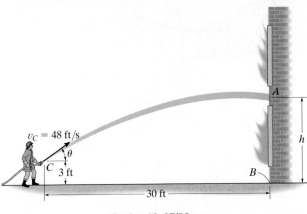

Probs. 12–97/98

12–99. Measurements of a shot recorded on a videotape during a basketball game are shown. The ball passed through the hoop even though it barely cleared the hands of the player B who attempted to block it. Neglecting the size of the ball, determine the magnitude v_A of its initial velocity and the height h of the ball when it passes over player B.

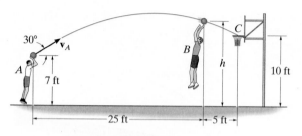

Prob. 12–99

*12–100. It is observed that the skier leaves the ramp A at an angle $\theta_A = 25°$ with the horizontal. If he strikes the ground at B, determine his initial speed v_A and the time of flight t_{AB}.

12–101. It is observed that the skier leaves the ramp A at an angle $\theta_A = 25°$ with the horizontal. If he strikes the ground at B, determine his initial speed v_A and the speed at which he strikes the ground.

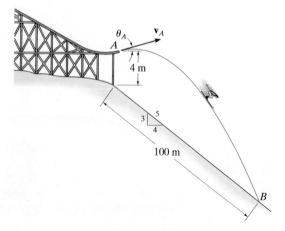

Probs. 12–100/101

12–102. A golf ball is struck with a velocity of 80 ft/s as shown. Determine the distance d to where it will land.

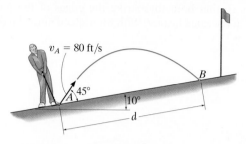

Prob. 12–102

12–103. The ball is thrown from the tower with a velocity of 20 ft/s as shown. Determine the x and y coordinates to where the ball strikes the slope. Also, determine the speed at which the ball hits the ground.

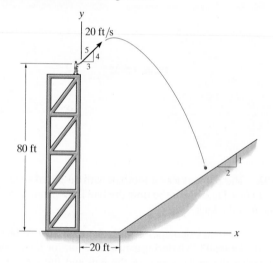

Prob. 12–103

*12–104. The projectile is launched with a velocity $\mathbf{v}_0$. Determine the range R, the maximum height h attained, and the time of flight. Express the results in terms of the angle θ and v_0. The acceleration due to gravity is g.

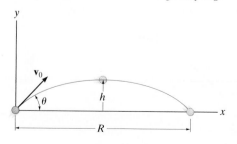

Prob. 12–104

12–105. Determine the horizontal velocity v_A of a tennis ball at A so that it just clears the net at B. Also, find the distance s where the ball strikes the ground.

*12–108.** The man at A wishes to throw two darts at the target at B so that they arrive at the *same time*. If each dart is thrown with a speed of 10 m/s, determine the angles θ_C and θ_D at which they should be thrown and the time between each throw. Note that the first dart must be thrown at θ_C $(>\theta_D)$, then the second dart is thrown at θ_D.

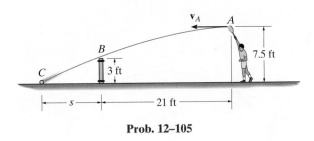

Prob. 12–105

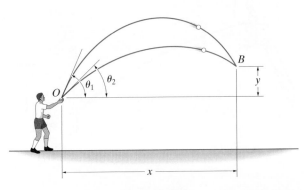

Prob. 12–108

12–106. The ball at A is kicked with a speed $v_A = 8$ ft/s and at an angle $\theta_A = 30°$. Determine the point (x, y) where it strikes the ground. Assume the ground has the shape of a parabola as shown.

12–107. The ball at A is kicked such that $\theta_A = 30°$. If it strikes the ground at B having coordinates $x = 15$ ft, $y = -9$ ft, determine the speed at which it is kicked and the speed at which it strikes the ground.

12–109. A boy throws a ball at O in the air with a speed v_0 at an angle θ_1. If he then throws another ball with the same speed v_0 at an angle $\theta_2 < \theta_1$, determine the time between the throws so that the balls collide in midair at B.

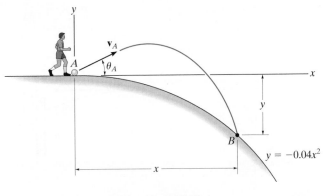

Probs. 12–106/107

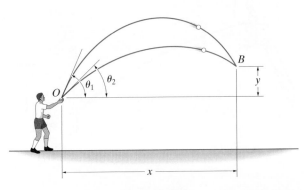

Prob. 12–109

12–110. Small packages traveling on the conveyor belt fall off into a l-m-long loading car. If the conveyor is running at a constant speed of $v_C = 2$ m/s, determine the smallest and largest distance R at which the end A of the car may be placed from the conveyor so that the packages enter the car.

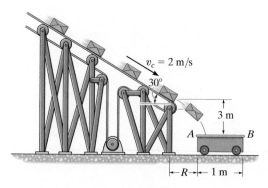

Prob. 12–110

12–111. The fireman wishes to direct the flow of water from his hose to the fire at B. Determine two possible angles θ_1 and θ_2 at which this can be done. Water flows from the hose at $v_A = 80$ ft/s.

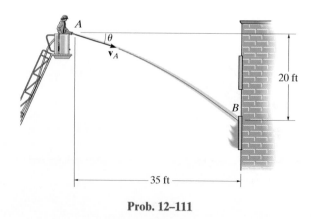

Prob. 12–111

*12–112. The baseball player A hits the baseball at $v_A = 40$ ft/s and $\theta_A = 60°$ from the horizontal. When the ball is directly overhead of player B he begins to run under it. Determine the constant speed at which B must run and the distance d in order to make the catch at the same elevation at which the ball was hit.

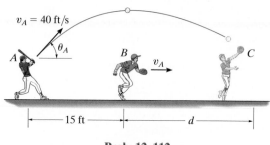

Prob. 12–112

12–113. The man stands 60 ft from the wall and throws a ball at it with a speed $v_0 = 50$ ft/s. Determine the angle θ at which he should release the ball so that it strikes the wall at the highest point possible. What is this height? The room has a ceiling height of 20 ft.

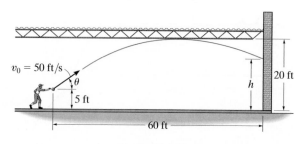

Prob. 12–113

12.7 Curvilinear Motion: Normal and Tangential Components

When the path along which a particle travels is *known*, then it is often convenient to describe the motion using n and t coordinate axes which act normal and tangent to the path, respectively, and at the instant considered have their *origin located at the particle*.

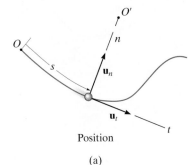

Position

(a)

Planar Motion. Consider the particle shown in Fig. 12–24*a*, which moves in a plane along a fixed curve, such that at a given instant it is at position s, measured from point O. We will now consider a coordinate system that has its origin on the curve, and at the instant considered this origin happens to *coincide* with the location of the particle. The t axis is *tangent* to the curve at the point and is positive in the direction of *increasing s*. We will designate this positive direction with the unit vector $\mathbf{u}_t$. A unique choice for the *normal axis* can be made by noting that geometrically the curve is constructed from a series of differential arc segments ds, Fig. 12–24*b*. Each segment ds is formed from the arc of an associated circle having a *radius of curvature* ρ (rho) and *center of curvature* O'. The normal axis n is perpendicular to the t axis with its positive sense directed *toward* the center of curvature O', Fig. 12–24*a*. This positive direction, which is *always* on the concave side of the curve, will be designated by the unit vector $\mathbf{u}_n$. The plane which contains the n and t axes is referred to as the embracing or *osculating plane*, and in this case it is fixed in the plane of motion.*

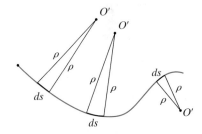

Radius of curvature

(b)

Velocity. Since the particle moves, s is a function of time. As indicated in Sec. 12.4, the particle's velocity $\mathbf{v}$ has a *direction* that is *always tangent to the path*, Fig. 12–24*c*, and a *magnitude* that is determined by taking the time derivative of the path function $s = s(t)$, i.e., $v = ds/dt$ (Eq. 12–8). Hence

$$\boxed{\mathbf{v} = v\mathbf{u}_t} \qquad (12\text{–}15)$$

where

$$\boxed{v = \dot{s}} \qquad (12\text{–}16)$$

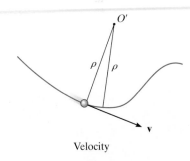

Velocity

(c)

Fig. 12–24

---

*The osculating plane may also be defined as the plane which has the greatest contact with the curve at a point. It is the limiting position of a plane contacting both the point and the arc segment ds. As noted above, the osculating plane is always coincident with a plane curve; however, each point on a three-dimensional curve has a unique osculating plane.

12

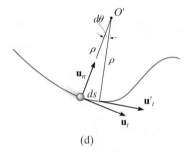

(d)

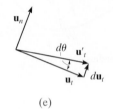

(e)

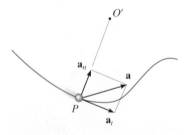

Acceleration

(f)

Fig. 12–24 (cont.)

Acceleration.   The acceleration of the particle is the time rate of change of the velocity. Thus,

$$\mathbf{a} = \dot{\mathbf{v}} = \dot{v}\mathbf{u}_t + v\dot{\mathbf{u}}_t \tag{12–17}$$

In order to determine the time derivative $\dot{\mathbf{u}}_t$, note that as the particle moves along the arc ds in time dt, $\mathbf{u}_t$ preserves its magnitude of unity; however, its *direction* changes, and becomes $\mathbf{u}'_t$, Fig. 12–24d. As shown in Fig. 12–24e, we require $\mathbf{u}'_t = \mathbf{u}_t + d\mathbf{u}_t$. Here $d\mathbf{u}_t$ stretches between the arrowheads of $\mathbf{u}_t$ and $\mathbf{u}'_t$, which lie on an infinitesimal arc of radius $u_t = 1$. Hence, $d\mathbf{u}_t$ has a *magnitude* of $du_t = (1)\,d\theta$, and its *direction* is defined by $\mathbf{u}_n$. Consequently, $d\mathbf{u}_t = d\theta\mathbf{u}_n$, and therefore the time derivative becomes $\dot{\mathbf{u}}_t = \dot{\theta}\mathbf{u}_n$. Since $ds = \rho\,d\theta$, Fig. 12–24d, then $\dot{\theta} = \dot{s}/\rho$, and therefore

$$\dot{\mathbf{u}}_t = \dot{\theta}\mathbf{u}_n = \frac{\dot{s}}{\rho}\mathbf{u}_n = \frac{v}{\rho}\mathbf{u}_n$$

Substituting into Eq. 12–17, $\mathbf{a}$ can be written as the sum of its two components,

$$\boxed{\mathbf{a} = a_t\mathbf{u}_t + a_n\mathbf{u}_n} \tag{12–18}$$

where

$$\boxed{a_t = \dot{v}} \quad \text{or} \quad \boxed{a_t\,ds = v\,dv} \tag{12–19}$$

and

$$\boxed{a_n = \frac{v^2}{\rho}} \tag{12–20}$$

These two mutually perpendicular components are shown in Fig. 12–24f. Therefore, the *magnitude* of acceleration is the positive value of

$$a = \sqrt{a_t^2 + a_n^2} \tag{12–21}$$

To better understand these results, consider the following two special cases of motion.

1. If the particle moves along a straight line, then $\rho \to \infty$ and from Eq. 12–20, $a_n = 0$. Thus $a = a_t = \dot{v}$, and we can conclude that the *tangential component of acceleration represents the time rate of change in the magnitude of the velocity.*

2. If the particle moves along a curve with a constant speed, then $a_t = \dot{v} = 0$ and $a = a_n = v^2/\rho$. Therefore, the *normal component of acceleration represents the time rate of change in the direction of the velocity.* Since $\mathbf{a}_n$ *always* acts towards the center of curvature, this component is sometimes referred to as the *centripetal* (or center seeking) *acceleration.*

As a result of these interpretations, a particle moving along the curved path in Fig. 12–25 will have accelerations directed as shown.

As the boy swings upward with a velocity $\mathbf{v}$, his motion can be analyzed using n–t coordinates. As he rises, the magnitude of his velocity (speed) is decreasing, and so a_t will be negative. The rate at which the direction of his velocity changes is a_n, which is always positive, that is, towards the center of rotation.

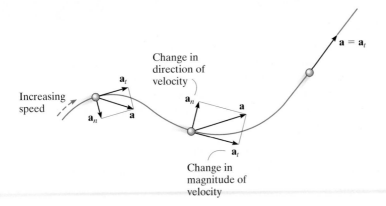

Fig. 12–25

Three-Dimensional Motion.

If the particle moves along a space curve, Fig. 12–26, then at a given instant the t axis is uniquely specified; however, an infinite number of straight lines can be constructed normal to the tangent axis. As in the case of planar motion, we will choose the positive n axis directed toward the path's center of curvature O'. This axis is referred to as the *principal normal* to the curve. With the n and t axes so defined, Eqs. 12–15 through 12–21 can be used to determine $\mathbf{v}$ and $\mathbf{a}$. Since $\mathbf{u}_t$ and $\mathbf{u}_n$ are always perpendicular to one another and lie in the osculating plane, for spatial motion a third unit vector, $\mathbf{u}_b$, defines the *binormal axis* b which is perpendicular to $\mathbf{u}_t$ and $\mathbf{u}_n$, Fig. 12–26.

Since the three unit vectors are related to one another by the vector cross product, e.g., $\mathbf{u}_b = \mathbf{u}_t \times \mathbf{u}_n$, Fig. 12–26, it may be possible to use this relation to establish the direction of one of the axes, if the directions of the other two are known. For example, no motion occurs in the $\mathbf{u}_b$ direction, and if this direction and $\mathbf{u}_t$ are known, then $\mathbf{u}_n$ can be determined, where in this case $\mathbf{u}_n = \mathbf{u}_b \times \mathbf{u}_t$, Fig. 12–26. Remember, though, that $\mathbf{u}_n$ is always on the concave side of the curve.

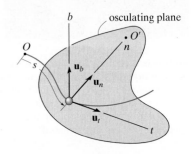

Fig. 12–26

Procedure for Analysis

Coordinate System.

- Provided the *path* of the particle is *known*, we can establish a set of n and t coordinates having a *fixed origin*, which is coincident with the particle at the instant considered.
- The positive tangent axis acts in the direction of motion and the positive normal axis is directed toward the path's center of curvature.

Velocity.

- The particle's *velocity* is always tangent to the path.
- The magnitude of velocity is found from the time derivative of the path function.

$$v = \dot{s}$$

Tangential Acceleration.

- The tangential component of acceleration is the result of the time rate of change in the *magnitude* of velocity. This component acts in the positive s direction if the particle's speed is increasing or in the opposite direction if the speed is decreasing.
- The relations between a_t, v, t and s are the same as for rectilinear motion, namely,

$$a_t = \dot{v} \qquad a_t\, ds = v\, dv$$

- If a_t is constant, $a_t = (a_t)_c$, the above equations, when integrated, yield

$$s = s_0 + v_0 t + \tfrac{1}{2}(a_t)_c t^2$$
$$v = v_0 + (a_t)_c t$$
$$v^2 = v_0^2 + 2(a_t)_c(s - s_0)$$

Normal Acceleration.

- The normal component of acceleration is the result of the time rate of change in the *direction* of the velocity. This component is *always* directed toward the center of curvature of the path, i.e., along the positive n axis.
- The magnitude of this component is determined from

$$a_n = \frac{v^2}{\rho}$$

- If the path is expressed as $y = f(x)$, the radius of curvature ρ at any point on the path is determined from the equation

$$\rho = \frac{[1 + (dy/dx)^2]^{3/2}}{|d^2y/dx^2|}$$

The derivation of this result is given in any standard calculus text.

Motorists traveling along this cloverleaf interchange experience a normal acceleration due to the change in direction of their velocity. A tangential component of acceleration occurs when the cars' speed is increased or decreased.

12

EXAMPLE 12.14

When the skier reaches point A along the parabolic path in Fig. 12–27a, he has a speed of 6 m/s which is increasing at 2 m/s². Determine the direction of his velocity and the direction and magnitude of his acceleration at this instant. Neglect the size of the skier in the calculation.

SOLUTION

Coordinate System. Although the path has been expressed in terms of its x and y coordinates, we can still establish the origin of the n, t axes at the fixed point A on the path and determine the components of $\mathbf{v}$ and $\mathbf{a}$ along these axes, Fig. 12–27a.

Velocity. By definition, the velocity is always directed tangent to the path. Since $y = \frac{1}{20}x^2$, $dy/dx = \frac{1}{10}x$, then at $x = 10$ m, $dy/dx = 1$. Hence, at A, $\mathbf{v}$ makes an angle of $\theta = \tan^{-1}1 = 45°$ with the x axis, Fig. 12–27b. Therefore,

$$v_A = 6 \text{ m/s} \quad 45° \, \nearrow \qquad\qquad \textit{Ans.}$$

The acceleration is determined from $\mathbf{a} = \dot{v}\mathbf{u}_t + (v^2/\rho)\mathbf{u}_n$. However, it is first necessary to determine the radius of curvature of the path at A (10 m, 5 m). Since $d^2y/dx^2 = \frac{1}{10}$, then

$$\rho = \frac{[1 + (dy/dx)^2]^{3/2}}{|d^2y/dx^2|} = \frac{[1 + (\frac{1}{10}x)^2]^{3/2}}{|\frac{1}{10}|}\bigg|_{x=10 \text{ m}} = 28.28 \text{ m}$$

The acceleration becomes

$$\mathbf{a}_A = \dot{v}\mathbf{u}_t + \frac{v^2}{\rho}\mathbf{u}_n$$

$$= 2\mathbf{u}_t + \frac{(6 \text{ m/s})^2}{28.28 \text{ m}}\mathbf{u}_n$$

$$= \{2\mathbf{u}_t + 1.273\mathbf{u}_n\} \text{ m/s}^2$$

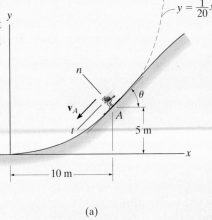

(a)

As shown in Fig. 12–27b,

$$a = \sqrt{(2 \text{ m/s}^2)^2 + (1.273 \text{ m/s}^2)^2} = 2.37 \text{ m/s}^2$$

$$\phi = \tan^{-1}\frac{2}{1.273} = 57.5°$$

Thus, $45° + 90° + 57.5° - 180° = 12.5°$ so that,

$$a = 2.37 \text{ m/s}^2 \quad 12.5° \, \nearrow \qquad\qquad \textit{Ans.}$$

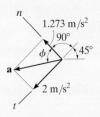

(b)

Fig. 12–27

NOTE: By using n, t coordinates, we were able to readily solve this problem through the use of Eq. 12–18, since it accounts for the separate changes in the magnitude and direction of $\mathbf{v}$.

EXAMPLE   12.15

A race car C travels around the horizontal circular track that has a radius of 300 ft, Fig. 12–28. If the car increases its speed at a constant rate of 7 ft/s^2, starting from rest, determine the time needed for it to reach an acceleration of 8 ft/s^2. What is its speed at this instant?

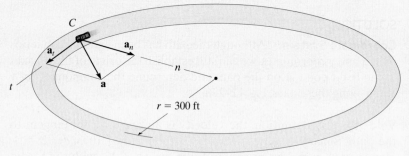

Fig. 12–28

SOLUTION

Coordinate System.   The origin of the n and t axes is coincident with the car at the instant considered. The t axis is in the direction of motion, and the positive n axis is directed toward the center of the circle. This coordinate system is selected since the path is known.

Acceleration.   The magnitude of acceleration can be related to its components using $a = \sqrt{a_t^2 + a_n^2}$. Here $a_t = 7$ ft/s^2. Since $a_n = v^2/\rho$, the velocity as a function of time must be determined first.

$$v = v_0 + (a_t)_c t$$
$$v = 0 + 7t$$

Thus

$$a_n = \frac{v^2}{\rho} = \frac{(7t)^2}{300} = 0.163t^2 \text{ ft/s}^2$$

The time needed for the acceleration to reach 8 ft/s^2 is therefore

$$a = \sqrt{a_t^2 + a_n^2}$$
$$8 \text{ ft/s}^2 = \sqrt{(7 \text{ ft/s}^2)^2 + (0.163t^2)^2}$$

Solving for the positive value of t yields

$$0.163t^2 = \sqrt{(8 \text{ ft/s}^2)^2 - (7 \text{ ft/s}^2)^2}$$
$$t = 4.87 \text{ s} \qquad\qquad Ans.$$

Velocity.   The speed at time $t = 4.87$ s is

$$v = 7t = 7(4.87) = 34.1 \text{ ft/s} \qquad\qquad Ans.$$

NOTE: Remember the velocity will always be tangent to the path, whereas the acceleration will be directed within the curvature of the path.

EXAMPLE 12.16

12

The boxes in Fig. 12–29a travel along the industrial conveyor. If a box as in Fig. 12–29b starts from rest at A and increases its speed such that $a_t = (0.2t)$ m/s^2, where t is in seconds, determine the magnitude of its acceleration when it arrives at point B.

SOLUTION

Coordinate System. The position of the box at any instant is defined from the fixed point A using the position or path coordinate s, Fig. 12–29b. The acceleration is to be determined at B, so the origin of the n, t axes is at this point.

(a)

Acceleration. To determine the acceleration components $a_t = \dot{v}$ and $a_n = v^2/\rho$, it is first necessary to formulate v and $\dot{v}$ so that they may be evaluated at B. Since $v_A = 0$ when $t = 0$, then

$$a_t = \dot{v} = 0.2t \qquad (1)$$

$$\int_0^v dv = \int_0^t 0.2t \, dt$$

$$v = 0.1t^2 \qquad (2)$$

The time needed for the box to reach point B can be determined by realizing that the position of B is $s_B = 3 + 2\pi(2)/4 = 6.142$ m, Fig. 12–29b, and since $s_A = 0$ when $t = 0$ we have

$$v = \frac{ds}{dt} = 0.1t^2$$

$$\int_0^{6.142 \text{ m}} ds = \int_0^{t_B} 0.1t^2 dt$$

$$6.142 \text{ m} = 0.0333 t_B^3$$

$$t_B = 5.690 \text{s}$$

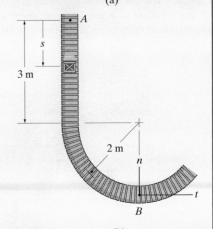

(b)

Substituting into Eqs. 1 and 2 yields

$$(a_B)_t = \dot{v}_B = 0.2(5.690) = 1.138 \text{ m/s}^2$$

$$v_B = 0.1(5.69)^2 = 3.238 \text{ m/s}$$

At B, $\rho_B = 2$ m, so that

$$(a_B)_n = \frac{v_B^2}{\rho_B} = \frac{(3.238 \text{ m/s})^2}{2 \text{ m}} = 5.242 \text{ m/s}^2$$

The magnitude of $\mathbf{a}_B$, Fig. 12–29c, is therefore

$$a_B = \sqrt{(1.138 \text{ m/s}^2)^2 + (5.242 \text{ m/s}^2)^2} = 5.36 \text{ m/s}^2 \quad Ans.$$

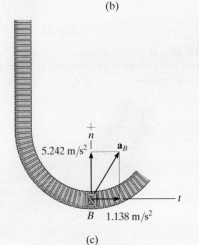

(c)

Fig. 12–29

FUNDAMENTAL PROBLEMS

F12–27. The boat is traveling along the circular path with a speed of $v = (0.0625t^2)$ m/s, where t is in seconds. Determine the magnitude of its acceleration when $t = 10$ s.

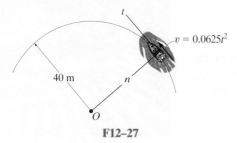

F12–27

F12–28. The car is traveling along the road with a speed of $v = (2\,s)$ m/s, where s is in meters. Determine the magnitude of its acceleration when $s = 10$ m.

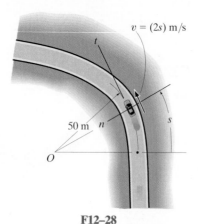

F12–28

F12–29. If the car decelerates uniformly along the curved road from 25 m/s at A to 15 m/s at C, determine the acceleration of the car at B.

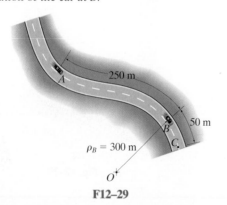

F12–29

F12–30. When $x = 10$ ft, the crate has a speed of 20 ft/s which is increasing at 6 ft/s². Determine the direction of the crate's velocity and the magnitude of the crate's acceleration at this instant.

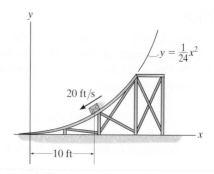

F12–30

F12–31. If the motorcycle has a deceleration of $a_t = -(0.001s)$ m/s² and its speed at position A is 25 m/s, determine the magnitude of its acceleration when it passes point B.

F12–31

F12–32. The car travels up the hill with a speed of $v = (0.2s)$ m/s, where s is in meters, measured from A. Determine the magnitude of its acceleration when it is at point $s = 50$ m, where $\rho = 500$ m.

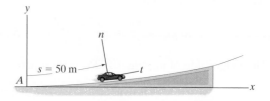

F12–32

PROBLEMS

12–114. A car is traveling along a circular curve that has a radius of 50 m. If its speed is 16 m/s and is increasing uniformly at 8 m/s^2, determine the magnitude of its acceleration at this instant.

12–115. Determine the maximum constant speed a race car can have if the acceleration of the car cannot exceed 7.5 m/s^2 while rounding a track having a radius of curvature of 200 m.

*12–116. A car moves along a circular track of radius 250 ft such that its speed for a short period of time $0 \le t \le 4$ s, is $v = 3(t + t^2)$ ft/s, where t is in seconds. Determine the magnitude of its acceleration when $t = 3$ s. How far has it traveled in $t = 3$ s?

12–117. A car travels along a horizontal circular curved road that has a radius of 600 m. If the speed is uniformly increased at a rate of 2000 km/h^2, determine the magnitude of the acceleration at the instant the speed of the car is 60 km/h.

12–118. The truck travels in a circular path having a radius of 50 m at a speed of $v = 4$ m/s. For a short distance from $s = 0$, its speed is increased by $\dot{v} = (0.05s)$ m/s^2, where s is in meters. Determine its speed and the magnitude of its acceleration when it has moved $s = 10$ m.

12–119. The automobile is originally at rest at $s = 0$. If its speed is increased by $\dot{v} = (0.05t^2)$ ft/s^2, where t is in seconds, determine the magnitudes of its velocity and acceleration when $t = 18$ s.

*12–120. The automobile is originally at rest $s = 0$. If it then starts to increase its speed at $\dot{v} = (0.05t^2)$ ft/s^2, where t is in seconds, determine the magnitudes of its velocity and acceleration at $s = 550$ ft.

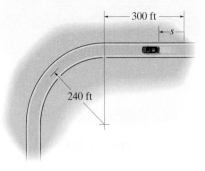

Probs. 12–119/120

12–121. When the roller coaster is at B, it has a speed of 25 m/s, which is increasing at $a_t = 3$ m/s^2. Determine the magnitude of the acceleration of the roller coaster at this instant and the direction angle it makes with the x axis.

12–122. If the roller coaster starts from rest at A and its speed increases at $a_t = (6 - 0.06s)$ m/s^2, determine the magnitude of its acceleration when it reaches B where $s_B = 40$ m.

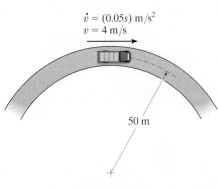

Prob. 12–118

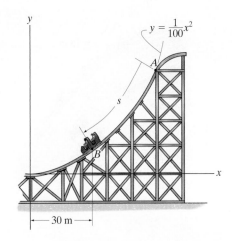

Probs. 12–121/122

12–123. The speedboat travels at a constant speed of 15 m/s while making a turn on a circular curve from A to B. If it takes 45 s to make the turn, determine the magnitude of the boat's acceleration during the turn.

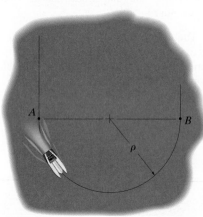

Prob. 12–123

12–125. The car passes point A with a speed of 25 m/s after which its speed is defined by $v = (25 - 0.15s)$ m/s. Determine the magnitude of the car's acceleration when it reaches point B, where $s = 51.5$ m and $x = 50$ m.

12–126. If the car passes point A with a speed of 20 m/s and begins to increase its speed at a constant rate of $a_t = 0.5$ m/s², determine the magnitude of the car's acceleration when $s = 100$ m and $x = 0$.

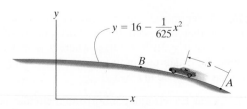

Probs. 12–125/126

*12–124. The car travels along the circular path such that its speed is increased by $a_t = (0.5e^t)$ m/s², where t is in seconds. Determine the magnitudes of its velocity and acceleration after the car has traveled $s = 18$ m starting from rest. Neglect the size of the car.

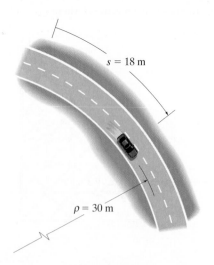

Prob. 12–124

12–127. A train is traveling with a constant speed of 14 m/s along the curved path. Determine the magnitude of the acceleration of the front of the train, B, at the instant it reaches point A ($y = 0$).

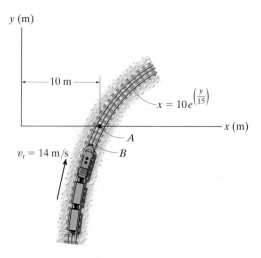

Prob. 12–127

*12–128. When a car starts to round a curved road with the radius of curvature of 600 ft, it is traveling at 75 ft/s. If the car's speed begins to decrease at a rate of $\dot{v} = (-0.06t^2)$ ft/s^2, determine the magnitude of the acceleration of the car when it has traveled a distance of $s = 700$ ft.

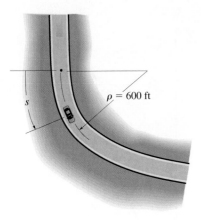

Prob. 12–128

12–129. When the motorcyclist is at A, he increases his speed along the vertical circular path at the rate of $\dot{v} = (0.3t)$ ft/s^2, where t is in seconds. If he starts from rest at A, determine the magnitudes of his velocity and acceleration when he reaches B.

12–130. When the motorcyclist is at A, he increases his speed along the vertical circular path at the rate of $\dot{v} = (0.04s)$ ft/s^2 where s is in ft. If he starts at $v_A = 2$ ft/s where $s = 0$ at A, determine the magnitude of his velocity when he reaches B. Also, what is his initial acceleration?

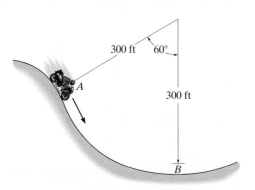

Probs. 12–129/130

12–131. At a given instant the train engine at E has a speed of 20 m/s and an acceleration of 14 m/s^2 acting in the direction shown. Determine the rate of increase in the train's speed and the radius of curvature ρ of the path.

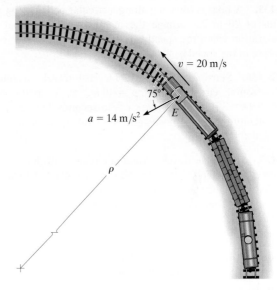

Prob. 12–131

*12–132. Car B turns such that its speed is increased by $(a_t)_B = (0.5e^t)$ m/s^2, where t is in seconds. If the car starts from rest when $\theta = 0°$, determine the magnitudes of its velocity and acceleration when the arm AB rotates $\theta = 30°$. Neglect the size of the car.

12–133. Car B turns such that its speed is increased by $(a_t)_B = (0.5e^t)$ m/s^2, where t is in seconds. If the car starts from rest when $\theta = 0°$, determine the magnitudes of its velocity and acceleration when $t = 2$ s. Neglect the size of the car.

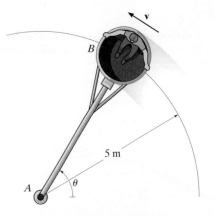

Probs. 12–132/133

12–134. A boat is traveling along a circular curve having a radius of 100 ft. If its speed at $t = 0$ is 15 ft/s and is increasing at $\dot{v} = (0.8t)$ ft/s^2, determine the magnitude of its acceleration at the instant $t = 5$ s.

12–135. A boat is traveling along a circular path having a radius of 20 m. Determine the magnitude of the boat's acceleration when the speed is $v = 5$ m/s and the rate of increase in the speed is $\dot{v} = 2$ m/s^2.

*■12–136. Starting from rest, a bicyclist travels around a horizontal circular path, $\rho = 10$ m, at a speed of $v = (0.09t^2 + 0.1t)$ m/s, where t is in seconds. Determine the magnitudes of his velocity and acceleration when he has traveled $s = 3$ m.

12–137. A particle travels around a circular path having a radius of 50 m. If it is initially traveling with a speed of 10 m/s and its speed then increases at a rate of $\dot{v} = (0.05\ v)$ m/s^2, determine the magnitude of the particle's acceleraton four seconds later.

12–138. When the bicycle passes point A, it has a speed of 6 m/s, which is increasing at the rate of $\dot{v} = 0.5$ m/s^2. Determine the magnitude of its acceleration when it is at point A.

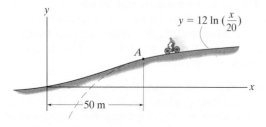

$$y = 12 \ln \left(\frac{x}{20}\right)$$

Prob. 12–138

12–139. The motorcycle is traveling at a constant speed of 60 km/h. Determine the magnitude of its acceleration when it is at point A.

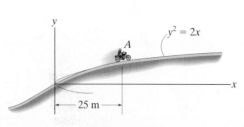

$$y^2 = 2x$$

Prob. 12–139

*12–140. The jet plane travels along the vertical parabolic path. When it is at point A it has a speed of 200 m/s, which is increasing at the rate of 0.8 m/s^2. Determine the magnitude of acceleration of the plane when it is at point A.

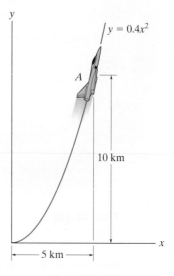

$$y = 0.4x^2$$

Prob. 12–140

12–141. The ball is ejected horizontally from the tube with a speed of 8 m/s. Find the equation of the path, $y = f(x)$, and then find the ball's velocity and the normal and tangential components of acceleration when $t = 0.25$ s.

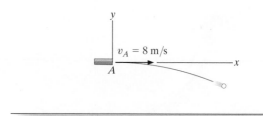

$$v_A = 8 \text{ m/s}$$

Prob. 12–141

12–142. A toboggan is traveling down along a curve which can be approximated by the parabola $y = 0.01x^2$. Determine the magnitude of its acceleration when it reaches point A, where its speed is $v_A = 10$ m/s, and it is increasing at the rate of $\dot{v}_A = 3$ m/s^2.

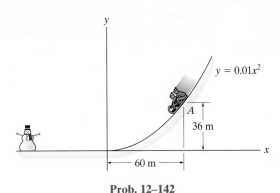

$y = 0.01x^2$

A

36 m

60 m

Prob. 12–142

12–143. A particle P moves along the curve $y = (x^2 - 4)$ m with a constant speed of 5 m/s. Determine the point on the curve where the maximum magnitude of acceleration occurs and compute its value.

*12–144. The Ferris wheel turns such that the speed of the passengers is increased by $\dot{v} = (4t)$ ft/s^2, where t is in seconds. If the wheel starts from rest when $\theta = 0°$, determine the magnitudes of the velocity and acceleration of the passengers when the wheel turns $\theta = 30°$.

40 ft

θ

Prob. 12–144

12–145. If the speed of the crate at A is 15 ft/s, which is increasing at a rate $\dot{v} = 3$ ft/s^2, determine the magnitude of the acceleration of the crate at this instant.

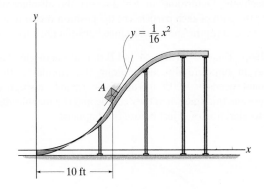

$y = \frac{1}{16}x^2$

A

10 ft

Prob. 12–145

12–146. The race car has an initial speed $v_A = 15$ m/s at A. If it increases its speed along the circular track at the rate $a_t = (0.4s)$ m/s^2, where s is in meters, determine the time needed for the car to travel 20 m. Take $\rho = 150$ m.

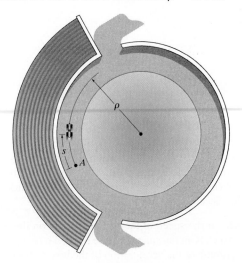

ρ

s

A

Prob. 12–146

12–147. A boy sits on a merry-go-round so that he is always located at $r = 8$ ft from the center of rotation. The merry-go-round is originally at rest, and then due to rotation the boy's speed is increased at 2 ft/s^2. Determine the time needed for his acceleration to become 4 ft/s^2.

*12–148. A particle travels along the path $y = a + bx + cx^2$, where a, b, c are constants. If the speed of the particle is constant, $v = v_0$, determine the x and y components of velocity and the normal component of acceleration when $x = 0$.

12–149. The two particles A and B start at the origin O and travel in opposite directions along the circular path at constant speeds $v_A = 0.7$ m/s and $v_B = 1.5$ m/s, respectively. Determine in $t = 2$ s, (a) the displacement along the path of each particle, (b) the position vector to each particle, and (c) the shortest distance between the particles.

12–150. The two particles A and B start at the origin O and travel in opposite directions along the circular path at constant speeds $v_A = 0.7$ m/s and $v_B = 1.5$ m/s, respectively. Determine the time when they collide and the magnitude of the acceleration of B just before this happens.

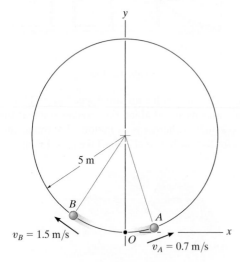

Probs. 12–149/150

12–151. The position of a particle traveling along a curved path is $s = (3t^3 - 4t^2 + 4)$ m, where t is in seconds. When $t = 2$ s, the particle is at a position on the path where the radius of curvature is 25 m. Determine the magnitude of the particle's acceleration at this instant.

*12–152. If the speed of the box at point A on the track is 30 ft/s which is increasing at the rate of $\dot{v} = 5$ ft/s², determine the magnitude of the acceleration of the box at this instant.

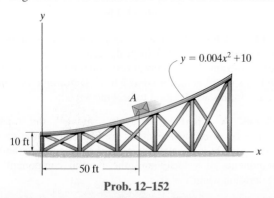

Prob. 12–152

■12–153. A go-cart moves along a circular track of radius 100 ft such that its speed for a short period of time, $0 \le t \le 4$ s, is $v = 60(1 - e^{-t^2})$ ft/s. Determine the magnitude of its acceleration when $t = 2$ s. How far has it traveled in $t = 2$ s? Use a numerical method to evaluate the integral.

■12–154. The ball is kicked with an initial speed $v_A = 8$ m/s at an angle $\theta_A = 40°$ with the horizontal. Find the equation of the path, $y = f(x)$, and then determine the ball's velocity and the normal and tangential components of its acceleration when $t = 0.25$ s.

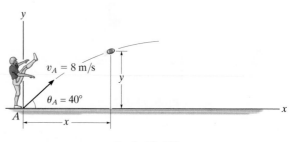

Prob. 12–154

12–155. The race car travels around the circular track with a speed of 16 m/s. When it reaches point A it increases its speed at $a_t = (\frac{4}{3} v^{1/4})$ m/s², where v is in m/s. Determine the magnitudes of the velocity and acceleration of the car when it reaches point B. Also, how much time is required for it to travel from A to B?

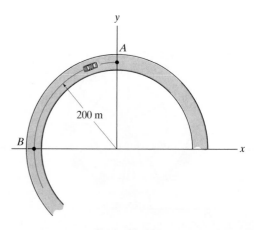

Prob. 12–155

*12–156. A particle P travels along an elliptical spiral path such that its position vector $\mathbf{r}$ is defined by $\mathbf{r} = \{2\cos(0.1t)\mathbf{i} + 1.5\sin(0.1t)\mathbf{j} + (2t)\mathbf{k}\}$ m, where t is in seconds and the arguments for the sine and cosine are given in radians. When $t = 8$ s, determine the coordinate direction angles α, β, and γ, which the binormal axis to the osculating plane makes with the x, y, and z axes. *Hint:* Solve for the velocity $\mathbf{v}_P$ and acceleration $\mathbf{a}_P$ of the particle in terms of their $\mathbf{i}$, $\mathbf{j}$, $\mathbf{k}$ components. The binormal is parallel to $\mathbf{v}_P \times \mathbf{a}_P$. Why?

12–157. The motion of a particle is defined by the equations $x = (2t + t^2)$ m and $y = (t^2)$ m, where t is in seconds. Determine the normal and tangential components of the particle's velocity and acceleration when $t = 2$ s.

12–158. The motorcycle travels along the elliptical track at a constant speed v. Determine the greatest magnitude of the acceleration if $a > b$.

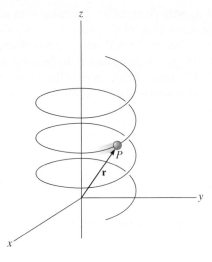

Prob. 12–156

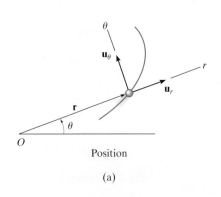

$$\frac{x^2}{a^2} + \frac{y^2}{b^2} = 1$$

Prob. 12–158

12.8   Curvilinear Motion: Cylindrical Components

Sometimes the motion of the particle is constrained on a path that is best described using cylindrical coordinates. If motion is restricted to the plane, then polar coordinates are used.

Polar Coordinates.   We can specify the location of the particle shown in Fig. 12–30a using a *radial coordinate r*, which extends outward from the fixed origin O to the particle, and a *transverse coordinate* θ, which is the counterclockwise angle between a fixed reference line and the r axis. The angle is generally measured in degrees or radians, where $1 \text{ rad} = 180°/\pi$. The positive directions of the r and θ coordinates are defined by the unit vectors $\mathbf{u}_r$ and $\mathbf{u}_\theta$, respectively. Here $\mathbf{u}_r$ is in the direction of increasing r when θ is held fixed, and $\mathbf{u}_\theta$ is in a direction of increasing θ when r is held fixed. Note that these directions are perpendicular to one another.

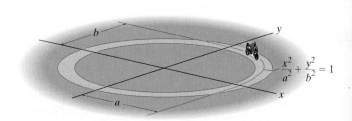

Position

(a)

Fig. 12–30

12

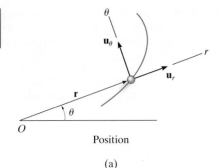

Position

(a)

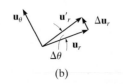

(b)

Position.   At any instant the position of the particle, Fig. 12–30a, is defined by the position vector

$$\mathbf{r} = r\mathbf{u}_r \qquad (12\text{–}22)$$

Velocity.   The instantaneous velocity **v** is obtained by taking the time derivative of **r**. Using a dot to represent the time derivative, we have

$$\mathbf{v} = \dot{\mathbf{r}} = \dot{r}\mathbf{u}_r + r\dot{\mathbf{u}}_r$$

To evaluate $\dot{\mathbf{u}}_r$, notice that $\mathbf{u}_r$ only changes its direction with respect to time, since by definition the magnitude of this vector is always one unit. Hence, during the time Δt, a change Δr will not cause a change in the direction of $\mathbf{u}_r$; however, a change $\Delta \theta$ will cause $\mathbf{u}_r$ to become $\mathbf{u}'_r$, where $\mathbf{u}'_r = \mathbf{u}_r + \Delta \mathbf{u}_r$, Fig. 12–30b. The time change in $\mathbf{u}_r$ is then $\Delta \mathbf{u}_r$. For small angles $\Delta \theta$ this vector has a magnitude $\Delta u_r \approx 1(\Delta \theta)$ and acts in the $\mathbf{u}_\theta$ direction. Therefore, $\Delta \mathbf{u}_r = \Delta \theta \mathbf{u}_\theta$, and so

$$\dot{\mathbf{u}}_r = \lim_{\Delta t \to 0} \frac{\Delta \mathbf{u}_r}{\Delta t} = \left(\lim_{\Delta t \to 0} \frac{\Delta \theta}{\Delta t} \right) \mathbf{u}_\theta$$

$$\dot{\mathbf{u}}_r = \dot{\theta}\mathbf{u}_\theta \qquad (12\text{–}23)$$

Substituting into the above equation, the velocity can be written in component form as

$$\boxed{\mathbf{v} = v_r \mathbf{u}_r + v_\theta \mathbf{u}_\theta} \qquad (12\text{–}24)$$

where

$$\boxed{\begin{aligned} v_r &= \dot{r} \\ v_\theta &= r\dot{\theta} \end{aligned}} \qquad (12\text{–}25)$$

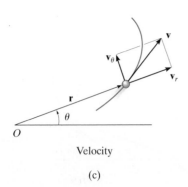

Velocity

(c)

Fig. 12–30 (cont.)

These components are shown graphically in Fig. 12–30c. The *radial component* v_r is a measure of the rate of increase or decrease in the length of the radial coordinate, i.e., $\dot{r}$; whereas the *transverse component* v_θ can be interpreted as the rate of motion along the circumference of a circle having a radius r. In particular, the term $\dot{\theta} = d\theta/dt$ is called the *angular velocity*, since it indicates the time rate of change of the angle θ. Common units used for this measurement are rad/s.

Since v_r and v_θ are mutually perpendicular, the *magnitude* of velocity or speed is simply the positive value of

$$v = \sqrt{(\dot{r})^2 + (r\dot{\theta})^2} \qquad (12\text{–}26)$$

and the *direction* of **v** is, of course, tangent to the path, Fig. 12–30c.

Acceleration.

Taking the time derivatives of Eq. 12–24, using Eqs. 12–25, we obtain the particle's instantaneous acceleration,

$$\mathbf{a} = \dot{\mathbf{v}} = \ddot{r}\mathbf{u}_r + \dot{r}\dot{\mathbf{u}}_r + \dot{r}\dot{\theta}\mathbf{u}_\theta + r\ddot{\theta}\mathbf{u}_\theta + r\dot{\theta}\dot{\mathbf{u}}_\theta$$

To evaluate $\dot{\mathbf{u}}_\theta$, it is necessary only to find the change in the direction of $\mathbf{u}_\theta$ since its magnitude is always unity. During the time Δt, a change Δr will not change the direction of $\mathbf{u}_\theta$, however, a change $\Delta\theta$ will cause $\mathbf{u}_\theta$ to become $\mathbf{u}'_\theta$, where $\mathbf{u}'_\theta = \mathbf{u}_\theta + \Delta\mathbf{u}_\theta$, Fig. 12–30d. The time change in $\mathbf{u}_\theta$ is thus $\Delta\mathbf{u}_\theta$. For small angles this vector has a magnitude $\Delta u_\theta \approx 1(\Delta\theta)$ and acts in the $-\mathbf{u}_r$ direction; i.e., $\Delta\mathbf{u}_\theta = -\Delta\theta\mathbf{u}_r$. Thus,

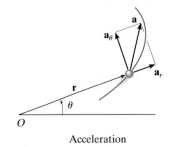

(d)

$$\dot{\mathbf{u}}_\theta = \lim_{\Delta t \to 0}\frac{\Delta\mathbf{u}_\theta}{\Delta t} = -\left(\lim_{\Delta t \to 0}\frac{\Delta\theta}{\Delta t}\right)\mathbf{u}_r$$

$$\dot{\mathbf{u}}_\theta = -\dot{\theta}\mathbf{u}_r \qquad (12\text{–}27)$$

Substituting this result and Eq. 12–23 into the above equation for $\mathbf{a}$, we can write the acceleration in component form as

$$\boxed{\mathbf{a} = a_r\mathbf{u}_r + a_\theta\mathbf{u}_\theta} \qquad (12\text{–}28)$$

where

$$\boxed{\begin{aligned} a_r &= \ddot{r} - r\dot{\theta}^2 \\ a_\theta &= r\ddot{\theta} + 2\dot{r}\dot{\theta} \end{aligned}} \qquad (12\text{–}29)$$

The term $\ddot{\theta} = d^2\theta/dt^2 = d/dt(d\theta/dt)$ is called the *angular acceleration* since it measures the change made in the angular velocity during an instant of time. Units for this measurement are rad/s².

Since $\mathbf{a}_r$ and $\mathbf{a}_\theta$ are always perpendicular, the *magnitude* of acceleration is simply the positive value of

$$a = \sqrt{(\ddot{r} - r\dot{\theta}^2)^2 + (r\ddot{\theta} + 2\dot{r}\dot{\theta})^2} \qquad (12\text{–}30)$$

Acceleration

(e)

The *direction* is determined from the vector addition of its two components. In general, $\mathbf{a}$ will *not* be tangent to the path, Fig. 12–30e.

12

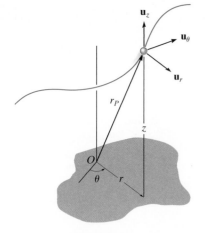

Fig. 12–31

The spiral motion of this girl can be followed by using cylindrical components. Here the radial coordinate r is constant, the transverse coordinate θ will increase with time as the girl rotates about the vertical, and her altitude z will decrease with time.

Cylindrical Coordinates.

If the particle moves along a space curve as shown in Fig. 12–31, then its location may be specified by the three *cylindrical coordinates, r, θ, z*. The z coordinate is identical to that used for rectangular coordinates. Since the unit vector defining its direction, $\mathbf{u}_z$, is constant, the time derivatives of this vector are zero, and therefore the position, velocity, and acceleration of the particle can be written in terms of its cylindrical coordinates as follows:

$$\mathbf{r}_P = r\mathbf{u}_r + z\mathbf{u}_z$$

$$\mathbf{v} = \dot{r}\mathbf{u}_r + r\dot{\theta}\mathbf{u}_\theta + \dot{z}\mathbf{u}_z \qquad (12\text{--}31)$$

$$\mathbf{a} = (\ddot{r} - r\dot{\theta}^2)\mathbf{u}_r + (r\ddot{\theta} + 2\dot{r}\dot{\theta})\mathbf{u}_\theta + \ddot{z}\mathbf{u}_z \qquad (12\text{--}32)$$

Time Derivatives.

The above equations require that we obtain the time derivatives $\dot{r}$, $\ddot{r}$, $\dot{\theta}$, and $\ddot{\theta}$ in order to evaluate the r and θ components of $\mathbf{v}$ and $\mathbf{a}$. Two types of problems generally occur:

1. If the polar coordinates are specified as time parametric equations, $r = r(t)$ and $\theta = \theta(t)$, then the time derivatives can be found directly.

2. If the time-parametric equations are not given, then the path $r = f(\theta)$ must be known. Using the chain rule of calculus we can then find the relation between $\dot{r}$ and $\dot{\theta}$, and between $\ddot{r}$ and $\ddot{\theta}$. Application of the chain rule, along with some examples, is explained in Appendix C.

Procedure for Analysis

Coordinate System.

- Polar coordinates are a suitable choice for solving problems when data regarding the angular motion of the radial coordinate r is given to describe the particle's motion. Also, some paths of motion can conveniently be described in terms of these coordinates.

- To use polar coordinates, the origin is established at a fixed point, and the radial line r is directed to the particle.

- The transverse coordinate θ is measured from a fixed reference line to the radial line.

Velocity and Acceleration.

- Once r and the four time derivatives $\dot{r}$, $\ddot{r}$, $\dot{\theta}$, and $\ddot{\theta}$ have been evaluated at the instant considered, their values can be substituted into Eqs. 12–25 and 12–29 to obtain the radial and transverse components of $\mathbf{v}$ and $\mathbf{a}$.

- If it is necessary to take the time derivatives of $r = f(\theta)$, then the chain rule of calculus must be used. See Appendix C.

- Motion in three dimensions requires a simple extension of the above procedure to include $\dot{z}$ and $\ddot{z}$.

EXAMPLE | 12.17

The amusement park ride shown in Fig. 12–32a consists of a chair that is rotating in a horizontal circular path of radius r such that the arm OB has an angular velocity $\dot{\theta}$ and angular acceleration $\ddot{\theta}$. Determine the radial and transverse components of velocity and acceleration of the passenger. Neglect his size in the calculation.

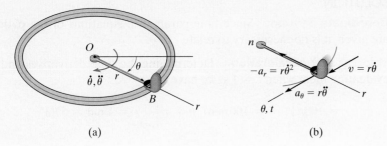

(a)   (b)

Fig. 12–32

SOLUTION

Coordinate System.   Since the angular motion of the arm is reported, polar coordinates are chosen for the solution, Fig. 12–32a. Here θ is not related to r, since the radius is constant for all θ.

Velocity and Acceleration.   It is first necessary to specify the first and second time derivatives of r and θ. Since r is *constant*, we have

$$r = r \qquad \dot{r} = 0 \qquad \ddot{r} = 0$$

Thus,

$$v_r = \dot{r} = 0 \qquad\qquad\qquad\qquad Ans.$$

$$v_\theta = r\dot{\theta} \qquad\qquad\qquad\qquad Ans.$$

$$a_r = \ddot{r} - r\dot{\theta}^2 = -r\dot{\theta}^2 \qquad\qquad Ans.$$

$$a_\theta = r\ddot{\theta} + 2\dot{r}\dot{\theta} = r\ddot{\theta} \qquad\qquad Ans.$$

These results are shown in Fig. 12–32b.

NOTE: The n, t axes are also shown in Fig. 12–32b, which in this special case of circular motion happen to be *collinear* with the r and θ axes, respectively. Since $v = v_\theta = v_t = r\dot{\theta}$, then by comparison,

$$-a_r = a_n = \frac{v^2}{\rho} = \frac{(r\dot{\theta})^2}{r} = r\dot{\theta}^2$$

$$a_\theta = a_t = \frac{dv}{dt} = \frac{d}{dt}(r\dot{\theta}) = \frac{dr}{dt}\dot{\theta} + r\frac{d\dot{\theta}}{dt} = 0 + r\ddot{\theta}$$

12

EXAMPLE | 12.18

The rod OA in Fig. 12–33a rotates in the horizontal plane such that $\theta = (t^3)$ rad. At the same time, the collar B is sliding outward along OA so that $r = (100t^2)$ mm. If in both cases t is in seconds, determine the velocity and acceleration of the collar when $t = 1$ s.

SOLUTION

Coordinate System. Since time-parametric equations of the path are given, it is not necessary to relate r to θ.

Velocity and Acceleration. Determining the time derivatives and evaluating them when $t = 1$ s, we have

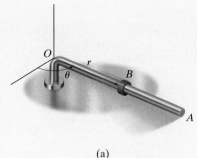

(a)

$$r = 100t^2 \Big|_{t=1\text{ s}} = 100\text{ mm} \quad \theta = t^3 \Big|_{t=1\text{ s}} = 1 \text{ rad} = 57.3°$$

$$\dot{r} = 200t \Big|_{t=1\text{ s}} = 200 \text{ mm/s} \quad \dot{\theta} = 3t^2 \Big|_{t=1\text{ s}} = 3 \text{ rad/s}$$

$$\ddot{r} = 200 \Big|_{t=1\text{ s}} = 200 \text{ mm/s}^2 \quad \ddot{\theta} = 6t \Big|_{t=1\text{ s}} = 6 \text{ rad/s}^2.$$

As shown in Fig. 12–33b,

$$\mathbf{v} = \dot{r}\mathbf{u}_r + r\dot{\theta}\mathbf{u}_\theta$$

$$= 200\mathbf{u}_r + 100(3)\mathbf{u}_\theta = \{200\mathbf{u}_r + 300\mathbf{u}_\theta\} \text{ mm/s}$$

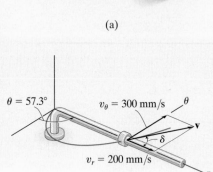

(b)

The magnitude of $\mathbf{v}$ is

$$v = \sqrt{(200)^2 + (300)^2} = 361 \text{ mm/s} \qquad Ans.$$

$$\delta = \tan^{-1}\left(\frac{300}{200}\right) = 56.3° \quad \delta + 57.3° = 114° \qquad Ans.$$

As shown in Fig. 12–33c,

$$\mathbf{a} = (\ddot{r} - r\dot{\theta}^2)\mathbf{u}_r + (r\ddot{\theta} + 2\dot{r}\dot{\theta})\mathbf{u}_\theta$$

$$= [200 - 100(3)^2]\mathbf{u}_r + [100(6) + 2(200)3]\mathbf{u}_\theta$$

$$= \{-700\mathbf{u}_r + 1800\mathbf{u}_\theta\} \text{ mm/s}^2$$

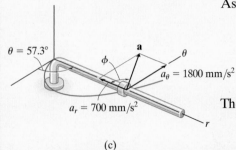

(c)

Fig. 12–33

The magnitude of $\mathbf{a}$ is

$$a = \sqrt{(-700)^2 + (1800)^2} = 1930 \text{ mm/s}^2 \qquad Ans.$$

$$\phi = \tan^{-1}\left(\frac{1800}{700}\right) = 68.7° \quad (180° - \phi) + 57.3° = 169° \qquad Ans.$$

NOTE: The velocity is tangent to the path; however, the acceleration is directed within the curvature of the path, as expected.

EXAMPLE | 12.19

The searchlight in Fig. 12–34a casts a spot of light along the face of a wall that is located 100 m from the searchlight. Determine the magnitudes of the velocity and acceleration at which the spot appears to travel across the wall at the instant $\theta = 45°$. The searchlight rotates at a constant rate of $\dot{\theta} = 4$ rad/s.

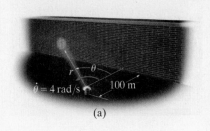

(a)

SOLUTION

Coordinate System. Polar coordinates will be used to solve this problem since the angular rate of the searchlight is given. To find the necessary time derivatives it is first necessary to relate r to θ. From Fig. 12–34a,

$$r = 100/\cos \theta = 100 \sec \theta$$

Velocity and Acceleration. Using the chain rule of calculus, noting that $d(\sec \theta) = \sec \theta \tan \theta \, d\theta$, and $d(\tan \theta) = \sec^2 \theta \, d\theta$, we have

$$\dot{r} = 100(\sec \theta \tan \theta)\dot{\theta}$$
$$\ddot{r} = 100(\sec \theta \tan \theta)\dot{\theta}(\tan \theta)\dot{\theta} + 100 \sec \theta(\sec^2 \theta)\dot{\theta}(\dot{\theta})$$
$$\quad + 100 \sec \theta \tan \theta(\ddot{\theta})$$
$$\quad = 100 \sec \theta \tan^2 \theta \, (\dot{\theta})^2 + 100 \sec^3 \theta \, (\dot{\theta})^2 + 100(\sec \theta \tan \theta)\ddot{\theta}$$

Since $\dot{\theta} = 4$ rad/s = constant, then $\ddot{\theta} = 0$, and the above equations, when $\theta = 45°$, become

$$r = 100 \sec 45° = 141.4$$
$$\dot{r} = 400 \sec 45° \tan 45° = 565.7$$
$$\ddot{r} = 1600 \,(\sec 45° \tan^2 45° + \sec^3 45°) = 6788.2$$

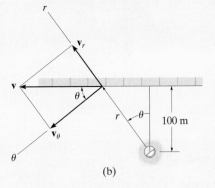

(b)

As shown in Fig. 12–34b,

$$\mathbf{v} = \dot{r}\mathbf{u}_r + r\dot{\theta}\mathbf{u}_\theta$$
$$\quad = 565.7\mathbf{u}_r + 141.4(4)\mathbf{u}_\theta$$
$$\quad = \{565.7\mathbf{u}_r + 565.7\mathbf{u}_\theta\} \text{ m/s}$$
$$v = \sqrt{v_r^2 + v_\theta^2} = \sqrt{(565.7)^2 + (565.7)^2}$$
$$\quad = 800 \text{ m/s} \qquad\qquad Ans.$$

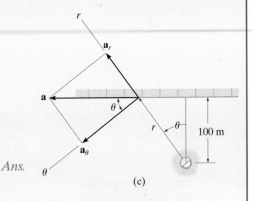

(c)

As shown in Fig. 12–34c,

$$\mathbf{a} = (\ddot{r} - r\dot{\theta}^2)\mathbf{u}_r + (r\ddot{\theta} + 2\dot{r}\dot{\theta})\mathbf{u}_\theta$$
$$\quad = [6788.2 - 141.4(4)^2]\mathbf{u}_r + [141.4(0) + 2(565.7)4]\mathbf{u}_\theta$$
$$\quad = \{4525.5\mathbf{u}_r + 4525.5\mathbf{u}_\theta\} \text{ m/s}^2$$
$$a = \sqrt{a_r^2 + a_\theta^2} = \sqrt{(4525.5)^2 + (4525.5)^2}$$
$$\quad = 6400 \text{ m/s}^2 \qquad\qquad Ans.$$

NOTE: It is also possible to find a without having to calculate $\ddot{r}$ (or a_r). As shown in Fig. 12–34d, since $a_\theta = 4525.5$ m/s², then by vector resolution, $a = 4525.5/\cos 45° = 6400$ m/s².

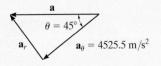

(d)

Fig. 12–34

12

EXAMPLE | 12.20

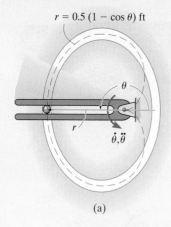

$r = 0.5 (1 - \cos \theta)$ ft

(a)

Due to the rotation of the forked rod, the ball in Fig. 12–35a travels around the slotted path, a portion of which is in the shape of a cardioid, $r = 0.5(1 - \cos \theta)$ ft, where θ is in radians. If the ball's velocity is $v = 4$ ft/s and its acceleration is $a = 30$ ft/s² at the instant $\theta = 180°$, determine the angular velocity $\dot{\theta}$ and angular acceleration $\ddot{\theta}$ of the fork.

SOLUTION

Coordinate System.   This path is most unusual, and mathematically it is best expressed using polar coordinates, as done here, rather than rectangular coordinates. Also, since $\dot{\theta}$ and $\ddot{\theta}$ must be determined, then r, θ coordinates are an obvious choice.

Velocity and Acceleration.   The time derivatives of r and θ can be determined using the chain rule.

$$r = 0.5(1 - \cos \theta)$$

$$\dot{r} = 0.5(\sin \theta)\dot{\theta}$$

$$\ddot{r} = 0.5(\cos \theta)\dot{\theta}(\dot{\theta}) + 0.5(\sin \theta)\ddot{\theta}$$

Evaluating these results at $\theta = 180°$, we have

$$r = 1 \text{ ft} \qquad \dot{r} = 0 \qquad \ddot{r} = -0.5\dot{\theta}^2$$

Since $v = 4$ ft/s, using Eq. 12–26 to determine $\dot{\theta}$ yields

$$v = \sqrt{(\dot{r})^2 + (r\dot{\theta})^2}$$

$$4 = \sqrt{(0)^2 + (1\dot{\theta})^2}$$

$$\dot{\theta} = 4 \text{ rad/s} \qquad\qquad Ans.$$

In a similar manner, $\ddot{\theta}$ can be found using Eq. 12–30.

$$a = \sqrt{(\ddot{r} - r\dot{\theta}^2)^2 + (r\ddot{\theta} + 2\dot{r}\dot{\theta})^2}$$

$$30 = \sqrt{[-0.5(4)^2 - 1(4)^2]^2 + [1\ddot{\theta} + 2(0)(4)]^2}$$

$$(30)^2 = (-24)^2 + \ddot{\theta}^2$$

$$\ddot{\theta} = 18 \text{ rad/s}^2 \qquad\qquad Ans.$$

Vectors **a** and **v** are shown in Fig. 12–35b.

NOTE: At this location, the θ and t (tangential) axes will coincide. The $+n$ (normal) axis is directed to the right, opposite to $+r$.

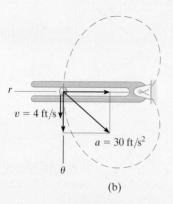

r

$v = 4$ ft/s

$a = 30$ ft/s²

θ

(b)

Fig. 12–35

FUNDAMENTAL PROBLEMS

F12–33. The car has a speed of 55 ft/s. Determine the angular velocity $\dot\theta$ of the radial line OA at this instant.

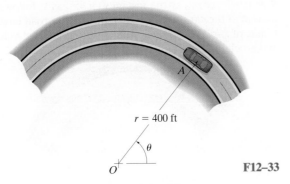

$r = 400$ ft

F12–33

F12–34. The platform is rotating about the vertical axis such that at any instant its angular position is $\theta = (4t^{3/2})$ rad, where t is in seconds. A ball rolls outward along the radial groove so that its position is $r = (0.1t^3)$ m, where t is in seconds. Determine the magnitudes of the velocity and acceleration of the ball when $t = 1.5$ s.

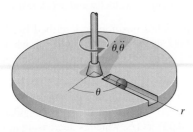

F12–34

F12–35. Peg P is driven by the fork link OA along the curved path described by $r = (2\theta)$ ft. At the instant $\theta = \pi/4$ rad, the angular velocity and angular acceleration of the link are $\dot\theta = 3$ rad/s and $\ddot\theta = 1$ rad/s². Determine the magnitude of the peg's acceleration at this instant.

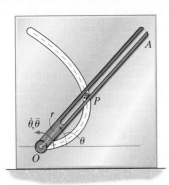

F12–35

F12–36. Peg P is driven by the forked link OA along the path described by $r = e^\theta$, where r is in meters. When $\theta = \frac{\pi}{4}$ rad, the link has an angular velocity and angular acceleration of $\dot\theta = 2$ rad/s and $\ddot\theta = 4$ rad/s². Determine the radial and transverse components of the peg's acceleration at this instant.

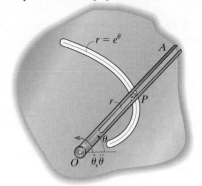

F12–36

F12–37. The collars are pin connected at B and are free to move along rod OA and the curved guide OC having the shape of a cardioid, $r = [0.2(1 + \cos\theta)]$ m. At $\theta = 30°$, the angular velocity of OA is $\dot\theta = 3$ rad/s. Determine the magnitude of the velocity of the collars at this point.

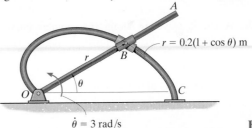

$\dot\theta = 3$ rad/s   **F12–37**

F12–38. At the instant $\theta = 45°$, the athlete is running with a constant speed of 2 m/s. Determine the angular velocity at which the camera must turn in order to follow the motion.

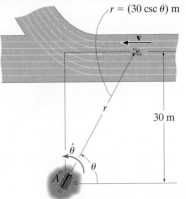

$r = (30 \csc\theta)$ m

30 m

F12–38

PROBLEMS

12–159. A particle is moving along a circular path having a radius of 4 in. such that its position as a function of time is given by $\theta = \cos 2t$, where θ is in radians and t is in seconds. Determine the magnitude of the acceleration of the particle when $\theta = 30°$.

*12–160. A particle travels around a limaçon, defined by the equation $r = b - a \cos \theta$, where a and b are constants. Determine the particle's radial and transverse components of velocity and acceleration as a function of θ and its time derivatives.

12–161. If a particle's position is described by the polar coordinates $r = 4(1 + \sin t)$ m and $\theta = (2e^{-t})$ rad, where t is in seconds and the argument for the sine is in radians, determine the radial and tangential components of the particle's velocity and acceleration when $t = 2$ s.

12–162. An airplane is flying in a straight line with a velocity of 200 mi/h and an acceleration of 3 mi/h². If the propeller has a diameter of 6 ft and is rotating at a constant angular rate of 120 rad/s, determine the magnitudes of velocity and acceleration of a particle located on the tip of the propeller.

12–163. A car is traveling along the circular curve of radius $r = 300$ ft. At the instant shown, its angular rate of rotation is $\dot{\theta} = 0.4$ rad/s, which is increasing at the rate of $\ddot{\theta} = 0.2$ rad/s². Determine the magnitudes of the car's velocity and acceleration at this instant.

*12–164. A radar gun at O rotates with the angular velocity of $\dot{\theta} = 0.1$ rad/s and angular acceleration of $\ddot{\theta} = 0.025$ rad/s², at the instant $\theta = 45°$, as it follows the motion of the car traveling along the circular road having a radius of $r = 200$ m. Determine the magnitudes of velocity and acceleration of the car at this instant.

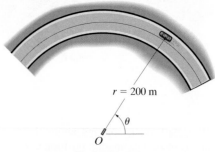

Prob. 12–164

12–165. If a particle moves along a path such that $r = (2 \cos t)$ ft and $\theta = (t/2)$ rad, where t is in seconds, plot the path $r = f(\theta)$ and determine the particle's radial and transverse components of velocity and acceleration.

12–166. If a particle's position is described by the polar coordinates $r = (2 \sin 2\theta)$ m and $\theta = (4t)$ rad, where t is in seconds, determine the radial and tangential components of its velocity and acceleration when $t = 1$ s.

12–167. The car travels along the circular curve having a radius $r = 400$ ft. At the instant shown, its angular rate of rotation is $\dot{\theta} = 0.025$ rad/s, which is decreasing at the rate $\ddot{\theta} = -0.008$ rad/s². Determine the radial and transverse components of the car's velocity and acceleration at this instant and sketch these components on the curve.

*12–168. The car travels along the circular curve of radius $r = 400$ ft with a constant speed of $v = 30$ ft/s. Determine the angular rate of rotation $\dot{\theta}$ of the radial line r and the magnitude of the car's acceleration.

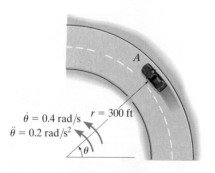

$\dot{\theta} = 0.4$ rad/s $r = 300$ ft
$\ddot{\theta} = 0.2$ rad/s²

θ

Prob. 12–163

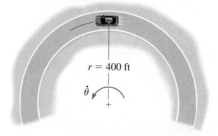

$r = 400$ ft

$\dot{\theta}$

Probs. 12–167/168

12–169. The time rate of change of acceleration is referred to as the *jerk*, which is often used as a means of measuring passenger discomfort. Calculate this vector, $\dot{\mathbf{a}}$, in terms of its cylindrical components, using Eq. 12–32.

*12–170. A particle is moving along a circular path having a radius of 6 in. such that its position as a function of time is given by $\theta = \sin 3t$, where θ and the argument for the sine are in radians, and t is in seconds. Determine the acceleration of the particle at $\theta = 30°$. The particle starts from rest at $\theta = 0°$.

12–171. The slotted link is pinned at O, and as a result of the constant angular velocity $\dot{\theta} = 3$ rad/s it drives the peg P for a short distance along the spiral guide $r = (0.4\,\theta)$ m, where θ is in radians. Determine the radial and transverse components of the velocity and acceleration of P at the instant $\theta = \pi/3$ rad.

12–172. Solve Prob. 12–171 if the slotted link has an angular acceleration $\ddot{\theta} = 8$ rad/s² when $\dot{\theta} = 3$ rad/s at $\theta = \pi/3$ rad.

12–173. The slotted link is pinned at O, and as a result of the constant angular velocity $\dot{\theta} = 3$ rad/s it drives the peg P for a short distance along the spiral guide $r = (0.4\,\theta)$ m, where θ is in radians. Determine the velocity and acceleration of the particle at the instant it leaves the slot in the link, i.e., when $r = 0.5$ m.

12–175. A particle P moves along the spiral path $r = (10/\theta)$ ft, where θ is in radians. If it maintains a constant speed of $v = 20$ ft/s, determine v_r and v_θ as functions of θ and evaluate each at $\theta = 1$ rad.

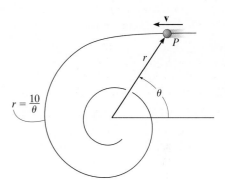

Prob. 12–175

*12–176. The driver of the car maintains a constant speed of 40 m/s. Determine the angular velocity of the camera tracking the car when $\theta = 15°$.

12–177. When $\theta = 15°$, the car has a speed of 50 m/s which is increasing at 6 m/s². Determine the angular velocity of the camera tracking the car at this instant.

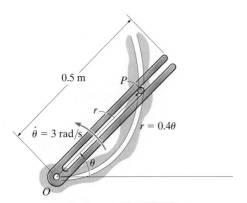

Probs. 12–171/172/173

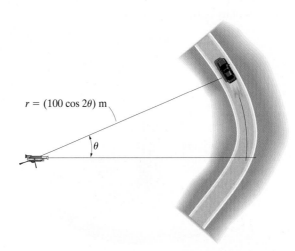

Probs. 12–176/177

12–174. A particle moves in the x–y plane such that its position is defined by $r = \{2t\mathbf{i} + 4t^2\mathbf{j}\}$ ft, where t is in seconds. Determine the radial and transverse components of the particle's velocity and acceleration when $t = 2$ s.

12–178. The small washer slides down the cord *OA*. When it is at the midpoint, its speed is 200 mm/s and its acceleration is 10 mm/s². Express the velocity and acceleration of the washer at this point in terms of its cylindrical components.

*12–180. Pin *P* is constrained to move along the curve defined by the lemniscate $r = (4 \sin 2\theta)$ ft. If the slotted arm *OA* rotates counterclockwise with a constant angular velocity of $\dot\theta = 1.5$ rad/s, determine the magnitudes of the velocity and acceleration of peg *P* when $\theta = 60°$.

12–181. Pin *P* is constrained to move along the curve defined by the lemniscate $r = (4 \sin 2\theta)$ ft. If the angular position of the slotted arm *OA* is defined by $\theta = (3t^{3/2})$ rad, determine the magnitudes of the velocity and acceleration of the pin *P* when $\theta = 60°$.

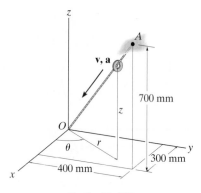

Prob. 12–178

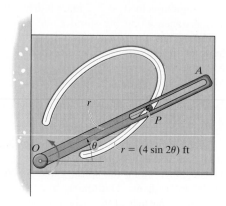

Probs. 12–180/181

12–179. A block moves outward along the slot in the platform with a speed of $\dot r = (4t)$ m/s, where *t* is in seconds. The platform rotates at a constant rate of 6 rad/s. If the block starts from rest at the center, determine the magnitudes of its velocity and acceleration when $t = 1$ s.

12–182. A cameraman standing at *A* is following the movement of a race car, *B*, which is traveling around a curved track at a constant speed of 30 m/s. Determine the angular rate $\dot\theta$ at which the man must turn in order to keep the camera directed on the car at the instant $\theta = 30°$.

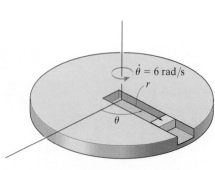

Prob. 12–179

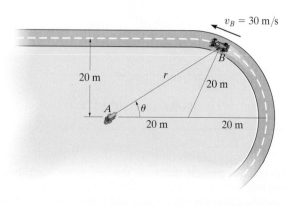

Prob. 12–182

12–183. The slotted arm AB drives pin C through the spiral groove described by the equation $r = a\theta$. If the angular velocity is constant at $\dot{\theta}$, determine the radial and transverse components of velocity and acceleration of the pin.

*12–184. The slotted arm AB drives pin C through the spiral groove described by the equation $r = (1.5\,\theta)$ ft, where θ is in radians. If the arm starts from rest when $\theta = 60°$ and is driven at an angular velocity of $\dot{\theta} = (4t)$ rad/s, where t is in seconds, determine the radial and transverse components of velocity and acceleration of the pin C when $t = 1$ s.

12–187. If the circular plate rotates clockwise with a constant angular velocity of $\dot{\theta} = 1.5$ rad/s, determine the magnitudes of the velocity and acceleration of the follower rod AB when $\theta = 2/3\pi$ rad.

*12–188. When $\theta = 2/3\pi$ rad, the angular velocity and angular acceleration of the circular plate are $\dot{\theta} = 1.5$ rad/s and $\ddot{\theta} = 3$ rad/s², respectively. Determine the magnitudes of the velocity and acceleration of the rod AB at this instant.

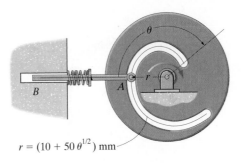

$r = (10 + 50\,\theta^{1/2})$ mm

Probs. 12–187/188

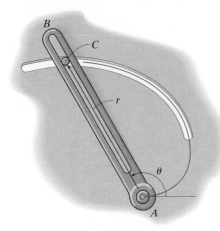

Probs. 12–183/184

12–185. If the slotted arm AB rotates counterclockwise with a constant angular velocity of $\dot{\theta} = 2$ rad/s, determine the magnitudes of the velocity and acceleration of peg P at $\theta = 30°$. The peg is constrained to move in the slots of the fixed bar CD and rotating bar AB.

12–186. The peg is constrained to move in the slots of the fixed bar CD and rotating bar AB. When $\theta = 30°$, the angular velocity and angular acceleration of arm AB are $\dot{\theta} = 2$ rad/s and $\ddot{\theta} = 3$ rad/s², respectively. Determine the magnitudes of the velocity and acceleration of the peg P at this instant.

12–189. The box slides down the helical ramp with a constant speed of $v = 2$ m/s. Determine the magnitude of its acceleration. The ramp descends a vertical distance of 1 m for every full revolution. The mean radius of the ramp is $r = 0.5$ m.

12–190. The box slides down the helical ramp such that $r = 0.5$ m, $\theta = (0.5t^3)$ rad, and $z = (2 - 0.2t^2)$ m, where t is in seconds. Determine the magnitudes of the velocity and acceleration of the box at the instant $\theta = 2\pi$ rad.

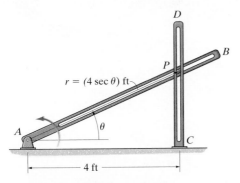

$r = (4 \sec \theta)$ ft

4 ft

Probs. 12–185/186

0.5 m

Probs. 12–189/190

12–191. For a *short distance* the train travels along a track having the shape of a spiral, $r = (1000/\theta)$ m, where θ is in radians. If it maintains a constant speed $v = 20$ m/s, determine the radial and transverse components of its velocity when $\theta = (9\pi/4)$ rad.

***12–192.** For a *short distance* the train travels along a track having the shape of a spiral, $r = (1000/\theta)$ m, where θ is in radians. If the angular rate is constant, $\dot{\theta} = 0.2$ rad/s, determine the radial and transverse components of its velocity and acceleration when $\theta = (9\pi/4)$ rad.

12–195. The arm of the robot has a fixed length of $r = 3$ ft and its grip A moves along the path $z = (3 \sin 4\theta)$ ft, where θ is in radians. If $\theta = (0.5t)$ rad, where t is in seconds, determine the magnitudes of the grip's velocity and acceleration when $t = 3$ s.

***12–196.** For a short time the arm of the robot is extending at a constant rate such that $\dot{r} = 1.5$ ft/s when $r = 3$ ft, $z = (4t^2)$ ft, and $\theta = 0.5t$ rad, where t is in seconds. Determine the magnitudes of the velocity and acceleration of the grip A when $t = 3$ s.

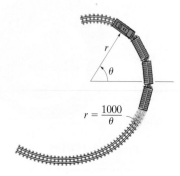

Probs. 12–191/192

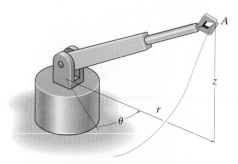

Probs. 12–195/196

12–193. A particle moves along an Archimedean spiral $r = (8\theta)$ ft, where θ is given in radians. If $\dot{\theta} = 4$ rad/s (constant), determine the radial and transverse components of the particle's velocity and acceleration at the instant $\theta = \pi/2$ rad. Sketch the curve and show the components on the curve.

12–194. Solve Prob. 12–193 if the particle has an angular acceleration $\ddot{\theta} = 5$ rad/s² when $\dot{\theta} = 4$ rad/s at $\theta = \pi/2$ rad.

12–197. The partial surface of the cam is that of a logarithmic spiral $r = (40e^{0.05\theta})$ mm, where θ is in radians. If the cam is rotating at a constant angular rate of $\dot{\theta} = 4$ rad/s, determine the magnitudes of the velocity and acceleration of the follower rod at the instant $\theta = 30°$.

12–198. Solve Prob. 12–197, if the cam has an angular acceleration of $\ddot{\theta} = 2$ rad/s² when its angular velocity is $\dot{\theta} = 4$ rad/s at $\theta = 30°$.

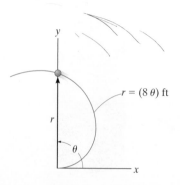

Probs. 12–193/194

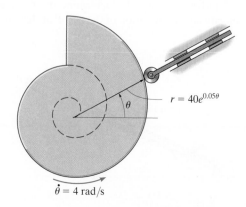

Probs. 12–197/198

12.9 Absolute Dependent Motion Analysis of Two Particles

In some types of problems the motion of one particle will *depend* on the corresponding motion of another particle. This dependency commonly occurs if the particles, here represented by blocks, are interconnected by inextensible cords which are wrapped around pulleys. For example, the movement of block A downward along the inclined plane in Fig. 12–36 will cause a corresponding movement of block B up the other incline. We can show this mathematically by first specifying the location of the blocks using *position coordinates* s_A and s_B. Note that each of the coordinate axes is (1) measured from a *fixed* point (O) or *fixed* datum line, (2) measured along each inclined plane *in the direction of motion* of each block, and (3) has a positive sense from the fixed datums to A and to B. If the total cord length is l_T, the two position coordinates are related by the equation

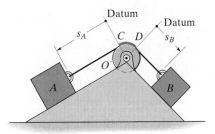

$$s_A + l_{CD} + s_B = l_T$$

Here l_{CD} is the length of the cord passing over arc CD. Taking the time derivative of this expression, realizing that l_{CD} and l_T remain *constant*, while s_A and s_B measure the segments of the cord that change in length, we have

$$\frac{ds_A}{dt} + \frac{ds_B}{dt} = 0 \quad \text{or} \quad v_B = -v_A$$

The negative sign indicates that when block A has a velocity downward, i.e., in the direction of positive s_A, it causes a corresponding upward velocity of block B; i.e., B moves in the negative s_B direction.

In a similar manner, time differentiation of the velocities yields the relation between the accelerations, i.e.,

$$a_B = -a_A$$

A more complicated example is shown in Fig. 12–37a. In this case, the position of block A is specified by s_A, and the position of the *end* of the cord from which block B is suspended is defined by s_B. As above, we have chosen position coordinates which (1) have their origin at fixed points or datums, (2) are measured in the direction of motion of each block, and (3) from the fixed datums are positive to the right for s_A and positive downward for s_B. During the motion, the length of the red colored segments of the cord in Fig. 12–37a remains constant. If l represents the total length of cord minus these segments, then the position coordinates can be related by the equation

$$2s_B + h + s_A = l$$

Since l and h are constant during the motion, the two time derivatives yield

$$2v_B = -v_A \quad 2a_B = -a_A$$

Hence, when B moves downward ($+s_B$), A moves to the left ($-s_A$) with twice the motion.

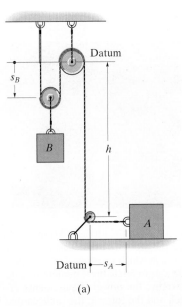

Fig. 12–36

(a)

Fig. 12–37

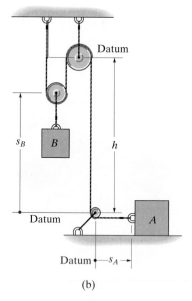

(b)

Fig. 12–37 (cont.)

This example can also be worked by defining the position of block B from the center of the bottom pulley (a fixed point), Fig. 12–37b. In this case

$$2(h - s_B) + h + s_A = l$$

Time differentiation yields

$$2v_B = v_A \qquad 2a_B = a_A$$

Here the signs are the same. Why?

The motion of the lift on this crane depends upon the motion of the cable connected to the winch which operates it. It is important to be able to relate these motions in order to determine the power requirements of the winch and the force in the cable caused by any accelerated motion.

Procedure for Analysis

The above method of relating the dependent motion of one particle to that of another can be performed using algebraic scalars or position coordinates provided each particle moves along a rectilinear path. When this is the case, only the magnitudes of the velocity and acceleration of the particles will change, not their line of direction.

Position-Coordinate Equation.

• Establish each position coordinate with an origin located at a *fixed* point or datum.

• It is *not necessary* that the *origin* be the *same* for each of the coordinates; however, it is *important* that each coordinate axis selected be directed along the *path of motion* of the particle.

• Using geometry or trigonometry, relate the position coordinates to the total length of the cord, l_T, or to that portion of cord, l, which *excludes* the segments that do not change length as the particles move—such as arc segments wrapped over pulleys.

• If a problem involves a *system* of two or more cords wrapped around pulleys, then the position of a point on one cord must be related to the position of a point on another cord using the above procedure. Separate equations are written for a fixed length of each cord of the system and the positions of the two particles are then related by these equations (see Examples 12.22 and 12.23).

Time Derivatives.

• Two successive time derivatives of the position-coordinate equations yield the required velocity and acceleration equations which relate the motions of the particles.

• The signs of the terms in these equations will be consistent with those that specify the positive and negative sense of the position coordinates.

EXAMPLE 12.21

Determine the speed of block A in Fig. 12–38 if block B has an upward speed of 6 ft/s.

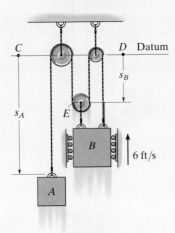

Fig. 12–38

SOLUTION

Position-Coordinate Equation. There is *one cord* in this system having segments which change length. Position coordinates s_A and s_B will be used since each is measured from a fixed point (C or D) and extends along each block's *path of motion*. In particular, s_B is directed to point E since motion of B and E is the *same*.

The red colored segments of the cord in Fig. 12–38 remain at a constant length and do not have to be considered as the blocks move. The remaining length of cord, l, is also constant and is related to the changing position coordinates s_A and s_B by the equation

$$s_A + 3s_B = l$$

Time Derivative. Taking the time derivative yields

$$v_A + 3v_B = 0$$

so that when $v_B = -6$ ft/s (upward),

$$v_A = 18 \text{ ft/s} \downarrow \qquad\qquad \textit{Ans.}$$

12

EXAMPLE | **12.22**

Determine the speed of A in Fig. 12–39 if B has an upward speed of 6 ft/s.

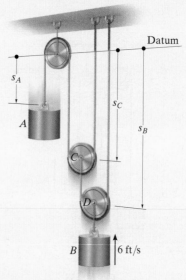

Fig. 12–39

SOLUTION

Position-Coordinate Equation. As shown, the positions of blocks A and B are defined using coordinates s_A and s_B. Since the system has *two cords* with segments that change length, it will be necessary to use a third coordinate, s_C, in order to relate s_A to s_B. In other words, the length of one of the cords can be expressed in terms of s_A and s_C, and the length of the other cord can be expressed in terms of s_B and s_C.

The red colored segments of the cords in Fig. 12–39 do not have to be considered in the analysis. Why? For the remaining cord lengths, say l_1 and l_2, we have

$$s_A + 2s_C = l_1 \qquad s_B + (s_B - s_C) = l_2$$

Time Derivative. Taking the time derivative of these equations yields

$$v_A + 2v_C = 0 \qquad 2v_B - v_C = 0$$

Eliminating v_C produces the relationship between the motions of each cylinder.

$$v_A + 4v_B = 0$$

so that when $v_B = -6$ ft/s (upward),

$$v_A = +24 \text{ ft/s} = 24 \text{ ft/s} \downarrow \qquad\qquad Ans.$$

EXAMPLE | 12.23

Determine the speed of block B in Fig. 12–40 if the end of the cord at A is pulled down with a speed of 2 m/s.

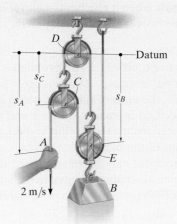

Fig. 12–40

SOLUTION

Position-Coordinate Equation.   The position of point A is defined by s_A, and the position of block B is specified by s_B since point E on the pulley will have the *same motion* as the block. Both coordinates are measured from a horizontal datum passing through the *fixed* pin at pulley D. Since the system consists of *two* cords, the coordinates s_A and s_B cannot be related directly. Instead, by establishing a third position coordinate, s_C, we can now express the length of one of the cords in terms of s_B and s_C, and the length of the other cord in terms of s_A, s_B, and s_C.

Excluding the red colored segments of the cords in Fig. 12–40, the remaining constant cord lengths l_1 and l_2 (along with the hook and link dimensions) can be expressed as

$$s_C + s_B = l_1$$
$$(s_A - s_C) + (s_B - s_C) + s_B = l_2$$

Time Derivative.   The time derivative of each equation gives

$$v_C + v_B = 0$$
$$v_A - 2v_C + 2v_B = 0$$

Eliminating v_C, we obtain

$$v_A + 4v_B = 0$$

so that when $v_A = 2$ m/s (downward),

$$v_B = -0.5 \text{ m/s} = 0.5 \text{ m/s} \uparrow \qquad\qquad Ans.$$

12

EXAMPLE | 12.24

A man at A is hoisting a safe S as shown in Fig. 12–41 by walking to the right with a constant velocity $v_A = 0.5$ m/s. Determine the velocity and acceleration of the safe when it reaches the elevation of 10 m. The rope is 30 m long and passes over a small pulley at D.

SOLUTION

Position-Coordinate Equation.   This problem is unlike the previous examples since rope segment DA changes *both direction and magnitude*. However, the ends of the rope, which define the positions of C and A, are specified by means of the x and y coordinates since they must be measured from a fixed point and *directed along the paths of motion* of the ends of the rope.

The x and y coordinates may be related since the rope has a fixed length $l = 30$ m, which at all times is equal to the length of segment DA plus CD. Using the Pythagorean theorem to determine l_{DA}, we have $l_{DA} = \sqrt{(15)^2 + x^2}$; also, $l_{CD} = 15 - y$. Hence,

$$l = l_{DA} + l_{CD}$$
$$30 = \sqrt{(15)^2 + x^2} + (15 - y)$$
$$y = \sqrt{225 + x^2} - 15 \qquad (1)$$

Time Derivatives.   Taking the time derivative, using the chain rule (see Appendix C), where $v_S = dy/dt$ and $v_A = dx/dt$, yields

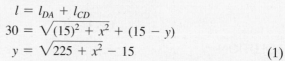

$$v_S = \frac{dy}{dt} = \left[\frac{1}{2} \frac{2x}{\sqrt{225 + x^2}} \right] \frac{dx}{dt}$$

$$= \frac{x}{\sqrt{225 + x^2}} v_A \qquad (2)$$

At $y = 10$ m, x is determined from Eq. 1, i.e., $x = 20$ m. Hence, from Eq. 2 with $v_A = 0.5$ m/s,

$$v_S = \frac{20}{\sqrt{225 + (20)^2}}(0.5) = 0.4 \text{ m/s} = 400 \text{ mm/s} \uparrow \qquad Ans.$$

The acceleration is determined by taking the time derivative of Eq. 2. Since v_A is constant, then $a_A = dv_A/dt = 0$, and we have

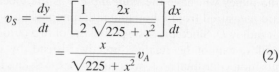

$$a_S = \frac{d^2y}{dt^2} = \left[\frac{-x(dx/dt)}{(225 + x^2)^{3/2}} \right] xv_A + \left[\frac{1}{\sqrt{225 + x^2}} \right] \left(\frac{dx}{dt} \right) v_A + \left[\frac{1}{\sqrt{225 + x^2}} \right] x \frac{dv_A}{dt} = \frac{225 v_A^2}{(225 + x^2)^{3/2}}$$

At $x = 20$ m, with $v_A = 0.5$ m/s, the acceleration becomes

$$a_S = \frac{225(0.5 \text{ m/s})^2}{[225 + (20 \text{ m})^2]^{3/2}} = 0.00360 \text{ m/s}^2 = 3.60 \text{ mm/s}^2 \uparrow \qquad Ans.$$

NOTE: The constant velocity at A causes the other end C of the rope to have an acceleration since $\mathbf{v}_A$ causes segment DA to change its direction as well as its length.

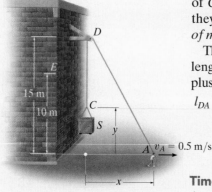

Fig. 12–41

12.10   Relative-Motion of Two Particles Using Translating Axes

Throughout this chapter the absolute motion of a particle has been determined using a single fixed reference frame. There are many cases, however, where the path of motion for a particle is complicated, so that it may be easier to analyze the motion in parts by using two or more frames of reference. For example, the motion of a particle located at the tip of an airplane propeller, while the plane is in flight, is more easily described if one observes first the motion of the airplane from a fixed reference and then superimposes (vectorially) the circular motion of the particle measured from a reference attached to the airplane.

In this section *translating frames of reference* will be considered for the analysis.

Position.   Consider particles A and B, which move along the arbitrary paths shown in Fig. 12–42. The *absolute position* of each particle, $\mathbf{r}_A$ and $\mathbf{r}_B$, is measured from the common origin O of the *fixed* x, y, z reference frame. The origin of a second frame of reference x', y', z' is attached to and moves with particle A. The axes of this frame are *only permitted to translate* relative to the fixed frame. The position of B measured relative to A is denoted by the *relative-position vector* $\mathbf{r}_{B/A}$. Using vector addition, the three vectors shown in Fig. 12–42 can be related by the equation

$$\mathbf{r}_B = \mathbf{r}_A + \mathbf{r}_{B/A} \qquad (12\text{–}33)$$

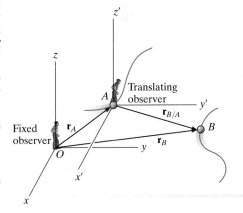

Fig. 12–42

Velocity.   An equation that relates the velocities of the particles is determined by taking the time derivative of the above equation; i.e.,

$$\mathbf{v}_B = \mathbf{v}_A + \mathbf{v}_{B/A} \qquad (12\text{–}34)$$

Here $\mathbf{v}_B = d\mathbf{r}_B/dt$ and $\mathbf{v}_A = d\mathbf{r}_A/dt$ refer to *absolute velocities*, since they are observed from the fixed frame; whereas the *relative velocity* $\mathbf{v}_{B/A} = d\mathbf{r}_{B/A}/dt$ is observed from the translating frame. It is important to note that since the x', y', z' axes translate, the *components* of $\mathbf{r}_{B/A}$ will *not* change direction and therefore the time derivative of these components will only have to account for the change in their magnitudes. Equation 12–34 therefore states that the velocity of B is equal to the velocity of A plus (vectorially) the velocity of "B with respect to A," as measured by the *translating observer* fixed in the x', y', z' reference frame.

12

Acceleration.   The time derivative of Eq. 12–34 yields a similar vector relation between the *absolute* and *relative accelerations* of particles A and B.

$$\mathbf{a}_B = \mathbf{a}_A + \mathbf{a}_{B/A}$$ (12–35)

Here $\mathbf{a}_{B/A}$ is the acceleration of B as seen by the observer located at A and translating with the x', y', z' reference frame.*

Procedure for Analysis

- When applying the relative velocity and acceleration equations, it is first necessary to specify the particle A that is the origin for the translating x', y', z' axes. Usually this point has a *known* velocity or acceleration.

- Since vector addition forms a triangle, there can be at most *two unknowns*, represented by the magnitudes and/or directions of the vector quantities.

- These unknowns can be solved for either graphically, using trigonometry (law of sines, law of cosines), or by resolving each of the three vectors into rectangular or Cartesian components, thereby generating a set of scalar equations.

The pilots of these jet planes flying close to one another must be aware of their relative positions and velocities at all times in order to avoid a collision.

---

* An easy way to remember the setup of these equations is to note the "cancellation" of the subscript A between the two terms, e.g., $\mathbf{a}_B = \mathbf{a}_{\cancel{A}} + \mathbf{a}_{B/\cancel{A}}$.

EXAMPLE   12.25

A train travels at a constant speed of 60 mi/h and crosses over a road as shown in Fig. 12–43*a*. If the automobile *A* is traveling at 45 mi/h along the road, determine the magnitude and direction of the velocity of the train relative to the automobile.

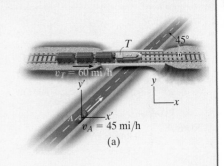

(a)

SOLUTION I

Vector Analysis. The relative velocity $\mathbf{v}_{T/A}$ is measured from the translating x', y' axes attached to the automobile, Fig. 12–43*a*. It is determined from $\mathbf{v}_T = \mathbf{v}_A + \mathbf{v}_{T/A}$. Since $\mathbf{v}_T$ and $\mathbf{v}_A$ are known in *both* magnitude and direction, the unknowns become the x and y components of $\mathbf{v}_{T/A}$. Using the x, y axes in Fig. 12–43*a*, we have

$$\mathbf{v}_T = \mathbf{v}_A + \mathbf{v}_{T/A}$$

$$60\mathbf{i} = (45\cos 45°\mathbf{i} + 45\sin 45°\mathbf{j}) + \mathbf{v}_{T/A}$$

$$\mathbf{v}_{T/A} = \{28.2\mathbf{i} - 31.8\mathbf{j}\}\ \text{mi/h}$$

The magnitude of $\mathbf{v}_{T/A}$ is thus

$$v_{T/A} = \sqrt{(28.2)^2 + (-31.8)^2} = 42.5\ \text{mi/h} \qquad \textit{Ans.}$$

From the direction of each component, Fig. 12–43*b*, the direction of $\mathbf{v}_{T/A}$ is

$$\tan\theta = \frac{(v_{T/A})_y}{(v_{T/A})_x} = \frac{31.8}{28.2}$$

$$\theta = 48.5° \quad c \qquad\qquad \textit{Ans.}$$

Note that the vector addition shown in Fig. 12–43*b* indicates the correct sense for $\mathbf{v}_{T/A}$. This figure anticipates the answer and can be used to check it.

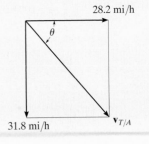

(b)

SOLUTION II

Scalar Analysis. The unknown components of $\mathbf{v}_{T/A}$ can also be determined by applying a scalar analysis. We will assume these components act in the *positive* x and y directions. Thus,

$$\mathbf{v}_T = \mathbf{v}_A + \mathbf{v}_{T/A}$$

$$\begin{bmatrix} 60\ \text{mi/h} \\ \rightarrow \end{bmatrix} = \begin{bmatrix} 45\ \text{mi/h} \\ a\ ^{45°} \end{bmatrix} + \begin{bmatrix} (v_{T/A})_x \\ \rightarrow \end{bmatrix} + \begin{bmatrix} (v_{T/A})_y \\ \uparrow \end{bmatrix}$$

Resolving each vector into its x and y components yields

$(\xrightarrow{+})$ $60 = 45\cos 45° + (v_{T/A})_x + 0$

$(+\uparrow)$ $0 = 45\sin 45° + 0 + (v_{T/A})_y$

Solving, we obtain the previous results,

$$(v_{T/A})_x = 28.2\ \text{mi/h} = 28.2\ \text{mi/h} \rightarrow$$

$$(v_{T/A})_y = -31.8\ \text{mi/h} = 31.8\ \text{mi/h} \downarrow$$

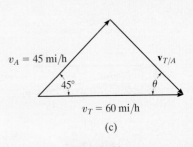

(c)

Fig. 12–43

EXAMPLE | 12.26

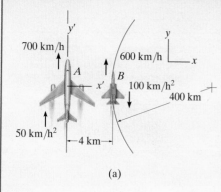

(a)

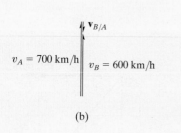

(b)

Plane A in Fig. 12–44a is flying along a straight-line path, whereas plane B is flying along a circular path having a radius of curvature of $\rho_B = 400$ km. Determine the velocity and acceleration of B as measured by the pilot of A.

SOLUTION

Velocity. The origin of the x *and* y axes are located at an arbitrary fixed point. Since the motion relative to plane A is to be determined, the *translating frame of reference* x', y' is attached to it, Fig. 12–44a. Applying the relative-velocity equation in scalar form since the velocity vectors of both planes are parallel at the instant shown, we have

$(+\uparrow)$
$$v_B = v_A + v_{B/A}$$
$$600 \text{ km/h} = 700 \text{ km/h} + v_{B/A}$$
$$v_{B/A} = -100 \text{ km/h} = 100 \text{ km/h} \downarrow \qquad Ans.$$

The vector addition is shown in Fig. 12–44b.

Acceleration. Plane B has both tangential and normal components of acceleration since it is flying along a *curved path*. From Eq. 12–20, the magnitude of the normal component is

$$(a_B)_n = \frac{v_B^2}{\rho} = \frac{(600 \text{ km/h})^2}{400 \text{ km}} = 900 \text{ km/h}^2$$

Applying the relative-acceleration equation gives

$$\mathbf{a}_B = \mathbf{a}_A + \mathbf{a}_{B/A}$$
$$900\mathbf{i} - 100\mathbf{j} = 50\mathbf{j} + \mathbf{a}_{B/A}$$

Thus,

$$\mathbf{a}_{B/A} = \{900\mathbf{i} - 150\mathbf{j}\} \text{ km/h}^2$$

From Fig. 12–44c, the magnitude and direction of $\mathbf{a}_{B/A}$ are therefore

$$a_{B/A} = 912 \text{ km/h}^2 \quad \theta = \tan^{-1}\frac{150}{900} = 9.46° \quad \text{⟍} \qquad Ans.$$

NOTE: The solution to this problem was possible using a translating frame of reference, since the pilot in plane A is "translating." Observation of the motion of plane A with respect to the pilot of plane B, however, must be obtained using a *rotating* set of axes attached to plane B. (This assumes, of course, that the pilot of B is fixed in the rotating frame, so he does not turn his eyes to follow the motion of A.) The analysis for this case is given in Example 16.21.

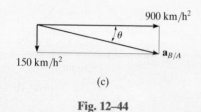

(c)

Fig. 12–44

EXAMPLE | 12.27

At the instant shown in Fig. 12–45a, cars A and B are traveling with speeds of 18 m/s and 12 m/s, respectively. Also at this instant, A has a decrease in speed of 2 m/s², and B has an increase in speed of 3 m/s². Determine the velocity and acceleration of B with respect to A.

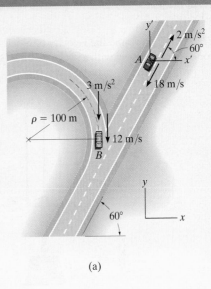

SOLUTION

Velocity. The fixed x, y axes are established at an arbitrary point on the ground and the translating x', y' axes are attached to car A, Fig. 12–45a. Why? The relative velocity is determined from $\mathbf{v}_B = \mathbf{v}_A + \mathbf{v}_{B/A}$. What are the two unknowns? Using a Cartesian vector analysis, we have

$$\mathbf{v}_B = \mathbf{v}_A + \mathbf{v}_{B/A}$$

$$-12\mathbf{j} = (-18\cos 60°\mathbf{i} - 18\sin 60°\mathbf{j}) + \mathbf{v}_{B/A}$$

$$\mathbf{v}_{B/A} = \{9\mathbf{i} + 3.588\mathbf{j}\} \text{ m/s}$$

(a)

Thus,

$$v_{B/A} = \sqrt{(9)^2 + (3.588)^2} = 9.69 \text{ m/s} \qquad Ans.$$

Noting that $\mathbf{v}_{B/A}$ has $+\mathbf{i}$ and $+\mathbf{j}$ components, Fig. 12–45b, its direction is

$$\tan \theta = \frac{(v_{B/A})_y}{(v_{B/A})_x} = \frac{3.588}{9}$$

$$\theta = 21.7° \qquad a \qquad Ans.$$

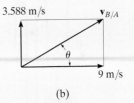

(b)

Acceleration. Car B has both tangential and normal components of acceleration. Why? The magnitude of the normal component is

$$(a_B)_n = \frac{v_B^2}{\rho} = \frac{(12 \text{ m/s})^2}{100 \text{ m}} = 1.440 \text{ m/s}^2$$

Applying the equation for relative acceleration yields

$$\mathbf{a}_B = \mathbf{a}_A + \mathbf{a}_{B/A}$$

$$(-1.440\mathbf{i} - 3\mathbf{j}) = (2\cos 60°\mathbf{i} + 2\sin 60°\mathbf{j}) + \mathbf{a}_{B/A}$$

$$\mathbf{a}_{B/A} = \{-2.440\mathbf{i} - 4.732\mathbf{j}\} \text{ m/s}^2$$

Here $\mathbf{a}_{B/A}$ has $-\mathbf{i}$ and $-\mathbf{j}$ components. Thus, from Fig. 12–45c,

$$a_{B/A} = \sqrt{(2.440)^2 + (4.732)^2} = 5.32 \text{ m/s}^2 \qquad Ans.$$

$$\tan \phi = \frac{(a_{B/A})_y}{(a_{B/A})_x} = \frac{4.732}{2.440}$$

$$\phi = 62.7° \qquad d \qquad Ans.$$

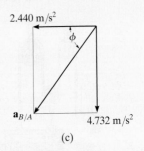

(c)

NOTE: Is it possible to obtain the relative acceleration of $\mathbf{a}_{A/B}$ using this method? Refer to the comment made at the end of Example 12.26.

Fig. 12–45

F12–39. Determine the velocity of block D if end A of the rope is pulled down with a speed of $v_A = 3$ m/s.

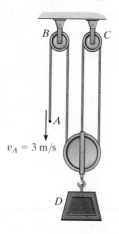

F12–39

F12–40. Determine the velocity of block A if end B of the rope is pulled down with a speed of 6 m/s.

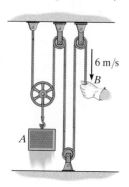

F12–40

F12–41. Determine the velocity of block A if end B of the rope is pulled down with a speed of 1.5 m/s.

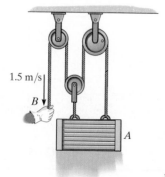

F12–41

F12–42. Determine the velocity of block A if end F of the rope is pulled down with a speed of $v_F = 3$ m/s.

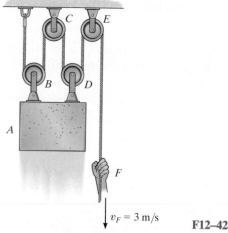

F12–42

F12–43. Determine the velocity of car A if point P on the cable has a speed of 4 m/s when the motor M winds the cable in.

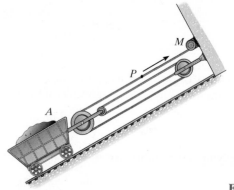

F12–43

F12–44. Determine the velocity of cylinder B if cylinder A moves downward with a speed of $v_A = 4$ ft/s.

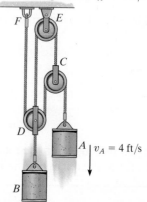

F12–44

F12–45. Car *A* is traveling with a constant speed of 80 km/h due north, while car *B* is traveling with a constant speed of 100 km/h due east. Determine the velocity of car *B* relative to car *A*.

F12–47. The boats *A* and *B* travel with constant speeds of $v_A = 15$ m/s and $v_B = 10$ m/s when they leave the pier at *O* at the same time. Determine the distance between them when $t = 4$ s.

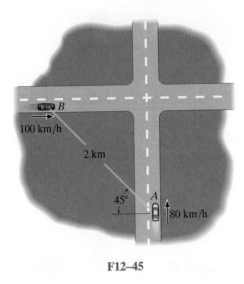

F12–45

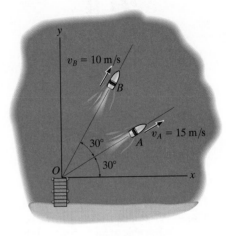

F12–47

F12–46. Two planes *A* and *B* are traveling with the constant velocities shown. Determine the magnitude and direction of the velocity of plane *B* relative to plane *A*.

F12–48. At the instant shown, cars *A* and *B* are traveling at the speeds shown. If *B* is accelerating at 1200 km/h^2 while *A* maintains a constant speed, determine the velocity and acceleration of *A* with respect to *B*.

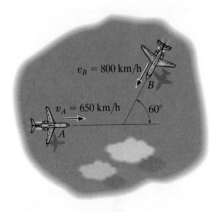

F12–46

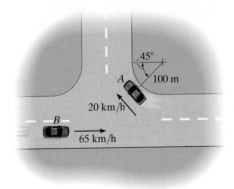

F12–48

PROBLEMS

12–199. If the end of the cable at A is pulled down with a speed of 2 m/s, determine the speed at which block B rises.

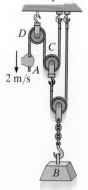

Prob. 12–199

*12–200. The motor at C pulls in the cable with an acceleration $a_C = (3t^2)\,\text{m/s}^2$, where t is in seconds. The motor at D draws in its cable at $a_D = 5\,\text{m/s}^2$. If both motors start at the same instant from rest when $d = 3$ m, determine (a) the time needed from $d = 0$, and (b) the velocities of blocks A and B when this occurs.

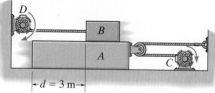

Prob. 12–200

12–201. The crate is being lifted up the inclined plane using the motor M and the rope and pulley arrangement shown. Determine the speed at which the cable must be taken up by the motor in order to move the crate up the plane with a constant speed of 4 ft/s.

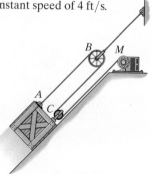

Prob. 12–201

12–202. Determine the time needed for the load at B to attain a speed of 8 m/s, starting from rest, if the cable is drawn into the motor with an acceleration of $0.2\,\text{m/s}^2$.

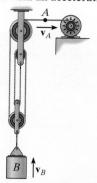

Prob. 12–202

12–203. Determine the displacement of the log if the truck at C pulls the cable 4 ft to the right.

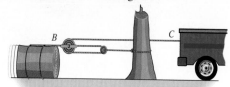

Prob. 12–203

*12–204. Determine the speed of cylinder A, if the rope is drawn toward the motor M at a constant rate of 10 m/s.

12–205. If the rope is drawn toward the motor M at a speed of $v_M = (5t^{3/2})\,\text{m/s}$, where t is in seconds, determine the speed of cylinder A when $t = 1$ s.

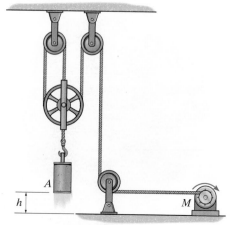

Probs. 12–204/205

12–206. If the hydraulic cylinder H draws in rod BC at 2 ft/s, determine the speed of slider A.

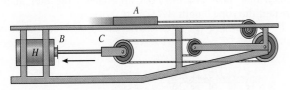

Prob. 12–206

12–207. If block A is moving downward with a speed of 4 ft/s while C is moving up at 2 ft/s, determine the velocity of block B.

*12–208. If block A is moving downward at 6 ft/s while block C is moving down at 18 ft/s, determine the speed of block B.

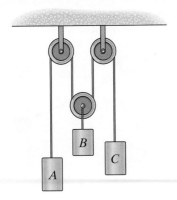

Probs. 12–207/208

12–209. Determine the displacement of block B if A is pulled down 4 ft.

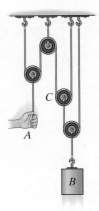

Prob. 12–209

12–210. The pulley arrangement shown is designed for hoisting materials. If BC *remains fixed* while the plunger P is pushed downward with a speed of 4 ft/s, determine the speed of the load at A.

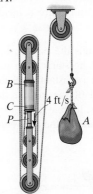

Prob. 12–210

12–211. Determine the speed of block A if the end of the rope is pulled down with a speed of 4 m/s.

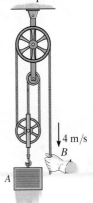

Prob. 12–211

*12–212. The cylinder C is being lifted using the cable and pulley system shown. If point A on the cable is being drawn toward the drum with a speed of 2 m/s, determine the velocity of the cylinder.

Prob. 12–212

12–213. The man pulls the boy up to the tree limb C by walking backward at a constant speed of 1.5 m/s. Determine the speed at which the boy is being lifted at the instant $x_A = 4$ m. Neglect the size of the limb. When $x_A = 0$, $y_B = 8$ m, so that A and B are coincident, i.e., the rope is 16 m long.

12–214. The man pulls the boy up to the tree limb C by walking backward. If he starts from rest when $x_A = 0$ and moves backward with a constant acceleration $a_A = 0.2$ m/s², determine the speed of the boy at the instant $y_B = 4$ m. Neglect the size of the limb. When $x_A = 0$, $y_B = 8$ m, so that A and B are coincident, i.e., the rope is 16 m long.

Probs. 12–213/214

12–215. The roller at A is moving upward with a velocity of $v_A = 3$ ft/s and has an acceleration of $a_A = 4$ ft/s² when $s_A = 4$ ft. Determine the velocity and acceleration of block B at this instant.

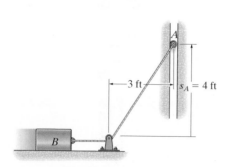

Prob. 12–215

*12–216. The girl at C stands near the edge of the pier and pulls in the rope *horizontally* at a constant speed of 6 ft/s. Determine how fast the boat approaches the pier at the instant the rope length AB is 50 ft.

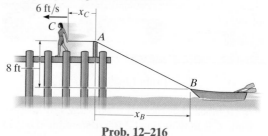

Prob. 12–216

12–217. The crate C is being lifted by moving the roller at A downward with a constant speed of $v_A = 2$ m/s along the guide. Determine the velocity and acceleration of the crate at the instant $s = 1$ m. When the roller is at B, the crate rests on the ground. Neglect the size of the pulley in the calculation. *Hint:* Relate the coordinates x_C and x_A using the problem geometry, then take the first and second time derivatives.

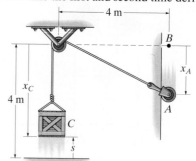

Prob. 12–217

12–218. The man can row the boat in still water with a speed of 5 m/s. If the river is flowing at 2 m/s, determine the speed of the boat and the angle θ he must direct the boat so that it travels from A to B.

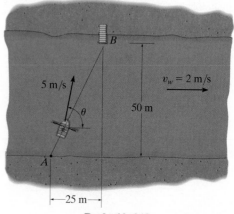

Prob. 12–218

12–219. Vertical motion of the load is produced by movement of the piston at A on the boom. Determine the distance the piston or pulley at C must move to the left in order to lift the load 2 ft. The cable is attached at B, passes over the pulley at C, then D, E, F, and again around E, and is attached at G.

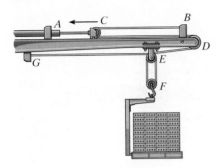

Prob. 12–219

*12–220. If block B is moving down with a velocity v_B and has an acceleration a_B, determine the velocity and acceleration of block A in terms of the parameters shown.

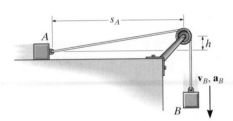

Prob. 12–220

12–221. Collars A and B are connected to the cord that passes over the small pulley at C. When A is located at D, B is 24 ft to the left of D. If A moves at a constant speed of 2 ft/s to the right, determine the speed of B when A is 4 ft to the right of D.

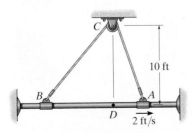

Prob. 12–221

12–222. Two planes, A and B, are flying at the same altitude. If their velocities are $v_A = 600$ km/h and $v_B = 500$ km/h such that the angle between their straight-line courses is $\theta = 75°$, determine the velocity of plane B with respect to plane A.

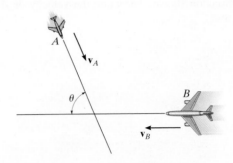

Prob. 12–222

12–223. At the instant shown, cars A and B are traveling at speeds of 55 mi/h and 40 mi/h, respectively. If B is increasing its speed by 1200 mi/h², while A maintains a constant speed, determine the velocity and acceleration of B with respect to A. Car B moves along a curve having a radius of curvature of 0.5 mi.

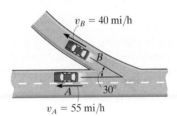

Prob. 12–223

*12–224. At the instant shown, car A travels along the straight portion of the road with a speed of 25 m/s. At this same instant car B travels along the circular portion of the road with a speed of 15 m/s. Determine the velocity of car B relative to car A.

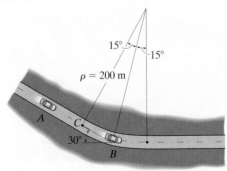

Prob. 12–224

12–225. An aircraft carrier is traveling forward with a velocity of 50 km/h. At the instant shown, the plane at *A* has just taken off and has attained a forward horizontal air speed of 200 km/h, measured from still water. If the plane at *B* is traveling along the runway of the carrier at 175 km/h in the direction shown, determine the velocity of *A* with respect to *B*.

*12–228. At the instant shown, the bicyclist at *A* is traveling at 7 m/s around the curve on the race track while increasing his speed at 0.5 m/s². The bicyclist at *B* is traveling at 8.5 m/s along the straight-a-way and increasing his speed at 0.7 m/s². Determine the relative velocity and relative acceleration of *A* with respect to *B* at this instant.

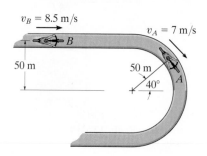

$v_B = 8.5$ m/s $v_A = 7$ m/s
50 m   50 m   40°

Prob. 12–228

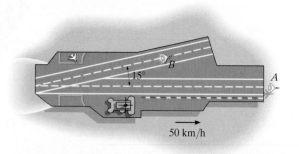

50 km/h

Prob. 12–225

12–226. A car is traveling north along a straight road at 50 km/h. An instrument in the car indicates that the wind is coming from the east. If the car's speed is 80 km/h, the instrument indicates that the wind is coming from the north-east. Determine the speed and direction of the wind.

12–227. Two boats leave the shore at the same time and travel in the directions shown. If $v_A = 20$ ft/s and $v_B = 15$ ft/s, determine the velocity of boat *A* with respect to boat *B*. How long after leaving the shore will the boats be 800 ft apart?

12–229. Cars *A* and *B* are traveling around the circular race track. At the instant shown, *A* has a speed of 90 ft/s and is increasing its speed at the rate of 15 ft/s², whereas *B* has a speed of 105 ft/s and is decreasing its speed at 25 ft/s². Determine the relative velocity and relative acceleration of car *A* with respect to car *B* at this instant.

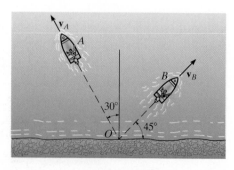

Prob. 12–227

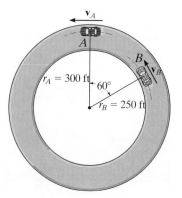

v_A
A
$r_A = 300$ ft 60°
B v_B
$r_B = 250$ ft

Prob. 12–229

12–230. The two cyclists A and B travel at the same constant speed v. Determine the speed of A with respect to B if A travels along the circular track, while B travels along the diameter of the circle.

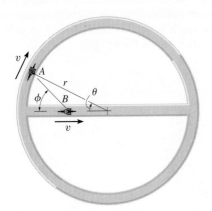

Prob. 12–230

12–231. At the instant shown, cars A and B travel at speeds of 70 mi/h and 50 mi/h, respectively. If B is increasing its speed by 1100 mi/h^2, while A maintains a constant speed, determine the velocity and acceleration of B with respect to A. Car B moves along a curve having a radius of curvature of 0.7 mi.

*12–232. At the instant shown, cars A and B travel at speeds of 70 mi/h and 50 mi/h, respectively. If B is decreasing its speed at 1400 mi/h^2 while A is increasing its speed at 800 mi/h^2, determine the acceleration of B with respect to A. Car B moves along a curve having a radius of curvature of 0.7 mi.

12–233. A passenger in an automobile observes that raindrops make an angle of 30° with the horizontal as the auto travels forward with a speed of 60 km/h. Compute the terminal (constant) velocity $\mathbf{v}_r$ of the rain if it is assumed to fall vertically.

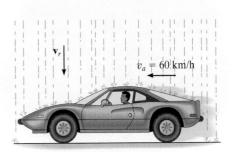

Prob. 12–233

12–234. A man can swim at 4 ft/s in still water. He wishes to cross the 40-ft-wide river to point B, 30 ft downstream. If the river flows with a velocity of 2 ft/s, determine the speed of the man and the time needed to make the crossing. *Note:* While in the water he must not direct himself toward point B to reach this point. Why?

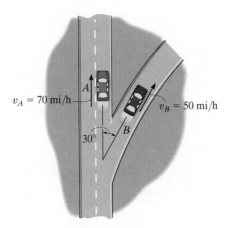

Probs. 12–231/232

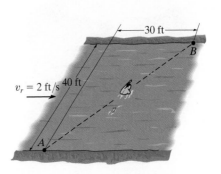

Prob. 12–234

12–235. The ship travels at a constant speed of $v_s = 20$ m/s and the wind is blowing at a speed of $v_w = 10$ m/s, as shown. Determine the magnitude and direction of the horizontal component of velocity of the smoke coming from the smoke stack as it appears to a passenger on the ship.

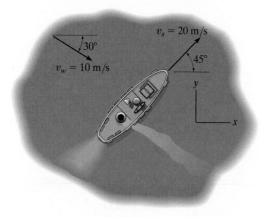

Prob. 12–235

*12–236. Car A travels along a straight road at a speed of 25 m/s while accelerating at 1.5 m/s². At this same instant car C is traveling along the straight road with a speed of 30 m/s while decelerating at 3 m/s². Determine the velocity and acceleration of car A relative to car C.

12–237. Car B is traveling along the curved road with a speed of 15 m/s while decreasing its speed at 2 m/s². At this same instant car C is traveling along the straight road with a speed of 30 m/s while decelerating at 3 m/s². Determine the velocity and acceleration of car B relative to car C.

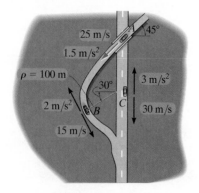

Probs. 12–236/237

12–238. At a given instant the football player at A throws a football C with a velocity of 20 m/s in the direction shown. Determine the constant speed at which the player at B must run so that he can catch the football at the same elevation at which it was thrown. Also calculate the relative velocity and relative acceleration of the football with respect to B at the instant the catch is made. Player B is 15 m away from A when A starts to throw the football.

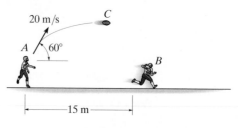

Prob. 12–238

12–239. Both boats A and B leave the shore at O at the same time. If A travels at v_A and B travels at v_B, write a general expression to determine the velocity of A with respect to B.

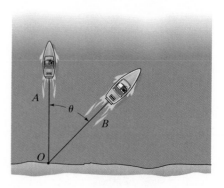

Prob. 12–239

CONCEPTUAL PROBLEMS

P12–1. If you measured the time it takes for the construction elevator to go from *A* to *B*, then *B* to *C*, and then *C* to *D*, and you also know the distance between each of the points, how could you determine the average velocity and average acceleration of the elevator as it ascends from *A* to *D*? Use numerical values to explain how this can be done.

P12–1

P12–2. If the sprinkler at *A* is 1 m from the ground, then scale the necessary measurements from the photo to determine the approximate velocity of the water jet as it flows from the nozzle of the sprinkler.

P12–2

P12–3. The basketball was thrown at an angle measured from the horizontal to the man's outstretched arm. If the basket is 3 m from the ground, make appropriate measurements in the photo and determine if the ball located as shown will pass through the basket.

P12–3

P12–4. The pilot tells you the wingspan of her plane and her constant airspeed. How would you determine the acceleration of the plane at the moment shown? Use numerical values and take any necessary measurements from the photo.

P12–4

CHAPTER REVIEW

Rectilinear Kinematics

Rectilinear kinematics refers to motion along a straight line. A position coordinate s specifies the location of the particle on the line, and the displacement Δs is the change in this position.

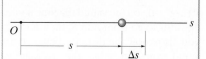

The average velocity is a vector quantity, defined as the displacement divided by the time interval.

$$v_{\text{avg}} = \frac{\Delta s}{\Delta t}$$

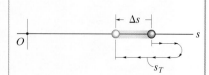

The average speed is a scalar, and is the total distance traveled divided by the time of travel.

$$(v_{\text{sp}})_{\text{avg}} = \frac{s_T}{\Delta t}$$

The time, position, velocity, and acceleration are related by three differential equations.

$$a = \frac{dv}{dt}, \qquad v = \frac{ds}{dt}, \qquad a\,ds = v\,dv$$

If the acceleration is known to be constant, then the differential equations relating time, position, velocity, and acceleration can be integrated.

$$v = v_0 + a_c t$$
$$s = s_0 + v_0 t + \tfrac{1}{2} a_c t^2$$
$$v^2 = v_0^2 + 2a_c(s - s_0)$$

Graphical Solutions

If the motion is erratic, then it can be described by a graph. If one of these graphs is given, then the others can be established using the differential relations between a, v, s, and t.

$$a = \frac{dv}{dt},$$
$$v = \frac{ds}{dt},$$
$$a\,ds = v\,dv$$

Curvilinear Motion, x, y, z

Curvilinear motion along the path can be resolved into rectilinear motion along the x, y, z axes. The equation of the path is used to relate the motion along each axis.

$$v_x = \dot{x} \qquad a_x = \dot{v}_x$$
$$v_y = \dot{y} \qquad a_y = \dot{v}_y$$
$$v_z = \dot{z} \qquad a_z = \dot{v}_z$$

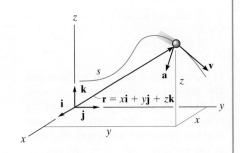

Projectile Motion

Free-flight motion of a projectile follows a parabolic path. It has a constant velocity in the horizontal direction, and a constant downward acceleration of $g = 9.81$ m/s^2 or 32.2 ft/s^2 in the vertical direction. Any two of the three equations for constant acceleration apply in the vertical direction, and in the horizontal direction only one equation applies.

$$(+\uparrow) \quad v_y = (v_0)_y + a_c t$$
$$(+\uparrow) \quad y = y_0 + (v_0)_y t + \tfrac{1}{2} a_c t^2$$
$$(+\uparrow) \quad v_y^2 = (v_0)_y^2 + 2a_c(y - y_0)$$
$$(\xrightarrow{+}) \quad x = x_0 + (v_0)_x t$$

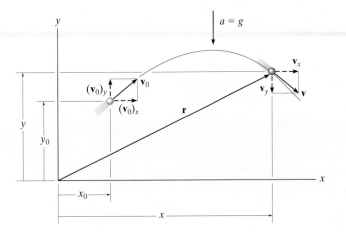

12

Curvilinear Motion n, t

If normal and tangential axes are used for the analysis, then **v** is always in the positive t direction.

The acceleration has two components. The tangential component, $\mathbf{a}_t$, accounts for the change in the magnitude of the velocity; a slowing down is in the negative t direction, and a speeding up is in the positive t direction. The normal component $\mathbf{a}_n$ accounts for the change in the direction of the velocity. This component is always in the positive n direction.

$$a_t = \dot{v} \quad \text{or} \quad a_t\, ds = v\, dv$$

$$a_n = \frac{v^2}{\rho}$$

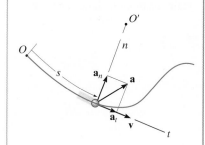

Curvilinear Motion r, θ

If the path of motion is expressed in polar coordinates, then the velocity and acceleration components can be related to the time derivatives of r and θ.

To apply the time-derivative equations, it is necessary to determine $r, \dot{r}, \ddot{r}, \dot{\theta}, \ddot{\theta}$ at the instant considered. If the path $r = f(\theta)$ is given, then the chain rule of calculus must be used to obtain time derivatives. (See Appendix C.)

Once the data are substituted into the equations, then the algebraic sign of the results will indicate the direction of the components of **v** or **a** along each axis.

$$v_r = \dot{r}$$

$$v_\theta = r\dot{\theta}$$

$$a_r = \ddot{r} - r\dot{\theta}^2$$

$$a_\theta = r\ddot{\theta} + 2\dot{r}\dot{\theta}$$

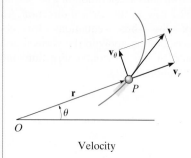

Velocity

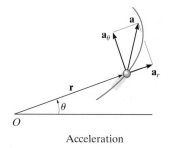

Acceleration

Absolute Dependent Motion of Two Particles

The dependent motion of blocks that are suspended from pulleys and cables can be related by the geometry of the system. This is done by first establishing position coordinates, measured from a fixed origin to each block. Each coordinate must be directed along the line of motion of a block.

Using geometry and/or trigonometry, the coordinates are then related to the cable length in order to formulate a position coordinate equation.

$$2s_B + h + s_A = l$$

The first time derivative of this equation gives a relationship between the velocities of the blocks, and a second time derivative gives the relation between their accelerations.

$$2v_B = -v_A$$

$$2a_B = -a_A$$

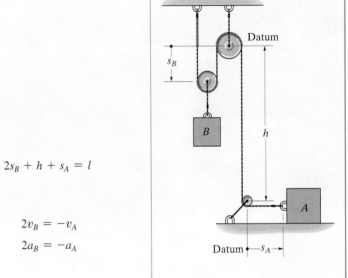

Relative-Motion Analysis Using Translating Axes

If two particles A and B undergo independent motions, then these motions can be related to their relative motion using a *translating set of axes* attached to one of the particles (A).

$$\mathbf{r}_B = \mathbf{r}_A + \mathbf{r}_{B/A}$$

For planar motion, each vector equation produces two scalar equations, one in the x, and the other in the y direction. For solution, the vectors can be expressed in Cartesian form, or the x and y scalar components can be written directly.

$$\mathbf{v}_B = \mathbf{v}_A + \mathbf{v}_{B/A}$$

$$\mathbf{a}_B = \mathbf{a}_A + \mathbf{a}_{B/A}$$

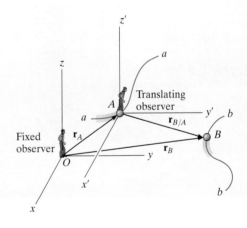

A car driving along this road will be subjected to forces that create both normal and tangential accelerations. In this chapter we will study how these forces are related to the accelerations they create.

Kinetics of a Particle: Force and Acceleration

CHAPTER OBJECTIVES

■ To state Newton's Second Law of Motion and to define mass and weight.

■ To analyze the accelerated motion of a particle using the equation of motion with different coordinate systems.

■ To investigate central-force motion and apply it to problems in space mechanics.

13.1 Newton's Second Law of Motion

Kinetics is a branch of dynamics that deals with the relationship between the change in motion of a body and the forces that cause this change. The basis for kinetics is Newton's second law, which states that when an *unbalanced force* acts on a particle, the particle will *accelerate* in the direction of the force with a magnitude that is proportional to the force.

This law can be verified experimentally by applying a known unbalanced force $\mathbf{F}$ to a particle, and then measuring the acceleration $\mathbf{a}$. Since the force and acceleration are directly proportional, the constant of proportionality, m, may be determined from the ratio $m = F/a$. This positive scalar m is called the *mass* of the particle. Being constant during any acceleration, m provides a quantitative measure of the resistance of the particle to a change in its velocity, that is its inertia.

If the mass of the particle is m, Newton's second law of motion may be written in mathematical form as

$$\mathbf{F} = m\mathbf{a}$$

The above equation, which is referred to as the *equation of motion*, is one of the most important formulations in mechanics.* As previously stated, its validity is based solely on *experimental evidence*. In 1905, however, Albert Einstein developed the theory of relativity and placed limitations on the use of Newton's second law for describing general particle motion. Through experiments it was proven that *time* is not an absolute quantity as assumed by Newton; and as a result, the equation of motion fails to predict the exact behavior of a particle, especially when the particle's speed approaches the speed of light (0.3 Gm/s). Developments of the theory of quantum mechanics by Erwin Schrödinger and others indicate further that conclusions drawn from using this equation are also invalid when particles are the size of an atom and move close to one another. For the most part, however, these requirements regarding particle speed and size are not encountered in engineering problems, so their effects will not be considered in this book.

Newton's Law of Gravitational Attraction.
Shortly after formulating his three laws of motion, Newton postulated a law governing the mutual attraction between any two particles. In mathematical form this law can be expressed as

$$F = G\frac{m_1 m_2}{r^2} \qquad (13\text{–}1)$$

where

$F =$ force of attraction between the two particles
$G =$ universal constant of gravitation; according to experimental evidence $G = 66.73(10^{-12})$ m^3/(kg · s^2)
$m_1, m_2 =$ mass of each of the two particles
$r =$ distance between the centers of the two particles

---

*Since m is constant, we can also write $\mathbf{F} = d(m\mathbf{v})/dt$, where $m\mathbf{v}$ is the particle's linear momentum. Here the unbalanced force acting on the particle is proportional to the time rate of change of the particle's linear momentum.

In the case of a particle located at or near the surface of the earth, the only gravitational force having any sizable magnitude is that between the earth and the particle. This force is termed the "weight" and, for our purpose, it will be the only gravitational force considered.

From Eq. 13–1, we can develop a general expression for finding the weight W of a particle having a mass $m_1 = m$. Let $m_2 = M_e$ be the mass of the earth and r the distance between the earth's center and the particle. Then, if $g = GM_e/r^2$, we have

$$W = mg$$

By comparison with $F = ma$, we term g the acceleration due to gravity. For most engineering calculations g is measured at a point on the surface of the earth at sea level, and at a latitude of 45°, which is considered the "standard location." Here the values $g = 9.81 \text{ m/s}^2 = 32.2 \text{ ft/s}^2$ will be used for calculations.

In the SI system the mass of the body is specified in kilograms, and the weight must be calculated using the above equation, Fig. 13–1a. Thus,

$$W = mg \text{ (N)} \quad (g = 9.81 \text{ m/s}^2) \qquad (13\text{–}2)$$

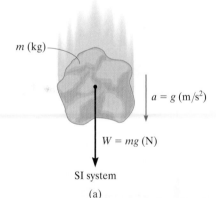

m (kg)

$a = g \text{ (m/s}^2)$

$W = mg \text{ (N)}$

SI system
(a)

As a result, a body of mass 1 kg has a weight of 9.81 N; a 2-kg body weighs 19.62 N; and so on.

In the FPS system the weight of the body is specified in pounds. The mass is measured in slugs, a term derived from "sluggish" which refers to the body's inertia. It must be calculated, Fig. 13–1b, using

$$m = \frac{W}{g} \text{ (slug)} \quad (g = 32.2 \text{ ft/s}^2) \qquad (13\text{–}3)$$

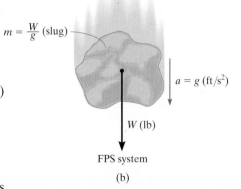

$m = \dfrac{W}{g}$ (slug)

$a = g \text{ (ft/s}^2)$

W (lb)

FPS system
(b)

Therefore, a body weighing 32.2 lb has a mass of 1 slug; a 64.4-lb body has a mass of 2 slugs; and so on.

Fig. 13–1

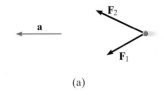

(a)

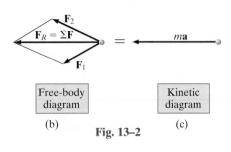

Free-body diagram

Kinetic diagram

(b)

Fig. 13–2

(c)

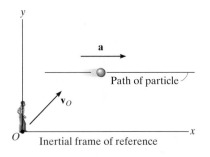

Inertial frame of reference

Fig. 13–3

13.2 The Equation of Motion

When more than one force acts on a particle, the resultant force is determined by a vector summation of all the forces; i.e., $\mathbf{F}_R = \Sigma\mathbf{F}$. For this more general case, the equation of motion may be written as

$$\Sigma\mathbf{F} = m\mathbf{a} \qquad (13\text{–}4)$$

To illustrate application of this equation, consider the particle shown in Fig. 13–2a, which has a mass m and is subjected to the action of two forces, $\mathbf{F}_1$ and $\mathbf{F}_2$. We can graphically account for the magnitude and direction of each force acting on the particle by drawing the particle's *free-body diagram*, Fig. 13–2b. Since the *resultant* of these forces *produces* the vector $m\mathbf{a}$, its magnitude and direction can be represented graphically on the *kinetic diagram*, shown in Fig. 13–2c.* The equal sign written between the diagrams symbolizes the *graphical* equivalency between the free-body diagram and the kinetic diagram; i.e., $\Sigma\mathbf{F} = m\mathbf{a}$.† In particular, note that if $\mathbf{F}_R = \Sigma\mathbf{F} = \mathbf{0}$, then the acceleration is also zero, so that the particle will either remain at *rest* or move along a straight-line path with *constant velocity*. Such are the conditions of *static equilibrium*, Newton's first law of motion.

Inertial Reference Frame. When applying the equation of motion, it is important that the acceleration of the particle be measured with respect to a reference frame that is *either fixed or translates with a constant velocity*. In this way, the observer will not accelerate and measurements of the particle's acceleration will be the *same* from *any reference* of this type. Such a frame of reference is commonly known as a *Newtonian* or *inertial reference frame*, Fig. 13–3.

When studying the motions of rockets and satellites, it is justifiable to consider the inertial reference frame as fixed to the stars, whereas dynamics problems concerned with motions on or near the surface of the earth may be solved by using an inertial frame which is assumed fixed to the earth. Even though the earth both rotates about its own axis and revolves about the sun, the accelerations created by these rotations are relatively small and so they can be neglected for most applications.

---

*Recall the free-body diagram considers the particle to be free of its surrounding supports and shows all the forces acting on the particle. The kinetic diagram pertains to the particle's motion as caused by the forces.

†The equation of motion can also be rewritten in the form $\Sigma\mathbf{F} - m\mathbf{a} = \mathbf{0}$. The vector $-m\mathbf{a}$ is referred to as the *inertia force vector*. If it is treated in the same way as a "force vector," then the state of "equilibrium" created is referred to as *dynamic equilibrium*. This method of application, which will not be used in this text, is often referred to as the *D'Alembert principle*, named after the French mathematician Jean le Rond d'Alembert.

We are all familiar with the sensation one feels when sitting in a car that is subjected to a forward acceleration. Often people think this is caused by a "force" which acts on them and tends to push them back in their seats; however, this is not the case. Instead, this sensation occurs due to their inertia or the resistance of their mass to a change in velocity.

Consider the passenger who is strapped to the seat of a rocket sled. Provided the sled is at rest or is moving with constant velocity, then no force is exerted on his back as shown on his free-body diagram.

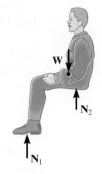

At rest or constant velocity

When the thrust of the rocket engine causes the sled to accelerate, then the seat upon which he is sitting exerts a force **F** on him which pushes him forward with the sled. In the photo, notice that the inertia of his head resists this change in motion (acceleration), and so his head moves back against the seat and his face, which is nonrigid, tends to distort backward.

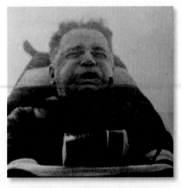
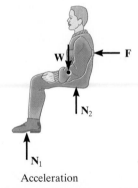

Acceleration

Upon deceleration the force of the seatbelt **F′** tends to pull his body to a stop, but his head leaves contact with the back of the seat and his face distorts forward, again due to his inertia or tendency to continue to move forward. No force is pulling him forward, although this is the sensation he receives.

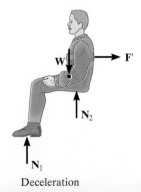

Deceleration

13.3   Equation of Motion for a System of Particles

The equation of motion will now be extended to include a system of particles isolated within an enclosed region in space, as shown in Fig. 13–4*a*. In particular, there is no restriction in the way the particles are connected, so the following analysis applies equally well to the motion of a solid, liquid, or gas system.

At the instant considered, the arbitrary *i*-th particle, having a mass m_i, is subjected to a system of internal forces and a resultant external force. The *internal force*, represented symbolically as $\mathbf{f}_i$, is the resultant of all the forces the other particles exert on the *i*th particle. The *resultant external force* $\mathbf{F}_i$ represents, for example, the effect of gravitational, electrical, magnetic, or contact forces between the *i*th particle and adjacent bodies or particles *not* included within the system.

The free-body and kinetic diagrams for the *i*th particle are shown in Fig. 13–4*b*. Applying the equation of motion,

$$\Sigma \mathbf{F} = m\mathbf{a}; \qquad\qquad \mathbf{F}_i + \mathbf{f}_i = m_i \mathbf{a}_i$$

When the equation of motion is applied to each of the other particles of the system, similar equations will result. And, if all these equations are added together *vectorially*, we obtain

$$\Sigma \mathbf{F}_i + \Sigma \mathbf{f}_i = \Sigma m_i \mathbf{a}_i$$

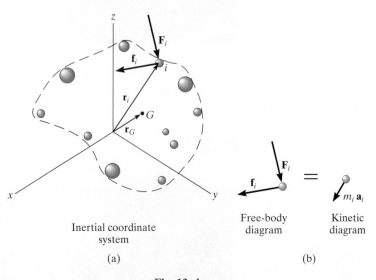

Inertial coordinate
system

(a)

Free-body
diagram

Kinetic
diagram

(b)

Fig. 13–4

The summation of the internal forces, if carried out, will equal zero, since internal forces between any two particles occur in equal but opposite collinear pairs. Consequently, only the sum of the external forces will remain, and therefore the equation of motion, written for the system of particles, becomes

$$\Sigma \mathbf{F}_i = \Sigma m_i \mathbf{a}_i \qquad (13\text{--}5)$$

If $\mathbf{r}_G$ is a position vector which locates the *center of mass* G of the particles, Fig. 13–4a, then by definition of the center of mass, $m\mathbf{r}_G = \Sigma m_i \mathbf{r}_i$, where $m = \Sigma m_i$ is the total mass of all the particles. Differentiating this equation twice with respect to time, assuming that no mass is entering or leaving the system, yields

$$m\mathbf{a}_G = \Sigma m_i \mathbf{a}_i$$

Substituting this result into Eq. 13–5, we obtain

$$\boxed{\Sigma \mathbf{F} = m\mathbf{a}_G} \qquad (13\text{--}6)$$

Hence, the sum of the external forces acting on the system of particles is equal to the total mass of the particles times the acceleration of its center of mass G. Since in reality all particles must have a finite size to possess mass, Eq. 13–6 justifies application of the equation of motion to a *body* that is represented as a single particle.

Important Points

- The equation of motion is based on experimental evidence and is valid only when applied within an inertial frame of reference.
- The equation of motion states that the *unbalanced force* on a particle causes it to *accelerate*.
- An inertial frame of reference does not rotate, rather its axes either translate with constant velocity or are at rest.
- Mass is a property of matter that provides a quantitative measure of its resistance to a change in velocity. It is an absolute quantity and so it does not change from one location to another.
- Weight is a force that is caused by the earth's gravitation. It is not absolute; rather it depends on the altitude of the mass from the earth's surface.

13.4   Equations of Motion: Rectangular Coordinates

When a particle moves relative to an inertial x, y, z frame of reference, the forces acting on the particle, as well as its acceleration, can be expressed in terms of their $\mathbf{i}, \mathbf{j}, \mathbf{k}$ components, Fig. 13–5. Applying the equation of motion, we have

$$\Sigma \mathbf{F} = m\mathbf{a}; \qquad \Sigma F_x \mathbf{i} + \Sigma F_y \mathbf{j} + \Sigma F_z \mathbf{k} = m(a_x \mathbf{i} + a_y \mathbf{j} + a_z \mathbf{k})$$

For this equation to be satisfied, the respective $\mathbf{i}, \mathbf{j}, \mathbf{k}$ components on the left side must equal the corresponding components on the right side. Consequently, we may write the following three scalar equations:

$$
\begin{aligned}
\Sigma F_x &= ma_x \\
\Sigma F_y &= ma_y \\
\Sigma F_z &= ma_z
\end{aligned}
\qquad (13\text{–}7)
$$

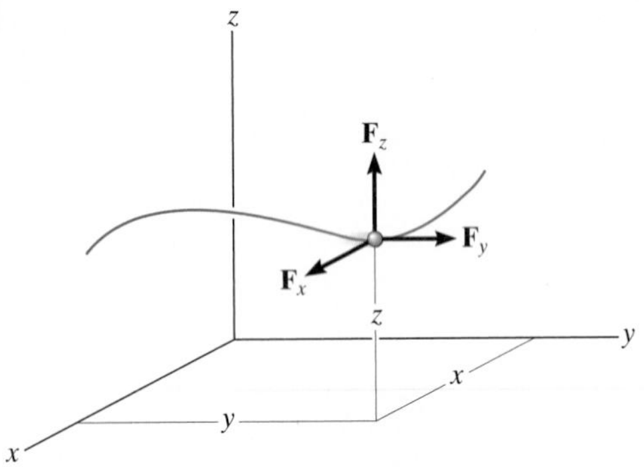

In particular, if the particle is constrained to move only in the x–y plane, then the first two of these equations are used to specify the motion.

Fig. 13–5

Procedure for Analysis

The equations of motion are used to solve problems which require a relationship between the forces acting on a particle and the accelerated motion they cause.

Free-Body Diagram.

- Select the inertial coordinate system. Most often, rectangular or x, y, z coordinates are chosen to analyze problems for which the particle has *rectilinear motion*.

- Once the coordinates are established, draw the particle's free-body diagram. Drawing this diagram is *very important* since it provides a graphical representation that accounts for *all the forces* ($\Sigma \mathbf{F}$) which act on the particle, and thereby makes it possible to resolve these forces into their x, y, z components.

- The direction and sense of the particle's acceleration $\mathbf{a}$ should also be established. If the sense is unknown, for mathematical convenience assume that the sense of each acceleration component acts in the *same direction* as its *positive* inertial coordinate axis.

- The acceleration may be represented as the $m\mathbf{a}$ vector on the kinetic diagram.*

- Identify the unknowns in the problem.

*It is a convention in this text always to use the kinetic diagram as a graphical aid when developing the proofs and theory. The particle's acceleration or its components will be shown as blue colored vectors near the free-body diagram in the examples.

Equations of Motion.

- If the forces can be resolved directly from the free-body diagram, apply the equations of motion in their scalar component form.

- If the geometry of the problem appears complicated, which often occurs in three dimensions, Cartesian vector analysis can be used for the solution.

- *Friction.* If a moving particle contacts a rough surface, it may be necessary to use the *frictional equation*, which relates the frictional and normal forces F_f and N acting at the surface of contact by using the coefficient of kinetic friction, i.e., $F_f = \mu_k N$. Remember that F_f always acts on the free-body diagram such that it opposes the motion of the particle relative to the surface it contacts. If the particle is *on the verge* of relative motion, then the coefficient of static friction should be used.

- *Spring.* If the particle is connected to an *elastic spring* having negligible mass, the spring force F_s can be related to the deformation of the spring by the equation $F_s = ks$. Here k is the spring's stiffness measured as a force per unit length, and s is the stretch or compression defined as the difference between the deformed length l and the undeformed length l_0, i.e., $s = l - l_0$.

Kinematics.

- If the velocity or position of the particle is to be found, it will be necessary to apply the necessary kinematic equations once the particle's acceleration is determined from $\Sigma \mathbf{F} = m\mathbf{a}$.

- If *acceleration* is a function of time, use $a = dv/dt$ and $v = ds/dt$ which, when integrated, yield the particle's velocity and position, respectively.

- If *acceleration* is a function of displacement, integrate $a\,ds = v\,dv$ to obtain the velocity as a function of position.

- If *acceleration is constant*, use $v = v_0 + a_c t$, $s = s_0 + v_0 t + \frac{1}{2}a_c t^2$, $v^2 = v_0^2 + 2a_c(s - s_0)$ to determine the velocity or position of the particle.

- If the problem involves the dependent motion of several particles, use the method outlined in Sec. 12.9 to relate their accelerations. In all cases, make sure the positive inertial coordinate directions used for writing the kinematic equations are the same as those used for writing the equations of motion; otherwise, simultaneous solution of the equations will result in errors.

- If the solution for an unknown vector component yields a negative scalar, it indicates that the component acts in the direction opposite to that which was assumed.

13

EXAMPLE 13.1

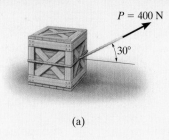

(a)

The 50-kg crate shown in Fig. 13–6a rests on a horizontal surface for which the coefficient of kinetic friction is $\mu_k = 0.3$. If the crate is subjected to a 400-N towing force as shown, determine the velocity of the crate in 3 s starting from rest.

SOLUTION

Using the equations of motion, we can relate the crate's acceleration to the force causing the motion. The crate's velocity can then be determined using kinematics.

Free-Body Diagram.   The weight of the crate is $W = mg = 50$ kg $(9.81$ m/s$^2) = 490.5$ N. As shown in Fig. 13–6b, the frictional force has a magnitude $F = \mu_k N_C$ and acts to the left, since it opposes the motion of the crate. The acceleration **a** is assumed to act horizontally, in the positive x direction. There are two unknowns, namely N_C and a.

Equations of Motion.   Using the data shown on the free-body diagram, we have

$$\xrightarrow{+} \Sigma F_x = ma_x; \quad 400 \cos 30° - 0.3N_C = 50a \tag{1}$$
$$+\uparrow \Sigma F_y = ma_y; \quad N_C - 490.5 + 400 \sin 30° = 0 \tag{2}$$

Solving Eq. 2 for N_C, substituting the result into Eq. 1, and solving for a yields

$$N_C = 290.5 \text{ N}$$
$$a = 5.185 \text{ m/s}^2$$

Kinematics.   Notice that the acceleration is *constant*, since the applied force **P** is constant. Since the initial velocity is zero, the velocity of the crate in 3 s is

$$(\xrightarrow{+}) \qquad v = v_0 + a_c t = 0 + 5.185(3)$$
$$= 15.6 \text{ m/s} \rightarrow \qquad\qquad Ans.$$

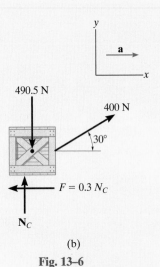

(b)

Fig. 13–6

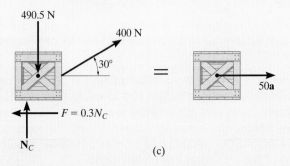

(c)

NOTE: We can also use the alternative procedure of drawing the crate's free-body *and* kinetic diagrams, Fig. 13–6c, prior to applying the equations of motion.

EXAMPLE 13.2

A 10-kg projectile is fired vertically upward from the ground, with an initial velocity of 50 m/s, Fig. 13–7a. Determine the maximum height to which it will travel if (a) atmospheric resistance is neglected; and (b) atmospheric resistance is measured as $F_D = (0.01v^2)$ N, where v is the speed of the projectile at any instant, measured in m/s.

(a)

SOLUTION

In both cases the known force on the projectile can be related to its acceleration using the equation of motion. Kinematics can then be used to relate the projectile's acceleration to its position.

Part (a) Free-Body Diagram. As shown in Fig. 13–7b, the projectile's weight is $W = mg = 10(9.81) = 98.1$ N. We will assume the unknown acceleration **a** acts upward in the *positive z* direction.

Equation of Motion.

$$+\uparrow \Sigma F_z = ma_z; \qquad -98.1 = 10a, \qquad a = -9.81 \text{ m/s}^2$$

The result indicates that the projectile, like every object having free-flight motion near the earth's surface, is subjected to a *constant* downward acceleration of 9.81 m/s².

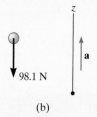

(b)

Kinematics. Initially, $z_0 = 0$ and $v_0 = 50$ m/s, and at the maximum height $z = h$, $v = 0$. Since the acceleration is *constant*, then

$$(+\uparrow) \qquad v^2 = v_0^2 + 2a_c(z - z_0)$$
$$0 = (50)^2 + 2(-9.81)(h - 0)$$
$$h = 127 \text{ m} \qquad\qquad Ans.$$

Part (b) Free-Body Diagram. Since the force $F_D = (0.01v^2)$ N tends to retard the upward motion of the projectile, it acts downward as shown on the free-body diagram, Fig. 13–7c.

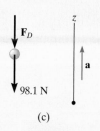

(c)

Fig. 13–7

Equation of Motion.

$$+\uparrow \Sigma F_z = ma_z; \quad -0.01v^2 - 98.1 = 10a, \quad a = -(0.001v^2 + 9.81)$$

Kinematics. Here the acceleration is *not constant* since F_D depends on the velocity. Since $a = f(v)$, we can relate a to position using

$$(+\uparrow)\, a\, dz = v\, dv; \qquad -(0.001v^2 + 9.81)\, dz = v\, dv$$

Separating the variables and integrating, realizing that initially $z_0 = 0$, $v_0 = 50$ m/s (positive upward), and at $z = h$, $v = 0$, we have

$$\int_0^h dz = -\int_{50 \text{ m/s}}^0 \frac{v\, dv}{0.001v^2 + 9.81} = -500 \ln(v^2 + 9810)\Big|_{50 \text{ m/s}}^0$$

$$h = 114 \text{ m} \qquad\qquad Ans.$$

NOTE: The answer indicates a lower elevation than that obtained in part (a) due to atmospheric resistance or drag.

EXAMPLE 13.3

The baggage truck A shown in the photo has a weight of 900 lb and tows a 550-lb cart B and a 325-lb cart C. For a short time the driving frictional force developed at the wheels of the truck is $F_A = (40t)$ lb, where t is in seconds. If the truck starts from rest, determine its speed in 2 seconds. Also, what is the horizontal force acting on the coupling between the truck and cart B at this instant? Neglect the size of the truck and carts.

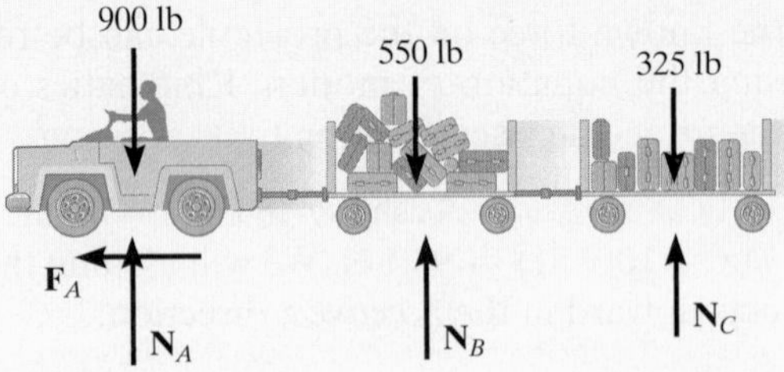

(a)

SOLUTION

Free-Body Diagram. As shown in Fig. 13–8a, it is the frictional driving force that gives both the truck and carts an acceleration. Here we have considered all three vehicles as a single system.

Equation of Motion. Only motion in the horizontal direction has to be considered.

$$\overset{+}{\leftarrow} \Sigma F_x = ma_x; \qquad 40t = \left(\frac{900 + 550 + 325}{32.2}\right)a$$

$$a = 0.7256t$$

Kinematics. Since the acceleration is a function of time, the velocity of the truck is obtained using $a = dv/dt$ with the initial condition that $v_0 = 0$ at $t = 0$. We have

$$\int_0^v dv = \int_0^{2\text{ s}} 0.7256t\,dt; \qquad v = 0.3628t^2\Big|_0^{2\text{ s}} = 1.45\text{ ft/s} \qquad \textit{Ans.}$$

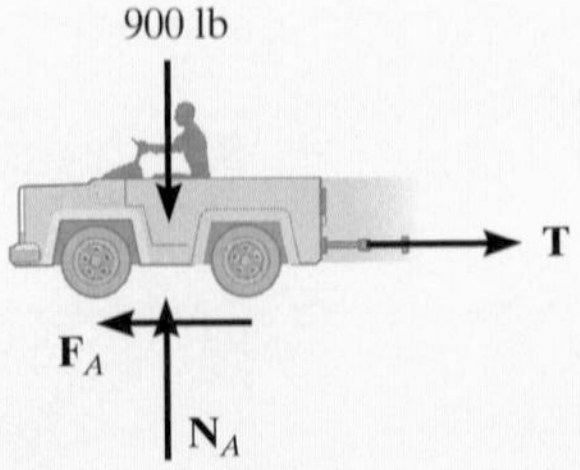

(b)

Fig. 13–8

Free-Body Diagram. In order to determine the force between the truck and cart B, we will consider a free-body diagram of the truck so that we can "expose" the coupling force **T** as external to the free-body diagram, Fig. 13–8b.

Equation of Motion. When $t = 2$ s, then

$$\overset{+}{\leftarrow} \Sigma F_x = ma_x: \qquad 40(2) - T = \left(\frac{900}{32.2}\right)[0.7256(2)]$$

$$T = 39.4\text{ lb} \qquad \textit{Ans.}$$

NOTE: Try and obtain this same result by considering a free-body diagram of carts B and C as a single system.

EXAMPLE | 13.4

A smooth 2-kg collar C, shown in Fig. 13–9a, is attached to a spring having a stiffness $k = 3$ N/m and an unstretched length of 0.75 m. If the collar is released from rest at A, determine its acceleration and the normal force of the rod on the collar at the instant $y = 1$ m.

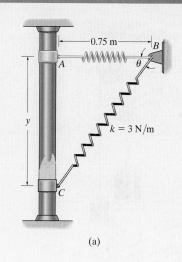

(a)

SOLUTION

Free-Body Diagram.   The free-body diagram of the collar when it is located at the arbitrary position y is shown in Fig. 13–9b. Furthermore, the collar is *assumed* to be accelerating so that "**a**" acts downward in the *positive y* direction. There are four unknowns, namely, N_C, F_s, a, and θ.

Equations of Motion.

$$\xrightarrow{+} \Sigma F_x = ma_x; \qquad -N_C + F_s \cos \theta = 0 \qquad (1)$$

$$+\downarrow \Sigma F_y = ma_y; \qquad 19.62 - F_s \sin \theta = 2a \qquad (2)$$

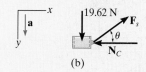

(b)

Fig. 13–9

From Eq. 2 it is seen that the acceleration depends on the magnitude and direction of the spring force. Solution for N_C and a is possible once F_s and θ are known.

The magnitude of the spring force is a function of the stretch s of the spring; i.e., $F_s = ks$. Here the unstretched length is $AB = 0.75$ m, Fig. 13–9a; therefore, $s = CB - AB = \sqrt{y^2 + (0.75)^2} - 0.75$. Since $k = 3$ N/m, then

$$F_s = ks = 3\left(\sqrt{y^2 + (0.75)^2} - 0.75 \right) \qquad (3)$$

From Fig. 13–9a, the angle θ is related to y by trigonometry.

$$\tan \theta = \frac{y}{0.75}$$

Substituting $y = 1$ m into Eqs. 3 and 4 yields $F_s = 1.50$ N and $\theta = 53.1°$. Substituting these results into Eqs. 1 and 2, we obtain

$$N_C = 0.900 \text{ N} \qquad\qquad Ans.$$

$$a = 9.21 \text{ m/s}^2 \downarrow \qquad\qquad Ans.$$

NOTE: This is not a case of constant acceleration, since the spring force changes both its magnitude and direction as the collar moves downward.

EXAMPLE | 13.5

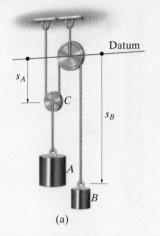

(a)

2T

(b)

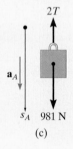

(c)

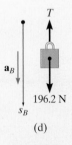

(d)

Fig. 13–10

The 100-kg block A shown in Fig. 13–10a is released from rest. If the masses of the pulleys and the cord are neglected, determine the velocity of the 20-kg block B in 2 s.

SOLUTION

Free-Body Diagrams.   Since the mass of the pulleys is *neglected*, then for pulley C, $ma = 0$ and we can apply $\Sigma F_y = 0$ as shown in Fig. 13–10b. The free-body diagrams for blocks A and B are shown in Fig. 13–10c and d, respectively. Notice that for A to remain stationary $T = 490.5$ N, whereas for B to remain static $T = 196.2$ N. Hence A will move down while B moves up. Although this is the case, we will assume both blocks accelerate downward, in the direction of $+s_A$ and $+s_B$. The three unknowns are T, a_A, and a_B.

Equations of Motion.   Block A,

$$+\downarrow \Sigma F_y = ma_y; \qquad\qquad 981 - 2T = 100a_A \qquad\qquad (1)$$

Block B,

$$+\downarrow \Sigma F_y = ma_y; \qquad\qquad 196.2 - T = 20a_B \qquad\qquad (2)$$

Kinematics.   The necessary third equation is obtained by relating a_A to a_B using a dependent motion analysis, discussed in Sec. 12.9. The coordinates s_A and s_B in Fig. 13–10a measure the positions of A and B from the fixed datum. It is seen that

$$2s_A + s_B = l$$

where l is constant and represents the total vertical length of cord. Differentiating this expression twice with respect to time yields

$$2a_A = -a_B \qquad\qquad (3)$$

Notice that when writing Eqs. 1 to 3, the *positive direction was always assumed downward*. It is very important to be *consistent* in this assumption since we are seeking a simultaneous solution of equations. The results are

$$T = 327.0 \text{ N}$$
$$a_A = 3.27 \text{ m/s}^2$$
$$a_B = -6.54 \text{ m/s}^2$$

Hence when block A accelerates *downward*, block B accelerates *upward* as expected. Since a_B is constant, the velocity of block B in 2 s is thus

$$(+\downarrow) \qquad\qquad v = v_0 + a_B t$$
$$= 0 + (-6.54)(2)$$
$$= -13.1 \text{ m/s} \qquad\qquad Ans.$$

The negative sign indicates that block B is moving upward.

FUNDAMENTAL PROBLEMS

F13–1. The motor winds in the cable with a constant acceleration, such that the 20-kg crate moves a distance $s = 6$ m in 3 s, starting from rest. Determine the tension developed in the cable. The coefficient of kinetic friction between the crate and the plane is $\mu_k = 0.3$.

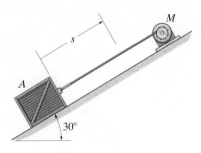

F13–1

F13–2. If motor M exerts a force of $F = (10t^2 + 100)$ N on the cable, where t is in seconds, determine the velocity of the 25-kg crate when $t = 4$ s. The coefficients of static and kinetic friction between the crate and the plane are $\mu_s = 0.3$ and $\mu_k = 0.25$, respectively. The crate is initially at rest.

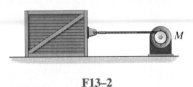

F13–2

F13–3. A spring of stiffness $k = 500$ N/m is mounted against the 10-kg block. If the block is subjected to the force of $F = 500$ N, determine its velocity at $s = 0.5$ m. When $s = 0$, the block is at rest and the spring is uncompressed. The contact surface is smooth.

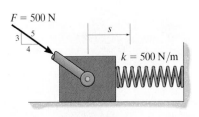

F13–3

F13–4. The 2-Mg car is being towed by a winch. If the winch exerts a force of $T = 100(s + 1)$ N on the cable, where s is the displacement of the car in meters, determine the speed of the car when $s = 10$ m, starting from rest. Neglect rolling resistance of the car.

F13–4

F13–5. The spring has a stiffness $k = 200$ N/m and is unstretched when the 25-kg block is at A. Determine the acceleration of the block when $s = 0.4$ m. The contact surface between the block and the plane is smooth.

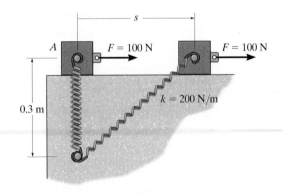

F13–5

F13–6. Block B rests upon a smooth surface. If the coefficients of static and kinetic friction between A and B are $\mu_s = 0.4$ and $\mu_k = 0.3$, respectively, determine the acceleration of each block if $P = 6$ lb.

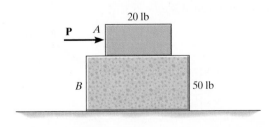

F13–6

PROBLEMS

13–1. The 6-lb particle is subjected to the action of its weight and forces $\mathbf{F}_1 = \{2\mathbf{i} + 6\mathbf{j} - 2t\mathbf{k}\}$ lb, $\mathbf{F}_2 = \{t^2\mathbf{i} - 4t\mathbf{j} - 1\mathbf{k}\}$ lb, and $\mathbf{F}_3 = \{-2t\mathbf{i}\}$ lb, where t is in seconds. Determine the distance the ball is from the origin 2 s after being released from rest.

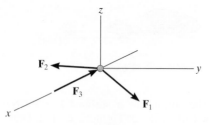

Prob. 13–1

13–2. The 10-lb block has an initial velocity of 10 ft/s on the smooth plane. If a force $F = (2.5t)$ lb, where t is in seconds, acts on the block for 3 s, determine the final velocity of the block and the distance the block travels during this time.

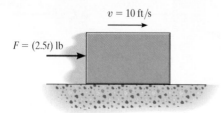

Prob. 13–2

13–3. If the coefficient of kinetic friction between the 50-kg crate and the ground is $\mu_k = 0.3$, determine the distance the crate travels and its velocity when $t = 3$ s. The crate starts from rest, and $P = 200$ N.

*13–4. If the 50-kg crate starts from rest and achieves a velocity of $v = 4$ m/s when it travels a distance of 5 m to the right, determine the magnitude of force $\mathbf{P}$ acting on the crate. The coefficient of kinetic friction between crate and the ground is $\mu_k = 0.3$.

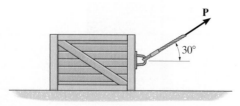

Probs. 13–3/4

13–5. The water-park ride consists of an 800-lb sled which slides from rest down the incline and then into the pool. If the frictional resistance on the incline is $F_r = 30$ lb, and in the pool for a short distance $F_r = 80$ lb, determine how fast the sled is traveling when $s = 5$ ft.

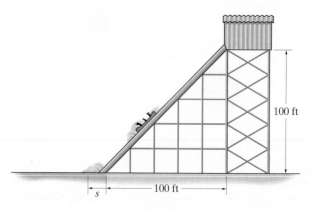

Prob. 13–5

13–6. If $P = 400$ N and the coefficient of kinetic friction between the 50-kg crate and the inclined plane is $\mu_k = 0.25$, determine the velocity of the crate after it travels 6 m up the plane. The crate starts from rest.

13–7. If the 50-kg crate starts from rest and travels a distance of 6 m up the plane in 4 s, determine the magnitude of force $\mathbf{P}$ acting on the crate. The coefficient of kinetic friction between the crate and the ground is $\mu_k = 0.25$.

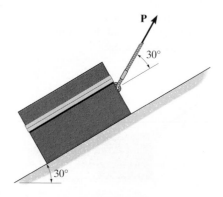

Probs. 13–6/7

*13–8. The speed of the 3500-lb sports car is plotted over the 30-s time period. Plot the variation of the traction force **F** needed to cause the motion.

13–11. The safe S has a weight of 200 lb and is supported by the rope and pulley arrangement shown. If the end of the rope is given to a boy B of weight 90 lb, determine his acceleration if in the confusion he doesn't let go of the rope. Neglect the mass of the pulleys and rope.

v (ft/s)

80

60

10   30 t (s)

Prob. 13–8

Prob. 13–11

13–9. The crate has a mass of 80 kg and is being towed by a chain which is always directed at 20° from the horizontal as shown. If the magnitude of **P** is increased until the crate begins to slide, determine the crate's initial acceleration if the coefficient of static friction is $\mu_s = 0.5$ and the coefficient of kinetic friction is $\mu_k = 0.3$.

13–10. The crate has a mass of 80 kg and is being towed by a chain which is always directed at 20° from the horizontal as shown. Determine the crate's acceleration in $t = 2$ s if the coefficient of static friction is $\mu_s = 0.4$, the coefficient of kinetic friction is $\mu_k = 0.3$, and the towing force is $P = (90t^2)$ N, where t is in seconds.

*13–12. The boy having a weight of 80 lb hangs uniformly from the bar. Determine the force in each of his arms in $t = 2$ s if the bar is moving upward with (a) a constant velocity of 3 ft/s, and (b) a speed of $v = (4t^2)$ ft/s, where t is in seconds.

P

20°

Probs. 13–9/10

Prob. 13–12

13–13. The bullet of mass m is given a velocity due to gas pressure caused by the burning of powder within the chamber of the gun. Assuming this pressure creates a force of $F = F_0 \sin(\pi t / t_0)$ on the bullet, determine the velocity of the bullet at any instant it is in the barrel. What is the bullet's maximum velocity? Also, determine the position of the bullet in the barrel as a function of time.

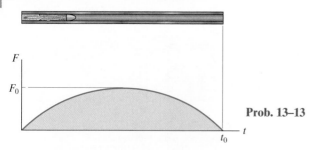

Prob. 13–13

13–14. The 2-Mg truck is traveling at 15 m/s when the brakes on all its wheels are applied, causing it to skid for a distance of 10 m before coming to rest. Determine the constant horizontal force developed in the coupling C, and the frictional force developed between the tires of the truck and the road during this time. The total mass of the boat and trailer is 1 Mg.

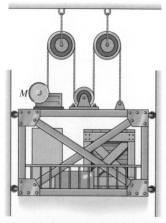

Prob. 13–14

13–15. A freight elevator, including its load, has a mass of 500 kg. It is prevented from rotating by the track and wheels mounted along its sides. When $t = 2$ s, the motor M draws in the cable with a speed of 6 m/s, *measured relative to the elevator.* If it starts from rest, determine the constant acceleration of the elevator and the tension in the cable. Neglect the mass of the pulleys, motor, and cables.

Prob. 13–15

*13–16. The man pushes on the 60-lb crate with a force **F**. The force is always directed down at 30° from the horizontal as shown, and its magnitude is increased until the crate begins to slide. Determine the crate's initial acceleration if the coefficient of static friction is $\mu_s = 0.6$ and the coefficient of kinetic friction is $\mu_k = 0.3$.

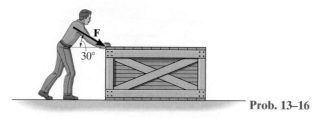

Prob. 13–16

13–17. The double inclined plane supports two blocks A and B, each having a weight of 10 lb. If the coefficient of kinetic friction between the blocks and the plane is $\mu_k = 0.1$, determine the acceleration of each block.

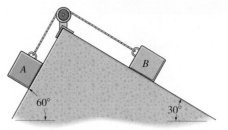

Prob. 13–17

13–18. A 40-lb suitcase slides from rest 20 ft down the smooth ramp. Determine the point where it strikes the ground at C. How long does it take to go from A to C?

13–19. Solve Prob. 13–18 if the suitcase has an initial velocity down the ramp of $v_A = 10$ ft/s and the coefficient of kinetic friction along AB is $\mu_k = 0.2$.

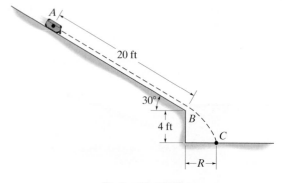

Probs. 13–18/19

*13–20.   The 400-kg mine car is hoisted up the incline using the cable and motor M. For a short time, the force in the cable is $F = (3200t^2)$ N, where t is in seconds. If the car has an initial velocity $v_1 = 2$ m/s when $t = 0$, determine its velocity when $t = 2$ s.

13–21.   The 400-kg mine car is hoisted up the incline using the cable and motor M. For a short time, the force in the cable is $F = (3200t^2)$ N, where t is in seconds. If the car has an initial velocity $v_1 = 2$ m/s at $s = 0$ and $t = 0$, determine the distance it moves up the plane when $t = 2$ s.

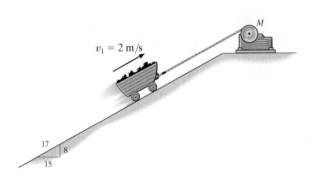

$v_1 = 2$ m/s

17
8
15

M

Probs. 13–20/21

13–22.   Determine the required mass of block A so that when it is released from rest it moves the 5-kg block B 0.75 m up along the smooth inclined plane in $t = 2$ s. Neglect the mass of the pulleys and cords.

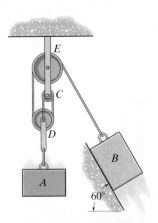

E

C

D

A

B

60°

Prob. 13–22

13–23.   The winding drum D is drawing in the cable at an accelerated rate of 5 m/s^2. Determine the cable tension if the suspended crate has a mass of 800 kg.

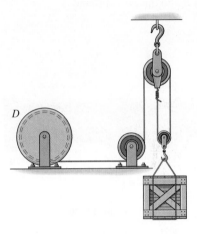

D

Prob. 13–23

*13–24.   If the motor draws in the cable at a rate of $v = (0.05s^{3/2})$ m/s, where s is in meters, determine the tension developed in the cable when $s = 10$ m. The crate has a mass of 20 kg, and the coefficient of kinetic friction between the crate and the ground is $\mu_k = 0.2$.

13–25.   If the motor draws in the cable at a rate of $v = (0.05t^2)$ m/s, where t is in seconds, determine the tension developed in the cable when $t = 5$ s. The crate has a mass of 20 kg and the coefficient of kinetic friction between the crate and the ground is $\mu_k = 0.2$.

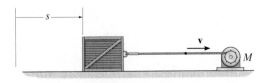

s

v

M

Probs. 13–24/25

13

13–26. The 2-kg shaft CA passes through a smooth journal bearing at B. Initially, the springs, which are coiled loosely around the shaft, are unstretched when no force is applied to the shaft. In this position $s = s' = 250$ mm and the shaft is at rest. If a horizontal force of $F = 5$ kN is applied, determine the speed of the shaft at the instant $s = 50$ mm, $s' = 450$ mm. The ends of the springs are attached to the bearing at B and the caps at C and A.

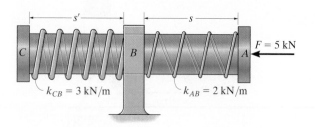

Prob. 13–26

13–27. The 30-lb crate is being hoisted upward with a constant acceleration of 6 ft/s². If the uniform beam AB has a weight of 200 lb, determine the components of reaction at A. Neglect the size and mass of the pulley at B. *Hint:* First find the tension in the cable, then analyze the forces in the beam using statics.

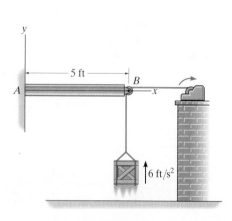

Prob. 13–27

*13–28. The driver attempts to tow the crate using a rope that has a tensile strength of 200 lb. If the crate is originally at rest and has a weight of 500 lb, determine the greatest acceleration it can have if the coefficient of static friction between the crate and the road is $\mu_s = 0.4$, and the coefficient of kinetic friction is $\mu_k = 0.3$.

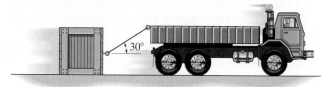

Prob. 13–28

13–29. The force exerted by the motor on the cable is shown in the graph. Determine the velocity of the 200-lb crate when $t = 2.5$ s.

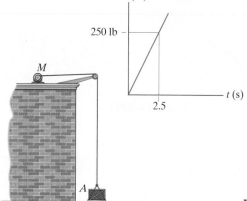

Prob. 13–29

13–30. The force of the motor M on the cable is shown in the graph. Determine the velocity of the 400-kg crate A when $t = 2$ s.

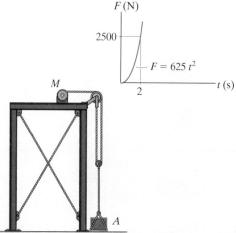

Prob. 13–30

13–31. The tractor is used to lift the 150-kg load B with the 24-m-long rope, boom, and pulley system. If the tractor travels to the right at a constant speed of 4 m/s, determine the tension in the rope when $s_A = 5$ m. When $s_A = 0, s_B = 0$.

*13–32. The tractor is used to lift the 150-kg load B with the 24-m-long rope, boom, and pulley system. If the tractor travels to the right with an acceleration of 3 m/s² and has a velocity of 4 m/s at the instant $s_A = 5$ m, determine the tension in the rope at this instant. When $s_A = 0, s_B = 0$.

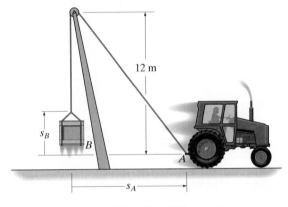

12 m

s_B

B

s_A

Probs. 13–31/32

13–33. Each of the three plates has a mass of 10 kg. If the coefficients of static and kinetic friction at each surface of contact are $\mu_s = 0.3$ and $\mu_k = 0.2$, respectively, determine the acceleration of each plate when the three horizontal forces are applied.

18 N

D

C   100 N

15 N

B

A

Prob. 13–33

13–34. Each of the two blocks has a mass m. The coefficient of kinetic friction at all surfaces of contact is μ. If a horizontal force $\mathbf{P}$ moves the bottom block, determine the acceleration of the bottom block in each case.

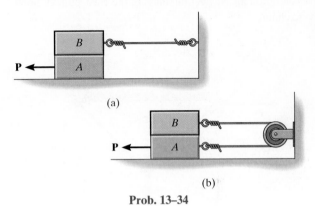

B

P   A

(a)

B

P   A

(b)

Prob. 13–34

13–35. The conveyor belt is moving at 4 m/s. If the coefficient of static friction between the conveyor and the 10-kg package B is $\mu_s = 0.2$, determine the shortest time the belt can stop so that the package does not slide on the belt.

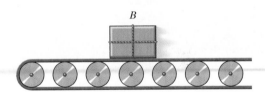

B

Prob. 13–35

*13–36. The 2-lb collar C fits loosely on the smooth shaft. If the spring is unstretched when $s = 0$ and the collar is given a velocity of 15 ft/s, determine the velocity of the collar when $s = 1$ ft.

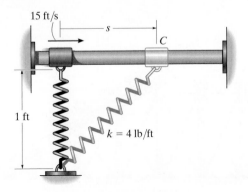

15 ft/s

s

C

1 ft

$k = 4$ lb/ft

Prob. 13–36

13

13–37. Cylinder B has a mass m and is hoisted using the cord and pulley system shown. Determine the magnitude of force $\mathbf{F}$ as a function of the cylinder's vertical position y so that when $\mathbf{F}$ is applied the cylinder rises with a constant acceleration $\mathbf{a}_B$. Neglect the mass of the cord, pulleys, hook and chain.

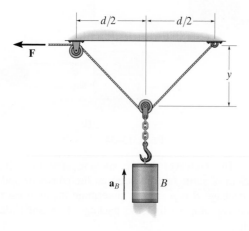

Prob. 13–37

13–38. The conveyor belt delivers each 12-kg crate to the ramp at A such that the crate's speed is $v_A = 2.5$ m/s, directed down *along* the ramp. If the coefficient of kinetic friction between each crate and the ramp is $\mu_k = 0.3$, determine the speed at which each crate slides off the ramp at B. Assume that no tipping occurs. Take $\theta = 30°$.

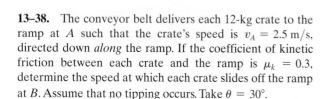

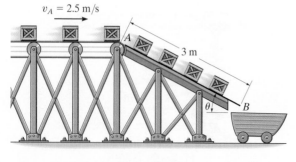

Prob. 13–38

13–39. An electron of mass m is discharged with an initial horizontal velocity of $\mathbf{v}_0$. If it is subjected to two fields of force for which $F_x = F_0$ and $F_y = 0.3F_0$, where F_0 is constant, determine the equation of the path, and the speed of the electron at any time t.

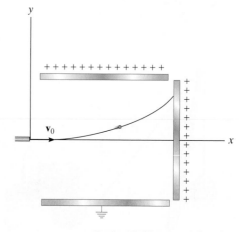

Prob. 13–39

*****13–40.** The engine of the van produces a constant driving traction force $\mathbf{F}$ at the wheels as it ascends the slope at a constant velocity $\mathbf{v}$. Determine the acceleration of the van when it passes point A and begins to travel on a level road, provided that it maintains the *same* traction force.

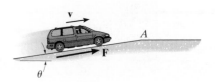

Prob. 13–40

13

13–41. The 2-kg collar C is free to slide along the smooth shaft AB. Determine the acceleration of collar C if (a) the shaft is fixed from moving, (b) collar A, which is fixed to shaft AB, moves downward at constant velocity along the vertical rod, and (c) collar A is subjected to a downward acceleration of 2 m/s^2. In all cases, the collar moves in the plane.

13–42. The 2-kg collar C is free to slide along the smooth shaft AB. Determine the acceleration of collar C if collar A is subjected to an upward acceleration of 4 m/s^2.

*13–44. When the blocks are released, determine their acceleration and the tension of the cable. Neglect the mass of the pulley.

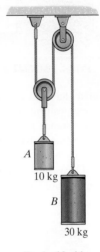

Prob. 13–44

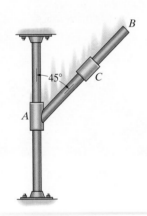

Probs. 13–41/42

13–43. The coefficient of static friction between the 200-kg crate and the flat bed of the truck is $\mu_s = 0.3$. Determine the shortest time for the truck to reach a speed of 60 km/h, starting from rest with constant acceleration, so that the crate does not slip.

13–45. If the force exerted on cable AB by the motor is $F = (100t^{3/2})$ N, where t is in seconds, determine the 50-kg crate's velocity when $t = 5$ s. The coefficients of static and kinetic friction between the crate and the ground are $\mu_s = 0.4$ and $\mu_k = 0.3$, respectively. Initially the crate is at rest.

Prob. 13–43

Prob. 13–45

13–46. Blocks A and B each have a mass m. Determine the largest horizontal force **P** which can be applied to B so that A will not move relative to B. All surfaces are smooth.

13–47. Blocks A and B each have a mass m. Determine the largest horizontal force **P** which can be applied to B so that A will not slip on B. The coefficient of static friction between A and B is μ_s. Neglect any friction between B and C.

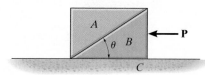

Probs. 13–46/47

*__13–48.__ A parachutist having a mass m opens his parachute from an at-rest position at a very high altitude. If the atmospheric drag resistance is $F_D = kv^2$, where k is a constant, determine his velocity when he has fallen for a time t. What is his velocity when he lands on the ground? This velocity is referred to as the *terminal velocity*, which is found by letting the time of fall $t \to \infty$.

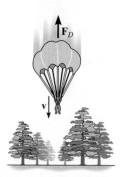

Prob. 13–48

13–49. The smooth block B of negligible size has a mass m and rests on the horizontal plane. If the board AC pushes on the block at an angle θ with a constant acceleration $\mathbf{a}_0$, determine the velocity of the block along the board and the distance s the block moves along the board as a function of time t. The block starts from rest when $s = 0, t = 0$.

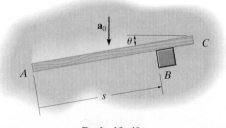

Prob. 13–49

13–50. A projectile of mass m is fired into a liquid at an angle θ_0 with an initial velocity v_0 as shown. If the liquid develops a frictional or drag resistance on the projectile which is proportional to its velocity, i.e., $F = kv$, where k is a constant, determine the x and y components of its position at any instant. Also, what is the maximum distance x_{max} that it travels?

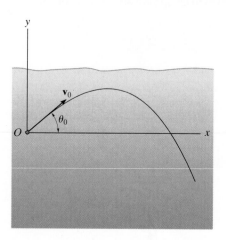

Prob. 13–50

13–51. The block A has a mass m_A and rests on the pan B, which has a mass m_B. Both are supported by a spring having a stiffness k that is attached to the bottom of the pan and to the ground. Determine the distance d the pan should be pushed down from the equilibrium position and then released from rest so that separation of the block will take place from the surface of the pan at the instant the spring becomes unstretched.

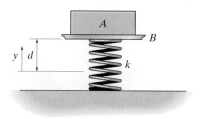

Prob. 13–51

13.5   Equations of Motion: Normal and Tangential Coordinates

When a particle moves along a curved path which is known, the equation of motion for the particle may be written in the tangential, normal, and binormal directions, Fig. 13–11. Note that there is no motion of the particle in the binormal direction, since the particle is constrained to move along the path. We have

$$\Sigma \mathbf{F} = m\mathbf{a}$$

$$\Sigma F_t \mathbf{u}_t + \Sigma F_n \mathbf{u}_n + \Sigma F_b \mathbf{u}_b = m\mathbf{a}_t + m\mathbf{a}_n$$

This equation is satisfied provided

$$\boxed{\begin{aligned} \Sigma F_t &= ma_t \\ \Sigma F_n &= ma_n \\ \Sigma F_b &= 0 \end{aligned}} \qquad (13\text{--}8)$$

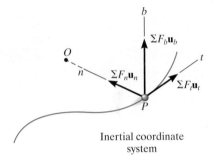

Inertial coordinate system

Fig. 13–11

Recall that $a_t \,(= dv/dt)$ represents the time rate of change in the magnitude of velocity. So if $\Sigma \mathbf{F}_t$ acts in the direction of motion, the particle's speed will increase, whereas if it acts in the opposite direction, the particle will slow down. Likewise, $a_n \,(= v^2/\rho)$ represents the time rate of change in the velocity's direction. It is caused by $\Sigma \mathbf{F}_n$, which *always* acts in the positive n direction, i.e., toward the path's center of curvature. From this reason it is often referred to as the *centripetal force*.

The centrifuge is used to subject a passenger to a very large normal acceleration caused by rapid rotation. Realize that this acceleration is *caused by* the unbalanced normal force exerted on the passenger by the seat of the centrifuge.

13

Procedure for Analysis

When a problem involves the motion of a particle along a *known curved path*, normal and tangential coordinates should be considered for the analysis since the acceleration components can be readily formulated. The method for applying the equations of motion, which relate the forces to the acceleration, has been outlined in the procedure given in Sec. 13.4. Specifically, for t, n, b coordinates it may be stated as follows:

Free-Body Diagram.

- Establish the inertial t, n, b coordinate system at the particle and draw the particle's free-body diagram.

- The particle's normal acceleration $\mathbf{a}_n$ *always* acts in the positive n direction.

- If the tangential acceleration $\mathbf{a}_t$ is unknown, assume it acts in the positive t direction.

- There is no acceleration in the b direction.

- Identify the unknowns in the problem.

Equations of Motion.

- Apply the equations of motion, Eq. 13–8.

Kinematics.

- Formulate the tangential and normal components of acceleration; i.e., $a_t = dv/dt$ or $a_t = v\, dv/ds$ and $a_n = v^2/\rho$.

- If the path is defined as $y = f(x)$, the radius of curvature at the point where the particle is located can be obtained from $\rho = [1 + (dy/dx)^2]^{3/2}/|d^2y/dx^2|$.

EXAMPLE | 13.6

Determine the banking angle θ for the race track so that the wheels of the racing cars shown in Fig. 13–12a will not have to depend upon friction to prevent any car from sliding up or down the track. Assume the cars have negligible size, a mass m, and travel around the curve of radius ρ with a constant speed v.

(a)

SOLUTION

Before looking at the following solution, give some thought as to why it should be solved using t, n, b coordinates.

Free-Body Diagram.   As shown in Fig. 13–12b, and as stated in the problem, no frictional force acts on the car. Here $\mathbf{N}_C$ represents the *resultant* of the ground on all four wheels. Since a_n can be calculated, the unknowns are N_C and θ.

Equations of Motion.   Using the n, b axes shown,

$$\xrightarrow{+} \Sigma F_n = ma_n; \qquad N_C \sin \theta = m\frac{v^2}{\rho} \qquad (1)$$

$$+\uparrow \Sigma F_b = 0; \qquad N_C \cos \theta - mg = 0 \qquad (2)$$

Eliminating N_C and m from these equations by dividing Eq. 1 by Eq. 2, we obtain

$$\tan \theta = \frac{v^2}{g\rho}$$

$$\theta = \tan^{-1}\left(\frac{v^2}{g\rho}\right) \qquad \qquad \textit{Ans.}$$

NOTE: The result is independent of the mass of the car. Also, a force summation in the tangential direction is of no consequence to the solution. If it were considered, then $a_t = dv/dt = 0$, since the car moves with *constant speed*. A further analysis of this problem is discussed in Prob. 21–53.

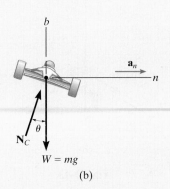

(b)

Fig. 13–12

13

EXAMPLE | 13.7

The 3-kg disk D is attached to the end of a cord as shown in Fig. 13–13a. The other end of the cord is attached to a ball-and-socket joint located at the center of a platform. If the platform rotates rapidly, and the disk is placed on it and released from rest as shown, determine the time it takes for the disk to reach a speed great enough to break the cord. The maximum tension the cord can sustain is 100 N, and the coefficient of kinetic friction between the disk and the platform is $\mu_k = 0.1$.

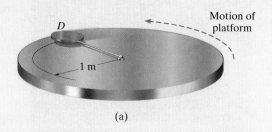

(a)

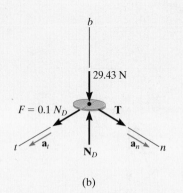

(b)

Fig. 13–13

SOLUTION

Free-Body Diagram. The frictional force has a magnitude $F = \mu_k N_D = 0.1N_D$ and a sense of direction that opposes the *relative motion* of the disk with respect to the platform. It is this force that gives the disk a tangential component of acceleration causing v to increase, thereby causing T to increase until it reaches 100 N. The weight of the disk is $W = 3(9.81) = 29.43$ N. Since a_n can be related to v, the unknowns are N_D, a_t, and v.

Equations of Motion.

$$\Sigma F_n = ma_n; \qquad\qquad T = 3\left(\frac{v^2}{1}\right) \qquad\qquad (1)$$

$$\Sigma F_t = ma_t; \qquad\qquad 0.1N_D = 3a_t \qquad\qquad (2)$$

$$\Sigma F_b = 0; \qquad\qquad N_D - 29.43 = 0 \qquad\qquad (3)$$

Setting $T = 100$ N, Eq. 1 can be solved for the critical speed v_{cr} of the disk needed to break the cord. Solving all the equations, we obtain

$$N_D = 29.43 \text{ N}$$

$$a_t = 0.981 \text{ m/s}^2$$

$$v_{cr} = 5.77 \text{ m/s}$$

Kinematics.   Since a_t is *constant*, the time needed to break the cord is

$$v_{cr} = v_0 + a_t t$$

$$5.77 = 0 + (0.981)t$$

$$t = 5.89 \text{ s} \qquad\qquad\qquad\qquad Ans.$$

EXAMPLE 13.8

Design of the ski jump shown in the photo requires knowing the type of forces that will be exerted on the skier and her approximate trajectory. If in this case the jump can be approximated by the parabola shown in Fig. 13–14a, determine the normal force on the 150-lb skier the instant she arrives at the end of the jump, point A, where her velocity is 65 ft/s. Also, what is her acceleration at this point?

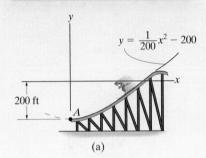

SOLUTION
Why consider using n, t coordinates to solve this problem?

Free-Body Diagram.   Since $dy/dx = x/100|_{x=0} = 0$, the slope at A is horizontal. The free-body diagram of the skier when she is at A is shown in Fig. 13–14b. Since the path is *curved*, there are two components of acceleration, $\mathbf{a}_n$ and $\mathbf{a}_t$. Since a_n can be calculated, the unknowns are a_t and N_A.

Equations of Motion.

$$+\uparrow \Sigma F_n = ma_n; \qquad N_A - 150 = \frac{150}{32.2}\left(\frac{(65)^2}{\rho}\right)$$

$$\xleftarrow{+} \Sigma F_t = ma_t; \qquad 0 = \frac{150}{32.2}a_t$$

The radius of curvature ρ for the path must be determined at point $A(0, -200$ ft$)$. Here $y = \frac{1}{200}x^2 - 200$, $dy/dx = \frac{1}{100}x$, $d^2y/dx^2 = \frac{1}{100}$, so that at $x = 0$,

$$\rho = \frac{[1 + (dy/dx)^2]^{3/2}}{|d^2y/dx^2|}\bigg|_{x=0} = \frac{[1 + (0)^2]^{3/2}}{\left|\frac{1}{100}\right|} = 100 \text{ ft}$$

Substituting this into Eq. 1 and solving for N_A, we obtain

$$N_A = 347 \text{ lb} \qquad\qquad Ans.$$

Kinematics.   From Eq. 2,

$$a_t = 0$$

Thus,

$$a_n = \frac{v^2}{\rho} = \frac{(65)^2}{100} = 42.2 \text{ ft/s}^2$$

$$a_A = a_n = 42.2 \text{ ft/s}^2 \uparrow \qquad\qquad Ans.$$

NOTE: Apply the equation of motion in the y direction and show that when the skier is in midair her acceleration is 32.2 ft/s².

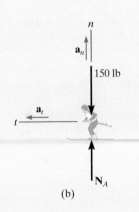

(a)

(b)

Fig. 13–14

EXAMPLE 13.9

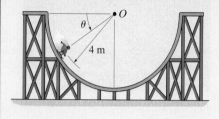

(a)

The 60-kg skateboarder in Fig. 13–15a coasts down the circular track. If he starts from rest when $\theta = 0°$, determine the magnitude of the normal reaction the track exerts on him when $\theta = 60°$. Neglect his size for the calculation.

SOLUTION

Free-Body Diagram. The free-body diagram of the skateboarder when he is at an *arbitrary position* θ is shown in Fig. 13–15b. At $\theta = 60°$ there are three unknowns, N_s, a_t, and a_n (or v).

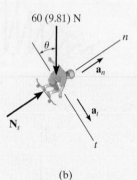

60 (9.81) N

(b)

Equations of Motion.

$$+\nearrow \Sigma F_n = ma_n; \quad N_s - [60(9.81)\text{N}]\sin\theta = (60\text{ kg})\left(\frac{v^2}{4\text{ m}}\right) \quad (1)$$

$$+\searrow \Sigma F_t = ma_t; \quad [60(9.81)\text{N}]\cos\theta = (60\text{ kg})\,a_t$$

$$a_t = 9.81\cos\theta$$

Kinematics. Since a_t is expressed in terms of θ, the equation $v\,dv = a_t\,ds$ must be used to determine the speed of the skateboarder when $\theta = 60°$. Using the geometric relation $s = \theta r$, where $ds = r\,d\theta = (4\text{ m})\,d\theta$, Fig. 13–15c, and the initial condition $v = 0$ at $\theta = 0°$, we have,

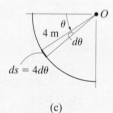

(c)

Fig. 13–15

$$v\,dv = a_t\,ds$$

$$\int_0^v v\,dv = \int_0^{60°} 9.81\cos\theta(4\,d\theta)$$

$$\left.\frac{v^2}{2}\right|_0^v = 39.24\sin\theta \Big|_0^{60°}$$

$$\frac{v^2}{2} - 0 = 39.24(\sin 60° - 0)$$

$$v^2 = 67.97 \text{ m}^2/\text{s}^2$$

Substituting this result and $\theta = 60°$ into Eq. (1), yields

$$N_s = 1529.23 \text{ N} = 1.53 \text{ kN} \qquad \textit{Ans.}$$

FUNDAMENTAL PROBLEMS

F13–7. The block rests at a distance of 2 m from the center of the platform. If the coefficient of static friction between the block and the platform is $\mu_s = 0.3$, determine the maximum speed which the block can attain before it begins to slip. Assume the angular motion of the disk is slowly increasing.

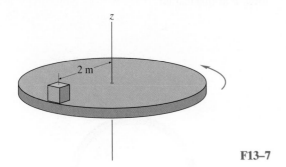

F13–7

F13–8. Determine the maximum speed that the jeep can travel over the crest of the hill and not lose contact with the road.

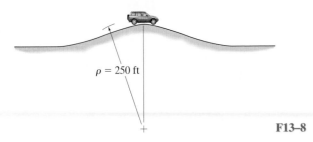

F13–8

F13–9. A pilot weighs 150 lb and is traveling at a constant speed of 120 ft/s. Determine the normal force he exerts on the seat of the plane when he is upside down at A. The loop has a radius of curvature of 400 ft.

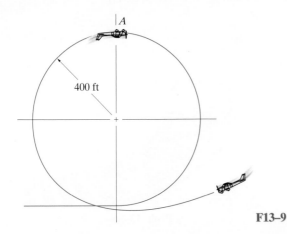

F13–9

F13–10. The sports car is traveling along a 30° banked road having a radius of curvature of $\rho = 500$ ft. If the coefficient of static friction between the tires and the road is $\mu_s = 0.2$, determine the maximum safe speed so no slipping occurs. Neglect the size of the car.

F13–10

F13–11. If the 10-kg ball has a velocity of 3 m/s when it is at the position A, along the vertical path, determine the tension in the cord and the increase in the speed of the ball at this position.

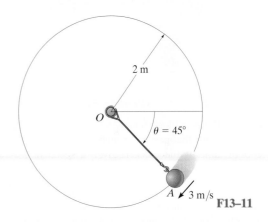

F13–11

F13–12. The motorcycle has a mass of 0.5 Mg and a negligible size. It passes point A traveling with a speed of 15 m/s, which is increasing at a constant rate of 1.5 m/s². Determine the resultant frictional force exerted by the road on the tires at this instant.

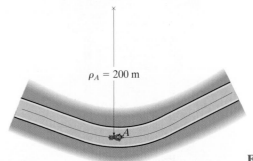

F13–12

PROBLEMS

*13–52. A girl, having a mass of 15 kg, sits motionless relative to the surface of a horizontal platform at a distance of $r = 5$ m from the platform's center. If the angular motion of the platform is *slowly* increased so that the girl's tangential component of acceleration can be neglected, determine the maximum speed which the girl will have before she begins to slip off the platform. The coefficient of static friction between the girl and the platform is $\mu = 0.2$.

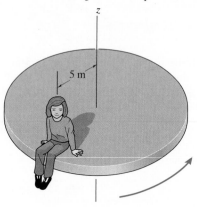

Prob. 13–52

13–53. The 2-kg block B and 15-kg cylinder A are connected to a light cord that passes through a hole in the center of the smooth table. If the block is given a speed of $v = 10$ m/s, determine the radius r of the circular path along which it travels.

13–54. The 2-kg block B and 15-kg cylinder A are connected to a light cord that passes through a hole in the center of the smooth table. If the block travels along a circular path of radius $r = 1.5$ m, determine the speed of the block.

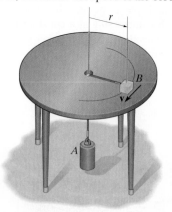

Probs. 13–53/54

13–55. The 5-kg collar A is sliding around a smooth vertical guide rod. At the instant shown, the speed of the collar is $v = 4$ m/s, which is increasing at 3 m/s². Determine the normal reaction of the guide rod on the collar, and force **P** at this instant.

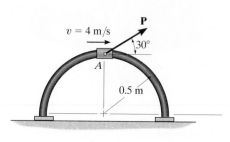

Prob. 13–55

*13–56. Cartons having a mass of 5 kg are required to move along the assembly line at a constant speed of 8 m/s. Determine the smallest radius of curvature, ρ, for the conveyor so the cartons do not slip. The coefficients of static and kinetic friction between a carton and the conveyor are $\mu_s = 0.7$ and $\mu_k = 0.5$, respectively.

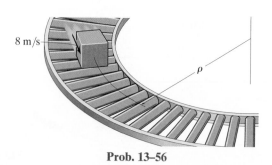

Prob. 13–56

13–57. The block B, having a mass of 0.2 kg, is attached to the vertex A of the right circular cone using a light cord. The cone is rotating at a constant angular rate about the z axis such that the block attains a speed of 0.5 m/s. At this speed, determine the tension in the cord and the reaction which the cone exerts on the block. Neglect the size of the block and the effect of friction.

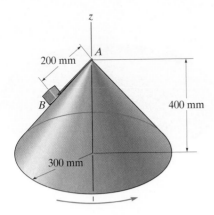

Prob. 13–57

13–58. The 2-kg spool S fits loosely on the inclined rod for which the coefficient of static friction is $\mu_s = 0.2$. If the spool is located 0.25 m from A, determine the minimum constant speed the spool can have so that it does not slip down the rod.

13–59. The 2-kg spool S fits loosely on the inclined rod for which the coefficient of static friction is $\mu_s = 0.2$. If the spool is located 0.25 m from A, determine the maximum constant speed the spool can have so that it does not slip up the rod.

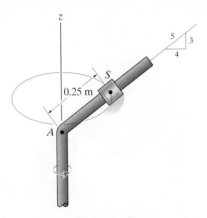

Probs. 13–58/59

13–60. At the instant $\theta = 60°$, the boy's center of mass G has a downward speed $v_G = 15$ ft/s. Determine the rate of increase in his speed and the tension in each of the two supporting cords of the swing at this instant. The boy has a weight of 60 lb. Neglect his size and the mass of the seat and cords.

13–61. At the instant $\theta = 60°$, the boy's center of mass G is momentarily at rest. Determine his speed and the tension in each of the two supporting cords of the swing when $\theta = 90°$. The boy has a weight of 60 lb. Neglect his size and the mass of the seat and cords.

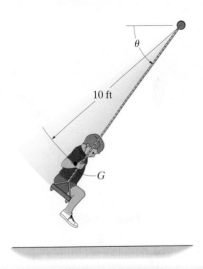

Probs. 13–60/61

13–62. The 10-lb suitcase slides down the curved ramp for which the coefficient of kinetic friction is $\mu_k = 0.2$. If at the instant it reaches point A it has a speed of 5 ft/s, determine the normal force on the suitcase and the rate of increase of its speed.

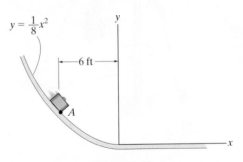

Prob. 13–62

13–63. The 150-lb man lies against the cushion for which the coefficient of static friction is $\mu_s = 0.5$. Determine the resultant normal and frictional forces the cushion exerts on him if, due to rotation about the z axis, he has a constant speed $v = 20$ ft/s. Neglect the size of the man. Take $\theta = 60°$.

*13–64. The 150-lb man lies against the cushion for which the coefficient of static friction is $\mu_s = 0.5$. If he rotates about the z axis with a constant speed $v = 30$ ft/s, determine the smallest angle θ of the cushion at which he will begin to slip off.

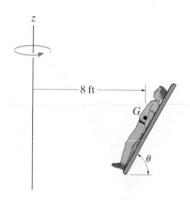

Probs. 13–63/64

13–65. Determine the constant speed of the passengers on the amusement-park ride if it is observed that the supporting cables are directed at $\theta = 30°$ from the vertical. Each chair including its passenger has a mass of 80 kg. Also, what are the components of force in the n, t, and b directions which the chair exerts on a 50-kg passenger during the motion?

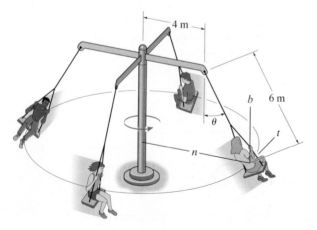

Prob. 13–65

13–66. The man has a mass of 80 kg and sits 3 m from the center of the rotating platform. Due to the rotation his speed is increased from rest by $\dot{v} = 0.4$ m/s². If the coefficient of static friction between his clothes and the platform is $\mu_s = 0.3$, determine the time required to cause him to slip.

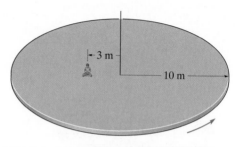

Prob. 13–66

13–67. The vehicle is designed to combine the feel of a motorcycle with the comfort and safety of an automobile. If the vehicle is traveling at a constant speed of 80 km/h along a circular curved road of radius 100 m, determine the tilt angle θ of the vehicle so that only a normal force from the seat acts on the driver. Neglect the size of the driver.

Prob. 13–67

*13–68. The 0.8-Mg car travels over the hill having the shape of a parabola. If the driver maintains a constant speed of 9 m/s, determine both the resultant normal force and the resultant frictional force that all the wheels of the car exert on the road at the instant it reaches point A. Neglect the size of the car.

13–69. The 0.8-Mg car travels over the hill having the shape of a parabola. When the car is at point A, it is traveling at 9 m/s and increasing its speed at 3 m/s². Determine both the resultant normal force and the resultant frictional force that all the wheels of the car exert on the road at this instant. Neglect the size of the car.

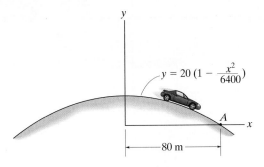

$$y = 20 \left(1 - \frac{x^2}{6400}\right)$$

Probs. 13–68/69

13–70. The package has a weight of 5 lb and slides down the chute. When it reaches the curved portion AB, it is traveling at 8 ft/s ($\theta = 0°$). If the chute is smooth, determine the speed of the package when it reaches the intermediate point C ($\theta = 30°$) and when it reaches the horizontal plane ($\theta = 45°$). Also, find the normal force on the package at C.

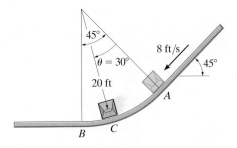

Prob. 13–70

13–71. If the ball has a mass of 30 kg and a speed $v = 4$ m/s at the instant it is at its lowest point, $\theta = 0°$, determine the tension in the cord at this instant. Also, determine the angle θ to which the ball swings at the instant it momentarily stops. Neglect the size of the ball.

*13–72. The ball has a mass of 30 kg and a speed $v = 4$ m/s at the instant it is at its lowest point, $\theta = 0°$. Determine the tension in the cord and the rate at which the ball's speed is decreasing at the instant $\theta = 20°$. Neglect the size of the ball.

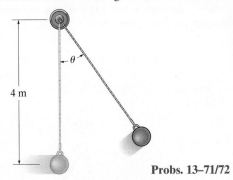

Probs. 13–71/72

13–73. Determine the maximum speed at which the car with mass m can pass over the top point A of the vertical curved road and still maintain contact with the road. If the car maintains this speed, what is the normal reaction the road exerts on the car when it passes the lowest point B on the road?

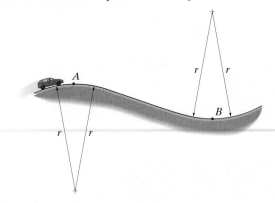

Prob. 13–73

13–74. If the crest of the hill has a radius of curvature $\rho = 200$ ft, determine the maximum constant speed at which the car can travel over it without leaving the surface of the road. Neglect the size of the car in the calculation. The car has a weight of 3500 lb.

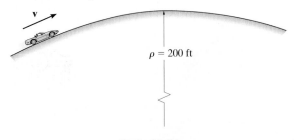

Prob. 13–74

13–75. Bobs A and B of mass m_A and m_B ($m_A > m_B$) are connected to an inextensible light string of length l that passes through the smooth ring at C. If bob B moves as a conical pendulum such that A is suspended a distance of h from C, determine the angle θ and the speed of bob B. Neglect the size of both bobs.

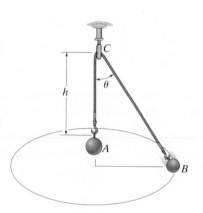

Prob. 13–75

*13–76. Prove that if the block is released from rest at point B of a smooth path of *arbitrary shape*, the speed it attains when it reaches point A is equal to the speed it attains when it falls freely through a distance h; i.e., $v = \sqrt{2gh}$.

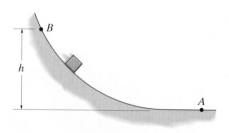

Prob. 13–76

13–77. The skier starts from rest at $A(10\text{ m}, 0)$ and descends the smooth slope, which may be approximated by a parabola. If she has a mass of 52 kg, determine the normal force the ground exerts on the skier at the instant she arrives at point B. Neglect the size of the skier. *Hint:* Use the result of Prob. 13–76.

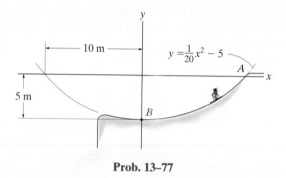

Prob. 13–77

13–78. A spring, having an unstretched length of 2 ft, has one end attached to the 10-lb ball. Determine the angle θ of the spring if the ball has a speed of 6 ft/s tangent to the horizontal circular path.

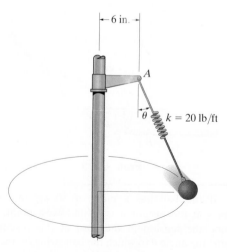

Prob. 13–78

13–79. The airplane, traveling at a constant speed of 50 m/s, is executing a horizontal turn. If the plane is banked at $\theta = 15°$, when the pilot experiences only a normal force on the seat of the plane, determine the radius of curvature ρ of the turn. Also, what is the normal force of the seat on the pilot if he has a mass of 70 kg.

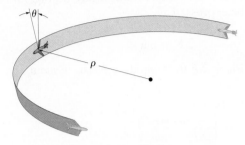

Prob. 13–79

*13–80.** A 5-Mg airplane is flying at a constant speed of 350 km/h along a horizontal circular path of radius $r = 3000$ m. Determine the uplift force **L** acting on the airplane and the banking angle θ. Neglect the size of the airplane.

13–81. A 5-Mg airplane is flying at a constant speed of 350 km/h along a horizontal circular path. If the banking angle $\theta = 15°$, determine the uplift force **L** acting on the airplane and the radius r of the circular path. Neglect the size of the airplane.

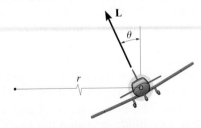

Probs. 13–80/81

13–82. The 800-kg motorbike travels with a constant speed of 80 km/h up the hill. Determine the normal force the surface exerts on its wheels when it reaches point A. Neglect its size.

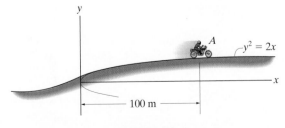

Prob. 13–82

13–83. The ball has a mass m and is attached to the cord of length l. The cord is tied at the top to a swivel and the ball is given a velocity $\mathbf{v}_0$. Show that the angle θ which the cord makes with the vertical as the ball travels around the circular path must satisfy the equation $\tan \theta \, \sin \theta = v_0^2/gl$. Neglect air resistance and the size of the ball.

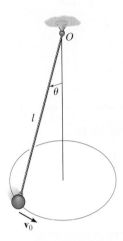

Prob. 13–83

*13–84.** The 5-lb collar slides on the smooth rod, so that when it is at A it has a speed of 10 ft/s. If the spring to which it is attached has an unstretched length of 3 ft and a stiffness of $k = 10$ lb/ft, determine the normal force on the collar and the magnitude of the acceleration of the collar at this instant.

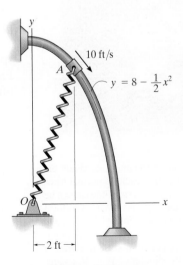

Prob. 13–84

13

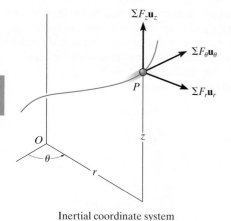

Inertial coordinate system

Fig. 13–16

13

13.6   Equations of Motion: Cylindrical Coordinates

When all the forces acting on a particle are resolved into cylindrical components, i.e., along the unit-vector directions $\mathbf{u}_r$, $\mathbf{u}_\theta$, $\mathbf{u}_z$, Fig. 13–16, the equation of motion can be expressed as

$$\Sigma \mathbf{F} = m\mathbf{a}$$

$$\Sigma F_r \mathbf{u}_r + \Sigma F_\theta \mathbf{u}_\theta + \Sigma F_z \mathbf{u}_z = ma_r \mathbf{u}_r + ma_\theta \mathbf{u}_\theta + ma_z \mathbf{u}_z$$

To satisfy this equation, we require

$$
\begin{aligned}
\Sigma F_r &= ma_r \\
\Sigma F_\theta &= ma_\theta \\
\Sigma F_z &= ma_z
\end{aligned}
\tag{13–9}
$$

If the particle is constrained to move only in the r–θ plane, then only the first two of Eq. 13–9 are used to specify the motion.

Tangential and Normal Forces.   The most straightforward type of problem involving cylindrical coordinates requires the determination of the resultant force components ΣF_r, ΣF_θ, ΣF_z which cause a particle to move with a *known* acceleration. If, however, the particle's accelerated motion is not completely specified at the given instant, then some information regarding the directions or magnitudes of the forces acting on the particle must be known or calculated in order to solve Eqs. 13–9. For example, the force $\mathbf{P}$ causes the particle in Fig. 13–17a to move along a path $r = f(\theta)$. The *normal force* $\mathbf{N}$ which the path exerts on the particle is always *perpendicular to the tangent of the path*, whereas the frictional force $\mathbf{F}$ always acts along the tangent in the opposite direction of motion. The *directions* of $\mathbf{N}$ and $\mathbf{F}$ can be specified relative to the radial coordinate by using the angle ψ (psi), Fig. 13–17b, which is defined between the *extended* radial line and the tangent to the curve.

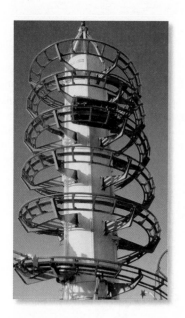

As the car of weight W descends the spiral track, the resultant normal force which the track exerts on the car can be represented by its three cylindrical components. $\mathbf{N}_r$, directed horizontally inward, creates a radial acceleration $-\mathbf{a}_r$, $\mathbf{N}_\theta$ creates a transverse acceleration $\mathbf{a}_\theta$, and the difference $\mathbf{W} - \mathbf{N}_z$ creates an azimuthal acceleration $-\mathbf{a}_z$.

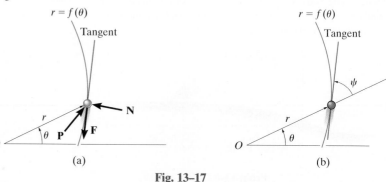

Fig. 13–17

This angle can be obtained by noting that when the particle is displaced a distance ds along the path, Fig. 13–17c, the component of displacement in the radial direction is dr and the component of displacement in the transverse direction is $r\,d\theta$. Since these two components are mutually perpendicular, the angle ψ can be determined from $\tan\psi = r\,d\theta/dr$, or

$$\tan\psi = \frac{r}{dr/d\theta} \qquad (13\text{–}10)$$

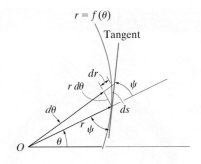

(c)

Fig. 13–17

If ψ is calculated as a positive quantity, it is measured from the *extended radial line* to the tangent in a counterclockwise sense or in the positive direction of θ. If it is negative, it is measured in the opposite direction to positive θ. For example, consider the cardioid $r = a(1 + \cos\theta)$, shown in Fig. 13–18. Because $dr/d\theta = -a\sin\theta$, then when $\theta = 30°$, $\tan\psi = a(1 + \cos 30°)/(-a\sin 30°) = -3.732$, or $\psi = -75°$, measured clockwise, opposite to $+\theta$ as shown in the figure.

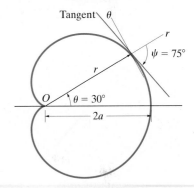

Fig. 13–18

Procedure for Analysis

Cylindrical or polar coordinates are a suitable choice for the analysis of a problem for which data regarding the angular motion of the radial line r are given, or in cases where the path can be conveniently expressed in terms of these coordinates. Once these coordinates have been established, the equations of motion can then be applied in order to relate the forces acting on the particle to its acceleration components. The method for doing this has been outlined in the procedure for analysis given in Sec. 13.4. The following is a summary of this procedure.

Free-Body Diagram.

- Establish the r, θ, z inertial coordinate system and draw the particle's free-body diagram.
- Assume that $\mathbf{a}_r$, $\mathbf{a}_\theta$, $\mathbf{a}_z$ act in the positive directions of r, θ, z if they are unknown.
- Identify all the unknowns in the problem.

Equations of Motion.

- Apply the equations of motion, Eq. 13–9.

Kinematics.

- Use the methods of Sec. 12.8 to determine r and the time derivatives $\dot{r}$, $\ddot{r}$, $\dot{\theta}$, $\ddot{\theta}$, $\ddot{z}$, and then evaluate the acceleration components $a_r = \ddot{r} - r\dot{\theta}^2$, $a_\theta = r\ddot{\theta} + 2\dot{r}\dot{\theta}$, $a_z = \ddot{z}$.
- If any of the acceleration components is computed as a negative quantity, it indicates that it acts in its negative coordinate direction.
- When taking the time derivatives of $r = f(\theta)$, it is very important to use the chain rule of calculus, which is discussed in Appendix C.

EXAMPLE | **13.10**

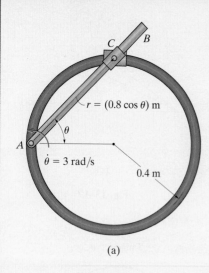

(a)

Fig. 13–19

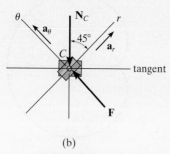

(b)

The smooth 0.5-kg double-collar in Fig. 13–19a can freely slide on arm *AB* and the circular guide rod. If the arm rotates with a constant angular velocity of $\dot{\theta} = 3$ rad/s, determine the force the arm exerts on the collar at the instant $\theta = 45°$. Motion is in the horizontal plane.

SOLUTION

Free-Body Diagram. The normal reaction $\mathbf{N}_C$ of the circular guide rod and the force $\mathbf{F}$ of arm *AB* act on the collar in the plane of motion, Fig. 13–19b. Note that $\mathbf{F}$ acts perpendicular to the axis of arm *AB*, that is, in the direction of the θ axis, while $\mathbf{N}_C$ acts perpendicular to the tangent of the circular path at $\theta = 45°$. The four unknowns are N_C, F, a_r, a_θ.

Equations of Motion.

$$+\nearrow \Sigma F_r = ma_r: \qquad -N_C \cos 45° = (0.5 \text{ kg}) a_r \qquad (1)$$

$$+\nwarrow \Sigma F_\theta = ma_\theta: \qquad F - N_C \sin 45° = (0.5 \text{ kg}) a_\theta \qquad (2)$$

Kinematics. Using the chain rule (see Appendix C), the first and second time derivatives of *r* when $\theta = 45°$, $\dot{\theta} = 3$ rad/s, $\ddot{\theta} = 0$, are

$$r = 0.8 \cos \theta = 0.8 \cos 45° = 0.5657 \text{ m}$$

$$\dot{r} = -0.8 \sin \theta \, \dot{\theta} = -0.8 \sin 45°(3) = -1.6971 \text{ m/s}$$

$$\ddot{r} = -0.8 \left[\sin \theta \, \ddot{\theta} + \cos \theta \, \dot{\theta}^2 \right]$$

$$= -0.8[\sin 45°(0) + \cos 45°(3^2)] = -5.091 \text{ m/s}^2$$

We have

$$a_r = \ddot{r} - r\dot{\theta}^2 = -5.091 \text{ m/s}^2 - (0.5657 \text{ m})(3 \text{ rad/s})^2 = -10.18 \text{ m/s}^2$$

$$a_\theta = r\ddot{\theta} + 2\dot{r}\dot{\theta} = (0.5657 \text{ m})(0) + 2(-1.6971 \text{ m/s})(3 \text{ rad/s})$$

$$= -10.18 \text{ m/s}^2$$

Substituting these results into Eqs. (1) and (2) and solving, we get

$$N_C = 7.20 \text{ N}$$

$$F = 0 \qquad \qquad Ans.$$

EXAMPLE 13.11

The smooth 2-kg cylinder C in Fig. 13–20a has a pin P through its center which passes through the slot in arm OA. If the arm is forced to rotate in the *vertical plane* at a constant rate $\dot{\theta} = 0.5$ rad/s, determine the force that the arm exerts on the peg at the instant $\theta = 60°$.

SOLUTION

Why is it a good idea to use polar coordinates to solve this problem?

Free-Body Diagram. The free-body diagram for the cylinder is shown in Fig. 13–20b. The force on the peg, $\mathbf{F}_P$, acts perpendicular to the slot in the arm. As usual, $\mathbf{a}_r$ and $\mathbf{a}_\theta$ are assumed to act in the directions of *positive r* and θ, respectively. Identify the four unknowns.

Equations of Motion. Using the data in Fig. 13–20b, we have

$$+\swarrow\Sigma F_r = ma_r; \quad 19.62 \sin\theta - N_C \sin\theta = 2a_r \quad (1)$$
$$+\searrow\Sigma F_\theta = ma_\theta; \quad 19.62 \cos\theta + F_P - N_C \cos\theta = 2a_\theta \quad (2)$$

Kinematics. From Fig. 13–20a, r can be related to θ by the equation

$$r = \frac{0.4}{\sin\theta} = 0.4 \csc\theta$$

Since $d(\csc\theta) = -(\csc\theta\cot\theta)\,d\theta$ and $d(\cot\theta) = -(\csc^2\theta)\,d\theta$, then r and the necessary time derivatives become

$$\dot{\theta} = 0.5 \quad r = 0.4\csc\theta$$
$$\ddot{\theta} = 0 \quad \dot{r} = -0.4(\csc\theta\cot\theta)\dot{\theta}$$
$$= -0.2\csc\theta\cot\theta$$
$$\ddot{r} = -0.2(-\csc\theta\cot\theta)(\dot{\theta})\cot\theta - 0.2\csc\theta(-\csc^2\theta)\dot{\theta}$$
$$= 0.1\csc\theta(\cot^2\theta + \csc^2\theta)$$

Evaluating these formulas at $\theta = 60°$, we get

$$\dot{\theta} = 0.5 \quad r = 0.462$$
$$\ddot{\theta} = 0 \quad \dot{r} = -0.133$$
$$\ddot{r} = 0.192$$
$$a_r = \ddot{r} - r\dot{\theta}^2 = 0.192 - 0.462(0.5)^2 = 0.0770$$
$$a_\theta = r\ddot{\theta} + 2\dot{r}\dot{\theta} = 0 + 2(-0.133)(0.5) = -0.133$$

Substituting these results into Eqs. 1 and 2 with $\theta = 60°$ and solving yields

$$N_C = 19.4 \text{ N} \qquad F_P = -0.356 \text{ N} \qquad \textit{Ans.}$$

The negative sign indicates that $\mathbf{F}_P$ acts opposite to the direction shown in Fig. 13–20b.

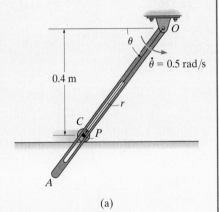

(a)

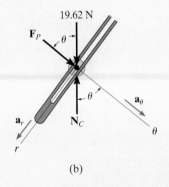

(b)

Fig. 13–20

13

EXAMPLE 13.12

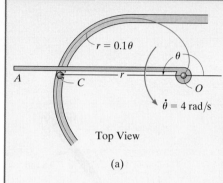

$r = 0.1\theta$

$\dot{\theta} = 4$ rad/s

Top View

(a)

A can C, having a mass of 0.5 kg, moves along a grooved horizontal slot shown in Fig. 13–21a. The slot is in the form of a spiral, which is defined by the equation $r = (0.1\theta)$ m, where θ is in radians. If the arm OA rotates with a constant rate $\dot{\theta} = 4$ rad/s in the horizontal plane, determine the force it exerts on the can at the instant $\theta = \pi$ rad. Neglect friction and the size of the can.

SOLUTION

Free-Body Diagram. The driving force $\mathbf{F}_C$ acts perpendicular to the arm OA, whereas the normal force of the wall of the slot on the can, N_C, acts perpendicular to the tangent to the curve at $\theta = \pi$ rad, Fig. 13–21b. As usual, $\mathbf{a}_r$ and $\mathbf{a}_\theta$ are assumed to act in the *positive directions* of r and θ, respectively. Since the path is specified, the angle ψ which the extended radial line r makes with the tangent, Fig. 13–21c, can be determined from Eq. 13–10. We have $r = 0.1\theta$, so that $dr/d\theta = 0.1$, and therefore

$$\tan \psi = \frac{r}{dr/d\theta} = \frac{0.1\theta}{0.1} = \theta$$

When $\theta = \pi$, $\psi = \tan^{-1}\pi = 72.3°$, so that $\phi = 90° - \psi = 17.7°$, as shown in Fig. 13–21$c$. Identify the four unknowns in Fig. 13–21b.

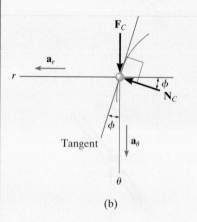

$\mathbf{F}_C$

$\mathbf{a}_r$

r

ϕ

$\mathbf{N}_C$

ϕ

Tangent

$\mathbf{a}_\theta$

θ

(b)

Equations of Motion. Using $\phi = 17.7°$ and the data shown in Fig. 13–21b, we have

$$\overset{+}{\rightarrow} \Sigma F_r = ma_r; \qquad\qquad N_C \cos 17.7° = 0.5a_r \qquad (1)$$

$$+\downarrow \Sigma F_\theta = ma_\theta; \qquad\qquad F_C - N_C \sin 17.7° = 0.5a_\theta \qquad (2)$$

Kinematics. The time derivatives of r and θ are

$$\dot{\theta} = 4 \text{ rad/s} \qquad\qquad r = 0.1\theta$$

$$\ddot{\theta} = 0 \qquad\qquad \dot{r} = 0.1\dot{\theta} = 0.1(4) = 0.4 \text{ m/s}$$

$$\ddot{r} = 0.1\ddot{\theta} = 0$$

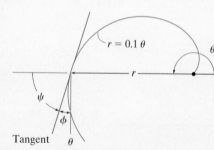

$r = 0.1\theta$

$\theta = \pi$

r

ψ

ϕ

Tangent

θ

(c)

Fig. 13–21

At the instant $\theta = \pi$ rad,

$$a_r = \ddot{r} - r\dot{\theta}^2 = 0 - 0.1(\pi)(4)^2 = -5.03 \text{ m/s}^2$$

$$a_\theta = r\ddot{\theta} + 2\dot{r}\dot{\theta} = 0 + 2(0.4)(4) = 3.20 \text{ m/s}^2$$

Substituting these results into Eqs. 1 and 2 and solving yields

$$N_C = -2.64 \text{ N}$$

$$F_C = 0.800 \text{ N} \qquad\qquad Ans.$$

What does the negative sign for N_C indicate?

FUNDAMENTAL PROBLEMS

F13–13. Determine the constant angular velocity $\dot{\theta}$ of the vertical shaft of the amusement ride if $\phi = 45°$. Neglect the mass of the cables and the size of the passengers.

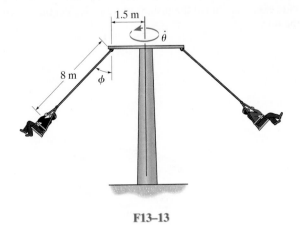

F13–13

F13–14. The 0.2-kg ball is blown through the smooth vertical circular tube whose shape is defined by $r = (0.6 \sin \theta)$ m, where θ is in radians. If $\theta = (\pi t^2)$ rad, where t is in seconds, determine the magnitude of force **F** exerted by the blower on the ball when $t = 0.5$ s.

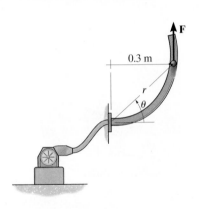

F13–14

F13–15. The 2-Mg car is traveling along the curved road described by $r = (50e^{2\theta})$ m, where θ is in radians. If a camera is located at A and it rotates with an angular velocity of $\dot{\theta} = 0.05$ rad/s and an angular acceleration of $\ddot{\theta} = 0.01$ rad/s^2 at the instant $\theta = \frac{\pi}{6}$ rad, determine the resultant friction force developed between the tires and the road at this instant.

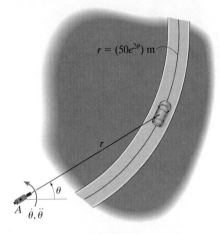

$r = (50e^{2\theta})$ m

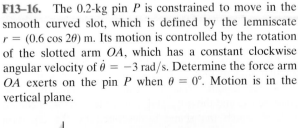

F13–15

F13–16. The 0.2-kg pin P is constrained to move in the smooth curved slot, which is defined by the lemniscate $r = (0.6 \cos 2\theta)$ m. Its motion is controlled by the rotation of the slotted arm OA, which has a constant clockwise angular velocity of $\dot{\theta} = -3$ rad/s. Determine the force arm OA exerts on the pin P when $\theta = 0°$. Motion is in the vertical plane.

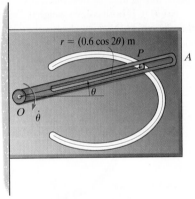

$r = (0.6 \cos 2\theta)$ m

F13–16

PROBLEMS

13–85. The spring-held follower AB has a weight of 0.75 lb and moves back and forth as its end rolls on the contoured surface of the cam, where $r = 0.2$ ft and $z = (0.1 \sin 2\theta)$ ft. If the cam is rotating at a constant rate of 6 rad/s, determine the force at the end A of the follower when $\theta = 45°$. In this position the spring is compressed 0.4 ft. Neglect friction at the bearing C.

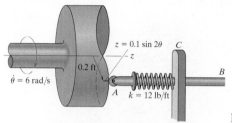

Prob. 13–85

13–86. Determine the magnitude of the resultant force acting on a 5-kg particle at the instant $t = 2$ s, if the particle is moving along a horizontal path defined by the equations $r = (2t + 10)$ m and $\theta = (1.5t^2 - 6t)$ rad, where t is in seconds.

13–87. The path of motion of a 5-lb particle in the horizontal plane is described in terms of polar coordinates as $r = (2t + 1)$ ft and $\theta = (0.5t^2 - t)$ rad, where t is in seconds. Determine the magnitude of the unbalanced force acting on the particle when $t = 2$ s.

*13–88. A particle, having a mass of 1.5 kg, moves along a path defined by the equations $r = (4 + 3t)$ m, $\theta = (t^2 + 2)$ rad, and $z = (6 - t^3)$ m, where t is in seconds. Determine the r, θ, and z components of force which the path exerts on the particle when $t = 2$ s.

13–89. Rod OA rotates counterclockwise with a constant angular velocity of $\dot\theta = 5$ rad/s. The double collar B is pin-connected together such that one collar slides over the rotating rod and the other slides over the *horizontal* curved rod, of which the shape is described by the equation $r = 1.5(2 - \cos\theta)$ ft. If both collars weigh 0.75 lb, determine the normal force which the curved rod exerts on one collar at the instant $\theta = 120°$. Neglect friction.

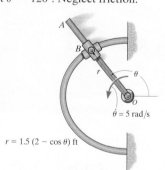

Prob. 13–89

13–90. The boy of mass 40 kg is sliding down the spiral slide at a constant speed such that his position, measured from the top of the chute, has components $r = 1.5$ m, $\theta = (0.7t)$ rad, and $z = (-0.5t)$ m, where t is in seconds. Determine the components of force $\mathbf{F}_r$, $\mathbf{F}_\theta$, and $\mathbf{F}_z$ which the slide exerts on him at the instant $t = 2$ s. Neglect the size of the boy.

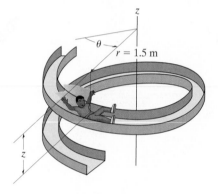

Prob. 13–90

13–91. The 0.5-lb particle is guided along the circular path using the slotted arm guide. If the arm has an angular velocity $\dot\theta = 4$ rad/s. and an angular acceleration $\ddot\theta = 8$ rad/s. at the instant $\theta = 30°$, determine the force of the guide on the particle. Motion occurs in the *horizontal plane*.

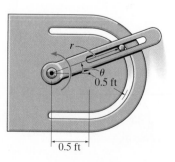

Prob. 13–91

*13–92. Using a forked rod, a smooth cylinder C having a mass of 0.5 kg is forced to move along the *vertical slotted* path $r = (0.5\theta)$ m, where θ is in radians. If the angular position of the arm is $\theta = (0.5t^2)$ rad, where t is in seconds, determine the force of the rod on the cylinder and the normal force of the slot on the cylinder at the instant $t = 2$ s. The cylinder is in contact with only *one edge* of the rod and slot at any instant.

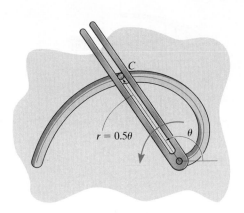

Prob. 13–92

13–93. If arm OA rotates with a constant clockwise angular velocity of $\dot{\theta} = 1.5$ rad/s. determine the force arm OA exerts on the smooth 4-lb cylinder B when $\theta = 45°$.

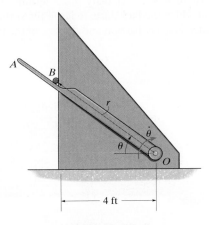

Prob. 13–93

13–94. The collar has a mass of 2 kg and travels along the smooth horizontal rod defined by the equiangular spiral $r = (e^{\theta})$ m, where θ is in radians. Determine the tangential force F and the normal force N acting on the collar when $\theta = 90°$, if the force F maintains a constant angular motion $\dot{\theta} = 2$ rad/s.

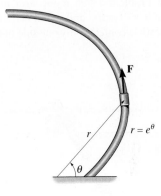

Prob. 13–94

13–95. The ball has a mass of 2 kg and a negligible size. It is originally traveling around the horizontal circular path of radius $r_0 = 0.5$ m such that the angular rate of rotation is $\dot{\theta}_0 = 1$ rad/s. If the attached cord ABC is drawn down through the hole at a constant speed of 0.2 m/s, determine the tension the cord exerts on the ball at the instant $r = 0.25$ m. Also, compute the angular velocity of the ball at this instant. Neglect the effects of friction between the ball and horizontal plane. *Hint:* First show that the equation of motion in the θ direction yields $a_{\theta} = r\ddot{\theta} + 2\dot{r}\dot{\theta} = (1/r)\,(d(r^2\dot{\theta}/dt)) = 0$. When integrated, $r^2\dot{\theta} = c$, where the constant c is determined from the problem data.

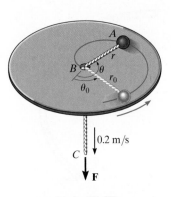

Prob. 13–95

*13–96. The particle has a mass of 0.5 kg and is confined to move along the smooth horizontal slot due to the rotation of the arm *OA*. Determine the force of the rod on the particle and the normal force of the slot on the particle when $\theta = 30°$. The rod is rotating with a constant angular velocity $\dot{\theta} = 2$ rad/s. Assume the particle contacts only one side of the slot at any instant.

13–97. Solve Problem 13–96 if the arm has an angular acceleration of $\ddot{\theta} = 3$ rad/s² and $\dot{\theta} = 2$ rad/s at this instant. Assume the particle contacts only one side of the slot at any instant.

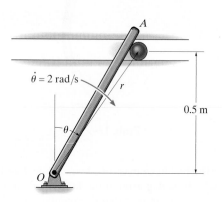

Probs. 13–96/97

13–98. The collar has a mass of 2 kg and travels along the smooth horizontal rod defined by the equiangular spiral $r = (e^\theta)$ m, where θ is in radians. Determine the tangential force F and the normal force N acting on the collar when $\theta = 45°$, if the force F maintains a constant angular motion $\dot{\theta} = 2$ rad/s.

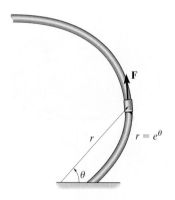

Prob. 13–98

13–99. For a short time, the 250-kg roller coaster car is traveling along the spiral track such that its position measured from the top of the track has components $r = 8$ m, $\theta = (0.1t + 0.5)$ rad, and $z = (-0.2t)$ m, where t is in seconds. Determine the magnitudes of the components of force which the track exerts on the car in the r, θ, and z directions at the instant $t = 2$ s. Neglect the size of the car.

Prob. 13–99

*13–100. The 0.5-lb ball is guided along the vertical circular path $r = 2r_c \cos\theta$ using the arm *OA*. If the arm has an angular velocity $\dot{\theta} = 0.4$ rad/s and an angular acceleration $\ddot{\theta} = 0.8$ rad/s² at the instant $\theta = 30°$, determine the force of the arm on the ball. Neglect friction and the size of the ball. Set $r_c = 0.4$ ft.

13–101. The ball of mass m is guided along the vertical circular path $r = 2r_c \cos\theta$ using the arm *OA*. If the arm has a constant angular velocity $\dot{\theta}_0$, determine the angle $\theta \le 45°$ at which the ball starts to leave the surface of the semicylinder. Neglect friction and the size of the ball.

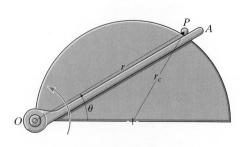

Probs. 13–100/101

13–102. Using a forked rod, a smooth cylinder P, having a mass of 0.4 kg, is forced to move along the *vertical slotted* path $r = (0.6\theta)$ m, where θ is in radians. If the cylinder has a constant speed of $v_C = 2$ m/s, determine the force of the rod and the normal force of the slot on the cylinder at the instant $\theta = \pi$ rad. Assume the cylinder is in contact with only *one edge* of the rod and slot at any instant. *Hint:* To obtain the time derivatives necessary to compute the cylinder's acceleration components a_r and a_θ, take the first and second time derivatives of $r = 0.6\theta$. Then, for further information, use Eq. 12–26 to determine $\dot{\theta}$. Also, take the time derivative of Eq. 12–26, noting that $\dot{v} = 0$ to determine $\ddot{\theta}$.

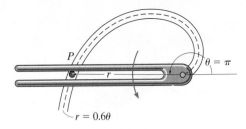

Prob. 13–102

13–103. A ride in an amusement park consists of a cart which is supported by small wheels. Initially the cart is traveling in a circular path of radius $r_0 = 16$ ft such that the angular rate of rotation is $\dot{\theta}_0 = 0.2$ rad/s. If the attached cable OC is drawn inward at a constant speed of $\dot{r} = -0.5$ ft/s, determine the tension it exerts on the cart at the instant $r = 4$ ft. The cart and its passengers have a total weight of 400 lb. Neglect the effects of friction. *Hint:* First show that the equation of motion in the θ direction yields $a_\theta = r\ddot{\theta} + 2\dot{r}\dot{\theta} = (1/r)\,d(r^2\dot{\theta})/dt = 0$. When integrated, $r^2\dot{\theta} = c$, where the constant c is determined from the problem data.

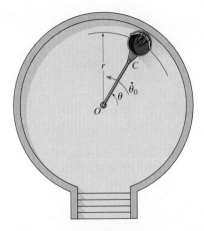

Prob. 13–103

***13–104.** The arm is rotating at a rate of $\dot{\theta} = 5$ rad/s when $\ddot{\theta} = 2$ rad/s^2 and $\theta = 90°$. Determine the normal force it must exert on the 0.5-kg particle if the particle is confined to move along the slotted path defined by the *horizontal* hyperbolic spiral $r\theta = 0.2$ m.

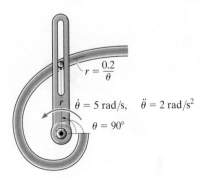

Prob. 13–104

13–105. The forked rod is used to move the smooth 2-lb particle around the horizontal path in the shape of a limaçon, $r = (2 + \cos\theta)$ ft. If at all times $\dot{\theta} = 0.5$ rad/s, determine the force which the rod exerts on the particle at the instant $\theta = 90°$. The fork and path contact the particle on only one side.

13–106. Solve Prob. 13–105 at the instant $\theta = 60°$.

13–107. The forked rod is used to move the smooth 2-lb particle around the horizontal path in the shape of a limaçon, $r = (2 + \cos\theta)$ ft. If $\theta = (0.5t^2)$ rad, where t is in seconds, determine the force which the rod exerts on the particle at the instant $t = 1$ s. The fork and path contact the particle on only one side.

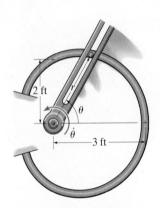

Probs. 13–105/106/107

*13-108. The collar, which has a weight of 3 lb, slides along the smooth rod lying in the *horizontal plane* and having the shape of a parabola $r = 4/(1 - \cos\theta)$, where θ is in radians and r is in feet. If the collar's angular rate is constant and equals $\dot\theta = 4$ rad/s, determine the tangential retarding force P needed to cause the motion and the normal force that the collar exerts on the rod at the instant $\theta = 90°$.

Prob. 13-108

13-109. The smooth particle has a mass of 80 g. It is attached to an elastic cord extending from O to P and due to the slotted arm guide moves along the *horizontal* circular path $r = (0.8 \sin\theta)$ m. If the cord has a stiffness $k = 30$ N/m and an unstretched length of 0.25 m, determine the force of the guide on the particle when $\theta = 60°$. The guide has a constant angular velocity $\dot\theta = 5$ rad/s.

13-110. Solve Prob. 13-109 if $\ddot\theta = 2$ rad/s^2 when $\dot\theta = 5$ rad/s and $\theta = 60°$.

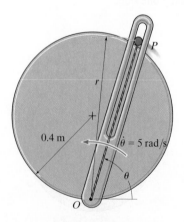

Probs. 13-109/110

13-111. A 0.2-kg spool slides down along a smooth rod. If the rod has a constant angular rate of rotation $\dot\theta = 2$ rad/s in the vertical plane, show that the equations of motion for the spool are $\ddot{r} - 4r - 9.81 \sin\theta = 0$ and $0.8\dot{r} + N_s - 1.962 \cos\theta = 0$, where N_s is the magnitude of the normal force of the rod on the spool. Using the methods of differential equations, it can be shown that the solution of the first of these equations is $r = C_1 e^{-2t} + C_2 e^{2t} - (9.81/8) \sin 2t$. If r, $\dot{r}$, and θ are zero when $t = 0$, evaluate the constants C_1 and C_2 determine r at the instant $\theta = \pi/4$ rad.

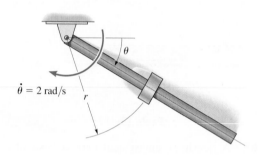

$\dot\theta = 2$ rad/s

Prob. 13-111

*13-112. The pilot of an airplane executes a vertical loop which in part follows the path of a cardioid, $r = 600(1 + \cos\theta)$ ft. If his speed at A $(\theta = 0°)$ is a constant $v_P = 80$ ft/s, determine the vertical force the seat belt must exert on him to hold him to his seat when the plane is upside down at A. He weighs 150 lb.

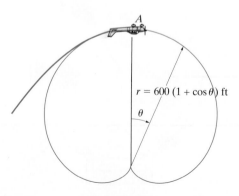

$r = 600 (1 + \cos\theta)$ ft

Prob. 13-112

*13.7   Central-Force Motion and Space Mechanics

If a particle is moving only under the influence of a force having a line of action which is always directed toward a fixed point, the motion is called *central-force motion*. This type of motion is commonly caused by electrostatic and gravitational forces.

In order to analyze the motion, we will consider the particle P shown in Fig. 13–22a, which has a mass m and is acted upon only by the central force $\mathbf{F}$. The free-body diagram for the particle is shown in Fig. 13–22b. Using polar coordinates (r, θ), the equations of motion, Eq. 13–9, become

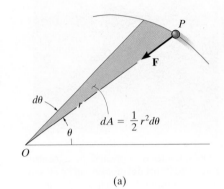

(a)

$$\Sigma F_r = ma_r; \qquad -F = m\left[\frac{d^2r}{dt^2} - r\left(\frac{d\theta}{dt}\right)^2\right] \qquad (13\text{–}11)$$

$$\Sigma F_\theta = ma_\theta; \qquad 0 = m\left(r\frac{d^2\theta}{dt^2} + 2\frac{dr}{dt}\frac{d\theta}{dt}\right)$$

The second of these equations may be written in the form

$$\frac{1}{r}\left[\frac{d}{dt}\left(r^2\frac{d\theta}{dt}\right)\right] = 0$$

so that integrating yields

$$r^2\frac{d\theta}{dt} = h \qquad (13\text{–}12)$$

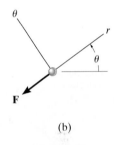

(b)

Fig. 13–22

Here h is the constant of integration.

From Fig. 13–22a notice that the shaded area described by the radius r, as r moves through an angle $d\theta$, is $dA = \frac{1}{2}r^2\,d\theta$. If the *areal velocity* is defined as

$$\frac{dA}{dt} = \frac{1}{2}r^2\frac{d\theta}{dt} = \frac{h}{2} \qquad (13\text{–}13)$$

then it is seen that the areal velocity for a particle subjected to central-force motion is *constant*. In other words, the particle will sweep out equal segments of area per unit of time as it travels along the path. To obtain the *path of motion*, $r = f(\theta)$, the independent variable t must be eliminated from Eqs. 13–11. Using the chain rule of calculus and Eq. 13–12, the time derivatives of Eqs. 13–11 may be replaced by

$$\frac{dr}{dt} = \frac{dr}{d\theta}\frac{d\theta}{dt} = \frac{h}{r^2}\frac{dr}{d\theta}$$

$$\frac{d^2r}{dt^2} = \frac{d}{dt}\left(\frac{h}{r^2}\frac{dr}{d\theta}\right) = \frac{d}{d\theta}\left(\frac{h}{r^2}\frac{dr}{d\theta}\right)\frac{d\theta}{dt} = \left[\frac{d}{d\theta}\left(\frac{h}{r^2}\frac{dr}{d\theta}\right)\right]\frac{h}{r^2}$$

This satellite is subjected to a central force and its orbital motion can be closely predicted using the equations developed in this section.

Substituting a new dependent variable (xi) $\xi = 1/r$ into the second equation, we have

$$\frac{d^2r}{dt^2} = -h^2\xi^2\frac{d^2\xi}{d\theta^2}$$

Also, the square of Eq. 13–12 becomes

$$\left(\frac{d\theta}{dt}\right)^2 = h^2\xi^4$$

Substituting these two equations into the first of Eqs. 13–11 yields

$$-h^2\xi^2\frac{d^2\xi}{d\theta^2} - h^2\xi^3 = -\frac{F}{m}$$

or

$$\frac{d^2\xi}{d\theta^2} + \xi = \frac{F}{mh^2\xi^2} \qquad (13\text{–}14)$$

This differential equation defines the path over which the particle travels when it is subjected to the central force **F**.*

For application, the force of gravitational attraction will be considered. Some common examples of central-force systems which depend on gravitation include the motion of the moon and artificial satellites about the earth, and the motion of the planets about the sun. As a typical problem in space mechanics, consider the trajectory of a space satellite or space vehicle launched into free-flight orbit with an initial velocity $\mathbf{v}_0$, Fig. 13–23. It will be assumed that this velocity is initially *parallel* to the tangent at the surface of the earth, as shown in the figure.† Just after the satellite is released into free flight, the only force acting on it is the gravitational force of the earth. (Gravitational attractions involving other bodies such as the moon or sun will be neglected, since for orbits close to the earth their effect is small in comparison with the earth's gravitation.) According to Newton's law of gravitation, force **F** will always act between the mass centers of the earth and the satellite, Fig. 13–23. From Eq. 13–1, this force of attraction has a magnitude of

$$F = G\frac{M_e m}{r^2}$$

where M_e and m represent the mass of the earth and the satellite, respectively, G is the gravitational constant, and r is the distance between

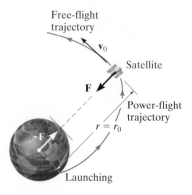

Free-flight trajectory

$\mathbf{v}_0$

Satellite

F

Power-flight trajectory

$r = r_0$

-**F**

Launching

Fig. 13–23

---

*In the derivation, **F** is considered positive when it is directed toward point O. If **F** is oppositely directed, the right side of Eq. 13–14 should be negative.

†The case where $\mathbf{v}_0$ acts at some initial angle θ to the tangent is best described using the conservation of angular momentum.

the mass centers. To obtain the orbital path, we set $\xi = 1/r$ in the foregoing equation and substitute the result into Eq. 13–14. We obtain

$$\frac{d^2\xi}{d\theta^2} + \xi = \frac{GM_e}{h^2} \qquad (13\text{–}15)$$

This second-order differential equation has constant coefficients and is nonhomogeneous. The solution is the sum of the complementary and particular solutions given by

$$\xi = \frac{1}{r} = C\cos(\theta - \phi) + \frac{GM_e}{h^2} \qquad (13\text{–}16)$$

This equation represents the *free-flight trajectory* of the satellite. It is the equation of a conic section expressed in terms of polar coordinates.

A geometric interpretation of Eq. 13–16 requires knowledge of the equation for a conic section. As shown in Fig. 13–24, a conic section is defined as the locus of a point P that moves in such a way that the ratio of its distance to a *focus*, or fixed point F, to its perpendicular distance to a fixed line DD called the *directrix*, is constant. This constant ratio will be denoted as e and is called the *eccentricity*. By definition

$$e = \frac{FP}{PA}$$

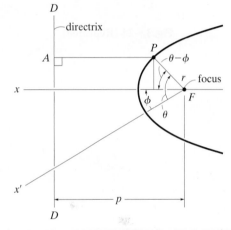

Fig. 13–24

From Fig. 13–24,

$$FP = r = e(PA) = e[p - r\cos(\theta - \phi)]$$

or

$$\frac{1}{r} = \frac{1}{p}\cos(\theta - \phi) + \frac{1}{ep}$$

Comparing this equation with Eq. 13–16, it is seen that the fixed distance from the focus to the directrix is

$$p = \frac{1}{C} \qquad (13\text{–}17)$$

And the eccentricity of the conic section for the trajectory is

$$e = \frac{Ch^2}{GM_e} \qquad (13\text{–}18)$$

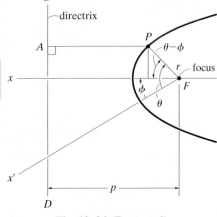

Fig. 13–24 (Repeated)

Provided the polar angle θ is measured from the x axis (an axis of symmetry since it is perpendicular to the directrix), the angle ϕ is zero, Fig. 13–24, and therefore Eq. 13–16 reduces to

$$\frac{1}{r} = C \cos \theta + \frac{GM_e}{h^2} \tag{13–19}$$

The constants h and C are determined from the data obtained for the position and velocity of the satellite at the end of the *power-flight trajectory*. For example, if the initial height or distance to the space vehicle is r_0, measured from the center of the earth, and its initial speed is v_0 at the beginning of its free flight, Fig. 13–25, then the constant h may be obtained from Eq. 13–12. When $\theta = \phi = 0°$, the velocity v_0 has no radial component; therefore, from Eq. 12–25, $v_0 = r_0(d\theta/dt)$, so that

$$h = r_0^2 \frac{d\theta}{dt}$$

or

$$h = r_0 v_0 \tag{13–20}$$

To determine C, use Eq. 13–19 with $\theta = 0°, r = r_0$, and substitute Eq. 13–20 for h:

$$C = \frac{1}{r_0}\left(1 - \frac{GM_e}{r_0 v_0^2}\right) \tag{13–21}$$

The equation for the free-flight trajectory therefore becomes

$$\frac{1}{r} = \frac{1}{r_0}\left(1 - \frac{GM_e}{r_0 v_0^2}\right)\cos \theta + \frac{GM_e}{r_0^2 v_0^2} \tag{13–22}$$

The type of path traveled by the satellite is determined from the value of the eccentricity of the conic section as given by Eq. 13–18. If

$e = 0$	free-flight trajectory is a circle
$e = 1$	free-flight trajectory is a parabola
$e < 1$	free-flight trajectory is an ellipse
$e > 1$	free-flight trajectory is a hyperbola

$$\tag{13–23}$$

Parabolic Path.   Each of these trajectories is shown in Fig. 13–25. From the curves it is seen that when the satellite follows a parabolic path, it is "on the border" of never returning to its initial starting point. The initial launch velocity, v_0, required for the satellite to follow a parabolic path is called the *escape velocity*. The speed, v_e, can be determined by using the second of Eqs. 13–23, $e = 1$, with Eqs. 13–18, 13–20, and 13–21. It is left as an exercise to show that

$$v_e = \sqrt{\frac{2GM_e}{r_0}} \qquad (13\text{–}24)$$

Circular Orbit.   The speed v_c required to launch a satellite into a *circular orbit* can be found using the first of Eqs. 13–23, $e = 0$. Since e is related to h and C, Eq. 13–18, C must be zero to satisfy this equation (from Eq. 13–20, h cannot be zero); and therefore, using Eq. 13–21, we have

$$v_c = \sqrt{\frac{GM_e}{r_0}} \qquad (13\text{–}25)$$

Provided r_0 represents a minimum height for launching, in which frictional resistance from the atmosphere is neglected, speeds at launch which are less than v_c will cause the satellite to reenter the earth's atmosphere and either burn up or crash, Fig. 13–25.

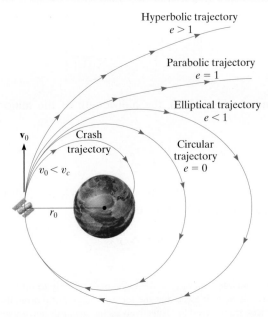

Fig. 13–25

13

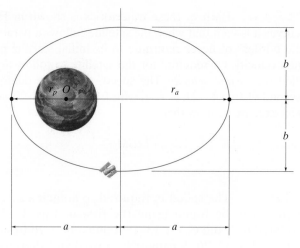

Fig. 13–26

Elliptical Orbit.   All the trajectories attained by planets and most satellites are elliptical, Fig. 13–26. For a satellite's orbit about the earth, the *minimum distance* from the orbit to the center of the earth O (which is located at one of the foci of the ellipse) is r_p and can be found using Eq. 13–22 with $\theta = 0°$. Therefore;

$$r_p = r_0 \qquad (13\text{–}26)$$

This minimum distance is called the *perigee* of the orbit. The *apogee* or maximum distance r_a can be found using Eq. 13–22 with $\theta = 180°$.* Thus,

$$r_a = \frac{r_0}{(2GM_e/r_0v_0^2) - 1} \qquad (13\text{–}27)$$

With reference to Fig. 13–26, the half-length of the major axis of the ellipse is

$$a = \frac{r_p + r_a}{2} \qquad (13\text{–}28)$$

Using analytical geometry, it can be shown that the half-length of the minor axis is determined from the equation

$$b = \sqrt{r_p r_a} \qquad (13\text{–}29)$$

---

*Actually, the terminology perigee and apogee pertains only to orbits about the *earth*. If any other heavenly body is located at the focus of an elliptical orbit, the minimum and maximum distances are referred to respectively as the *periapsis* and *apoapsis* of the orbit.

Furthermore, by direct integration, the area of an ellipse is

$$A = \pi ab = \frac{\pi}{2}(r_p + r_a)\sqrt{r_p r_a} \qquad (13\text{-}30)$$

The areal velocity has been defined by Eq. 13–13, $dA/dt = h/2$. Integrating yields $A = hT/2$, where T is the *period* of time required to make one orbital revolution. From Eq. 13–30, the period is

$$T = \frac{\pi}{h}(r_p + r_a)\sqrt{r_p r_a} \qquad (13\text{-}31)$$

In addition to predicting the orbital trajectory of earth satellites, the theory developed in this section is valid, to a surprisingly close approximation, at predicting the actual motion of the planets traveling around the sun. In this case the mass of the sun, M_s, should be substituted for M_e when the appropriate formulas are used.

The fact that the planets do indeed follow elliptic orbits about the sun was discovered by the German astronomer Johannes Kepler in the early seventeenth century. His discovery was made *before* Newton had developed the laws of motion and the law of gravitation, and so at the time it provided important proof as to the validity of these laws. Kepler's laws, developed after 20 years of planetary observation, are summarized as follows:

1. Every planet travels in its orbit such that the line joining it to the center of the sun sweeps over equal areas in equal intervals of time, whatever the line's length.

2. The orbit of every planet is an ellipse with the sun placed at one of its foci.

3. The square of the period of any planet is directly proportional to the cube of the major axis of its orbit.

A mathematical statement of the first and second laws is given by Eqs. 13–13 and 13–22, respectively. The third law can be shown from Eq. 13–31 using Eqs. 13–19, 13–28, and 13–29. (See Prob. 13–117.)

EXAMPLE | 13.13

A satellite is launched 600 km from the surface of the earth, with an initial velocity of 30 Mm/h acting parallel to the tangent at the surface of the earth, Fig. 13–27. Assuming that the radius of the earth is 6378 km and that its mass is $5.976(10^{24})$ kg, determine (a) the eccentricity of the orbital path, and (b) the velocity of the satellite at apogee.

SOLUTION

Part (a). The eccentricity of the orbit is obtained using Eq. 13–18. The constants h and C are first determined from Eq. 13–20 and 13–21. Since

$$r_p = r_0 = 6378 \text{ km} + 600 \text{ km} = 6.978(10^6) \text{ m}$$

$$v_0 = 30 \text{ Mm/h} = 8333.3 \text{ m/s}$$

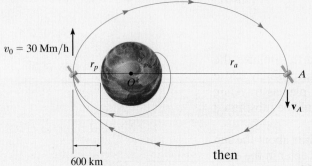

$v_0 = 30$ Mm/h

600 km

Fig. 13–27

then

$$h = r_p v_0 = 6.978(10^6)(8333.3) = 58.15(10^9) \text{ m}^2/\text{s}$$

$$C = \frac{1}{r_p}\left(1 - \frac{GM_e}{r_p v_0^2}\right)$$

$$= \frac{1}{6.978(10^6)}\left\{1 - \frac{66.73(10^{-12})[5.976(10^{24})]}{6.978(10^6)(8333.3)^2}\right\} = 25.4(10^{-9}) \text{ m}^{-1}$$

Hence,

$$e = \frac{Ch^2}{GM_e} = \frac{2.54(10^{-8})[58.15(10^9)]^2}{66.73(10^{-12})[5.976(10^{24})]} = 0.215 < 1 \quad \textit{Ans.}$$

From Eqs. 13–23, observe that the orbit is an *ellipse*.

Part (b). If the satellite were launched at the apogee A shown in Fig. 13–27, with a velocity $\mathbf{v}_A$, the same orbit would be maintained provided

$$h = r_p v_0 = r_a v_A = 58.15(10^9) \text{ m}^2/\text{s}$$

Using Eq. 13–27, we have

$$r_a = \frac{r_p}{\dfrac{2GM_e}{r_p v_0^2} - 1} = \frac{6.978(10^6)}{\dfrac{2[66.73(10^{-12})][5.976(10^{24})]}{6.978(10^6)(8333.3)^2} - 1} = 10.804(10^6) \text{ m}$$

Thus,

$$v_A = \frac{58.15(10^9)}{10.804(10^6)} = 5382.2 \text{ m/s} = 19.4 \text{ Mm/h} \quad \textit{Ans.}$$

NOTE: The farther the satellite is from the earth, the slower it moves, which is to be expected since h is constant.

PROBLEMS

In the following problems, except where otherwise indicated, assume that the radius of the earth is 6378 km, the earth's mass is $5.976(10^{24})$ kg, the mass of the sun is $1.99(10^{30})$ kg, and the gravitational constant is $G = 66.73(10^{-12})$ m^3/(kg·s^2).

13–113. The earth has an orbit with eccentricity $e = 0.0821$ around the sun. Knowing that the earth's minimum distance from the sun is $151.3(10^6)$ km, find the speed at which the earth travels when it is at this distance. Determine the equation in polar coordinates which describes the earth's orbit about the sun.

13–114. A communications satellite is in a circular orbit above the earth such that it always remains directly over a point on the earth's surface. As a result, the period of the satellite must equal the rotation of the earth, which is approximately 24 hours. Determine the satellite's altitude h above the earth's surface and its orbital speed.

13–115. The speed of a satellite launched into a circular orbit about the earth is given by Eq. 13–25. Determine the speed of a satellite launched parallel to the surface of the earth so that it travels in a circular orbit 800 km from the earth's surface.

*13–116. The rocket is in circular orbit about the earth at an altitude of $h = 4$ Mm. Determine the minimum increment in speed it must have in order to escape the earth's gravitational field.

13–117. Prove Kepler's third law of motion. *Hint:* Use Eqs. 13–19, 13–28, 13–29, and 13–31.

13–118. The satellite is moving in an elliptical orbit with an eccentricity $e = 0.25$. Determine its speed when it is at its maximum distance A and minimum distance B from the earth.

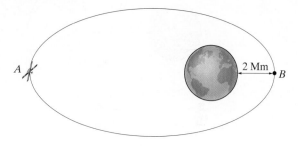

Prob. 13–118

13–119. The elliptical orbit of a satellite orbiting the earth has an eccentricity of $e = 0.45$. If the satellite has an altitude of 6 Mm at perigee, determine the velocity of the satellite at apogee and the period.

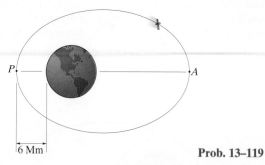

Prob. 13–119

*13–120. Determine the constant speed of satellite S so that it circles the earth with an orbit of radius $r = 15$ Mm. *Hint:* Use Eq. 13–1.

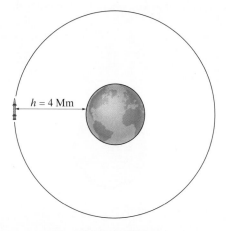

Prob. 13–116

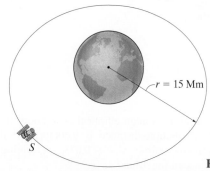

Prob. 13–120

13–121. The rocket is in free flight along an elliptical trajectory $A'A$. The planet has no atmosphere, and its mass is 0.70 times that of the earth. If the rocket has an apoapsis and periapsis as shown in the figure, determine the speed of the rocket when it is at point A.

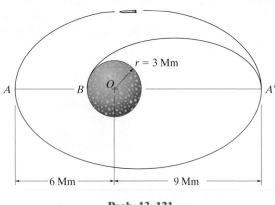

Prob. 13–121

*13–124.** An elliptical path of a satellite has an eccentricity $e = 0.130$. If it has a speed of 15 Mm/h when it is at perigee, P, determine its speed when it arrives at apogee, A. Also, how far is it from the earth's surface when it is at A?

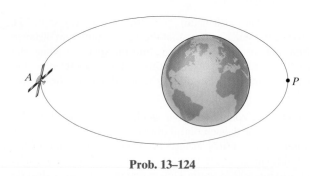

Prob. 13–124

13–122. A satellite S travels in a circular orbit around the earth. A rocket is located at the apogee of its elliptical orbit for which $e = 0.58$. Determine the sudden change in speed that must occur at A so that the rocket can enter the satellite's orbit while in free flight along the blue elliptical trajectory. When it arrives at B, determine the sudden adjustment in speed that must be given to the rocket in order to maintain the circular orbit.

13–125. A satellite is launched with an initial velocity $v_0 = 2500$ mi/h parallel to the surface of the earth. Determine the required altitude (or range of altitudes) above the earth's surface for launching if the free-flight trajectory is to be (a) circular, (b) parabolic, (c) elliptical, and (d) hyperbolic. Take $G = 34.4(10^{-9})(\text{lb} \cdot \text{ft}^2)/\text{slug}^2$, $M_e = 409(10^{21})$ slug, the earth's radius $r_e = 3960$ mi, and 1 mi = 5280 ft.

13–126. A probe has a circular orbit around a planet of radius R and mass M. If the radius of the orbit is nR and the explorer is traveling with a constant speed v_0, determine the angle θ at which it lands on the surface of the planet B when its speed is reduced to kv_0, where $k < 1$ at point A.

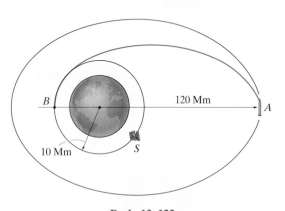

Prob. 13–122

13–123. An asteroid is in an elliptical orbit about the sun such that its perihelion distance is $9.30(10^9)$ km. If the eccentricity of the orbit is $e = 0.073$, determine the aphelion distance of the orbit.

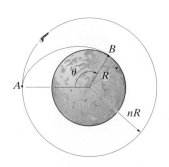

Prob. 13–126

13–127. Upon completion of the moon exploration mission, the command module, which was originally in a circular orbit as shown, is given a boost so that it escapes from the moon's gravitational field. Determine the necessary increase in velocity so that the command module follows a parabolic trajectory. The mass of the moon is $0.01230 M_e$.

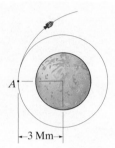

Prob. 13–127

*13–128.** The rocket is traveling in a free-flight elliptical orbit about the earth such that $e = 0.76$ and its perigee is 9 Mm as shown. Determine its speed when it is at point B. Also determine the sudden decrease in speed the rocket must experience at A in order to travel in a circular orbit about the earth.

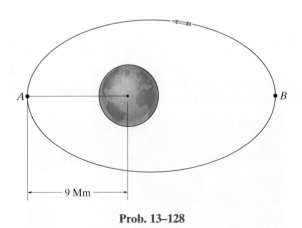

Prob. 13–128

13–129. A rocket is in circular orbit about the earth at an altitude above the earth's surface of $h = 4$ Mm. Determine the minimum increment in speed it must have in order to escape the earth's gravitational field.

13–130. The satellite is an elliptical orbit having an eccentricity of $e = 0.15$. If its velocity at perigee is $v_P = 15$ Mm/h, determine its velocity at apogee A and the period of the satellite.

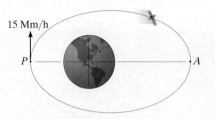

Prob. 13–130

13–131. A rocket is in a free-flight elliptical orbit about the earth such that the eccentricity of its orbit is e and its perigee is r_0. Determine the minimum increment of speed it should have in order to escape the earth's gravitational field when it is at this point along its orbit.

*13–132.** The rocket shown is originally in a circular orbit 6 Mm above the surface of the earth. It is required that it travel in another circular orbit having an altitude of 14 Mm. To do this, the rocket is given a short pulse of power at A so that it travels in free flight along the gray elliptical path from the first orbit to the second orbit. Determine the necessary speed it must have at A just after the power pulse, and the time required to get to the outer orbit along the path AA'. What adjustment in speed must be made at A' to maintain the second circular orbit?

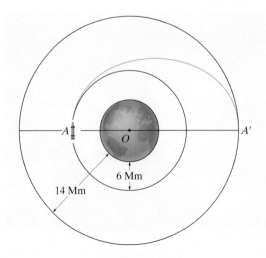

Prob. 13–132

13

CONCEPTUAL PROBLEMS

P13–1. If the box is released from rest at *A*, use numerical values to show how you would estimate the time for it to arrive at *B*. Also, list the assumptions for your analysis.

P13–1

P13–2. The tugboat has a known mass and its propeller provides a known maximum thrust. When the tug is fully powered you observe the time it takes for the tug to reach a speed of known value starting from rest. Show how you could determine the mass of the barge. Neglect the drag force of the water on the tug. Use numerical values to explain your answer.

P13–2

P13–3. Determine the smallest speed of each car *A* and *B* so that the passengers do not lose contact with the seat while the arms turn at a constant rate. What is the largest normal force of the seat on each passenger? Use numerical values to explain your answer.

P13–3

P13–4. Each car is pin connected at its ends to the rim of the wheel which turns at a constant speed. Using numerical values, show how to determine the resultant force the seat exerts on the passenger located in the top car *A*. The passengers are seated towards the center of the wheel. Also, list the assumptions for your analysis.

P13–4

CHAPTER REVIEW

Kinetics

Kinetics is the study of the relation between forces and the acceleration they cause. This relation is based on Newton's second law of motion, expressed mathematically as $\Sigma \mathbf{F} = m\mathbf{a}$.

Before applying the equation of motion, it is important to first draw the particle's *free-body diagram* in order to account for all of the forces that act on the particle. Graphically, this diagram is equal to the *kinetic diagram*, which shows the result of the forces, that is, the $m\mathbf{a}$ vector.

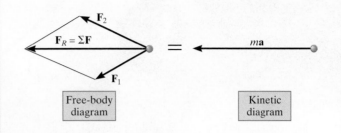

Free-body diagram   Kinetic diagram

Inertial Coordinate Systems

When applying the equation of motion, it is important to measure the acceleration from an inertial coordinate system. This system has axes that do not rotate but are either fixed or translate with a constant velocity. Various types of inertial coordinate systems can be used to apply $\Sigma \mathbf{F} = m\mathbf{a}$ in component form.

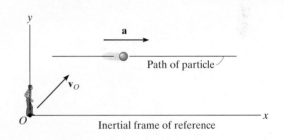

Inertial frame of reference

Rectangular x, y, z axes are used to describe rectilinear motion along each of the axes.

$$\Sigma F_x = ma_x, \ \Sigma F_y = ma_y, \ \Sigma F_z = ma_z$$

Normal and tangential n, t axes are often used when the path is known. Recall that $\mathbf{a}_n$ is always directed in the $+n$ direction. It indicates the change in the velocity direction. Also recall that $\mathbf{a}_t$ is tangent to the path. It indicates the change in the velocity magnitude.

$$\Sigma F_t = ma_t, \ \Sigma F_n = ma_n, \ \Sigma F_b = 0$$
$$a_t = dv/dt \quad \text{or} \quad a_t = v\,dv/ds$$

$$a_n = v^2/\rho \quad \text{where} \quad \rho = \frac{[1 + (dy/dx)^2]^{3/2}}{|d^2y/dx^2|}$$

Cylindrical coordinates are useful when angular motion of the radial line r is specified or when the path can conveniently be described with these coordinates.

$$\Sigma F_r = m(\ddot{r} - r\dot{\theta}^2)$$
$$\Sigma F_\theta = m(r\ddot{\theta} + 2\dot{r}\dot{\theta})$$
$$\Sigma F_z = m\ddot{z}$$

Central-Force Motion

When a single force acts upon a particle, such as during the free-flight trajectory of a satellite in a gravitational field, then the motion is referred to as central-force motion. The orbit depends upon the eccentricity e; and as a result, the trajectory can either be circular, parabolic, elliptical, or hyperbolic.

Chapter 14

As the woman falls, her energy will have to be absorbed by the bungee cord.
The principles of work and energy can be used to predict the motion.

Kinetics of a Particle: Work and Energy

CHAPTER OBJECTIVES

■ To develop the principle of work and energy and apply it to solve problems that involve force, velocity, and displacement.

■ To study problems that involve power and efficiency.

■ To introduce the concept of a conservative force and apply the theorem of conservation of energy to solve kinetic problems.

14.1 The Work of a Force

In this chapter, we will analyze motion of a particle using the concepts of work and energy. The resulting equation will be useful for solving problems that involve force, velocity, and displacement. Before we do this, however, we must first define the work of a force. Specifically, a force $\mathbf{F}$ will do *work* on a particle only when the particle undergoes a *displacement in the direction of the force*. For example, if the force $\mathbf{F}$ in Fig. 14–1 causes the particle to move along the path s from position $\mathbf{r}$ to a new position $\mathbf{r}'$, the displacement is then $d\mathbf{r} = \mathbf{r}' - \mathbf{r}$. The magnitude of $d\mathbf{r}$ is ds, the length of the differential segment along the path. If the angle between the tails of $d\mathbf{r}$ and $\mathbf{F}$ is θ, Fig. 14–1, then the work done by $\mathbf{F}$ is a *scalar quantity*, defined by

$$dU = F \, ds \cos \theta$$

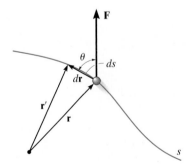

Fig. 14–1

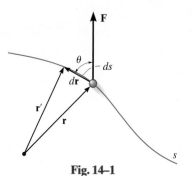

Fig. 14–1

By definition of the dot product (see Eq. B–14) this equation can also be written as

$$dU = \mathbf{F} \cdot d\mathbf{r}$$

This result may be interpreted in one of two ways: either as the product of F and the component of displacement $ds \cos \theta$ in the direction of the force, or as the product of ds and the component of force, $F \cos \theta$, in the direction of displacement. Note that if $0° \le \theta < 90°$, then the force component and the displacement have the *same sense* so that the work is *positive;* whereas if $90° < \theta \le 180°$, these vectors will have *opposite sense*, and therefore the work is *negative*. Also, $dU = 0$ if the force is *perpendicular* to displacement, since $\cos 90° = 0$, or if the force is applied at a *fixed point*, in which case the displacement is zero.

The unit of work in SI units is the joule (J), which is the amount of work done by a one-newton force when it moves through a distance of one meter in the direction of the force ($1\,\text{J} = 1\,\text{N} \cdot \text{m}$). In the FPS system, work is measured in units of foot-pounds ($\text{ft} \cdot \text{lb}$), which is the work done by a one-pound force acting through a distance of one foot in the direction of the force.*

Work of a Variable Force.

If the particle acted upon by the force $\mathbf{F}$ undergoes a finite displacement along its path from $\mathbf{r}_1$ to $\mathbf{r}_2$ or s_1 to s_2, Fig. 14–2a, the work of force $\mathbf{F}$ is determined by integration. Provided $\mathbf{F}$ and θ can be expressed as a function of position, then

$$U_{1-2} = \int_{\mathbf{r}_1}^{\mathbf{r}_2} \mathbf{F} \cdot d\mathbf{r} = \int_{s_1}^{s_2} F \cos \theta \, ds \qquad (14\text{–}1)$$

Sometimes, this relation may be obtained by using experimental data to plot a graph of $F \cos \theta$ vs. s. Then the area under this graph bounded by s_1 and s_2 represents the total work, Fig. 14–2b.

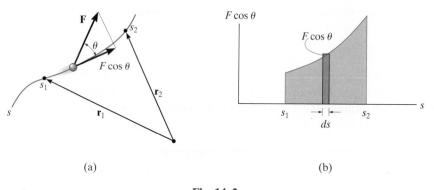

(a)   (b)

Fig. 14–2

*By convention, the units for the moment of a force or torque are written as $\text{lb} \cdot \text{ft}$, to distinguish them from those used to signify work, $\text{ft} \cdot \text{lb}$.

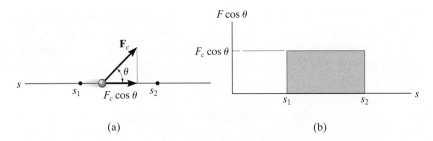

Fig. 14–3

Work of a Constant Force Moving Along a Straight Line.

If the force $\mathbf{F}_c$ has a constant magnitude and acts at a constant angle θ from its straight-line path, Fig. 14–3a, then the component of $\mathbf{F}_c$ in the direction of displacement is always $F_c \cos \theta$. The work done by $\mathbf{F}_c$ when the particle is displaced from s_1 to s_2 is determined from Eq. 14–1, in which case

$$U_{1-2} = F_c \cos \theta \int_{s_1}^{s_2} ds$$

or

$$\boxed{U_{1-2} = F_c \cos \theta (s_2 - s_1)} \qquad (14\text{–}2)$$

Here the work of $\mathbf{F}_c$ represents the *area of the rectangle* in Fig. 14–3b.

Work of a Weight.

Consider a particle of weight $\mathbf{W}$, which moves up along the path s shown in Fig. 14–4 from position s_1 to position s_2. At an intermediate point, the displacement $d\mathbf{r} = dx\mathbf{i} + dy\mathbf{j} + dz\mathbf{k}$. Since $\mathbf{W} = -W\mathbf{j}$, applying Eq. 14–1 we have

$$U_{1-2} = \int \mathbf{F} \cdot d\mathbf{r} = \int_{\mathbf{r}_1}^{\mathbf{r}_2} (-W\mathbf{j}) \cdot (dx\mathbf{i} + dy\mathbf{j} + dz\mathbf{k})$$

$$= \int_{y_1}^{y_2} -W \, dy = -W(y_2 - y_1)$$

or

$$\boxed{U_{1-2} = -W \Delta y} \qquad (14\text{–}3)$$

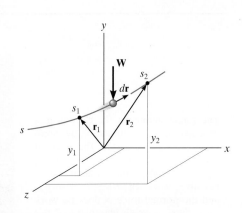

Fig. 14–4

Thus, the work is independent of the path and is equal to the magnitude of the particle's weight times its vertical displacement. In the case shown in Fig. 14–4 the work is *negative*, since W is downward and Δy is upward. Note, however, that if the particle is displaced *downward* $(-\Delta y)$, the work of the weight is *positive*. Why?

Work of a Spring Force.   If an elastic spring is elongated a distance ds, Fig. 14–5a, then the work done by the force that acts on the attached particle is $dU = -F_s ds = -ks\,ds$. The work is *negative* since $\mathbf{F}_s$ acts in the opposite sense to ds. If the particle displaces from s_1 to s_2, the work of $\mathbf{F}_s$ is then

$$U_{1-2} = \int_{s_1}^{s_2} F_s\,ds = \int_{s_1}^{s_2} -ks\,ds$$

$$\boxed{U_{1-2} = -\left(\tfrac{1}{2}ks_2^2 - \tfrac{1}{2}ks_1^2\right)} \qquad (14\text{–}4)$$

This work represents the trapezoidal area under the line $F_s = ks$, Fig. 14–5b.

A mistake in sign can be avoided when applying this equation if one simply notes the direction of the spring force acting on the particle and compares it with the sense of direction of displacement of the particle— if both are in the *same sense, positive work* results; if they are *opposite* to one another, the *work is negative.*

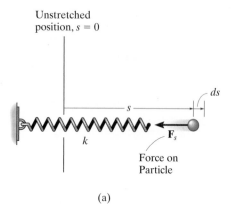

Unstretched position, $s = 0$

k

$\mathbf{F}_s$

ds

Force on Particle

(a)

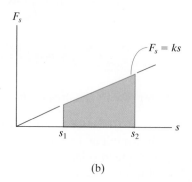

F_s

$F_s = ks$

s_1 s_2

s

(b)

Fig. 14–5

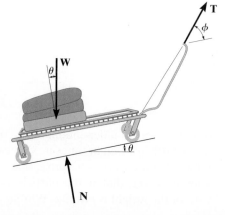

The forces acting on the cart, as it is pulled a distance s up the incline, are shown on its free-body diagram. The constant towing force $\mathbf{T}$ does positive work of $U_T = (T\cos\phi)s$, the weight does negative work of $U_W = -(W\sin\theta)s$, and the normal force $\mathbf{N}$ does no work since there is no displacement of this force along its line of action.

EXAMPLE | 14.1

The 10-kg block shown in Fig. 14–6a rests on the smooth incline. If the spring is originally stretched 0.5 m, determine the total work done by all the forces acting on the block when a horizontal force $P = 400$ N pushes the block up the plane $s = 2$ m.

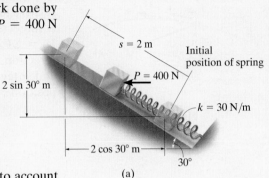

(a)

(b)

Fig. 14–6

SOLUTION

First the free-body diagram of the block is drawn in order to account for all the forces that act on the block, Fig. 14–6b.

Horizontal Force P. Since this force is *constant*, the work is determined using Eq. 14–2. The result can be calculated as the force times the component of displacement in the direction of the force; i.e.,

$$U_P = 400 \text{ N } (2 \text{ m } \cos 30°) = 692.8 \text{ J}$$

or the displacement times the component of force in the direction of displacement, i.e.,

$$U_P = 400 \text{ N } \cos 30°(2 \text{ m}) = 692.8 \text{ J}$$

Spring Force F_s. In the initial position the spring is stretched $s_1 = 0.5$ m and in the final position it is stretched $s_2 = 0.5$ m + 2 m = 2.5 m. We require the work to be negative since the force and displacement are opposite to each other. The work of $\mathbf{F}_s$ is thus

$$U_s = -\left[\tfrac{1}{2}(30 \text{ N/m})(2.5 \text{ m})^2 - \tfrac{1}{2}(30 \text{ N/m})(0.5 \text{ m})^2\right] = -90 \text{ J}$$

Weight W. Since the weight acts in the opposite sense to its vertical displacement, the work is negative; i.e.,

$$U_W = -(98.1 \text{ N}) (2 \text{ m } \sin 30°) = -98.1 \text{ J}$$

Note that it is also possible to consider the component of weight in the direction of displacement; i.e.,

$$U_W = -(98.1 \sin 30° \text{ N}) (2 \text{ m}) = -98.1 \text{ J}$$

Normal Force N_B. This force does *no work* since it is *always* perpendicular to the displacement.

Total Work. The work of all the forces when the block is displaced 2 m is therefore

$$U_T = 692.8 \text{ J} - 90 \text{ J} - 98.1 \text{ J} = 505 \text{ J} \qquad \textit{Ans.}$$

Fig. 14–7

14.2   Principle of Work and Energy

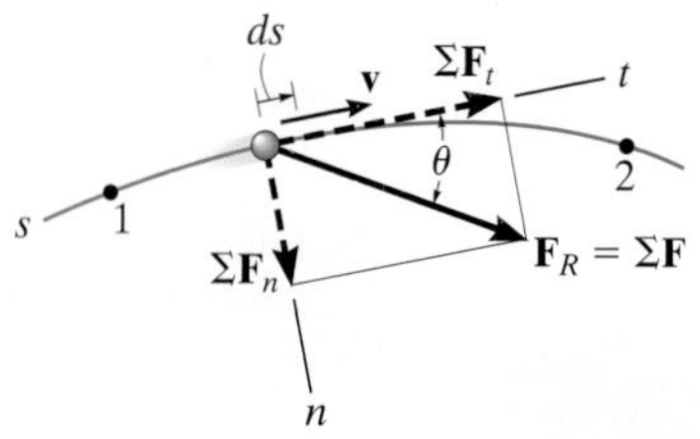

Consider the particle in Fig. 14–7, which is located on the path defined relative to an inertial coordinate system. If the particle has a mass m and is subjected to a system of external forces represented by the resultant $\mathbf{F}_R = \Sigma\mathbf{F}$, then the equation of motion for the particle in the tangential direction is $\Sigma F_t = ma_t$. Applying the kinematic equation $a_t = v\,dv/ds$ and integrating both sides, assuming initially that the particle has a position $s = s_1$ and a speed $v = v_1$, and later at $s = s_2$, $v = v_2$, we have

$$\Sigma \int_{s_1}^{s_2} F_t\,ds = \int_{v_1}^{v_2} mv\,dv$$

$$\Sigma \int_{s_1}^{s_2} F_t\,ds = \tfrac{1}{2}mv_2^2 - \tfrac{1}{2}mv_1^2 \tag{14–5}$$

From Fig. 14–7, note that $\Sigma F_t = \Sigma F\cos\theta$, and since work is defined from Eq. 14–1, the final result can be written as

$$\Sigma U_{1-2} = \tfrac{1}{2}mv_2^2 - \tfrac{1}{2}mv_1^2 \tag{14–6}$$

This equation represents the *principle of work and energy* for the particle. The term on the left is the sum of the work done by *all* the forces acting on the particle as the particle moves from point 1 to point 2. The two terms on the right side, which are of the form $T = \tfrac{1}{2}mv^2$, define the particle's final and initial *kinetic energy*, respectively. Like work, kinetic energy is a *scalar* and has units of joules (J) and ft · lb. However, unlike work, which can be either positive or negative, the kinetic energy is *always positive*, regardless of the direction of motion of the particle.

When Eq. 14–6 is applied, it is often expressed in the form

$$T_1 + \Sigma U_{1-2} = T_2 \tag{14–7}$$

which states that the particle's initial kinetic energy plus the work done by all the forces acting on the particle as it moves from its initial to its final position is equal to the particle's final kinetic energy.

As noted from the derivation, the principle of work and energy represents an integrated form of $\Sigma F_t = ma_t$, obtained by using the kinematic equation $a_t = v\,dv/ds$. As a result, this principle will provide a convenient *substitution* for $\Sigma F_t = ma_t$ when solving those types of kinetic problems which involve *force*, *velocity*, and *displacement* since these quantities are involved in Eq. 14–7. For application, it is suggested that the following procedure be used.

Procedure for Analysis

Work (Free-Body Diagram).

- Establish the inertial coordinate system and draw a free-body diagram of the particle in order to account for all the forces that do work on the particle as it moves along its path.

Principle of Work and Energy.

- Apply the principle of work and energy, $T_1 + \Sigma U_{1-2} = T_2$.

- The kinetic energy at the initial and final points is *always positive*, since it involves the speed squared $\left(T = \frac{1}{2}mv^2\right)$.

- A force does work when it moves through a displacement in the direction of the force.

- Work is *positive* when the force component is in the *same sense of direction* as its displacement, otherwise it is negative.

- Forces that are functions of displacement must be integrated to obtain the work. Graphically, the work is equal to the area under the force-displacement curve.

- The work of a weight is the product of the weight magnitude and the vertical displacement, $U_W = \pm Wy$. It is positive when the weight moves downwards.

- The work of a spring is of the form $U_s = \frac{1}{2}ks^2$, where k is the spring stiffness and s is the stretch or compression of the spring.

Numerical application of this procedure is illustrated in the examples following Sec. 14.3.

If an oncoming car strikes these crash barrels, the car's kinetic energy will be transformed into work, which causes the barrels, and to some extent the car, to be deformed. By knowing the amount of energy that can be absorbed by each barrel it is possible to design a crash cushion such as this.

14.3 Principle of Work and Energy for a System of Particles

The principle of work and energy can be extended to include a system of particles isolated within an enclosed region of space as shown in Fig. 14–8. Here the arbitrary ith particle, having a mass m_i, is subjected to a resultant external force $\mathbf{F}_i$ and a resultant internal force $\mathbf{f}_i$ which all the other particles exert on the ith particle. If we apply the principle of work and energy to this and each of the other particles in the system, then since work and energy are scalar quantities, the equations can be summed algebraically, which gives

$$\Sigma T_1 + \Sigma U_{1-2} = \Sigma T_2 \qquad (14\text{–}8)$$

In this case, the initial kinetic energy of the system plus the work done by all the external and internal forces acting on the system is equal to the final kinetic energy of the system.

If the system represents a *translating rigid body*, or a series of connected translating bodies, then all the particles in each body will undergo the *same displacement*. Therefore, the work of all the internal forces will occur in equal but opposite collinear pairs and so it will cancel out. On the other hand, if the body is assumed to be *nonrigid*, the particles of the body may be displaced along *different paths*, and some of the energy due to force interactions would be given off and lost as heat or stored in the body if permanent deformations occur. We will discuss these effects briefly at the end of this section and in Sec. 15.4. Throughout this text, however, the principle of work and energy will be applied to problems where direct accountability of such energy losses does not have to be considered.

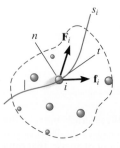

Inertial coordinate system

Fig. 14–8

Work of Friction Caused by Sliding.

A special class of problems will now be investigated which requires a careful application of Eq. 14–8. These problems involve cases where a body slides over the surface of another body in the presence of friction. Consider, for example, a block which is translating a distance s over a rough surface as shown in Fig. 14–9a. If the applied force **P** just balances the *resultant* frictional force $\mu_k N$, Fig. 14–9b, then due to equilibrium a constant velocity **v** is maintained, and one would expect Eq. 14–8 to be applied as follows:

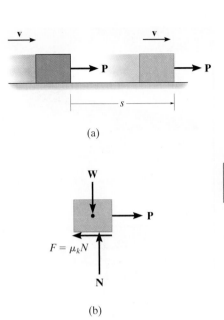

(a)

$$\tfrac{1}{2}mv^2 + Ps - \mu_k Ns = \tfrac{1}{2}mv^2$$

(b)

Indeed this equation is satisfied if $P = \mu_k N$; however, as one realizes from experience, the sliding motion will *generate heat*, a form of energy which seems not to be accounted for in the work-energy equation. In order to explain this paradox and thereby more closely represent the nature of friction, we should actually model the block so that the surfaces of contact are *deformable* (nonrigid).* Recall that the rough portions at the bottom of the block act as "teeth," and when the block slides these teeth *deform slightly* and either break off or vibrate as they pull away from "teeth" at the contacting surface, Fig. 14–9c. As a result, frictional forces that act on the block at these points are displaced slightly, due to the localized deformations, and later they are replaced by other frictional forces as other points of contact are made. At any instant, the *resultant* **F** of all these frictional forces remains essentially constant, i.e., $\mu_k N$; however, due to the many *localized deformations*, the actual displacement s' of $\mu_k N$ is *not* the same as the displacement s of the applied force **P**. Instead, s' will be *less* than s ($s' < s$), and therefore the *external work* done by the resultant frictional force will be $\mu_k Ns'$ and not $\mu_k Ns$. The remaining amount of work, $\mu_k N(s - s')$, manifests itself as an increase in *internal energy*, which in fact causes the block's temperature to rise.

(c)

Fig. 14–9

In summary then, Eq. 14–8 can be applied to problems involving sliding friction; however, it should be fully realized that the work of the resultant frictional force is not represented by $\mu_k Ns$; instead, this term represents *both* the external work of friction ($\mu_k Ns'$) *and* internal work $[\mu_k N(s - s')]$ which is converted into various forms of internal energy, such as heat.†

---

*See Chapter 8 of *Engineering Mechanics: Statics.*
†See B. A. Sherwood and W. H. Bernard, "Work and Heat Transfer in the Presence of Sliding Friction," *Am. J. Phys.* 52, 1001 (1984).

EXAMPLE | 14.2

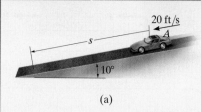

(a)

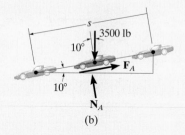

(b)

Fig. 14–10

The 3500-lb automobile shown in Fig. 14–10a travels down the 10° inclined road at a speed of 20 ft/s. If the driver jams on the brakes, causing his wheels to lock, determine how far s the tires skid on the road. The coefficient of kinetic friction between the wheels and the road is $\mu_k = 0.5$.

SOLUTION

This problem can be solved using the principle of work and energy, since it involves force, velocity, and displacement.

Work (Free-Body Diagram). As shown in Fig. 14–10b, the normal force N_A does no work since it never undergoes displacement along its line of action. The weight, 3500 lb, is displaced $s \sin 10°$ and does positive work. Why? The frictional force F_A does both external and internal work when it undergoes a displacement s. This work is negative since it is in the opposite sense of direction to the displacement. Applying the equation of equilibrium normal to the road, we have

$$+\nwarrow\Sigma F_n = 0; \qquad N_A - 3500\cos 10° \text{ lb} = 0 \qquad N_A = 3446.8 \text{ lb}$$

Thus,

$$F_A = \mu_k N_A = 0.5\,(3446.8 \text{ lb}) = 1723.4 \text{ lb}$$

Principle of Work and Energy.

$$T_1 + \Sigma U_{1-2} = T_2$$

$$\frac{1}{2}\left(\frac{3500 \text{ lb}}{32.2 \text{ ft/s}^2}\right)(20 \text{ ft/s})^2 + 3500 \text{ lb}(s \sin 10°) - (1723.4 \text{ lb})s = 0$$

Solving for s yields

$$s = 19.5 \text{ ft} \qquad\qquad Ans.$$

NOTE: If this problem is solved by using the equation of motion, *two steps* are involved. First, from the free-body diagram, Fig. 14–10b, the equation of motion is applied along the incline. This yields

$$+\swarrow\Sigma F_s = ma_s; \qquad 3500\sin 10° \text{ lb} - 1723.4 \text{ lb} = \frac{3500 \text{ lb}}{32.2 \text{ ft/s}^2}a$$

$$a = -10.3 \text{ ft/s}^2$$

Then, since a is constant, we have

$$\left(+\swarrow\right) \quad v^2 = v_0^2 + 2a_c(s - s_0);$$

$$(0)^2 = (20 \text{ ft/s})^2 + 2(-10.3 \text{ ft/s}^2)(s - 0)$$

$$s = 19.5 \text{ ft} \qquad\qquad Ans.$$

EXAMPLE   14.3

For a short time the crane in Fig. 14–11a lifts the 2.50-Mg beam with a force of $F = (28 + 3s^2)$ kN. Determine the speed of the beam when it has risen $s = 3$ m. Also, how much time does it take to attain this height starting from rest?

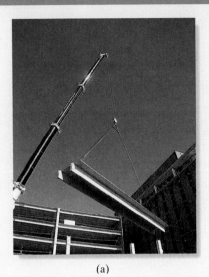

(a)

SOLUTION
We can solve part of this problem using the principle of work and energy since it involves force, velocity, and displacement. Kinematics must be used to determine the time. Note that at $s = 0$, $F = 28(10^3)$N $> W = 2.50(10^3)(9.81)$N, so motion will occur.

Work (Free-Body Diagram).   As shown on the free-body diagram, Fig. 14–11b, the lifting force **F** does positive work, which must be determined by integration since this force is a variable. Also, the weight is constant and will do negative work since the displacement is upwards.

Principles of Work and Energy.

$$T_1 + \Sigma U_{1-2} = T_2$$

$$0 + \int_0^s (28 + 3s^2)(10^3)\, ds - (2.50)(10^3)(9.81)s = \tfrac{1}{2}(2.50)(10^3)v^2$$

$$28(10^3)s + (10^3)s^3 - 24.525(10^3)s = 1.25(10^3)v^2$$

$$v = (2.78s + 0.8s^3)^{\frac{1}{2}} \qquad (1)$$

When $s = 3$ m,

$$v = 5.47 \text{ m/s} \qquad \textit{Ans.}$$

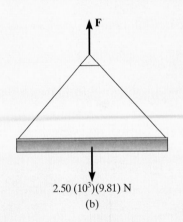

2.50 $(10^3)(9.81)$ N
(b)

Fig. 14–11

Kinematics.   Since we were able to express the velocity as a function of displacement, the time can be determined using $v = ds/dt$. In this case,

$$(2.78s + 0.8s^3)^{\frac{1}{2}} = \frac{ds}{dt}$$

$$t = \int_0^3 \frac{ds}{(2.78s + 0.8s^3)^{\frac{1}{2}}}$$

The integration can be performed numerically using a pocket calculator. The result is

$$t = 1.79 \text{ s} \qquad \textit{Ans.}$$

NOTE: The acceleration of the beam can be determined by integrating Eq. (1) using $v\, dv = a\, ds$, or more directly, by applying the equation of motion, $\Sigma F = ma$.

EXAMPLE 14.4

The platform P, shown in Fig. 14–12a, has negligible mass and is tied down so that the 0.4-m-long cords keep a 1-m-long spring compressed 0.6 m when *nothing* is on the platform. If a 2-kg block is placed on the platform and released from rest after the platform is pushed down 0.1 m, Fig. 14–12b, determine the maximum height h the block rises in the air, measured from the ground.

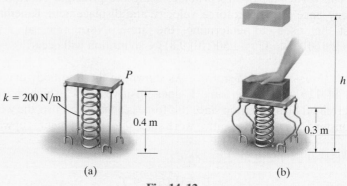

(a)   (b)

Fig. 14–12

SOLUTION

Work (Free-Body Diagram). Since the block is released from rest and later reaches its maximum height, the initial and final velocities are zero. The free-body diagram of the block when it is still in contact with the platform is shown in Fig. 14–12c. Note that the weight does negative work and the spring force does positive work. Why? In particular, the *initial compression* in the spring is $s_1 = 0.6$ m $+ 0.1$ m $= 0.7$ m. Due to the cords, the spring's *final compression* is $s_2 = 0.6$ m (after the block leaves the platform). The bottom of the block rises from a height of $(0.4$ m $- 0.1$ m$) = 0.3$ m to a final height h.

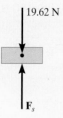

(c)

Principle of Work and Energy.

$$T_1 + \Sigma U_{1\text{-}2} = T_2$$

$$\tfrac{1}{2}mv_1^2 + \left\{ -\left(\tfrac{1}{2}ks_2^2 - \tfrac{1}{2}ks_1^2\right) - W\,\Delta y \right\} = \tfrac{1}{2}mv_2^2$$

Note that here $s_1 = 0.7$ m $> s_2 = 0.6$ m and so the work of the spring as determined from Eq. 14–4 will indeed be positive once the calculation is made. Thus,

$$0 + \left\{ -\left[\tfrac{1}{2}(200 \text{ N/m})(0.6 \text{ m})^2 - \tfrac{1}{2}(200 \text{ N/m})(0.7 \text{ m})^2\right] \right.$$

$$\left. - (19.62 \text{ N})[h - (0.3 \text{ m})] \right\} = 0$$

Solving yields

$$h = 0.963 \text{ m} \qquad \qquad Ans.$$

EXAMPLE 14.5

The 40-kg boy in Fig. 14–13a slides down the smooth water slide. If he starts from rest at A, determine his speed when he reaches B and the normal reaction the slide exerts on the boy at this position.

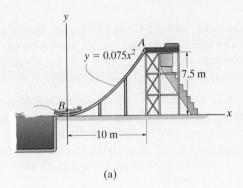

(a)

SOLUTION

Work (Free-Body Diagram). As shown on the free-body diagram, Fig. 14–13b, there are two forces acting on the boy as he goes down the slide. Note that the normal force does no work.

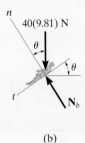

(b)

Principle of Work and Energy.

$$T_A + \Sigma U_{A-B} = T_B$$

$$0 + (40(9.81)\text{N})\,(7.5\text{ m}) = \tfrac{1}{2}(40\text{ kg})v_B^2$$

$$v_B = 12.13\text{ m/s} = 12.1\text{ m/s} \qquad \textit{Ans.}$$

Equation of Motion. Referring to the free-body diagram of the boy when he is at B, Fig. 14–13c, the normal reaction N_B can now be obtained by applying the equation of motion along the n axis. Here the radius of curvature of the path is

$$\rho_B = \frac{\left[1 + \left(\dfrac{dy}{dx}\right)^2\right]^{3/2}}{|d^2y/dx^2|} = \left.\frac{\left[1 + (0.15x)^2\right]^{3/2}}{|0.15|}\right|_{x=0} = 6.667\text{ m}$$

Thus,

$$+\uparrow \Sigma F_n = ma_n; \qquad N_B - 40(9.81)\text{ N} = 40\text{ kg}\left(\frac{(12.13\text{ m/s})^2}{6.667\text{ m}}\right)$$

$$N_B = 1275.3\text{ N} = 1.28\text{ kN} \qquad \textit{Ans.}$$

(c)

Fig. 14–13

EXAMPLE | 14.6

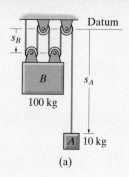

(a)

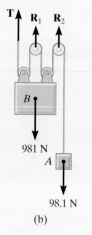

(b)

Fig. 14–14

Blocks A and B shown in Fig. 14–14a have a mass of 10 kg and 100 kg, respectively. Determine the distance B travels when it is released from rest to the point where its speed becomes 2 m/s.

SOLUTION
This problem may be solved by considering the blocks separately and applying the principle of work and energy to each block. However, the work of the (unknown) cable tension can be eliminated from the analysis by considering blocks A and B together as a *single system*.

Work (Free-Body Diagram).   As shown on the free-body diagram of the system, Fig. 14–14b, the cable force $\mathbf{T}$ and reactions $\mathbf{R}_1$ and $\mathbf{R}_2$ do *no work*, since these forces represent the reactions at the supports and consequently they do not move while the blocks are displaced. The weights both do positive work if we *assume* both move downward, in the positive sense of direction of s_A and s_B.

Principle of Work and Energy.   Realizing the blocks are released from rest, we have

$$\Sigma T_1 + \Sigma U_{1-2} = \Sigma T_2$$

$$\left\{\tfrac{1}{2}m_A(v_A)_1^2 + \tfrac{1}{2}m_B(v_B)_1^2\right\} + \left\{W_A\,\Delta s_A + W_B\,\Delta s_B\right\} =$$

$$\left\{\tfrac{1}{2}m_A(v_A)_2^2 + \tfrac{1}{2}m_B(v_B)_2^2\right\}$$

$$\{0 + 0\} + \{98.1 \text{ N }(\Delta s_A) + 981 \text{ N }(\Delta s_B)\} =$$

$$\left\{\tfrac{1}{2}(10 \text{ kg})(v_A)_2^2 + \tfrac{1}{2}(100 \text{ kg})(2 \text{ m/s})^2\right\} \qquad (1)$$

Kinematics.   Using methods of kinematics, as discussed in Sec. 12.9, it may be seen from Fig. 14–14a that the total length l of all the vertical segments of cable may be expressed in terms of the position coordinates s_A and s_B as

$$s_A + 4s_B = l$$

Hence, a change in position yields the displacement equation

$$\Delta s_A + 4\,\Delta s_B = 0$$
$$\Delta s_A = -4\,\Delta s_B$$

Here we see that a downward displacement of one block produces an upward displacement of the other block. Note that Δs_A and Δs_B must have the *same* sign convention in both Eqs. 1 and 2. Taking the time derivative yields

$$v_A = -4v_B = -4(2 \text{ m/s}) = -8 \text{ m/s} \qquad (2)$$

Retaining the negative sign in Eq. 2 and substituting into Eq. 1 yields

$$\Delta s_B = 0.883 \text{ m} \downarrow \qquad\qquad Ans.$$

FUNDAMENTAL PROBLEMS

F14–1. The spring is placed between the wall and the 10-kg block. If the block is subjected to a force of $F = 500$ N, determine its velocity when $s = 0.5$ m. When $s = 0$, the block is at rest and the spring is uncompressed. The contact surface is smooth.

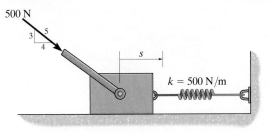

500 N

3
5
4

s

$k = 500$ N/m

F14–1

F14–2. If the motor exerts a constant force of 300 N on the cable, determine the speed of the 20-kg crate when it travels $s = 10$ m up the plane, starting from rest. The coefficient of kinetic friction between the crate and the plane is $\mu_k = 0.3$.

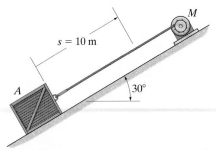

M

$s = 10$ m

A

30°

F14–2

F14–3. If the motor exerts a force of $F = (600 + 2s^2)$ N on the cable, determine the speed of the 100-kg crate when it rises to $s = 15$ m. The crate is initially at rest on the ground.

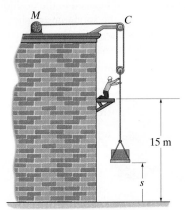

M

C

15 m

s

F14–3

F14–4. The 1.8-Mg dragster is traveling at 125 m/s when the engine is shut off and the parachute is released. If the drag force of the parachute can be approximated by the graph, determine the speed of the dragster when it has traveled 400 m.

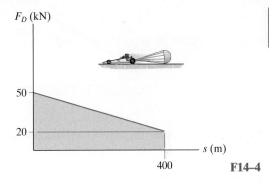

F_D (kN)

50

20

400

s (m)

F14–4

F14–5. When $s = 0.6$ m, the spring is unstretched and the 10-kg block has a speed of 5 m/s down the smooth plane. Determine the distance s when the block stops.

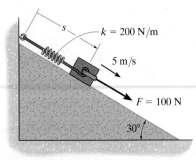

s

$k = 200$ N/m

5 m/s

$F = 100$ N

30°

F14–5

F14–6. The 5-lb collar is pulled by a cord that passes around a small peg at C. If the cord is subjected to a constant force of $F = 10$ lb, and the collar is at rest when it is at A, determine its speed when it reaches B. Neglect friction.

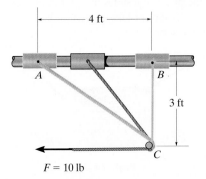

4 ft

A

B

3 ft

C

$F = 10$ lb

F14–6

PROBLEMS

14–1. The 20-kg crate is subjected to a force having a constant direction and a magnitude $F = 100$ N. When $s = 15$ m, the crate is moving to the right with a speed of 8 m/s. Determine its speed when $s = 25$ m. The coefficient of kinetic friction between the crate and the ground is $\mu_k = 0.25$.

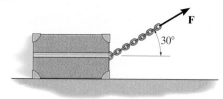

Prob. 14–1

14–2. For protection, the barrel barrier is placed in front of the bridge pier. If the relation between the force and deflection of the barrier is $F = (90(10^3)x^{1/2})$ lb, where x is in ft, determine the car's maximum penetration in the barrier. The car has a weight of 4000 lb and it is traveling with a speed of 75 ft/s just before it hits the barrier.

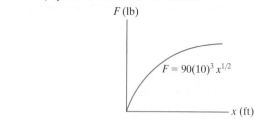

F (lb)

$F = 90(10)^3\, x^{1/2}$

x (ft)

Prob. 14–2

14–3. The crate, which has a mass of 100 kg, is subjected to the action of the two forces. If it is originally at rest, determine the distance it slides in order to attain a speed of 6 m/s. The coefficient of kinetic friction between the crate and the surface is $\mu_k = 0.2$.

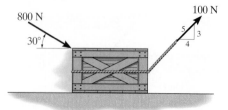

Prob. 14–3

*14–4. The 2-kg block is subjected to a force having a constant direction and a magnitude $F = (300/(1 + s))$ N, where s is in meters. When $s = 4$ m, the block is moving to the left with a speed of 8 m/s. Determine its speed when $s = 12$ m. The coefficient of kinetic friction between the block and the ground is $\mu_k = 0.25$.

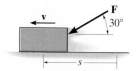

Prob. 14–4

14–5. When a 7-kg projectile is fired from a cannon barrel that has a length of 2 m, the explosive force exerted on the projectile, while it is in the barrel, varies in the manner shown. Determine the approximate muzzle velocity of the projectile at the instant it leaves the barrel. Neglect the effects of friction inside the barrel and assume the barrel is horizontal.

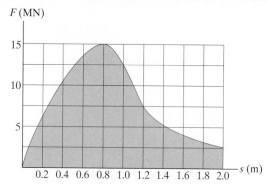

Prob. 14–5

14–6. The spring in the toy gun has an unstretched length of 100 mm. It is compressed and locked in the position shown. When the trigger is pulled, the spring unstretches 12.5 mm, and the 20-g ball moves along the barrel. Determine the speed of the ball when it leaves the gun. Neglect friction.

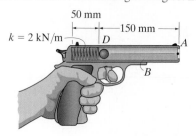

Prob. 14–6

14–7.   As indicated by the derivation, the principle of work and energy is valid for observers in *any* inertial reference frame. Show that this is so, by considering the 10-kg block which rests on the smooth surface and is subjected to a horizontal force of 6 N. If observer A is in a *fixed* frame x, determine the final speed of the block if it has an initial speed of 5 m/s and travels 10 m, both directed to the right and measured from the fixed frame. Compare the result with that obtained by an observer B, attached to the x' axis and moving at a constant velocity of 2 m/s relative to A. *Hint:* The distance the block travels will first have to be computed for observer B before applying the principle of work and energy.

14–10.   The 2-Mg car has a velocity of $v_1 = 100\,\text{km/h}$ when the driver sees an obstacle in front of the car. If it takes 0.75 s for him to react and lock the brakes, causing the car to skid, determine the distance the car travels before it stops. The coefficient of kinetic friction between the tires and the road is $\mu_k = 0.25$.

14–11.   The 2-Mg car has a velocity of $v_1 = 100\,\text{km/h}$ when the driver sees an obstacle in front of the car. It takes 0.75 s for him to react and lock the brakes, causing the car to skid. If the car stops when it has traveled a distance of 175 m, determine the coefficient of kinetic friction between the tires and the road.

Probs. 14–10/11

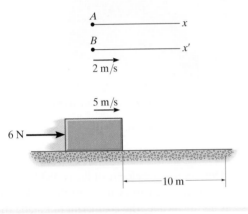

Prob. 14–7

*14–12.**   Design considerations for the bumper B on the 5-Mg train car require use of a nonlinear spring having the load-deflection characteristics shown in the graph. Select the proper value of k so that the maximum deflection of the spring is limited to 0.2 m when the car, traveling at 4 m/s, strikes the rigid stop. Neglect the mass of the car wheels.

*14–8.**   If the 50-kg crate is subjected to a force of $P = 200$ N, determine its speed when it has traveled 15 m starting from rest. The coefficient of kinetic friction between the crate and the ground is $\mu_k = 0.3$.

14–9.   If the 50-kg crate starts from rest and attains a speed of 6 m/s when it has traveled a distance of 15 m, determine the force **P** acting on the crate. The coefficient of kinetic friction between the crate and the ground is $\mu_k = 0.3$.

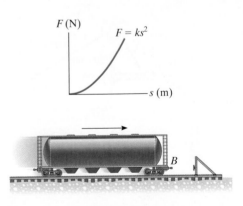

Prob. 14–12

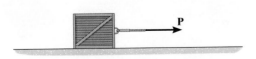

Probs. 14–8/9

14–13. The 2-lb brick slides down a smooth roof, such that when it is at A it has a velocity of 5 ft/s. Determine the speed of the brick just before it leaves the surface at B, the distance d from the wall to where it strikes the ground, and the speed at which it his the ground.

14–15. The crash cushion for a highway barrier consists of a nest of barrels filled with an impact-absorbing material. The barrier stopping force is measured versus the vehicle penetration into the barrier. Determine the distance a car having a weight of 4000 lb will penetrate the barrier if it is originally traveling at 55 ft/s when it strikes the first barrel.

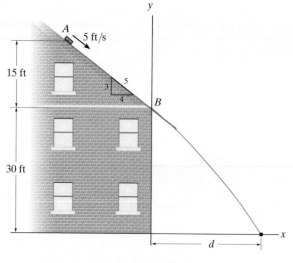

Prob. 14–13

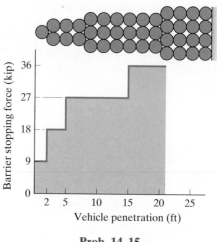

Prob. 14–15

14–14. If the cord is subjected to a constant force of $F = 300$ N and the 15-kg smooth collar starts from rest at A, determine the velocity of the collar when it reaches point B. Neglect the size of the pulley.

*14–16. Determine the velocity of the 60-lb block A if the two blocks are released from rest and the 40-lb block B moves 2 ft up the incline. The coefficient of kinetic friction between both blocks and the inclined planes is $\mu_k = 0.10$.

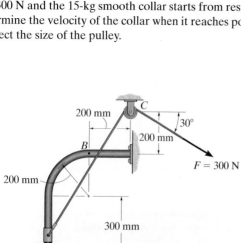

Prob. 14–14

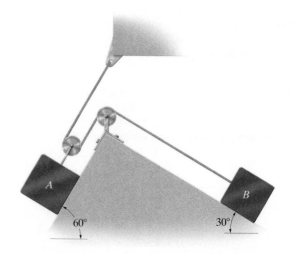

Prob. 14–16

14–17. If the cord is subjected to a constant force of $F = 30$ lb and the smooth 10-lb collar starts from rest at A, determine its speed when it passes point B. Neglect the size of pulley C.

14–19. If the 10-lb block passes point A on the smooth track with a speed of $v_A = 5$ ft/s, determine the normal reaction on the block when it reaches point B.

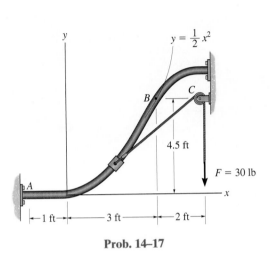

Prob. 14–17

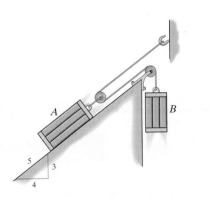

Prob. 14–19

14–18. The two blocks A and B have weights $W_A = 60$ lb and $W_B = 10$ lb. If the kinetic coefficient of friction between the incline and block A is $\mu_k = 0.2$, determine the speed of A after it moves 3 ft down the plane starting from rest. Neglect the mass of the cord and pulleys.

***14–20.** The steel ingot has a mass of 1800 kg. It travels along the conveyor at a speed $v = 0.5$ m/s when it collides with the "nested" spring assembly. Determine the maximum deflection in each spring needed to stop the motion of the ingot. Take $k_A = 5$ kN/m, $k_B = 3$ kN/m.

14–21. The steel ingot has a mass of 1800 kg. It travels along the conveyor at a speed $v = 0.5$ m/s when it collides with the "nested" spring assembly. If the stiffness of the outer spring is $k_A = 5$ kN/m, determine the required stiffness k_B of the inner spring so that the motion of the ingot is stopped at the moment the front, C, of the ingot is 0.3 m from the wall.

Prob. 14–18

Probs. 14–20/21

14–22. The 25-lb block has an initial speed of $v_0 = 10$ ft/s when it is midway between springs A and B. After striking spring B, it rebounds and slides across the horizontal plane toward spring A, etc. If the coefficient of kinetic friction between the plane and the block is $\mu_k = 0.4$, determine the total distance traveled by the block before it comes to rest.

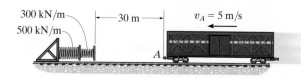

Prob. 14–22

14–23. The train car has a mass of 10 Mg and is traveling at 5 m/s when it reaches A. If the rolling resistance is 1/100 of the weight of the car, determine the compression of each spring when the car is momentarily brought to rest.

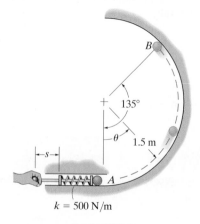

Prob. 14–23

*14–24.** The 0.5-kg ball is fired up the smooth vertical circular track using the spring plunger. The plunger keeps the spring compressed 0.08 m when $s = 0$. Determine how far s it must be pulled back and released so that the ball will begin to leave the track when $\theta = 135°$.

14–25. The skier starts from rest at A and travels down the ramp. If friction and air resistance can be neglected, determine his speed v_B when he reaches B. Also, find the distance s to where he strikes the ground at C, if he makes the jump traveling horizontally at B. Neglect the skier's size. He has a mass of 70 kg.

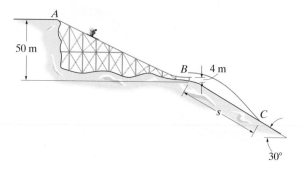

Prob. 14–25

14–26. The catapulting mechanism is used to propel the 10-kg slider A to the right along the smooth track. The propelling action is obtained by drawing the pulley attached to rod BC rapidly to the left by means of a piston P. If the piston applies a constant force $F = 20$ kN to rod BC such that it moves it 0.2 m, determine the speed attained by the slider if it was originally at rest. Neglect the mass of the pulleys, cable, piston, and rod BC.

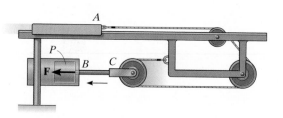

Prob. 14–26

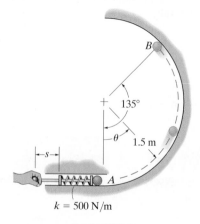

Prob. 14–24

14–27. Block *A* has a weight of 60 lb and block *B* has a weight of 10 lb. Determine the distance *A* must descend from rest before it obtains a speed of 8 ft/s. Also, what is the tension in the cord supporting block *A*? Neglect the mass of the cord and pulleys.

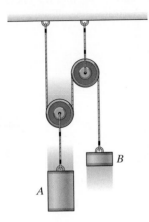

Prob. 14–27

*14–28.** The cyclist travels to point *A*, pedaling until he reaches a speed $v_A = 4$ m/s. He then coasts freely up the curved surface. Determine how high he reaches up the surface before he comes to a stop. Also, what are the resultant normal force on the surface at this point and his acceleration? The total mass of the bike and man is 75 kg. Neglect friction, the mass of the wheels, and the size of the bicycle.

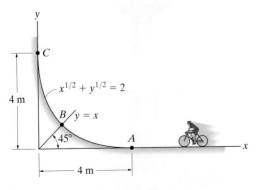

Prob. 14–28

14–29. The collar has a mass of 20 kg and slides along the smooth rod. Two springs are attached to it and the ends of the rod as shown. If each spring has an uncompressed length of 1 m and the collar has a speed of 2 m/s when $s = 0$, determine the maximum compression of each spring due to the back-and-forth (oscillating) motion of the collar.

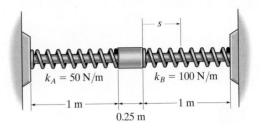

Prob. 14–29

14–30. The 30-lb box *A* is released from rest and slides down along the smooth ramp and onto the surface of a cart. If the cart is *prevented from moving*, determine the distance *s* from the end of the cart to where the box stops. The coefficient of kinetic friction between the cart and the box is $\mu_k = 0.6$.

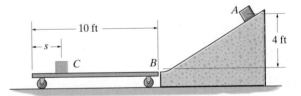

Prob. 14–30

14–31. Marbles having a mass of 5 g are dropped from rest at *A* through the smooth glass tube and accumulate in the can at *C*. Determine the placement *R* of the can from the end of the tube and the speed at which the marbles fall into the can. Neglect the size of the can.

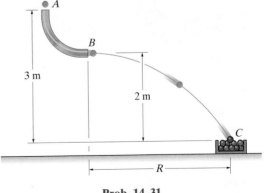

Prob. 14–31

*14–32. The cyclist travels to point A, pedaling until he reaches a speed $v_A = 8$ m/s. He then coasts freely up the curved surface. Determine the normal force he exerts on the surface when he reaches point B. The total mass of the bike and man is 75 kg. Neglect friction, the mass of the wheels, and the size of the bicycle.

14–34. The spring bumper is used to arrest the motion of the 4-lb block, which is sliding toward it at $v = 9$ ft/s. As shown, the spring is confined by the plate P and wall using cables so that its length is 1.5 ft. If the stiffness of the spring is $k = 50$ lb/ft, determine the required unstretched length of the spring so that the plate is not displaced more than 0.2 ft after the block collides into it. Neglect friction, the mass of the plate and spring, and the energy loss between the plate and block during the collision.

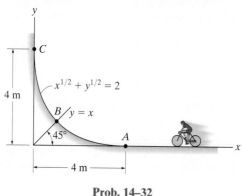

Prob. 14–32

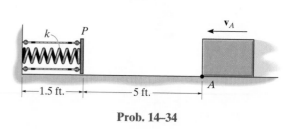

Prob. 14–34

14–33. The man at the window A wishes to throw the 30-kg sack on the ground. To do this he allows it to swing from rest at B to point C, when he releases the cord at $\theta = 30°$. Determine the speed at which it strikes the ground and the distance R.

14–35. The collar has a mass of 20 kg and is supported on the smooth rod. The attached springs are undeformed when $d = 0.5$ m. Determine the speed of the collar after the applied force $F = 100$ N causes it to be displaced so that $d = 0.3$ m. When $d = 0.5$ m the collar is at rest.

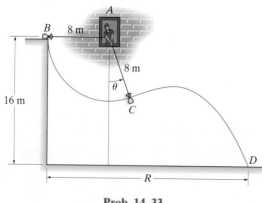

Prob. 14–33

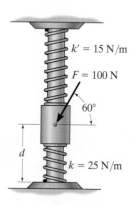

Prob. 14–35

*14–36. If the force exerted by the motor M on the cable is 250 N, determine the speed of the 100-kg crate when it is hoisted to $s = 3$ m. The crate is at rest when $s = 0$.

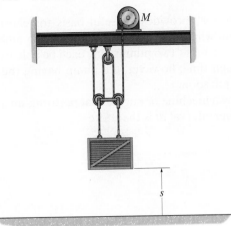

Prob. 14–36

14–37. If the track is to be designed so that the passengers of the roller coaster do not experience a normal force equal to zero or more than 4 times their weight, determine the limiting heights h_A and h_C so that this does not occur. The roller coaster starts from rest at position A. Neglect friction.

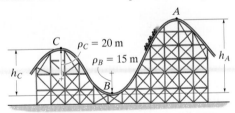

Prob. 14–37

14–38. The 150-lb skater passes point A with a speed of 6 ft/s. Determine his speed when he reaches point B and the normal force exerted on him by the track at this point. Neglect friction.

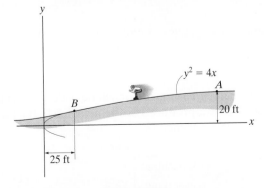

Prob. 14–38

14–39. The 8-kg cylinder A and 3-kg cylinder B are released from rest. Determine the speed of A after it has moved 2 m starting from rest. Neglect the mass of the cord and pulleys.

*14–40. Cylinder A has a mass of 3 kg and cylinder B has a mass of 8 kg. Determine the speed of A after it has moved 2 m starting from rest. Neglect the mass of the cord and pulleys.

Probs. 14–39/40

14–41. A 2-lb block rests on the smooth semicylindrical surface. An elastic cord having a stiffness $k = 2$ lb/ft is attached to the block at B and to the base of the semicylinder at point C. If the block is released from rest at A ($\theta = 0°$), determine the unstretched length of the cord so the block begins to leave the semicylinder at the instant $\theta = 45°$. Neglect the size of the block.

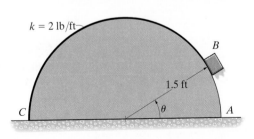

Prob. 14–41

14

14.4   Power and Efficiency

Power.   The term "power" provides a useful basis for chosing the type of motor or machine which is required to do a certain amount of work in a given time. For example, two pumps may each be able to empty a reservoir if given enough time; however, the pump having the larger power will complete the job sooner.

The *power* generated by a machine or engine that performs an amount of work dU within the time interval dt is therefore

$$P = \frac{dU}{dt} \tag{14–9}$$

If the work dU is expressed as $dU = \mathbf{F} \cdot d\mathbf{r}$, then

$$P = \frac{dU}{dt} = \frac{\mathbf{F} \cdot d\mathbf{r}}{dt} = \mathbf{F} \cdot \frac{d\mathbf{r}}{dt}$$

or

$$P = \mathbf{F} \cdot \mathbf{v} \tag{14–10}$$

Hence, power is a *scalar*, where in this formulation $\mathbf{v}$ represents the velocity of the particle which is acted upon by the force $\mathbf{F}$.

The basic units of power used in the SI and FPS systems are the watt (W) and horsepower (hp), respectively. These units are defined as

$$1 \text{ W} = 1 \text{ J/s} = 1 \text{ N} \cdot \text{m/s}$$

$$1 \text{ hp} = 550 \text{ ft} \cdot \text{lb/s}$$

For conversion between the two systems of units, 1 hp = 746 W.

The power output of this locomotive comes from the driving frictional force $\mathbf{F}$ developed at its wheels. It is this force that overcomes the frictional resistance of the cars in tow and is able to lift the weight of the train up the grade.

Efficiency.   The *mechanical efficiency* of a machine is defined as the ratio of the output of useful power produced by the machine to the input of power supplied to the machine. Hence,

$$\varepsilon = \frac{\text{power output}}{\text{power input}} \tag{14–11}$$

If energy supplied to the machine occurs during the *same time interval* at which it is drawn, then the efficiency may also be expressed in terms of the ratio

$$\varepsilon = \frac{\text{energy output}}{\text{energy input}}$$ (14–12)

Since machines consist of a series of moving parts, frictional forces will always be developed within the machine, and as a result, extra energy or power is needed to overcome these forces. Consequently, power output will be less than power input and so *the efficiency of a machine is always less than 1.*

The power supplied to a body can be determined using the following procedure.

Procedure for Analysis

- First determine the external force **F** acting on the body which causes the motion. This force is usually developed by a machine or engine placed either within or external to the body.

- If the body is accelerating, it may be necessary to draw its free-body diagram and apply the equation of motion ($\Sigma \mathbf{F} = m\mathbf{a}$) to determine **F**.

- Once **F** and the velocity **v** of the particle where **F** is applied have been found, the power is determined by multiplying the force magnitude with the component of velocity acting in the direction of **F**, (i.e., $P = \mathbf{F} \cdot \mathbf{v} = Fv \cos \theta$).

- In some problems the power may be found by calculating the work done by **F** per unit of time ($P_{\text{avg}} = \Delta U / \Delta t$,).

The power requirements of this elevator depend upon the vertical force **F** that acts on the elevator and causes it to move upwards. If the velocity of the elevator is **v**, then the power output is $P = \mathbf{F} \cdot \mathbf{v}$.

EXAMPLE | **14.7**

The man in Fig. 14–15a pushes on the 50-kg crate with a force of $F = 150$ N. Determine the power supplied by the man when $t = 4$ s. The coefficient of kinetic friction between the floor and the crate is $\mu_k = 0.2$. Initially the create is at rest.

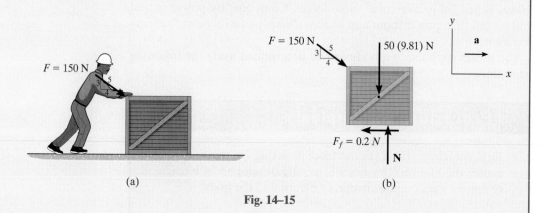

(a)   (b)

Fig. 14–15

SOLUTION

To determine the power developed by the man, the velocity of the 150-N force must be obtained first. The free-body diagram of the crate is shown in Fig. 14–15b. Applying the equation of motion,

$$+\uparrow \Sigma F_y = ma_y; \qquad N - \left(\tfrac{3}{5}\right)150 \text{ N} - 50(9.81) \text{ N} = 0$$

$$N = 580.5 \text{ N}$$

$$\xrightarrow{+} \Sigma F_x = ma_x; \qquad \left(\tfrac{4}{5}\right)150 \text{ N} - 0.2(580.5 \text{ N}) = (50 \text{ kg})a$$

$$a = 0.078 \text{ m/s}^2$$

The velocity of the crate when $t = 4$ s is therefore

$$(\xrightarrow{+}) \qquad\qquad v = v_0 + a_c t$$

$$v = 0 + (0.078 \text{ m/s}^2)(4 \text{ s}) = 0.312 \text{ m/s}$$

The power supplied to the crate by the man when $t = 4$ s is therefore

$$P = \mathbf{F} \cdot \mathbf{v} = F_x v = \left(\tfrac{4}{5}\right)(150 \text{ N})(0.312 \text{ m/s})$$

$$= 37.4 \text{ W} \qquad\qquad\qquad Ans.$$

EXAMPLE 14.8

The motor M of the hoist shown in Fig. 14–16a lifts the 75-lb crate C so that the acceleration of point P is 4 ft/s². Determine the power that must be supplied to the motor at the instant P has a velocity of 2 ft/s. Neglect the mass of the pulley and cable and take $\varepsilon = 0.85$.

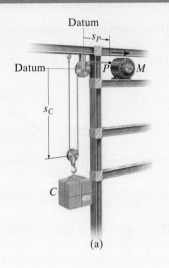

Datum

(a)

SOLUTION

In order to find the power output of the motor, it is first necessary to determine the tension in the cable since this force is developed by the motor.

From the free-body diagram, Fig. 14–16b, we have

$$+\downarrow \ \Sigma F_y = ma_y; \quad -2T + 75 \text{ lb} = \frac{75 \text{ lb}}{32.2 \text{ ft/s}^2} a_c \quad (1)$$

The acceleration of the crate can be obtained by using kinematics to relate it to the known acceleration of point P, Fig. 14–16a. Using the methods of absolute dependent motion, the coordinates s_C and s_P can be related to a constant portion of cable length l which is changing in the vertical and horizontal directions. We have $2s_C + s_P = l$. Taking the second time derivative of this equation yields

$$2a_C = -a_P \quad (2)$$

Since $a_P = +4$ ft/s², then $a_C = -(4 \text{ ft/s}^2)/2 = -2 \text{ ft/s}^2$. What does the negative sign indicate? Substituting this result into Eq. 1 and *retaining* the negative sign since the acceleration in *both* Eq. 1 and Eq. 2 was considered positive downward, we have

$$-2T + 75 \text{ lb} = \left(\frac{75 \text{ lb}}{32.2 \text{ ft/s}^2}\right)(-2 \text{ ft/s}^2)$$

$$T = 39.83 \text{ lb}$$

(b)

Fig. 14–16

The power output, measured in units of horsepower, required to draw the cable in at a rate of 2 ft/s is therefore

$$P = \mathbf{T} \cdot \mathbf{v} = (39.83 \text{ lb})(2 \text{ ft/s})[1 \text{ hp}/(550 \text{ ft} \cdot \text{lb/s})]$$

$$= 0.1448 \text{ hp}$$

This *power output* requires that the motor provide a *power input* of

$$\text{power input} = \frac{1}{\varepsilon}(\text{power output})$$

$$= \frac{1}{0.85}(0.1448 \text{ hp}) = 0.170 \text{ hp} \qquad \textit{Ans.}$$

NOTE: Since the velocity of the crate is constantly changing, the power requirement is *instantaneous*.

FUNDAMENTAL PROBLEMS

F14–7. If the contact surface between the 20-kg block and the ground is smooth, determine the power of force **F** when $t = 4$ s. Initially, the block is at rest.

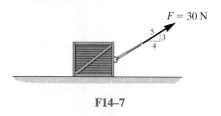

$F = 30$ N

F14–7

F14–8. If $F = (10\,s)$ N, where s is in meters, and the contact surface between the block and the ground is smooth, determine the power of force **F** when $s = 5$ m. When $s = 0$, the 20-kg block is moving at $v = 1$ m/s.

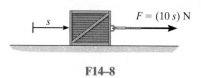

s

$F = (10\,s)$ N

F14–8

F14–9. If the motor winds in the cable with a constant speed of $v = 3$ ft/s, determine the power supplied to the motor. The load weighs 100 lb and the efficiency of the motor is $\varepsilon = 0.8$. Neglect the mass of the pulleys.

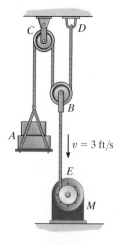

C D

B

A

$v = 3$ ft/s

E

M

F14–9

F14–10. The coefficient of kinetic friction between the 20-kg block and the inclined plane is $\mu_k = 0.2$. If the block is traveling up the inclined plane with a constant velocity $v = 5$ m/s, determine the power of force **F**.

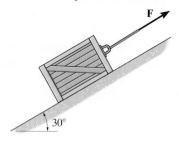

F

30°

F14–10

F14–11. If the 50-kg load A is hoisted by motor M so that the load has a constant velocity of 1.5 m/s, determine the power input to the motor, which operates at an efficiency $\varepsilon = 0.8$.

M

1.5 m/s

A

F14–11

F14–12. At the instant shown, point P on the cable has a velocity $v_P = 12$ m/s, which is increasing at a rate of $a_P = 6$ m/s^2. Determine the power input to motor M at this instant if it operates with an efficiency $\varepsilon = 0.8$. The mass of block A is 50 kg.

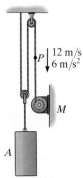

P 12 m/s   6 m/s^2

M

A

F14–12

PROBLEMS

14–42.   The jeep has a weight of 2500 lb and an engine which transmits a power of 100 hp to *all* the wheels. Assuming the wheels do not slip on the ground, determine the angle θ of the largest incline the jeep can climb at a constant speed $v = 30$ ft/s.

Prob. 14–42

14–43.   Determine the power input for a motor necessary to lift 300 lb at a constant rate of 5 ft/s. The efficiency of the motor is $\varepsilon = 0.65$.

*14–44.   An automobile having a mass of 2 Mg travels up a 7° slope at a constant speed of $v = 100$ km/h. If mechanical friction and wind resistance are neglected, determine the power developed by the engine if the automobile has an efficiency $\varepsilon = 0.65$.

Prob. 14–44

14–45.   The Milkin Aircraft Co. manufactures a turbojet engine that is placed in a plane having a weight of 13000 lb. If the engine develops a constant thrust of 5200 lb, determine the power output of the plane when it is just ready to take off with a speed of 600 mi/h.

14–46.   To dramatize the loss of energy in an automobile, consider a car having a weight of 5000 lb that is traveling at 35 mi/h. If the car is brought to a stop, determine how long a 100-W light bulb must burn to expend the same amount of energy. (1 mi = 5280 ft.)

14–47.   The escalator steps move with a constant speed of 0.6 m/s. If the steps are 125 mm high and 250 mm in length, determine the power of a motor needed to lift an average mass of 150 kg per step. There are 32 steps.

*14–48.   If the escalator in Prob. 14–46 is not moving, determine the constant speed at which a man having a mass of 80 kg must walk up the steps to generate 100 W of power—the same amount that is needed to power a standard light bulb.

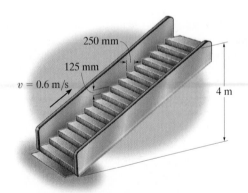

Probs. 14–47/48

14–49.   The 2-Mg car increases its speed uniformly from rest to 25 m/s in 30 s up the inclined road. Determine the maximum power that must be supplied by the engine, which operates with an efficiency of $\varepsilon = 0.8$. Also, find the average power supplied by the engine.

Prob. 14–49

14–50. The crate has a mass of 150 kg and rests on a surface for which the coefficients of static and kinetic friction are $\mu_s = 0.3$ and $\mu_k = 0.2$, respectively. If the motor M supplies a cable force of $F = (8t^2 + 20)$ N, where t is in seconds, determine the power output developed by the motor when $t = 5$ s.

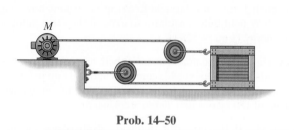

Prob. 14–50

14–51. The 50-kg crate is hoisted up the 30° incline by the pulley system and motor M. If the crate starts from rest and by constant acceleration attains a speed of 4 m/s after traveling 8 m along the plane, determine the power that must be supplied to the motor at this instant. Neglect friction along the plane. The motor has an efficiency of $\varepsilon = 0.74$.

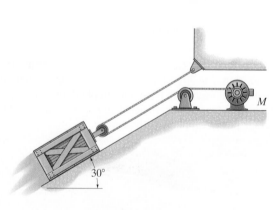

Prob. 14–51

*14–52. The 50-lb load is hoisted by the pulley system and motor M. If the motor exerts a constant force of 30 lb on the cable, determine the power that must be supplied to the motor if the load has been hoisted $s = 10$ ft starting from rest. The motor has an efficiency of $\varepsilon = 0.76$.

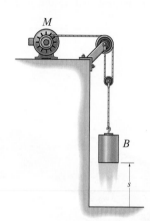

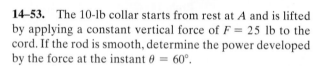

Prob. 14–52

14–53. The 10-lb collar starts from rest at A and is lifted by applying a constant vertical force of $F = 25$ lb to the cord. If the rod is smooth, determine the power developed by the force at the instant $\theta = 60°$.

14–54. The 10-lb collar starts from rest at A and is lifted with a constant speed of 2 ft/s along the smooth rod. Determine the power developed by the force $\mathbf{F}$ at the instant shown.

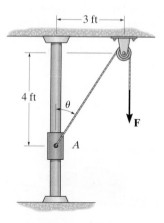

Probs. 14–53/54

14–55. The elevator E and its freight have a total mass of 400 kg. Hoisting is provided by the motor M and the 60-kg block C. If the motor has an efficiency of $\varepsilon = 0.6$, determine the power that must be supplied to the motor when the elevator is hoisted upward at a constant speed of $v_E = 4$ m/s.

14–58. The block has a mass of 150 kg and rests on a surface for which the coefficients of static and kinetic friction are $\mu_s = 0.5$ and $\mu_k = 0.4$, respectively. If a force $F = (60t^2)$ N, where t is in seconds, is applied to the cable, determine the power developed by the force when $t = 5$ s. *Hint:* First determine the time needed for the force to cause motion.

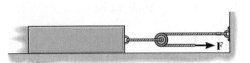

Prob. 14–58

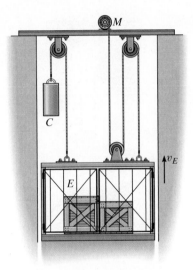

Prob. 14–55

14–59. The rocket sled has a mass of 4 Mg and travels from rest along the horizontal track for which the coefficient of kinetic friction is $\mu_k = 0.20$. If the engine provides a constant thrust $T = 150$ kN, determine the power output of the engine as a function of time. Neglect the loss of fuel mass and air resistance.

*14–56. The sports car has a mass of 2.3 Mg, and while it is traveling at 28 m/s the driver causes it to accelerate at 5 m/s². If the drag resistance on the car due to the wind is $F_D = (0.3v^2)$ N, where v is the velocity in m/s, determine the power supplied to the engine at this instant. The engine has a running efficiency of $\varepsilon = 0.68$.

14–57. The sports car has a mass of 2.3 Mg and accelerates at 6 m/s², starting from rest. If the drag resistance on the car due to the wind is $F_D = (10v)$ N, where v is the velocity in m/s, determine the power supplied to the engine when $t = 5$ s. The engine has a running efficiency of $\varepsilon = 0.68$.

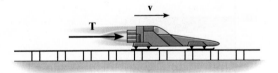

Prob. 14–59

Probs. 14–56/57

*14–60. A loaded truck weighs $16(10^3)$ lb and accelerates uniformly on a level road from 15 ft/s to 30 ft/s during 4 s. If the frictional resistance to motion is 325 lb, determine the maximum power that must be delivered to the wheels.

14–61. If the jet on the dragster supplies a constant thrust of $T = 20$ kN, determine the power generated by the jet as a function of time. Neglect drag and rolling resistance, and the loss of fuel. The dragster has a mass of 1 Mg and starts from rest.

*14–64. The 500-kg elevator starts from rest and travels upward with a constant acceleration $a_c = 2$ m/s². Determine the power output of the motor M when $t = 3$ s. Neglect the mass of the pulleys and cable.

Prob. 14–61

14–62. An athlete pushes against an exercise machine with a force that varies with time as shown in the first graph. Also, the velocity of the athlete's arm acting in the same direction as the force varies with time as shown in the second graph. Determine the power applied as a function of time and the work done in $t = 0.3$ s.

14–63. An athlete pushes against an exercise machine with a force that varies with time as shown in the first graph. Also, the velocity of the athlete's arm acting in the same direction as the force varies with time as shown in the second graph. Determine the maximum power developed during the 0.3-second time period.

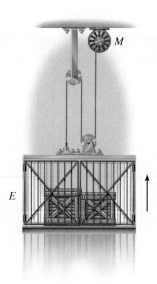

Prob. 14–64

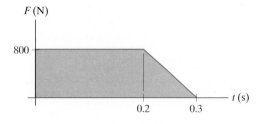

14–65. The 50-lb block rests on the rough surface for which the coefficient of kinetic friction is $\mu_k = 0.2$. A force $F = (40 + s^2)$ lb, where s is in ft, acts on the block in the direction shown. If the spring is originally unstretched $(s = 0)$ and the block is at rest, determine the power developed by the force the instant the block has moved $s = 1.5$ ft.

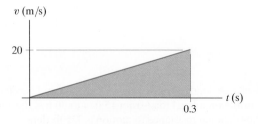

Probs. 14–62/63

Prob. 14–65

14.5 Conservative Forces and Potential Energy

Conservative Force.   If the work of a force is *independent of the path* and depends only on the force's initial and final positions on the path, then we can classify this force as a *conservative force*. Examples of conservative forces are the weight of a particle and the force developed by a spring. The work done by the weight depends *only* on the *vertical displacement* of the weight, and the work done by a spring force depends *only* on the spring's *elongation* or *compression*.

In contrast to a conservative force, consider the force of friction exerted *on a sliding object* by a fixed surface. The work done by the frictional force *depends on the path*—the longer the path, the greater the work. Consequently, *frictional forces are nonconservative*. The work is dissipated from the body in the form of heat.

Energy.   Energy is defined as the capacity for doing work. For example, if a particle is originally at rest, then the principle of work and energy states that $\Sigma U_{1 \to 2} = T_2$. In other words, the kinetic energy is equal to the work that must be done on the particle to bring it from a state of rest to a speed v. Thus, the *kinetic energy* is a measure of the particle's *capacity to do work*, which is associated with the *motion* of the particle. When energy comes from the *position* of the particle, measured from a fixed datum or reference plane, it is called potential energy. Thus, *potential energy* is a measure of the amount of work a conservative force will do when it moves from a given position to the datum. In mechanics, the potential energy created by gravity (weight) and an elastic spring is important.

Gravitational Potential Energy.   If a particle is located a distance y *above* an arbitrarily selected datum, as shown in Fig. 14–17, the particle's weight **W** has positive *gravitational potential energy*, V_g, since **W** has the capacity of doing positive work when the particle is moved back down to the datum. Likewise, if the particle is located a distance y *below* the datum, V_g is negative since the weight does negative work when the particle is moved back up to the datum. At the datum $V_g = 0$.

In general, if y is *positive upward*, the gravitational potential energy of the particle of weight W is*

$$V_g = Wy \tag{14–13}$$

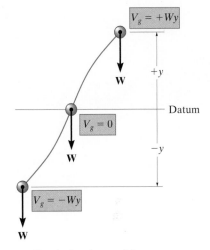

Gravitational potential energy

Fig. 14–17

14

---

*Here the weight is assumed to be *constant*. This assumption is suitable for small differences in elevation Δy. If the elevation change is significant, however, a variation of weight with elevation must be taken into account (see Prob. 14–82).

Elastic Potential Energy.

When an elastic spring is elongated or compressed a distance s from its unstretched position, elastic potential energy V_e can be stored in the spring. This energy is

$$V_e = +\tfrac{1}{2}ks^2 \qquad (14\text{--}14)$$

Here V_e is *always positive* since, in the deformed position, the force of the spring has the *capacity* or "potential" for always doing positive work on the particle when the spring is returned to its unstretched position, Fig. 14–18.

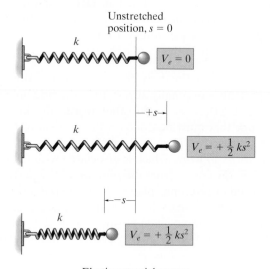

Elastic potential energy

Fig. 14–18

The weight of the sacks resting on this platform causes potential energy to be stored in the supporting springs. As each sack is removed, the platform will *rise* slightly since some of the potential energy within the springs will be transformed into an increase in gravitational potential energy of the remaining sacks. Such a device is useful for removing the sacks without having to bend over to pick them up as they are unloaded.

Potential Function.   In the general case, if a particle is subjected to both gravitational and elastic forces, the particle's potential energy can be expressed as a *potential function*, which is the algebraic sum

$$V = V_g + V_e \qquad (14\text{–}15)$$

Measurement of V depends on the location of the particle with respect to a selected datum in accordance with Eqs. 14–13 and 14–14.

The work done by a conservative force in moving the particle from one point to another point is measured by the *difference* of this function, i.e.,

$$U_{1-2} = V_1 - V_2 \qquad (14\text{–}16)$$

For example, the potential function for a particle of weight W suspended from a spring can be expressed in terms of its position, s, measured from a datum located at the unstretched length of the spring, Fig. 14–19. We have

$$V = V_g + V_e$$
$$= -Ws + \tfrac{1}{2}ks^2$$

If the particle moves from s_1 to a lower position s_2, then applying Eq. 14–16 it can be seen that the work of **W** and $\mathbf{F}_s$ is

$$U_{1-2} = V_1 - V_2 = \left(-Ws_1 + \tfrac{1}{2}ks_1^2\right) - \left(-Ws_2 + \tfrac{1}{2}ks_2^2\right)$$
$$= W(s_2 - s_1) - \left(\tfrac{1}{2}ks_2^2 - \tfrac{1}{2}ks_1^2\right)$$

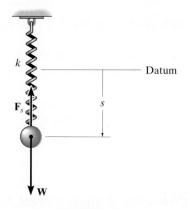

Fig. 14–19

When the displacement along the path is infinitesimal, i.e., from point (x, y, z) to $(x + dx, y + dy, z + dz)$, Eq. 14–16 becomes

$$dU = V(x, y, z) - V(x + dx, y + dy, z + dz)$$
$$= -dV(x, y, z) \tag{14-17}$$

If we represent both the force and its displacement as Cartesian vectors, then the work can also be expressed as

$$dU = \mathbf{F} \cdot d\mathbf{r} = (F_x\mathbf{i} + F_y\mathbf{j} + F_z\mathbf{k}) \cdot (dx\mathbf{i} + dy\mathbf{j} + dz\mathbf{k})$$
$$= F_x\,dx + F_y\,dy + F_z\,dz$$

Substituting this result into Eq. 14–17 and expressing the differential $dV(x, y, z)$ in terms of its partial derivatives yields

$$F_x\,dx + F_y\,dy + F_z\,dz = -\left(\frac{\partial V}{\partial x}dx + \frac{\partial V}{\partial y}dy + \frac{\partial V}{\partial z}dz\right)$$

Since changes in x, y, and z are all independent of one another, this equation is satisfied provided

$$F_x = -\frac{\partial V}{\partial x}, \qquad F_y = -\frac{\partial V}{\partial y}, \qquad F_z = -\frac{\partial V}{\partial z} \tag{14-18}$$

Thus,

$$\mathbf{F} = -\frac{\partial V}{\partial x}\mathbf{i} - \frac{\partial V}{\partial y}\mathbf{j} - \frac{\partial V}{\partial z}\mathbf{k}$$

$$= -\left(\frac{\partial}{\partial x}\mathbf{i} + \frac{\partial}{\partial y}\mathbf{j} + \frac{\partial}{\partial z}\mathbf{k}\right)V$$

or

$$\mathbf{F} = -\nabla V \tag{14-19}$$

where ∇ (del) represents the vector operator $\nabla = (\partial/\partial x)\mathbf{i} + (\partial/\partial y)\mathbf{j} + (\partial/\partial z)\mathbf{k}$.

Equation 14–19 relates a force $\mathbf{F}$ to its potential function V and thereby provides a mathematical criterion for proving that $\mathbf{F}$ is conservative. For example, the gravitational potential function for a weight located a distance y above a datum is $V_g = Wy$. To prove that $\mathbf{W}$ is conservative, it is necessary to show that it satisfies Eq. 14–18 (or Eq. 14–19), in which case

$$F_y = -\frac{\partial V}{\partial y}; \qquad F_y = -\frac{\partial}{\partial y}(Wy) = -W$$

The negative sign indicates that $\mathbf{W}$ acts downward, opposite to positive y, which is upward.

14.6   Conservation of Energy

When a particle is acted upon by a system of *both* conservative and nonconservative forces, the portion of the work done by the *conservative forces* can be written in terms of the difference in their potential energies using Eq. 14–16, i.e., $(\Sigma U_{1-2})_{\text{cons.}} = V_1 - V_2$. As a result, the principle of work and energy can be written as

$$T_1 + V_1 + (\Sigma U_{1-2})_{\text{noncons.}} = T_2 + V_2 \qquad (14\text{–}20)$$

Here $(\Sigma U_{1-2})_{\text{noncons.}}$ represents the work of the nonconservative forces acting on the particle. If *only conservative forces* do work then we have

$$\boxed{T_1 + V_1 = T_2 + V_2} \qquad (14\text{–}21)$$

This equation is referred to as the *conservation of mechanical energy* or simply the *conservation of energy*. It states that during the motion the sum of the particle's kinetic and potential energies remains *constant*. For this to occur, kinetic energy must be transformed into potential energy, and vice versa. For example, if a ball of weight **W** is dropped from a height h above the ground (datum), Fig. 14–20, the potential energy of the ball is maximum before it is dropped, at which time its kinetic energy is zero. The total mechanical energy of the ball in its initial position is thus

$$E = T_1 + V_1 = 0 + Wh = Wh$$

When the ball has fallen a distance $h/2$, its speed can be determined by using $v^2 = v_0^2 + 2a_c(y - y_0)$, which yields $v = \sqrt{2g(h/2)} = \sqrt{gh}$. The energy of the ball at the mid-height position is therefore

$$E = T_2 + V_2 = \frac{1}{2}\frac{W}{g}(\sqrt{gh})^2 + W\left(\frac{h}{2}\right) = Wh$$

Just before the ball strikes the ground, its potential energy is zero and its speed is $v = \sqrt{2gh}$. Here, again, the total energy of the ball is

$$E = T_3 + V_3 = \frac{1}{2}\frac{W}{g}(\sqrt{2gh})^2 + 0 = Wh$$

Note that when the ball comes in contact with the ground, it deforms somewhat, and provided the ground is hard enough, the ball will rebound off the surface, reaching a new height h', which will be *less* than the height h from which it was first released. Neglecting air friction, the difference in height accounts for an energy loss, $E_l = W(h - h')$, which occurs during the collision. Portions of this loss produce noise, localized deformation of the ball and ground, and heat.

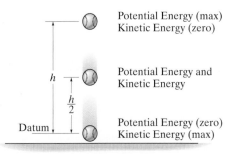

Potential Energy (max)
Kinetic Energy (zero)

Potential Energy and
Kinetic Energy

Potential Energy (zero)
Kinetic Energy (max)

Datum

Fig. 14–20

System of Particles.   If a system of particles is *subjected only to conservative forces*, then an equation similar to Eq. 14–21 can be written for the particles. Applying the ideas of the preceding discussion, Eq. 14–8 ($\Sigma T_1 + \Sigma U_{1-2} = \Sigma T_2$) becomes

$$\Sigma T_1 + \Sigma V_1 = \Sigma T_2 + \Sigma V_2 \qquad\qquad (14\text{–}22)$$

Here, the sum of the system's initial kinetic and potential energies is equal to the sum of the system's final kinetic and potential energies. In other words, $\Sigma T + \Sigma V = $ const.

Procedure for Analysis

The conservation of energy equation can be used to solve problems involving *velocity, displacement*, and *conservative force systems*. It is generally *easier to apply* than the principle of work and energy because this equation requires specifying the particle's kinetic and potential energies at only *two points* along the path, rather than determining the work when the particle moves through a *displacement*. For application it is suggested that the following procedure be used.

Potential Energy.

- Draw two diagrams showing the particle located at its initial and final points along the path.

- If the particle is subjected to a vertical displacement, establish the fixed horizontal datum from which to measure the particle's gravitational potential energy V_g.

- Data pertaining to the elevation y of the particle from the datum and the stretch or compression s of any connecting springs can be determined from the geometry associated with the two diagrams.

- Recall $V_g = Wy$, where y is positive upward from the datum and negative downward from the datum; also for a spring, $V_e = \frac{1}{2}ks^2$, which is *always positive*.

Conservation of Energy.

- Apply the equation $T_1 + V_1 = T_2 + V_2$.

- When determining the kinetic energy, $T = \frac{1}{2}mv^2$, remember that the particle's speed v must be measured from an inertial reference frame.

EXAMPLE | 14.9

The gantry structure in the photo is used to test the response of an airplane during a crash. As shown in Fig. 14–21a, the plane, having a mass of 8 Mg, is hoisted back until $\theta = 60°$, and then the pull-back cable AC is released when the plane is at rest. Determine the speed of the plane just before it crashes into the ground, $\theta = 15°$. Also, what is the maximum tension developed in the supporting cable during the motion? Neglect the size of the airplane and the effect of lift caused by the wings during the motion.

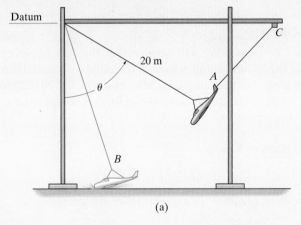

(a)

SOLUTION

Since the force of the cable does *no work* on the plane, it must be obtained using the equation of motion. First, however, we must determine the plane's speed at B.

Potential Energy.   For convenience, the datum has been established at the top of the gantry, Fig. 14–21a.

Conservation of Energy.

$$T_A + V_A = T_B + V_B$$

$$0 - 8000 \text{ kg } (9.81 \text{ m/s}^2)(20 \cos 60° \text{ m}) =$$

$$\tfrac{1}{2}(8000 \text{ kg})v_B^2 - 8000 \text{ kg } (9.81 \text{ m/s}^2)(20 \cos 15° \text{ m})$$

$$v_B = 13.52 \text{ m/s} = 13.5 \text{ m/s} \qquad Ans.$$

Equation of Motion.   From the free-body diagram when the plane is at B, Fig. 14–21b, we have

$$+\nwarrow \ \Sigma F_n = ma_n;$$

$$T - (8000(9.81) \text{ N}) \cos 15° = (8000 \text{ kg})\frac{(13.52 \text{ m/s})^2}{20 \text{ m}}$$

$$T = 149 \text{ kN} \qquad Ans.$$

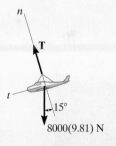

(b)

Fig. 14–21

EXAMPLE 14.10

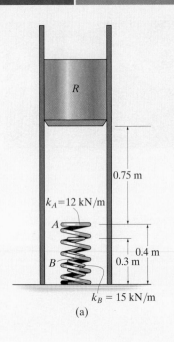

$k_A = 12$ kN/m

0.75 m

0.4 m
0.3 m

$k_B = 15$ kN/m

(a)

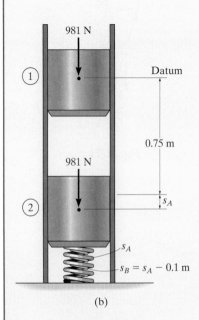

981 N

981 N

Datum

0.75 m

s_A

s_A

$s_B = s_A - 0.1$ m

(b)

Fig. 14–22

The ram R shown in Fig. 14–22a has a mass of 100 kg and is released from rest 0.75 m from the top of a spring, A, that has a stiffness $k_A = 12$ kN/m. If a second spring B, having a stiffness $k_B = 15$ kN/m, is "nested" in A, determine the maximum displacement of A needed to stop the downward motion of the ram. The unstretched length of each spring is indicated in the figure. Neglect the mass of the springs.

SOLUTION

Potential Energy.   We will *assume* that the ram compresses *both* springs at the instant it comes to rest. The datum is located through the center of gravity of the ram at its initial position, Fig. 14–22b. When the kinetic energy is reduced to zero ($v_2 = 0$), A is compressed a distance s_A and B compresses $s_B = s_A - 0.1$ m.

Conservation of Energy.

$$T_1 + V_1 = T_2 + V_2$$

$$0 + 0 = 0 + \left\{ \tfrac{1}{2}k_A s_A^2 + \tfrac{1}{2}k_B(s_A - 0.1)^2 - Wh \right\}$$

$$0 + 0 = 0 + \left\{ \tfrac{1}{2}(12\,000 \text{ N/m})s_A^2 + \tfrac{1}{2}(15\,000 \text{ N/m})(s_A - 0.1 \text{ m})^2 \right.$$
$$\left. - 981 \text{ N} (0.75 \text{ m} + s_A) \right\}$$

Rearranging the terms,

$$13\,500 s_A^2 - 2481 s_A - 660.75 = 0$$

Using the quadratic formula and solving for the positive root, we have

$$s_A = 0.331 \text{ m} \qquad\qquad Ans.$$

Since $s_B = 0.331$ m $- 0.1$ m $= 0.231$ m,   which is positive, the assumption that *both* springs are compressed by the ram is correct.

NOTE: The second root, $s_A = -0.148$ m, does not represent the physical situation. Since positive s is measured downward, the negative sign indicates that spring A would have to be "extended" by an amount of 0.148 m to stop the ram.

EXAMPLE 14.11

A smooth 2-kg collar, shown in Fig. 14–23a, fits loosely on the vertical shaft. If the spring is unstretched when the collar is in the position A, determine the speed at which the collar is moving when $y = 1$ m, if (a) it is released from rest at A, and (b) it is released at A with an *upward* velocity $v_A = 2$ m/s.

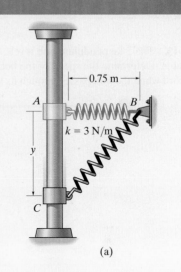

(a)

SOLUTION

Part (a) Potential Energy. For convenience, the datum is established through AB, Fig. 14–23b. When the collar is at C, the gravitational potential energy is $-(mg)y$, since the collar is *below* the datum, and the elastic potential energy is $\frac{1}{2}ks_{CB}^2$. Here $s_{CB} = 0.5$ m, which represents the *stretch* in the spring as shown in the figure.

Conservation of Energy.

$$T_A + V_A = T_C + V_C$$

$$0 + 0 = \tfrac{1}{2}mv_C^2 + \left\{\tfrac{1}{2}ks_{CB}^2 - mgy\right\}$$

$$0 + 0 = \left\{\tfrac{1}{2}(2\ \text{kg})v_C^2\right\} + \left\{\tfrac{1}{2}(3\ \text{N/m})(0.5\ \text{m})^2 - 2(9.81)\ \text{N}\,(1\ \text{m})\right\}$$

$$v_C = 4.39\ \text{m/s} \downarrow \qquad\qquad Ans.$$

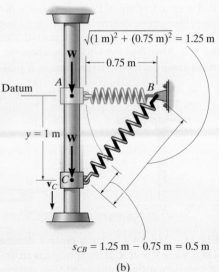

$$\sqrt{(1\ \text{m})^2 + (0.75\ \text{m})^2} = 1.25\ \text{m}$$

$$s_{CB} = 1.25\ \text{m} - 0.75\ \text{m} = 0.5\ \text{m}$$

(b)

Fig. 14–23

This problem can also be solved by using the equation of motion or the principle of work and energy. Note that for *both* of these methods the variation of the magnitude and direction of the spring force must be taken into account (see Example 13.4). Here, however, the above solution is clearly advantageous since the calculations depend *only* on data calculated at the initial and final points of the path.

Part (b) Conservation of Energy. If $v_A = 2$ m/s, using the data in Fig. 14–23b, we have

$$T_A + V_A = T_C + V_C$$

$$\tfrac{1}{2}mv_A^2 + 0 = \tfrac{1}{2}mv_C^2 + \left\{\tfrac{1}{2}ks_{CB}^2 - mgy\right\}$$

$$\tfrac{1}{2}(2\ \text{kg})(2\ \text{m/s})^2 + 0 = \tfrac{1}{2}(2\ \text{kg})v_C^2 + \left\{\tfrac{1}{2}(3\ \text{N/m})(0.5\ \text{m})^2\right.$$

$$\left. - 2(9.81)\ \text{N}\,(1\ \text{m})\right\}$$

$$v_C = 4.82\ \text{m/s} \downarrow \qquad\qquad Ans.$$

NOTE: The kinetic energy of the collar depends only on the *magnitude* of velocity, and therefore it is immaterial if the collar is moving up or down at 2 m/s when released at A.

FUNDAMENTAL PROBLEMS

14

F14–13. The 2-kg pendulum bob is released from rest when it is at A. Determine the speed of the bob and the tension in the cord when the bob passes through its lowest position, B.

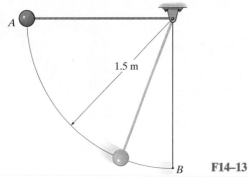

F14–13

F14–14. The 2-kg package leaves the conveyor belt at A with a speed of $v_A = 1$ m/s and slides down the smooth ramp. Determine the required speed of the conveyor belt at B so that the package can be delivered without slipping on the belt. Also, find the normal reaction the curved portion of the ramp exerts on the package at B if $\rho_B = 2$ m.

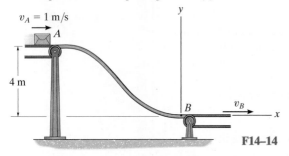

F14–14

F14–15. The 2-kg collar is given a downward velocity of 4 m/s when it is at A. If the spring has an unstretched length of 1 m and a stiffness of $k = 30$ N/m, determine the velocity of the collar at $s = 1$ m.

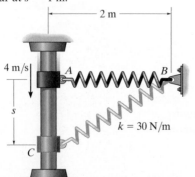

F14–15

F14–16. The 5-lb collar is released from rest at A and travels along the frictionless guide. Determine the speed of the collar when it strikes the stop B. The spring has an unstretched length of 0.5 ft.

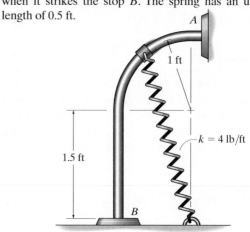

F14–16

F14–17. The 75-lb block is released from rest 5 ft above the plate. Determine the compression of each spring when the block momentarily comes to rest after striking the plate. Neglect the mass of the plate. The springs are initially unstretched.

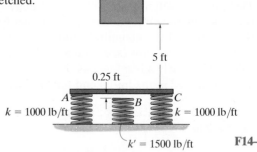

F14–17

F14–18. The 4-kg collar C has a velocity of $v_A = 2$ m/s when it is at A. If the guide rod is smooth, determine the speed of the collar when it is at B. The spring has an unstretched length of $l_0 = 0.2$ m.

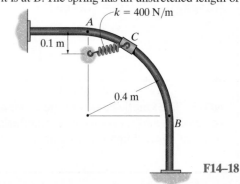

F14–18

PROBLEMS

14–66. The girl has a mass of 40 kg and center of mass at G. If she is swinging to a maximum height defined by $\theta = 60°$, determine the force developed along each of the four supporting posts such as AB at the instant $\theta = 0°$. The swing is centrally located between the posts.

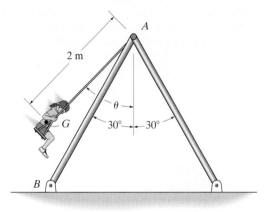

2 m

A

θ

G 30°   30°

B

Prob. 14–66

14–67. Two equal-length springs are "nested" together in order to form a shock absorber. If it is designed to arrest the motion of a 2-kg mass that is dropped $s = 0.5$ m above the top of the springs from an at-rest position, and the maximum compression of the springs is to be 0.2 m, determine the required stiffness of the inner spring, k_B, if the outer spring has a stiffness $k_A = 400$ N/m.

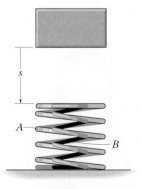

s

A

B

Prob. 14–67

*14–68. The collar has a weight of 8 lb. If it is pushed down so as to compress the spring 2 ft and then released from rest ($h = 0$), determine its speed when it is displaced $h = 4.5$ ft. The spring is not attached to the collar. Neglect friction.

14–69. The collar has a weight of 8 lb. If it is released from rest at a height of $h = 2$ ft from the top of the uncompressed spring, determine the speed of the collar after it falls and compresses the spring 0.3 ft.

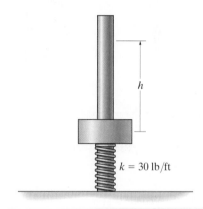

h

$k = 30$ lb/ft

Probs. 14–68/69

14–70. The 2-kg ball of negligible size is fired from point A with an initial velocity of 10 m/s up the smooth inclined plane. Determine the distance from point C to where it hits the horizontal surface at D. Also, what is its velocity when it strikes the surface?

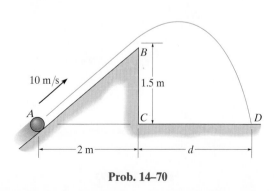

B

10 m/s   1.5 m

A C D

2 m d

Prob. 14–70

14–71. The ride at an amusement park consists of a gondola which is lifted to a height of 120 ft at A. If it is released from rest and falls along the parabolic track, determine the speed at the instant $y = 20$ ft. Also determine the normal reaction of the tracks on the gondola at this instant. The gondola and passenger have a total weight of 500 lb. Neglect the effects of friction and the mass of the wheels.

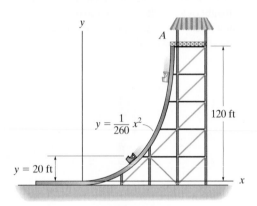

Prob. 14–71

*14–72. The 2-kg collar is attached to a spring that has an unstretched length of 3 m. If the collar is drawn to point B and released from rest, determine its speed when it arrives at point A.

14–73. The 2-kg collar is attached to a spring that has an unstretched length of 2 m. If the collar is drawn to point B and released from rest, determine its speed when it arrives at point A.

14–74. The 0.5-lb ball is shot from the spring device shown. The spring has a stiffness $k = 10$ lb/in. and the four cords C and plate P keep the spring compressed 2 in. when no load is on the plate. The plate is pushed back 3 in. from its initial position. If it is then released from rest, determine the speed of the ball when it travels 30 in. up the smooth plane.

14–75. The 0.5-lb ball is shot from the spring device shown. Determine the smallest stiffness k which is required to shoot the ball a maximum distance of 30 in. up the smooth plane after the spring is pushed back 3 in. and the ball is released from rest. The four cords C and plate P keep the spring compressed 2 in. when no load is on the plate.

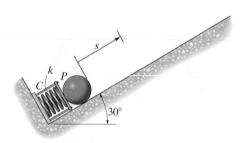

Probs. 14–74/75

*14–76. The roller coaster car having a mass m is released from rest at point A. If the track is to be designed so that the car does not leave it at B, determine the required height h. Also, find the speed of the car when it reaches point C. Neglect friction.

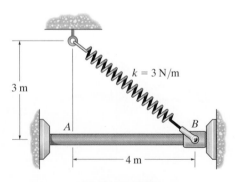

Probs. 14–72/73

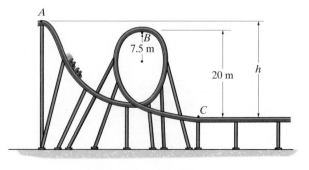

Prob. 14–76

14–77. A 750-mm-long spring is compressed and confined by the plate P, which can slide freely along the vertical 600-mm-long rods. The 40-kg block is given a speed of v = 5 m/s when it is h = 2 m above the plate. Determine how far the plate moves downwards when the block momentarily stops after striking it. Neglect the mass of the plate.

14–79. The block has a weight of 1.5 lb and slides along the smooth chute AB. It is released from rest at A, which has coordinates of A(5 ft, 0, 10 ft). Determine the speed at which it slides off at B, which has coordinates of B(0, 8 ft, 0).

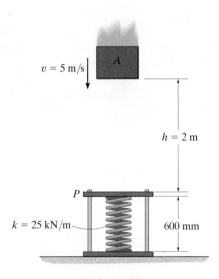

Prob. 14–77

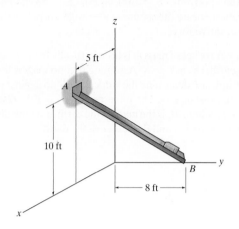

Prob. 14–79

14–78. The 2-lb block is given an initial velocity of 20 ft/s when it is at A. If the spring has an unstretched length of 2 ft and a stiffness of k = 100 lb/ft, determine the velocity of the block when s = 1 ft.

***14–80.** Each of the two elastic rubber bands of the slingshot has an unstretched length of 200 mm. If they are pulled back to the position shown and released from rest, determine the speed of the 25-g pellet just after the rubber bands become unstretched. Neglect the mass of the rubber bands. Each rubber band has a stiffness of k = 50 N/m.

14–81. Each of the two elastic rubber bands of the slingshot has an unstretched length of 200 mm. If they are pulled back to the position shown and released from rest, determine the maximum height the 25-g pellet will reach if it is fired vertically upward. Neglect the mass of the rubber bands and the change in elevation of the pellet while it is constrained by the rubber bands. Each rubber band has a stiffness k = 50 N/m.

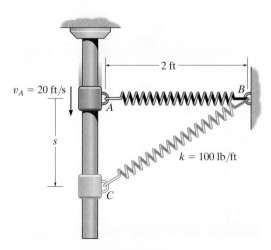

Prob. 14–78

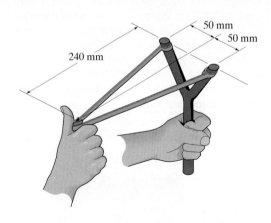

Probs. 14–80/81

14–82. If the mass of the earth is M_e, show that the gravitational potential energy of a body of mass m located a distance r from the center of the earth is $V_g = -GM_em/r$. Recall that the gravitational force acting between the earth and the body is $F = G(M_em/r^2)$, Eq. 13–1. For the calculation, locate the datum at $r \to \infty$. Also, prove that F is a conservative force.

14–83. A rocket of mass m is fired vertically from the surface of the earth, i.e., at $r = r_1$. Assuming that no mass is lost as it travels upward, determine the work it must do against gravity to reach a distance r_2. The force of gravity is $F = GM_em/r^2$ (Eq. 13–1), where M_e is the mass of the earth and r the distance between the rocket and the center of the earth.

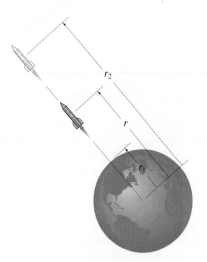

Probs. 14–82/83

*14–84. The firing mechanism of a pinball machine consists of a plunger P having a mass of 0.25 kg and a spring of stiffness $k = 300$ N/m. When $s = 0$, the spring is compressed 50 mm. If the arm is pulled back such that $s = 100$ mm and released, determine the speed of the 0.3-kg pinball B *just before* the plunger strikes the stop, i.e., $s = 0$. Assume all surfaces of contact to be smooth. The ball moves in the horizontal plane. Neglect friction, the mass of the spring, and the rolling motion of the ball.

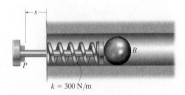

Prob. 14–84

14–85. A 60-kg satellite travels in free flight along an elliptical orbit such that at A, where $r_A = 20$ Mm, it has a speed $v_A = 40$ Mm/h. What is the speed of the satellite when it reaches point B, where $r_B = 80$ Mm? *Hint:* See Prob. 14–82, where $M_e = 5.976(10^{24})$ kg and $G = 66.73(10^{-12})$ m^3/(kg·s^2).

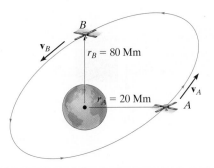

Prob. 14–85

14–86. Just for fun, two 150-lb engineering students A and B intend to jump off the bridge from rest using an elastic cord (bungee cord) having a stiffness $k = 80$ lb/ft. They wish to just reach the surface of the river, when A, attached to the cord, lets go of B at the instant they touch the water. Determine the proper unstretched length of the cord to do the stunt, and calculate the maximum acceleration of student A and the maximum height he reaches above the water after the rebound. From your results, comment on the feasibility of doing this stunt.

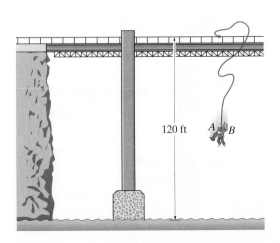

Prob. 14–86

14–87. The 20-lb collar slides along the smooth rod. If the collar is released from rest at A, determine its speed when it passes point B. The spring has an unstretched length of 3 ft.

Prob. 14–87

*14–88.** Two equal-length springs having a stiffness $k_A = 300$ N/m and $k_B = 200$ N/m are "nested" together in order to form a shock absorber. If a 2-kg block is dropped from an at-rest position 0.6 m above the top of the springs, determine their deformation when the block momentarily stops.

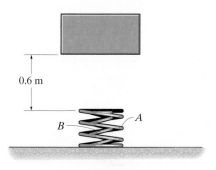

Prob. 14–88

14–89. When the 6-kg box reaches point A it has a speed of $v_A = 2$ m/s. Determine the angle θ at which it leaves the smooth circular ramp and the distance s to where it falls into the cart. Neglect friction.

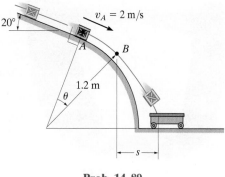

Prob. 14–89

14–90. The Raptor is an outside loop roller coaster in which riders are belted into seats resembling ski-lift chairs. Determine the minimum speed v_0 at which the cars should coast down from the top of the hill, so that passengers can just make the loop without leaving contact with their seats. Neglect friction, the size of the car and passenger, and assume each passenger and car has a mass m.

14–91. The Raptor is an outside loop roller coaster in which riders are belted into seats resembling ski-lift chairs. If the cars travel at $v_0 = 4$ m/s when they are at the top of the hill, determine their speed when they are at the top of the loop and the reaction of the 70-kg passenger on his seat at this instant. The car has a mass of 50 kg. Take $h = 12$ m, $\rho = 5$ m. Neglect friction and the size of the car and passenger.

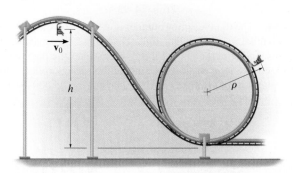

Probs. 14–90/91

*14–92.   The 75-kg man bungee jumps off the bridge at *A* with an initial downward speed of 1.5 m/s. Determine the required unstretched length of the elastic cord to which he is attached in order that he stops momentarily just above the surface of the water. The stiffness of the elastic cord is $k = 80$ N/m. Neglect the size of the man.

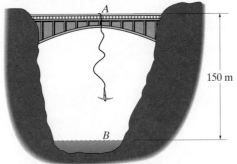

Prob. 14–92

14–93.   The 10-kg sphere *C* is released from rest when $\theta = 0°$ and the tension in the spring is 100 N. Determine the speed of the sphere at the instant $\theta = 90°$. Neglect the mass of rod *AB* and the size of the sphere.

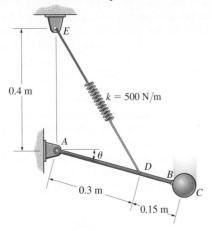

Prob. 14–93

14–94.   The double-spring bumper is used to stop the 1500-lb steel billet in the rolling mill. Determine the maximum displacement of the plate *A* if the billet strikes the plate with a speed of 8 ft/s. Neglect the mass of the springs, rollers and the plates *A* and *B*. Take $k_1 = 3000$ lb/ft, $k_2 = 45\,000$ lb/ft.

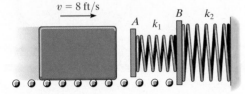

Prob. 14–94

14–95.   The 2-lb box has a velocity of 5 ft/s when it begins to slide down the smooth inclined surface at *A*. Determine the point *C* (*x, y*) where it strikes the lower incline.

*14–96.   The 2-lb box has a velocity of 5 ft/s when it begins to slide down the smooth inclined surface at *A*. Determine its speed just before hitting the surface at *C* and the time to travel from *A* to *C*. The coordinates of point *C* are $x = 17.66$ ft, and $y = 8.832$ ft.

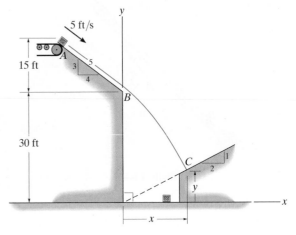

Probs. 14–95/96

14–97.   A pan of negligible mass is attached to two identical springs of stiffness $k = 250$ N/m. If a 10-kg box is dropped from a height of 0.5 m above the pan, determine the maximum vertical displacement *d*. Initially each spring has a tension of 50 N.

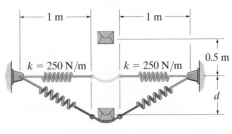

Prob. 14–97

CONCEPTUAL PROBLEMS

P14–1. The roller coaster is momentarily at rest at *A*. Determine the approximate normal force it exerts on the track at *B*. Also determine its approximate acceleration at this point. Use numerical data, and take scaled measurements from the photo with a known height at *A*.

P14–1

P14–2. As the large ring rotates, the operator can apply a breaking mechanism that binds the cars to the ring, which then allows the cars to rotate with the ring. Assuming the passengers are not belted into the cars, determine the smallest speed of the ring (cars) so that no passenger will fall out. When should the operator release the brake so that the cars can achieve their greatest speed as they slide freely on the ring? Estimate the greatest normal force of the seat on a passenger when this speed is reached. Use numerical values to explain your answer.

P14–2

P14–3. The woman pulls the water balloon launcher back, stretching each of the four elastic cords. Estimate the maximum height and the maximum range of a ball placed within the container if it is released from the position shown. Use numerical values and any necessary measurements from the photo. Assume the unstretched length and stiffness of each cord is known.

P14–3

P14–4. The girl is momentarily at rest in the position shown. If the unstretched length and stiffness of each of the two elastic cords is known, determine approximately how far the girl descends before she again becomes momentarily at rest. Use numerical values and take any necessary measurements from the photo.

P14–4

14

CHAPTER REVIEW

Work of a Force

A force does work when it undergoes a displacement along its line of action. If the force varies with the displacement, then the work is $U = \int F \cos \theta \, ds$.

Graphically, this represents the area under the F–s diagram.

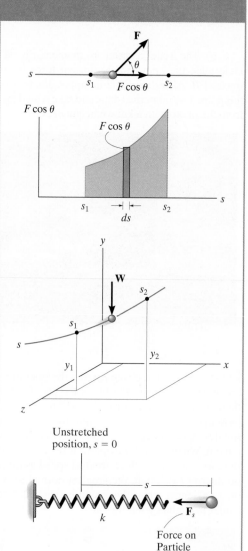

If the force is constant, then for a displacement Δs in the direction of the force, $U = F_c \, \Delta s$. A typical example of this case is the work of a weight, $U = -W \, \Delta y$. Here, Δy is the vertical displacement.

The work done by a spring force, $F = ks$, depends upon the stretch or compression s of the spring.

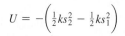

$$U = -\left(\tfrac{1}{2} k s_2^2 - \tfrac{1}{2} k s_1^2 \right)$$

Unstretched position, $s = 0$

Force on Particle

The Principle of Work and Energy

If the equation of motion in the tangential direction, $\Sigma F_t = ma_t$, is combined with the kinematic equation, $a_t \, ds = v \, dv$, we obtain the principle of work and energy. This equation states that the initial kinetic energy T, plus the work done ΣU_{1-2} is equal to the final kinetic energy.

$$T_1 + \Sigma U_{1-2} = T_2$$

The principle of work and energy is useful for solving problems that involve force, velocity, and displacement. For application, the free-body diagram of the particle should be drawn in order to identify the forces that do work.

Power and Efficiency

Power is the time rate of doing work. For application, the force **F** creating the power and its velocity **v** must be specified.

$$P = \frac{dU}{dt}$$

$$P = \mathbf{F} \cdot \mathbf{v}$$

Efficiency represents the ratio of power output to power input. Due to frictional losses, it is always less than one.

$$\varepsilon = \frac{\text{power output}}{\text{power input}}$$

Conservation of Energy

A conservative force does work that is independent of its path. Two examples are the weight of a particle and the spring force.

Friction is a nonconservative force since the work depends upon the length of the path. The longer the path, the more work done.

The work done by a conservative force depends upon its position relative to a datum. When this work is referenced from a datum, it is called potential energy. For a weight, it is $V_g = \pm Wy$, and for a spring it is $V_e = +\frac{1}{2}ks^2$.

Mechanical energy consists of kinetic energy T and gravitational and elastic potential energies V. According to the conservation of energy, this sum is constant and has the same value at any position on the path. If only gravitational and spring forces cause motion of the particle, then the conservation-of-energy equation can be used to solve problems involving these conservative forces, displacement, and velocity.

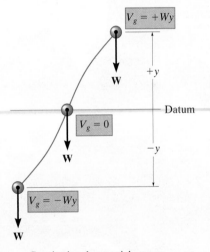

Gravitational potential energy

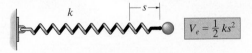

Elastic potential energy

$$T_1 + V_1 = T_2 + V_2$$

Chapter 15

The design of the bumper cars used for this amusement park ride requires knowledge of the principles of impulse and momentum.

Kinetics of a Particle: Impulse and Momentum

CHAPTER OBJECTIVES

■ To develop the principle of linear impulse and momentum for a particle and apply it to solve problems that involve force, velocity, and time.

■ To study the conservation of linear momentum for particles.

■ To analyze the mechanics of impact.

■ To introduce the concept of angular impulse and momentum.

■ To solve problems involving steady fluid streams and propulsion with variable mass.

15.1 Principle of Linear Impulse and Momentum

In this section we will integrate the equation of motion with respect to time and thereby obtain the principle of impulse and momentum. The resulting equation will be useful for solving problems involving force, velocity, and time.

Using kinematics, the equation of motion for a particle of mass m can be written as

$$\Sigma \mathbf{F} = m\mathbf{a} = m\frac{d\mathbf{v}}{dt} \tag{15–1}$$

where $\mathbf{a}$ and $\mathbf{v}$ are both measured from an inertial frame of reference. Rearranging the terms and integrating between the limits $\mathbf{v} = \mathbf{v}_1$ at $t = t_1$ and $\mathbf{v} = \mathbf{v}_2$ at $t = t_2$, we have

$$\Sigma \int_{t_1}^{t_2} \mathbf{F}dt = m \int_{\mathbf{v}_1}^{\mathbf{v}_2} d\mathbf{v}$$

The impulse tool is used to remove the dent in the trailer fender. To do so its end is first screwed into a hole drilled in the fender, then the weight is gripped and jerked upwards, striking the stop ring. The impulse developed is transferred along the shaft of the tool and pulls suddenly on the dent.

or

$$\Sigma \int_{t_1}^{t_2} \mathbf{F}dt = m\mathbf{v}_2 - m\mathbf{v}_1 \qquad (15\text{--}2)$$

This equation is referred to as the *principle of linear impulse and momentum*. From the derivation it can be seen that it is simply a time integration of the equation of motion. It provides a *direct means* of obtaining the particle's final velocity $\mathbf{v}_2$ after a specified time period when the particle's initial velocity is known and the forces acting on the particle are either constant or can be expressed as functions of time. By comparison, if $\mathbf{v}_2$ was determined using the equation of motion, a two-step process would be necessary; i.e., apply $\Sigma\mathbf{F} = m\mathbf{a}$ to obtain $\mathbf{a}$, then integrate $\mathbf{a} = d\mathbf{v}/dt$ to obtain $\mathbf{v}_2$.

Linear Momentum.   Each of the two vectors of the form $\mathbf{L} = m\mathbf{v}$ in Eq. 15–2 is referred to as the particle's linear momentum. Since m is a positive scalar, the linear-momentum vector has the same direction as $\mathbf{v}$, and its magnitude mv has units of mass times velocity, e.g., $\text{kg} \cdot \text{m/s}$, or $\text{slug} \cdot \text{ft/s}$.

Linear Impulse.   The integral $\mathbf{I} = \int \mathbf{F}\, dt$ in Eq. 15–2 is referred to as the *linear impulse*. This term is a vector quantity which measures the effect of a force during the time the force acts. Since time is a positive scalar, the impulse acts in the same direction as the force, and its magnitude has units of force times time, e.g., $\text{N} \cdot \text{s}$ or $\text{lb} \cdot \text{s}$.*

If the force is expressed as a function of time, the impulse can be determined by direct evaluation of the integral. In particular, if the force is constant in both magnitude and direction, the resulting impulse becomes

$$\mathbf{I} = \int_{t_1}^{t_2}\mathbf{F}_c\, dt = \mathbf{F}_c(t_2 - t_1).$$

Graphically the magnitude of the impulse can be represented by the shaded area under the curve of force versus time, Fig. 15–1. A constant force creates the shaded rectangular area shown in Fig. 15–2.

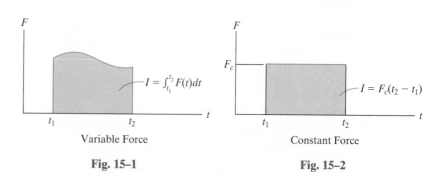

Variable Force	Constant Force
Fig. 15–1	**Fig. 15–2**

*Although the units for impulse and momentum are defined differently, it can be shown that Eq. 15–2 is dimensionally homogeneous.

Principle of Linear Impulse and Momentum. For problem solving, Eq. 15–2 will be rewritten in the form

$$ m\mathbf{v}_1 + \Sigma \int_{t_1}^{t_2} \mathbf{F}\, dt = m\mathbf{v}_2 \qquad (15\text{--}3) $$

which states that the initial momentum of the particle at time t_1 plus the sum of all the impulses applied to the particle from t_1 to t_2 is equivalent to the final momentum of the particle at time t_2. These three terms are illustrated graphically on the *impulse and momentum diagrams* shown in Fig. 15–3. The two *momentum diagrams* are simply outlined shapes of the particle which indicate the direction and magnitude of the particle's initial and final momenta, $m\mathbf{v}_1$ and $m\mathbf{v}_2$. Similar to the free-body diagram, the *impulse diagram* is an outlined shape of the particle showing all the impulses that act on the particle when it is located at some intermediate point along its path.

If each of the vectors in Eq. 15–3 is resolved into its x, y, z components, we can write the following three scalar equations of linear impulse and momentum.

Many types of sports, such as baseball, require application of the principle of linear impulse and momentum.

$$ m(v_x)_1 + \Sigma \int_{t_1}^{t_2} F_x\, dt = m(v_x)_2 $$

$$ m(v_y)_1 + \Sigma \int_{t_1}^{t_2} F_y\, dt = m(v_y)_2 \qquad (15\text{--}4) $$

$$ m(v_z)_1 + \Sigma \int_{t_1}^{t_2} F_z\, dt = m(v_z)_2 $$

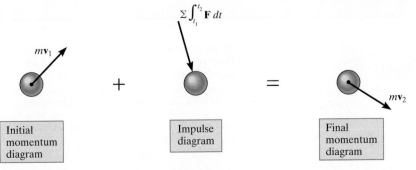

$$ \Sigma \int_{t_1}^{t_2} \mathbf{F}\, dt $$

$m\mathbf{v}_1$

$+$

$=$

$m\mathbf{v}_2$

Initial momentum diagram

Impulse diagram

Final momentum diagram

Fig. 15–3

15

Procedure for Analysis

The principle of linear impulse and momentum is used to solve problems involving *force, time*, and *velocity*, since these terms are involved in the formulation. For application it is suggested that the following procedure be used.*

Free-Body Diagram.

- Establish the x, y, z inertial frame of reference and draw the particle's free-body diagram in order to account for all the forces that produce impulses on the particle.

- The direction and sense of the particle's initial and final velocities should be established.

- If a vector is unknown, assume that the sense of its components is in the direction of the positive inertial coordinate(s).

- As an alternative procedure, draw the impulse and momentum diagrams for the particle as discussed in reference to Fig. 15–3.

Principle of Impulse and Momentum.

- In accordance with the established coordinate system, apply the principle of linear impulse and momentum, $m\mathbf{v}_1 + \Sigma \int_{t_1}^{t_2} \mathbf{F}\, dt = m\mathbf{v}_2$. If motion occurs in the x–y plane, the two scalar component equations can be formulated by either resolving the vector components of $\mathbf{F}$ from the free-body diagram, or by using the data on the impulse and momentum diagrams.

- Realize that every force acting on the particle's free-body diagram will create an impulse, even though some of these forces will do no work.

- Forces that are functions of time must be integrated to obtain the impulse. Graphically, the impulse is equal to the area under the force–time curve.

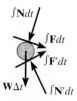

As the wheels of the pitching machine rotate, they apply frictional impulses to the ball, thereby giving it a linear momentum. These impulses are shown on the impulse diagram. Here both the frictional and normal impulses vary with time. By comparison, the weight impulse is constant and is very small since the time Δt the ball is in contact with the wheels is very small.

*This procedure will be followed when developing the proofs and theory in the text.

EXAMPLE | 15.1

The 100-kg crate shown in Fig. 15–4a is originally at rest on the smooth horizontal surface. If a towing force of 200 N, acting at an angle of 45°, is applied for 10 s, determine the final velocity and the normal force which the surface exerts on the crate during this time interval.

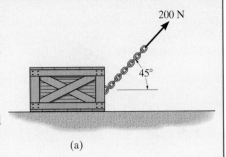

(a)

SOLUTION

This problem can be solved using the principle of impulse and momentum since it involves force, velocity, and time.

Free-Body Diagram. See Fig. 15–4b. Since all the forces acting are *constant*, the impulses are simply the product of the force magnitude and 10 s $[\mathbf{I} = \mathbf{F}_c(t_2 - t_1)]$. Note the alternative procedure of drawing the crate's impulse and momentum diagrams, Fig. 15–4c.

Principle of Impulse and Momentum. Applying Eqs. 15–4 yields

$(\xrightarrow{+})$
$$m(v_x)_1 + \Sigma \int_{t_1}^{t_2} F_x \, dt = m(v_x)_2$$

$$0 + 200 \text{ N} \cos 45°(10 \text{ s}) = (100 \text{ kg})v_2$$

$$v_2 = 14.1 \text{ m/s} \qquad Ans.$$

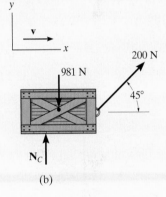

(b)

$(+\uparrow)$
$$m(v_y)_1 + \Sigma \int_{t_1}^{t_2} F_y \, dt = m(v_y)_2$$

$$0 + N_C(10 \text{ s}) - 981 \text{ N}(10 \text{ s}) + 200 \text{ N} \sin 45°(10 \text{ s}) = 0$$

$$N_C = 840 \text{ N} \qquad Ans.$$

NOTE: Since no motion occurs in the y direction, direct application of the equilibrium equation $\Sigma F_y = 0$ gives the same result for N_C. Try to solve the problem by first applying $\Sigma F_x = ma_x$, then $v = v_0 + a_c t$.

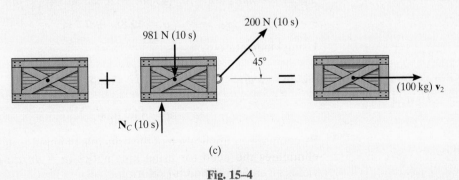

(c)

Fig. 15–4

EXAMPLE | 15.2

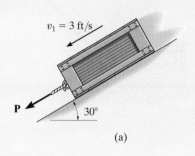

$v_1 = 3$ ft/s

P

30°

(a)

15

The 50-lb crate shown in Fig. 15–5a is acted upon by a force having a variable magnitude $P = (20t)$ lb, where t is in seconds. Determine the crate's velocity 2 s after **P** has been applied. The initial velocity is $v_1 = 3$ ft/s down the plane, and the coefficient of kinetic friction between the crate and the plane is $\mu_k = 0.3$.

SOLUTION

Free-Body Diagram.   See Fig. 15–5b. Since the magnitude of force $P = 20t$ varies with time, the impulse it creates must be determined by integrating over the 2-s time interval.

Principle of Impulse and Momentum.   Applying Eqs. 15–4 in the x direction, we have

$$(+\swarrow) \qquad\qquad m(v_x)_1 + \Sigma \int_{t_1}^{t_2} F_x \, dt = m(v_x)_2$$

$$\frac{50 \text{ lb}}{32.2 \text{ ft/s}^2}(3 \text{ ft/s}) + \int_0^{2\text{ s}} 20t \, dt - 0.3N_C(2\text{ s}) + (50 \text{ lb}) \sin 30°(2\text{ s}) = \frac{50 \text{ lb}}{32.2 \text{ ft/s}^2}v_2$$

$$4.658 + 40 - 0.6N_C + 50 = 1.553v_2$$

The equation of equilibrium can be applied in the y direction. Why?

$$+\nwarrow\Sigma F_y = 0; \qquad\qquad N_C - 50 \cos 30° \text{ lb} = 0$$

Solving,

$$N_C = 43.30 \text{ lb}$$

$$v_2 = 44.2 \text{ ft/s} \swarrow \qquad\qquad Ans.$$

NOTE: We can also solve this problem using the equation of motion. From Fig. 15–5b,

$$+\swarrow\Sigma F_x = ma_x; \quad 20t - 0.3(43.30) + 50 \sin 30° = \frac{50}{32.2}a$$

$$a = 12.88t + 7.734$$

Using kinematics

$$+\swarrow dv = a \, dt; \qquad \int_{3\text{ ft/s}}^{v} dv = \int_0^{2\text{ s}} (12.88t + 7.734)dt$$

$$v = 44.2 \text{ ft/s} \qquad\qquad Ans.$$

By comparison, application of the principle of impulse and momentum eliminates the need for using kinematics ($a = dv/dt$) and thereby yields an easier method for solution.

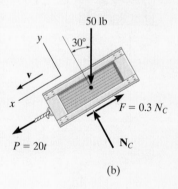

50 lb

30°

y

v

x

$F = 0.3 N_C$

$P = 20t$

N_C

(b)

Fig. 15–5

EXAMPLE 15.3

Blocks A and B shown in Fig. 15–6a have a mass of 3 kg and 5 kg, respectively. If the system is released from rest, determine the velocity of block B in 6 s. Neglect the mass of the pulleys and cord.

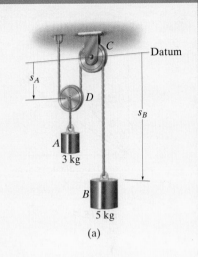

Datum

(a)

SOLUTION

Free-Body Diagram. See Fig. 15–6b. Since the weight of each block is constant, the cord tensions will also be constant. Furthermore, since the mass of pulley D is neglected, the cord tension $T_A = 2T_B$. Note that the blocks are both assumed to be moving downward in the positive coordinate directions, s_A and s_B.

Principle of Impulse and Momentum.

Block A:

$(+\downarrow)$ $$m(v_A)_1 + \Sigma \int_{t_1}^{t_2} F_y\, dt = m(v_A)_2$$

$$0 - 2T_B(6\text{ s}) + 3(9.81)\text{ N}(6\text{ s}) = (3\text{ kg})(v_A)_2 \qquad (1)$$

Block B:

$(+\downarrow)$ $$m(v_B)_1 + \Sigma \int_{t_1}^{t_2} F_y\, dt = m(v_B)_2$$

$$0 + 5(9.81)\text{ N}(6\text{ s}) - T_B(6\text{ s}) = (5\text{ kg})(v_B)_2 \qquad (2)$$

$T_A = 2T_B$

Kinematics.   Since the blocks are subjected to dependent motion, the velocity of A can be related to that of B by using the kinematic analysis discussed in Sec. 12–9. A horizontal datum is established through the fixed point at C, Fig. 15–6a, and the position coordinates, s_A and s_B, are related to the constant total length l of the vertical segments of the cord by the equation

$$2s_A + s_B = l$$

Taking the time derivative yields

$$2v_A = -v_B \qquad (3)$$

As indicated by the negative sign, when B moves downward A moves upward. Substituting this result into Eq. 1 and solving Eqs. 1 and 2 yields

$$(v_B)_2 = 35.8 \text{ m/s} \downarrow \qquad Ans.$$

$$T_B = 19.2 \text{ N}$$

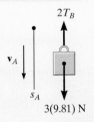

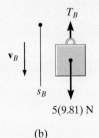

(b)

NOTE: Realize that the *positive* (downward) direction for $\mathbf{v}_A$ and $\mathbf{v}_B$ is *consistent* in Figs. 15–6a and 15–6b and in Eqs. 1 to 3. This is important since we are seeking a simultaneous solution of equations.

Fig. 15–6

15

15.2   Principle of Linear Impulse and Momentum for a System of Particles

The principle of linear impulse and momentum for a system of particles moving relative to an inertial reference, Fig. 15–7, is obtained from the equation of motion applied to all the particles in the system, i.e.,

$$\Sigma \mathbf{F}_i = \Sigma m_i \frac{d\mathbf{v}_i}{dt} \qquad (15\text{--}5)$$

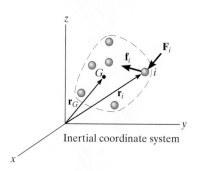

Inertial coordinate system

Fig. 15–7

The term on the left side represents only the sum of the *external forces* acting on the particles. Recall that the internal forces $\mathbf{f}_i$ acting between particles do not appear with this summation, since by Newton's third law they occur in equal but opposite collinear pairs and therefore cancel out. Multiplying both sides of Eq. 15–5 by dt and integrating between the limits $t = t_1, \mathbf{v}_i = (\mathbf{v}_i)_1$ and $t = t_2$, $\mathbf{v}_i = (\mathbf{v}_i)_2$ yields

$$\Sigma m_i(\mathbf{v}_i)_1 + \Sigma \int_{t_1}^{t_2} \mathbf{F}_i \, dt = \Sigma m_i(\mathbf{v}_i)_2 \qquad (15\text{--}6)$$

This equation states that the initial linear momenta of the system plus the impulses of all the *external forces* acting on the system from t_1 to t_2 is equal to the system's final linear momenta.

Since the location of the mass center G of the system is determined from $m\mathbf{r}_G = \Sigma m_i\mathbf{r}_i$, where $m = \Sigma m_i$ is the total mass of all the particles, Fig. 15–7, then taking the time derivative, we have

$$m\mathbf{v}_G = \Sigma m_i\mathbf{v}_i$$

which states that the total linear momentum of the system of particles is equivalent to the linear momentum of a "fictitious" aggregate particle of mass $m = \Sigma m_i$ moving with the velocity of the mass center of the system. Substituting into Eq. 15–6 yields

$$m(\mathbf{v}_G)_1 + \Sigma \int_{t_1}^{t_2} \mathbf{F}_i \, dt = m(\mathbf{v}_G)_2 \qquad (15\text{--}7)$$

Here the initial linear momentum of the aggregate particle plus the external impulses acting on the system of particles from t_1 to t_2 is equal to the aggregate particle's final linear momentum. As a result, the above equation justifies application of the principle of linear impulse and momentum to a system of particles that compose a rigid body.

FUNDAMENTAL PROBLEMS

F15–1. The 0.5-kg ball strikes the rough ground and rebounds with the velocities shown. Determine the magnitude of the impulse the ground exerts on the ball. Assume that the ball does not slip when it strikes the ground, and neglect the size of the ball and the impulse produced by the weight of the ball.

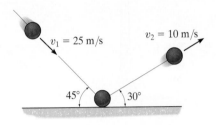

F15–1

F15–2. If the coefficient of kinetic friction between the 150-lb crate and the ground is $\mu_k = 0.2$, determine the speed of the crate when $t = 4$ s. The crate starts from rest and is towed by the 100-lb force.

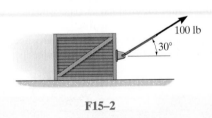

F15–2

F15–3. The motor exerts a force of $F = (20t^2)$ N on the cable, where t is in seconds. Determine the speed of the 25-kg crate when $t = 4$ s. The coefficients of static and kinetic friction between the crate and the plane are $\mu_s = 0.3$ and $\mu_k = 0.25$, respectively.

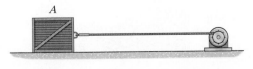

F15–3

F15–4. The wheels of the 1.5-Mg car generate the traction force $\mathbf{F}$ described by the graph. If the car starts from rest, determine its speed when $t = 6$ s.

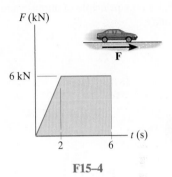

F15–4

F15–5. The 2.5-Mg four-wheel-drive SUV tows the 1.5-Mg trailer. The traction force developed at the wheels is $F_D = 9$ kN. Determine the speed of the truck in 20 s, starting from rest. Also, determine the tension developed in the coupling between the SUV and the trailer. Neglect the mass of the wheels.

F15–5

F15–6. The 10-lb block A attains a velocity of 1 ft/s in 5 seconds, starting from rest. Determine the tension in the cord and the coefficient of kinetic friction between block A and the horizontal plane. Neglect the weight of the pulley. Block B has a weight of 8 lb.

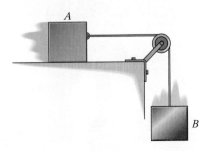

F15–6

15

PROBLEMS

15–1. A 2-lb ball is thrown in the direction shown with an initial speed $v_A = 18$ ft/s. Determine the time needed for it to reach its highest point B and the speed at which it is traveling at B. Use the principle of impulse and momentum for the solution.

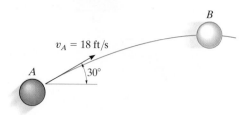

Prob. 15–1

15–2. A 20-lb block slides down a 30° inclined plane with an initial velocity of 2 ft/s. Determine the velocity of the block in 3 s if the coefficient of kinetic friction between the block and the plane is $\mu_k = 0.25$.

15–3. A 5-lb block is given an initial velocity of 10 ft/s up a 45° smooth slope. Determine the time it will take to travel up the slope before it stops.

***15–4.** The 180-lb iron worker is secured by a fall-arrest system consisting of a harness and lanyard AB, which is fixed to the beam. If the lanyard has a slack of 4 ft, determine the average impulsive force developed in the lanyard if he happens to fall 4 feet. Neglect his size in the calculation and assume the impulse takes place in 0.6 seconds.

Prob. 15–4

15–5. A man hits the 50-g golf ball such that it leaves the tee at an angle of 40° with the horizontal and strikes the ground at the same elevation a distance of 20 m away. Determine the impulse of the club C on the ball. Neglect the impulse caused by the ball's weight while the club is striking the ball.

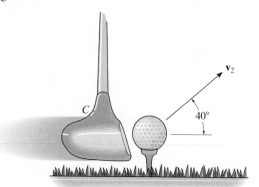

Prob. 15–5

15–6. A train consists of a 50-Mg engine and three cars, each having a mass of 30 Mg. If it takes 80 s for the train to increase its speed uniformly to 40 km/h, starting from rest, determine the force T developed at the coupling between the engine E and the first car A. The wheels of the engine provide a resultant frictional tractive force $\mathbf{F}$ which gives the train forward motion, whereas the car wheels roll freely. Also, determine F acting on the engine wheels.

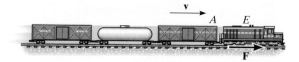

Prob. 15–6

15–7. Crates A and B weigh 100 lb and 50 lb, respectively. If they start from rest, determine their speed when $t = 5$ s. Also, find the force exerted by crate A on crate B during the motion. The coefficient of kinetic friction between the crates and the ground is $\mu_k = 0.25$.

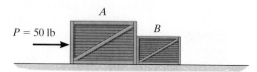

Prob. 15–7

*15–8. If the jets exert a vertical thrust of $T = (500t^{3/2})$ N, where t is in seconds, determine the man's speed when $t = 3$ s. The total mass of the man and the jet suit is 100 kg. Neglect the loss of mass due to the fuel consumed during the lift which begins from rest on the ground.

Prob. 15–8

15–9. Under a constant thrust of $T = 40$ kN, the 1.5-Mg dragster reaches its maximum speed of 125 m/s in 8 s starting from rest. Determine the average drag resistance $\mathbf{F}_D$ during this period of time.

$T = 40$ kN

$\mathbf{F}_D$

Prob. 15–9

15–10. The 50-kg crate is pulled by the constant force $\mathbf{P}$. If the crate starts from rest and achieves a speed of 10 m/s in 5 s, determine the magnitude of $\mathbf{P}$. The coefficient of kinetic friction between the crate and the ground is $\mu_k = 0.2$.

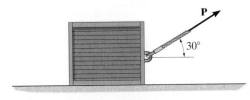

$\mathbf{P}$

$30°$

Prob. 15–10

15–11. When the 5-kg block is 6 m from the wall, it is sliding at $v_1 = 14$ m/s. If the coefficient of kinetic friction between the block and the horizontal plane is $\mu_k = 0.3$, determine the impulse of the wall on the block necessary to stop the block. Neglect the friction impulse acting on the block during the collision.

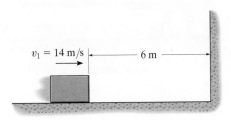

$v_1 = 14$ m/s   6 m

Prob. 15–11

*15–12. For a short period of time, the frictional driving force acting on the wheels of the 2.5-Mg van is $F_D = (600t^2)$ N, where t is in seconds. If the van has a speed of 20 km/h when $t = 0$, determine its speed when $t = 5$ s.

$\mathbf{F}_D$

Prob. 15–12

15–13. The 2.5-Mg van is traveling with a speed of 100 km/h when the brakes are applied and all four wheels lock. If the speed decreases to 40 km/h in 5 s, determine the coefficient of kinetic friction between the tires and the road.

Prob. 15–13

15

15–14. The force acting on a projectile having a mass m as it passes horizontally through the barrel of the cannon is $F = C \sin (\pi t/t')$. Determine the projectile's velocity when $t = t'$. If the projectile reaches the end of the barrel at this instant, determine the length s.

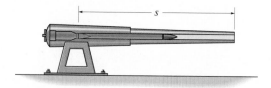

Prob. 15–14

15–15. During operation the breaker hammer develops on the concrete surface a force which is indicated in the graph. To achieve this the 2-lb spike S is fired from rest into the surface at 200 ft/s. Determine the speed of the spike just after rebounding.

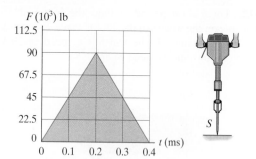

Prob. 15–15

*15–16. The twitch in a muscle of the arm develops a force which can be measured as a function of time as shown in the graph. If the effective contraction of the muscle lasts for a time t_0, determine the impulse developed by the muscle.

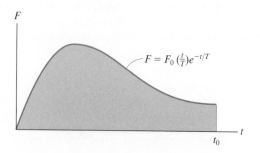

$$F = F_0 \left(\frac{t}{T}\right)e^{-t/T}$$

Prob. 15–16

15–17. A hammer head H having a weight of 0.25 lb is moving vertically downward at 40 ft/s when it strikes the head of a nail of negligible mass and drives it into a block of wood. Find the impulse on the nail if it is assumed that the grip at A is loose, the handle has a negligible mass, and the hammer stays in contact with the nail while it comes to rest. Neglect the impulse caused by the weight of the hammer head during contact with the nail.

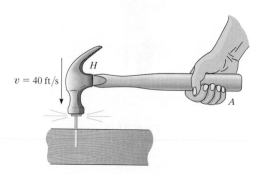

$v = 40$ ft/s

Prob. 15–17

15–18. The 40-kg slider block is moving to the right with a speed of 1.5 m/s when it is acted upon by the forces $\mathbf{F}_1$ and $\mathbf{F}_2$. If these loadings vary in the manner shown on the graph, determine the speed of the block at $t = 6$ s. Neglect friction and the mass of the pulleys and cords.

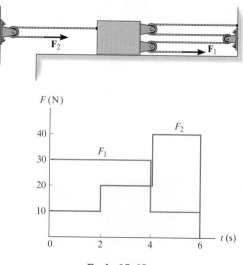

Prob. 15–18

15–19.   Determine the velocity of each block 2 s after the blocks are released from rest. Neglect the mass of the pulleys and cord.

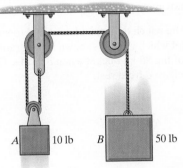

Prob. 15–19

*****15–20.**   The particle P is acted upon by its weight of 3 lb and forces $\mathbf{F}_1$ and $\mathbf{F}_2$, where t is in seconds. If the particle orginally has a velocity of $\mathbf{v}_1 = \{3\mathbf{i} + 1\mathbf{j} + 6\mathbf{k}\}$ ft/s, determine its speed after 2 s.

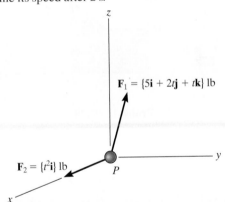

Prob. 15–20

15–21.   If it takes 35 s for the 50-Mg tugboat to increase its speed uniformly to 25 km/h, starting from rest, determine the force of the rope on the tugboat. The propeller provides the propulsion force $\mathbf{F}$ which gives the tugboat forward motion, whereas the barge moves freely. Also, determine F acting on the tugboat. The barge has a mass of 75 Mg.

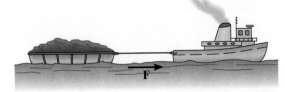

Prob. 15–21

15-22.   If the force T exerted on the cable by the motor M is indicated by the graph, determine the speed of the 500-lb crate when $t = 4$ s, starting from rest. The coefficients of static and kinetic friction are $\mu_s = 0.3$ and $\mu_k = 0.25$, respectively.

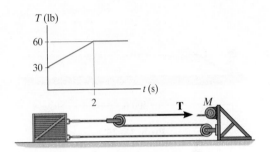

Prob. 15–22

15–23.   The 5-kg block is moving downward at $v_1 = 2$ m/s when it is 8 m from the sandy surface. Determine the impulse of the sand on the block necessary to stop its motion. Neglect the distance the block dents into the sand and assume the block does not rebound. Neglect the weight of the block during the impact with the sand.

*****15–24.**   The 5-kg block is falling downward at $v_1 = 2$ m/s when it is 8 m from the sandy surface. Determine the average impulsive force acting on the block by the sand if the motion of the block is stopped in 0.9 s once the block strikes the sand. Neglect the distance the block dents into the sand and assume the block does not rebound. Neglect the weight of the block during the impact with the sand.

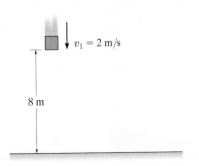

Probs. 15–23/24

15–25. The 0.1-lb golf ball is struck by the club and then travels along the trajectory shown. Determine the average impulsive force the club imparts on the ball if the club maintains contact with the ball for 0.5 ms.

15–27. The winch delivers a horizontal towing force $\mathbf{F}$ to its cable at A which varies as shown in the graph. Determine the speed of the 70-kg bucket when $t = 18$ s. Originally the bucket is moving upward at $v_1 = 3$ m/s.

*15–28. The winch delivers a horizontal towing force $\mathbf{F}$ to its cable at A which varies as shown in the graph. Determine the speed of the 80-kg bucket when $t = 24$ s. Originally the bucket is released from rest.

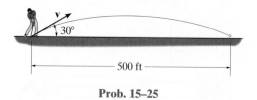

Prob. 15–25

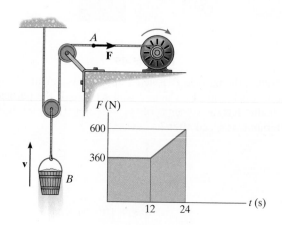

Probs. 15–27/28

15–26. As indicated by the derivation, the principle of impulse and momentum is valid for observers in *any* inertial reference frame. Show that this is so, by considering the 10-kg block which rests on the smooth surface and is subjected to a horizontal force of 6 N. If observer A is in a *fixed* frame x, determine the final speed of the block in 4 s if it has an initial speed of 5 m/s measured from the fixed frame. Compare the result with that obtained by an observer B, attached to the x' axis that moves at a constant velocity of 2 m/s relative to A.

15–29. The train consists of a 30-Mg engine E, and cars A, B, and C, which have a mass of 15 Mg, 10 Mg, and 8 Mg, respectively. If the tracks provide a traction force of $F = 30$ kN on the engine wheels, determine the speed of the train when $t = 30$ s, starting from rest. Also, find the horizontal coupling force at D between the engine E and car A. Neglect rolling resistance.

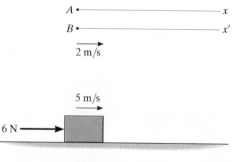

Prob. 15–26

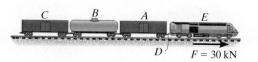

Prob. 15–29

15–30. The crate B and cylinder A have a mass of 200 kg and 75 kg, respectively. If the system is released from rest, determine the speed of the crate and cylinder when $t = 3$ s. Neglect the mass of the pulleys.

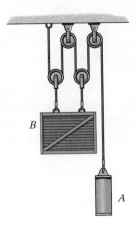

B

A

Prob. 15–30

15–31. Block A weighs 10 lb and block B weighs 3 lb. If B is moving downward with a velocity $(v_B)_1 = 3$ ft/s at $t = 0$, determine the velocity of A when $t = 1$ s. Assume that the horizontal plane is smooth. Neglect the mass of the pulleys and cords.

*15–32. Block A weighs 10 lb and block B weighs 3 lb. If B is moving downward with a velocity $(v_B)_1 = 3$ ft/s at $t = 0$, determine the velocity of A when $t = 1$ s. The coefficient of kinetic friction between the horizontal plane and block A is $\mu_A = 0.15$.

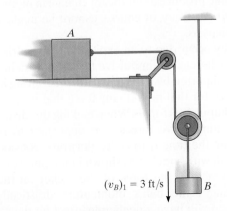

A

$(v_B)_1 = 3$ ft/s   B

Probs. 15–31/32

15–33. The log has a mass of 500 kg and rests on the ground for which the coefficients of static and kinetic friction are $\mu_s = 0.5$ and $\mu_k = 0.4$, respectively. The winch delivers a horizontal towing force T to its cable at A which varies as shown in the graph. Determine the speed of the log when $t = 5$ s. Originally the tension in the cable is zero. *Hint:* First determine the force needed to begin moving the log.

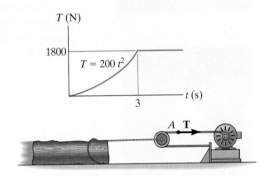

T (N)

1800

$T = 200\ t^2$

3   t (s)

$A\ \mathbf{T}$

Prob. 15–33

15–34. The 50-kg block is hoisted up the incline using the cable and motor arrangement shown. The coefficient of kinetic friction between the block and the surface is $\mu_k = 0.4$. If the block is initially moving up the plane at $v_0 = 2$ m/s, and at this instant ($t = 0$) the motor develops a tension in the cord of $T = (300 + 120\sqrt{t})$ N, where t is in seconds, determine the velocity of the block when $t = 2$ s.

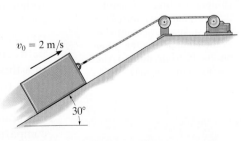

$v_0 = 2$ m/s

30°

Prob. 15–34

15

The hammer in the top photo applies an impulsive force to the stake. During this extremely short time of contact the weight of the stake can be considered nonimpulsive, and provided the stake is driven into soft ground, the impulse of the ground acting on the stake can also be considered nonimpulsive. By contrast, if the stake is used in a concrete chipper to break concrete, then two impulsive forces act on the stake: one at its top due to the chipper and the other on its bottom due to the rigidity of the concrete.

15.3   Conservation of Linear Momentum for a System of Particles

When the sum of the *external impulses* acting on a system of particles is *zero*, Eq. 15–6 reduces to a simplified form, namely,

$$\Sigma m_i(\mathbf{v}_i)_1 = \Sigma m_i(\mathbf{v}_i)_2 \qquad (15\text{–}8)$$

This equation is referred to as the *conservation of linear momentum*. It states that the total linear momentum for a system of particles remains constant during the time period t_1 to t_2. Substituting $m\mathbf{v}_G = \Sigma m_i \mathbf{v}_i$ into Eq. 15–8, we can also write

$$(\mathbf{v}_G)_1 = (\mathbf{v}_G)_2 \qquad (15\text{–}9)$$

which indicates that the velocity $\mathbf{v}_G$ of the mass center for the system of particles does not change if no external impulses are applied to the system.

The conservation of linear momentum is often applied when particles collide or interact. For application, a careful study of the free-body diagram for the *entire* system of particles should be made in order to identify the forces which create either external or internal impulses and thereby determine in what direction(s) linear momentum is conserved. As stated earlier, the *internal impulses* for the system will always cancel out, since they occur in equal but opposite collinear pairs. If the time period over which the motion is studied is *very short*, some of the external impulses may also be neglected or considered approximately equal to zero. The forces causing these negligible impulses are called *nonimpulsive forces*. By comparison, forces which are very large and act for a very short period of time produce a significant change in momentum and are called *impulsive forces*. They, of course, cannot be neglected in the impulse–momentum analysis.

Impulsive forces normally occur due to an explosion or the striking of one body against another, whereas nonimpulsive forces may include the weight of a body, the force imparted by a slightly deformed spring having a relatively small stiffness, or for that matter, any force that is very small compared to other larger (impulsive) forces. When making this distinction between impulsive and nonimpulsive forces, it is important to realize that this only applies during the time t_1 to t_2. To illustrate, consider the effect of striking a tennis ball with a racket as shown in the photo. During the *very short* time of interaction, the force of the racket on the ball is impulsive since it changes the ball's momentum drastically. By comparison, the ball's weight will have a negligible effect on the change in momentum, and therefore it is nonimpulsive. Consequently, it can be neglected from an impulse–momentum analysis during this time. If an impulse–momentum analysis is considered during the much longer time of flight after the racket–ball interaction, then the impulse of the ball's weight is important since it, along with air resistance, causes the change in the momentum of the ball.

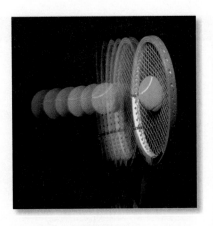

Procedure for Analysis

Generally, the principle of linear impulse and momentum or the conservation of linear momentum is applied to a *system of particles* in order to determine the final velocities of the particles *just after* the time period considered. By applying this principle to the entire system, the internal impulses acting within the system, which may be unknown, are *eliminated* from the analysis. For application it is suggested that the following procedure be used.

Free-Body Diagram.
- Establish the *x, y, z* inertial frame of reference and draw the free-body diagram for each particle of the system in order to identify the internal and external forces.

- The conservation of linear momentum applies to the system in a direction which either has no external forces or the forces can be considered nonimpulsive.

- Establish the direction and sense of the particles' initial and final velocities. If the sense is unknown, assume it is along a positive inertial coordinate axis.

- As an alternative procedure, draw the impulse and momentum diagrams for each particle of the system.

Momentum Equations.
- Apply the principle of linear impulse and momentum or the conservation of linear momentum in the appropriate directions.

- If it is necessary to determine the *internal impulse* $\int F \, dt$ acting on only one particle of a system, then the particle must be *isolated* (free-body diagram), and the principle of linear impulse and momentum must be applied *to this particle*.

- After the impulse is calculated, and provided the time Δt for which the impulse acts is known, then the *average impulsive force* F_{avg} can be determined from $F_{avg} = \int F \, dt / \Delta t$.

15

EXAMPLE 15.4

The 15-Mg boxcar A is coasting at 1.5 m/s on the horizontal track when it encounters a 12-Mg tank car B coasting at 0.75 m/s toward it as shown in Fig. 15–8a. If the cars collide and couple together, determine (a) the speed of both cars just after the coupling, and (b) the average force between them if the coupling takes place in 0.8 s.

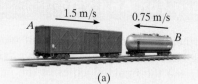

(a)

SOLUTION

Part (a) Free-Body Diagram.* Here we have considered *both* cars as a single system, Fig. 15–8b. By inspection, momentum is conserved in the x direction since the coupling force **F** is *internal* to the system and will therefore cancel out. It is assumed both cars, when coupled, move at v_2 in the positive x direction.

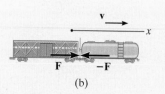

(b)

Conservation of Linear Momentum.

$$(\xrightarrow{+})\qquad m_A(v_A)_1 + m_B(v_B)_1 = (m_A + m_B)v_2$$

$$(15\,000 \text{ kg})(1.5 \text{ m/s}) - 12\,000 \text{ kg}(0.75 \text{ m/s}) = (27\,000 \text{ kg})v_2$$

$$v_2 = 0.5 \text{ m/s} \rightarrow \qquad Ans.$$

Part (b). The average (impulsive) coupling force, $\mathbf{F}_{avg}$, can be determined by applying the principle of linear momentum to *either one* of the cars.

Free-Body Diagram. As shown in Fig. 15–8c, by isolating the boxcar the coupling force is *external* to the car.

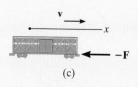

(c)

Fig. 15–8

Principle of Impulse and Momentum. Since $\int F\,dt = F_{avg}\,\Delta t = F_{avg}(0.8 \text{ s})$, we have

$$(\xrightarrow{+})\qquad m_A(v_A)_1 + \Sigma \int F\,dt = m_A v_2$$

$$(15\,000 \text{ kg})(1.5 \text{ m/s}) - F_{avg}(0.8 \text{ s}) = (15\,000 \text{ kg})(0.5 \text{ m/s})$$

$$F_{avg} = 18.8 \text{ kN} \qquad Ans.$$

NOTE: Solution was possible here since the boxcar's final velocity was obtained in Part (a). Try solving for F_{avg} by applying the principle of impulse and momentum to the tank car.

*Only horizontal forces are shown on the free-body diagram.

EXAMPLE 15.5

The bumper cars A and B in Fig. 15–9a each have a mass of 150 kg and are coasting with the velocities shown before they freely collide head on. If no energy is lost during the collision, determine their velocities after collision.

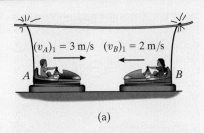

(a)

SOLUTION

Free-Body Diagram.   The cars will be considered as a single system. The free-body diagram is shown in Fig. 15–9b.

Conservation of Momentum.

$(\pm)$ $\qquad m_A(v_A)_1 + m_B(v_B)_1 = m_A(v_A)_2 + m_B(v_B)_2$

$(150 \text{ kg})(3 \text{ m/s}) + (150 \text{ kg})(-2 \text{ m/s}) = (150 \text{ kg})(v_A)_2 + (150 \text{ kg})(v_B)_2$

$$(v_A)_2 = 1 - (v_B)_2 \qquad\qquad (1)$$

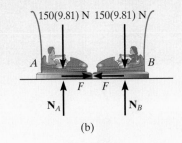

(b)

Fig. 15–9

Conservation of Energy.   Since no energy is lost, the conservation of energy theorem gives

$$T_1 + V_1 = T_2 + V_2$$

$$\frac{1}{2}m_A(v_A)_1^2 + \frac{1}{2}m_B(v_B)_1^2 + 0 = \frac{1}{2}m_A(v_A)_2^2 + \frac{1}{2}m_B(v_B)_2^2 + 0$$

$$\frac{1}{2}(150 \text{ kg})(3 \text{ m/s})^2 + \frac{1}{2}(150 \text{ kg})(2 \text{ m/s})^2 + 0 = \frac{1}{2}(150 \text{ kg})(v_A)_2^2$$

$$+ \frac{1}{2}(150 \text{ kg})(v_B)_2^2 + 0$$

$$(v_A)_2^2 + (v_B)_2^2 = 13 \qquad\qquad (2)$$

Substituting Eq. (1) into (2) and simplifying, we get

$$(v_B)_2^2 - (v_B)_2 - 6 = 0$$

Solving for the two roots,

$$(v_B)_2 = 3 \text{ m/s} \qquad \text{and} \qquad (v_B)_2 = -2 \text{ m/s}$$

Since $(v_B)_2 = -2$ m/s refers to the velocity of B just *before* collision, then the velocity of B just after the collision must be

$$(v_B)_2 = 3 \text{ m/s} \rightarrow \qquad\qquad Ans.$$

Substituting this result into Eq. (1), we obtain

$$(v_A)_2 = 1 - 3 \text{ m/s} = -2 \text{ m/s} = 2 \text{ m/s} \leftarrow \qquad Ans.$$

EXAMPLE 15.6

An 800-kg rigid pile shown in Fig. 15–10a is driven into the ground using a 300-kg hammer. The hammer falls from rest at a height $y_0 = 0.5$ m and strikes the top of the pile. Determine the impulse which the pile exerts on the hammer if the pile is surrounded entirely by loose sand so that after striking, the hammer does *not* rebound off the pile.

SOLUTION

Conservation of Energy. The velocity at which the hammer strikes the pile can be determined using the conservation of energy equation applied to the hammer. With the datum at the top of the pile, Fig. 15–10a, we have

$$T_0 + V_0 = T_1 + V_1$$

$$\frac{1}{2}m_H(v_H)_0^2 + W_H y_0 = \frac{1}{2}m_H(v_H)_1^2 + W_H y_1$$

$$0 + 300(9.81)\,\text{N}(0.5\,\text{m}) = \frac{1}{2}(300\,\text{kg})(v_H)_1^2 + 0$$

$$(v_H)_1 = 3.132\,\text{m/s}$$

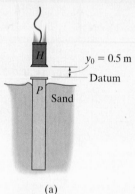

(a)

Free-Body Diagram. From the physical aspects of the problem, the free-body diagram of the hammer and pile, Fig. 15–10b, indicates that during the *short time* from *just before* to *just after* the *collision*, the weights of the hammer and pile and the resistance force $\mathbf{F}_s$ of the sand are all *nonimpulsive*. The impulsive force $\mathbf{R}$ is internal to the system and therefore cancels. Consequently, momentum is conserved in the vertical direction during this short time.

Conservation of Momentum. Since the hammer does not rebound off the pile just after collision, then $(v_H)_2 = (v_P)_2 = v_2$.

$$(+\downarrow) \qquad m_H(v_H)_1 + m_P(v_P)_1 = m_H v_2 + m_P v_2$$

$$(300\,\text{kg})(3.132\,\text{m/s}) + 0 = (300\,\text{kg})v_2 + (800\,\text{kg})v_2$$

$$v_2 = 0.8542\,\text{m/s}$$

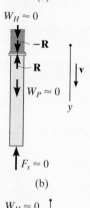

(b)

(c)

Fig. 15–10

Principle of Impulse and Momentum. The impulse which the pile imparts to the hammer can now be determined since v_2 is known. From the free-body diagram for the hammer, Fig. 15–10c, we have

$$(+\downarrow) \qquad m_H(v_H)_1 + \Sigma \int_{t_1}^{t_2} F_y\,dt = m_H v_2$$

$$(300\,\text{kg})(3.132\,\text{m/s}) - \int R\,dt = (300\,\text{kg})(0.8542\,\text{m/s})$$

$$\int R\,dt = 683\,\text{N} \cdot \text{s} \qquad \qquad Ans.$$

NOTE: The equal but opposite impulse acts on the pile. Try finding this impulse by applying the principle of impulse and momentum to the pile.

EXAMPLE | 15.7

The 80-kg man can throw the 20-kg box horizontally at 4 m/s when standing on the ground. If instead he firmly stands in the 120-kg boat and throws the box, as shown in the photo, determine how far the boat will move in three seconds. Neglect water resistance.

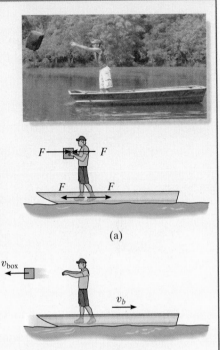

SOLUTION

Free-Body Diagram.   If the man, boat, and box are considered as a single system, the horizontal forces between the man and the boat and the man and the box become internal to the system, Fig. 15–11a, and so linear momentum will be conserved along the x axis.

Conservation of Momentum.   When writing the conservation of momentum equation, it is *important* that the velocities be measured from the same inertial coordinate system, assumed here to be fixed. From this coordinate system, we will assume that the boat and man go to the right while the box goes to the left, as shown in Fig. 15–11b.

Applying the conservation of linear momentum to the man, boat, box system,

$(\stackrel{+}{\rightarrow})$

$$0 + 0 + 0 = (m_m + m_b)\, v_b - m_{box}\, v_{box}$$

$$0 = (80 \text{ kg} + 120 \text{ kg})\, v_b - (20 \text{ kg})\, v_{box}$$

$$v_{box} = 10\, v_b \qquad (1)$$

(a)

(b)

Fig. 15–11

Kinematics.   Since the velocity of the box *relative* to the man (and boat), $v_{box/b}$, is known, then v_b can also be related to v_{box} using the relative velocity equation.

$(\stackrel{+}{\rightarrow})$

$$v_{box} = v_b + v_{box/b}$$

$$-v_{box} = v_b - 4 \text{ m/s} \qquad (2)$$

Solving Eqs. (1) and (2),

$$v_{box} = 3.64 \text{ m/s} \leftarrow$$

$$v_b = 0.3636 \text{ m/s} \rightarrow$$

The displacement of the boat in three seconds is therefore

$$s_b = v_b t = (0.3636 \text{ m/s})(3 \text{ s}) = 1.09 \text{ m} \qquad \textit{Ans.}$$

15

EXAMPLE | 15.8

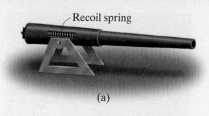

Recoil spring

(a)

The 1200-lb cannon shown in Fig. 15–12a fires an 8-lb projectile with a muzzle velocity of 1500 ft/s measured relative to the cannon. If firing takes place in 0.03 s, determine the recoil velocity of the cannon just after firing. The cannon support is fixed to the ground, and the horizontal recoil of the cannon is absorbed by two springs.

SOLUTION

Part (a) Free-Body Diagram.*   As shown in Fig. 15–12b, we will consider the projectile and cannon as a single system, since the impulsive forces, $\mathbf{F}$ and $-\mathbf{F}$, between the cannon and projectile are *internal* to the system and will therefore cancel from the analysis. Furthermore, during the time $\Delta t = 0.03$ s, the two recoil springs which are attached to the support each exert a *nonimpulsive force* $\mathbf{F}_s$ on the cannon. This is because Δt is very short, so that during this time the cannon only moves through a *very small* distance s. Consequently, $F_s = ks \approx 0$, where k is the spring's stiffness, which is also considered to be relatively small. Hence it can be concluded that momentum for the system is conserved in the *horizontal direction*.

Conservation of Linear Momentum.

$$(\overset{+}{\rightarrow})\qquad m_c(v_c)_1 + m_p(v_p)_1 = -m_c(v_c)_2 + m_p(v_p)_2$$

$$0 + 0 = -\left(\frac{1200\ \text{lb}}{32.2\ \text{ft/s}^2}\right)(v_c)_2 + \left(\frac{8\ \text{lb}}{32.2\ \text{ft/s}^2}\right)(v_p)_2$$

$$(v_p)_2 = 150\,(v_c)_2 \qquad\qquad (1)$$

These unknown velocities are measured by a *fixed* observer. As in Example 15–7, they can also be related using the relative velocity equation.

$$\overset{+}{\rightarrow}\qquad (v_p)_2 = (v_c)_2 + v_{p/c}$$
$$(v_p)_2 = -(v_c)_2 + 1500\ \text{ft/s} \qquad\qquad (2)$$

Solving Eqs. (1) and (2) yields

$$(v_c)_2 = 9.93\ \text{ft/s} \qquad\qquad\qquad Ans.$$
$$(v_p)_2 = 1490\ \text{ft/s}$$

Apply the principle of impulse and momentum to the projectile (or the cannon) and show that the average impulsive force on the projectile is 12.3 kip.

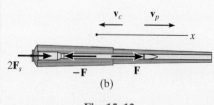

v_c v_p

x

$2\mathbf{F}_s$
$-\mathbf{F}$ $\mathbf{F}$
(b)

Fig. 15–12

NOTE: If the cannon is firmly fixed to its support (no springs), the reactive force of the support on the cannon must be considered as an external impulse to the system, since the support would allow no movement of the cannon. In this case momentum is *not* conserved.

*Only horizontal forces are shown on the free-body diagram.

FUNDAMENTAL PROBLEMS

F15–7. The freight cars A and B have a mass of 20 Mg and 15 Mg, respectively. Determine the velocity of A after collision if the cars collide and rebound, such that B moves to the right with a speed of 2 m/s. If A and B are in contact for 0.5 s, find the average impulsive force which acts between them.

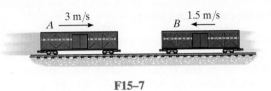

F15–7

F15–8. The cart and package have a mass of 20 kg and 5 kg, respectively. If the cart has a smooth surface and it is initially at rest, while the velocity of the package is as shown, determine the final common velocity of the cart and package after the impact.

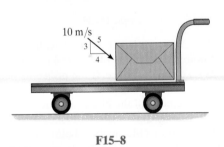

F15–8

F15–9. The 5-kg block A has an initial speed of 5 m/s as it slides down the smooth ramp, after which it collides with the stationary block B of mass 8 kg. If the two blocks couple together after collision, determine their common velocity immediately after collision.

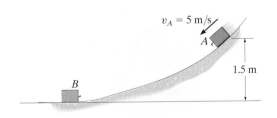

F15–9

F15–10. The spring is fixed to block A and block B is pressed against the spring. If the spring is compressed $s = 200$ mm and then the blocks are released, determine their velocity at the instant block B loses contact with the spring. The masses of blocks A and B are 10 kg and 15 kg, respectively.

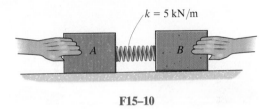

F15–10

F15–11. Blocks A and B have a mass of 15 kg and 10 kg, respectively. If A is stationary and B has a velocity of 15 m/s just before collision, and the blocks couple together after impact, determine the maximum compression of the spring.

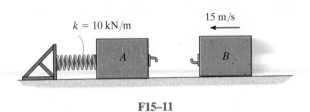

F15–11

F15–12. The cannon and support without a projectile have a mass of 250 kg. If a 20-kg projectile is fired from the cannon with a velocity of 400 m/s, measured *relative* to the cannon, determine the speed of the projectile as it leaves the barrel of the cannon. Neglect rolling resistance.

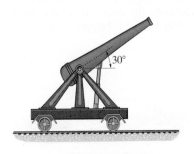

F15–12

15

PROBLEMS

15–35. The bus B has a weight of 15 000 lb and is traveling to the right at 5 ft/s. Meanwhile a 3000-lb car A is traveling at 4 ft/s to the left. If the vehicles crash head-on and become entangled, determine their common velocity just after the collision. Assume that the vehicles are free to roll during collision.

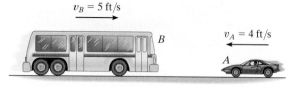

$v_B = 5$ ft/s

$v_A = 4$ ft/s

Prob. 15–35

*15–36. The 50-kg boy jumps on the 5-kg skateboard with a horizontal velocity of 5 m/s. Determine the distance s the boy reaches up the inclined plane before momentarily coming to rest. Neglect the skateboard's rolling resistance.

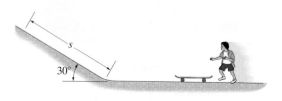

s

$30°$

Prob. 15–36

15–37. The 2.5-Mg pickup truck is towing the 1.5-Mg car using a cable as shown. If the car is initially at rest and the truck is coasting with a velocity of 30 km/h when the cable is slack, determine the common velocity of the truck and the car just after the cable becomes taut. Also, find the loss of energy.

30 km/h

Prob. 15–37

15–38. A railroad car having a mass of 15 Mg is coasting at 1.5 m/s on a horizontal track. At the same time another car having a mass of 12 Mg is coasting at 0.75 m/s in the opposite direction. If the cars meet and couple together, determine the speed of both cars just after the coupling. Find the difference between the total kinetic energy before and after coupling has occurred, and explain qualitatively what happened to this energy.

15–39. The car A has a weight of 4500 lb and is traveling to the right at 3 ft/s. Meanwhile a 3000-lb car B is traveling at 6 ft/s to the left. If the cars crash head-on and become entangled, determine their common velocity just after the collision. Assume that the brakes are not applied during collision.

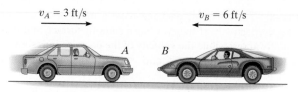

$v_A = 3$ ft/s $v_B = 6$ ft/s

A B

Prob. 15–39

*15–40. The 200-g projectile is fired with a velocity of 900 m/s towards the center of the 15-kg wooden block, which rests on a rough surface. If the projectile penetrates and emerges from the block with a velocity of 300 m/s, determine the velocity of the block just after the projectile emerges. How long does the block slide on the rough surface, after the projectile emerges, before it comes to rest again? The coefficient of kinetic friction between the surface and the block is $\mu_k = 0.2$.

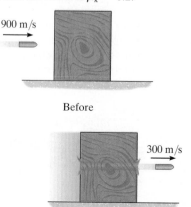

900 m/s

Before

300 m/s

After   Prob. 15–40

15–41. The block has a mass of 50 kg and rests on the surface of the cart having a mass of 75 kg. If the spring which is attached to the cart and not the block is compressed 0.2 m and the system is released from rest, determine the speed of the block relative to the *ground* after the spring becomes undeformed. Neglect the mass of the cart's wheels and the spring in the calculation. Also neglect friction. Take $k = 300$ N/m.

15–42. The block has a mass of 50 kg and rests on the surface of the cart having a mass of 75 kg. If the spring which is attached to the cart and not the block is compressed 0.2 m and the system is released from rest, determine the speed of the block with respect to the *cart* after the spring becomes undeformed. Neglect the mass of the wheels and the spring in the calculation. Also neglect friction. Take $k = 300$ N/m.

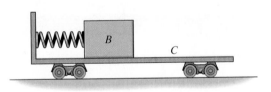

Probs. 15–41/42

15–43. The three freight cars A, B, and C have masses of 10 Mg, 5 Mg, and 20 Mg, respectively. They are traveling along the track with the velocities shown. Car A collides with car B first, followed by car C. If the three cars couple together after collision, determine the common velocity of the cars after the two collisions have taken place.

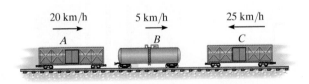

Prob. 15–43

*****15–44.** Two men A and B, each having a weight of 160 lb, stand on the stationary 200-lb cart. Each then runs with a speed of 3 ft/s measured relative to the cart. Determine the final speed of the cart if (a) A runs and jumps off, then B runs and jumps off the same end, and (b) both run at the same time and jump off at the same time. Neglect the mass of the wheels and assume the jumps are made horizontally.

Prob. 15–44

15–45. The block of mass m travels at v_1 in the direction θ_1 shown at the top of the smooth slope. Determine its speed v_2 and its direction θ_2 when it reaches the bottom.

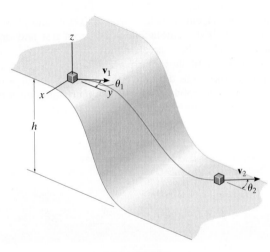

Prob. 15–45

15–46. The barge *B* weighs 30 000 lb and supports an automobile weighing 3000 lb. If the barge is not tied to the pier *P* and someone drives the automobile to the other side of the barge for unloading, determine how far the barge moves away from the pier. Neglect the resistance of the water.

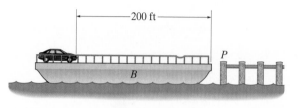

Prob. 15–46

15–47. The 30-Mg freight car *A* and 15-Mg freight car *B* are moving towards each other with the velocities shown. Determine the maximum compression of the spring mounted on car *A*. Neglect rolling resistance.

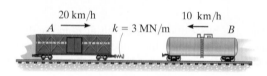

Prob. 15–47

*15–48.** The barge weighs 45 000 lb and supports two automobiles *A* and *B*, which weigh 4000 lb and 3000 lb, respectively. If the automobiles start from rest and drive towards each other, accelerating at $a_A = 4$ ft/s^2 and $a_B = 8$ ft/s^2 until they reach a constant speed of 6 ft/s relative to the barge, determine the speed of the barge just before the automobiles collide. How much time does this take? Originally the barge is at rest. Neglect water resistance.

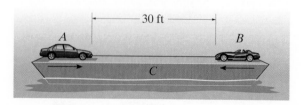

Prob. 15–48

15–49. The man *M* weighs 150 lb and jumps onto the boat *B* which has a weight of 200 lb. If he has a horizontal component of velocity *relative to the boat* of 3 ft/s, just before he enters the boat, and the boat is traveling $v_B = 2$ ft/s away from the pier when he makes the jump, determine the resulting velocity of the man and boat.

15–50. The man *M* weighs 150 lb and jumps onto the boat *B* which is originally at rest. If he has a horizontal component of velocity of 3 ft/s just before he enters the boat, determine the weight of the boat if it has a velocity of 2 ft/s once the man enters it.

Probs. 15–49/50

15–51. The 20-kg package has a speed of 1.5 m/s when it is delivered to the smooth ramp. After sliding down the ramp it lands onto a 10-kg cart as shown. Determine the speed of the cart and package after the package stops sliding on the cart.

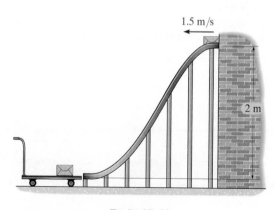

Prob. 15–51

*15–52. The free-rolling ramp has a mass of 40 kg. A 10-kg crate is released from rest at A and slides down 3.5 m to point B. If the surface of the ramp is smooth, determine the ramp's speed when the crate reaches B. Also, what is the velocity of the crate?

15–55. A tugboat T having a mass of 19 Mg is tied to a barge B having a mass of 75 Mg. If the rope is "elastic" such that it has a stiffness $k = 600$ kN/m, determine the maximum stretch in the rope during the initial towing. Originally both the tugboat and barge are moving in the same direction with speeds $(v_T)_1 = 15$ km/h and $(v_B)_1 = 10$ km/h, respectively. Neglect the resistance of the water.

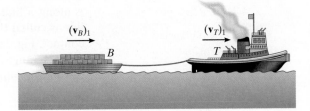

Prob. 15–55

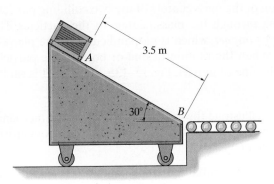

Prob. 15–52

*15–56. Two boxes A and B, each having a weight of 160 lb, sit on the 500-lb conveyor which is free to roll on the ground. If the belt starts from rest and begins to run with a speed of 3 ft/s, determine the final speed of the conveyor if (a) the boxes are not stacked and A falls off then B falls off, and (b) A is stacked on top of B and both fall off together.

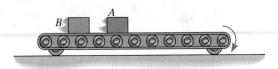

Prob. 15–56

15–53. The 80-lb boy and 60-lb girl walk towards each other with a constant speed on the 300-lb cart. If their velocities, measured relative to the cart, are 3 ft/s to the right and 2 ft/s to the left, respectively, determine the velocities of the boy and girl during the motion. Also, find the distance the cart has traveled at the instant the boy and girl meet.

15–54. The 80-lb boy and 60-lb girl walk towards each other with constant speed on the 300-lb cart. If their velocities measured relative to the cart are 3 ft/s to the right and 2 ft/s to the left, respectively, determine the velocity of the cart while they are walking.

15–57. The 10-kg block is held at rest on the smooth inclined plane by the stop block at A. If the 10-g bullet is traveling at 300 m/s when it becomes embedded in the 10-kg block, determine the distance the block will slide up along the plane before momentarily stopping.

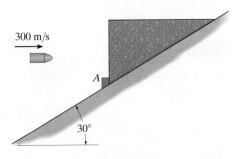

Prob. 15–57

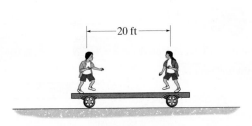

Probs. 15–53/54

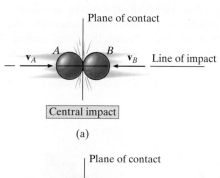

Central impact

(a)

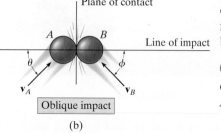

Oblique impact

(b)

Fig. 15–13

15.4 Impact

Impact occurs when two bodies collide with each other during a very *short* period of time, causing relatively large (impulsive) forces to be exerted between the bodies. The striking of a hammer on a nail, or a golf club on a ball, are common examples of impact loadings.

In general, there are two types of impact. *Central impact* occurs when the direction of motion of the mass centers of the two colliding particles is along a line passing through the mass centers of the particles. This line is called the *line of impact*, which is perpendicular to the plane of contact, Fig. 15–13a. When the motion of one or both of the particles make an angle with the line of impact, Fig. 15–13b, the impact is said to be *oblique impact*.

Central Impact.   To illustrate the method for analyzing the mechanics of impact, consider the case involving the central impact of the two particles *A* and *B* shown in Fig. 15–14.

- The particles have the initial momenta shown in Fig. 15–14a. Provided $(v_A)_1 > (v_B)_1$, collision will eventually occur.

- During the collision the particles must be thought of as *deformable* or nonrigid. The particles will undergo a *period of deformation* such that they exert an equal but opposite deformation impulse $\int \mathbf{P}\, dt$ on each other, Fig. 15–14b.

- Only at the instant of *maximum deformation* will both particles move with a common velocity $\mathbf{v}$, since their relative motion is zero, Fig. 15–14c.

- Afterward a *period of restitution* occurs, in which case the particles will either return to their original shape or remain permanently deformed. The equal but opposite *restitution impulse* $\int \mathbf{R}\, dt$ pushes the particles apart from one another, Fig. 15–14d. In reality, the physical properties of any two bodies are such that the deformation impulse will *always be greater* than that of restitution, i.e., $\int P\, dt > \int R\, dt$.

- Just after separation the particles will have the final momenta shown in Fig. 15–14e, where $(v_B)_2 > (v_A)_2$.

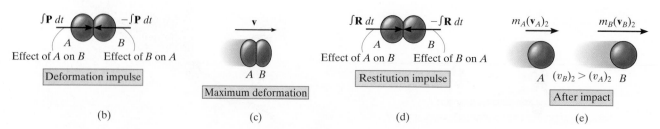

Fig. 15–14

In most problems the initial velocities of the particles will be *known*, and it will be necessary to determine their final velocities $(v_A)_2$ and $(v_B)_2$. In this regard, *momentum* for the *system of particles* is *conserved* since during collision the internal impulses of deformation and restitution *cancel*. Hence, referring to Fig. 15–14a and Fig. 15–14e we require

$$(\stackrel{+}{\rightarrow}) \qquad m_A(v_A)_1 + m_B(v_B)_1 = m_A(v_A)_2 + m_B(v_B)_2 \qquad (15\text{--}10)$$

In order to obtain a second equation necessary to solve for $(v_A)_2$ and $(v_B)_2$, we must apply the principle of impulse and momentum to *each* *particle*. For example, during the deformation phase for particle A, Figs. 15–14a, 15–14b, and 15–14c, we have

$$(\stackrel{+}{\rightarrow}) \qquad m_A(v_A)_1 - \int P \, dt = m_A v$$

For the restitution phase, Figs. 15–14c, 15–14d, and 15–14e,

$$(\stackrel{+}{\rightarrow}) \qquad m_A v - \int R \, dt = m_A(v_A)_2$$

The ratio of the restitution impulse to the deformation impulse is called the *coefficient of restitution*, e. From the above equations, this value for particle A is

$$e = \frac{\int R \, dt}{\int P \, dt} = \frac{v - (v_A)_2}{(v_A)_1 - v}$$

In a similar manner, we can establish e by considering particle B, Fig. 15–14. This yields

$$e = \frac{\int R \, dt}{\int P \, dt} = \frac{(v_B)_2 - v}{v - (v_B)_1}$$

If the unknown v is eliminated from the above two equations, the coefficient of restitution can be expressed in terms of the particles' initial and final velocities as

$$(\stackrel{+}{\rightarrow}) \qquad \boxed{e = \frac{(v_B)_2 - (v_A)_2}{(v_A)_1 - (v_B)_1}} \qquad (15\text{--}11)$$

The quality of a manufactured tennis ball is measured by the height of its bounce, which can be related to its coefficient of restitution. Using the mechanics of oblique impact, engineers can design a separation device to remove substandard tennis balls from a production line.

Provided a value for e is specified, Eqs. 15–10 and 15–11 can be solved simultaneously to obtain $(v_A)_2$ and $(v_B)_2$. In doing so, however, it is important to carefully establish a sign convention for defining the positive direction for both $\mathbf{v}_A$ and $\mathbf{v}_B$ and then use it *consistently* when writing *both* equations. As noted from the application shown, and indicated symbolically by the arrow in parentheses, we have defined the positive direction to the right when referring to the motions of both A and B. Consequently, if a negative value results from the solution of either $(v_A)_2$ or $(v_B)_2$, it indicates motion is to the left.

Coefficient of Restitution.

From Figs. 15–14*a* and 15–14*e*, it is seen that Eq. 15–11 states that e is equal to the ratio of the relative velocity of the particles' separation *just after impact*, $(v_B)_2 - (v_A)_2$, to the relative velocity of the particles' approach *just before impact*, $(v_A)_1 - (v_B)_1$. By measuring these relative velocities experimentally, it has been found that e varies appreciably with impact velocity as well as with the size and shape of the colliding bodies. For these reasons the coefficient of restitution is reliable only when used with data which closely approximate the conditions which were known to exist when measurements of it were made. In general e has a value between zero and one, and one should be aware of the physical meaning of these two limits.

Elastic Impact ($e = 1$).

If the collision between the two particles is *perfectly elastic*, the deformation impulse $\left(\int \mathbf{P} \, dt \right)$ is equal and opposite to the restitution impulse $\left(\int \mathbf{R} \, dt \right)$. Although in reality this can never be achieved, $e = 1$ for an elastic collision.

The mechanics of pool depends upon application of the conservation of momentum and the coefficient of restitution.

Plastic Impact ($e = 0$).

The impact is said to be *inelastic or plastic* when $e = 0$. In this case there is no restitution impulse $\left(\int \mathbf{R} \, dt = \mathbf{0} \right)$, so that after collision both particles couple or stick *together* and move with a common velocity.

From the above derivation it should be evident that the principle of work and energy cannot be used for the analysis of impact problems since it is not possible to know how the *internal forces* of deformation and restitution vary or displace during the collision. By knowing the particle's velocities before and after collision, however, the energy loss during collision can be calculated on the basis of the difference in the particle's kinetic energy. This energy loss, $\Sigma U_{1-2} = \Sigma T_2 - \Sigma T_1$, occurs because some of the initial kinetic energy of the particle is transformed into thermal energy as well as creating sound and localized deformation of the material when the collision occurs. In particular, if the impact is *perfectly elastic*, no energy is lost in the collision; whereas if the collision is *plastic*, the energy lost during collision is a maximum.

Procedure for Analysis (Central Impact)

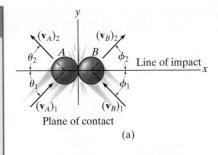

In most cases the *final velocities* of two smooth particles are to be determined *just after* they are subjected to direct central impact. Provided the coefficient of restitution, the mass of each particle, and each particle's initial velocity *just before* impact are known, the solution to this problem can be obtained using the following two equations:

- The conservation of momentum applies to the system of particles, $\Sigma m v_1 = \Sigma m v_2$.
- The coefficient of restitution, $e = [(v_B)_2 - (v_A)_2]/[(v_A)_1 - (v_B)_1]$, relates the relative velocities of the particles along the line of impact, just before and just after collision.

When applying these two equations, the sense of an unknown velocity can be assumed. If the solution yields a negative magnitude, the velocity acts in the opposite sense.

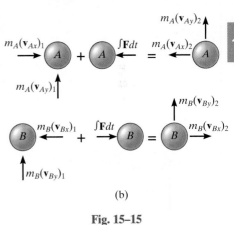

(b)

Fig. 15–15

Oblique Impact.

When oblique impact occurs between two smooth particles, the particles move away from each other with velocities having unknown directions as well as unknown magnitudes. Provided the initial velocities are known, then four unknowns are present in the problem. As shown in Fig. 15–15*a*, these unknowns may be represented either as $(v_A)_2$, $(v_B)_2$, θ_2, and ϕ_2, or as the x and y components of the final velocities.

Procedure for Analysis (Oblique Impact)

If the y axis is established within the plane of contact and the x axis along the line of impact, the impulsive forces of deformation and restitution act *only in the x direction*, Fig. 15–15*b*. By resolving the velocity or momentum vectors into components along the x and y axes, Fig. 15–15*b*, it is then possible to write four independent scalar equations in order to determine $(v_{Ax})_2$, $(v_{Ay})_2$, $(v_{Bx})_2$, and $(v_{By})_2$.

- Momentum of the system is conserved *along the line of impact*, x axis, so that $\Sigma m(v_x)_1 = \Sigma m(v_x)_2$.
- The coefficient of restitution, $e = [(v_{Bx})_2 - (v_{Ax})_2]/[(v_{Ax})_1 - (v_{Bx})_1]$, relates the relative-velocity *components* of the particles *along the line of impact* (x axis).
- If these two equations are solved simultaneously, we obtain $(v_{Ax})_2$ and $(v_{Bx})_2$.
- Momentum of particle A is conserved along the y axis, perpendicular to the line of impact, since no impulse acts on particle A in this direction. As a result $m_A(v_{Ay})_1 = m_A(v_{Ay})_2$ or $(v_{Ay})_1 = (v_{Ay})_2$
- Momentum of particle B is conserved along the y axis, perpendicular to the line of impact, since no impulse acts on particle B in this direction. Consequently $(v_{By})_1 = (v_{By})_2$.

Application of these four equations is illustrated in Example 15.11.

EXAMPLE   15.9

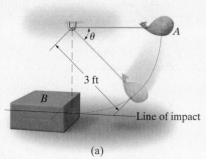

(a)

15

The bag A, having a weight of 6 lb, is released from rest at the position $\theta = 0°$, as shown in Fig. 15–16a. After falling to $\theta = 90°$, it strikes an 18-lb box B. If the coefficient of restitution between the bag and box is $e = 0.5$, determine the velocities of the bag and box just after impact. What is the loss of energy during collision?

SOLUTION

This problem involves central impact. Why? Before analyzing the mechanics of the impact, however, it is first necessary to obtain the velocity of the bag *just before* it strikes the box.

Conservation of Energy.   With the datum at $\theta = 0°$, Fig. 15–16b, we have

$$T_0 + V_0 = T_1 + V_1$$

$$0 + 0 = \frac{1}{2}\left(\frac{6\ \text{lb}}{32.2\ \text{ft/s}^2}\right)(v_A)_1^2 - 6\ \text{lb}(3\ \text{ft}); (v_A)_1 = 13.90\ \text{ft/s}$$

Conservation of Momentum.   After impact we will assume A and B travel to the left. Applying the conservation of momentum to the system, Fig. 15–16c, we have

$$(\pm)\qquad m_B(v_B)_1 + m_A(v_A)_1 = m_B(v_B)_2 + m_A(v_A)_2$$

$$0 + \left(\frac{6\ \text{lb}}{32.2\ \text{ft/s}^2}\right)(13.90\ \text{ft/s}) = \left(\frac{18\ \text{lb}}{32.2\ \text{ft/s}^2}\right)(v_B)_2 + \left(\frac{6\ \text{lb}}{32.2\ \text{ft/s}^2}\right)(v_A)_2$$

$$(v_A)_2 = 13.90 - 3(v_B)_2 \qquad (1)$$

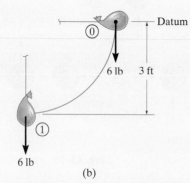

(b)

Coefficient of Restitution.   Realizing that for separation to occur after collision $(v_B)_2 > (v_A)_2$, Fig. 15–16c, we have

$$(\pm)\qquad e = \frac{(v_B)_2 - (v_A)_2}{(v_A)_1 - (v_B)_1}; \qquad 0.5 = \frac{(v_B)_2 - (v_A)_2}{13.90\ \text{ft/s} - 0}$$

$$(v_A)_2 = (v_B)_2 - 6.950 \qquad (2)$$

Solving Eqs. 1 and 2 simultaneously yields

$$(v_A)_2 = -1.74\ \text{ft/s} = 1.74\ \text{ft/s} \rightarrow \quad \text{and} \quad (v_B)_2 = 5.21\ \text{ft/s} \leftarrow \quad Ans.$$

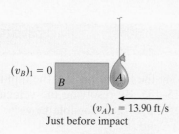

$(v_B)_1 = 0$

$(v_A)_1 = 13.90\ \text{ft/s}$

Just before impact

Loss of Energy.   Applying the principle of work and energy to the bag and box just before and just after collision, we have

$$\Sigma U_{1-2} = T_2 - T_1;$$

$$\Sigma U_{1-2} = \left[\frac{1}{2}\left(\frac{18\ \text{lb}}{32.2\ \text{ft/s}^2}\right)(5.21\ \text{ft/s})^2 + \frac{1}{2}\left(\frac{6\ \text{lb}}{32.2\ \text{ft/s}^2}\right)(1.74\ \text{ft/s})^2\right]$$

$$- \left[\frac{1}{2}\left(\frac{6\ \text{lb}}{32.2\ \text{ft/s}^2}\right)(13.9\ \text{ft/s})^2\right]$$

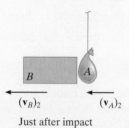

$(v_B)_2$ $(v_A)_2$

Just after impact

(c)

Fig. 15–16

$$\Sigma U_{1-2} = -10.1\ \text{ft}\cdot\text{lb} \qquad Ans.$$

NOTE: The energy loss occurs due to inelastic deformation during the collision.

EXAMPLE | 15.10

Ball *B* shown in Fig. 15–17a has a mass of 1.5 kg and is suspended from the ceiling by a 1-m-long elastic cord. If the cord is *stretched* downward 0.25 m and the ball is released from rest, determine how far the cord stretches after the ball rebounds from the ceiling. The stiffness of the cord is $k = 800$ N/m, and the coefficient of restitution between the ball and ceiling is $e = 0.8$. The ball makes a central impact with the ceiling.

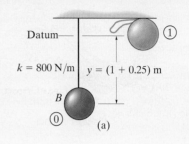
(a)

SOLUTION

First we must obtain the velocity of the ball *just before* it strikes the ceiling using energy methods, then consider the impulse and momentum between the ball and ceiling, and finally again use energy methods to determine the stretch in the cord.

Conservation of Energy. With the datum located as shown in Fig. 15–17a, realizing that initially $y = y_0 = (1 + 0.25)$ m $= 1.25$ m, we have

$$T_0 + V_0 = T_1 + V_1$$
$$\tfrac{1}{2}m(v_B)_0^2 - W_B y_0 + \tfrac{1}{2}ks^2 = \tfrac{1}{2}m(v_B)_1^2 + 0$$
$$0 - 1.5(9.81)\text{N}(1.25 \text{ m}) + \tfrac{1}{2}(800 \text{ N/m})(0.25 \text{ m})^2 = \tfrac{1}{2}(1.5 \text{ kg})(v_B)_1^2$$
$$(v_B)_1 = 2.968 \text{ m/s} \uparrow$$

The interaction of the ball with the ceiling will now be considered using the principles of impact.* Since an unknown portion of the mass of the ceiling is involved in the impact, the conservation of momentum for the ball–ceiling system will not be written. The "velocity" of this portion of ceiling is zero since it (or the earth) are assumed to remain at rest *both* before and after impact.

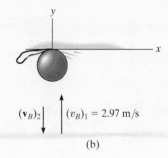
(b)

Coefficient of Restitution. Fig. 15–17b.

$$(+\uparrow) \quad e = \frac{(v_B)_2 - (v_A)_2}{(v_A)_1 - (v_B)_1}; \qquad 0.8 = \frac{(v_B)_2 - 0}{0 - 2.968 \text{ m/s}}$$
$$(v_B)_2 = -2.374 \text{ m/s} = 2.374 \text{ m/s} \downarrow$$

Conservation of Energy. The maximum stretch s_3 in the cord can be determined by again applying the conservation of energy equation to the ball just after collision. Assuming that $y = y_3 = (1 + s_3)$ m, Fig. 15–17c, then

$$T_2 + V_2 = T_3 + V_3$$
$$\tfrac{1}{2}m(v_B)_2^2 + 0 = \tfrac{1}{2}m(v_B)_3^2 - W_B y_3 + \tfrac{1}{2}ks_3^2$$
$$\tfrac{1}{2}(1.5 \text{ kg})(2.37 \text{ m/s})^2 = 0 - 9.81(1.5) \text{ N}(1 \text{ m} + s_3) + \tfrac{1}{2}(800 \text{ N/m})s_3^2$$
$$400s_3^2 - 14.715s_3 - 18.94 = 0$$

Solving this quadratic equation for the positive root yields

$$s_3 = 0.237 \text{ m} = 237 \text{ mm} \qquad \qquad Ans.$$

*The weight of the ball is considered a nonimpulsive force.

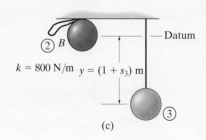
(c)

Fig. 15–17

EXAMPLE | **15.11**

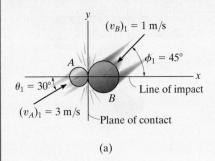

(a)

Two smooth disks A and B, having a mass of 1 kg and 2 kg, respectively, collide with the velocities shown in Fig. 15–18a. If the coefficient of restitution for the disks is $e = 0.75$, determine the x and y components of the final velocity of each disk just after collision.

SOLUTION
This problem involves *oblique impact*. Why? In order to solve it, we have established the x and y axes along the line of impact and the plane of contact, respectively, Fig. 15–18a.

Resolving each of the initial velocities into x and y components, we have

$$(v_{Ax})_1 = 3 \cos 30° = 2.598 \text{ m/s} \qquad (v_{Ay})_1 = 3 \sin 30° = 1.50 \text{ m/s}$$
$$(v_{Bx})_1 = -1 \cos 45° = -0.7071 \text{ m/s} \quad (v_{By})_1 = -1 \sin 45° = -0.7071 \text{ m/s}$$

The four unknown velocity components after collision are *assumed to act in the positive directions*, Fig. 15–18b. Since the impact occurs in the x direction (line of impact), the conservation of momentum for *both* disks can be applied in this direction. Why?

Conservation of "x" Momentum. In reference to the momentum diagrams, we have

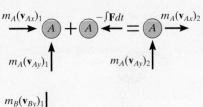

$$(\stackrel{+}{\rightarrow}) \qquad m_A(v_{Ax})_1 + m_B(v_{Bx})_1 = m_A(v_{Ax})_2 + m_B(v_{Bx})_2$$
$$1 \text{ kg}(2.598 \text{ m/s}) + 2 \text{ kg}(-0.707 \text{ m/s}) = 1 \text{ kg}(v_{Ax})_2 + 2 \text{ kg}(v_{Bx})_2$$
$$(v_{Ax})_2 + 2(v_{Bx})_2 = 1.184 \tag{1}$$

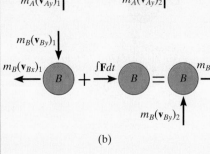

(b)

Coefficient of Restitution (x).

$$(\stackrel{+}{\rightarrow}) \qquad e = \frac{(v_{Bx})_2 - (v_{Ax})_2}{(v_{Ax})_1 - (v_{Bx})_1}; \quad 0.75 = \frac{(v_{Bx})_2 - (v_{Ax})_2}{2.598 \text{ m/s} - (-0.7071 \text{ m/s})}$$
$$(v_{Bx})_2 - (v_{Ax})_2 = 2.482 \tag{2}$$

Solving Eqs. 1 and 2 for $(v_{Ax})_2$ and $(v_{Bx})_2$ yields
$$(v_{Ax})_2 = -1.26 \text{ m/s} = 1.26 \text{ m/s} \leftarrow \quad (v_{Bx})_2 = 1.22 \text{ m/s} \rightarrow \qquad Ans.$$

Conservation of "y" Momentum. The momentum of *each disk* is *conserved* in the y direction (plane of contact), since the disks are smooth and therefore *no* external impulse acts in this direction. From Fig. 15–18b,

$$(+\uparrow) \; m_A(v_{Ay})_1 = m_A(v_{Ay})_2; \quad (v_{Ay})_2 = 1.50 \text{ m/s} \uparrow \qquad Ans.$$
$$(+\uparrow) \; m_B(v_{By})_1 = m_B(v_{By})_2; \quad (v_{By})_2 = -0.707 \text{ m/s} = 0.707 \text{ m/s} \downarrow \; Ans.$$

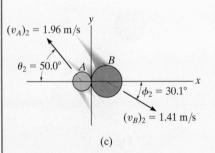

(c)

Fig. 15–18

NOTE: Show that when the velocity components are summed vectorially, one obtains the results shown in Fig. 15–18c.

FUNDAMENTAL PROBLEMS

F15–13. Determine the coefficient of restitution e between ball A and ball B. The velocities of A and B before and after the collision are shown.

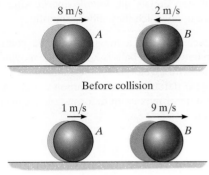

Before collision

After collision

F15–13

F15–14. The 15-Mg tank car A and 25-Mg freight car B travel towards each other with the velocities shown. If the coefficient of restitution between the bumpers is $e = 0.6$, determine the velocity of each car just after the collision.

F15–14

F15–15. The 30-lb package A has a speed of 5 ft/s when it enters the smooth ramp. As it slides down the ramp, it strikes the 80-lb package B which is initially at rest. If the coefficient of restitution between A and B is $e = 0.6$, determine the velocity of B just after the impact.

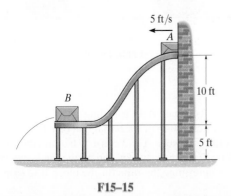

F15–15

F15–16. The ball strikes the smooth wall with a velocity of $(v_b)_1 = 20$ m/s. If the coefficient of restitution between the ball and the wall is $e = 0.75$, determine the velocity of the ball just after the impact.

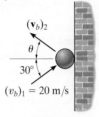

F15–16

F15–17. Disk A weighs 2 lb and slides on the smooth horizontal plane with a velocity of 3 ft/s. Disk B weighs 11 lb and is initially at rest. If after impact A has a velocity of 1 ft/s, parallel to the positive x axis, determine the speed of disk B after impact.

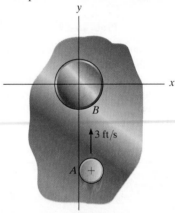

F15–17

F15–18. Two disks A and B each have a weight of 2 lb and the initial velocities shown just before they collide. If the coefficient of restitution is $e = 0.5$, determine their speeds just after impact.

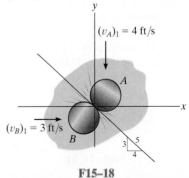

F15–18

PROBLEMS

15–58. A ball having a mass of 200 g is released from rest at a height of 400 mm above a very large fixed metal surface. If the ball rebounds to a height of 325 mm above the surface, determine the coefficient of restitution between the ball and the surface.

15–59. The 5-Mg truck and 2-Mg car are traveling with the free-rolling velocities shown just before they collide. After the collision, the car moves with a velocity of 15 km/h to the right *relative* to the truck. Determine the coefficient of restitution between the truck and car and the loss of energy due to the collision.

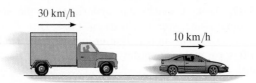

30 km/h

10 km/h

Prob. 15–59

*15–60. Disk A has a mass of 2 kg and is sliding forward on the *smooth* surface with a velocity $(v_A)_1 = 5$ m/s when it strikes the 4-kg disk B, which is sliding towards A at $(v_B)_1 = 2$ m/s, with direct central impact. If the coefficient of restitution between the disks is $e = 0.4$, compute the velocities of A and B just after collision.

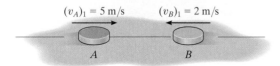

$(v_A)_1 = 5$ m/s $(v_B)_1 = 2$ m/s

A   B

Prob. 15–60

15–61. Block A has a mass of 3 kg and is sliding on a rough horizontal surface with a velocity $(v_A)_1 = 2$ m/s when it makes a direct collision with block B, which has a mass of 2 kg and is originally at rest. If the collision is perfectly elastic ($e = 1$), determine the velocity of each block just after collision and the distance between the blocks when they stop sliding. The coefficient of kinetic friction between the blocks and the plane is $\mu_k = 0.3$.

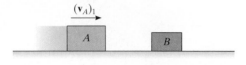

$(v_A)_1$

A   B

Prob. 15–61

15–62. If two disks A and B have the same mass and are subjected to direct central impact such that the collision is perfectly elastic ($e = 1$), prove that the kinetic energy before collision equals the kinetic energy after collision. The surface upon which they slide is smooth.

15–63. Each ball has a mass m and the coefficient of restitution between the balls is e. If they are moving towards one another with a velocity v, determine their speeds after collision. Also, determine their common velocity when they reach the state of maximum deformation. Neglect the size of each ball.

v v

A   B

Prob. 15–63

*15–64. The three balls each have a mass m. If A has a speed v just before a direct collision with B, determine the speed of C after collision. The coefficient of restitution between each pair of balls is e. Neglect the size of each ball.

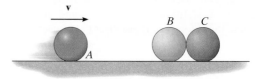

v

B   C

A

Prob. 15–64

15–65. A 1-lb ball A is traveling horizontally at 20 ft/s when it strikes a 10-lb block B that is at rest. If the coefficient of restitution between A and B is $e = 0.6$, and the coefficient of kinetic friction between the plane and the block is $\mu_k = 0.4$, determine the time for the block B to stop sliding.

15–66. If the girl throws the ball with a horizontal velocity of $v_A = 8$ ft/s, determine the distance d so that the ball bounces once on the smooth surface and then lands in the cup at C. Take $e = 0.8$.

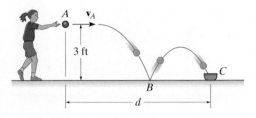

Prob. 15–66

15–67. The three balls each weigh 0.5 lb and have a coefficient of restitution of $e = 0.85$. If ball A is released from rest and strikes ball B and then ball B strikes ball C, determine the velocity of each ball after the second collision has occurred. The balls slide without friction.

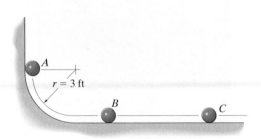

Prob. 15–67

*15–68. The girl throws the ball with a horizontal velocity of $v_1 = 8$ ft/s. If the coefficient of restitution between the ball and the ground is $e = 0.8$, determine (a) the velocity of the ball just after it rebounds from the ground and (b) the maximum height to which the ball rises after the first bounce.

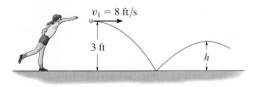

Prob. 15–68

15–69. A 300-g ball is kicked with a velocity of $v_A = 25$ m/s at point A as shown. If the coefficient of restitution between the ball and the field is $e = 0.4$, determine the magnitude and direction θ of the velocity of the rebounding ball at B.

Prob. 15–69

15–70. Two smooth spheres A and B each have a mass m. If A is given a velocity of v_0, while sphere B is at rest, determine the velocity of B just after it strikes the wall. The coefficient of restitution for any collision is e.

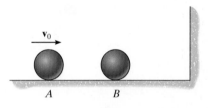

Prob. 15–70

15–71. It was observed that a tennis ball when served horizontally 7.5 ft above the ground strikes the smooth ground at B 20 ft away. Determine the initial velocity $\mathbf{v}_A$ of the ball and the velocity $\mathbf{v}_B$ (and θ) of the ball just after it strikes the court at B. Take $e = 0.7$.

*15–72. The tennis ball is struck with a horizontal velocity $\mathbf{v}_A$, strikes the smooth ground at B, and bounces upward at $\theta = 30°$. Determine the initial velocity $\mathbf{v}_A$, the final velocity $\mathbf{v}_B$, and the coefficient of restitution between the ball and the ground.

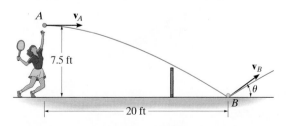

Probs. 15–71/72

15–73. The 1 lb ball is dropped from rest and falls a distance of 4 ft before striking the smooth plane at A. If $e = 0.8$, determine the distance d to where it again strikes the plane at B.

15–74. The 1 lb ball is dropped from rest and falls a distance of 4 ft before striking the smooth plane at A. If it rebounds and in $t = 0.5$ s again strikes the plane at B, determine the coefficient of restitution e between the ball and the plane. Also, what is the distance d?

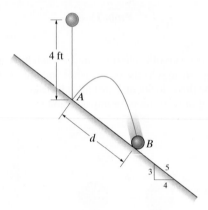

Probs. 15–73/74

15–75. The 1-kg ball is dropped from rest at point A, 2 m above the smooth plane. If the coefficient of restitution between the ball and the plane is $e = 0.6$, determine the distance d where the ball again strikes the plane.

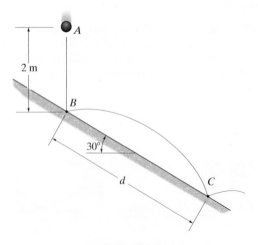

Prob. 15–75

*15–76. A ball of mass m is dropped vertically from a height h_0 above the ground. If it rebounds to a height of h_1, determine the coefficient of restitution between the ball and the ground.

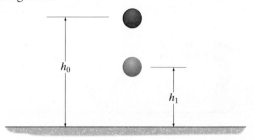

Prob. 15–76

15–77. The cue ball A is given an initial velocity $(v_A)_1 = 5$ m/s. If it makes a direct collision with ball B ($e = 0.8$), determine the velocity of B and the angle θ just after it rebounds from the cushion at C ($e' = 0.6$). Each ball has a mass of 0.4 kg. Neglect their size.

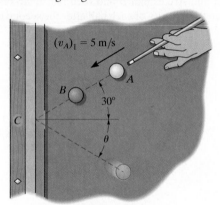

Prob. 15–77

15–78. Using a slingshot, the boy fires the 0.2-lb marble at the concrete wall, striking it at B. If the coefficient of restitution between the marble and the wall is $e = 0.5$, determine the speed of the marble after it rebounds from the wall.

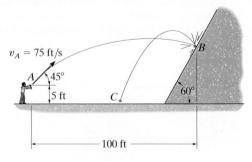

Prob. 15–78

15–79. The sphere of mass m falls and strikes the triangular block with a vertical velocity v. If the block rests on a smooth surface and has a mass $3\,m$, determine its velocity just after the collision. The coefficient of restitution is e.

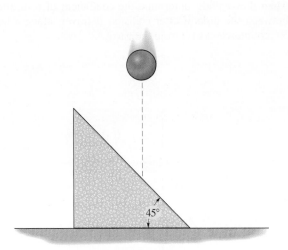

Prob. 15–79

15–81. The girl throws the 0.5-kg ball toward the wall with an initial velocity $v_A = 10\,\text{m/s}$. Determine (a) the velocity at which it strikes the wall at B, (b) the velocity at which it rebounds from the wall if the coefficient of restitution $e = 0.5$, and (c) the distance s from the wall to where it strikes the ground at C.

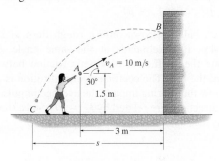

Prob. 15–81

15–82. The 20-lb box slides on the surface for which $\mu_k = 0.3$. The box has a velocity $v = 15\,\text{ft/s}$ when it is 2 ft from the plate. If it strikes the smooth plate, which has a weight of 10 lb and is held in position by an unstretched spring of stiffness $k = 400\,\text{lb/ft}$, determine the maximum compression imparted to the spring. Take $e = 0.8$ between the box and the plate. Assume that the plate slides smoothly.

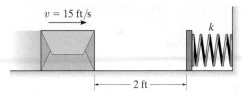

Prob. 15–82

15–83. Before a cranberry can make it to your dinner plate, it must pass a bouncing test which rates its quality. If cranberries having an $e \geq 0.8$ are to be accepted, determine the dimensions d and h for the barrier so that when a cranberry falls from rest at A it strikes the incline at B and bounces over the barrier at C.

*15–80. Block A, having a mass m, is released from rest, falls a distance h and strikes the plate B having a mass $2\,m$. If the coefficient of restitution between A and B is e, determine the velocity of the plate just after collision. The spring has a stiffness k.

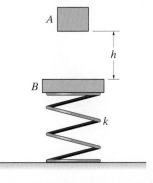

Prob. 15–80

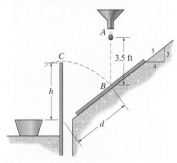

Prob. 15–83

15

*15–84.   A ball is thrown onto a rough floor at an angle θ. If it rebounds at an angle ϕ and the coefficient of kinetic friction is μ, determine the coefficient of restitution e. Neglect the size of the ball. *Hint:* Show that during impact, the average impulses in the x and y directions are related by $I_x = \mu I_y$. Since the time of impact is the same, $F_x \, \Delta t = \mu F_y \, \Delta t$ or $F_x = \mu F_y$.

15–85.   A ball is thrown onto a rough floor at an angle of $\theta = 45°$. If it rebounds at the same angle $\phi = 45°$, determine the coefficient of kinetic friction between the floor and the ball. The coefficient of restitution is $e = 0.6$. *Hint:* Show that during impact, the average impulses in the x and y directions are related by $I_x = \mu I_y$. Since the time of impact is the same, $F_x \, \Delta t = \mu F_y \, \Delta t$ or $F_x = \mu F_y$.

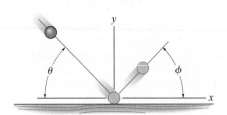

Probs. 15–84/85

15–86.   The "stone" A used in the sport of curling slides over the ice track and strikes another "stone" B as shown. If each "stone" is smooth and has a weight of 47 lb, and the coefficient of restitution between the "stones" is $e = 0.8$, determine their speeds just after collision. Initially A has a velocity of 8 ft/s and B is at rest. Neglect friction.

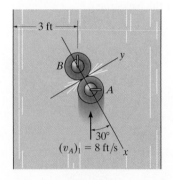

Prob. 15–86

15–87.   Two smooth disks A and B each have a mass of 0.5 kg. If both disks are moving with the velocities shown when they collide, determine their final velocities just after collision. The coefficient of restitution is $e = 0.75$.

*15–88.   Two smooth disks A and B each have a mass of 0.5 kg. If both disks are moving with the velocities shown when they collide, determine the coefficient of restitution between the disks if after collision B travels along a line, $30°$ counterclockwise from the y axis.

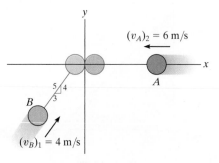

Probs. 15–87/88

15–89.   Two smooth disks A and B have the initial velocities shown just before they collide at O. If they have masses $m_A = 8$ kg and $m_B = 6$ kg, determine their speeds just after impact. The coefficient of restitution is $e = 0.5$.

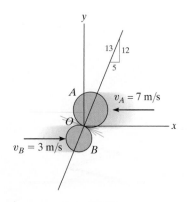

Prob. 15–89

15–90.   If disk A is sliding along the tangent to disk B and strikes B with a velocity $\mathbf{v}$, determine the velocity of B after the collision and compute the loss of kinetic energy during the collision. Neglect friction. Disk B is originally at rest. The coefficient of restitution is e, and each disk has the same size and mass m.

*15–92.   Two smooth coins A and B, each having the same mass, slide on a smooth surface with the motion shown. Determine the speed of each coin after collision if they move off along the dashed paths. *Hint:* Since the line of impact has not been defined, apply the conservation of momentum along the x and y axes, respectively.

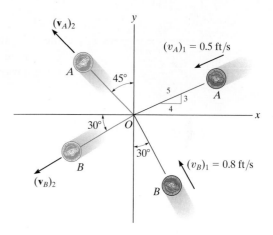

Prob. 15–92

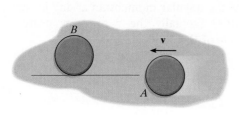

Prob. 15–90

15–91.   Two disks A and B weigh 2 lb and 5 lb, respectively. If they are sliding on the smooth horizontal plane with the velocities shown, determine their velocities just after impact. The coefficient of restitution between the disks is $e = 0.6$.

15–93.   Disks A and B have a mass of 15 kg and 10 kg, respectively. If they are sliding on a smooth horizontal plane with the velocities shown, determine their speeds just after impact. The coefficient of restitution between them is $e = 0.8$.

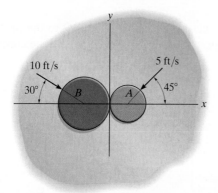

Prob. 15–91

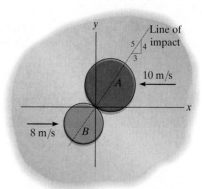

Prob. 15–93

15

15.5 Angular Momentum

The *angular momentum* of a particle about point O is defined as the "moment" of the particle's linear momentum about O. Since this concept is analogous to finding the moment of a force about a point, the angular momentum, $\mathbf{H}_O$, is sometimes referred to as the *moment of momentum*.

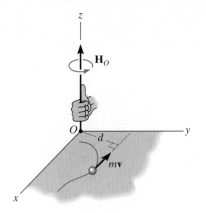

Fig. 15–19

Scalar Formulation.   If a particle moves along a curve lying in the x–y plane, Fig. 15–19, the angular momentum at any instant can be determined about point O (actually the z axis) by using a scalar formulation. The *magnitude* of $\mathbf{H}_O$ is

$$(H_O)_z = (d)(mv) \tag{15–12}$$

Here d is the moment arm or perpendicular distance from O to the line of action of $m\mathbf{v}$. Common units for $(H_O)_z$ are $\text{kg} \cdot \text{m}^2/\text{s}$ or $\text{slug} \cdot \text{ft}^2/\text{s}$. The *direction* of $\mathbf{H}_O$ is defined by the right-hand rule. As shown, the curl of the fingers of the right hand indicates the sense of rotation of $m\mathbf{v}$ about O, so that in this case the thumb (or $\mathbf{H}_O$) is directed perpendicular to the x–y plane along the $+z$ axis.

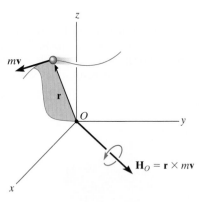

Fig. 15–20

Vector Formulation.   If the particle moves along a space curve, Fig. 15–20, the vector cross product can be used to determine the *angular momentum* about O. In this case

$$\mathbf{H}_O = \mathbf{r} \times m\mathbf{v} \tag{15–13}$$

Here $\mathbf{r}$ denotes a position vector drawn from point O to the particle. As shown in the figure, $\mathbf{H}_O$ is *perpendicular* to the shaded plane containing $\mathbf{r}$ and $m\mathbf{v}$.

In order to evaluate the cross product, $\mathbf{r}$ and $m\mathbf{v}$ should be expressed in terms of their Cartesian components, so that the angular momentum can be determined by evaluating the determinant:

$$\mathbf{H}_O = \begin{vmatrix} \mathbf{i} & \mathbf{j} & \mathbf{k} \\ r_x & r_y & r_z \\ mv_x & mv_y & mv_z \end{vmatrix} \tag{15–14}$$

15.6   Relation Between Moment of a Force and Angular Momentum

The moments about point O of all the forces acting on the particle in Fig. 15–21a can be related to the particle's angular momentum by applying the equation of motion. If the mass of the particle is constant, we may write

$$\Sigma \mathbf{F} = m\dot{\mathbf{v}}$$

The moments of the forces about point O can be obtained by performing a cross-product multiplication of each side of this equation by the position vector $\mathbf{r}$, which is measured from the x, y, z inertial frame of reference. We have

$$\Sigma \mathbf{M}_O = \mathbf{r} \times \Sigma \mathbf{F} = \mathbf{r} \times m\dot{\mathbf{v}}$$

From Appendix B, the derivative of $\mathbf{r} \times m\mathbf{v}$ can be written as

$$\dot{\mathbf{H}}_O = \frac{d}{dt}(\mathbf{r} \times m\mathbf{v}) = \dot{\mathbf{r}} \times m\mathbf{v} + \mathbf{r} \times m\dot{\mathbf{v}}$$

The first term on the right side, $\dot{\mathbf{r}} \times m\mathbf{v} = m(\dot{\mathbf{r}} \times \dot{\mathbf{r}}) = \mathbf{0}$, since the cross product of a vector with itself is zero. Hence, the above equation becomes

$$\boxed{\Sigma \mathbf{M}_O = \dot{\mathbf{H}}_O} \qquad (15\text{--}15)$$

which states that *the resultant moment about point O of all the forces acting on the particle is equal to the time rate of change of the particle's angular momentum about point O.* This result is similar to Eq. 15–1, i.e.,

$$\boxed{\Sigma \mathbf{F} = \dot{\mathbf{L}}} \qquad (15\text{--}16)$$

Here $\mathbf{L} = m\mathbf{v}$, so that *the resultant force acting on the particle is equal to the time rate of change of the particle's linear momentum.*

From the derivations, it is seen that Eqs. 15–15 and 15–16 are actually another way of stating Newton's second law of motion. In other sections of this book it will be shown that these equations have many practical applications when extended and applied to problems involving either a system of particles or a rigid body.

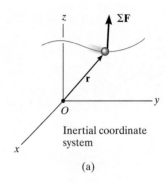

Inertial coordinate system

(a)

Fig. 15–21

15

System of Particles.   An equation having the same form as Eq. 15–15 may be derived for the system of particles shown in Fig. 15–21*b*. The forces acting on the arbitrary *i*th particle of the system consist of a resultant *external force* $\mathbf{F}_i$ and a resultant *internal force* $\mathbf{f}_i$. Expressing the moments of these forces about point O, using the form of Eq. 15–15, we have

$$(\mathbf{r}_i \times \mathbf{F}_i) + (\mathbf{r}_i \times \mathbf{f}_i) = (\dot{\mathbf{H}}_i)_O$$

Here $(\dot{\mathbf{H}}_i)_O$ is the time rate of change in the angular momentum of the *i*th particle about O. Similar equations can be written for each of the other particles of the system. When the results are summed vectorially, the result is

$$\Sigma(\mathbf{r}_i \times \mathbf{F}_i) + \Sigma(\mathbf{r}_i \times \mathbf{f}_i) = \Sigma(\dot{\mathbf{H}}_i)_O$$

The second term is zero since the internal forces occur in equal but opposite collinear pairs, and hence the moment of each pair about point O is zero. Dropping the index notation, the above equation can be written in a simplified form as

$$\Sigma\mathbf{M}_O = \dot{\mathbf{H}}_O \tag{15–17}$$

which states that *the sum of the moments about point O of all the external forces acting on a system of particles is equal to the time rate of change of the total angular momentum of the system about point O.* Although O has been chosen here as the origin of coordinates, it actually can represent any *fixed point* in the inertial frame of reference.

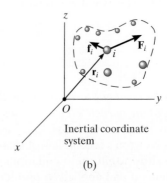

Inertial coordinate system

(b)

Fig. 15–21 (cont.)

EXAMPLE | 15.12

The box shown in Fig. 15–22a has a mass m and travels down the smooth circular ramp such that when it is at the angle θ it has a speed v. Determine its angular momentum about point O at this instant and the rate of increase in its speed, i.e., a_t.

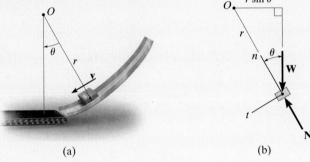

(a)   (b)

Fig. 15–22

SOLUTION

Since $\mathbf{v}$ is tangent to the path, applying Eq. 15–12 the angular momentum is

$$H_O = rmv \, \circlearrowleft \qquad\qquad Ans.$$

The rate of increase in its speed (dv/dt) can be found by applying Eq. 15–15. From the free-body diagram of the box, Fig. 15–22b, it can be seen that only the weight $W = mg$ contributes a moment about point O. We have

$$\zeta + \Sigma M_O = \dot{H}_O; \qquad mg(r \sin \theta) = \frac{d}{dt}(rmv)$$

Since r and m are constant,

$$mgr \sin \theta = rm \frac{dv}{dt}$$

$$\frac{dv}{dt} = g \sin \theta \qquad\qquad Ans.$$

NOTE: This same result can, of course, be obtained from the equation of motion applied in the tangential direction, Fig. 15–22b, i.e.,

$$+\swarrow \Sigma F_t = ma_t; \qquad mg \sin \theta = m\left(\frac{dv}{dt}\right)$$

$$\frac{dv}{dt} = g \sin \theta \qquad\qquad Ans.$$

15.7   Principle of Angular Impulse and Momentum

Principle of Angular Impulse and Momentum.   If Eq. 15–15 is rewritten in the form $\Sigma \mathbf{M}_O\, dt = d\mathbf{H}_O$ and integrated, assuming that at time $t = t_1$, $\mathbf{H}_O = (\mathbf{H}_O)_1$ and at time $t = t_2$, $\mathbf{H}_O = (\mathbf{H}_O)_2$, we have

$$\Sigma \int_{t_1}^{t_2} \mathbf{M}_O\, dt = (\mathbf{H}_O)_2 - (\mathbf{H}_O)_1$$

or

$$(\mathbf{H}_O)_1 + \Sigma \int_{t_1}^{t_2} \mathbf{M}_O\, dt = (\mathbf{H}_O)_2 \qquad (15\text{–}18)$$

This equation is referred to as the *principle of angular impulse and momentum*. The initial and final angular momenta $(\mathbf{H}_O)_1$ and $(\mathbf{H}_O)_2$ are defined as the moment of the linear momentum of the particle $(\mathbf{H}_O = \mathbf{r} \times m\mathbf{v})$ at the instants t_1 and t_2, respectively. The second term on the left side, $\Sigma \int \mathbf{M}_O\, dt$, is called the *angular impulse*. It is determined by integrating, with respect to time, the moments of all the forces acting on the particle over the time period t_1 to t_2. Since the moment of a force about point O is $\mathbf{M}_O = \mathbf{r} \times \mathbf{F}$, the angular impulse may be expressed in vector form as

$$\text{angular impulse} = \int_{t_1}^{t_2} \mathbf{M}_O\, dt = \int_{t_1}^{t_2} (\mathbf{r} \times \mathbf{F})\, dt \qquad (15\text{–}19)$$

Here $\mathbf{r}$ is a position vector which extends from point O to any point on the line of action of $\mathbf{F}$.

In a similar manner, using Eq. 15–18, the principle of angular impulse and momentum for a system of particles may be written as

$$\Sigma(\mathbf{H}_O)_1 + \Sigma \int_{t_1}^{t_2} \mathbf{M}_O\, dt = \Sigma(\mathbf{H}_O)_2 \qquad (15\text{–}20)$$

Here the first and third terms represent the angular momenta of all the particles $[\Sigma \mathbf{H}_O = \Sigma(\mathbf{r}_i \times m\mathbf{v}_i)]$ at the instants t_1 and t_2. The second term is the sum of the angular impulses given to all the particles from t_1 to t_2. Recall that these impulses are created only by the moments of the external forces acting on the system where, for the ith particle, $\mathbf{M}_O = \mathbf{r}_i \times \mathbf{F}_i$.

Vector Formulation.   Using impulse and momentum principles, it is therefore possible to write two equations which define the particle's motion, namely, Eqs. 15–3 and Eqs. 15–18, restated as

$$
\begin{aligned}
m\mathbf{v}_1 + \Sigma \int_{t_1}^{t_2} \mathbf{F}\, dt &= m\mathbf{v}_2 \\[2mm]
(\mathbf{H}_O)_1 + \Sigma \int_{t_1}^{t_2} \mathbf{M}_O\, dt &= (\mathbf{H}_O)_2
\end{aligned}
\tag{15–21}
$$

Scalar Formulation.   In general, the above equations can be expressed in x, y, z component form, yielding a total of six scalar equations. If the particle is confined to move in the x–y plane, three scalar equations can be written to express the motion, namely,

$$
\begin{aligned}
m(v_x)_1 + \Sigma \int_{t_1}^{t_2} F_x\, dt &= m(v_x)_2 \\[2mm]
m(v_y)_1 + \Sigma \int_{t_1}^{t_2} F_y\, dt &= m(v_y)_2 \\[2mm]
(H_O)_1 + \Sigma \int_{t_1}^{t_2} M_O\, dt &= (H_O)_2
\end{aligned}
\tag{15–22}
$$

The first two of these equations represent the principle of linear impulse and momentum in the x and y directions, which has been discussed in Sec. 15–1, and the third equation represents the principle of angular impulse and momentum about the z axis.

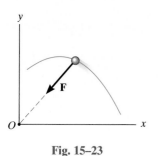

Fig. 15–23

Conservation of Angular Momentum.

When the angular impulses acting on a particle are all zero during the time t_1 to t_2, Eq. 15–18 reduces to the following simplified form:

$$(\mathbf{H}_O)_1 = (\mathbf{H}_O)_2 \qquad (15\text{–}23)$$

This equation is known as the *conservation of angular momentum*. It states that from t_1 to t_2 the particle's angular momentum remains constant. Obviously, if no external impulse is applied to the particle, both linear and angular momentum will be conserved. In some cases, however, the particle's angular momentum will be conserved and linear momentum may not. An example of this occurs when the particle is subjected *only* to a *central force* (see Sec. 13–7). As shown in Fig. 15–23, the impulsive central force $\mathbf{F}$ is always directed toward point O as the particle moves along the path. Hence, the angular impulse (moment) created by $\mathbf{F}$ about the z axis is always zero, and therefore angular momentum of the particle is conserved about this axis.

From Eq. 15–20, we can also write the conservation of angular momentum for a system of particles as

$$\Sigma(\mathbf{H}_O)_1 = \Sigma(\mathbf{H}_O)_2 \qquad (15\text{–}24)$$

In this case the summation must include the angular momenta of all particles in the system.

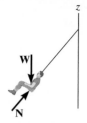

Provided air resistance is neglected, the passengers on this amusement-park ride are subjected to a conservation of angular momentum about the z axis of rotation. As shown on the free-body diagram, the line of action of the normal force $\mathbf{N}$ of the seat on the passenger passes through this axis, and the passenger's weight $\mathbf{W}$ is parallel to it. Thus, no angular impulse acts around the z axis.

Procedure for Analysis

When applying the principles of angular impulse and momentum, or the conservation of angular momentum, it is suggested that the following procedure be used.

Free-Body Diagram.

- Draw the particle's free-body diagram in order to determine any axis about which angular momentum may be conserved. For this to occur, the moments of all the forces (or impulses) must either be parallel or pass through the axis so as to create zero moment throughout the time period t_1 to t_2.

- The direction and sense of the particle's initial and final velocities should also be established.

- An alternative procedure would be to draw the impulse and momentum diagrams for the particle.

Momentum Equations.

- Apply the principle of angular impulse and momentum, $(\mathbf{H}_O)_1 + \Sigma \int_{t_1}^{t_2} \mathbf{M}_O \, dt = (\mathbf{H}_O)_2$, or if appropriate, the conservation of angular momentum, $(\mathbf{H}_O)_1 = (\mathbf{H}_O)_2$.

EXAMPLE | **15.13**

The 1.5-Mg car travels along the circular road as shown in Fig. 15–24a. If the traction force of the wheels on the road is $F = (150t^2)$ N, where t is in seconds, determine the speed of the car when $t = 5$ s. The car initially travels with a speed of 5 m/s. Neglect the size of the car.

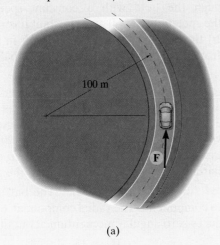

(a)

Free-Body Diagram.   The free-body diagram of the car is shown in Fig. 15–24b. If we apply the principle of angular impulse and momentum about the z axis, then the angular impulse created by the weight, normal force, and radial frictional force will be eliminated since they act parallel to the axis or pass through it.

Principle of Angular Impulse and Momentum.

$$(H_z)_1 + \Sigma \int_{t_1}^{t_2} M_z \, dt = (H_z)_2$$

$$rm_c(v_c)_1 + \int_{t_1}^{t_2} rF \, dt = rm_c(v_c)_2$$

$$(100 \text{ m})(1500 \text{ kg})(5 \text{ m/s}) + \int_0^{5 \text{ s}} (100 \text{ m})[(150t^2) \text{ N}] \, dt$$

$$= (100 \text{ m})(1500 \text{ kg})(v_c)_2$$

$$750(10^3) + 5000t^3 \Big|_0^{5 \text{ s}} = 150(10^3)(v_c)_2$$

$$(v_c)_2 = 9.17 \text{ m/s} \qquad Ans.$$

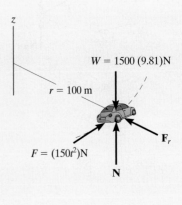

(b)

Fig. 15–24

EXAMPLE 15.14

(a)

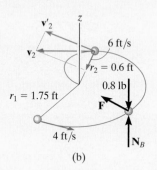

(b)

Fig. 15–25

The 0.8-lb ball B, shown in Fig. 15–25a, is attached to a cord which passes through a hole at A in a smooth table. When the ball is $r_1 = 1.75$ ft from the hole, it is rotating around in a circle such that its speed is $v_1 = 4$ ft/s. By applying the force $\mathbf{F}$ the cord is pulled downward through the hole with a constant speed $v_c = 6$ ft/s. Determine (a) the speed of the ball at the instant it is $r_2 = 0.6$ ft from the hole, and (b) the amount of work done by $\mathbf{F}$ in shortening the radial distance from r_1 to r_2. Neglect the size of the ball.

SOLUTION

Part (a) Free-Body Diagram. As the ball moves from r_1 to r_2, Fig. 15–25b, the cord force $\mathbf{F}$ on the ball always passes through the z axis, and the weight and $\mathbf{N}_B$ are parallel to it. Hence the moments, or angular impulses created by these forces, are all *zero* about this axis. Therefore, angular momentum is conserved about the z axis.

Conservation of Angular Momentum. The ball's velocity $\mathbf{v}_2$ is resolved into two components. The radial component, 6 ft/s, is known; however, it produces zero angular momentum about the z axis. Thus,

$$\mathbf{H}_1 = \mathbf{H}_2$$

$$r_1 m_B v_1 = r_2 m_B v_2'$$

$$1.75 \text{ ft} \left(\frac{0.8 \text{ lb}}{32.2 \text{ ft/s}^2} \right) 4 \text{ ft/s} = 0.6 \text{ ft} \left(\frac{0.8 \text{ lb}}{32.2 \text{ ft/s}^2} \right) v_2'$$

$$v_2' = 11.67 \text{ ft/s}$$

The speed of the ball is thus

$$v_2 = \sqrt{(11.67 \text{ ft/s})^2 + (6 \text{ ft/s})^2}$$

$$= 13.1 \text{ ft/s}$$

Part (b). The only force that does work on the ball is $\mathbf{F}$. (The normal force and weight do not move vertically.) The initial and final kinetic energies of the ball can be determined so that from the principle of work and energy we have

$$T_1 + \Sigma U_{1-2} = T_2$$

$$\frac{1}{2} \left(\frac{0.8 \text{ lb}}{32.2 \text{ ft/s}^2} \right) (4 \text{ ft/s})^2 + U_F = \frac{1}{2} \left(\frac{0.8 \text{ lb}}{32.2 \text{ ft/s}^2} \right) (13.1 \text{ ft/s})^2$$

$$U_F = 1.94 \text{ ft} \cdot \text{lb} \qquad \qquad Ans.$$

NOTE: The force F is not constant because the normal component of acceleration, $a_n = v^2/r$, changes as r changes.

EXAMPLE | 15.15

The 2-kg disk shown in Fig. 15–26a rests on a smooth horizontal surface and is attached to an elastic cord that has a stiffness $k_c = 20 \text{ N/m}$ and is initially unstretched. If the disk is given a velocity $(v_D)_1 = 1.5 \text{ m/s}$, perpendicular to the cord, determine the rate at which the cord is being stretched and the speed of the disk at the instant the cord is stretched 0.2 m.

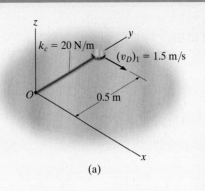

(a)

SOLUTION

Free-Body Diagram.   After the disk has been launched, it slides along the path shown in Fig. 15–26b. By inspection, angular momentum about point O (or the z axis) is *conserved*, since none of the forces produce an angular impulse about this axis. Also, when the distance is 0.7 m, only the transverse component $(v'_D)_2$ produces angular momentum of the disk about O.

Conservation of Angular Momentum.   The component $(v'_D)_2$ can be obtained by applying the conservation of angular momentum about O (the z axis).

$$(H_O)_1 = (H_O)_2$$

$$r_1 m_D (v_D)_1 = r_2 m_D (v'_D)_2$$

$$0.5 \text{ m } (2 \text{ kg})(1.5 \text{ m/s}) = 0.7 \text{ m}(2 \text{ kg})(v'_D)_2$$

$$(v'_D)_2 = 1.071 \text{ m/s}$$

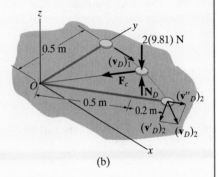

(b)

Fig. 15–26

Conservation of Energy.   The speed of the disk can be obtained by applying the conservation of energy equation at the point where the disk was launched and at the point where the cord is stretched 0.2 m.

$$T_1 + V_1 = T_2 + V_2$$

$$\tfrac{1}{2} m_D (v_D)_1^2 + \tfrac{1}{2} k x_1^2 = \tfrac{1}{2} m_D (v_D)_2^2 + \tfrac{1}{2} k x_2^2$$

$$\tfrac{1}{2}(2 \text{ kg})(1.5 \text{ m/s})^2 + 0 = \tfrac{1}{2}(2 \text{ kg})(v_D)_2^2 + \tfrac{1}{2}(20 \text{ N/m})(0.2 \text{ m})^2$$

$$(v_D)_2 = 1.360 \text{ m/s} = 1.36 \text{ m/s} \qquad Ans.$$

Having determined $(v_D)_2$ and its component $(v'_D)_2$, the rate of stretch of the cord, or radial component, $(v''_D)_2$ is determined from the Pythagorean theorem,

$$(v''_D)_2 = \sqrt{(v_D)_2^2 - (v'_D)_2^2}$$

$$= \sqrt{(1.360 \text{ m/s})^2 - (1.071 \text{ m/s})^2}$$

$$= 0.838 \text{ m/s} \qquad Ans.$$

FUNDAMENTAL PROBLEMS

F15–19. The 2-kg particle A has the velocity shown. Determine its angular momentum $\mathbf{H}_O$ about point O.

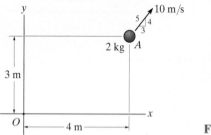

F15–19

F15–20. The 2-kg particle A has the velocity shown. Determine its angular momentum $\mathbf{H}_P$ about point P.

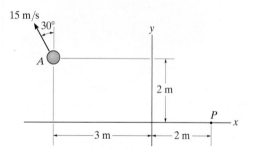

F15–20

F15–21. Initially the 5-kg block is rotating with a constant speed of 2 m/s around the circular path centered at O on the smooth horizontal plane. If a constant tangential force $F = 5$ N is applied to the block, determine its speed when $t = 3$ s. Neglect the size of the block.

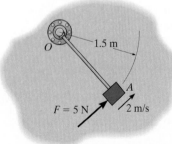

F15–21

F15–22. The 5-kg block is rotating around the circular path centered at O on the smooth horizontal plane when it is subjected to the force $F = (10t)$ N, where t is in seconds. If the block starts from rest, determine its speed when $t = 4$ s. Neglect the size of the block. The force maintains the same constant angle tangent to the path.

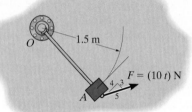

F15–22

F15–23. The 2-kg sphere is attached to the light rigid rod, which rotates in the *horizontal plane* centered at O. If the system is subjected to a couple moment $M = (0.9t^2)$ N $\cdot$ m, where t is in seconds, determine the speed of the sphere at the instant $t = 5$ s starting from rest.

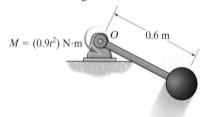

F15–23

F15–24. Two identical 10-kg spheres are attached to the light rigid rod, which rotates in the horizontal plane centered at pin O. If the spheres are subjected to tangential forces of $P = 10$ N, and the rod is subjected to a couple moment $M = (8t)$ N $\cdot$ m, where t is in seconds, determine the speed of the spheres at the instant $t = 4$ s. The system starts from rest. Neglect the size of the spheres.

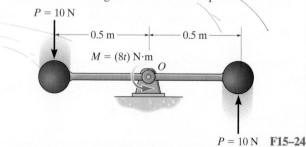

$P = 10$ N   **F15–24**

PROBLEMS

15–94. Determine the angular momentum $\mathbf{H}_O$ of the particle about point O.

15–97. Determine the total angular momentum $\mathbf{H}_O$ for the system of three particles about point O. All the particles are moving in the x–y plane.

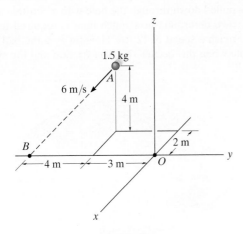

Prob. 15–94

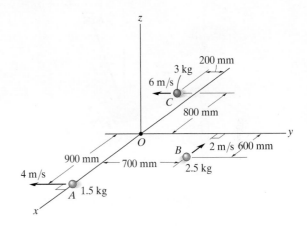

Prob. 15–97

15–95. Determine the angular momentum $\mathbf{H}_O$ of the particle about point O.

*15–96.** Determine the angular momentum $\mathbf{H}_P$ of the particle about point P.

15–98. Determine the angular momentum $\mathbf{H}_O$ of each of the two particles about point O. Use a scalar solution.

15–99. Determine the angular momentum $\mathbf{H}_P$ of each of the two particles about point P. Use a scalar solution.

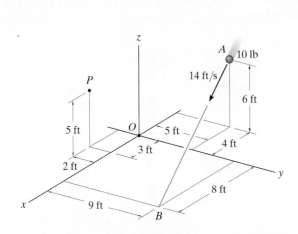

Probs. 15–95/96

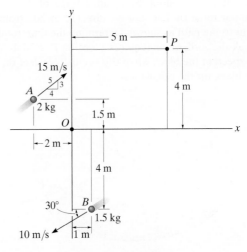

Probs. 15–98/99

*15–100. The small cylinder C has a mass of 10 kg and is attached to the end of a rod whose mass may be neglected. If the frame is subjected to a couple $M = (8t^2 + 5)$ N·m, where t is in seconds, and the cylinder is subjected to a force of 60 N, which is always directed as shown, determine the speed of the cylinder when $t = 2$ s. The cylinder has a speed $v_0 = 2$ m/s when $t = 0$.

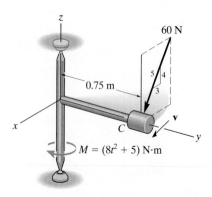

Prob. 15–100

15–101. The 10-lb block rests on a surface for which $\mu_k = 0.5$. It is acted upon by a radial force of 2 lb and a horizontal force of 7 lb, always directed at 30° from the tangent to the path as shown. If the block is initially moving in a circular path with a speed $v_1 = 2$ ft/s at the instant the forces are applied, determine the time required before the tension in cord AB becomes 20 lb. Neglect the size of the block for the calculation.

15–102. The 10-lb block is originally at rest on the smooth surface. It is acted upon by a radial force of 2 lb and a horizontal force of 7 lb, always directed at 30° from the tangent to the path as shown. Determine the time required to break the cord, which requires a tension $T = 30$ lb. What is the speed of the block when this occurs? Neglect the size of the block for the calculation.

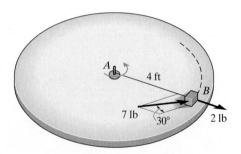

Probs. 15–101/102

15–103. A 4-lb ball B is traveling around in a circle of radius $r_1 = 3$ ft with a speed $(v_B)_1 = 6$ ft/s. If the attached cord is pulled down through the hole with a constant speed $v_r = 2$ ft/s, determine the ball's speed at the instant $r_2 = 2$ ft. How much work has to be done to pull down the cord? Neglect friction and the size of the ball.

*15–104. A 4-lb ball B is traveling around in a circle of radius $r_1 = 3$ ft with a speed $(v_B)_1 = 6$ ft/s. If the attached cord is pulled down through the hole with a constant speed $v_r = 2$ ft/s, determine how much time is required for the ball to reach a speed of 12 ft/s. How far r_2 is the ball from the hole when this occurs? Neglect friction and the size of the ball.

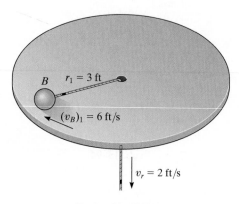

Probs. 15–103/104

15–105. The four 5-lb spheres are rigidly attached to the crossbar frame having a negligible weight. If a couple moment $M = (0.5t + 0.8)$ lb·ft, where t is in seconds, is applied as shown, determine the speed of each of the spheres in 4 seconds starting from rest. Neglect the size of the spheres.

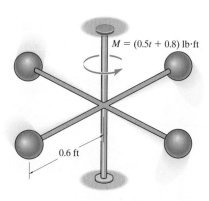

Prob. 15–105

15–106. A small particle having a mass m is placed inside the semicircular tube. The particle is placed at the position shown and released. Apply the principle of angular momentum about point O ($\Sigma M_O = \dot{H}_O$), and show that the motion of the particle is governed by the differential equation $\ddot{\theta} + (g/R) \sin\theta = 0$.

*15–108. A child having a mass of 50 kg holds her legs up as shown as she swings downward from rest at $\theta_1 = 30°$. Her center of mass is located at point G_1. When she is at the bottom position $\theta = 0°$, she *suddenly* lets her legs come down, shifting her center of mass to position G_2. Determine her speed in the upswing due to this sudden movement and the angle θ_2 to which she swings before momentarily coming to rest. Treat the child's body as a particle.

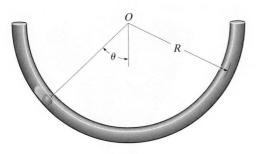

Prob. 15–106

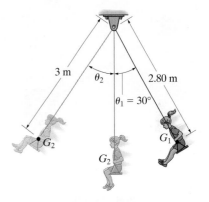

Prob. 15–108

15–107. The ball B has a weight of 5 lb and is originally rotating in a circle. As shown, the cord AB has a length of 3 ft and passes through the hole A, which is 2 ft above the plane of motion. If 1.5 ft of cord is pulled through the hole, determine the speed of the ball when it moves in a circular path at C.

15–109. The 150-lb car of an amusement park ride is connected to a rotating telescopic boom. When $r = 15$ ft, the car is moving on a horizontal circular path with a speed of 30 ft/s. If the boom is shortened at a rate of 3 ft/s, determine the speed of the car when $r = 10$ ft. Also, find the work done by the axial force $\mathbf{F}$ along the boom. Neglect the size of the car and the mass of the boom.

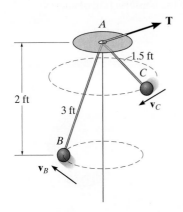

Prob. 15–107

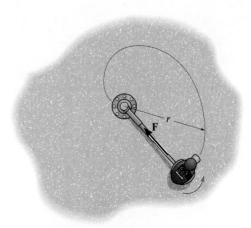

Prob. 15–109

15–110. An amusement park ride consists of a car which is attached to the cable OA. The car rotates in a horizontal circular path and is brought to a speed $v_1 = 4$ ft/s when $r = 12$ ft. The cable is then pulled in at the constant rate of 0.5 ft/s. Determine the speed of the car in 3 s.

*15–112. A small block having a mass of 0.1 kg is given a horizontal velocity of $v_1 = 0.4$ m/s when $r_1 = 500$ mm. It slides along the smooth conical surface. Determine the distance h it must descend for it to reach a speed of $v_2 = 2$ m/s. Also, what is the angle of descent θ, that is, the angle measured from the horizontal to the tangent of the path?

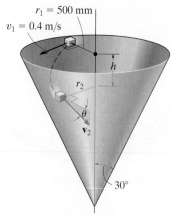

Prob. 15–112

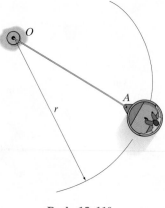

Prob. 15–110

15–111. The 800-lb roller-coaster car starts from rest on the track having the shape of a cylindrical helix. If the helix descends 8 ft for every one revolution, determine the speed of the car in $t = 4$ s. Also, how far has the car descended in this time? Neglect friction and the size of the car.

15–113. An earth satellite of mass 700 kg is launched into a free-flight trajectory about the earth with an initial speed of $v_A = 10$ km/s when the distance from the center of the earth is $r_A = 15$ Mm. If the launch angle at this position is $\phi_A = 70°$, determine the speed v_B of the satellite and its closest distance r_B from the center of the earth. The earth has a mass $M_e = 5.976(10^{24})$ kg. *Hint:* Under these conditions, the satellite is subjected only to the earth's gravitational force, $F = GM_e m_s/r^2$, Eq. 13–1. For part of the solution, use the conservation of energy.

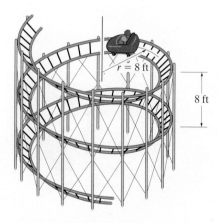

Prob. 15–111

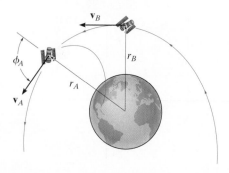

Prob. 15–113

15.8   Steady Flow of a Fluid Stream

Up to this point we have restricted our study of impulse and momentum principles to a system of particles contained within a *closed volume*. In this section, however, we will apply the principle of impulse and momentum to the steady mass flow of fluid particles entering into and then out of a *control volume*. This volume is defined as a region in space where fluid particles can flow into or out of a region. The size and shape of the control volume is frequently made to coincide with the solid boundaries and openings of a pipe, turbine, or pump. Provided the flow of the fluid into the control volume is equal to the flow out, then the flow can be classified as *steady flow*.

Principle of Impulse and Momentum.   Consider the steady flow of a fluid stream in Fig. 15–27a that passes through a pipe. The region within the pipe and its openings will be taken as the control volume. As shown, the fluid flows into and out of the control volume with velocities $\mathbf{v}_A$ and $\mathbf{v}_B$, respectively. The change in the direction of the fluid flow within the control volume is caused by an impulse produced by the resultant external force exerted on the control surface by the wall of the pipe. This resultant force can be determined by applying the principle of impulse and momentum to the control volume.

(a)

Fig. 15–27

The conveyor belt must supply frictional forces to the gravel that falls upon it in order to change the momentum of the gravel stream, so that it begins to travel along the belt.

The air on one side of this fan is essentially at rest, and as it passes through the blades its momentum is increased. To change the momentum of the air flow in this manner, the blades must exert a horizontal thrust on the air stream. As the blades turn faster, the equal but opposite thrust of the air on the blades could overcome the rolling resistance of the wheels on the ground and begin to move the frame of the fan.

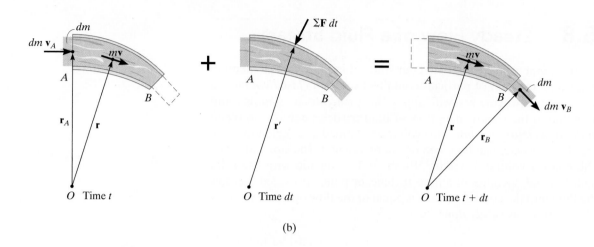

(b)

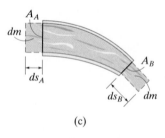

(c)

Fig. 15–27 (cont.)

As indicated in Fig. 15–27b, a small amount of fluid having a mass dm is about to enter the control volume through opening A with a velocity of $\mathbf{v}_A$ at time t. Since the flow is considered steady, at time $t + dt$, the same amount of fluid will leave the control volume through opening B with a velocity $\mathbf{v}_B$. The momenta of the fluid entering and leaving the control volume are therefore $dm\,\mathbf{v}_A$ and $dm\,\mathbf{v}_B$, respectively. Also, during the time dt, the momentum of the fluid mass within the control volume remains constant and is denoted as $m\mathbf{v}$. As shown on the center diagram, the resultant external force exerted on the control volume produces the impulse $\Sigma\mathbf{F}\,dt$. If we apply the principle of linear impulse and momentum, we have

$$dm\,\mathbf{v}_A + m\mathbf{v} + \Sigma\mathbf{F}\,dt = dm\,\mathbf{v}_B + m\mathbf{v}$$

If $\mathbf{r}$, $\mathbf{r}_A$, $\mathbf{r}_B$ are position vectors measured from point O to the geometric centers of the control volume and the openings at A and B, Fig. 15–27b, then the principle of angular impulse and momentum about O becomes

$$\mathbf{r}_A \times dm\,\mathbf{v}_A + \mathbf{r} \times m\mathbf{v} + \mathbf{r}' \times \Sigma\mathbf{F}\,dt = \mathbf{r} \times m\mathbf{v} + \mathbf{r}_B \times dm\,\mathbf{v}_B$$

Dividing both sides of the above two equations by dt and simplifying, we get

$$\Sigma\mathbf{F} = \frac{dm}{dt}(\mathbf{v}_B - \mathbf{v}_A) \tag{15–25}$$

$$\Sigma\mathbf{M}_O = \frac{dm}{dt}(\mathbf{r}_B \times \mathbf{v}_B - \mathbf{r}_A \times \mathbf{v}_A) \tag{15–26}$$

The term dm/dt is called the *mass flow*. It indicates the constant amount of fluid which flows either into or out of the control volume per unit of time. If the cross-sectional areas and densities of the fluid at the entrance A are A_A, ρ_A and at exit B, A_B, ρ_B, Fig. 15–27c, then for an incompressible fluid, the *continuity of mass* requires $dm = \rho dV = \rho_A(ds_A A_A) = \rho_B(ds_B A_B)$. Hence, during the time dt, since $v_A = ds_A/dt$ and $v_B = ds_B/dt$, we have $dm/dt = \rho_A v_A A_A = \rho_B v_B A_B$ or in general,

$$\boxed{\frac{dm}{dt} = \rho v A = \rho Q} \qquad (15\text{–}27)$$

The term $Q = vA$ measures the volume of fluid flow per unit of time and is referred to as the *discharge* or the *volumetric flow*.

Procedure for Analysis

Problems involving steady flow can be solved using the following procedure.

Kinematic Diagram.

- Identify the control volume. If it is *moving*, a *kinematic diagram* may be helpful for determining the entrance and exit velocities of the fluid flowing into and out of its openings since a *relative-motion analysis* of velocity will be involved.

- The measurement of velocities v_A and v_B must be made by an observer fixed in an inertial frame of reference.

- Once the velocity of the fluid flowing into the control volume is determined, the mass flow is calculated using Eq. 15–27.

Free-Body Diagram.

- Draw the free-body diagram of the control volume in order to establish the forces $\Sigma\mathbf{F}$ that act on it. These forces will include the support reactions, the weight of all solid parts and the fluid contained within the control volume, and the static gauge pressure forces of the fluid on the entrance and exit sections.* The gauge pressure is the pressure measured above atmospheric pressure, and so if an opening is exposed to the atmosphere, the gauge pressure there will be zero.

Equations of Steady Flow.

- Apply the equations of steady flow, Eq. 15–25 and 15–26, using the appropriate components of velocity and force shown on the kinematic and free-body diagrams.

* In the SI system, pressure is measured using the pascal (Pa), where $1\text{Pa} = 1\text{ N/m}^2$.

EXAMPLE 15.16

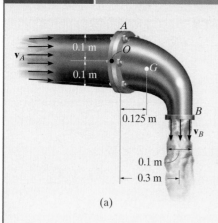

(a)

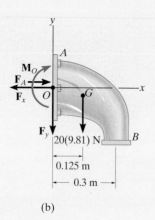

(b)

Fig. 15–28

Determine the components of reaction which the fixed pipe joint at A exerts on the elbow in Fig. 15–28a, if water flowing through the pipe is subjected to a static gauge pressure of 100 kPa at A. The discharge at B is $Q_B = 0.2 \text{ m}^3/\text{s}$. Water has a density $\rho_w = 1000 \text{ kg/m}^3$, and the water-filled elbow has a mass of 20 kg and center of mass at G.

SOLUTION

We will consider the control volume to be the outer surface of the elbow. Using a fixed inertial coordinate system, the velocity of flow at A and B and the mass flow rate can be obtained from Eq. 15–27. Since the density of water is constant, $Q_B = Q_A = Q$. Hence,

$$\frac{dm}{dt} = \rho_w Q = (1000 \text{ kg/m}^3)(0.2 \text{ m}^3/\text{s}) = 200 \text{ kg/s}$$

$$v_B = \frac{Q}{A_B} = \frac{0.2 \text{ m}^3/\text{s}}{\pi(0.05 \text{ m})^2} = 25.46 \text{ m/s} \downarrow$$

$$v_A = \frac{Q}{A_A} = \frac{0.2 \text{ m}^3/\text{s}}{\pi(0.1 \text{ m})^2} = 6.37 \text{ m/s} \rightarrow$$

Free-Body Diagram.   As shown on the free-body diagram of the control volume (elbow) Fig. 15–28b, the *fixed* connection at A exerts a resultant couple moment $\mathbf{M}_O$ and force components $\mathbf{F}_x$ and $\mathbf{F}_y$ on the elbow. Due to the static pressure of water in the pipe, the pressure force acting on the open control surface at A is $F_A = p_A A_A$. Since 1 kPa = 1000 N/m²,

$$F_A = p_A A_A = [100(10^3) \text{ N/m}^2][\pi(0.1 \text{ m})^2] = 3141.6 \text{ N}$$

There is no static pressure acting at B, since the water is discharged at atmospheric pressure; i.e., the pressure measured by a gauge at B is equal to zero, $p_B = 0$.

Equations of Steady Flow.

$$\xrightarrow{+} \Sigma F_x = \frac{dm}{dt}(v_{Bx} - v_{Ax}); \quad -F_x + 3141.6 \text{ N} = 200 \text{ kg/s}(0 - 6.37 \text{ m/s})$$

$$F_x = 4.41 \text{ kN} \qquad\qquad Ans.$$

$$+\uparrow \Sigma F_y = \frac{dm}{dt}(v_{By} - v_{Ay}); -F_y - 20(9.81) \text{ N} = 200 \text{ kg/s}(-25.46 \text{ m/s} - 0)$$

$$F_y = 4.90 \text{ kN} \qquad\qquad Ans.$$

If moments are summed about point O, Fig. 15–28b, then $\mathbf{F}_x$, $\mathbf{F}_y$, and the static pressure $\mathbf{F}_A$ are eliminated, as well as the moment of momentum of the water entering at A, Fig. 15–28a. Hence,

$$\zeta + \Sigma M_O = \frac{dm}{dt}(d_{OB} v_B - d_{OA} v_A)$$

$$M_O + 20(9.81) \text{ N } (0.125 \text{ m}) = 200 \text{ kg/s}[(0.3 \text{ m})(25.46 \text{ m/s}) - 0]$$

$$M_O = 1.50 \text{ kN} \cdot \text{m} \qquad\qquad Ans.$$

EXAMPLE 15.17

A 2-in.-diameter water jet having a velocity of 25 ft/s impinges upon a single moving blade, Fig. 15–29a. If the blade moves with a constant velocity of 5 ft/s away from the jet, determine the horizontal and vertical components of force which the blade is exerting on the water. What power does the water generate on the blade? Water has a specific weight of $\gamma_w = 62.4 \text{ lb/ft}^3$.

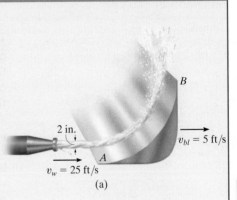

(a)

SOLUTION

Kinematic Diagram.   Here the control volume will be the stream of water on the blade. From a fixed inertial coordinate system, Fig. 15–29b, the rate at which water enters the control volume at A is

$$\mathbf{v}_A = \{25\mathbf{i}\} \text{ ft/s}$$

The *relative-flow velocity* within the control volume is $\mathbf{v}_{w/cv} = \mathbf{v}_w - \mathbf{v}_{cv} = 25\mathbf{i} - 5\mathbf{i} = \{20\mathbf{i}\} \text{ ft/s}$. Since the control volume is moving with a velocity of $\mathbf{v}_{cv} = \{5\mathbf{i}\} \text{ ft/s}$, the velocity of flow at B measured from the fixed x, y axes is the vector sum, shown in Fig. 15–29b. Here,

$$\mathbf{v}_B = \mathbf{v}_{cv} + \mathbf{v}_{w/cv}$$

$$= \{5\mathbf{i} + 20\mathbf{j}\} \text{ ft/s}$$

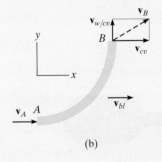

(b)

Thus, the mass flow of water *onto* the control volume that undergoes a momentum change is

$$\frac{dm}{dt} = \rho_w(v_{w/cv})A_A = \left(\frac{62.4}{32.2}\right)(20)\left[\pi\left(\frac{1}{12}\right)^2\right] = 0.8456 \text{ slug/s}$$

Free-Body Diagram.   The free-body diagram of the control volume is shown in Fig. 15–29c. The weight of the water will be neglected in the calculation, since this force will be small compared to the reactive components $\mathbf{F}_x$ and $\mathbf{F}_y$.

Equations of Steady Flow.

$$\Sigma \mathbf{F} = \frac{dm}{dt}(\mathbf{v}_B - \mathbf{v}_A)$$

$$-F_x\mathbf{i} + F_y\mathbf{j} = 0.8456(5\mathbf{i} + 20\mathbf{j} - 25\mathbf{i})$$

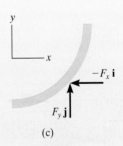

(c)

Fig. 15–29

Equating the respective $\mathbf{i}$ and $\mathbf{j}$ components gives

$$F_x = 0.8456(20) = 16.9 \text{ lb} \leftarrow \qquad Ans.$$
$$F_y = 0.8456(20) = 16.9 \text{ lb} \uparrow \qquad Ans.$$

The water exerts equal but opposite forces on the blade.

Since the water force which causes the blade to move forward horizontally with a velocity of 5 ft/s is $F_x = 16.9$ lb, then from Eq. 14–10 the power is

$$P = \mathbf{F} \cdot \mathbf{v}; \qquad P = \frac{16.9 \text{ lb}(5 \text{ ft/s})}{550 \text{ hp}/(\text{ft} \cdot \text{lb/s})} = 0.154 \text{ hp}$$

15

*15.9   Propulsion with Variable Mass

A Control Volume That Loses Mass. Consider a device such as a rocket which at an instant of time has a mass m and is moving forward with a velocity $\mathbf{v}$, Fig. 15–30a. At this same instant the amount of mass m_e is expelled from the device with a mass flow velocity $\mathbf{v}_e$. For the analysis, the control volume will include *both the mass m of the device and the expelled mass m_e*. The impulse and momentum diagrams for the control volume are shown in Fig. 15–30b. During the time dt, its velocity is increased from $\mathbf{v}$ to $\mathbf{v} + d\mathbf{v}$ since an amount of mass dm_e has been ejected and thereby gained in the exhaust. This increase in forward velocity, however, does not change the velocity $\mathbf{v}_e$ of the expelled mass, as seen by a fixed observer, since this mass moves with a constant velocity once it has been ejected. The impulses are created by $\Sigma\mathbf{F}_{cv}$, which represents the resultant of all the external forces, such as drag or weight, that *act on the control volume* in the direction of motion. This force resultant *does not include* the force which causes the control volume to move forward, since this force (called a *thrust*) is *internal to the control volume*; that is, the thrust acts with equal magnitude but opposite direction on the mass m of the device and the expelled exhaust mass m_e.* Applying the principle of impulse and momentum to the control volume, Fig. 15–30b, we have

$$(\overset{+}{\rightarrow})\quad mv - m_e v_e + \Sigma F_{cv}\, dt = (m - dm_e)(v + dv) - (m_e + dm_e)v_e$$

or

$$\Sigma F_{cv}\, dt = -v\, dm_e + m\, dv - dm_e\, dv - v_e\, dm_e$$

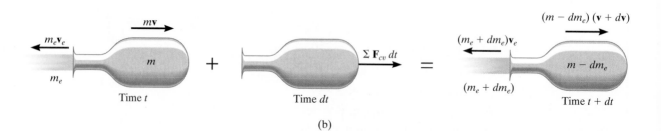

Fig. 15–30

*$\Sigma\mathbf{F}$ represents the external resultant force *acting on the control volume*, which is different from $\mathbf{F}$, the resultant force acting only on the device.

Without loss of accuracy, the third term on the right side may be neglected since it is a "second-order" differential. Dividing by dt gives

$$\Sigma F_{cv} = m\frac{dv}{dt} - (v + v_e)\frac{dm_e}{dt}$$

The velocity of the device as seen by an observer moving with the particles of the ejected mass is $v_{D/e} = (v + v_e)$, and so the final result can be written as

$$\Sigma F_{cv} = m\frac{dv}{dt} - v_{D/e}\frac{dm_e}{dt} \qquad (15\text{--}28)$$

Here the term dm_e/dt represents the rate at which mass is being ejected.

To illustrate an application of Eq. 15–28, consider the rocket shown in Fig. 15–31, which has a weight $\mathbf{W}$ and is moving upward against an atmospheric drag force $\mathbf{F}_D$. The control volume to be considered consists of the mass of the rocket and the mass of ejected gas m_e. Applying Eq. 15–28 gives

$(+\uparrow) \qquad -F_D - W = \frac{W}{g}\frac{dv}{dt} - v_{D/e}\frac{dm_e}{dt}$

The last term of this equation represents the *thrust* $\mathbf{T}$ which the engine exhaust exerts on the rocket, Fig. 15–31. Recognizing that $dv/dt = a$, we can therefore write

$(+\uparrow) \qquad T - F_D - W = \frac{W}{g}a$

If a free-body diagram of the rocket is drawn, it becomes obvious that this equation represents an application of $\Sigma F = ma$ for the rocket.

A Control Volume That Gains Mass.

A device such as a scoop or a shovel may gain mass as it moves forward. For example, the device shown in Fig. 15–32a has a mass m and moves forward with a velocity $\mathbf{v}$. At this instant, the device is collecting a particle stream of mass m_i. The flow velocity $\mathbf{v}_i$ of this injected mass is constant and independent of the velocity $\mathbf{v}$ such that $v > v_i$. The control volume to be considered here includes both the mass of the device and the mass of the injected particles.

Fig. 15–31

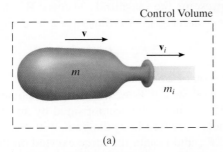

(a)

Fig. 15–32

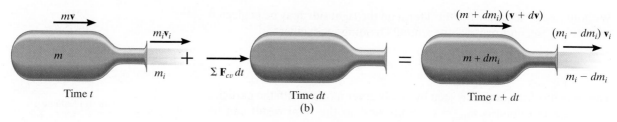

Time t Time dt Time $t + dt$
                  (b)

Fig. 15–32 (cont.)

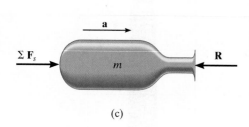

(c)

The impulse and momentum diagrams are shown in Fig. 15–32b. Along with an increase in mass dm_i gained by the device, there is an assumed increase in velocity $d\mathbf{v}$ during the time interval dt. This increase is caused by the impulse created by $\Sigma \mathbf{F}_{cv}$, the resultant of all the external forces *acting on the control volume* in the direction of motion. The force summation does not include the retarding force of the injected mass acting on the device. Why? Applying the principle of impulse and momentum to the control volume, we have

$$(\overset{+}{\rightarrow}) \qquad mv + m_i v_i + \Sigma F_{cv}\, dt = (m + dm_i)(v + dv) + (m_i - dm_i)v_i$$

Using the same procedure as in the previous case, we may write this equation as

$$\Sigma F_{cv} = m\frac{dv}{dt} + (v - v_i)\frac{dm_i}{dt}$$

Since the velocity of the device as seen by an observer moving with the particles of the injected mass is $v_{D/i} = (v - v_i)$, the final result can be written as

$$\boxed{\Sigma F_{cv} = m\frac{dv}{dt} + v_{D/i}\frac{dm_i}{dt}} \qquad (15\text{--}29)$$

where dm_i/dt is the rate of mass injected into the device. The last term in this equation represents the magnitude of force $\mathbf{R}$, which the injected mass *exerts on the device*, Fig. 15–32c. Since $dv/dt = a$, Eq. 15–29 becomes

$$\Sigma F_{cv} - R = ma$$

This is the application of $\Sigma \mathbf{F} = m\mathbf{a}$.

As in the case of steady flow, problems which are solved using Eqs. 15–28 and 15–29 should be accompanied by an identified control volume and the necessary free-body diagram. With this diagram one can then determine ΣF_{cv} and isolate the force exerted on the device by the particle stream.

The scraper box behind this tractor represents a device that gains mass. If the tractor maintains a constant velocity v, then $dv/dt = 0$ and, because the soil is originally at rest, $v_{D/i} = v$. Applying Eq. 15–29, the horizontal towing force on the scraper box is then $T = 0 + v(dm/dt)$, where dm/dt is the rate of soil accumulated in the box.

EXAMPLE | 15.18

The initial combined mass of a rocket and its fuel is m_0. A total mass m_f of fuel is consumed at a constant rate of $dm_e/dt = c$ and expelled at a constant speed of u relative to the rocket. Determine the maximum velocity of the rocket, i.e., at the instant the fuel runs out. Neglect the change in the rocket's weight with altitude and the drag resistance of the air. The rocket is fired vertically from rest.

SOLUTION

Since the rocket loses mass as it moves upward, Eq. 15–28 can be used for the solution. The only *external force* acting on the *control volume* consisting of the rocket and a portion of the expelled mass is the weight **W**, Fig. 15–33. Hence,

$$+\uparrow \Sigma F_{cv} = m\frac{dv}{dt} - v_{D/e}\frac{dm_e}{dt}; \qquad -W = m\frac{dv}{dt} - uc \qquad (1)$$

The rocket's velocity is obtained by integrating this equation.

At any given instant t during the flight, the mass of the rocket can be expressed as $m = m_0 - (dm_e/dt)t = m_0 - ct$. Since $W = mg$, Eq. 1 becomes

$$-(m_0 - ct)g = (m_0 - ct)\frac{dv}{dt} - uc$$

Separating the variables and integrating, realizing that $v = 0$ at $t = 0$, we have

$$\int_0^v dv = \int_0^t \left(\frac{uc}{m_0 - ct} - g\right) dt$$

$$v = -u\ln(m_0 - ct) - gt\Big|_0^t = u\ln\left(\frac{m_0}{m_0 - ct}\right) - gt \qquad (2)$$

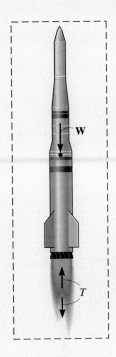

Note that liftoff requires the first term on the right to be greater than the second during the initial phase of motion. The time t' needed to consume all the fuel is

$$m_f = \left(\frac{dm_e}{dt}\right)t' = ct'$$

Hence,

$$t' = m_f/c$$

Substituting into Eq. 2 yields

$$v_{max} = u\ln\left(\frac{m_0}{m_0 - m_f}\right) - \frac{gm_f}{c} \qquad \qquad Ans.$$

EXAMPLE | 15.19

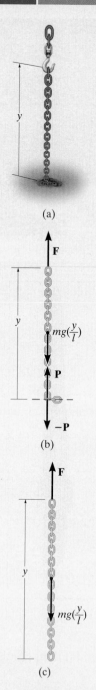

(a)

(b)

(c)

Fig. 15–34

A chain of length l, Fig. 15–34a, has a mass m. Determine the magnitude of force $\mathbf{F}$ required to (a) raise the chain with a constant speed v_c, starting from rest when $y = 0$; and (b) lower the chain with a constant speed v_c, starting from rest when $y = l$.

SOLUTION

Part (a).   As the chain is raised, all the suspended links are given a sudden downward impulse by each added link which is lifted off the ground. Thus, the *suspended portion* of the chain may be considered as a device which is *gaining mass*. The control volume to be considered is the length of chain y which is suspended by $\mathbf{F}$ at any instant, including the next link which is about to be added but is still at rest, Fig. 15–34b. The forces acting on the control volume *exclude* the internal forces $\mathbf{P}$ and $-\mathbf{P}$, which act between the added link and the suspended portion of the chain. Hence, $\Sigma F_{cv} = F - mg(y/l)$.

To apply Eq. 15–29, it is also necessary to find the rate at which mass is being added to the system. The velocity $\mathbf{v}_c$ of the chain is equivalent to $\mathbf{v}_{D/i}$. Why? Since v_c is constant, $dv_c/dt = 0$ and $dy/dt = v_c$. Integrating, using the initial condition that $y = 0$ when $t = 0$, gives $y = v_c t$. Thus, the mass of the control volume at any instant is $m_{cv} = m(y/l) = m(v_c t/l)$, and therefore the *rate* at which mass is *added* to the suspended chain is

$$\frac{dm_i}{dt} = m\left(\frac{v_c}{l}\right)$$

Applying Eq. 15–29 using this data, we have

$$+\uparrow \Sigma F_{cv} = m\frac{dv_c}{dt} + v_{D/i}\frac{dm_i}{dt}$$

$$F - mg\left(\frac{y}{l}\right) = 0 + v_c m\left(\frac{v_c}{l}\right)$$

Hence,

$$F = (m/l)(gy + v_c^2) \qquad \textit{Ans.}$$

Part (b).   When the chain is being lowered, the links which are expelled (given zero velocity) *do not* impart an impulse to the *remaining* suspended links. Why? Thus, the control volume in Part (*a*) will not be considered. Instead, the equation of motion will be used to obtain the solution. At time t the portion of chain still off the floor is y. The free-body diagram for a suspended portion of the chain is shown in Fig. 15–34c. Thus,

$$+\uparrow \Sigma F = ma; \qquad F - mg\left(\frac{y}{l}\right) = 0$$

$$F = mg\left(\frac{y}{l}\right) \qquad \textit{Ans.}$$

PROBLEMS

15–114. The fire boat discharges two streams of seawater, each at a flow of 0.25 m³/s and with a nozzle velocity of 50 m/s. Determine the tension developed in the anchor chain, needed to secure the boat. The density of seawater is $\rho_{sw} = 1020 \text{ kg/m}^3$.

*15–116. The 200-kg boat is powered by the fan which develops a slipstream having a diameter of 0.75 m. If the fan ejects air with a speed of 14 m/s, measured relative to the boat, determine the initial acceleration of the boat if it is initially at rest. Assume that air has a constant density of $\rho_w = 1.22 \text{ kg/m}^3$ and that the entering air is essentially at rest. Neglect the drag resistance of the water.

Prob. 15–116

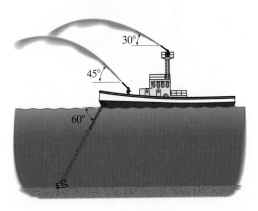

Prob. 15–114

15–117. The chute is used to divert the flow of water, $Q = 0.6 \text{ m}^3/\text{s}$. If the water has a cross-sectional area of 0.05 m^2, determine the force components at the pin D and roller C necessary for equilibrium. Neglect the weight of the chute and weight of the water on the chute, $\rho_w = 1 \text{ Mg/m}^3$.

15–115. A jet of water having a cross-sectional area of 4 in^2 strikes the fixed blade with a speed of 25 ft/s. Determine the horizontal and vertical components of force which the blade exerts on the water, $\gamma_w = 62.4 \text{ lb/ft}^3$.

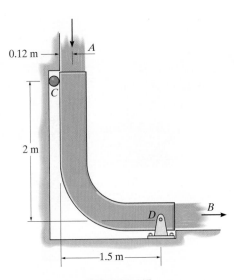

Prob. 15–115

Prob. 15–117

15–118. The buckets on the *Pelton wheel* are subjected to a 2-in-diameter jet of water, which has a velocity of 150 ft/s. If each bucket is traveling at 95 ft/s when the water strikes it, determine the power developed by the bucket. $\gamma_w = 62.4$ lb/ft^3.

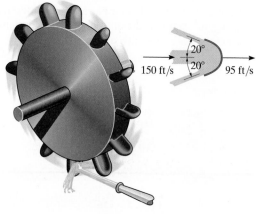

Prob. 15–118

15–119. The blade divides the jet of water having a diameter of 3 in. If one-fourth of the water flows downward while the other three-fourths flows upwards, and the total flow is $Q = 0.5$ ft^3/s, determine the horizontal and vertical components of force exerted on the blade by the jet, $\gamma_w = 62.4$ lb/ft^3.

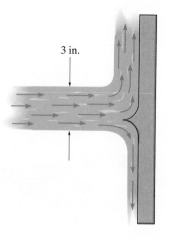

3 in.

Prob. 15–119

*15–120. The fan draws air through a vent with a speed of 12 ft/s. If the cross-sectional area of the vent is 2 ft^2, determine the horizontal thrust on the blade. The specific weight of the air is $\gamma_a = 0.076$ lb/ft^3.

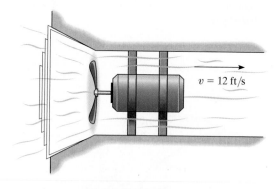

$v = 12$ ft/s

Prob. 15–120

15–121. The gauge pressure of water at C is 40 lb/in^2. If water flows out of the pipe at A and B with velocities $v_A = 12$ ft/s and $v_B = 25$ ft/s, determine the horizontal and vertical components of force necessary to hold the pipe assembly in equilibrium. Neglect the weight of water within the pipe and the weight of the pipe. The pipe has a diameter of 0.75 in. at C, and at A and B the diameter is 0.5 in. $\gamma_w = 62.4$ lb/ft^3.

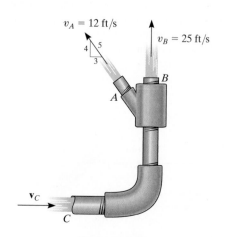

$v_A = 12$ ft/s

$v_B = 25$ ft/s

Prob. 15–121

15–122. A speedboat is powered by the jet drive shown. Seawater is drawn into the pump housing at the rate of 20 ft³/s through a 6-in.-diameter intake A. An impeller accelerates the water flow and forces it out horizontally through a 4-in.-diameter nozzle B. Determine the horizontal and vertical components of thrust exerted on the speedboat. The specific weight of seawater is $\gamma_{sw} = 64.3$ lb/ft³.

15–125. Water is discharged from a nozzle with a velocity of 12 m/s and strikes the blade mounted on the 20-kg cart. Determine the tension developed in the cord, needed to hold the cart stationary, and the normal reaction of the wheels on the cart. The nozzle has a diameter of 50 mm and the density of water is $\rho_w = 1000$ kg/mg³.

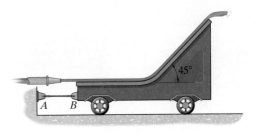

Prob. 15–125

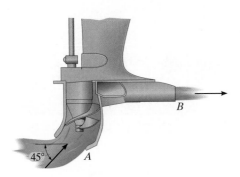

Prob. 15–122

15–123. A plow located on the front of a locomotive scoops up snow at the rate of 10 ft³/s and stores it in the train. If the locomotive is traveling at a constant speed of 12 ft/s, determine the resistance to motion caused by the shoveling. The specific weight of snow is $\gamma_s = 6$ lb/ft³.

***15–124.** The boat has a mass of 180 kg and is traveling forward on a river with a constant velocity of 70 km/h, measured *relative* to the river. The river is flowing in the opposite direction at 5 km/h. If a tube is placed in the water, as shown, and it collects 40 kg of water in the boat in 80 s, determine the horizontal thrust T on the tube that is required to overcome the resistance due to the water collection and yet maintain the constant speed of the boat. $\rho_w = 1$ Mg/m³.

15–126. Water is flowing from the 150-mm-diameter fire hydrant with a velocity $v_B = 15$ m/s. Determine the horizontal and vertical components of force and the moment developed at the base joint A, if the static (gauge) pressure at A is 50 kPa. The diameter of the fire hydrant at A is 200 mm. $\rho_w = 1$ Mg/m³.

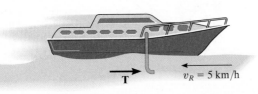

Prob. 15–124

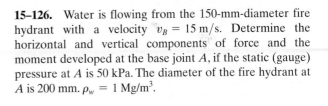

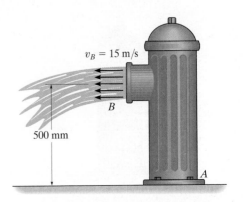

Prob. 15–126

15–127. A coil of heavy open chain is used to reduce the stopping distance of a sled that has a mass M and travels at a speed of v_0. Determine the required mass per unit length of the chain needed to slow down the sled to $(1/2)v_0$ within a distance $x = s$ if the sled is hooked to the chain at $x = 0$. Neglect friction between the chain and the ground.

*15–128. The car is used to scoop up water that is lying in a trough at the tracks. Determine the force needed to pull the car forward at constant velocity $\mathbf{v}$ for each of the three cases. The scoop has a cross-sectional area A and the density of water is ρ_w.

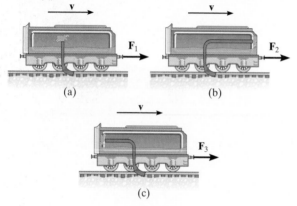

(a)

(b)

(c)

Prob. 15–128

15–129. The water flow enters below the hydrant at C at the rate of 0.75 m³/s. It is then divided equally between the two outlets at A and B. If the gauge pressure at C is 300 kPa, determine the horizontal and vertical force reactions and the moment reaction on the fixed support at C. The diameter of the two outlets at A and B is 75 mm, and the diameter of the inlet pipe at C is 150 mm. The density of water is $\rho_w = 1000$ kg/m³. Neglect the mass of the contained water and the hydrant.

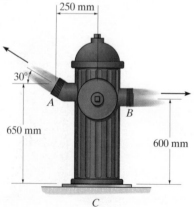

Prob. 15–129

15–130. The mini hovercraft is designed so that air is drawn in at a constant rate of 20 m³/s by the fan blade and channeled to provide a vertical thrust $\mathbf{F}$, just sufficient to lift the hovercraft off the water, while the remaining air is channeled to produce a horizontal thrust $\mathbf{T}$ on the hovercraft. If the air is discharged horizontally at 200 m/s and vertically at 800 m/s, determine the thrust $\mathbf{T}$ produced. The hovercraft and its passenger have a total mass of 1.5 Mg, and the density of the air is $\rho_a = 1.20$ kg/m³.

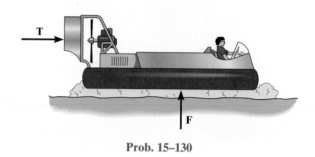

Prob. 15–130

15–131. Sand is discharged from the silo at A at a rate of 50 kg/s with a vertical velocity of 10 m/s onto the conveyor belt, which is moving with a constant velocity of 1.5 m/s. If the conveyor system and the sand on it have a total mass of 750 kg and center of mass at point G, determine the horizontal and vertical components of reaction at the pin support B and roller support A. Neglect the thickness of the conveyor.

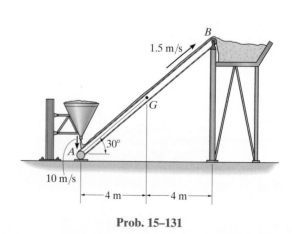

Prob. 15–131

*15–132. The fan blows air at 6000 ft³/min. If the fan has a weight of 30 lb and a center of gravity at G, determine the smallest diameter d of its base so that it will not tip over. The specific weight of air is $\gamma = 0.076$ lb/ft³.

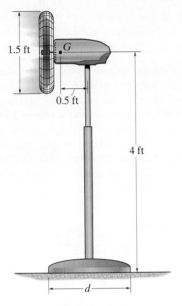

1.5 ft

G

0.5 ft

4 ft

d

Prob. 15–132

15–133. The bend is connected to the pipe at flanges A and B as shown. If the diameter of the pipe is 1 ft and it carries a discharge of 50 ft³/s, determine the horizontal and vertical components of force reaction and the moment reaction exerted at the fixed base D of the support. The total weight of the bend and the water within it is 500 lb, with a mass center at point G. The gauge pressure of the water at the flanges at A and B are 15 psi and 12 psi, respectively. Assume that no force is transferred to the flanges at A and B. The specific weight of water is $\gamma_w = 62.4$ lb/ft³.

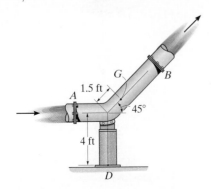

G

B

1.5 ft

A

45°

4 ft

D

Prob. 15–133

15–134. Each of the two stages A and B of the rocket has a mass of 2 Mg when their fuel tanks are empty. They each carry 500 kg of fuel and are capable of consuming it at a rate of 50 kg/s and eject it with a constant velocity of 2500 m/s, measured with respect to the rocket. The rocket is launched vertically from rest by first igniting stage B. Then stage A is ignited immediately after all the fuel in B is consumed and A has separated from B. Determine the maximum velocity of stage A. Neglect drag resistance and the variation of the rocket's weight with altitude.

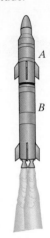

A

B

Prob. 15–134

15–135. A power lawn mower hovers very close over the ground. This is done by drawing air in at a speed of 6 m/s through an intake unit A, which has a cross-sectional area of $A_A = 0.25$ m², and then discharging it at the ground, B, where the cross-sectional area is $A_B = 0.35$ m². If air at A is subjected only to atmospheric pressure, determine the air pressure which the lawn mower exerts on the ground when the weight of the mower is freely supported and no load is placed on the handle. The mower has a mass of 15 kg with center of mass at G. Assume that air has a constant density of $\rho_a = 1.22$ kg/m³.

$\mathbf{v}_A$

A

G

B

Prob. 15–135

15

*15–136. The 12-ft-long open-link chain has 2 ft of its end suspended from the hole as shown. If a force of $P = 10$ lb is applied to its end and the chain is released from rest, determine the velocity of the chain at the instant the entire chain has been extended. The chain has a weight of 2 lb/ft.

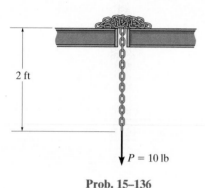

2 ft

$P = 10$ lb

Prob. 15–136

15–138. The second stage of the two-stage rocket weighs 2000 lb (empty) and is launched from the first stage with a velocity of 3000 mi/h. The fuel in the second stage weighs 1000 lb. If it is consumed at the rate of 50 lb/s and ejected with a relative velocity of 8000 ft/s, determine the acceleration of the second stage just after the engine is fired. What is the rocket's acceleration just before all the fuel is consumed? Neglect the effect of gravitation.

15–139. The missile weighs 40 000 lb. The constant thrust provided by the turbojet engine is $T = 15\,000$ lb. Additional thrust is provided by *two* rocket boosters B. The propellant in each booster is burned at a constant rate of 150 lb/s, with a relative exhaust velocity of 3000 ft/s. If the mass of the propellant lost by the turbojet engine can be neglected, determine the velocity of the missile after the 4-s burn time of the boosters. The initial velocity of the missile is 300 mi/h.

T

B

Prob. 15–139

15–137. A chain of mass m_0 per unit length is loosely coiled on the floor. If one end is subjected to a constant force **P** when $y = 0$, determine the velocity of the chain as a function of y.

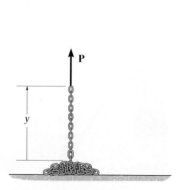

P

y

Prob. 15–137

*15–140. The 10-Mg helicopter carries a bucket containing 500 kg of water, which is used to fight fires. If it hovers over the land in a fixed position and then releases 50 kg/s of water at 10 m/s, measured relative to the helicopter, determine the initial upward acceleration the helicopter experiences as the water is being released.

a

Prob. 15–140

15–141. The earthmover initially carries 10 m^3 of sand having a density of 1520 kg/m^3. The sand is unloaded horizontally through a 2.5 m^2 dumping port P at a rate of 900 kg/s measured relative to the port. Determine the resultant tractive force $\mathbf{F}$ at its front wheels if the acceleration of the earthmover is 0.1 m/s^2 when half the sand is dumped. When empty, the earthmover has a mass of 30 Mg. Neglect any resistance to forward motion and the mass of the wheels. The rear wheels are free to roll.

Prob. 15–141

15–142. The 12-Mg jet airplane has a constant speed of 950 km/h when it is flying along a horizontal straight line. Air enters the intake scoops S at the rate of $50 \text{ m}^3/\text{s}$. If the engine burns fuel at the rate of 0.4 kg/s and the gas (air and fuel) is exhausted relative to the plane with a speed of 450 m/s, determine the resultant drag force exerted on the plane by air resistance. Assume that air has a constant density of 1.22 kg/m^3. *Hint:* Since mass both enters and exits the plane, Eqs. 15–28 and 15–29 must be combined to yield

$$\Sigma F_s = m\frac{dv}{dt} - v_{D/e}\frac{dm_e}{dt} + v_{D/i}\frac{dm_i}{dt}.$$

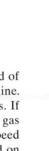

15–143. The jet is traveling at a speed of 500 mi/h, $30°$ with the horizontal. If the fuel is being spent at 3 lb/s, and the engine takes in air at 400 lb/s, whereas the exhaust gas (air and fuel) has a relative speed of $32\,800 \text{ ft/s}$, determine the acceleration of the plane at this instant. The drag resistance of the air is $F_D = (0.7v^2) \text{ lb}$, where the speed is measured in ft/s. The jet has a weight of $15\,000 \text{ lb}$. *Hint:* See Prob. 15–142.

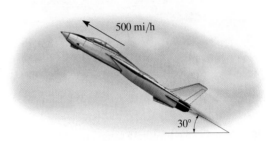

Prob. 15–143

*15–144. A four-engine commercial jumbo jet is cruising at a constant speed of 800 km/h in level flight when all four engines are in operation. Each of the engines is capable of discharging combustion gases with a velocity of 775 m/s relative to the plane. If during a test two of the engines, one on each side of the plane, are shut off, determine the new cruising speed of the jet. Assume that air resistance (drag) is proportional to the square of the speed, that is, $F_D = cv^2$, where c is a constant to be determined. Neglect the loss of mass due to fuel consumption.

$v = 950 \text{ km/h}$

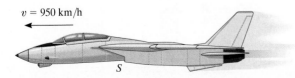

Prob. 15–142

Prob. 15–144

15–145. The car has a mass m_0 and is used to tow the smooth chain having a total length l and a mass per unit of length m'. If the chain is originally piled up, determine the tractive force F that must be supplied by the rear wheels of the car, necessary to maintain a constant speed v while the chain is being drawn out.

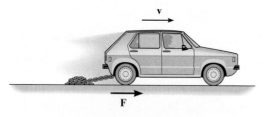

Prob. 15–145

15–146. A rocket has an empty weight of 500 lb and carries 300 lb of fuel. If the fuel is burned at the rate of 1.5 lb/s and ejected with a velocity of 4400 ft/s relative to the rocket, determine the maximum speed attained by the rocket starting from rest. Neglect the effect of gravitation on the rocket.

15–147. Determine the magnitude of force **F** as a function of time, which must be applied to the end of the cord at A to raise the hook H with a constant speed $v = 0.4$ m/s. Initially the chain is at rest on the ground. Neglect the mass of the cord and the hook. The chain has a mass of 2 kg/m.

*15–148. The truck has a mass of 50 Mg when empty. When it is unloading $5\ m^3$ of sand at a constant rate of $0.8\ m^3/s$, the sand flows out the back at a speed of 7 m/s, measured relative to the truck, in the direction shown. If the truck is free to roll, determine its initial acceleration just as the load begins to empty. Neglect the mass of the wheels and any frictional resistance to motion. The density of sand is $\rho_s = 1520\ kg/m^3$.

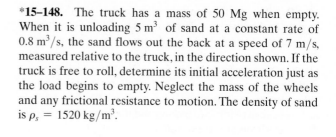

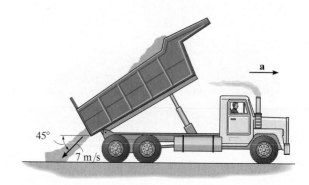

Prob. 15–148

15–149. The chain has a total length $L < d$ and a mass per unit length of m'. If a portion h of the chain is suspended over the table and released, determine the velocity of its end A as a function of its position y. Neglect friction.

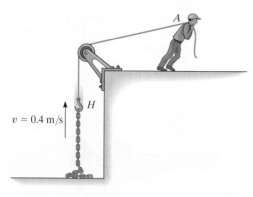

Prob. 15–147

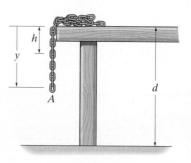

Prob. 15–149

CONCEPTUAL PROBLEMS

P15–1. The ball travels to the left when it is struck by the bat. If the ball then moves horizontally to the right, determine which measurements you could make in order to determine the net impulse given to the ball. Use numerical values to give an example of how this can be done.

P15–1

P15–2. The steel wrecking "ball" is suspended from the boom using an old rubber tire A. The crane operator lifts the ball then allows it to drop freely to break up the concrete. Explain, using appropriate numerical data, why it is a good idea to use the rubber tire for this work.

P15–2

P15–3. The train engine on the left, A, is at rest, and the one on the right, B, is coasting to the left. If the engines are identical, use numerical values to show how to determine the maximum compression in each of the spring bumpers that are mounted in the front of the engines. Each engine is free to roll.

P15–3

P15–4. Three train cars each have the same mass and are rolling freely when they strike the fixed bumper. Legs AB and BC on the bumper are pin connected at their ends and the angle BAC is 30° and BCA is 60°. Compare the average impulse in each leg needed to stop the motion if the cars have no bumper and if the cars have a spring bumper. Use appropriate numerical values to explain your answer.

P15–4

15

15

CHAPTER REVIEW

Impulse

An impulse is defined as the product of force and time. Graphically it represents the area under the F–t diagram. If the force is constant, then the impulse becomes $I = F_c(t_2 - t_1)$.

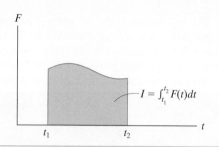

$$I = \int_{t_1}^{t_2} F(t)\,dt$$

Principle of Impulse and Momentum

When the equation of motion, $\Sigma \mathbf{F} = m\mathbf{a}$, and the kinematic equation, $a = dv/dt$, are combined, we obtain the principle of impulse and momentum. This is a vector equation that can be resolved into rectangular components and used to solve problems that involve force, velocity, and time. For application, the free-body diagram should be drawn in order to account for all the impulses that act on the particle.

$$m\mathbf{v}_1 + \Sigma \int_{t_1}^{t_2} \mathbf{F}\,dt = m\mathbf{v}_2$$

Conservation of Linear Momentum

If the principle of impulse and momentum is applied to a *system of particles*, then the collisions between the particles produce internal impulses that are equal, opposite, and collinear, and therefore cancel from the equation. Furthermore, if an external impulse is small, that is, the force is small and the time is short, then the impulse can be classified as nonimpulsive and can be neglected. Consequently, momentum for the system of particles is conserved.

The conservation-of-momentum equation is useful for finding the final velocity of a particle when internal impulses are exerted between two particles and the initial velocities of the particles is known. If the internal impulse is to be determined, then one of the particles is isolated and the principle of impulse and momentum is applied to this particle.

$$\Sigma m_i(\mathbf{v}_i)_1 = \Sigma m_i(\mathbf{v}_i)_2$$

Impact

When two particles A and B have a direct impact, the internal impulse between them is equal, opposite, and collinear. Consequently the conservation of momentum for this system applies along the line of impact.

$$m_A(v_A)_1 + m_B(v_B)_1 = m_A(v_A)_2 + m_B(v_B)_2$$

If the final velocities are unknown, a second equation is needed for solution. We must use the coefficient of restitution, e. This experimentally determined coefficient depends upon the physical properties of the colliding particles. It can be expressed as the ratio of their relative velocity after collision to their relative velocity before collision. If the collision is elastic, no energy is lost and $e = 1$. For a plastic collision $e = 0$.

$$e = \frac{(v_B)_2 - (v_A)_2}{(v_A)_1 - (v_B)_1}$$

If the impact is oblique, then the conservation of momentum for the system and the coefficient-of-restitution equation apply along the line of impact. Also, conservation of momentum for each particle applies perpendicular to this line (plane of contact) because no impulse acts on the particles in this direction.

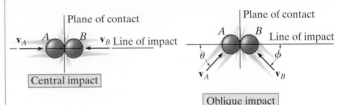

Central impact

Oblique impact

Principle of Angular Impulse and Momentum

The moment of the linear momentum about an axis (z) is called the angular momentum.

$$(H_O)_z = (d)(mv)$$

The principle of angular impulse and momentum is often used to eliminate unknown impulses by summing the moments about an axis through which the lines of action of these impulses produce no moment. For this reason, a free-body diagram should accompany the solution.

$$(\mathbf{H}_O)_1 + \Sigma \int_{t_1}^{t_2} \mathbf{M}_O \, dt = (\mathbf{H}_O)_2$$

Steady Fluid Streams

Impulse-and-momentum methods are often used to determine the forces that a device exerts on the mass flow of a fluid—liquid or gas. To do so, a free-body diagram of the fluid mass in contact with the device is drawn in order to identify these forces. Also, the velocity of the fluid as it flows into and out of a control volume for the device is calculated. The equations of steady flow involve summing the forces and the moments to determine these reactions.

$$\Sigma \mathbf{F} = \frac{dm}{dt}(\mathbf{v}_B - \mathbf{v}_A)$$

$$\Sigma \mathbf{M}_O = \frac{dm}{dt}(\mathbf{r}_B \times \mathbf{v}_B - \mathbf{r}_A \times \mathbf{v}_A)$$

Propulsion with Variable Mass

Some devices, such as a rocket, lose mass as they are propelled forward. Others gain mass, such as a shovel. We can account for this mass loss or gain by applying the principle of impulse and momentum to a control volume for the device. From this equation, the force exerted on the device by the mass flow can then be determined.

$$\Sigma F_{cv} = m\frac{dv}{dt} - v_{D/e}\frac{dm_e}{dt}$$

Loses Mass

$$\Sigma F_{cv} = m\frac{dv}{dt} + v_{D/i}\frac{dm_i}{dt}$$

Gains Mass

Review

1

Kinematics and Kinetics of a Particle

The topics and problems presented in Chapters 12 though 15 have all been *categorized* in order to provide a *clear focus* for learning the various problem-solving principles involved. In engineering practice, however, it is most important to be able to *identify* an appropriate method for the solution of a particular problem. In this regard, one must fully understand the limitations and use of the equations of dynamics, and be able to recognize which equations and principles to use for the problem's solution. For these reasons, we will now summarize the equations and principles of particle dynamics and provide the opportunity for applying them to a variety of problems.

Kinematics. Problems in kinematics require a study of the geometry of motion, and do not account for the forces causing the motion. When the equations of kinematics are applied, one should clearly establish a fixed origin and select an appropriate coordinate system used to define the position of the particle. Once the positive direction of each coordinate axis is established, then the directions of the components of position, velocity, and acceleration can be determined from the algebraic sign of their numerical quantities.

Rectilinear Motion. *Variable Acceleration.* If a mathematical (or graphical) relation is established between *any two* of the *four* variables s, v, a, and t, then a *third* variable can be determined by using one of the following equations which relates all three variables.

$$v = \frac{ds}{dt} \qquad a = \frac{dv}{dt} \qquad a\, ds = v\, dv$$

Constant Acceleration. Be *absolutely* certain that the acceleration is constant when using the following equations:

$$s = s_0 + v_0 t + \tfrac{1}{2} a_c t^2 \qquad v = v_0 + a_c t \qquad v^2 = v_0^2 + 2a_c(s - s_0)$$

Curvilinear Motion. *x, y, z Coordinates.* These coordinates are often used when the motion can be resolved into rectangular components. They are also useful for studying projectile motion since the acceleration of the projectile is *always* downward.

$$v_x = \dot{x} \qquad a_x = \dot{v}_x$$
$$v_y = \dot{y} \qquad a_y = \dot{v}_y$$
$$v_z = \dot{z} \qquad a_z = \dot{v}_z$$

n, t, b Coordinates. These coordinates are particularly advantageous for studying the particle's *acceleration* along a known path. This is because the t and n components of $\mathbf{a}$ represent the separate changes in the magnitude and direction of the velocity, respectively, and these components can be readily formulated.

$$v = \dot{s}$$

$$a_t = \dot{v} = v\frac{dv}{ds}$$

$$a_n = \frac{v^2}{\rho}$$

where

$$\rho = \left| \frac{[1 + (dy/dx)^2]^{3/2}}{d^2y/dx^2} \right|$$

when the path $y = f(x)$ is given.

r, θ, z Coordinates. These coordinates are used when data regarding the angular motion of the radial coordinate r is given to describe the particle's motion. Also, some paths of motion can conveniently be described using these coordinates.

$$v_r = \dot{r} \qquad a_r = \ddot{r} - r\dot{\theta}^2$$
$$v_\theta = r\dot{\theta} \qquad a_\theta = r\ddot{\theta} + 2\dot{r}\dot{\theta}$$
$$v_z = \dot{z} \qquad a_z = \ddot{z}$$

Relative Motion.   If the origin of a *translating* coordinate system is established at particle A, then for particle B,

$$\mathbf{r}_B = \mathbf{r}_A + \mathbf{r}_{B/A}$$
$$\mathbf{v}_B = \mathbf{v}_A + \mathbf{v}_{B/A}$$
$$\mathbf{a}_B = \mathbf{a}_A + \mathbf{a}_{B/A}$$

Here the relative motion is measured by an observer fixed in the translating coordinate system.

Kinetics.   Problems in kinetics involve the analysis of forces which cause the motion. When applying the equations of kinetics, it is absolutely necessary that measurements of the motion be made from an *inertial coordinate system*, i.e., one that does not rotate and is either fixed or translates with constant velocity. If a problem requires *simultaneous solution* of the equations of kinetics and kinematics, then it is important that the coordinate systems selected for writing each of the equations define the *positive directions* of the axes in the *same* manner.

Equations of Motion.   These equations are used to solve for the particle's acceleration or the forces causing the motion. If they are used to determine a particle's position, velocity, or time of motion, then kinematics will also have to be considered to complete the solution. Before applying the equations of motion, always draw a free-body diagram to identify all the forces acting on the particle. Also, establish the direction of the particle's acceleration or its components. (A kinetic diagram may accompany the solution in order to graphically account for the $m\mathbf{a}$ vector.)

$$\Sigma F_x = ma_x \qquad \Sigma F_n = ma_n \qquad \Sigma F_r = ma_r$$
$$\Sigma F_y = ma_y \qquad \Sigma F_t = ma_t \qquad \Sigma F_\theta = ma_\theta$$
$$\Sigma F_z = ma_z \qquad \Sigma F_b = 0 \qquad \Sigma F_z = ma_z$$

Work and Energy.   The equation of work and energy represents an integrated form of the tangential equation of motion, $\Sigma F_t = ma_t$, combined with kinematics ($a_t\, ds = v\, dv$). *It is used to solve problems involving force, velocity, and displacement.* Before applying this equation, *always draw a free-body diagram* in order to identify the forces which do work on the particle.

$$T_1 + \Sigma U_{1-2} = T_2$$

where

$$T = \tfrac{1}{2}mv^2 \qquad\qquad \text{(kinetic energy)}$$

$$U_F = \int_{s_1}^{s_2} F\cos\theta\, ds \qquad \text{(work of a variable force)}$$

$$U_{F_c} = F_c\cos\theta(s_2 - s_1) \qquad \text{(work of a constant force)}$$
$$U_W = -W\,\Delta y \qquad\qquad \text{(work of a weight)}$$
$$U_s = -\left(\tfrac{1}{2}ks_2^2 - \tfrac{1}{2}ks_1^2\right) \qquad \text{(work of an elastic spring)}$$

If the forces acting on the particle are *conservative forces,* i.e., those that *do not* cause a dissipation of energy, such as friction, then apply the conservation of energy equation. This equation is easier to use than the equation of work and energy since it applies at only *two points* on the path and *does not* require calculation of the work done by a force as the particle moves along the path.

$$T_1 + V_1 = T_2 + V_2$$

where $V = V_g + V_e$ and

$$V_g = Wy \qquad \text{(gravitational potential energy)}$$
$$V_e = \tfrac{1}{2}ks^2 \qquad \text{(elastic potential energy)}$$

If the *power* developed by a force is to be determined, use

$$P = \frac{dU}{dt} = \mathbf{F}\cdot\mathbf{v}$$

where $\mathbf{v}$ is the velocity of the particle acted upon by the force $\mathbf{F}$.

Impulse and Momentum. The equation of *linear impulse and momentum* is an integrated form of the equation of motion, $\Sigma\mathbf{F} = m\mathbf{a}$, combined with kinematics ($\mathbf{a} = d\mathbf{v}/dt$). *It is used to solve problems involving force, velocity, and time.* Before applying this equation, one should *always draw the free-body diagram,* in order to identify all the forces that cause impulses on the particle. From the diagram the impulsive and nonimpulsive forces should be identified. Recall that the nonimpulsive forces can be neglected in the analysis during the time of impact. Also, establish the direction of the particle's velocity just before and just after the impulses are applied. As an alternative procedure, impulse and momentum diagrams may accompany the solution in order to graphically account for the terms in the equation.

$$m\mathbf{v}_1 + \Sigma\int_{t_1}^{t_2} \mathbf{F}\, dt = m\mathbf{v}_2$$

If several particles are involved in the problem, consider applying the *conservation of momentum* to the system in order to eliminate the internal impulses from the analysis. This can be done in a specified direction, provided no external impulses act on the particles in that direction.

$$\Sigma m\mathbf{v}_1 = \Sigma m\mathbf{v}_2$$

If the problem involves impact and the coefficient of restitution e is given, then apply the following equation.

$$e = \frac{(v_B)_2 - (v_A)_2}{(v_A)_1 - (v_B)_1} \qquad \text{(along line of impact)}$$

Remember that during impact the principle of work and energy cannot be used, since the particles deform and therefore the work due to the internal forces will be unknown. The principle of work and energy can be used, however, to determine the energy loss during the collision once the particle's initial and final velocities are determined.

The *principle of angular impulse and momentum* and the *conservation of angular momentum* can be applied about an axis in order to *eliminate* some of the unknown impulses acting on the particle during the period when its motion is studied. Investigation of the particle's free-body diagram (or the impulse diagram) will aid in choosing the axis for application.

$$(\mathbf{H}_O)_1 + \Sigma\int_{t_1}^{t_2} \mathbf{M}_O dt = (\mathbf{H}_O)_2$$

$$(\mathbf{H}_O)_1 = (\mathbf{H}_O)_2$$

The following problems provide an opportunity for applying the above concepts. They are presented in *random order* so that practice may be gained in identifying the various types of problems and developing the skills necessary for their solution.

REVIEW PROBLEMS

R1–1. An automobile is traveling with a *constant speed* along a horizontal circular curve that has a radius $\rho = 750$ ft. If the magnitude of acceleration is $a = 8$ ft/s^2, determine the speed at which the automobile is traveling.

R1–2. Block B rests on a smooth surface. If the coefficients of friction between A and B are $\mu_s = 0.4$ and $\mu_k = 0.3$, determine the acceleration of each block if (a) $F = 6$ lb, and (b) $F = 50$ lb.

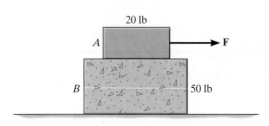

Prob. R1–2

R1–3. The small 2-lb collar starting from rest at A slides down along the smooth rod. During the motion, the collar is acted upon by a force $\mathbf{F} = \{10\mathbf{i} + 6y\mathbf{j} + 2z\mathbf{k}\}$ lb, where x, y, z are in feet. Determine the collar's speed when it strikes the wall at B.

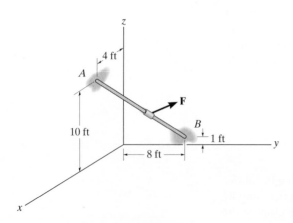

Prob. R1–3

*R1–4. The automobile travels from a parking deck down along a cylindrical spiral ramp at a constant speed of $v = 1.5$ m/s. If the ramp descends a distance of 12 m for every full revolution, $\theta = 2\pi$ rad, determine the magnitude of the car's acceleration as it moves along the ramp, $r = 10$ m. *Hint:* For part of the solution, note that the tangent to the ramp at any point is at an angle of $\phi = \tan^{-1}(12/[2\pi(10)]) = 10.81°$ from the horizontal. Use this to determine the velocity components v_θ and v_z, which in turn are used to determine $\dot{\theta}$ and $\dot{z}$.

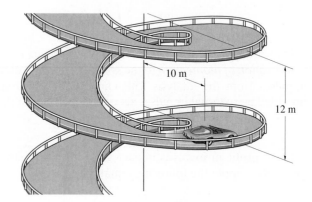

Prob. R1–4

R1–5. A rifle has a mass of 2.5 kg. If it is loosely gripped and a 1.5-g bullet is fired from it with a horizontal muzzle velocity of 1400 m/s, determine the recoil velocity of the rifle just after firing.

R1–6. If a 150-lb crate is released from rest at A, determine its speed after it slides 30 ft down the plane. The coefficient of kinetic friction between the crate and plane is $\mu_k = 0.3$.

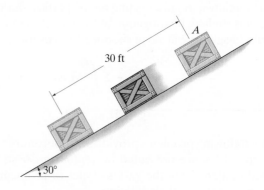

Prob. R1–6

R1–7. The van is traveling at 20 km/h when a coupling of the trailer at A fails. If the trailer has a mass of 250 kg and coasts 45 m before coming to rest, determine the constant horizontal force F created by rolling friction which causes the trailer to stop.

R1–10. Packages having a mass of 6 kg slide down a smooth chute and land horizontally with a speed of 3 m/s on the surface of a conveyor belt. If the coefficient of kinetic friction between the belt and a package is $\mu_k = 0.2$, determine the time needed to bring the package to rest on the belt if the belt is moving in the same direction as the package with a speed $v = 1$ m/s.

R1

Prob. R1–7

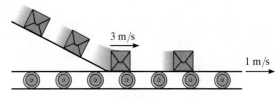

Prob. R1–10

*R1–8. The position of a particle along a straight line is given by $s = (t^3 - 9t^2 + 15t)$ ft, where t is in seconds. Determine its maximum acceleration and maximum velocity during the time interval $0 \le t \le 10$ s.

R1–9. The spool, which has a mass of 4 kg, slides along the rotating rod. At the instant shown, the angular rate of rotation of the rod is $\dot{\theta} = 6$ rad/s and this rotation is increasing at $\ddot{\theta} = 2$ rad/s². At this same instant, the spool has a velocity of 3 m/s and an acceleration of 1 m/s², both measured relative to the rod and directed away from the center O when $r = 0.5$ m. Determine the radial frictional force and the normal force, both exerted by the rod on the spool at this instant.

R1–11. A 20-kg block is originally at rest on a horizontal surface for which the coefficient of static friction is $\mu_s = 0.6$ and the coefficient of kinetic friction is $\mu_k = 0.5$. If a horizontal force F is applied such that it varies with time as shown, determine the speed of the block in 10 s. *Hint:* First determine the time needed to overcome friction and start the block moving.

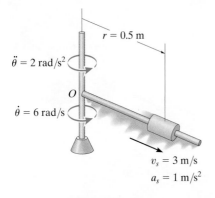

Prob. R1–9

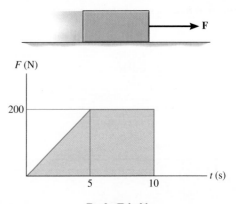

Prob. R1–11

*R1–12.   The 6-lb ball is fired from a tube by a spring having a stiffness $k = 20$ lb/in. Determine how far the spring must be compressed to fire the ball from the compressed position to a height of 8 ft, at which point it has a velocity of 6 ft/s.

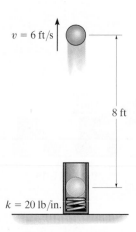

$v = 6$ ft/s

8 ft

$k = 20$ lb/in.

Prob. R1–12

R1–13.   A train car, having a mass of 25 Mg, travels up a 10° incline with a constant speed of 80 km/h. Determine the power required to overcome the force of gravity.

R1–14.   The rocket sled has a mass of 4 Mg and travels from rest along the smooth horizontal track such that it maintains a constant power output of 450 kW. Neglect the loss of fuel mass and air resistance, and determine how far it must travel to reach a speed of $v = 60$ m/s.

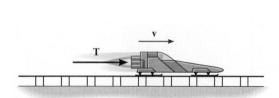

v

T

Prob. R1–14

■R1–15.   A projectile, initially at the origin, moves vertically downward along a straight-line path through a fluid medium such that its velocity is defined as $v = 3(8e^{-t} + t)^{1/2}$ m/s, where t is in seconds. Plot the position s of the projectile during the first 2 s. Use the Runge-Kutta method to evaluate s with incremental values of $h = 0.25$ s.

*R1–16.   The chain has a mass of 3 kg/m. If the coefficient of kinetic friction between the chain and the plane is $\mu_k = 0.2$, determine the velocity at which the end A will pass point B when the chain is released from rest.

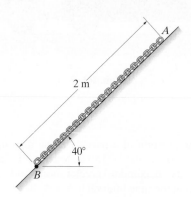

A

2 m

40°

B

Prob. R1–16

R1–17.   The motor M pulls in its attached rope with an acceleration $a_p = 6$ m/s². Determine the towing force exerted by M on the rope in order to move the 50-kg crate up the inclined plane. The coefficient of kinetic friction between the crate and the plane is $\mu_k = 0.3$. Neglect the mass of the pulleys and rope.

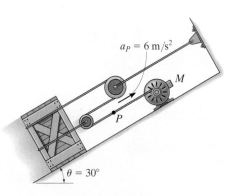

$a_P = 6$ m/s²

M

P

$\theta = 30°$

Prob. R1–17

R1–18. The drinking fountain is designed such that the nozzle is located from the edge of the basin as shown. Determine the maximum and minimum speed at which water can be ejected from the nozzle so that it does not splash over the sides of the basin at B and C.

R1–21. The ping-pong ball has a mass of 2 g. If it is struck with the velocity shown, determine how high h it rises above the end of the smooth table after the rebound. Take $e = 0.8$.

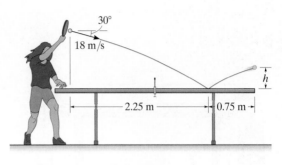

Prob. R1–21

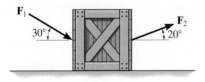

Prob. R1–18

R1–22. A sports car can accelerate at 6 m/s² and decelerate at 8 m/s². If the maximum speed it can attain is 60 m/s, determine the shortest time it takes to travel 900 m starting from rest and then stopping when s = 900 m.

R1–23. A 2-kg particle rests on a smooth horizontal plane and is acted upon by forces $F_x = 0$ and $F_y = 3$ N. If $x = 0$, $y = 0$, $v_x = 6$ m/s, and $v_y = 2$ m/s when $t = 0$, determine the equation $y = f(x)$ which describes the path.

R1–19. The 100-kg crate is subjected to the action of two forces, $F_1 = 800$ N and $F_2 = 1.5$ kN, as shown. If it is originally at rest, determine the distance it slides in order to attain a speed of 6 m/s. The coefficient of kinetic friction between the crate and the surface is $\mu_k = 0.2$.

*R1–24. A skier starts from rest at A (30 ft, 0) and descends the smooth slope, which may be approximated by a parabola. If she has a weight of 120 lb, determine the normal force she exerts on the ground at the instant she arrives at point B.

Prob. R1–19

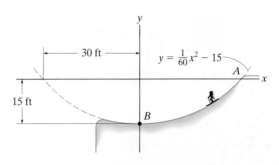

*R1–20. If a particle has an initial velocity $v_0 = 12$ ft/s to the right, and a constant acceleration of 2 ft/s² to the left, determine the particle's displacement in 10 s. Originally $s_0 = 0$.

Prob. R1–24

R1

R1–25. The 20-lb block B rests on the surface of a table for which the coefficient of kinetic friction is $\mu_k = 0.1$. Determine the speed of the 10-lb block A after it has moved downward 2 ft from rest. Neglect the mass of the pulleys and cords.

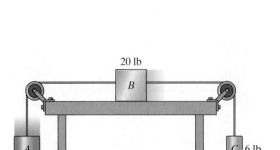

20 lb

B

A

10 lb

C 6 lb

Prob. R1–25

R1–26. At a given instant the 10-lb block A is moving downward with a speed of 6 ft/s. Determine its speed 2 s later. Block B has a weight of 4 lb, and the coefficient of kinetic friction between it and the horizontal plane is $\mu_k = 0.2$. Neglect the mass of the pulleys and cord.

B

A

Prob. R1–26

R1–27. Two smooth billiard balls A and B have an equal mass of $m = 200$ g. If A strikes B with a velocity of $(v_A)_1 = 2$ m/s as shown, determine their final velocities just after collision. Ball B is originally at rest and the coefficient of restitution is $e = 0.75$.

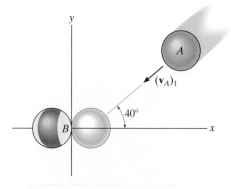

y

A

$(v_A)_1$

$40°$

B

x

Prob. R1–27

*R1–28. A crate has a weight of 1500 lb. If it is pulled along the ground at a constant speed for a distance of 20 ft, and the towing cable makes an angle of 15° with the horizontal, determine the tension in the cable and the work done by the towing force. The coefficient of kinetic friction between the crate and the ground is $\mu_k = 0.55$.

R1–29. The collar of negligible size has a mass of 0.25 kg and is attached to a spring having an unstretched length of 100 mm. If the collar is released from rest at A and travels along the smooth guide, determine its speed just before it strikes B.

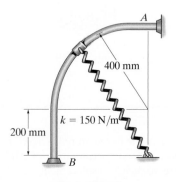

A

400 mm

$k = 150$ N/m

200 mm

B

Prob. R1–29

R1–30. Determine the tension developed in the two cords and the acceleration of each block. Neglect the mass of the pulleys and cords. *Hint:* Since the system consists of *two* cords, relate the motion of block *A* to *C*, and of block *B* to *C*. Then, by elimination, relate the motion of *A* to *B*.

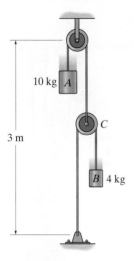

10 kg *A*

C

3 m

B 4 kg

Prob. R1–30

R1–31. The baggage truck *A* has a mass of 800 kg and is used to pull each of the 300-kg cars. Determine the tension in the couplings at *B* and *C* if the tractive force **F** on the truck is *F* = 480 N. What is the speed of the truck when *t* = 2 s, starting from the rest? The car wheels are free to roll. Neglect the mass of the wheels.

***R1–32.** The baggage truck *A* has a mass of 800 kg and is used to pull each of the 300-kg cars. If the tractive force **F** on the truck is *F* = 480 N, determine the initial acceleration of the truck. What is the acceleration of the truck if the coupling at *C* suddenly fails? The car wheels are free to roll. Neglect the mass of the wheels.

Probs. R1–31/32

■**R1–33.** Packages having a mass of 2.5 kg ride on the surface of the conveyor belt. If the belt starts from rest and with constant acceleration increases to a speed of 0.75 m/s in 2 s, determine the maximum angle of tilt, θ, so that none of the packages slip on the inclined surface *AB* of the belt. The coefficient of static friction between the belt and each package is $\mu_s = 0.3$. At what angle ϕ do the packages first begin to slip off the surface of the belt if the belt is moving at a constant speed of 0.75 m/s?

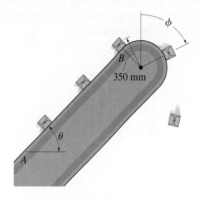

ϕ

B

350 mm

θ

A

Prob. R1–33

R1–34. A particle travels in a straight line such that for a short time $2 \text{ s} \leq t \leq 6 \text{ s}$ its motion is described by $v = (4/a)$ ft/s where *a* is in ft/s^2. If $v = 6$ ft/s when $t = 2$ s, determine the particle's acceleration when $t = 3$ s.

R1–35. The blocks *A* and *B* weigh 10 and 30 lb, respectively. They are connected together by a light cord and ride in the frictionless grooves. Determine the speed of each block after block *A* moves 6 ft up along the plane. The blocks are released from rest.

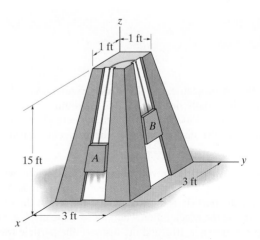

z

1 ft

1 ft

B

15 ft

A

y

3 ft

3 ft

x

Prob. R1–35

*R1–36. A motorcycle starts from rest at $t = 0$ and travels along a straight road with a constant acceleration of 6 ft/s² until it reaches a speed of 50 ft/s. Afterwards it maintains this speed. Also, when $t = 0$, a car located 6000 ft down the road is traveling toward the motocycle at a constant speed of 30 ft/s. Determine the time and the distance traveled by the motorcycle when they pass each other.

R1–37. The 5-lb ball, attached to the cord, is struck by the boy. Determine the smallest speed he must impart to the ball so that it will swing around in a vertical circle, without causing the cord to become slack.

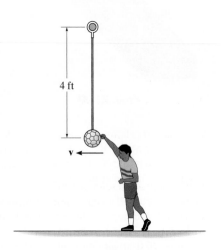

4 ft

v ←

Prob. R1–37

R1–38. A projectile, initially at the origin, moves along a straight-line path through a fluid medium such that its velocity is $v = 1800(1 - e^{-0.3t})$ mm/s where t is in seconds. Determine the displacement of the projectile during the first 3 s.

R1–39. A particle travels along a straight line with a velocity $v = (12 - 3t^2)$ m/s, where t is in seconds. When $t = 1$ s, the particle is located 10 m to the left of the origin. Determine the acceleration when $t = 4$ s, the displacement from $t = 0$ to $t = 10$ s, and the distance the particle travels during this time period.

*R1–40. A particle is moving along a circular path of 2-m radius such that its position as a function of time is given by $\theta = (5t^2)$ rad, where t is in seconds. Determine the magnitude of the particle's acceleration when $\theta = 30°$. The particle starts from rest when $\theta = 0°$.

R1–41. If the end of the cable at A is pulled down with a speed of 2 m/s, determine the speed at which block B rises.

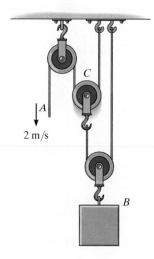

C

A

2 m/s

B

Prob. R1–41

R1–42. The bottle rests at a distance of 3 ft from the center of the horizontal platform. If the coefficient of static friction between the bottle and the platform is $\mu_s = 0.3$, determine the maximum speed that the bottle can attain before slipping. Assume the angular motion of the platform is slowly increasing.

R1–43. Work Prob. R1–42 assuming that the platform starts rotating from rest so that the speed of the bottle is increased at 2 ft/s².

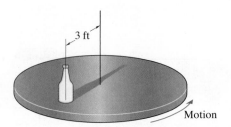

3 ft

Motion

Probs. R1–42/43

*R1–44. A 3-lb block, initially at rest at point A, slides along the smooth parabolic surface. Determine the normal force acting on the block when it reaches B. Neglect the size of the block.

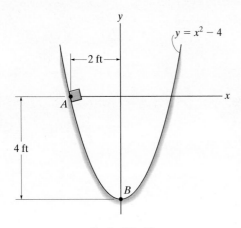

Prob. R1–44

R1–45. A car starts from rest and moves along a straight line with an acceleration of $a = (3s^{-1/3})$ m/s², where s is in meters. Determine the car's acceleration when $t = 4$ s.

R1–46. A particle travels along a curve defined by the equation $s = (t^3 - 3t^2 + 2t)$ m, where t is in seconds. Draw the s–t, v–t, and a–t graphs for the particle for $0 \le t \le 3$ s.

R1–47. The crate, having a weight of 50 lb, is hoisted by the pulley system and motor M. If the crate starts from rest and, by constant acceleration, attains a speed of 12 ft/s after rising 10 ft, determine the power that must be supplied to the motor at the instant $s = 10$ ft. The motor has an efficiency $\varepsilon = 0.74$.

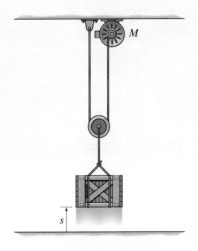

Prob. R1–47

*R1–48. The block has a mass of 0.5 kg and moves within the smooth vertical slot. If the block starts from rest when the *attached* spring is in the unstretched position at A, determine the *constant* vertical force F which must be applied to the cord so that the block attains a speed $v_B = 2.5$ m/s when it reaches B; $s_B = 0.15$ m. Neglect the mass of the cord and pulley.

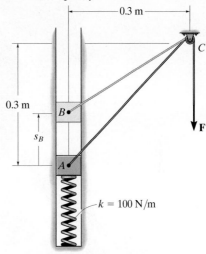

Prob. R1–48

R1–49. A ball having a mass of 200 g is released from rest at a height of 400 mm above a very large fixed metal surface. If the ball rebounds to a height of 325 mm above the surface, determine the coefficient of restitution between the ball and the surface.

R1–50. Determine the speed of block B if the end of the cable at C is pulled downward with a speed of 10 ft/s. What is the relative velocity of the block with respect to C.

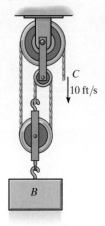

Prob. R1–50

Chapter **16**

Kinematics is important for the design of the mechanism used on this dump truck.

Planar Kinematics of a Rigid Body

CHAPTER OBJECTIVES

- To classify the various types of rigid-body planar motion.
- To investigate rigid-body translation and angular motion about a fixed axis.
- To study planar motion using an absolute motion analysis.
- To provide a relative motion analysis of velocity and acceleration using a translating frame of reference.
- To show how to find the instantaneous center of zero velocity and determine the velocity of a point on a body using this method.
- To provide a relative-motion analysis of velocity and acceleration using a rotating frame of reference.

16.1 Planar Rigid-Body Motion

In this chapter, the planar kinematics of a rigid body will be discussed. This study is important for the design of gears, cams, and mechanisms used for many mechanical operations. Once the kinematics is thoroughly understood, then we can apply the equations of motion, which relate the forces on the body to the body's motion.

The *planar motion* of a body occurs when all the particles of a rigid body move along paths which are equidistant from a fixed plane. There are three types of rigid body planar motion. In order of increasing complexity, they are

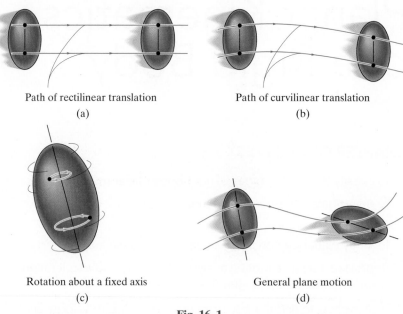

Path of rectilinear translation
(a)

Path of curvilinear translation
(b)

Rotation about a fixed axis
(c)

General plane motion
(d)

Fig. 16–1

- *Translation.*   This type of motion occurs when a line in the body remains parallel to its original orientation throughout the motion. When the paths of motion for any two points on the body are parallel lines, the motion is called *rectilinear translation*, Fig. 16–1*a*. If the paths of motion are along curved lines, the motion is called *curvilinear translation*, Fig. 16–1*b*.

- *Rotation about a fixed axis.*   When a rigid body rotates about a fixed axis, all the particles of the body, except those which lie on the axis of rotation, move along circular paths, Fig. 16–1*c*.

- *General plane motion.*   When a body is subjected to general plane motion, it undergoes a combination of translation *and* rotation, Fig. 16–1*d*. The translation occurs within a reference plane, and the rotation occurs about an axis perpendicular to the reference plane.

In the following sections we will consider each of these motions in detail. Examples of bodies undergoing these motions are shown in Fig. 16–2.

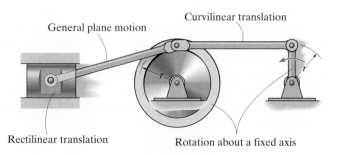

General plane motion

Curvilinear translation

Rectilinear translation

Rotation about a fixed axis

Fig. 16–2

16.2   Translation

Consider a rigid body which is subjected to either rectilinear or curvilinear translation in the x–y plane, Fig. 16–3.

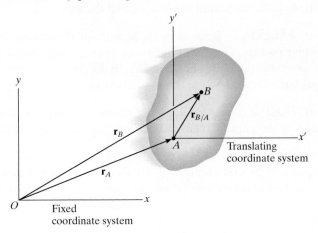

Fig. 16–3

Position.   The locations of points A and B on the body are defined with respect to fixed x, y reference frame using *position vectors* $\mathbf{r}_A$ and $\mathbf{r}_B$. The translating x', y' coordinate system is *fixed in the body* and has its origin at A, hereafter referred to as the *base point*. The position of B with respect to A is denoted by the *relative-position vector* $\mathbf{r}_{B/A}$ ("$\mathbf{r}$ of B with respect to A"). By vector addition,

$$\mathbf{r}_B = \mathbf{r}_A + \mathbf{r}_{B/A}$$

Velocity.   A relation between the instantaneous velocities of A and B is obtained by taking the time derivative of this equation, which yields $\mathbf{v}_B = \mathbf{v}_A + d\mathbf{r}_{B/A}/dt$. Here $\mathbf{v}_A$ and $\mathbf{v}_B$ denote *absolute velocities* since these vectors are measured with respect to the x, y axes. The term $d\mathbf{r}_{B/A}/dt = \mathbf{0}$, since the *magnitude* of $\mathbf{r}_{B/A}$ is *constant* by definition of a rigid body, and because the body is translating the *direction* of $\mathbf{r}_{B/A}$ is also *constant*. Therefore,

$$\mathbf{v}_B = \mathbf{v}_A$$

Acceleration.   Taking the time derivative of the velocity equation yields a similar relationship between the instantaneous accelerations of A and B:

$$\mathbf{a}_B = \mathbf{a}_A$$

The above two equations indicate that *all points in a rigid body subjected to either rectilinear or curvilinear translation move with the same velocity and acceleration*. As a result, the kinematics of particle motion, discussed in Chapter 12, can also be used to specify the kinematics of points located in a translating rigid body.

Riders on this amusement ride are subjected to curvilinear translation, since the vehicle moves in a circular path yet it always remains in an upright position.

16

(a)

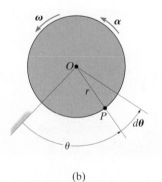

(b)

Fig. 16–4

16.3   Rotation about a Fixed Axis

When a body rotates about a fixed axis, any point P located in the body travels along a *circular path*. To study this motion it is first necessary to discuss the angular motion of the body about the axis.

Angular Motion.   Since a point is without dimension, it cannot have angular motion. *Only lines or bodies undergo angular motion.* For example, consider the body shown in Fig. 16–4a and the angular motion of a radial line r located within the shaded plane.

Angular Position.   At the instant shown, the *angular position* of r is defined by the angle θ, measured from a *fixed* reference line to r.

Angular Displacement.   The change in the angular position, which can be measured as a differential $d\boldsymbol{\theta}$, is called the *angular displacement*.* This vector has a *magnitude* of $d\boldsymbol{\theta}$, measured in degrees, radians, or revolutions, where 1 rev = 2π rad. Since motion is about a *fixed axis*, the direction of $d\boldsymbol{\theta}$ is *always* along this axis. Specifically, the *direction* is determined by the right-hand rule; that is, the fingers of the right hand are curled with the sense of rotation, so that in this case the thumb, or $d\boldsymbol{\theta}$, points upward, Fig. 16–4a. In two dimensions, as shown by the top view of the shaded plane, Fig. 16–4b, both θ and $d\theta$ are counterclockwise, and so the thumb points outward from the page.

Angular Velocity.   The time rate of change in the angular position is called the *angular velocity* $\boldsymbol{\omega}$ (omega). Since $d\boldsymbol{\theta}$ occurs during an instant of time dt, then,

$(\zeta+)$

$$\omega = \frac{d\theta}{dt} \qquad (16\text{–}1)$$

This vector has a *magnitude* which is often measured in rad/s. It is expressed here in scalar form since its *direction* is also along the axis of rotation, Fig. 16–4a. When indicating the angular motion in the shaded plane, Fig. 16–4b, we can refer to the sense of rotation as clockwise or counterclockwise. Here we have *arbitrarily* chosen counterclockwise rotations as *positive* and indicated this by the curl shown in parentheses next to Eq. 16–1. Realize, however, that the directional sense of $\boldsymbol{\omega}$ is actually outward from the page.

---

*It is shown in Sec. 20.1 that finite rotations or finite angular displacements are *not* vector quantities, although differential rotations $d\boldsymbol{\theta}$ are vectors.

Angular Acceleration.

The *angular acceleration* $\boldsymbol{\alpha}$ (alpha) measures the time rate of change of the angular velocity. The *magnitude* of this vector is

$(\zeta+)$

$$\alpha = \frac{d\omega}{dt}$$

$(16\text{–}2)$

Using Eq. 16–1, it is also possible to express α as

$(\zeta+)$

$$\alpha = \frac{d^2\theta}{dt^2}$$

$(16\text{–}3)$

The gears used in the operation of a crane all rotate about fixed axes. Engineers must be able to relate their angular motions in order to properly design this gear system.

The line of action of $\boldsymbol{\alpha}$ is the same as that for $\boldsymbol{\omega}$, Fig. 16–4a; however, its sense of *direction* depends on whether $\boldsymbol{\omega}$ is increasing or decreasing. If $\boldsymbol{\omega}$ is decreasing, then $\boldsymbol{\alpha}$ is called an *angular deceleration* and therefore has a sense of direction which is opposite to $\boldsymbol{\omega}$.

By eliminating dt from Eqs. 16–1 and 16–2, we obtain a differential relation between the angular acceleration, angular velocity, and angular displacement, namely,

$(\zeta+)$

$$\alpha \, d\theta = \omega \, d\omega$$

$(16\text{–}4)$

The similarity between the differential relations for angular motion and those developed for rectilinear motion of a particle ($v = ds/dt$, $a = dv/dt$, and $a \, ds = v \, dv$) should be apparent.

Constant Angular Acceleration.

If the angular acceleration of the body is *constant*, $\boldsymbol{\alpha} = \boldsymbol{\alpha}_c$, then Eqs. 16–1, 16–2, and 16–4, when integrated, yield a set of formulas which relate the body's angular velocity, angular position, and time. These equations are similar to Eqs. 12–4 to 12–6 used for rectilinear motion. The results are

$(\zeta+)$

$(\zeta+)$

$(\zeta+)$

$$\omega = \omega_0 + \alpha_c t$$ $(16\text{–}5)$

$$\theta = \theta_0 + \omega_0 t + \tfrac{1}{2}\alpha_c t^2$$ $(16\text{–}6)$

$$\omega^2 = \omega_0^2 + 2\alpha_c(\theta - \theta_0)$$ $(16\text{–}7)$

Constant Angular Acceleration

Here θ_0 and ω_0 are the initial values of the body's angular position and angular velocity, respectively.

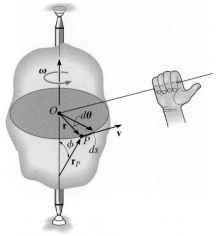

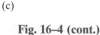

(c)

Fig. 16–4 (cont.)

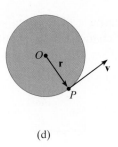

(d)

Motion of Point P.

As the rigid body in Fig. 16–4c rotates, point P travels along a *circular path* of radius r with center at point O. This path is contained within the shaded plane shown in top view, Fig. 16–4d.

Position and Displacement.

The position of P is defined by the position vector $\mathbf{r}$, which extends from O to P. If the body rotates $d\theta$ then P will displace $ds = r\,d\theta$.

Velocity.

The velocity of P has a magnitude which can be found by dividing $ds = r\,d\theta$ by dt so that

$$v = \omega r \tag{16–8}$$

As shown in Figs. 16–4c and 16–4d, the *direction* of $\mathbf{v}$ is *tangent* to the circular path.

Both the magnitude and direction of $\mathbf{v}$ can also be accounted for by using the cross product of $\boldsymbol{\omega}$ and $\mathbf{r}_P$ (see Appendix B). Here, $\mathbf{r}_P$ is directed from *any point* on the axis of rotation to point P, Fig. 16–4c. We have

$$\mathbf{v} = \boldsymbol{\omega} \times \mathbf{r}_P \tag{16–9}$$

The order of the vectors in this formulation is important, since the cross product is not commutative, i.e., $\boldsymbol{\omega} \times \mathbf{r}_P \neq \mathbf{r}_P \times \boldsymbol{\omega}$. Notice in Fig. 16–4c how the correct direction of $\mathbf{v}$ is established by the right-hand rule. The fingers of the right hand are curled from $\boldsymbol{\omega}$ toward $\mathbf{r}_P$ ($\boldsymbol{\omega}$ "cross" $\mathbf{r}_P$). The thumb indicates the correct direction of $\mathbf{v}$, which is tangent to the path in the direction of motion. From Eq. B–8, the magnitude of $\mathbf{v}$ in Eq. 16–9 is $v = \omega r_P \sin\phi$, and since $r = r_P \sin\phi$, Fig. 16–4c, then $v = \omega r$, which agrees with Eq. 16–8. As a special case, the position vector $\mathbf{r}$ can be chosen for $\mathbf{r}_P$. Here $\mathbf{r}$ lies in the plane of motion and again the velocity of point P is

$$\mathbf{v} = \boldsymbol{\omega} \times \mathbf{r} \tag{16–10}$$

Acceleration.   The acceleration of P can be expressed in terms of its normal and tangential components. Since $a_t = dv/dt$ and $a_n = v^2/\rho$, where $\rho = r$, $v = \omega r$, and $\alpha = d\omega/dt$, we have

$$a_t = \alpha r \qquad\qquad (16\text{–}11)$$

$$a_n = \omega^2 r \qquad\qquad (16\text{–}12)$$

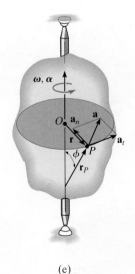

(e)

The *tangential component of acceleration*, Figs. 16–4e and 16–4f, represents the time rate of change in the velocity's magnitude. If the speed of P is increasing, then $\mathbf{a}_t$ acts in the same direction as $\mathbf{v}$; if the speed is decreasing, $\mathbf{a}_t$ acts in the opposite direction of $\mathbf{v}$; and finally, if the speed is constant, $\mathbf{a}_t$ is zero.

The *normal component of acceleration* represents the time rate of change in the velocity's direction. The *direction* of $\mathbf{a}_n$ is always toward O, the center of the circular path, Figs. 16–4e and 16–4f.

Like the velocity, the acceleration of point P can be expressed in terms of the vector cross product. Taking the time derivative of Eq. 16–9 we have

$$\mathbf{a} = \frac{d\mathbf{v}}{dt} = \frac{d\boldsymbol{\omega}}{dt} \times \mathbf{r}_P + \boldsymbol{\omega} \times \frac{d\mathbf{r}_P}{dt}$$

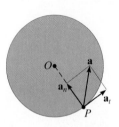

(f)

Fig. 16–4 (cont.)

Recalling that $\boldsymbol{\alpha} = d\boldsymbol{\omega}/dt$, and using Eq. 16–9 ($d\mathbf{r}_P/dt = \mathbf{v} = \boldsymbol{\omega} \times \mathbf{r}_P$), yields

$$\mathbf{a} = \boldsymbol{\alpha} \times \mathbf{r}_P + \boldsymbol{\omega} \times (\boldsymbol{\omega} \times \mathbf{r}_P) \qquad\qquad (16\text{–}13)$$

From the definition of the cross product, the first term on the right has a magnitude $a_t = \alpha r_P \sin \phi = \alpha r$, and by the right-hand rule, $\boldsymbol{\alpha} \times \mathbf{r}_P$ is in the direction of $\mathbf{a}_t$, Fig. 16–4e. Likewise, the second term has a magnitude $a_n = \omega^2 r_P \sin \phi = \omega^2 r$, and applying the right-hand rule twice, first to determine the result $\mathbf{v}_P = \boldsymbol{\omega} \times \mathbf{r}_P$ then $\boldsymbol{\omega} \times \mathbf{v}_P$, it can be seen that this result is in the same direction as $\mathbf{a}_n$, shown in Fig. 16–4e. Noting that this is also the *same* direction as $-\mathbf{r}$, which lies in the plane of motion, we can express $\mathbf{a}_n$ in a much simpler form as $\mathbf{a}_n = -\omega^2 \mathbf{r}$. Hence, Eq. 16–13 can be identified by its two components as

$$\begin{aligned} \mathbf{a} &= \mathbf{a}_t + \mathbf{a}_n \\ &= \boldsymbol{\alpha} \times \mathbf{r} - \omega^2 \mathbf{r} \end{aligned} \qquad\qquad (16\text{–}14)$$

Since $\mathbf{a}_t$ and $\mathbf{a}_n$ are perpendicular to one another, if needed the magnitude of acceleration can be determined from the Pythagorean theorem; namely, $a = \sqrt{a_n^2 + a_t^2}$, Fig. 16–4f.

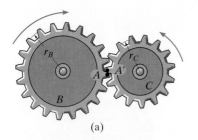

(a)

If two rotating bodies contact one another, then the *points in contact* move along *different circular paths*, and the velocity and *tangential components* of acceleration of the points will be the *same*: however, the *normal components* of acceleration will *not* be the same. For example, consider the two meshed gears in Fig. 16–5a. Point A is located on gear B and a coincident point A' is located on gear C. Due to the rotational motion, $\mathbf{v}_A = \mathbf{v}_{A'}$, Fig. 16–5b, and as a result, $\omega_B r_B = \omega_C r_C$ or $\omega_B = \omega_C(r_C/r_B)$. Also, from Fig. 16–5c, $(\mathbf{a}_A)_t = (\mathbf{a}_{A'})_t$, so that $\alpha_B = \alpha_C(r_C/r_B)$; however, since both points follow different circular paths, $(\mathbf{a}_A)_n \neq (\mathbf{a}_{A'})_n$ and therefore, as shown, $\mathbf{a}_A \neq \mathbf{a}_{A'}$.

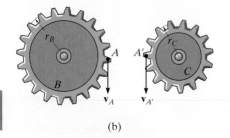

(b)

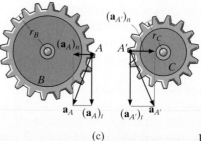

(c)

Fig. 16–5

Important Points

- A body can undergo two types of translation. During rectilinear translation all points follow parallel straight-line paths, and during curvilinear translation the points follow curved paths that are the same shape.

- All the points on a translating body move with the same velocity and acceleration.

- Points located on a body that rotates about a fixed axis follow circular paths.

- The relation $\alpha\,d\theta = \omega\,d\omega$ is derived from $\alpha = d\omega/dt$ and $\omega = d\theta/dt$ by eliminating dt.

- Once angular motions ω and α are known, the velocity and acceleration of any point on the body can be determined.

- The velocity always acts tangent to the path of motion.

- The acceleration has two components. The tangential acceleration measures the rate of change in the magnitude of the velocity and can be determined from $a_t = \alpha r$. The normal acceleration measures the rate of change in the direction of the velocity and can be determined from $a_n = \omega^2 r$.

Procedure for Analysis

The velocity and acceleration of a point located on a rigid body that is rotating about a fixed axis can be determined using the following procedure.

Angular Motion.

- Establish the positive sense of rotation about the axis of rotation and show it alongside each kinematic equation as it is applied.

- If a relation is known between any *two* of the four variables α, ω, θ, and t, then a third variable can be obtained by using one of the following kinematic equations which relates all three variables.

$$\omega = \frac{d\theta}{dt} \qquad \alpha = \frac{d\omega}{dt} \qquad \alpha\, d\theta = \omega\, d\omega$$

- If the body's angular acceleration is *constant*, then the following equations can be used:

$$\omega = \omega_0 + \alpha_c t$$
$$\theta = \theta_0 + \omega_0 t + \tfrac{1}{2}\alpha_c t^2$$
$$\omega^2 = \omega_0^2 + 2\alpha_c(\theta - \theta_0)$$

- Once the solution is obtained, the sense of θ, ω, and α is determined from the algebraic signs of their numerical quantities.

Motion of Point P.

- In most cases the velocity of P and its two components of acceleration can be determined from the scalar equations

$$v = \omega r$$
$$a_t = \alpha r$$
$$a_n = \omega^2 r$$

- If the geometry of the problem is difficult to visualize, the following vector equations should be used:

$$\mathbf{v} = \boldsymbol{\omega} \times \mathbf{r}_P = \boldsymbol{\omega} \times \mathbf{r}$$
$$\mathbf{a}_t = \boldsymbol{\alpha} \times \mathbf{r}_P = \boldsymbol{\alpha} \times \mathbf{r}$$
$$\mathbf{a}_n = \boldsymbol{\omega} \times (\boldsymbol{\omega} \times \mathbf{r}_P) = -\omega^2 \mathbf{r}$$

- Here $\mathbf{r}_P$ is directed from any point on the axis of rotation to point P, whereas $\mathbf{r}$ lies in the plane of motion of P. Either of these vectors, along with $\boldsymbol{\omega}$ and $\boldsymbol{\alpha}$, should be expressed in terms of its $\mathbf{i}$, $\mathbf{j}$, $\mathbf{k}$ components, and, if necessary, the cross products determined using a determinant expansion (see Eq. B–12).

16

EXAMPLE | 16.1

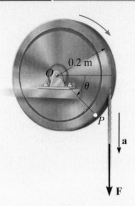

Fig. 16–6

A cord is wrapped around a wheel in Fig. 16–6, which is initially at rest when $\theta = 0$. If a force is applied to the cord and gives it an acceleration $a = (4t)$ m/s², where t is in seconds, determine, as a function of time, (a) the angular velocity of the wheel, and (b) the angular position of line OP in radians.

SOLUTION

Part (a). The wheel is subjected to rotation about a fixed axis passing through point O. Thus, point P on the wheel has motion about a circular path, and the acceleration of this point has *both* tangential and normal components. The tangential component is $(a_P)_t = (4t)$ m/s², since the cord is wrapped around the wheel and moves *tangent* to it. Hence the angular acceleration of the wheel is

$(\circlearrowleft +)$ $$(a_P)_t = \alpha r$$

$$(4t) \text{ m/s}^2 = \alpha(0.2 \text{ m})$$

$$\alpha = (20t) \text{ rad/s}^2 \, \circlearrowright$$

Using this result, the wheel's angular velocity ω can now be determined from $\alpha = d\omega/dt$, since this equation relates α, t, and ω. Integrating, with the initial condition that $\omega = 0$ when $t = 0$, yields

$(\circlearrowleft +)$ $$\alpha = \frac{d\omega}{dt} = (20t) \text{ rad/s}^2$$

$$\int_0^\omega d\omega = \int_0^t 20t \, dt$$

$$\omega = 10t^2 \text{ rad/s} \, \circlearrowright \qquad\qquad Ans.$$

Part (b). Using this result, the angular position θ of OP can be found from $\omega = d\theta/dt$, since this equation relates θ, ω, and t. Integrating, with the initial condition $\theta = 0$ when $t = 0$, we have

$(\circlearrowleft +)$ $$\frac{d\theta}{dt} = \omega = (10t^2) \text{ rad/s}$$

$$\int_0^\theta d\theta = \int_0^t 10t^2 \, dt$$

$$\theta = 3.33t^3 \text{ rad} \qquad\qquad Ans.$$

NOTE: We cannot use the equation of constant angular acceleration, since α is a function of time.

EXAMPLE | 16.2

The motor shown in the photo is used to turn a wheel and attached blower contained within the housing. The details are shown in Fig. 16–7a. If the pulley A connected to the motor begins to rotate from rest with a constant angular acceleration of $\alpha_A = 2 \text{ rad/s}^2$, determine the magnitudes of the velocity and acceleration of point P on the wheel, after the pulley has turned two revolutions. Assume the transmission belt does not slip on the pulley and wheel.

SOLUTION

Angular Motion.   First we will convert the two revolutions to radians. Since there are 2π rad in one revolution, then

$$\theta_A = 2 \text{ rev} \left(\frac{2\pi \text{ rad}}{1 \text{ rev}} \right) = 12.57 \text{ rad}$$

Since α_A is constant, the angular velocity of pulley A is therefore

$(\zeta +)$
$$\omega^2 = \omega_0^2 + 2\alpha_c(\theta - \theta_0)$$
$$\omega_A^2 = 0 + 2(2 \text{ rad/s}^2)(12.57 \text{ rad} - 0)$$
$$\omega_A = 7.090 \text{ rad/s}$$

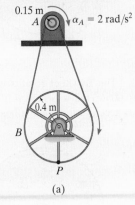

The belt has the same speed and tangential component of acceleration as it passes over the pulley and wheel. Thus,

$$v = \omega_A r_A = \omega_B r_B; \quad 7.090 \text{ rad/s} (0.15 \text{ m}) = \omega_B (0.4 \text{ m})$$
$$\omega_B = 2.659 \text{ rad/s}$$
$$a_t = \alpha_A r_A = \alpha_B r_B; \quad 2 \text{ rad/s}^2 (0.15 \text{ m}) = \alpha_B (0.4 \text{ m})$$
$$\alpha_B = 0.750 \text{ rad/s}^2$$

Motion of P.   As shown on the kinematic diagram in Fig. 16–7b, we have

$$v_P = \omega_B r_B = 2.659 \text{ rad/s} (0.4 \text{ m}) = 1.06 \text{ m/s} \qquad Ans.$$
$$(a_P)_t = \alpha_B r_B = 0.750 \text{ rad/s}^2 (0.4 \text{ m}) = 0.3 \text{ m/s}^2$$
$$(a_P)_n = \omega_B^2 r_B = (2.659 \text{ rad/s})^2 (0.4 \text{ m}) = 2.827 \text{ m/s}^2$$

Thus

$$a_P = \sqrt{(0.3 \text{ m/s}^2)^2 + (2.827 \text{ m/s}^2)^2} = 2.84 \text{ m/s}^2 \qquad Ans.$$

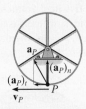

Fig. 16–7

FUNDAMENTAL PROBLEMS

F16–1. When the gear rotates 20 revolutions, it achieves an angular velocity of $\omega = 30$ rad/s, starting from rest. Determine its constant angular acceleration and the time required.

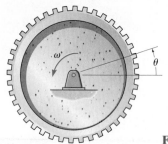

F16–1

F16–2. The flywheel rotates with an angular velocity of $\omega = (0.005\theta^2)$ rad/s, where θ is in radians. Determine the angular acceleration when it has rotated 20 revolutions.

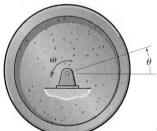

F16–2

F16–3. The flywheel rotates with an angular velocity of $\omega = (4\,\theta^{1/2})$ rad/s, where θ is in radians. Determine the time it takes to achieve an angular velocity of $\omega = 150$ rad/s. When $t = 0, \theta = 1$ rad.

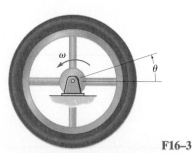

F16–3

F16–4. The bucket is hoisted by the rope that wraps around a drum wheel. If the angular displacement of the wheel is $\theta = (0.5t^3 + 15t)$ rad, where t is in seconds, determine the velocity and acceleration of the bucket when $t = 3$ s.

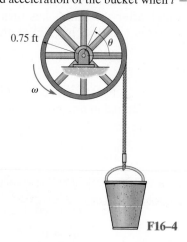

0.75 ft

F16–4

F16–5. A wheel has an angular acceleration of $\alpha = (0.5\,\theta)$ rad/s^2, where θ is in radians. Determine the magnitude of the velocity and acceleration of a point P located on its rim after the wheel has rotated 2 revolutions. The wheel has a radius of 0.2 m and starts at $\omega_0 = 2$ rad/s.

F16–6. For a short period of time, the motor turns gear A with a constant angular acceleration of $\alpha_A = 4.5$ rad/s^2, starting from rest. Determine the velocity of the cylinder and the distance it travels in three seconds. The cord is wrapped around pulley D which is rigidly attached to gear B.

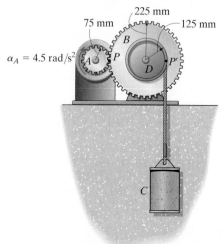

225 mm
75 mm
125 mm
$\alpha_A = 4.5$ rad/s^2
A P B D P'
C

F16–6

PROBLEMS

16–1. The angular velocity of the disk is defined by $\omega = (5t^2 + 2)$ rad/s, where t is in seconds. Determine the magnitudes of the velocity and acceleration of point A on the disk when $t = 0.5$ s.

Prob. 16–1

16–2. A flywheel has its angular speed increased uniformly from 15 rad/s to 60 rad/s in 80 s. If the diameter of the wheel is 2 ft, determine the magnitudes of the normal and tangential components of acceleration of a point on the rim of the wheel when $t = 80$ s, and the total distance the point travels during the time period.

16–3. The disk is originally rotating at $\omega_0 = 8$ rad/s. If it is subjected to a constant angular acceleration of $\alpha = 6$ rad/s², determine the magnitudes of the velocity and the n and t components of acceleration of point A at the instant $t = 0.5$ s.

*16–4. The disk is originally rotating at $\omega_0 = 8$ rad/s. If it is subjected to a constant angular acceleration of $\alpha = 6$ rad/s², determine the magnitudes of the velocity and the n and t components of acceleration of point B just after the wheel undergoes 2 revolutions.

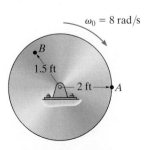

Probs. 16–3/4

16–5. Initially the motor on the circular saw turns its drive shaft at $\omega = (20t^{2/3})$ rad/s, where t is in seconds. If the radii of gears A and B are 0.25 in. and 1 in., respectively, determine the magnitudes of the velocity and acceleration of a tooth C on the saw blade after the drive shaft rotates $\theta = 5$ rad starting from rest.

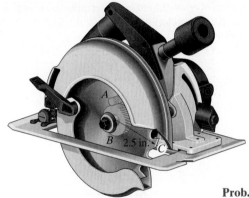

Prob. 16–5

16–6. A wheel has an initial clockwise angular velocity of 10 rad/s and a constant angular acceleration of 3 rad/s². Determine the number of revolutions it must undergo to acquire a clockwise angular velocity of 15 rad/s. What time is required?

16–7. If gear A rotates with a constant angular acceleration of $\alpha_A = 90$ rad/s², starting from rest, determine the time required for gear D to attain an angular velocity of 600 rpm. Also, find the number of revolutions of gear D to attain this angular velocity. Gears A, B, C, and D have radii of 15 mm, 50 mm, 25 mm, and 75 mm, respectively.

*16–8. If gear A rotates with an angular velocity of $\omega_A = (\theta_A + 1)$ rad/s, where θ_A is the angular displacement of gear A, measured in radians, determine the angular acceleration of gear D when $\theta_A = 3$ rad, starting from rest. Gears A, B, C, and D have radii of 15 mm, 50 mm, 25 mm, and 75 mm, respectively.

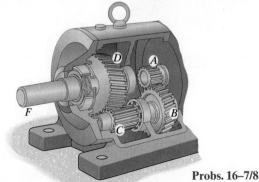

Probs. 16–7/8

16–9. The vertical-axis wind turbine consists of two blades that have a parabolic shape. If the blades are originally at rest and begin to turn with a constant angular acceleration of $\alpha_c = 0.5$ rad/s², determine the magnitude of the velocity and acceleration of points A and B on the blade after the blade has rotated through two revolutions.

16–10. The vertical-axis windturbine consists of two blades that have a parabolic shape. If the blades are originally at rest and begin to turn with a constant angular acceleration of $\alpha_c = 0.5$ rad/s², determine the magnitude of the velocity and acceleration of points A and B on the blade when $t = 4$ s.

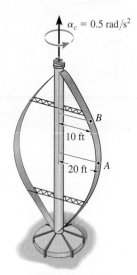

Probs. 16–9/10

16–11. If the angular velocity of the drum is increased uniformly from 6 rad/s when $t = 0$ to 12 rad/s when $t = 5$ s, determine the magnitudes of the velocity and acceleration of points A and B on the belt when $t = 1$ s. At this instant the points are located as shown.

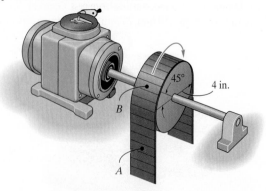

Prob. 16–11

*16–12.** A motor gives gear A an angular acceleration of $\alpha_A = (0.25\theta^3 + 0.5)$ rad/s², where θ is in radians. If this gear is initially turning at $(\omega_A)_0 = 20$ rad/s, determine the angular velocity of gear B after A undergoes an angular displacement of 10 rev.

16–13. A motor gives gear A an angular acceleration of $\alpha_A = (4t^3)$ rad/s², where t is in seconds. If this gear is initially turning at $(\omega_A)_0 = 20$ rad/s, determine the angular velocity of gear B when $t = 2$ s.

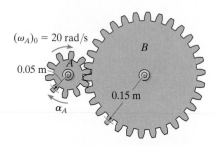

Probs. 16–12/13

16–14. The disk starts from rest and is given an angular acceleration $\alpha = (2t^2)$ rad/s², where t is in seconds. Determine the angular velocity of the disk and its angular displacement when $t = 4$ s.

16–15. The disk starts from rest and is given an angular acceleration $\alpha = (5t^{1/2})$ rad/s², where t is in seconds. Determine the magnitudes of the normal and tangential components of acceleration of a point P on the rim of the disk when $t = 2$ s.

*16–16.** The disk starts at $\omega_0 = 1$ rad/s when $\theta = 0$, and is given an angular acceleration $\alpha = (0.3\theta)$ rad/s², where θ is in radians. Determine the magnitudes of the normal and tangential components of acceleration of a point P on the rim of the disk when $\theta = 1$ rev.

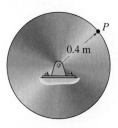

Probs. 16–14/15/16

16–17. Starting at $(\omega_A)_0 = 3$ rad/s, when $\theta = 0$, $s = 0$, pulley A is given an angular acceleration $\alpha = (0.6\theta)$ rad/s^2, where θ is in radians. Determine the speed of block B when it has risen $s = 0.5$ m. The pulley has an inner hub D which is fixed to C and turns with it.

16–18. Starting from rest when $s = 0$, pulley A is given a constant angular acceleration $\alpha_c = 6$ rad/s^2. Determine the speed of block B when it has risen $s = 6$ m. The pulley has an inner hub D which is fixed to C and turns with it.

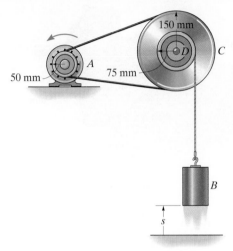

Probs. 16–17/18

16–19. The vacuum cleaner's armature shaft S rotates with an angular acceleration of $\alpha = 4\omega^{3/4}$ rad/s^2, where ω is in rad/s. Determine the brush's angular velocity when $t = 4$ s, starting from $\omega_0 = 1$ rad/s, at $\theta = 0$. The radii of the shaft and the brush are 0.25 in. and 1 in., respectively. Neglect the thickness of the drive belt.

Prob. 16–19

***16–20.** The operation of "reverse" for a three-speed automotive transmission is illustrated schematically in the figure. If the crank shaft G is turning with an angular speed of 60 rad/s, determine the angular speed of the drive shaft H. Each of the gears rotates about a fixed axis. Note that gears A and B, C and D, E and F are in mesh. The radii of each of these gears are reported in the figure.

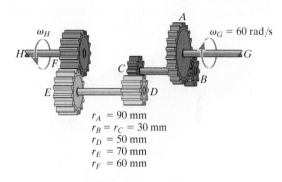

$r_A = 90$ mm
$r_B = r_C = 30$ mm
$r_D = 50$ mm
$r_E = 70$ mm
$r_F = 60$ mm

Prob. 16–20

16–21. A motor gives disk A a clockwise angular acceleration of $\alpha_A = (0.6t^2 + 0.75)$ rad/s^2, where t is in seconds. If the initial angular velocity of the disk is $\omega_0 = 6$ rad/s, determine the magnitudes of the velocity and acceleration of block B when $t = 2$ s.

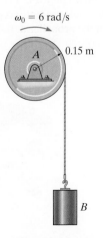

$\omega_0 = 6$ rad/s

0.15 m

Prob. 16–21

16

16–22. For a short time the motor turns gear A with an angular acceleration of $\alpha_A = (30t^{1/2})\,\text{rad/s}^2$, where t is in seconds. Determine the angular velocity of gear D when $t = 5$ s, starting from rest. Gear A is initially at rest. The radii of gears A, B, C, and D are $r_A = 25$ mm, $r_B = 100$ mm, $r_C = 40$ mm, and $r_D = 100$ mm, respectively.

16–23. The motor turns gear A so that its angular velocity increases uniformly from zero to 3000 rev/min after the shaft turns 200 rev. Determine the angular velocity of gear D when $t = 3$ s. The radii of gears A, B, C, and D are $r_A = 25$ mm, $r_B = 100$ mm, $r_C = 40$ mm, and $r_D = 100$ mm, respectively.

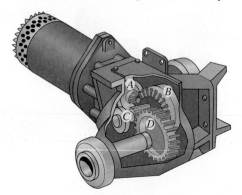

Probs. 16–22/23

***16–24.** The gear A on the drive shaft of the outboard motor has a radius $r_A = 0.5$ in. and the meshed pinion gear B on the propeller shaft has a radius $r_B = 1.2$ in. Determine the angular velocity of the propeller in $t = 1.5$ s, if the drive shaft rotates with an angular acceleration $\alpha = (400t^3)\,\text{rad/s}^2$, where t is in seconds. The propeller is originally at rest and the motor frame does not move.

16–25. For the outboard motor in Prob. 16–24, determine the magnitude of the velocity and acceleration of point P located on the tip of the propeller at the instant $t = 0.75$ s.

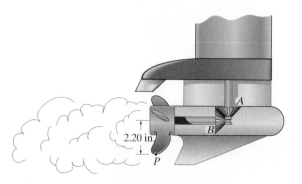

Probs. 16–24/25

16–26. The pinion gear A on the motor shaft is given a constant angular acceleration $\alpha = 3\,\text{rad/s}^2$. If the gears A and B have the dimensions shown, determine the angular velocity and angular displacement of the output shaft C, when $t = 2$ s starting from rest. The shaft is fixed to B and turns with it.

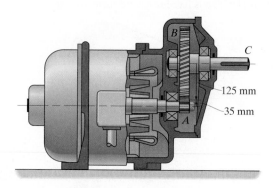

Prob. 16–26

16–27. For a short time, gear A of the automobile starter rotates with an angular acceleration of $\alpha_A = (450t^2 + 60)\,\text{rad/s}^2$, where t is in seconds. Determine the angular velocity and angular displacement of gear B when $t = 2$ s, starting from rest. The radii of gears A and B are 10 mm and 25 mm, respectively.

***16–28.** For a short time, gear A of the automobile starter rotates with an angular acceleration of $\alpha_A = (50\omega^{1/2})\,\text{rad/s}^2$, where ω is in rad/s. Determine the angular velocity of gear B when $t = 1$ s. Orginally $(\omega_A)_0 = 1$ rad/s when $t = 0$. The radii of gears A and B are 10 mm and 25 mm, respectively.

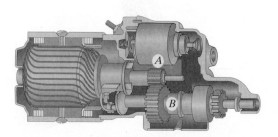

Probs. 16–27/28

16–29. A mill in a textile plant uses the belt-and-pulley arrangement shown to transmit power. When $t = 0$ an electric motor is turning pulley A with an angular velocity of $\omega_A = 5$ rad/s. If this pulley is subjected to a constant counterclockwise angular acceleration of $\alpha_A = 2$ rad/s^2, determine the angular velocity of pulley B after B turns 6 revolutions. The hub at D is rigidly *connected* to pulley C and turns with it.

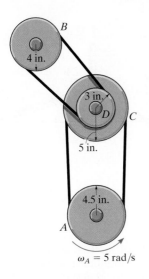

Prob. 16–29

16–30. A tape having a thickness s wraps around the wheel which is turning at a constant rate ω. Assuming the unwrapped portion of tape remains horizontal, determine the acceleration of point P of the unwrapped tape when the radius of the wrapped tape is r. *Hint:* Since $v_P = \omega r$, take the time derivative and note that $dr/dt = \omega(s/2\pi)$.

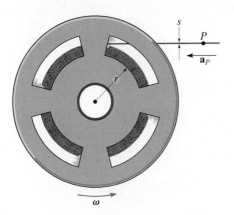

Prob. 16–30

16–31. Due to the screw at E, the actuator provides linear motion to the arm at F when the motor turns the gear at A. If the gears have the radii listed in the figure, and the screw at E has a pitch $p = 2$ mm, determine the speed at F when the motor turns A at $\omega_A = 20$ rad/s. *Hint:* The screw pitch indicates the amount of advance of the screw for each full revolution.

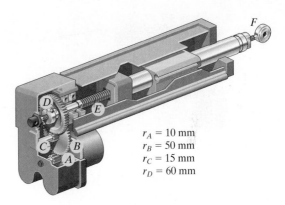

$r_A = 10$ mm
$r_B = 50$ mm
$r_C = 15$ mm
$r_D = 60$ mm

Prob. 16–31

*16–32. The driving belt is twisted so that pulley B rotates in the opposite direction to that of drive wheel A. If A has a constant angular acceleration of $\alpha_A = 30$ rad/s^2, determine the tangential and normal components of acceleration of a point located at the rim of B when $t = 3$ s, starting from rest.

16–33. The driving belt is twisted so that pulley B rotates in the opposite direction to that of drive wheel A. If the angular displacement of A is $\theta_A = (5t^3 + 10t^2)$ rad, where t is in seconds, determine the angular velocity and angular acceleration of B when $t = 3$ s.

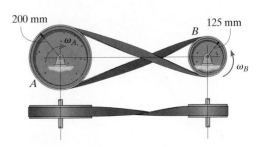

200 mm   125 mm

Probs. 16–32/33

16–34. The rope of diameter d is wrapped around the tapered drum which has the dimensions shown. If the drum is rotating at a constant rate of ω, determine the upward acceleration of the block. Neglect the small horizontal displacement of the block.

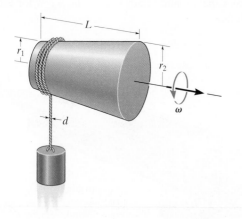

Prob. 16–34

16–35. If the shaft and plate rotates with a constant angular velocity of $\omega = 14$ rad/s, determine the velocity and acceleration of point C located on the corner of the plate at the instant shown. Express the result in Cartesian vector form.

*16–36. At the instant shown, the shaft and plate rotates with an angular velocity of $\omega = 14$ rad/s and angular acceleration of $\alpha = 7$ rad/s². Determine the velocity and acceleration of point D located on the corner of the plate at this instant. Express the result in Cartesian vector form.

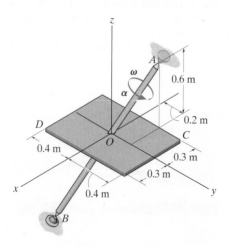

Probs. 16–35/36

16–37. The rod assembly is supported by ball-and-socket joints at A and B. At the instant shown it is rotating about the y axis with an angular velocity $\omega = 5$ rad/s and has an angular acceleration $\alpha = 8$ rad/s². Determine the magnitudes of the velocity and acceleration of point C at this instant. Solve the problem using Cartesian vectors and Eqs. 16–9 and 16–13.

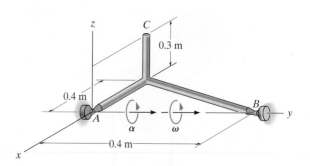

Prob. 16–37

16–38. Rotation of the robotic arm occurs due to linear movement of the hydraulic cylinders A and B. If this motion causes the gear at D to rotate clockwise at 5 rad/s, determine the magnitude of velocity and acceleration of the part C held by the grips of the arm.

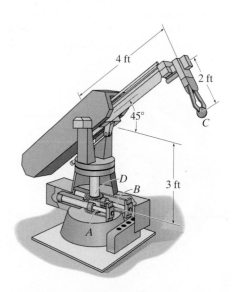

Prob. 16–38

16.4 Absolute Motion Analysis

A body subjected to *general plane motion* undergoes a *simultaneous* translation and rotation. If the body is represented by a thin slab, the slab translates in the plane of the slab and rotates about an axis perpendicular to this plane. The motion can be completely specified by knowing *both* the angular rotation of a line fixed in the body and the motion of a point on the body. One way to relate these motions is to use a rectilinear position coordinate s to locate the point along its path and an angular position coordinate θ to specify the orientation of the line. The two coordinates are then related using the geometry of the problem. By *direct application* of the time-differential equations $v = ds/dt$, $a = dv/dt$, $\omega = d\theta/dt$, and $\alpha = d\omega/dt$, the *motion* of the point and the *angular motion* of the line can then be related. This procedure is similar to that used to solve dependent motion problems involving pulleys, Sec. 12.9. In some cases, this same procedure may be used to relate the motion of one body, undergoing either rotation about a fixed axis or translation, to that of a connected body undergoing general plane motion.

The dumping bin on the truck rotates about a fixed axis passing through the pin at A. It is operated by the extension of the hydraulic cylinder BC. The angular position of the bin can be specified using the angular position coordinate θ, and the position of point C on the bin is specified using the rectilinear position coordinate s. Since a and b are fixed lengths, then the two coordinates can be related by the cosine law, $s = \sqrt{a^2 + b^2 - 2ab \cos \theta}$. The time derivative of this equation relates the speed at which the hydraulic cylinder extends to the angular velocity of the bin.

Procedure for Analysis

The velocity and acceleration of a point P undergoing rectilinear motion can be related to the angular velocity and angular acceleration of a line contained within a body using the following procedure.

Position Coordinate Equation.

• Locate point P on the body using a position coordinate s, which is measured from a *fixed origin* and is *directed along the straight-line path of motion* of point P.

• Measure from a fixed reference line the angular position θ of a line lying in the body.

• From the dimensions of the body, relate s to θ, $s = f(\theta)$, using geometry and/or trigonometry.

Time Derivatives.

• Take the first derivative of $s = f(\theta)$ with respect to time to get a relation between v and ω.

• Take the second time derivative to get a relation between a and α.

• In each case the chain rule of calculus must be used when taking the time derivatives of the position coordinate equation. See Appendix C.

EXAMPLE   16.3

The end of rod R shown in Fig. 16–8 maintains contact with the cam by means of a spring. If the cam rotates about an axis passing through point O with an angular acceleration $\boldsymbol{\alpha}$ and angular velocity $\boldsymbol{\omega}$, determine the velocity and acceleration of the rod when the cam is in the arbitrary position θ.

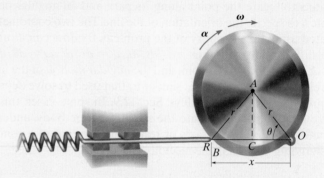

Fig. 16–8

SOLUTION

Position Coordinate Equation.   Coordinates θ and x are chosen in order to relate the *rotational motion* of the line segment OA on the cam to the *rectilinear translation* of the rod. These coordinates are measured from the *fixed point* O and can be related to each other using trigonometry. Since $OC = CB = r \cos \theta$, Fig. 16–8, then

$$x = 2r \cos \theta$$

Time Derivatives.   Using the chain rule of calculus, we have

$$\frac{dx}{dt} = -2r(\sin \theta)\frac{d\theta}{dt}$$

$$v = -2r\omega \sin \theta \qquad\qquad Ans.$$

$$\frac{dv}{dt} = -2r\left(\frac{d\omega}{dt}\right)\sin \theta - 2r\omega(\cos \theta)\frac{d\theta}{dt}$$

$$a = -2r(\alpha \sin \theta + \omega^2 \cos \theta) \qquad\qquad Ans.$$

NOTE: The negative signs indicate that v and a are opposite to the direction of positive x. This seems reasonable when you visualize the motion.

EXAMPLE | 16.4

At a given instant, the cylinder of radius r, shown in Fig. 16–9, has an angular velocity $\boldsymbol{\omega}$ and angular acceleration $\boldsymbol{\alpha}$. Determine the velocity and acceleration of its center G if the cylinder rolls without slipping.

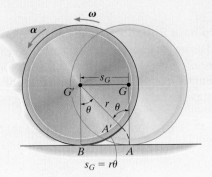

Fig. 16–9

SOLUTION

Position Coordinate Equation. The cylinder undergoes general plane motion since it simultaneously translates and rotates. By inspection, point G moves in a *straight line* to the left, from G to G', as the cylinder rolls, Fig. 16–9. Consequently its new position G' will be specified by the *horizontal* position coordinate s_G, which is measured from G to G'. Also, as the cylinder rolls (without slipping), the arc length $A'B$ on the rim which was in contact with the ground from A to B, is equivalent to s_G. Consequently, the motion requires the radial line GA to rotate θ to the position $G'A'$. Since the arc $A'B = r\theta$, then G travels a distance

$$s_G = r\theta$$

Time Derivatives. Taking successive time derivatives of this equation, realizing that r is constant, $\omega = d\theta/dt$, and $\alpha = d\omega/dt$, gives the necessary relationships:

$$s_G = r\theta$$

$$v_G = r\omega \qquad\qquad\qquad Ans.$$

$$a_G = r\alpha \qquad\qquad\qquad Ans.$$

NOTE: Remember that these relationships are valid only if the cylinder (disk, wheel, ball, etc.) rolls *without* slipping.

EXAMPLE 16.5

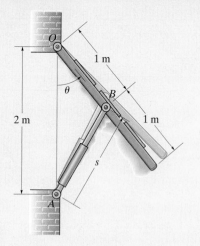

Fig. 16–10

The large window in Fig. 16–10 is opened using a hydraulic cylinder AB. If the cylinder extends at a constant rate of 0.5 m/s, determine the angular velocity and angular acceleration of the window at the instant $\theta = 30°$.

SOLUTION

Position Coordinate Equation. The angular motion of the window can be obtained using the coordinate θ, whereas the extension or motion *along the hydraulic cylinder* is defined using a coordinate s, which measures its length from the fixed point A to the moving point B. These coordinates can be related using the law of cosines, namely,

$$s^2 = (2 \text{ m})^2 + (1 \text{ m})^2 - 2(2 \text{ m})(1 \text{ m}) \cos \theta$$

$$s^2 = 5 - 4 \cos \theta \tag{1}$$

When $\theta = 30°$,

$$s = 1.239 \text{ m}$$

Time Derivatives. Taking the time derivatives of Eq. 1, we have

$$2s\frac{ds}{dt} = 0 - 4(-\sin \theta)\frac{d\theta}{dt}$$

$$s(v_s) = 2(\sin \theta)\omega \tag{2}$$

Since $v_s = 0.5$ m/s, then at $\theta = 30°$,

$$(1.239 \text{ m})(0.5 \text{ m/s}) = 2 \sin 30°\omega$$

$$\omega = 0.6197 \text{ rad/s} = 0.620 \text{ rad/s} \qquad \textit{Ans.}$$

Taking the time derivative of Eq. 2 yields

$$\frac{ds}{dt}v_s + s\frac{dv_s}{dt} = 2(\cos \theta)\frac{d\theta}{dt}\omega + 2(\sin \theta)\frac{d\omega}{dt}$$

$$v_s^2 + sa_s = 2(\cos \theta)\omega^2 + 2(\sin \theta)\alpha$$

Since $a_s = dv_s/dt = 0$, then

$$(0.5 \text{ m/s})^2 + 0 = 2 \cos 30°(0.6197 \text{ rad/s})^2 + 2 \sin 30°\alpha$$

$$\alpha = -0.415 \text{ rad/s}^2 \qquad \textit{Ans.}$$

Because the result is negative, it indicates the window has an angular deceleration.

PROBLEMS

16–39. The bar DC rotates uniformly about the shaft at D with a constant angular velocity $\boldsymbol{\omega}$. Determine the velocity and acceleration of the bar AB, which is confined by the guides to move vertically.

16–41. At the instant $\theta = 50°$, the slotted guide is moving upward with an acceleration of 3 m/s^2 and a velocity of 2 m/s. Determine the angular acceleration and angular velocity of link AB at this instant. *Note:* The upward motion of the guide is in the negative y direction.

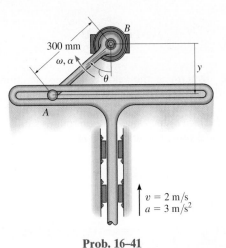

Prob. 16–41

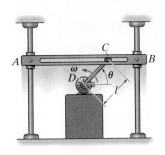

Prob. 16–39

16–42. The mechanism shown is known as a Nuremberg scissors. If the hook at C moves with a constant velocity of $\mathbf{v}$, determine the velocity and acceleration of collar A as a function of θ. The collar slides freely along the vertical guide.

*16–40. The mechanism is used to convert the constant circular motion ω of rod AB into translating motion of rod CD and the attached vertical slot. Determine the velocity and acceleration of CD for any angle θ of AB.

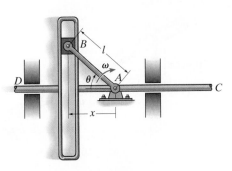

Prob. 16–40

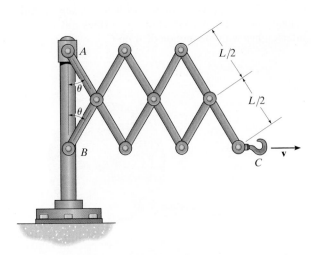

Prob. 16–42

16–43. The crankshaft AB is rotating at a constant angular velocity of $\omega = 150$ rad/s. Determine the velocity of the piston P at the instant $\theta = 30°$.

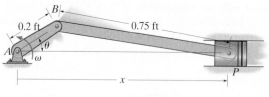

Prob. 16–43

*****16–44.** Determine the velocity and acceleration of the follower rod CD as a function of θ when the contact between the cam and follower is along the straight region AB on the face of the cam. The cam rotates with a constant counterclockwise angular velocity ω.

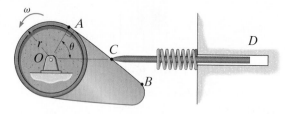

Prob. 16–44

16–45. Determine the velocity of rod R for any angle θ of the cam C if the cam rotates with a constant angular velocity ω. The pin connection at O does not cause an interference with the motion of A on C.

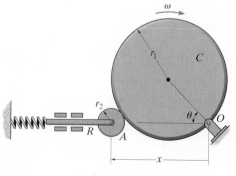

Prob. 16–45

16–46. The bridge girder G of a bascule bridge is raised and lowered using the drive mechanism shown. If the hydraulic cylinder AB shortens at a constant rate of 0.15 m/s, determine the angular velocity of the bridge girder at the instant $\theta = 60°$.

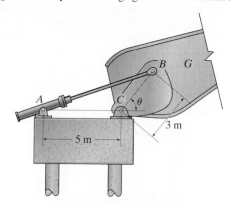

Prob. 16–46

16–47. The circular cam of radius r is rotating clockwise with a constant angular velocity ω about the pin at O, which is at an eccentric distance e from the center of the cam. Determine the velocity and acceleration of the follower rod A as a function of θ.

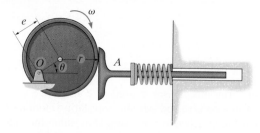

Prob. 16–47

*****16–48.** Peg B mounted on hydraulic cylinder BD slides freely along the slot in link AC. If the hydraulic cylinder extends at a constant rate of 0.5 m/s, determine the angular velocity and angular acceleration of the link at the instant $\theta = 45°$.

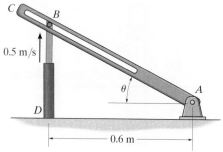

Prob. 16–48

16–49. Bar *AB* rotates uniformly about the fixed pin *A* with a constant angular velocity ω. Determine the velocity and acceleration of block *C*, at the instant $\theta = 60°$.

16–51. The bar is confined to move along the vertical and inclined planes. If the velocity of the roller at *A* is $v_A = 6$ ft/s when $\theta = 45°$, determine the bar's angular velocity and the velocity of roller *B* at this instant.

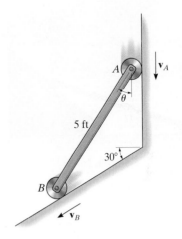

Prob. 16–51

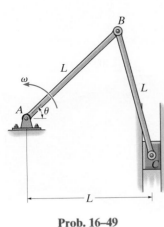

Prob. 16–49

*16–52.** Arm *AB* has an angular velocity of ω and an angular acceleration of α. If no slipping occurs between the disk and the fixed curved surface, determine the angular velocity and angular acceleration of the disk.

16–50. The block moves to the left with a constant velocity v_0. Determine the angular velocity and angular acceleration of the bar as a function of θ.

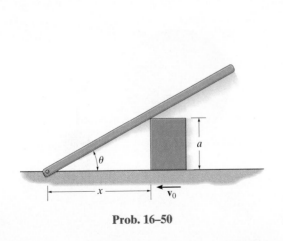

Prob. 16–50

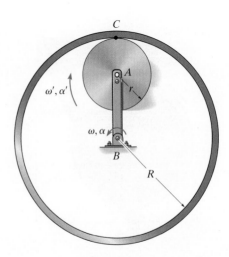

Prob. 16–52

16–53. If the wedge moves to the left with a constant velocity **v**, determine the angular velocity of the rod as a function of θ.

16–55. The Geneva wheel A provides intermittent rotary motion ω_A for continuous motion $\omega_D = 2 \text{ rad/s}$ of disk D. By choosing $d = 100\sqrt{2}$ mm, the wheel has zero angular velocity at the instant pin B enters or leaves one of the four slots. Determine the magnitude of the angular velocity ω_A of the Geneva wheel at any angle θ for which pin B is in contact with the slot.

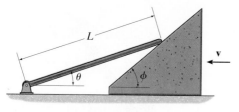

Prob. 16–53

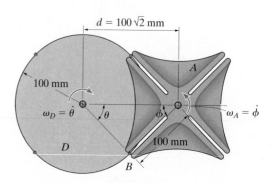

Prob. 16–55

16–54. The slotted yoke is pinned at A while end B is used to move the ram R horizontally. If the disk rotates with a constant angular velocity ω, determine the velocity and acceleration of the ram. The crank pin C is fixed to the disk and turns with it.

*16–56. At the instant shown, the disk is rotating with an angular velocity of ω and has an angular acceleration of α. Determine the velocity and acceleration of cylinder B at this instant. Neglect the size of the pulley at C.

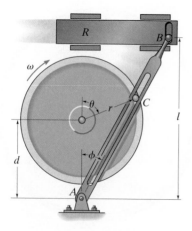

Prob. 16–54

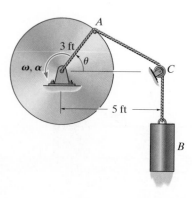

Prob. 16–56

16.5   Relative-Motion Analysis: Velocity

The general plane motion of a rigid body can be described as a *combination* of translation and rotation. To view these "component" motions *separately* we will use a *relative-motion analysis* involving two sets of coordinate axes. The x, y coordinate system is fixed and measures the *absolute* position of two points A and B on the body, here represented as a bar, Fig. 16–11a. The origin of the x', y' coordinate system will be attached to the selected "base point" A, which generally has a *known* motion. The axes of this coordinate system *translate* with respect to the fixed frame but do not rotate with the bar.

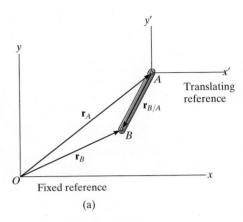

Fig. 16–11

Position.   The position vector $\mathbf{r}_A$ in Fig. 16–11a specifies the location of the "base point" A, and the relative-position vector $\mathbf{r}_{B/A}$ locates point B with respect to point A. By vector addition, the *position* of B is then

$$\mathbf{r}_B = \mathbf{r}_A + \mathbf{r}_{B/A}$$

Displacement.   During an instant of time dt, points A and B undergo displacements $d\mathbf{r}_A$ and $d\mathbf{r}_B$ as shown in Fig. 16–11b. If we consider the general plane motion by its component parts then the *entire bar* first *translates* by an amount $d\mathbf{r}_A$ so that A, the base point, moves to its *final position* and point B moves to B', Fig. 16–11c. The bar is then *rotated* about A by an amount $d\theta$ so that B' undergoes a *relative displacement* $d\mathbf{r}_{B/A}$ and thus moves to its final position B. Due to the rotation about A, $dr_{B/A} = r_{B/A}\, d\theta$, and the displacement of B is

$$d\mathbf{r}_B = d\mathbf{r}_A + d\mathbf{r}_{B/A}$$

due to rotation about A
due to translation of A
due to translation and rotation

16

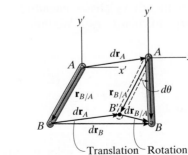

Time t   Time $t + dt$

General plane
motion

(b)

Translation   Rotation

(c)

As slider block *A* moves horizontally to the left with a velocity $\mathbf{v}_A$, it causes crank *CB* to rotate counterclockwise, such that $\mathbf{v}_B$ is directed tangent to its circular path, i.e., upward to the left. The connecting rod *AB* is subjected to general plane motion, and at the instant shown it has an angular velocity $\boldsymbol{\omega}$.

Velocity. To determine the relation between the velocities of points *A* and *B*, it is necessary to take the time derivative of the position equation, or simply divide the displacement equation by *dt*. This yields

$$\frac{d\mathbf{r}_B}{dt} = \frac{d\mathbf{r}_A}{dt} + \frac{d\mathbf{r}_{B/A}}{dt}$$

The terms $d\mathbf{r}_B/dt = \mathbf{v}_B$ and $d\mathbf{r}_A/dt = \mathbf{v}_A$ are measured with respect to the fixed *x, y* axes and represent the *absolute velocities* of points *A* and *B*, respectively. Since the relative displacement is caused by a rotation, the magnitude of the third term is $d\mathbf{r}_{B/A}/dt = r_{B/A}\, d\theta/dt = r_{B/A}\dot{\theta} = r_{B/A}\omega$, where ω is the angular velocity of the body at the instant considered. We will denote this term as the *relative velocity* $\mathbf{v}_{B/A}$, since it represents the velocity of *B* with respect to *A* as measured by an observer fixed to the translating *x′, y′* axes. In other words, *the bar appears to move as if it were rotating with an angular velocity $\boldsymbol{\omega}$ about the z′ axis passing through A*. Consequently, $\mathbf{v}_{B/A}$ has a magnitude of $v_{B/A} = \omega r_{B/A}$ and a *direction* which is perpendicular to $\mathbf{r}_{B/A}$. We therefore have

$$\boxed{\mathbf{v}_B = \mathbf{v}_A + \mathbf{v}_{B/A}} \tag{16–15}$$

where

$$\mathbf{v}_B = \text{velocity of point } B$$
$$\mathbf{v}_A = \text{velocity of the base point } A$$
$$\mathbf{v}_{B/A} = \text{velocity of } B \text{ with respect to } A$$

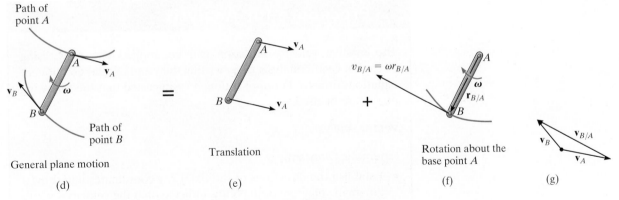

General plane motion

(d)

Translation

(e)

Rotation about the
base point A

(f)

(g)

Fig. 16–11 (cont.)

What this equation states is that the velocity of B, Fig. 16–11d, is determined by considering the entire bar to translate with a velocity of $\mathbf{v}_A$, Fig. 16–11e, and rotate about A with an angular velocity $\boldsymbol{\omega}$, Fig. 16–11f. Vector addition of these two effects, applied to B, yields $\mathbf{v}_B$, as shown in Fig. 16–11g.

Since the relative velocity $\mathbf{v}_{B/A}$ represents the effect of *circular motion*, about A, this term can be expressed by the cross product $\mathbf{v}_{B/A} = \boldsymbol{\omega} \times \mathbf{r}_{B/A}$, Eq. 16–9. Hence, for application using Cartesian vector analysis, we can also write Eq. 16–15 as

$$\mathbf{v}_B = \mathbf{v}_A + \boldsymbol{\omega} \times \mathbf{r}_{B/A} \qquad (16\text{–}16)$$

where

$$
\begin{aligned}
\mathbf{v}_B &= \text{velocity of } B \\
\mathbf{v}_A &= \text{velocity of the base point } A \\
\boldsymbol{\omega} &= \text{angular velocity of the body} \\
\mathbf{r}_{B/A} &= \text{position vector directed from } A \text{ to } B
\end{aligned}
$$

The velocity equation 16–15 or 16–16 may be used in a practical manner to study the general plane motion of a rigid body which is either pin connected to or in contact with other moving bodies. When applying this equation, points A and B should generally be selected as points on the body which are pin-connected to other bodies, or as points in contact with adjacent bodies which have a *known motion*. For example, point A on link AB in Fig. 16–12a must move along a horizontal path, whereas point B moves on a circular path. The *directions* of $\mathbf{v}_A$ and $\mathbf{v}_B$ can therefore be established since they are always *tangent* to their paths of motion, Fig. 16–12b. In the case of the wheel in Fig. 16–13, which rolls *without slipping*, point A on the wheel can be selected at the ground. Here A (momentarily) has zero velocity since the ground does not move. Furthermore, the center of the wheel, B, moves along a horizontal path so that $\mathbf{v}_B$ is horizontal.

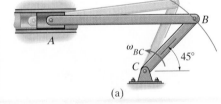

(a)

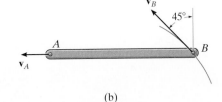

(b)

Fig. 16–12

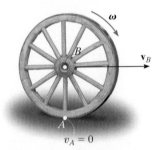

Fig. 16–13

Procedure for Analysis

The relative velocity equation can be applied either by using Cartesian vector analysis, or by writing the x and y scalar component equations directly. For application, it is suggested that the following procedure be used.

Vector Analysis

Kinematic Diagram.

• Establish the directions of the fixed x, y coordinates and draw a kinematic diagram of the body. Indicate on it the velocities $\mathbf{v}_A$, $\mathbf{v}_B$ of points A and B, the angular velocity $\boldsymbol{\omega}$, and the relative-position vector $\mathbf{r}_{B/A}$.

• If the magnitudes of $\mathbf{v}_A$, $\mathbf{v}_B$, or $\boldsymbol{\omega}$ are unknown, the sense of direction of these vectors can be assumed.

Velocity Equation.

• To apply $\mathbf{v}_B = \mathbf{v}_A + \boldsymbol{\omega} \times \mathbf{r}_{B/A}$, express the vectors in Cartesian vector form and substitute them into the equation. Evaluate the cross product and then equate the respective $\mathbf{i}$ and $\mathbf{j}$ components to obtain two scalar equations.

• If the solution yields a *negative* answer for an *unknown* magnitude, it indicates the sense of direction of the vector is opposite to that shown on the kinematic diagram.

Scalar Analysis

Kinematic Diagram.

• If the velocity equation is to be applied in scalar form, then the magnitude and direction of the relative velocity $\mathbf{v}_{B/A}$ must be established. Draw a kinematic diagram such as shown in Fig. 16–11g, which shows the relative motion. Since the body is considered to be "pinned" momentarily at the base point A, the magnitude of $\mathbf{v}_{B/A}$ is $v_{B/A} = \omega r_{B/A}$. The sense of direction of $\mathbf{v}_{B/A}$ is always perpendicular to $\mathbf{r}_{B/A}$ in accordance with the rotational motion $\boldsymbol{\omega}$ of the body.*

Velocity Equation.

• Write Eq. 16–15 in symbolic form, $\mathbf{v}_B = \mathbf{v}_A + \mathbf{v}_{B/A}$, and underneath each of the terms represent the vectors graphically by showing their magnitudes and directions. The scalar equations are determined from the x and y components of these vectors.

*The notation $\mathbf{v}_B = \mathbf{v}_A + \mathbf{v}_{B/A\text{(pin)}}$ may be helpful in recalling that A is "pinned."

EXAMPLE 16.6

The link shown in Fig. 16–14a is guided by two blocks at A and B, which move in the fixed slots. If the velocity of A is 2 m/s downward, determine the velocity of B at the instant $\theta = 45°$.

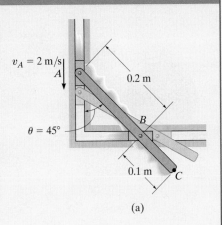

(a)

SOLUTION I (VECTOR ANALYSIS)

Kinematic Diagram.   Since points A and B are restricted to move along the fixed slots and $\mathbf{v}_A$ is directed downward, then velocity $\mathbf{v}_B$ must be directed horizontally to the right, Fig. 16–14b. This motion causes the link to rotate counterclockwise; that is, by the right-hand rule the angular velocity $\boldsymbol{\omega}$ is directed outward, perpendicular to the plane of motion.

Velocity Equation.   Expressing each of the vectors in Fig. 16–14b in terms of their $\mathbf{i}, \mathbf{j}, \mathbf{k}$ components and applying Eq. 16–16 to A, the base point, and B, we have

$$\mathbf{v}_B = \mathbf{v}_A + \boldsymbol{\omega} \times \mathbf{r}_{B/A}$$
$$v_B\mathbf{i} = -2\mathbf{j} + [\omega\mathbf{k} \times (0.2 \sin 45°\mathbf{i} - 0.2 \cos 45°\mathbf{j})]$$
$$v_B\mathbf{i} = -2\mathbf{j} + 0.2\omega \sin 45°\mathbf{j} + 0.2\omega \cos 45°\mathbf{i}$$

Equating the **i** and **j** components gives

$$v_B = 0.2\omega \cos 45° \quad 0 = -2 + 0.2\omega \sin 45°$$

Thus,

$$\omega = 14.1 \text{ rad/s} \circlearrowleft \qquad v_B = 2 \text{ m/s} \rightarrow \qquad Ans.$$

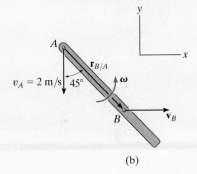

(b)

SOLUTION II (SCALAR ANALYSIS)

The kinematic diagram of the relative "circular motion" which produces $\mathbf{v}_{B/A}$ is shown in Fig. 16–14c. Here $v_{B/A} = \omega(0.2 \text{ m})$.
   Thus,

$$v_B = v_A + v_{B/A}$$

$$\begin{bmatrix} v_B \\ \rightarrow \end{bmatrix} = \begin{bmatrix} 2 \text{ m/s} \\ \downarrow \end{bmatrix} + \begin{bmatrix} \omega(0.2 \text{ m}) \\ \measuredangle 45° \end{bmatrix}$$

$(\xrightarrow{+})$ $\qquad v_B = 0 + \omega(0.2) \cos 45°$

$(+\uparrow)$ $\qquad 0 = -2 + \omega(0.2) \sin 45°$

   The solution produces the above results.
   It should be emphasized that these results are *valid only* at the instant $\theta = 45°$. A recalculation for $\theta = 44°$ yields $v_B = 2.07$ m/s and $\omega = 14.4$ rad/s; whereas when $\theta = 46°$, $v_B = 1.93$ m/s and $\omega = 13.9$ rad/s, etc.

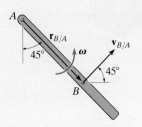

Relative motion

(c)

Fig. 16–14

NOTE: Since v_A and ω are *known*, the velocity of any other point on the link can be determined. As an exercise, see if you can apply Eq. 16–16 to points A and C or to points B and C and show that when $\theta = 45°$, $v_C = 3.16$ m/s, directed at an angle of 18.4° up from the horizontal.

EXAMPLE 16.7

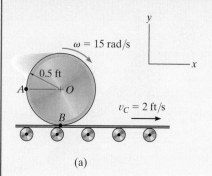

(a)

The cylinder shown in Fig. 16–15a rolls without slipping on the surface of a conveyor belt which is moving at 2 ft/s. Determine the velocity of point A. The cylinder has a clockwise angular velocity $\omega = 15$ rad/s at the instant shown.

SOLUTION I (VECTOR ANALYSIS)

Kinematic Diagram. Since no slipping occurs, point B on the cylinder has the same velocity as the conveyor, Fig. 16–15b. Also, the angular velocity of the cylinder is known, so we can apply the velocity equation to B, the base point, and A to determine $\mathbf{v}_A$.

Velocity Equation

$$\mathbf{v}_A = \mathbf{v}_B + \boldsymbol{\omega} \times \mathbf{r}_{A/B}$$

$$(v_A)_x\mathbf{i} + (v_A)_y\mathbf{j} = 2\mathbf{i} + (-15\mathbf{k}) \times (-0.5\mathbf{i} + 0.5\mathbf{j})$$

$$(v_A)_x\mathbf{i} + (v_A)_y\mathbf{j} = 2\mathbf{i} + 7.50\mathbf{j} + 7.50\mathbf{i}$$

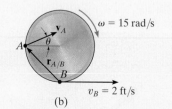

(b)

so that

$$(v_A)_x = 2 + 7.50 = 9.50 \text{ ft/s} \tag{1}$$

$$(v_A)_y = 7.50 \text{ ft/s} \tag{2}$$

Thus,

$$v_A = \sqrt{(9.50)^2 + (7.50)^2} = 12.1 \text{ ft/s} \qquad Ans.$$

$$\theta = \tan^{-1}\frac{7.50}{9.50} = 38.3° \; \measuredangle \qquad Ans.$$

SOLUTION II (SCALAR ANALYSIS)

As an alternative procedure, the scalar components of $\mathbf{v}_A = \mathbf{v}_B + \mathbf{v}_{A/B}$ can be obtained directly. From the kinematic diagram showing the relative "circular" motion which produces $\mathbf{v}_{A/B}$, Fig. 16–15c, we have

$$v_{A/B} = \omega r_{A/B} = (15 \text{ rad/s})\left(\frac{0.5 \text{ ft}}{\cos 45°}\right) = 10.6 \text{ ft/s}$$

Thus,

$$\mathbf{v}_A = \mathbf{v}_B + \mathbf{v}_{A/B}$$

$$\begin{bmatrix} (v_A)_x \\ \rightarrow \end{bmatrix} + \begin{bmatrix} (v_A)_y \\ \uparrow \end{bmatrix} = \begin{bmatrix} 2 \text{ ft/s} \\ \rightarrow \end{bmatrix} + \begin{bmatrix} 10.6 \text{ ft/s} \\ \measuredangle 45° \end{bmatrix}$$

Relative motion

(c)

Fig. 16–15

Equating the x and y components gives the same results as before, namely,

$$(\overset{+}{\rightarrow}) \qquad (v_A)_x = 2 + 10.6\cos 45° = 9.50 \text{ ft/s}$$

$$(+\uparrow) \qquad (v_A)_y = 0 + 10.6\sin 45° = 7.50 \text{ ft/s}$$

EXAMPLE | 16.8

The collar C in Fig. 16–16a is moving downward with a velocity of 2 m/s. Determine the angular velocity of CB at this instant.

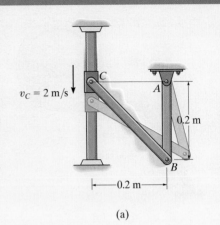

(a)

SOLUTION I (VECTOR ANALYSIS)

Kinematic Diagram. The downward motion of C causes B to move to the right along a curved path. Also, CB and AB rotate counterclockwise.

Velocity Equation. Link CB (general plane motion): See Fig. 16–16b.

$$\mathbf{v}_B = \mathbf{v}_C + \boldsymbol{\omega}_{CB} \times \mathbf{r}_{B/C}$$
$$v_B\mathbf{i} = -2\mathbf{j} + \omega_{CB}\mathbf{k} \times (0.2\mathbf{i} - 0.2\mathbf{j})$$
$$v_B\mathbf{i} = -2\mathbf{j} + 0.2\omega_{CB}\mathbf{j} + 0.2\omega_{CB}\mathbf{i}$$

$$v_B = 0.2\omega_{CB} \qquad (1)$$
$$0 = -2 + 0.2\omega_{CB} \qquad (2)$$
$$\omega_{CB} = 10 \text{ rad/s} \; \circlearrowright \qquad \textit{Ans.}$$
$$v_B = 2 \text{ m/s} \rightarrow$$

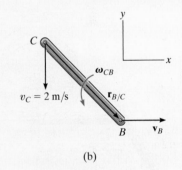

(b)

SOLUTION II (SCALAR ANALYSIS)

The scalar component equations of $\mathbf{v}_B = \mathbf{v}_C + \mathbf{v}_{B/C}$ can be obtained directly. The kinematic diagram in Fig. 16–16c shows the relative "circular" motion which produces $\mathbf{v}_{B/C}$. We have

$$\mathbf{v}_B = \mathbf{v}_C + \mathbf{v}_{B/C}$$
$$\begin{bmatrix} v_B \\ \rightarrow \end{bmatrix} = \begin{bmatrix} 2 \text{ m/s} \\ \downarrow \end{bmatrix} + \begin{bmatrix} \omega_{CB}\left(0.2\sqrt{2} \text{ m}\right) \\ \angle 45° \end{bmatrix}$$

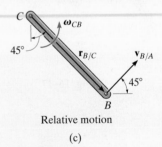

Relative motion

(c)

Resolving these vectors in the x and y directions yields

$$(\xrightarrow{+}) \qquad v_B = 0 + \omega_{CB}\left(0.2\sqrt{2} \cos 45°\right)$$
$$(+\uparrow) \qquad 0 = -2 + \omega_{CB}\left(0.2\sqrt{2} \sin 45°\right)$$

which is the same as Eqs. 1 and 2.

NOTE: Since link AB rotates about a fixed axis and v_B is known, Fig. 16–16d, its angular velocity is found from $v_B = \omega_{AB}r_{AB}$ or 2 m/s $= \omega_{AB}(0.2$ m$)$, $\omega_{AB} = 10$ rad/s.

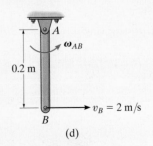

(d)

Fig. 16–16

16

EXAMPLE | 16.9

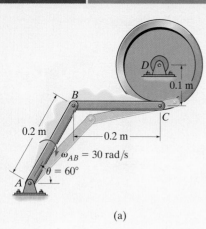

(a)

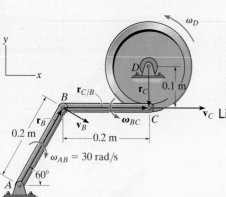

(b)

Fig. 16–17

The bar AB of the linkage shown in Fig. 16–17a has a clockwise angular velocity of 30 rad/s when $\theta = 60°$. Determine the angular velocities of member BC and the wheel at this instant.

SOLUTION (VECTOR ANALYSIS)

Kinematic Diagram.   By inspection, the velocities of points B and C are defined by the rotation of link AB and the wheel about their fixed axes. The position vectors and the angular velocity of each member are shown on the kinematic diagram in Fig. 16–17b. To solve, we will write the appropriate kinematic equation for each member.

Velocity Equation.   Link AB (rotation about a fixed axis):

$$\mathbf{v}_B = \boldsymbol{\omega}_{AB} \times \mathbf{r}_B$$

$$= (-30\mathbf{k}) \times (0.2 \cos 60°\mathbf{i} + 0.2 \sin 60°\mathbf{j})$$

$$= \{5.20\mathbf{i} - 3.0\mathbf{j}\} \text{ m/s}$$

Link BC (general plane motion):

$$\mathbf{v}_C = \mathbf{v}_B + \boldsymbol{\omega}_{BC} \times \mathbf{r}_{C/B}$$

$$v_C\mathbf{i} = 5.20\mathbf{i} - 3.0\mathbf{j} + (\omega_{BC}\mathbf{k}) \times (0.2\mathbf{i})$$

$$v_C\mathbf{i} = 5.20\mathbf{i} + (0.2\omega_{BC} - 3.0)\mathbf{j}$$

$$v_C = 5.20 \text{ m/s}$$

$$0 = 0.2\omega_{BC} - 3.0$$

$$\omega_{BC} = 15 \text{ rad/s} \circlearrowright \qquad\qquad\qquad Ans.$$

Wheel (rotation about a fixed axis):

$$\mathbf{v}_C = \boldsymbol{\omega}_D \times \mathbf{r}_C$$

$$5.20\mathbf{i} = (\omega_D\mathbf{k}) \times (-0.1\mathbf{j})$$

$$5.20 = 0.1\omega_D$$

$$\omega_D = 52.0 \text{ rad/s} \circlearrowright \qquad\qquad\qquad Ans.$$

FUNDAMENTAL PROBLEMS

F16–7. If roller *A* moves to the right with a constant velocity of $v_A = 3$ m/s, determine the angular velocity of the link and the velocity of roller *B* at the instant $\theta = 30°$.

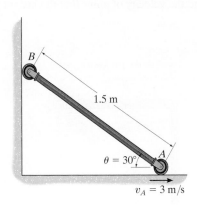

F16–7

F16–8. The wheel rolls without slipping with an angular velocity of $\omega = 10$ rad/s. Determine the magnitude of the velocity of point *B* at the instant shown.

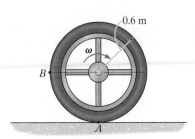

F16–8

F16–9. Determine the angular velocity of the spool. The cable wraps around the inner core, and the spool does not slip on the platform *P*.

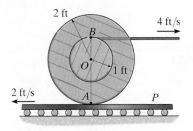

F16–9

F16–10. If crank *OA* rotates with an angular velocity of $\omega = 12$ rad/s, determine the velocity of piston *B* and the angular velocity of rod *AB* at the instant shown.

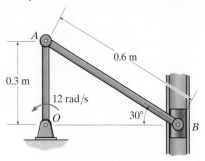

F16–10

F16–11. If rod *AB* slides along the horizontal slot with a velocity of 60 ft/s, determine the angular velocity of link *BC* at the instant shown.

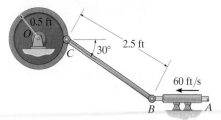

F16–11

F16–12. End *A* of the link has a velocity of $v_A = 3$ m/s. Determine the velocity of the peg at *B* at this instant. The peg is constrained to move along the slot.

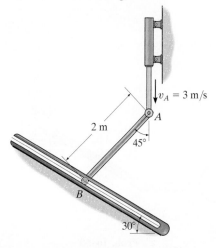

F16–12

16

PROBLEMS

16–57. If h and θ are known, and the speed of A and B is $v_A = v_B = v$, determine the angular velocity $\boldsymbol{\omega}$ of the body and the direction ϕ of $\mathbf{v}_B$.

16–59. The velocity of the slider block C is 4 ft/s up the inclined groove. Determine the angular velocity of links AB and BC and the velocity of point B at the instant shown.

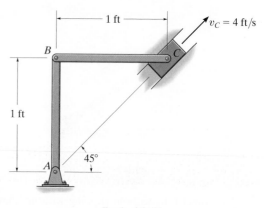

Prob. 16–59

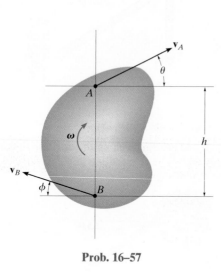

Prob. 16–57

*16–60. The epicyclic gear train consists of the sun gear A which is in mesh with the planet gear B. This gear has an inner hub C which is fixed to B and in mesh with the fixed ring gear R. If the connecting link DE pinned to B and C is rotating at $\omega_{DE} = 18$ rad/s about the pin at E, determine the angular velocities of the planet and sun gears.

16–58. If the block at C is moving downward at 4 ft/s, determine the angular velocity of bar AB at the instant shown.

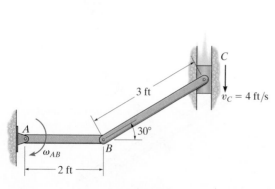

Prob. 16–58

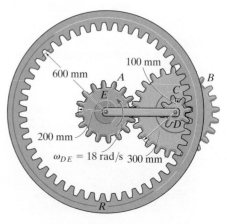

Prob. 16–60

16–61. Determine the velocity of the block at the instant $\theta = 60°$, if link AB is rotating at 4 rad/s.

16–63. If the angular velocity of link AB is $\omega_{AB} = 3$ rad/s, determine the velocity of the block at C and the angular velocity of the connecting link CB at the instant $\theta = 45°$ and $\phi = 30°$.

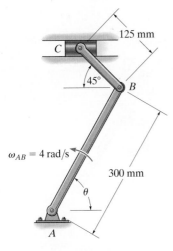

Prob. 16–61

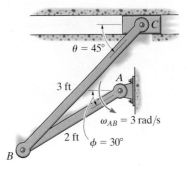

Prob. 16–63

16–62. If the flywheel is rotating with an angular velocity of $\omega_A = 6$ rad/s, determine the angular velocity of rod BC at the instant shown.

*16–64.** The pinion gear A rolls on the fixed gear rack B with an angular velocity $\omega = 4$ rad/s. Determine the velocity of the gear rack C.

16–65. The pinion gear rolls on the gear racks. If B is moving to the right at 8 ft/s and C is moving to the left at 4 ft/s, determine the angular velocity of the pinion gear and the velocity of its center A.

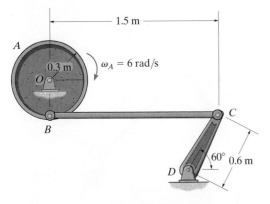

Prob. 16–62

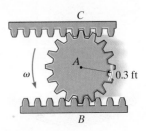

Probs. 16–64/65

16–66. Determine the angular velocity of the gear and the velocity of its center O at the instant shown.

16–67. Determine the velocity of point A on the rim of the gear at the instant shown.

16–69. If the gear rotates with an angular velocity of $\omega = 10 \text{ rad/s}$ and the gear rack moves at $v_C = 5 \text{ m/s}$, determine the velocity of the slider block A at the instant shown.

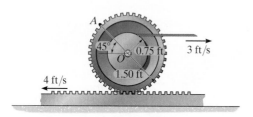

Probs. 16–66/67

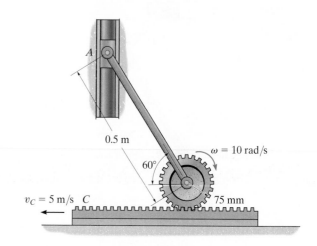

Prob. 16–69

***16–68.** Part of an automatic transmission consists of a *fixed* ring gear R, three equal planet gears P, the sun gear S, and the planet carrier C, which is shaded. If the sun gear is rotating at $\omega_S = 6 \text{ rad/s}$, determine the angular velocity ω_C of the *planet carrier*. Note that C is pin-connected to the center of each of the planet gears.

16–70. If the slider block C is moving at $v_C = 3 \text{ m/s}$, determine the angular velocity of BC and the crank AB at the instant shown.

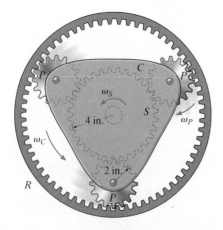

Prob. 16–68

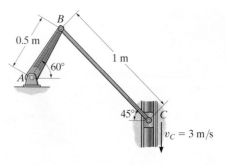

Prob. 16–70

16–71. The two-cylinder engine is designed so that the pistons are connected to the crankshaft BE using a master rod ABC and articulated rod AD. If the crankshaft is rotating at $\omega = 30\ \text{rad/s}$, determine the velocities of the pistons C and D at the instant shown.

16–74. If crank AB rotates with a constant angular velocity of $\omega_{AB} = 6\ \text{rad/s}$, determine the angular velocity of rod BC and the velocity of the slider block at the instant shown. The rod is in a horizontal position.

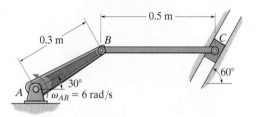

Prob. 16–74

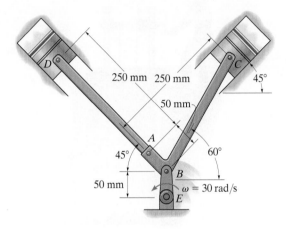

Prob. 16–71

*16–72.** Determine the velocity of the center O of the spool when the cable is pulled to the right with a velocity of $\mathbf{v}$. The spool rolls without slipping.

16–73. Determine the velocity of point A on the outer rim of the spool at the instant shown when the cable is pulled to the right with a velocity of $\mathbf{v}$. The spool rolls without slipping.

16–75. If the slider block A is moving downward at $v_A = 4\ \text{m/s}$, determine the velocity of block B at the instant shown.

*16–76.** If the slider block A is moving downward at $v_A = 4\ \text{m/s}$, determine the velocity of block C at the instant shown.

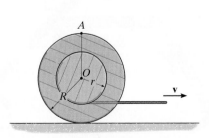

Probs. 16–72/73

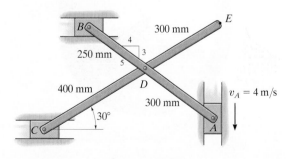

Probs. 16–75/76

16–77. The planetary gear system is used in an automatic transmission for an automobile. By locking or releasing certain gears, it has the advantage of operating the car at different speeds. Consider the case where the ring gear R is held fixed, $\omega_R = 0$, and the sun gear S is rotating at $\omega_S = 5$ rad/s. Determine the angular velocity of each of the planet gears P and shaft A.

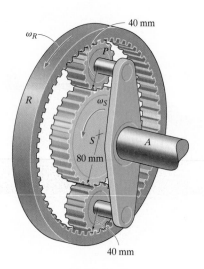

Prob. 16–77

16–78. If the ring gear D rotates counterclockwise with an angular velocity of $\omega_D = 5$ rad/s while link AB rotates clockwise with an angular velocity of $\omega_{AB} = 10$ rad/s, determine the angular velocity of gear C.

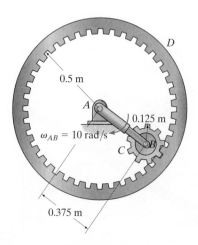

Prob. 16–78

16–79. The differential drum operates in such a manner that the rope is unwound from the small drum B and wound up on the large drum A. If the radii of the large and small drums are R and r, respectively, and for the pulley it is $(R + r)/2$, determine the speed at which the bucket C rises if the man rotates the handle with a constant angular velocity of ω. Neglect the thickness of the rope.

Prob. 16–79

*16–80. Mechanical toy animals often use a walking mechanism as shown idealized in the figure. If the driving crank AB is propelled by a spring motor such that $\omega_{AB} = 5$ rad/s, determine the velocity of the rear foot E at the instant shown. Although not part of this problem, the upper end of the foreleg has a slotted guide which is constrained by the fixed pin at G.

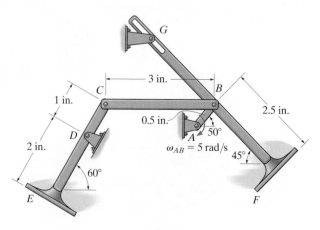

Prob. 16–80

16.6   Instantaneous Center of Zero Velocity

The velocity of any point B located on a rigid body can be obtained in a very direct way by choosing the base point A to be a point that has *zero velocity* at the instant considered. In this case, $\mathbf{v}_A = \mathbf{0}$, and therefore the velocity equation, $\mathbf{v}_B = \mathbf{v}_A + \boldsymbol{\omega} \times \mathbf{r}_{B/A}$, becomes $\mathbf{v}_B = \boldsymbol{\omega} \times \mathbf{r}_{B/A}$. For a body having general plane motion, point A so chosen is called the *instantaneous center of zero velocity (IC)*, and it lies on the *instantaneous axis of zero velocity*. This axis is always perpendicular to the plane of motion, and the intersection of the axis with this plane defines the location of the *IC*. Since point A coincides with the *IC*, then $\mathbf{v}_B = \boldsymbol{\omega} \times \mathbf{r}_{B/IC}$ and so point B moves momentarily about the *IC* in a *circular path*; in other words, the body appears to rotate about the instantaneous axis. The *magnitude* of $\mathbf{v}_B$ is simply $v_B = \omega r_{B/IC}$, where ω is the angular velocity of the body. Due to the circular motion, the *direction* of $\mathbf{v}_B$ must always be *perpendicular* to $\mathbf{r}_{B/IC}$.

For example, the *IC* for the bicycle wheel in Fig. 16–18 is at the contact point with the ground. There the spokes are somewhat visible, whereas at the top of the wheel they become blurred. If one imagines that the wheel is momentarily pinned at this point, the velocities of various points can be found using $v = \omega r$. Here the radial distances shown in the photo, Fig. 16–18, must be determined from the geometry of the wheel.

16

Fig. 16–18

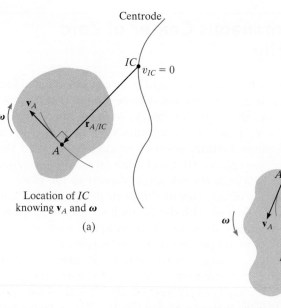

Location of *IC*
knowing $\mathbf{v}_A$ and $\boldsymbol{\omega}$

(a)

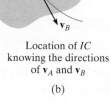

Location of *IC*
knowing the directions
of $\mathbf{v}_A$ and $\mathbf{v}_B$

(b)

Fig. 16–19

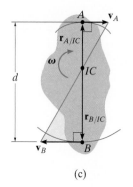

(c)

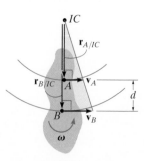

Location of *IC*
knowing $\mathbf{v}_A$ and $\mathbf{v}_B$

(d)

Location of the *IC*. To locate the *IC* we can use the fact that the *velocity* of a point on the body is *always perpendicular* to the *relative-position vector* directed from the *IC* to the point. Several possibilities exist:

- *The velocity $\mathbf{v}_A$ of a point A on the body and the angular velocity $\boldsymbol{\omega}$ of the body are known,* Fig. 16–19a. In this case, the *IC* is located along the line drawn perpendicular to $\mathbf{v}_A$ at A, such that the distance from A to the *IC* is $r_{A/IC} = v_A/\omega$. Note that the *IC* lies up and to the right of A since $\mathbf{v}_A$ must cause a clockwise angular velocity $\boldsymbol{\omega}$ about the *IC*.

- *The lines of action of two nonparallel velocities $\mathbf{v}_A$ and $\mathbf{v}_B$ are known,* Fig. 16–19b. Construct at points A and B line segments that are perpendicular to $\mathbf{v}_A$ and $\mathbf{v}_B$. Extending these perpendiculars to their *point of intersection* as shown locates the *IC* at the instant considered.

- *The magnitude and direction of two parallel velocities $\mathbf{v}_A$ and $\mathbf{v}_B$ are known.* Here the location of the *IC* is determined by proportional triangles. Examples are shown in Fig. 16–19c and d. In both cases $r_{A/IC} = v_A/\omega$ and $r_{B/IC} = v_B/\omega$. If d is a known distance between points A and B, then in Fig. 16–19c, $r_{A/IC} + r_{B/IC} = d$ and in Fig. 16–19d, $r_{B/IC} - r_{A/IC} = d$.

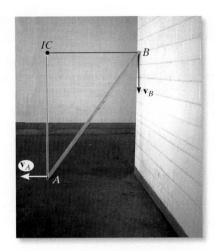

As the board slides downward to the left it is subjected to general plane motion. Since the directions of the velocities of its ends A and B are known, the IC is located as shown. At this instant the board will momentarily rotate about this point. Draw the board in several other positions and establish the IC for each case.

Realize that the point chosen as the instantaneous center of zero velocity for the body *can only be used at the instant considered* since the body changes its position from one instant to the next. The locus of points which define the location of the IC during the body's motion is called a *centrode*, Fig. 16–19a, and so each point on the centrode acts as the IC for the body only for an instant.

Although the IC may be conveniently used to determine the velocity of any point in a body, it generally *does not have zero acceleration* and therefore it *should not* be used for finding the accelerations of points in a body.

Procedure for Analysis

The velocity of a point on a body which is subjected to general plane motion can be determined with reference to its instantaneous center of zero velocity provided the location of the IC is first established using one of the three methods described above.

- As shown on the kinematic diagram in Fig. 16–20, the body is imagined as "extended and pinned" at the IC so that, at the instant considered, it rotates about this pin with its angular velocity $\boldsymbol{\omega}$.

- The *magnitude* of velocity for each of the arbitrary points A, B, and C on the body can be determined by using the equation $v = \omega r$, where r is the radial distance from the IC to each point.

- The line of action of each velocity vector $\mathbf{v}$ is *perpendicular* to its associated radial line $\mathbf{r}$, and the velocity has a *sense of direction* which tends to move the point in a manner consistent with the angular rotation $\boldsymbol{\omega}$ of the radial line, Fig. 16–20.

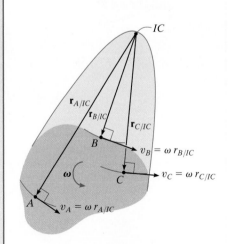

Fig. 16–20

EXAMPLE | **16.10**

Show how to determine the location of the instantaneous center of zero velocity for (a) member BC shown in Fig. 16–21a; and (b) the link CB shown in Fig. 16–21c.

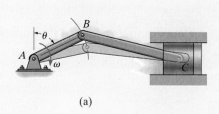

(a)

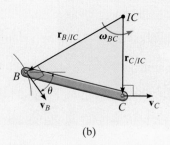

(b)

SOLUTION

Part (a). As shown in Fig. 16–21a, point B moves in a circular path such that $\mathbf{v}_B$ is perpendicular to AB. Therefore, it acts at an angle θ from the horizontal as shown in Fig. 16–21b. The motion of point B causes the piston to move forward *horizontally* with a velocity $\mathbf{v}_C$. When lines are drawn perpendicular to $\mathbf{v}_B$ and $\mathbf{v}_C$, Fig. 16–21b, they intersect at the IC.

Part (b). Points B and C follow circular paths of motion since links AB and DC are each subjected to rotation about a fixed axis, Fig. 16–21c. Since the velocity is always tangent to the path, at the instant considered, $\mathbf{v}_C$ on rod DC and $\mathbf{v}_B$ on rod AB are both directed vertically downward, along the axis of link CB, Fig. 16–21d. Radial lines drawn perpendicular to these two velocities form parallel lines which intersect at "infinity;" i.e., $r_{C/IC} \to \infty$ and $r_{B/IC} \to \infty$. Thus, $\omega_{CB} = (v_C / r_{C/IC}) \to 0$. As a result, link CB momentarily *translates*. An instant later, however, CB will move to a tilted position, causing the IC to move to some finite location.

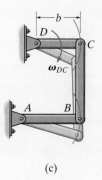

(c)

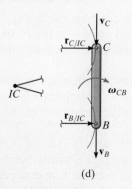

(d)

Fig. 16–21

EXAMPLE | 16.11

Block D shown in Fig. 16–22a moves with a speed of 3 m/s. Determine the angular velocities of links BD and AB, at the instant shown.

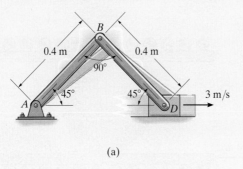

(a)

SOLUTION

As D moves to the right, it causes AB to rotate clockwise about point A. Hence, $\mathbf{v}_B$ is directed perpendicular to AB. The instantaneous center of zero velocity for BD is located at the intersection of the line segments drawn perpendicular to $\mathbf{v}_B$ and $\mathbf{v}_D$, Fig. 16–22b. From the geometry,

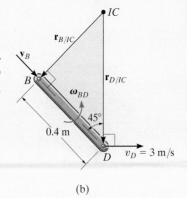

$$r_{B/IC} = 0.4 \tan 45° \text{ m} = 0.4 \text{ m}$$

$$r_{D/IC} = \frac{0.4 \text{ m}}{\cos 45°} = 0.5657 \text{ m}$$

(b)

Since the magnitude of $\mathbf{v}_D$ is known, the angular velocity of link BD is

$$\omega_{BD} = \frac{v_D}{r_{D/IC}} = \frac{3 \text{ m/s}}{0.5657 \text{ m}} = 5.30 \text{ rad/s} \circlearrowright \qquad Ans.$$

The velocity of B is therefore

$$v_B = \omega_{BD}(r_{B/IC}) = 5.30 \text{ rad/s } (0.4 \text{ m}) = 2.12 \text{ m/s} \quad \diagdown 45°$$

From Fig. 16–22c, the angular velocity of AB is

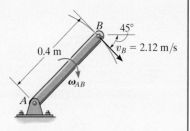

$$\omega_{AB} = \frac{v_B}{r_{B/A}} = \frac{2.12 \text{ m/s}}{0.4 \text{ m}} = 5.30 \text{ rad/s} \circlearrowleft \qquad Ans.$$

NOTE: Try and solve this problem by applying $\mathbf{v}_D = \mathbf{v}_B + \mathbf{v}_{D/B}$ to member BD.

(c)

Fig. 16–22

EXAMPLE   16.12

The cylinder shown in Fig. 16–23a rolls without slipping between the two moving plates E and D. Determine the angular velocity of the cylinder and the velocity of its center C.

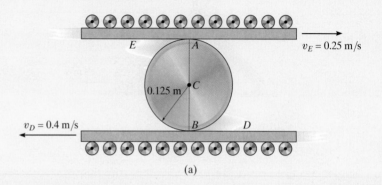

(a)

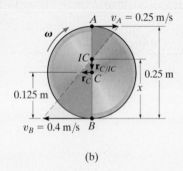

(b)

Fig. 16–23

SOLUTION
Since no slipping occurs, the contact points A and B on the cylinder have the same velocities as the plates E and D, respectively. Furthermore, the velocities $\mathbf{v}_A$ and $\mathbf{v}_B$ are *parallel*, so that by the proportionality of right triangles the IC is located at a point on line AB, Fig. 16–23b. Assuming this point to be a distance x from B, we have

$$v_B = \omega x; \qquad\qquad 0.4 \text{ m/s} = \omega x$$
$$v_A = \omega(0.25 \text{ m} - x); \qquad 0.25 \text{ m/s} = \omega(0.25 \text{ m} - x)$$

Dividing one equation into the other eliminates ω and yields

$$0.4(0.25 - x) = 0.25x$$

$$x = \frac{0.1}{0.65} = 0.1538 \text{ m}$$

Hence, the angular velocity of the cylinder is

$$\omega = \frac{v_B}{x} = \frac{0.4 \text{ m/s}}{0.1538 \text{ m}} = 2.60 \text{ rad/s} \circlearrowright \qquad\qquad \textit{Ans.}$$

The velocity of point C is therefore

$$v_C = \omega r_{C/IC} = 2.60 \text{ rad/s} (0.1538 \text{ m} - 0.125 \text{ m})$$
$$= 0.0750 \text{ m/s} \leftarrow \qquad\qquad \textit{Ans.}$$

EXAMPLE | 16.13

The crankshaft AB turns with a clockwise angular velocity of 10 rad/s, Fig. 16–24a. Determine the velocity of the piston at the instant shown.

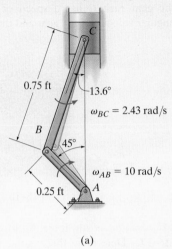

(a)

SOLUTION

The crankshaft rotates about a fixed axis, and so the velocity of point B is

$$v_B = 10 \text{ rad/s} (0.25 \text{ ft}) = 2.50 \text{ ft/s} \ \angle 45°$$

Since the directions of the velocities of B and C are known, then the location of the IC for the connecting rod BC is at the intersection of the lines extended from these points, perpendicular to $\mathbf{v}_B$ and $\mathbf{v}_C$, Fig. 16–24b. The magnitudes of $\mathbf{r}_{B/IC}$ and $\mathbf{r}_{C/IC}$ can be obtained from the geometry of the triangle and the law of sines, i.e.,

$$\frac{0.75 \text{ ft}}{\sin 45°} = \frac{r_{B/IC}}{\sin 76.4°}$$

$$r_{B/IC} = 1.031 \text{ ft}$$

$$\frac{0.75 \text{ ft}}{\sin 45°} = \frac{r_{C/IC}}{\sin 58.6°}$$

$$r_{C/IC} = 0.9056 \text{ ft}$$

The rotational sense of $\boldsymbol{\omega}_{BC}$ must be the same as the rotation caused by $\mathbf{v}_B$ about the IC, which is counterclockwise. Therefore,

$$\omega_{BC} = \frac{v_B}{r_{B/IC}} = \frac{2.5 \text{ ft/s}}{1.031 \text{ ft}} = 2.425 \text{ rad/s}$$

Using this result, the velocity of the piston is

$$v_C = \omega_{BC} r_{C/IC} = (2.425 \text{ rad/s})(0.9056 \text{ ft}) = 2.20 \text{ ft/s} \quad Ans.$$

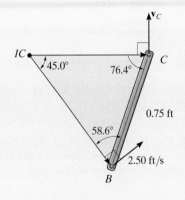

(b)

Fig. 16–24

16

FUNDAMENTAL PROBLEMS

F16–13. Determine the angular velocity of the rod and the velocity of point C at the instant shown.

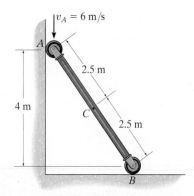

F16–13

F16–14. Determine the angular velocity of link BC and velocity of the piston C at the instant shown.

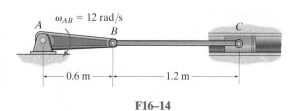

F16–14

F16–15. If the center O of the wheel is moving with a speed of $v_O = 6$ m/s, determine the velocity of point A on the wheel. The gear rack B is fixed.

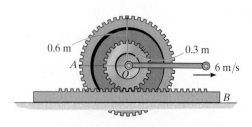

F16–15

F16–16. If cable AB is unwound with a speed of 3 m/s, and the gear rack C has a speed of 1.5 m/s, determine the angular velocity of the gear and the velocity of its center O.

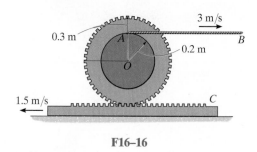

F16–16

F16–17. Determine the angular velocity of link BC and the velocity of the piston C at the instant shown.

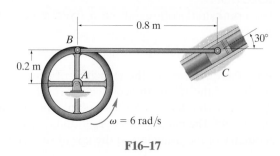

F16–17

F16–18. Determine the angular velocity of links BC and CD at the instant shown.

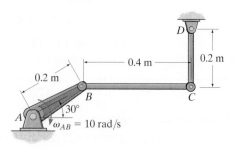

F16–18

PROBLEMS

16–81. In each case show graphically how to locate the instantaneous center of zero velocity of link AB. Assume the geometry is known.

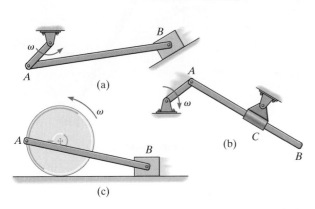

(a)

(b)

(c)

Prob. 16–81

16–83. At the instant shown, the disk is rotating at $\omega = 4$ rad/s. Determine the velocities of points A, B, and C.

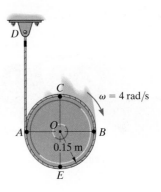

Prob. 16–83

16–82. Determine the angular velocity of link AB at the instant shown if block C is moving upward at 12 in/s.

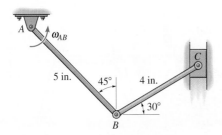

Prob. 16–82

*16–84. If link CD has an angular velocity of $\omega_{CD} = 6$ rad/s, determine the velocity of point B on link BC and the angular velocity of link AB at the instant shown.

16–85. If link CD has an angular velocity of $\omega_{CD} = 6$ rad/s, determine the velocity of point E on link BC and the angular velocity of link AB at the instant shown.

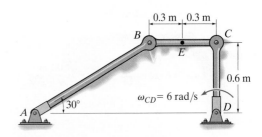

Probs. 16–84/85

16

16–86. At the instant shown, the truck travels to the right at 3 m/s, while the pipe rolls counterclockwise at $\omega = 6$ rad/s without slipping at B. Determine the velocity of the pipe's center G.

*16–88. If link AB is rotating at $\omega_{AB} = 6$ rad/s, determine the angular velocities of links BC and CD at the instant $\theta = 60°$.

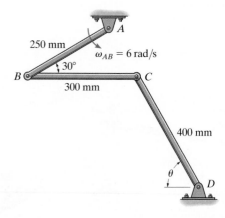

Prob. 16–88

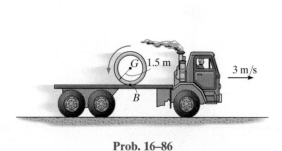

Prob. 16–86

16–89. The oil pumping unit consists of a walking beam AB, connecting rod BC, and crank CD. If the crank rotates at a constant rate of 6 rad/s, determine the speed of the rod hanger H at the instant shown. *Hint:* Point B follows a circular path about point E and therefore the velocity of B is *not* vertical.

16–87. If crank AB is rotating with an angular velocity of $\omega_{AB} = 6$ rad/s, determine the velocity of the center O of the gear at the instant shown.

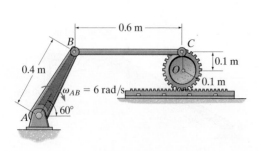

Prob. 16–87

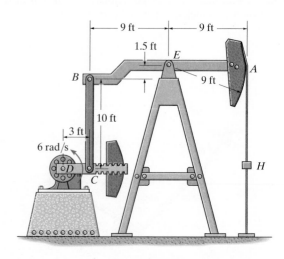

Prob. 16–89

16–90.   Due to slipping, points *A* and *B* on the rim of the disk have the velocities shown. Determine the velocities of the center point *C* and point *D* at this instant.

16–91.   Due to slipping, points *A* and *B* on the rim of the disk have the velocities shown. Determine the velocities of the center point *C* and point *E* at this instant.

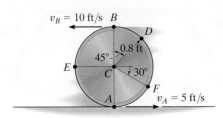

Probs. 16–90/91

*16–92.   Knowing that the angular velocity of link *AB* is $\omega_{AB} = 4$ rad/s, determine the velocity of the collar at *C* and the angular velocity of link *CB* at the instant shown. Link *CB* is horizontal at this instant.

16–93.   If the collar at *C* is moving downward to the left at $v_C = 8$ m/s, determine the angular velocity of link *AB* at the instant shown.

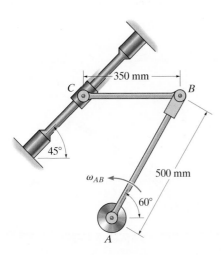

Probs. 16–92/93

16–94.   If the roller is given a velocity of $v_A = 6$ ft/s to the right, determine the angular velocity of the rod and the velocity of *C* at the instant shown.

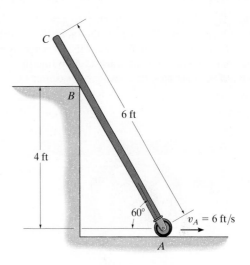

Prob. 16–94

16–95.   As the car travels forward at 80 ft/s on a wet road, due to slipping, the rear wheels have an angular velocity $\omega = 100$ rad/s. Determine the speeds of points *A*, *B*, and *C* caused by the motion.

Prob. 16–95

*16–96.   Determine the angular velocity of the double-tooth gear and the velocity of point *C* on the gear.

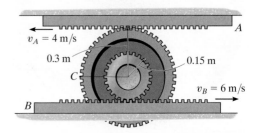

Prob. 16–96

16–97. The wheel is rigidly attached to gear A, which is in mesh with gear racks D and E. If D has a velocity of $v_D = 6$ ft/s to the right and the wheel rolls on track C without slipping, determine the velocity of gear rack E.

16–98. The wheel is rigidly attached to gear A, which is in mesh with gear racks D and E. If the racks have a velocity of $v_D = 6$ ft/s and $v_E = 10$ ft/s, show that it is necessary for the wheel to slip on the fixed track C. Also find the angular velocity of the gear and the velocity of its center O.

*16–100. The similar links AB and CD rotate about the fixed pins at A and C. If AB has an angular velocity $\omega_{AB} = 8$ rad/s, determine the angular velocity of BDP and the velocity of point P.

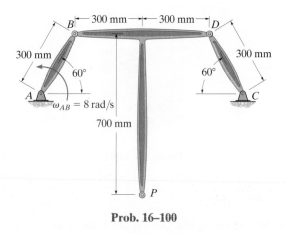

Prob. 16–100

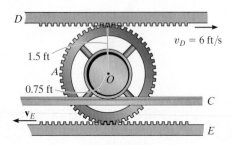

Probs. 16–97/98

16–99. The epicyclic gear train is driven by the rotating link DE, which has an angular velocity $\omega_{DE} = 5$ rad/s. If the ring gear F is fixed, determine the angular velocities of gears A, B, and C.

16–101. If rod AB is rotating with an angular velocity $\omega_{AB} = 3$ rad/s, determine the angular velocity of rod BC at the instant shown.

16–102. If rod AB is rotating with an angular velocity $\omega_{AB} = 3$ rad/s, determine the angular velocity of rod CD at the instant shown.

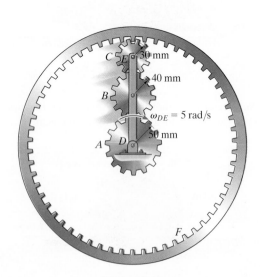

Prob. 16–99

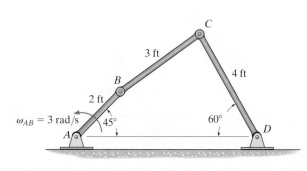

Probs. 16–101/102

16.7   Relative-Motion Analysis: Acceleration

An equation that relates the accelerations of two points on a bar (rigid body) subjected to general plane motion may be determined by differentiating $\mathbf{v}_B = \mathbf{v}_A + \mathbf{v}_{B/A}$ with respect to time. This yields

$$\frac{d\mathbf{v}_B}{dt} = \frac{d\mathbf{v}_A}{dt} + \frac{d\mathbf{v}_{B/A}}{dt}$$

The terms $d\mathbf{v}_B/dt = \mathbf{a}_B$ and $d\mathbf{v}_A/dt = \mathbf{a}_A$ are measured with respect to a set of *fixed x, y axes* and represent the *absolute accelerations* of points B and A. The last term represents the acceleration of B with respect to A as measured by an observer fixed to translating x', y' axes which have their origin at the base point A. In Sec. 16.5 it was shown that to this observer point B appears to move along a *circular arc* that has a radius of curvature $r_{B/A}$. Consequently, $\mathbf{a}_{B/A}$ can be expressed in terms of its tangential and normal components; i.e., $\mathbf{a}_{B/A} = (\mathbf{a}_{B/A})_t + (\mathbf{a}_{B/A})_n$, where $(a_{B/A})_t = \alpha r_{B/A}$ and $(a_{B/A})_n = \omega^2 r_{B/A}$. Hence, the relative-acceleration equation can be written in the form

$$\boxed{\mathbf{a}_B = \mathbf{a}_A + (\mathbf{a}_{B/A})_t + (\mathbf{a}_{B/A})_n} \qquad (16\text{--}17)$$

where

$\mathbf{a}_B =$ acceleration of point B

$\mathbf{a}_A =$ acceleration of point A

$(\mathbf{a}_{B/A})_t =$ tangential acceleration component of B with respect to A. The *magnitude* is $(a_{B/A})_t = \alpha r_{B/A}$, and the *direction* is perpendicular to $\mathbf{r}_{B/A}$.

$(\mathbf{a}_{B/A})_n =$ normal acceleration component of B with respect to A. The *magnitude* is $(a_{B/A})_n = \omega^2 r_{B/A}$, and the *direction* is always from B towards A.

The terms in Eq. 16–17 are represented graphically in Fig. 16–25. Here it is seen that at a given instant the acceleration of B, Fig. 16–25a, is determined by considering the bar to translate with an acceleration $\mathbf{a}_A$, Fig. 16–25b, and simultaneously rotate about the base point A with an instantaneous angular velocity $\boldsymbol{\omega}$ and angular acceleration $\boldsymbol{\alpha}$, Fig. 16–25c. Vector addition of these two effects, applied to B, yields $\mathbf{a}_B$, as shown in Fig. 16–25d. It should be noted from Fig. 16–25a that since points A and B move along *curved paths*, the accelerations of these points will have *both* tangential and normal components. (Recall that the acceleration of a point is *tangent to the path only* when the path is *rectilinear* or when it is an inflection point on a curve.)

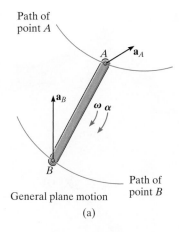

Path of point A

Path of point B

General plane motion

(a)

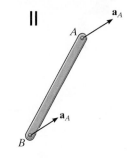

Translation

(b)

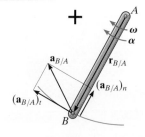

Rotation about the base point A

(c)

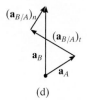

(d)

Fig. 16–25

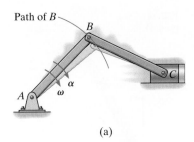

Path of B

(a)

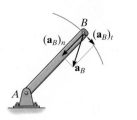

(b)

Fig. 16–26

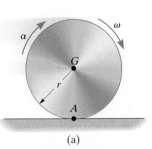

(a)

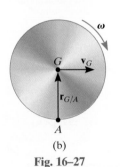

(b)

Fig. 16–27

Since the relative-acceleration components represent the effect of *circular motion* observed from translating axes having their origin at the base point A, these terms can be expressed as $(\mathbf{a}_{B/A})_t = \boldsymbol{\alpha} \times \mathbf{r}_{B/A}$ and $(\mathbf{a}_{B/A})_n = -\omega^2 \mathbf{r}_{B/A}$, Eq. 16–14. Hence, Eq. 16–17 becomes

$$\mathbf{a}_B = \mathbf{a}_A + \boldsymbol{\alpha} \times \mathbf{r}_{B/A} - \omega^2 \mathbf{r}_{B/A} \qquad (16\text{–}18)$$

where

$\mathbf{a}_B$ = acceleration of point B

$\mathbf{a}_A$ = acceleration of the base point A

$\boldsymbol{\alpha}$ = angular acceleration of the body

ω = angular velocity of the body

$\mathbf{r}_{B/A}$ = position vector directed from A to B

If Eq. 16–17 or 16–18 is applied in a practical manner to study the accelerated motion of a rigid body which is *pin connected* to two other bodies, it should be realized that points which are *coincident at the pin* move with the *same acceleration*, since the path of motion over which they travel is the *same*. For example, point B lying on either rod BA or BC of the crank mechanism shown in Fig. 16–26a has the same acceleration, since the rods are pin connected at B. Here the motion of B is along a *circular path*, so that $\mathbf{a}_B$ can be expressed in terms of its tangential and normal components. At the other end of rod BC point C moves along a *straight-lined path*, which is defined by the piston. Hence, $\mathbf{a}_C$ is horizontal, Fig. 16–26b.

Finally, consider a disk that rolls without slipping as shown in Fig. 16–27a. As a result, $v_A = 0$ and so from the kinematic diagram in Fig. 16–27b, the velocity of the mass center G is

$$\mathbf{v}_G = \mathbf{v}_A + \boldsymbol{\omega} \times \mathbf{r}_{G/A} = \mathbf{0} + (-\omega\mathbf{k}) \times (r\mathbf{j})$$

So that

$$v_G = \omega r \qquad (16\text{–}19)$$

This same result can also be determined using the IC method where point A is the *IC*.

Since G moves along a *straight line*, its acceleration in this case can be determined from the time derivative of its velocity.

$$\frac{dv_G}{dt} = \frac{d\omega}{dt} r$$

$$a_G = \alpha r \qquad (16\text{–}20)$$

These two important results were also obtained in Example 16–4. They apply as well to any circular object, such as a ball, gear, wheel, etc, that *rolls without slipping*.

Procedure for Analysis

The relative acceleration equation can be applied between any two points A and B on a body either by using a Cartesian vector analysis, or by writing the x and y scalar component equations directly.

Velocity Analysis.

- Determine the angular velocity $\boldsymbol{\omega}$ of the body by using a velocity analysis as discussed in Sec. 16.5 or 16.6. Also, determine the velocities $\mathbf{v}_A$ and $\mathbf{v}_B$ of points A and B if these points move along *curved paths*.

Vector Analysis
Kinematic Diagram.

- Establish the directions of the fixed x, y coordinates and draw the kinematic diagram of the body. Indicate on it $\mathbf{a}_A$, $\mathbf{a}_B$, $\boldsymbol{\omega}$, $\boldsymbol{\alpha}$, and $\mathbf{r}_{B/A}$.

- If points A and B move along *curved paths*, then their accelerations should be indicated in terms of their tangential and normal components, i.e., $\mathbf{a}_A = (\mathbf{a}_A)_t + (\mathbf{a}_A)_n$ and $\mathbf{a}_B = (\mathbf{a}_B)_t + (\mathbf{a}_B)_n$.

Acceleration Equation.

- To apply $\mathbf{a}_B = \mathbf{a}_A + \boldsymbol{\alpha} \times \mathbf{r}_{B/A} - \omega^2 \mathbf{r}_{B/A}$, express the vectors in Cartesian vector form and substitute them into the equation. Evaluate the cross product and then equate the respective $\mathbf{i}$ and $\mathbf{j}$ components to obtain two scalar equations.

- If the solution yields a *negative* answer for an *unknown* magnitude, it indicates that the sense of direction of the vector is opposite to that shown on the kinematic diagram.

Scalar Analysis
Kinematic Diagram.

- If the acceleration equation is applied in scalar form, then the magnitudes and directions of the relative-acceleration components $(\mathbf{a}_{B/A})_t$ and $(\mathbf{a}_{B/A})_n$ must be established. To do this draw a kinematic diagram such as shown in Fig. 16–25c. Since the body is considered to be momentarily "pinned" at the base point A, the *magnitudes* of these components are $(a_{B/A})_t = \alpha r_{B/A}$ and $(a_{B/A})_n = \omega^2 r_{B/A}$. Their *sense of direction* is established from the diagram such that $(\mathbf{a}_{B/A})_t$ acts perpendicular to $\mathbf{r}_{B/A}$, in accordance with the rotational motion $\boldsymbol{\alpha}$ of the body, and $(\mathbf{a}_{B/A})_n$ is directed from B towards A.*

Acceleration Equation.

- Represent the vectors in $\mathbf{a}_B = \mathbf{a}_A + (\mathbf{a}_{B/A})_t + (\mathbf{a}_{B/A})_n$ graphically by showing their magnitudes and directions underneath each term. The scalar equations are determined from the x and y components of these vectors.

*The notation $\mathbf{a}_B = \mathbf{a}_A + (\mathbf{a}_{B/A(\text{pin})})_t + (\mathbf{a}_{B/A(\text{pin})})_n$ may be helpful in recalling that A is assumed to be pinned.

The mechanism for a window is shown. Here CA rotates about a fixed axis through C, and AB undergoes general plane motion. Since point A moves along a curved path it has two components of acceleration, whereas point B moves along a straight track and the direction of its acceleration is specified.

16

EXAMPLE | 16.14

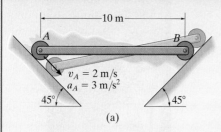

(a)

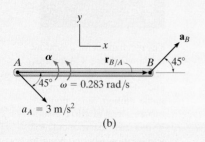

(b)

The rod AB shown in Fig. 16–28a is confined to move along the inclined planes at A and B. If point A has an acceleration of 3 m/s^2 and a velocity of 2 m/s, both directed down the plane at the instant the rod is horizontal, determine the angular acceleration of the rod at this instant.

SOLUTION I (VECTOR ANALYSIS)
We will apply the acceleration equation to points A and B on the rod. To do so it is first necessary to determine the angular velocity of the rod. Show that it is $\omega = 0.283 \text{ rad/s} \circlearrowright$ using either the velocity equation or the method of instantaneous centers.

Kinematic Diagram. Since points A and B both move along straight-line paths, they have *no* components of acceleration normal to the paths. There are two unknowns in Fig. 16–28b, namely, a_B and α.

Acceleration Equation.

$$\mathbf{a}_B = \mathbf{a}_A + \boldsymbol{\alpha} \times \mathbf{r}_{B/A} - \omega^2 \mathbf{r}_{B/A}$$

$$a_B \cos 45°\mathbf{i} + a_B \sin 45°\mathbf{j} = 3 \cos 45°\mathbf{i} - 3 \sin 45°\mathbf{j} + (\alpha\mathbf{k}) \times (10\mathbf{i}) - (0.283)^2(10\mathbf{i})$$

Carrying out the cross product and equating the $\mathbf{i}$ and $\mathbf{j}$ components yields

$$a_B \cos 45° = 3 \cos 45° - (0.283)^2(10) \qquad (1)$$

$$a_B \sin 45° = -3 \sin 45° + \alpha(10) \qquad (2)$$

Solving, we have

$$a_B = 1.87 \text{ m/s}^2 \measuredangle 45°$$

$$\alpha = 0.344 \text{ rad/s}^2 \circlearrowright \qquad \qquad Ans.$$

SOLUTION II (SCALAR ANALYSIS)
From the kinematic diagram, showing the relative-acceleration components $(\mathbf{a}_{B/A})_t$ and $(\mathbf{a}_{B/A})_n$, Fig. 16–28c, we have

$$\mathbf{a}_B = \mathbf{a}_A + (\mathbf{a}_{B/A})_t + (\mathbf{a}_{B/A})_n$$

$$\begin{bmatrix} a_B \\ \measuredangle 45° \end{bmatrix} = \begin{bmatrix} 3 \text{ m/s}^2 \\ \diagdown 45° \end{bmatrix} + \begin{bmatrix} \alpha(10 \text{ m}) \\ \uparrow \end{bmatrix} + \begin{bmatrix} (0.283 \text{ rad/s})^2(10 \text{ m}) \\ \leftarrow \end{bmatrix}$$

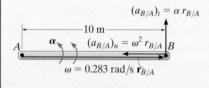

(c)

Fig. 16–28

Equating the x and y components yields Eqs. 1 and 2, and the solution proceeds as before.

EXAMPLE | 16.15

The disk rolls without slipping and has the angular motion shown in Fig. 16–29a. Determine the acceleration of point A at this instant.

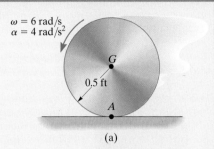

(a)

SOLUTION I (VECTOR ANALYSIS)

Kinematic Diagram.   Since no slipping occurs, applying Eq. 16–20,

$$a_G = \alpha r = (4 \text{ rad/s}^2)(0.5 \text{ ft}) = 2 \text{ ft/s}^2$$

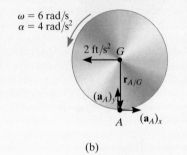

(b)

Acceleration Equation.

We will apply the acceleration equation to points G and A, Fig. 16–29b,

$$\mathbf{a}_A = \mathbf{a}_G + \boldsymbol{\alpha} \times \mathbf{r}_{A/G} - \omega^2 \mathbf{r}_{A/G}$$
$$\mathbf{a}_A = -2\mathbf{i} + (4\mathbf{k}) \times (-0.5\mathbf{j}) - (6)^2(-0.5\mathbf{j})$$
$$= \{18\mathbf{j}\} \text{ ft/s}^2$$

SOLUTION II (SCALAR ANALYSIS)

Using the result for $a_G = 2$ ft/s² determined above, and from the kinematic diagram, showing the relative motion $\mathbf{a}_{A/G}$, Fig. 16–29c, we have

$$\mathbf{a}_A = \mathbf{a}_G + (\mathbf{a}_{A/G})_x + (\mathbf{a}_{A/G})_y$$

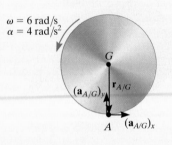

(c)

Fig. 16–29

$$\begin{bmatrix} (a_A)_x \\ \rightarrow \end{bmatrix} + \begin{bmatrix} (a_A)_y \\ \uparrow \end{bmatrix} = \begin{bmatrix} 2 \text{ ft/s}^2 \\ \leftarrow \end{bmatrix} + \begin{bmatrix} (4 \text{ rad/s}^2)(0.5 \text{ ft}) \\ \rightarrow \end{bmatrix} + \begin{bmatrix} (6 \text{ rad/s})^2(0.5 \text{ ft}) \\ \uparrow \end{bmatrix}$$

$\xrightarrow{+}$ $(a_A)_x = -2 + 2 = 0$

$+\uparrow$ $(a_A)_y = 18 \text{ ft/s}^2$

Therefore,

$$a_A = \sqrt{(0)^2 + (18 \text{ ft/s}^2)^2} = 18 \text{ ft/s}^2 \qquad \qquad Ans.$$

NOTE: The fact that $a_A = 18$ ft/s² indicates that the instantaneous center of zero velocity, point A, is *not* a point of zero acceleration.

EXAMPLE | **16.16**

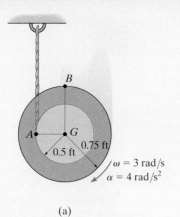

(a)

16

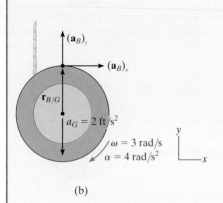

(b)

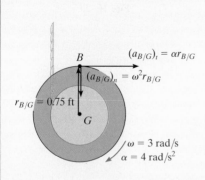

(c)

Fig. 16–30

The spool shown in Fig. 16–30a unravels from the cord, such that at the instant shown it has an angular velocity of 3 rad/s and an angular acceleration of 4 rad/s^2. Determine the acceleration of point B.

SOLUTION I (VECTOR ANALYSIS)

The spool "appears" to be rolling downward without slipping at point A. Therefore, we can use the results of Eq. 16–20 to determine the acceleration of point G, i.e.,

$$a_G = \alpha r = (4 \text{ rad/s}^2)(0.5 \text{ ft}) = 2 \text{ ft/s}^2$$

We will apply the acceleration equation to points G and B.

Kinematic Diagram. Point B moves along a *curved path* having an *unknown* radius of curvature.* Its acceleration will be represented by its unknown x and y components as shown in Fig. 16–30b.

Acceleration Equation.

$$\mathbf{a}_B = \mathbf{a}_G + \boldsymbol{\alpha} \times \mathbf{r}_{B/G} - \omega^2 \mathbf{r}_{B/G}$$

$$(a_B)_x \mathbf{i} + (a_B)_y \mathbf{j} = -2\mathbf{j} + (-4\mathbf{k}) \times (0.75\mathbf{j}) - (3)^2(0.75\mathbf{j})$$

Equating the $\mathbf{i}$ and $\mathbf{j}$ terms, the component equations are

$$(a_B)_x = 4(0.75) = 3 \text{ ft/s}^2 \rightarrow \qquad (1)$$

$$(a_B)_y = -2 - 6.75 = -8.75 \text{ ft/s}^2 = 8.75 \text{ ft/s}^2 \downarrow \qquad (2)$$

The magnitude and direction of $\mathbf{a}_B$ are therefore

$$a_B = \sqrt{(3)^2 + (8.75)^2} = 9.25 \text{ ft/s}^2 \qquad \textit{Ans.}$$

$$\theta = \tan^{-1}\frac{8.75}{3} = 71.1° \ \searrow \qquad \textit{Ans.}$$

SOLUTION II (SCALAR ANALYSIS)

This problem may be solved by writing the scalar component equations directly. The kinematic diagram in Fig. 16–30c shows the relative-acceleration components $(\mathbf{a}_{B/G})_t$ and $(\mathbf{a}_{B/G})_n$. Thus,

$$\mathbf{a}_B = \mathbf{a}_G + (\mathbf{a}_{B/G})_t + (\mathbf{a}_{B/G})_n$$

$$\begin{bmatrix} (a_B)_x \\ \rightarrow \end{bmatrix} + \begin{bmatrix} (a_B)_y \\ \uparrow \end{bmatrix}$$

$$= \begin{bmatrix} 2 \text{ ft/s}^2 \\ \downarrow \end{bmatrix} + \begin{bmatrix} 4 \text{ rad/s}^2 \, (0.75 \text{ ft}) \\ \rightarrow \end{bmatrix} + \begin{bmatrix} (3 \text{ rad/s})^2(0.75 \text{ ft}) \\ \downarrow \end{bmatrix}$$

The x and y components yield Eqs. 1 and 2 above.

*Realize that the path's radius of curvature ρ is not equal to the radius of the spool since the spool is not rotating about point G. Furthermore, ρ is not defined as the distance from A (IC) to B, since the location of the IC depends only on the velocity of a point and not the geometry of its path.

EXAMPLE | 16.17

The collar C in Fig. 16–31a moves downward with an acceleration of 1 m/s^2. At the instant shown, it has a speed of 2 m/s which gives links CB and AB an angular velocity $\omega_{AB} = \omega_{CB} = 10$ rad/s. (See Example 16.8.) Determine the angular accelerations of CB and AB at this instant.

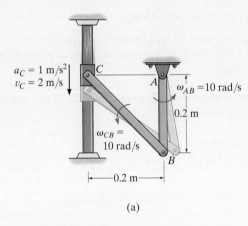

(a)

SOLUTION (VECTOR ANALYSIS)

Kinematic Diagram. The kinematic diagrams of *both* links AB and CB are shown in Fig. 16–31b. To solve, we will apply the appropriate kinematic equation to each link.

Acceleration Equation.

Link *AB* (rotation about a fixed axis):

$$\mathbf{a}_B = \boldsymbol{\alpha}_{AB} \times \mathbf{r}_B - \omega_{AB}^2 \mathbf{r}_B$$
$$\mathbf{a}_B = (\alpha_{AB}\mathbf{k}) \times (-0.2\mathbf{j}) - (10)^2(-0.2\mathbf{j})$$
$$\mathbf{a}_B = 0.2\alpha_{AB}\mathbf{i} + 20\mathbf{j}$$

Note that $\mathbf{a}_B$ has n and t components since it moves along a *circular path.*

Link *BC* (general plane motion): Using the result for $\mathbf{a}_B$ and applying Eq. 16–18, we have

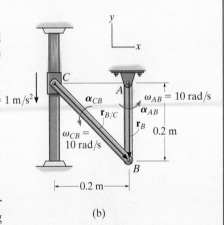

(b)

Fig. 16–31

$$\mathbf{a}_B = \mathbf{a}_C + \boldsymbol{\alpha}_{CB} \times \mathbf{r}_{B/C} - \omega_{CB}^2 \mathbf{r}_{B/C}$$
$$0.2\alpha_{AB}\mathbf{i} + 20\mathbf{j} = -1\mathbf{j} + (\alpha_{CB}\mathbf{k}) \times (0.2\mathbf{i} - 0.2\mathbf{j}) - (10)^2(0.2\mathbf{i} - 0.2\mathbf{j})$$
$$0.2\alpha_{AB}\mathbf{i} + 20\mathbf{j} = -1\mathbf{j} + 0.2\alpha_{CB}\mathbf{j} + 0.2\alpha_{CB}\mathbf{i} - 20\mathbf{i} + 20\mathbf{j}$$

Thus,

$$0.2\alpha_{AB} = 0.2\alpha_{CB} - 20$$
$$20 = -1 + 0.2\alpha_{CB} + 20$$

Solving,

$$\alpha_{CB} = 5 \text{ rad/s}^2 \ \circlearrowright \qquad Ans.$$
$$\alpha_{AB} = -95 \text{ rad/s}^2 = 95 \text{ rad/s}^2 \ \circlearrowleft \qquad Ans.$$

EXAMPLE 16.18

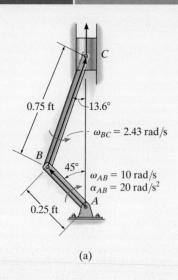

(a)

The crankshaft AB turns with a clockwise angular acceleration of 20 rad/s^2, Fig. 16–32a. Determine the acceleration of the piston at the instant AB is in the position shown. At this instant $\omega_{AB} = 10 \text{ rad/s}$ and $\omega_{BC} = 2.43 \text{ rad/s}$. (See Example 16.13.)

SOLUTION (VECTOR ANALYSIS)

Kinematic Diagram. The kinematic diagrams for both AB and BC are shown in Fig. 16–32b. Here $\mathbf{a}_C$ is vertical since C moves along a straight-line path.

Acceleration Equation. Expressing each of the position vectors in Cartesian vector form

$$\mathbf{r}_B = \{-0.25 \sin 45°\mathbf{i} + 0.25 \cos 45°\mathbf{j}\} \text{ ft} = \{-0.177\mathbf{i} + 0.177\mathbf{j}\} \text{ ft}$$

$$\mathbf{r}_{C/B} = \{0.75 \sin 13.6°\mathbf{i} + 0.75 \cos 13.6°\mathbf{j}\} \text{ ft} = \{0.177\mathbf{i} + 0.729\mathbf{j}\} \text{ ft}$$

Crankshaft AB (rotation about a fixed axis):

$$\mathbf{a}_B = \boldsymbol{\alpha}_{AB} \times \mathbf{r}_B - \omega_{AB}^2\mathbf{r}_B$$

$$= (-20\mathbf{k}) \times (-0.177\mathbf{i} + 0.177\mathbf{j}) - (10)^2(-0.177\mathbf{i} + 0.177\mathbf{j})$$

$$= \{21.21\mathbf{i} - 14.14\mathbf{j}\} \text{ ft/s}^2$$

Connecting Rod BC (general plane motion): Using the result for $\mathbf{a}_B$ and noting that $\mathbf{a}_C$ is in the vertical direction, we have

$$\mathbf{a}_C = \mathbf{a}_B + \boldsymbol{\alpha}_{BC} \times \mathbf{r}_{C/B} - \omega_{BC}^2\mathbf{r}_{C/B}$$

$$a_C\mathbf{j} = 21.21\mathbf{i} - 14.14\mathbf{j} + (\alpha_{BC}\mathbf{k}) \times (0.177\mathbf{i} + 0.729\mathbf{j}) - (2.43)^2(0.177\mathbf{i} + 0.729\mathbf{j})$$

$$a_C\mathbf{j} = 21.21\mathbf{i} - 14.14\mathbf{j} + 0.177\alpha_{BC}\mathbf{j} - 0.729\alpha_{BC}\mathbf{i} - 1.04\mathbf{i} - 4.30\mathbf{j}$$

$$0 = 20.17 - 0.729\alpha_{BC}$$

$$a_C = 0.177\alpha_{BC} - 18.45$$

Solving yields

$$\alpha_{BC} = 27.7 \text{ rad/s}^2 \quad \circlearrowright$$

$$a_C = -13.5 \text{ ft/s}^2 \qquad Ans.$$

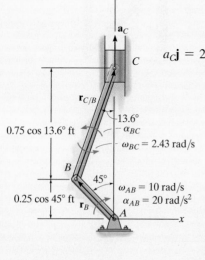

(b)

Fig. 16–32

NOTE: Since the piston is moving upward, the negative sign for a_C indicates that the piston is decelerating, i.e., $\mathbf{a}_C = \{-13.5\mathbf{j}\} \text{ ft/s}^2$. This causes the speed of the piston to decrease until AB becomes vertical, at which time the piston is momentarily at rest.

FUNDAMENTAL PROBLEMS

F16–19. At the instant shown, end A of the rod has the velocity and acceleration shown. Determine the angular acceleration of the rod and acceleration of end B of the rod.

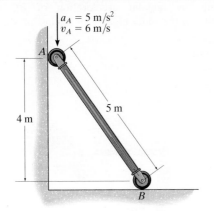

F16–19

F16–20. The gear rolls on the fixed rack with an angular velocity of $\omega = 12$ rad/s and angular acceleration of $\alpha = 6$ rad/s². Determine the acceleration of point A.

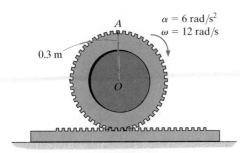

F16–20

F16–21. The gear rolls on the fixed rack B. At the instant shown, the center O of the gear moves with a velocity of $v_O = 6$ m/s and acceleration of $a_O = 3$ m/s². Determine the angular acceleration of the gear and acceleration of point A at this instant.

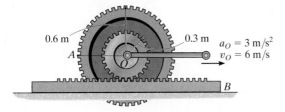

F16–21

F16–22. At the instant shown, cable AB has a velocity of 3 m/s and acceleration of 1.5 m/s², while the gear rack has a velocity of 1.5 m/s and acceleration of 0.75 m/s². Determine the angular acceleration of the gear at this instant.

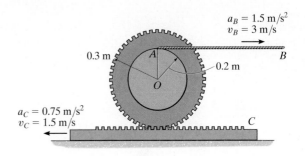

F16–22

F16–23. At the instant shown, the wheel rotates with an angular velocity of $\omega = 12$ rad/s and an angular acceleration of $\alpha = 6$ rad/s². Determine the angular acceleration of link BC and the acceleration of piston C at this instant.

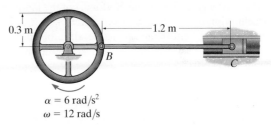

F16–23

F16–24. At the instant shown, wheel A rotates with an angular velocity of $\omega = 6$ rad/s and an angular acceleration of $\alpha = 3$ rad/s². Determine the angular acceleration of link BC and the acceleration of piston C.

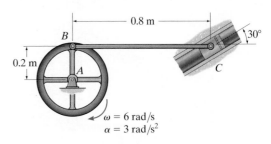

F16–24

16

PROBLEMS

16–103. At a given instant the top end A of the bar has the velocity and acceleration shown. Determine the acceleration of the bottom B and the bar's angular acceleration at this instant.

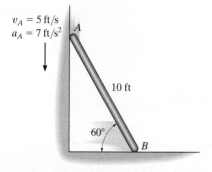

$v_A = 5$ ft/s
$a_A = 7$ ft/s^2

10 ft

60°

Prob. 16–103

*16–104. At a given instant the bottom A of the ladder has an acceleration $a_A = 4$ ft/s^2 and velocity $v_A = 6$ ft/s, both acting to the left. Determine the acceleration of the top of the ladder, B, and the ladder's angular acceleration at this same instant.

16–105. At a given instant the top B of the ladder has an acceleration $a_B = 2$ ft/s^2 and a velocity of $v_B = 4$ ft/s, both acting downward. Determine the acceleration of the bottom A of the ladder, and the ladder's angular acceleration at this instant.

16–106. Crank AB is rotating with an angular velocity of $\omega_{AB} = 5$ rad/s and an angular acceleration of $\alpha_{AB} = 6$ rad/s^2. Determine the angular acceleration of BC and the acceleration of the slider block C at the instant shown.

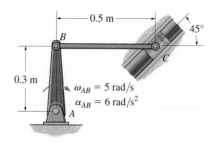

0.5 m

45°

0.3 m

$\omega_{AB} = 5$ rad/s
$\alpha_{AB} = 6$ rad/s^2

Prob. 16–106

16–107. At a given instant, the slider block A has the velocity and deceleration shown. Determine the acceleration of block B and the angular acceleration of the link at this instant.

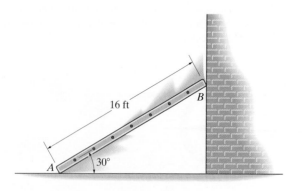

16 ft

30°

Probs. 16–104/105

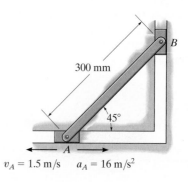

300 mm

45°

$v_A = 1.5$ m/s $a_A = 16$ m/s^2

Prob. 16–107

*16–108. As the cord unravels from the cylinder, the cylinder has an angular acceleration of $\alpha = 4 \text{ rad/s}^2$ and an angular velocity of $\omega = 2 \text{ rad/s}$ at the instant shown. Determine the accelerations of points A and B at this instant.

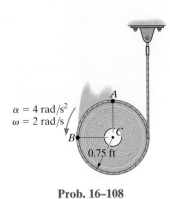

α = 4 rad/s²
ω = 2 rad/s
0.75 ft

Prob. 16–108

16–110. At a given instant the wheel is rotating with the angular motions shown. Determine the acceleration of the collar at A at this instant.

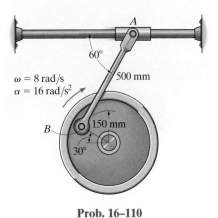

A
60°
$\omega = 8 \text{ rad/s}$
$\alpha = 16 \text{ rad/s}^2$
500 mm
B
150 mm
30°

Prob. 16–110

16

16–109. The hydraulic cylinder is extending with a velocity of $v_C = 3 \text{ ft/s}$ and an acceleration of $a_C = 1.5 \text{ ft/s}^2$. Determine the angular acceleration of links BC and AB at the instant shown.

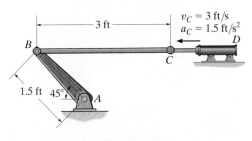

3 ft
$v_C = 3 \text{ ft/s}$
$a_C = 1.5 \text{ ft/s}^2$
B
D
C
1.5 ft   45° A

Prob. 16–109

16–111. Crank AB rotates with the angular velocity and angular acceleration shown. Determine the acceleration of the slider block C at the instant shown.

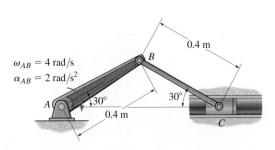

0.4 m
$\omega_{AB} = 4 \text{ rad/s}$
$\alpha_{AB} = 2 \text{ rad/s}^2$
B
A 30°
0.4 m
30°
C

Prob. 16–111

*16–112. The wheel is moving to the right such that it has an angular velocity $\omega = 2$ rad/s and angular acceleration $\alpha = 4$ rad/s^2 at the instant shown. If it does not slip at A, determine the acceleration of point B.

16–115. A cord is wrapped around the inner spool of the gear. If it is pulled with a constant velocity $\mathbf{v}$, determine the velocities and accelerations of points A and B. The gear rolls on the fixed gear rack.

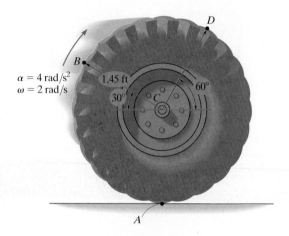

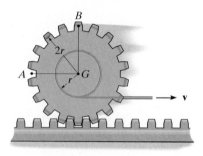

Prob. 16–115

Prob. 16–112

*16–116. At a given instant, the gear racks have the velocities and accelerations shown. Determine the acceleration of point A.

16–113. The disk is moving to the left such that it has an angular acceleration $\alpha = 8$ rad/s^2 and angular velocity $\omega = 3$ rad/s at the instant shown. If it does not slip at A, determine the acceleration of point B.

16–117. At a given instant, the gear racks have the velocities and accelerations shown. Determine the acceleration of point B.

16–114. The disk is moving to the left such that it has an angular acceleration $\alpha = 8$ rad/s^2 and angular velocity $\omega = 3$ rad/s at the instant shown. If it does not slip at A, determine the acceleration of point D.

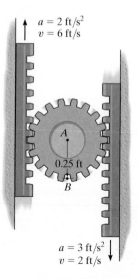

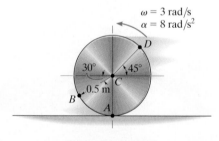

Probs. 16–113/114

Probs. 16–116/117

16–118. At a given instant gears A and B have the angular motions shown. Determine the angular acceleration of gear C and the acceleration of its center point D at this instant. Note that the inner hub of gear C is in mesh with gear A and its outer rim is in mesh with gear B.

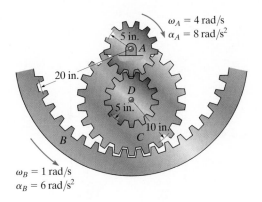

$\omega_A = 4$ rad/s
$\alpha_A = 8$ rad/s²
5 in.
20 in.
D
5 in.
10 in.
C
B
$\omega_B = 1$ rad/s
$\alpha_B = 6$ rad/s²

Prob. 16–118

16–119. The wheel rolls without slipping such that at the instant shown it has an angular velocity ω and angular acceleration α. Determine the velocity and acceleration of point B on the rod at this instant.

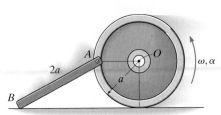

A
$2a$
O
ω, α
a
B

Prob. 16–119

*16–120. The center O of the gear and the gear rack P move with the velocities and accelerations shown. Determine the angular acceleration of the gear and the acceleration of point B located at the rim of the gear at the instant shown.

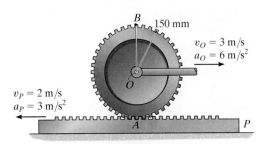

B 150 mm
$v_O = 3$ m/s
$a_O = 6$ m/s²
O
$v_P = 2$ m/s
$a_P = 3$ m/s²
A P

Prob. 16–120

16–121. The tied crank and gear mechanism gives rocking motion to crank AC, necessary for the operation of a printing press. If link DE has the angular motion shown, determine the respective angular velocities of gear F and crank AC at this instant, and the angular acceleration of crank AC.

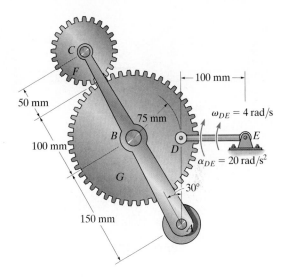

C
F
100 mm
50 mm
75 mm
$\omega_{DE} = 4$ rad/s
B
100 mm
D
E
$\alpha_{DE} = 20$ rad/s²
G
30°
150 mm
A

Prob. 16–121

16–122. Pulley A rotates with the angular velocity and angular acceleration shown. Determine the angular acceleration of pulley B at the instant shown.

16–123. Pulley A rotates with the angular velocity and angular acceleration shown. Determine the acceleration of block E at the instant shown.

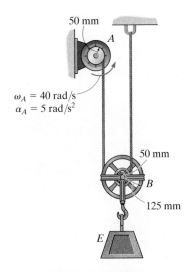

50 mm
A
$\omega_A = 40$ rad/s
$\alpha_A = 5$ rad/s²
50 mm
B
125 mm
E

Probs. 16–122/123

16

*16–124. At a given instant, the gear has the angular motion shown. Determine the accelerations of points A and B on the link and the link's angular acceleration at this instant.

16–126. At a given instant, the cables supporting the pipe have the motions shown. Determine the angular velocity and angular acceleration of the pipe and the velocity and acceleration of point B located on the pipe.

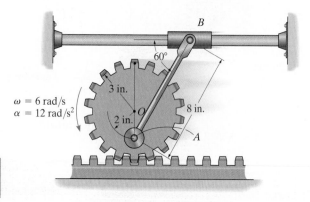

Prob. 16–124

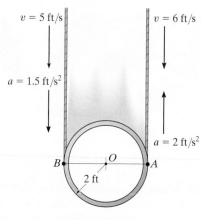

Prob. 16–126

16–125. The ends of the bar AB are confined to move along the paths shown. At a given instant, A has a velocity of $v_A = 4$ ft/s and an acceleration of $a_A = 7$ ft/s^2. Determine the angular velocity and angular acceleration of AB at this instant.

16–127. The slider block moves with a velocity of $v_B = 5$ ft/s and an acceleration of $a_B = 3$ ft/s^2. Determine the angular acceleration of rod AB at the instant shown.

*16–128. The slider block moves with a velocity of $v_B = 5$ ft/s and an acceleration of $a_B = 3$ ft/s^2. Determine the acceleration of A at the instant shown.

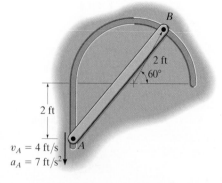

Prob. 16–125

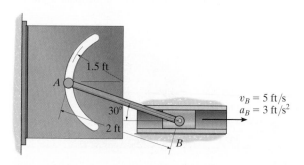

Probs. 16–127/128

16.8   Relative-Motion Analysis using Rotating Axes

In the previous sections the relative-motion analysis for velocity and acceleration was described using a translating coordinate system. This type of analysis is useful for determining the motion of points on the *same* rigid body, or the motion of points located on several pin-connected bodies. In some problems, however, rigid bodies (mechanisms) are constructed such that *sliding* will occur at their connections. The kinematic analysis for such cases is best performed if the motion is analyzed using a coordinate system which both *translates* and *rotates*. Furthermore, this frame of reference is useful for analyzing the motions of two points on a mechanism which are *not* located in the *same* body and for specifying the kinematics of particle motion when the particle moves along a rotating path.

In the following analysis two equations will be developed which relate the velocity and acceleration of two points, one of which is the origin of a moving frame of reference subjected to both a translation and a rotation in the plane.*

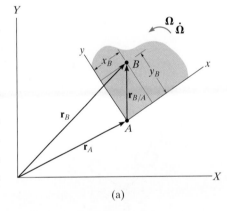

Position.   Consider the two points A and B shown in Fig. 16–33a. Their location is specified by the position vectors $\mathbf{r}_A$ and $\mathbf{r}_B$, which are measured with respect to the fixed X, Y, Z coordinate system. As shown in the figure, the "base point" A represents the origin of the x, y, z coordinate system, which is assumed to be both translating and rotating with respect to the X, Y, Z system. The position of B with respect to A is specified by the relative-position vector $\mathbf{r}_{B/A}$. The components of this vector may be expressed either in terms of unit vectors along the X, Y axes, i.e., $\mathbf{I}$ and $\mathbf{J}$, or by unit vectors along the x, y axes, i.e., $\mathbf{i}$ and $\mathbf{j}$. For the development which follows, $\mathbf{r}_{B/A}$ will be measured with respect to the moving x, y frame of reference. Thus, if B has coordinates (x_B, y_B), Fig. 16–33a, then

$$\mathbf{r}_{B/A} = x_B\mathbf{i} + y_B\mathbf{j}$$

(a)

Fig. 16–33

Using vector addition, the three position vectors in Fig. 16–33a are related by the equation

$$\boxed{\mathbf{r}_B = \mathbf{r}_A + \mathbf{r}_{B/A}} \qquad (16\text{–}21)$$

At the instant considered, point A has a velocity $\mathbf{v}_A$ and an acceleration $\mathbf{a}_A$, while the angular velocity and angular acceleration of the x, y axes are $\mathbf{\Omega}$ (omega) and $\dot{\mathbf{\Omega}} = d\mathbf{\Omega}/dt$, respectively.

---

*The more general, three-dimensional motion of the points is developed in Sec. 20.4.

Velocity.   The velocity of point B is determined by taking the time derivative of Eq. 16–21, which yields

$$\mathbf{v}_B = \mathbf{v}_A + \frac{d\mathbf{r}_{B/A}}{dt} \tag{16-22}$$

The last term in this equation is evaluated as follows:

$$\frac{d\mathbf{r}_{B/A}}{dt} = \frac{d}{dt}(x_B\mathbf{i} + y_B\mathbf{j})$$

$$= \frac{dx_B}{dt}\mathbf{i} + x_B\frac{d\mathbf{i}}{dt} + \frac{dy_B}{dt}\mathbf{j} + y_B\frac{d\mathbf{j}}{dt}$$

$$= \left(\frac{dx_B}{dt}\mathbf{i} + \frac{dy_B}{dt}\mathbf{j}\right) + \left(x_B\frac{d\mathbf{i}}{dt} + y_B\frac{d\mathbf{j}}{dt}\right) \tag{16-23}$$

The two terms in the first set of parentheses represent the components of velocity of point B as measured by an observer attached to the moving x, y, z coordinate system. These terms will be denoted by vector $(\mathbf{v}_{B/A})_{xyz}$. In the second set of parentheses the instantaneous time rate of change of the unit vectors $\mathbf{i}$ and $\mathbf{j}$ is measured by an observer located in the fixed X, Y, Z coordinate system. These changes, $d\mathbf{i}$ and $d\mathbf{j}$, are due *only* to the *rotation $d\theta$* of the x, y, z axes, causing $\mathbf{i}$ to become $\mathbf{i}' = \mathbf{i} + d\mathbf{i}$ and $\mathbf{j}$ to become $\mathbf{j}' = \mathbf{j} + d\mathbf{j}$, Fig. 16–33b. As shown, the *magnitudes* of both $d\mathbf{i}$ and $d\mathbf{j}$ equal $1\, d\theta$, since $i = i' = j = j' = 1$. The *direction* of $d\mathbf{i}$ is defined by $+\mathbf{j}$, since $d\mathbf{i}$ is tangent to the path described by the arrowhead of $\mathbf{i}$ in the limit as $\Delta t \to dt$. Likewise, $d\mathbf{j}$ acts in the $-\mathbf{i}$ direction, Fig. 16–33b. Hence,

$$\frac{d\mathbf{i}}{dt} = \frac{d\theta}{dt}(\mathbf{j}) = \Omega\mathbf{j} \qquad \frac{d\mathbf{j}}{dt} = \frac{d\theta}{dt}(-\mathbf{i}) = -\Omega\mathbf{i}$$

Viewing the axes in three dimensions, Fig. 16–33c, and noting that $\mathbf{\Omega} = \Omega\mathbf{k}$, we can express the above derivatives in terms of the cross product as

$$\frac{d\mathbf{i}}{dt} = \mathbf{\Omega} \times \mathbf{i} \qquad \frac{d\mathbf{j}}{dt} = \mathbf{\Omega} \times \mathbf{j} \tag{16-24}$$

Substituting these results into Eq. 16–23 and using the distributive property of the vector cross product, we obtain

$$\frac{d\mathbf{r}_{B/A}}{dt} = (\mathbf{v}_{B/A})_{xyz} + \mathbf{\Omega} \times (x_B\mathbf{i} + y_B\mathbf{j}) = (\mathbf{v}_{B/A})_{xyz} + \mathbf{\Omega} \times \mathbf{r}_{B/A} \tag{16-25}$$

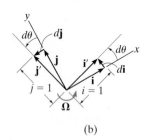

(b)

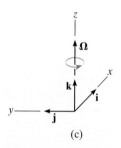

(c)

Fig. 16–33 (cont.)

Hence, Eq. 16–22 becomes

$$\mathbf{v}_B = \mathbf{v}_A + \mathbf{\Omega} \times \mathbf{r}_{B/A} + (\mathbf{v}_{B/A})_{xyz} \qquad (16\text{–}26)$$

where

$\mathbf{v}_B$ = velocity of B, measured from the X, Y, Z reference

$\mathbf{v}_A$ = velocity of the origin A of the x, y, z reference, measured from the X, Y, Z reference

$(\mathbf{v}_{B/A})_{xyz}$ = velocity of "B with respect to A," as measured by an observer attached to the rotating x, y, z reference

$\mathbf{\Omega}$ = angular velocity of the x, y, z reference, measured from the X, Y, Z reference

$\mathbf{r}_{B/A}$ = position of B with respect to A

16

Comparing Eq. 16–26 with Eq. 16–16 ($\mathbf{v}_B = \mathbf{v}_A + \mathbf{\Omega} \times \mathbf{r}_{B/A}$), which is valid for a translating frame of reference, it can be seen that the only difference between these two equations is represented by the term $(\mathbf{v}_{B/A})_{xyz}$.

When applying Eq. 16–26 it is often useful to understand what each of the terms represents. In order of appearance, they are as follows:

$\mathbf{v}_B$ $\begin{cases} \text{absolute velocity of } B \end{cases}$ $\begin{cases} \text{motion of } B \text{ observed} \\ \text{from the } X, Y, Z \text{ frame} \end{cases}$

(equals)

$\mathbf{v}_A$ $\begin{cases} \text{absolute velocity of the} \\ \text{origin of } x, y, z \text{ frame} \end{cases}$

(plus)

$\begin{cases} \text{motion of } x, y, z \text{ frame} \\ \text{observed from the} \\ X, Y, Z \text{ frame} \end{cases}$

$\mathbf{\Omega} \times \mathbf{r}_{B/A}$ $\begin{cases} \text{angular velocity effect caused} \\ \text{by rotation of } x, y, z \text{ frame} \end{cases}$

(plus)

$(\mathbf{v}_{B/A})_{xyz}$ $\begin{cases} \text{velocity of } B \\ \text{with respect to } A \end{cases}$ $\begin{cases} \text{motion of } B \text{ observed} \\ \text{from the } x, y, z \text{ frame} \end{cases}$

Acceleration.   The acceleration of B, observed from the X, Y, Z coordinate system, may be expressed in terms of its motion measured with respect to the rotating system of coordinates by taking the time derivative of Eq. 16–26.

$$\frac{d\mathbf{v}_B}{dt} = \frac{d\mathbf{v}_A}{dt} + \frac{d\boldsymbol{\Omega}}{dt} \times \mathbf{r}_{B/A} + \boldsymbol{\Omega} \times \frac{d\mathbf{r}_{B/A}}{dt} + \frac{d(\mathbf{v}_{B/A})_{xyz}}{dt}$$

$$\mathbf{a}_B = \mathbf{a}_A + \dot{\boldsymbol{\Omega}} \times \mathbf{r}_{B/A} + \boldsymbol{\Omega} \times \frac{d\mathbf{r}_{B/A}}{dt} + \frac{d(\mathbf{v}_{B/A})_{xyz}}{dt} \qquad (16\text{–}27)$$

Here $\dot{\boldsymbol{\Omega}} = d\boldsymbol{\Omega}/dt$ is the angular acceleration of the x, y, z coordinate system. Since $\boldsymbol{\Omega}$ is always perpendicular to the plane of motion, then $\dot{\boldsymbol{\Omega}}$ measures *only the change in magnitude* of $\boldsymbol{\Omega}$. The derivative $d\mathbf{r}_{B/A}/dt$ is defined by Eq. 16–25, so that

$$\boldsymbol{\Omega} \times \frac{d\mathbf{r}_{B/A}}{dt} = \boldsymbol{\Omega} \times (\mathbf{v}_{B/A})_{xyz} + \boldsymbol{\Omega} \times (\boldsymbol{\Omega} \times \mathbf{r}_{B/A}) \qquad (16\text{–}28)$$

Finding the time derivative of $(\mathbf{v}_{B/A})_{xyz} = (v_{B/A})_x \mathbf{i} + (v_{B/A})_y \mathbf{j}$,

$$\frac{d(\mathbf{v}_{B/A})_{xyz}}{dt} = \left[\frac{d(v_{B/A})_x}{dt} \mathbf{i} + \frac{d(v_{B/A})_y}{dt} \mathbf{j} \right] + \left[(v_{B/A})_x \frac{d\mathbf{i}}{dt} + (v_{B/A})_y \frac{d\mathbf{j}}{dt} \right]$$

The two terms in the first set of brackets represent the components of acceleration of point B as measured by an observer attached to the rotating coordinate system. These terms will be denoted by $(\mathbf{a}_{B/A})_{xyz}$. The terms in the second set of brackets can be simplified using Eqs. 16–24.

$$\frac{d(\mathbf{v}_{B/A})_{xyz}}{dt} = (\mathbf{a}_{B/A})_{xyz} + \boldsymbol{\Omega} \times (\mathbf{v}_{B/A})_{xyz}$$

Substituting this and Eq. 16–28 into Eq. 16–27 and rearranging terms,

$$\boxed{\mathbf{a}_B = \mathbf{a}_A + \dot{\boldsymbol{\Omega}} \times \mathbf{r}_{B/A} + \boldsymbol{\Omega} \times (\boldsymbol{\Omega} \times \mathbf{r}_{B/A}) + 2\boldsymbol{\Omega} \times (\mathbf{v}_{B/A})_{xyz} + (\mathbf{a}_{B/A})_{xyz}}$$

$$(16\text{–}29)$$

where

$$\mathbf{a}_B = \text{acceleration of } B, \text{ measured from the } X, Y, Z \text{ reference}$$

$$\mathbf{a}_A = \text{acceleration of the origin } A \text{ of the } x, y, z \text{ reference, measured from the } X, Y, Z \text{ reference}$$

$$(\mathbf{a}_{B/A})_{xyz}, (\mathbf{v}_{B/A})_{xyz} = \text{acceleration and velocity of } B \text{ with respect to } A, \text{ as measured by an observer attached to the rotating } x, y, z \text{ reference}$$

$$\dot{\boldsymbol{\Omega}}, \boldsymbol{\Omega} = \text{angular acceleration and angular velocity of the } x, y, z \text{ reference, measured from the } X, Y, Z \text{ reference}$$

$$\mathbf{r}_{B/A} = \text{position of } B \text{ with respect to } A$$

If Eq. 16–29 is compared with Eq. 16–18, written in the form $\mathbf{a}_B = \mathbf{a}_A + \dot{\boldsymbol{\Omega}} \times \mathbf{r}_{B/A} + \boldsymbol{\Omega} \times (\boldsymbol{\Omega} \times \mathbf{r}_{B/A})$, which is valid for a translating frame of reference, it can be seen that the difference between these two equations is represented by the terms $2\boldsymbol{\Omega} \times (\mathbf{v}_{B/A})_{xyz}$ and $(\mathbf{a}_{B/A})_{xyz}$. In particular, $2\boldsymbol{\Omega} \times (\mathbf{v}_{B/A})_{xyz}$ is called the *Coriolis acceleration*, named after the French engineer G. C. Coriolis, who was the first to determine it. This term represents the difference in the acceleration of B as measured from nonrotating and rotating x, y, z axes. As indicated by the vector cross product, the Coriolis acceleration will *always* be perpendicular to both $\boldsymbol{\Omega}$ and $(\mathbf{v}_{B/A})_{xyz}$. It is an important component of the acceleration which must be considered whenever rotating reference frames are used. This often occurs, for example, when studying the accelerations and forces which act on rockets, long-range projectiles, or other bodies having motions whose measurements are significantly affected by the rotation of the earth.

The following interpretation of the terms in Eq. 16–29 may be useful when applying this equation to the solution of problems.

$\mathbf{a}_B$	$\left\{\begin{array}{l}\text{absolute acceleration of } B\end{array}\right.$	$\left.\begin{array}{l}\text{motion of } B \text{ observed} \\ \text{from the } X, Y, Z \text{ frame}\end{array}\right\}$
	(equals)	
$\mathbf{a}_A$	$\left\{\begin{array}{l}\text{absolute acceleration of the} \\ \text{origin of } x, y, z \text{ frame}\end{array}\right.$	
	(plus)	
		motion of
$\dot{\boldsymbol{\Omega}} \times \mathbf{r}_{B/A}$	$\left\{\begin{array}{l}\text{angular acceleration effect} \\ \text{caused by rotation of } x, y, z \\ \text{frame}\end{array}\right.$	x, y, z frame observed from the X, Y, Z frame
	(plus)	
$\boldsymbol{\Omega} \times (\boldsymbol{\Omega} \times \mathbf{r}_{B/A})$	$\left\{\begin{array}{l}\text{angular velocity effect caused} \\ \text{by rotation of } x, y, z \text{ frame}\end{array}\right.$	
	(plus)	
$2\boldsymbol{\Omega} \times (\mathbf{v}_{B/A})_{xyz}$	$\left\{\begin{array}{l}\text{combined effect of } B \text{ moving} \\ \text{relative to } x, y, z \text{ coordinates} \\ \text{and rotation of } x, y, z \text{ frame}\end{array}\right.$	$\left.\begin{array}{l}\text{interacting motion}\end{array}\right\}$
	(plus)	
$(\mathbf{a}_{B/A})_{xyz}$	$\left\{\begin{array}{l}\text{acceleration of } B \text{ with} \\ \text{respect to } A\end{array}\right.$	$\left.\begin{array}{l}\text{motion of } B \text{ observed} \\ \text{from the } x, y, z \text{ frame}\end{array}\right\}$

16

Procedure for Analysis

Equations 16–26 and 16–29 can be applied to the solution of problems involving the planar motion of particles or rigid bodies using the following procedure.

Coordinate Axes.

- Choose an appropriate location for the origin and proper orientation of the axes for both fixed X, Y, Z and moving x, y, z reference frames.

- Most often solutions are easily obtained if at the instant considered:
  1. the origins are coincident
  2. the corresponding axes are collinear
  3. the corresponding axes are parallel

- The moving frame should be selected fixed to the body or device along which the relative motion occurs.

Kinematic Equations.

- After defining the origin A of the moving reference and specifying the moving point B, Eqs. 16–26 and 16–29 should be written in symbolic form

$$\mathbf{v}_B = \mathbf{v}_A + \boldsymbol{\Omega} \times \mathbf{r}_{B/A} + (\mathbf{v}_{B/A})_{xyz}$$
$$\mathbf{a}_B = \mathbf{a}_A + \dot{\boldsymbol{\Omega}} \times \mathbf{r}_{B/A} + \boldsymbol{\Omega} \times (\boldsymbol{\Omega} \times \mathbf{r}_{B/A}) + 2\boldsymbol{\Omega} \times (\mathbf{v}_{B/A})_{xyz} + (\mathbf{a}_{B/A})_{xyz}$$

- The Cartesian components of all these vectors may be expressed along either the X, Y, Z axes or the x, y, z axes. The choice is arbitrary provided a consistent set of unit vectors is used.

- Motion of the moving reference is expressed by $\mathbf{v}_A$, $\mathbf{a}_A$, $\boldsymbol{\Omega}$, and $\dot{\boldsymbol{\Omega}}$; and motion of B with respect to the moving reference is expressed by $\mathbf{r}_{B/A}$, $(\mathbf{v}_{B/A})_{xyz}$, and $(\mathbf{a}_{B/A})_{xyz}$.

The rotation of the dumping bin of the truck about point C is operated by the extension of the hydraulic cylinder AB. To determine the rotation of the bin due to this extension, we can use the equations of relative motion and fix the x, y axes to the cylinder so that the relative motion of the cylinder's extension occurs along the y axis.

EXAMPLE   16.19

At the instant $\theta = 60°$, the rod in Fig. 16–34 has an angular velocity of 3 rad/s and an angular acceleration of 2 rad/s². At this same instant, collar C travels outward along the rod such that when $x = 0.2$ m the velocity is 2 m/s and the acceleration is 3 m/s², both measured relative to the rod. Determine the Coriolis acceleration and the velocity and acceleration of the collar at this instant.

SOLUTION

Coordinate Axes.   The origin of both coordinate systems is located at point O, Fig. 16–34. Since motion of the collar is reported relative to the rod, the moving x, y, z frame of reference is *attached* to the rod.

Kinematic Equations.

$$\mathbf{v}_C = \mathbf{v}_O + \boldsymbol{\Omega} \times \mathbf{r}_{C/O} + (\mathbf{v}_{C/O})_{xyz} \tag{1}$$

$$\mathbf{a}_C = \mathbf{a}_O + \dot{\boldsymbol{\Omega}} \times \mathbf{r}_{C/O} + \boldsymbol{\Omega} \times (\boldsymbol{\Omega} \times \mathbf{r}_{C/O}) + 2\boldsymbol{\Omega} \times (\mathbf{v}_{C/O})_{xyz} + (\mathbf{a}_{C/O})_{xyz} \tag{2}$$

It will be simpler to express the data in terms of $\mathbf{i}, \mathbf{j}, \mathbf{k}$ component vectors rather than $\mathbf{I}, \mathbf{J}, \mathbf{K}$ components. Hence,

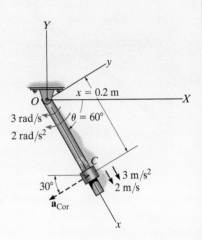

Fig. 16–34

Motion of moving reference	Motion of C with respect to moving reference
$\mathbf{v}_O = \mathbf{0}$	$\mathbf{r}_{C/O} = \{0.2\mathbf{i}\}$ m
$\mathbf{a}_O = \mathbf{0}$	$(\mathbf{v}_{C/O})_{xyz} = \{2\mathbf{i}\}$ m/s
$\boldsymbol{\Omega} = \{-3\mathbf{k}\}$ rad/s	$(\mathbf{a}_{C/O})_{xyz} = \{3\mathbf{i}\}$ m/s²
$\dot{\boldsymbol{\Omega}} = \{-2\mathbf{k}\}$ rad/s²	

The Coriolis acceleration is defined as

$$\mathbf{a}_{Cor} = 2\boldsymbol{\Omega} \times (\mathbf{v}_{C/O})_{xyz} = 2(-3\mathbf{k}) \times (2\mathbf{i}) = \{-12\mathbf{j}\} \text{ m/s}^2 \qquad Ans.$$

This vector is shown dashed in Fig. 16–34. If desired, it may be resolved into $\mathbf{I}, \mathbf{J}$ components acting along the X and Y axes, respectively.

The velocity and acceleration of the collar are determined by substituting the data into Eqs. 1 and 2 and evaluating the cross products, which yields

$$\mathbf{v}_C = \mathbf{v}_O + \boldsymbol{\Omega} \times \mathbf{r}_{C/O} + (\mathbf{v}_{C/O})_{xyz}$$

$$= 0 + (-3\mathbf{k}) \times (0.2\mathbf{i}) + 2\mathbf{i}$$

$$= \{2\mathbf{i} - 0.6\mathbf{j}\} \text{ m/s} \qquad Ans.$$

$$\mathbf{a}_C = \mathbf{a}_O + \dot{\boldsymbol{\Omega}} \times \mathbf{r}_{C/O} + \boldsymbol{\Omega} \times (\boldsymbol{\Omega} \times \mathbf{r}_{C/O}) + 2\boldsymbol{\Omega} \times (\mathbf{v}_{C/O})_{xyz} + (\mathbf{a}_{C/O})_{xyz}$$

$$= 0 + (-2\mathbf{k}) \times (0.2\mathbf{i}) + (-3\mathbf{k}) \times [(-3\mathbf{k}) \times (0.2\mathbf{i})] + 2(-3\mathbf{k}) \times (2\mathbf{i}) + 3\mathbf{i}$$

$$= 0 - 0.4\mathbf{j} - 1.80\mathbf{i} - 12\mathbf{j} + 3\mathbf{i}$$

$$= \{1.20\mathbf{i} - 12.4\mathbf{j}\} \text{ m/s}^2 \qquad Ans.$$

16

EXAMPLE | 16.20

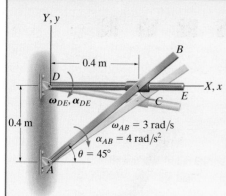

Fig. 16–35

Rod AB, shown in Fig. 16–35, rotates clockwise such that it has an angular velocity $\omega_{AB} = 3$ rad/s and angular acceleration $\alpha_{AB} = 4$ rad/s^2 when $\theta = 45°$. Determine the angular motion of rod DE at this instant. The collar at C is pin connected to AB and slides over rod DE.

SOLUTION

Coordinate Axes.   The origin of both the fixed and moving frames of reference is located at D, Fig. 16–35. Furthermore, the x, y, z reference is attached to and rotates with rod DE so that the relative motion of the collar is easy to follow.

Kinematic Equations.

$$\mathbf{v}_C = \mathbf{v}_D + \boldsymbol{\Omega} \times \mathbf{r}_{C/D} + (\mathbf{v}_{C/D})_{xyz} \tag{1}$$

$$\mathbf{a}_C = \mathbf{a}_D + \dot{\boldsymbol{\Omega}} \times \mathbf{r}_{C/D} + \boldsymbol{\Omega} \times (\boldsymbol{\Omega} \times \mathbf{r}_{C/D}) + 2\boldsymbol{\Omega} \times (\mathbf{v}_{C/D})_{xyz} + (\mathbf{a}_{C/D})_{xyz} \tag{2}$$

All vectors will be expressed in terms of $\mathbf{i}, \mathbf{j}, \mathbf{k}$ components.

Motion of moving reference	Motion of C with respect to moving reference
$\mathbf{v}_D = 0$	$\mathbf{r}_{C/D} = \{0.4\mathbf{i}\}$ m
$\mathbf{a}_D = 0$	$(\mathbf{v}_{C/D})_{xyz} = (v_{C/D})_{xyz}\mathbf{i}$
$\boldsymbol{\Omega} = -\omega_{DE}\mathbf{k}$	$(\mathbf{a}_{C/D})_{xyz} = (a_{C/D})_{xyz}\mathbf{i}$
$\dot{\boldsymbol{\Omega}} = -\alpha_{DE}\mathbf{k}$	

Motion of C: Since the collar moves along a *circular path* of radius AC, its velocity and acceleration can be determined using Eqs. 16–9 and 16–14.

$$\mathbf{v}_C = \boldsymbol{\omega}_{AB} \times \mathbf{r}_{C/A} = (-3\mathbf{k}) \times (0.4\mathbf{i} + 0.4\mathbf{j}) = \{1.2\mathbf{i} - 1.2\mathbf{j}\} \text{ m/s}$$

$$\mathbf{a}_C = \boldsymbol{\alpha}_{AB} \times \mathbf{r}_{C/A} - \omega_{AB}^2\mathbf{r}_{C/A}$$

$$= (-4\mathbf{k}) \times (0.4\mathbf{i} + 0.4\mathbf{j}) - (3)^2(0.4\mathbf{i} + 0.4\mathbf{j}) = \{-2\mathbf{i} - 5.2\mathbf{j}\} \text{ m/s}^2$$

Substituting the data into Eqs. 1 and 2, we have

$$\mathbf{v}_C = \mathbf{v}_D + \boldsymbol{\Omega} \times \mathbf{r}_{C/D} + (\mathbf{v}_{C/D})_{xyz}$$

$$1.2\mathbf{i} - 1.2\mathbf{j} = 0 + (-\omega_{DE}\mathbf{k}) \times (0.4\mathbf{i}) + (v_{C/D})_{xyz}\mathbf{i}$$

$$1.2\mathbf{i} - 1.2\mathbf{j} = 0 - 0.4\omega_{DE}\mathbf{j} + (v_{C/D})_{xyz}\mathbf{i}$$

$$(v_{C/D})_{xyz} = 1.2 \text{ m/s}$$

$$\omega_{DE} = 3 \text{ rad/s} \circlearrowright \qquad \qquad \textit{Ans.}$$

$$\mathbf{a}_C = \mathbf{a}_D + \dot{\boldsymbol{\Omega}} \times \mathbf{r}_{C/D} + \boldsymbol{\Omega} \times (\boldsymbol{\Omega} \times \mathbf{r}_{C/D}) + 2\boldsymbol{\Omega} \times (\mathbf{v}_{C/D})_{xyz} + (\mathbf{a}_{C/D})_{xyz}$$

$$-2\mathbf{i} - 5.2\mathbf{j} = 0 + (-\alpha_{DE}\mathbf{k}) \times (0.4\mathbf{i}) + (-3\mathbf{k}) \times [(-3\mathbf{k}) \times (0.4\mathbf{i})]$$

$$+ 2(-3\mathbf{k}) \times (1.2\mathbf{i}) + (a_{C/D})_{xyz}\mathbf{i}$$

$$-2\mathbf{i} - 5.2\mathbf{j} = -0.4\alpha_{DE}\mathbf{j} - 3.6\mathbf{i} - 7.2\mathbf{j} + (a_{C/D})_{xyz}\mathbf{i}$$

$$(a_{C/D})_{xyz} = 1.6 \text{ m/s}^2$$

$$\alpha_{DE} = -5 \text{ rad/s}^2 = 5 \text{ rad/s}^2 \circlearrowright \qquad \qquad \textit{Ans.}$$

EXAMPLE | 16.21

Planes A and B fly at the same elevation and have the motions shown in Fig. 16–36. Determine the velocity and acceleration of A as measured by the pilot of B.

SOLUTION

Coordinate Axes. Since the relative motion of A with respect to the pilot in B is being sought, the x, y, z axes are attached to plane B, Fig. 16–36. At the *instant* considered, the origin B coincides with the origin of the fixed X, Y, Z frame.

Kinematic Equations.

$$\mathbf{v}_A = \mathbf{v}_B + \boldsymbol{\Omega} \times \mathbf{r}_{A/B} + (\mathbf{v}_{A/B})_{xyz} \tag{1}$$

$$\mathbf{a}_A = \mathbf{a}_B + \dot{\boldsymbol{\Omega}} \times \mathbf{r}_{A/B} + \boldsymbol{\Omega} \times (\boldsymbol{\Omega} \times \mathbf{r}_{A/B}) + 2\boldsymbol{\Omega} \times (\mathbf{v}_{A/B})_{xyz} + (\mathbf{a}_{A/B})_{xyz} \tag{2}$$

Motion of Moving Reference:

$$\mathbf{v}_B = \{600\mathbf{j}\} \text{ km/h}$$

$$(a_B)_n = \frac{v_B^2}{\rho} = \frac{(600)^2}{400} = 900 \text{ km/h}^2$$

$$\mathbf{a}_B = (\mathbf{a}_B)_n + (\mathbf{a}_B)_t = \{900\mathbf{i} - 100\mathbf{j}\} \text{ km/h}^2$$

$$\boldsymbol{\Omega} = \frac{v_B}{\rho} = \frac{600 \text{ km/h}}{400 \text{ km}} = 1.5 \text{ rad/h} \, \text{↻} \qquad \boldsymbol{\Omega} = \{-1.5\mathbf{k}\} \text{ rad/h}$$

$$\dot{\boldsymbol{\Omega}} = \frac{(a_B)_t}{\rho} = \frac{100 \text{ km/h}^2}{400 \text{ km}} = 0.25 \text{ rad/h}^2 \, \text{↻} \qquad \dot{\boldsymbol{\Omega}} = \{0.25\mathbf{k}\} \text{ rad/h}^2$$

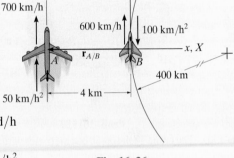

Fig. 16–36

16

Motion of A with Respect to Moving Reference:

$$\mathbf{r}_{A/B} = \{-4\mathbf{i}\} \text{ km} \quad (\mathbf{v}_{A/B})_{xyz} = ? \quad (\mathbf{a}_{A/B})_{xyz} = ?$$

Substituting the data into Eqs. 1 and 2, realizing that $\mathbf{v}_A = \{700\mathbf{j}\}$ km/h and $\mathbf{a}_A = \{50\mathbf{j}\}$ km/h^2, we have

$$\mathbf{v}_A = \mathbf{v}_B + \boldsymbol{\Omega} \times \mathbf{r}_{A/B} + (\mathbf{v}_{A/B})_{xyz}$$

$$700\mathbf{j} = 600\mathbf{j} + (-1.5\mathbf{k}) \times (-4\mathbf{i}) + (\mathbf{v}_{A/B})_{xyz}$$

$$(\mathbf{v}_{A/B})_{xyz} = \{94\mathbf{j}\} \text{ km/h} \qquad\qquad Ans.$$

$$\mathbf{a}_A = \mathbf{a}_B + \dot{\boldsymbol{\Omega}} \times \mathbf{r}_{A/B} + \boldsymbol{\Omega} \times (\boldsymbol{\Omega} \times \mathbf{r}_{A/B}) + 2\boldsymbol{\Omega} \times (\mathbf{v}_{A/B})_{xyz} + (\mathbf{a}_{A/B})_{xyz}$$

$$50\mathbf{j} = (900\mathbf{i} - 100\mathbf{j}) + (0.25\mathbf{k}) \times (-4\mathbf{i})$$

$$+ (-1.5\mathbf{k}) \times [(-1.5\mathbf{k}) \times (-4\mathbf{i})] + 2(-1.5\mathbf{k}) \times (94\mathbf{j}) + (\mathbf{a}_{A/B})_{xyz}$$

$$(\mathbf{a}_{A/B})_{xyz} = \{-1191\mathbf{i} + 151\mathbf{j}\} \text{ km/h}^2 \qquad\qquad Ans.$$

NOTE: The solution of this problem should be compared with that of Example 12.26, where it is seen that $(v_{B/A})_{xyz} \neq (v_{A/B})_{xyz}$ and $(a_{B/A})_{xyz} \neq (a_{A/B})_{xyz}$.

PROBLEMS

16–129. Ball C moves along the slot from A to B with a speed of 3 ft/s, which is increasing at 1.5 ft/s², both measured relative to the circular plate. At this same instant the plate rotates with the angular velocity and angular deceleration shown. Determine the velocity and acceleration of the ball at this instant.

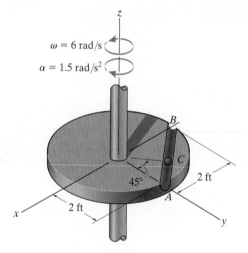

Prob. 16–129

16–130. The crane's telescopic boom rotates with the angular velocity and angular acceleration shown. At the same instant, the boom is extending with a constant speed of 0.5 ft/s, measured relative to the boom. Determine the magnitudes of the velocity and acceleration of point B at this instant.

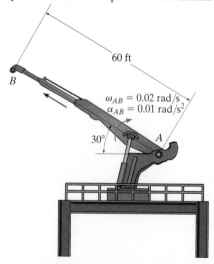

Prob. 16–130

16–131. While the swing bridge is closing with a constant rotation of 0.5 rad/s, a man runs along the roadway at a constant speed of 5 ft/s relative to the roadway. Determine his velocity and acceleration at the instant $d = 15$ ft.

*16–132. While the swing bridge is closing with a constant rotation of 0.5 rad/s, a man runs along the roadway such that when $d = 10$ ft he is running outward from the center at 5 ft/s with an acceleration of 2 ft/s², both measured relative to the roadway. Determine his velocity and acceleration at this instant.

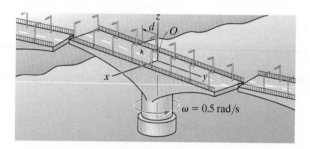

Probs. 16–131/132

16–133. Collar C moves along rod BA with a velocity of 3 m/s and an acceleration of 0.5 m/s², both directed from B towards A and measured relative to the rod. At the same instant, rod AB rotates with the angular velocity and angular acceleration shown. Determine the collar's velocity and acceleration at this instant.

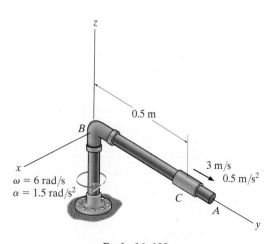

Prob. 16–133

16–134. Block A, which is attached to a cord, moves along the slot of a horizontal forked rod. At the instant shown, the cord is pulled down through the hole at O with an acceleration of 4 m/s² and its velocity is 2 m/s. Determine the acceleration of the block at this instant. The rod rotates about O with a constant angular velocity $\omega = 4$ rad/s.

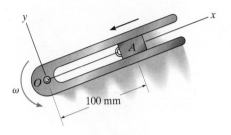

Prob. 16–134

16–135. A girl stands at A on a platform which is rotating with a constant angular velocity $\omega = 0.5$ rad/s. If she walks at a constant speed of $v = 0.75$ m/s measured relative to the platform, determine her acceleration (a) when she reaches point D in going along the path $ADC, d = 1$ m; and (b) when she reaches point B if she follows the path ABC, $r = 3$ m.

*16–136. A girl stands at A on a platform which is rotating with an angular acceleration $\alpha = 0.2$ rad/s² and at the instant shown has an angular velocity $\omega = 0.5$ rad/s. If she walks at a constant speed $v = 0.75$ m/s measured relative to the platform, determine her acceleration (a) when she reaches point D in going along the path $ADC, d = 1$ m; and (b) when she reaches point B if she follows the path ABC, $r = 3$ m.

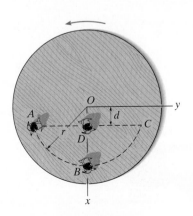

Probs. 16–135/136

16–137. At the instant shown, rod AB has an angular velocity $\omega_{AB} = 3$ rad/s and an angular acceleration $\alpha_{AB} = 5$ rad/s². Determine the angular velocity and angular acceleration of rod CD at this instant. The collar at C is pin-connected to CD and slides over AB.

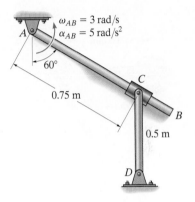

Prob. 16–137

16–138. Collar B moves to the left with a speed of 5 m/s, which is increasing at a constant rate of 1.5 m/s², relative to the hoop, while the hoop rotates with the angular velocity and angular acceleration shown. Determine the magnitudes of the velocity and acceleration of the collar at this instant.

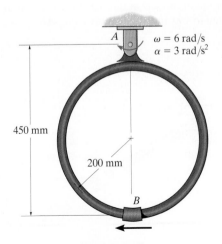

Prob. 16–138

16–139. Block *B* of the mechanism is confined to move within the slot member *CD*. If *AB* is rotating at a constant rate of $\omega_{AB} = 3$ rad/s, determine the angular velocity and angular acceleration of member *CD* at the instant shown.

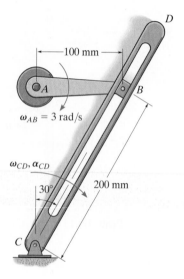

Prob. 16–139

16–141. The "quick-return" mechanism consists of a crank *AB*, slider block *B*, and slotted link *CD*. If the crank has the angular motion shown, determine the angular motion of the slotted link at this instant.

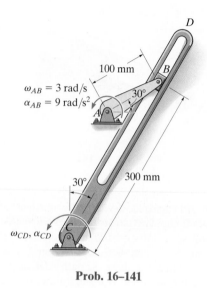

Prob. 16–141

*16–140. At the instant shown rod *AB* has an angular velocity $\omega_{AB} = 4$ rad/s and an angular acceleration $\alpha_{AB} = 2$ rad/s^2. Determine the angular velocity and angular acceleration of rod *CD* at this instant. The collar at *C* is pin connected to *CD* and slides freely along *AB*.

16–142. At the instant shown, the robotic arm *AB* is rotating counterclockwise at $\omega = 5$ rad/s and has an angular acceleration $\alpha = 2$ rad/s^2. Simultaneously, the grip *BC* is rotating counterclockwise at $\omega' = 6$ rad/s and $\alpha' = 2$ rad/s^2, both measured relative to a *fixed* reference. Determine the velocity and acceleration of the object held at the grip *C*.

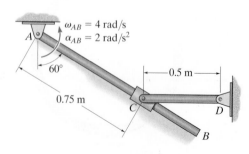

Prob. 16–140

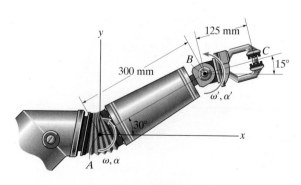

Prob. 16–142

16–143. Peg B on the gear slides freely along the slot in link AB. If the gear's center O moves with the velocity and acceleration shown, determine the angular velocity and angular acceleration of the link at this instant.

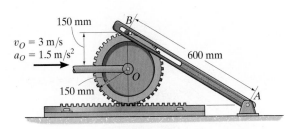

Prob. 16–143

*16–144. The cars on the amusement-park ride rotate around the axle at A with a constant angular velocity $\omega_{A/f} = 2$ rad/s, measured relative to the frame AB. At the same time the frame rotates around the main axle support at B with a constant angular velocity $\omega_f = 1$ rad/s. Determine the velocity and acceleration of the passenger at C at the instant shown.

16–145. The cars on the amusement-park ride rotate around the axle at A with a constant angular velocity $\omega_{A/f} = 2$ rad/s, measured relative to the frame AB. At the same time the frame rotates around the main axle support at B with a constant angular velocity $\omega_f = 1$ rad/s. Determine the velocity and acceleration of the passenger at D at the instant shown.

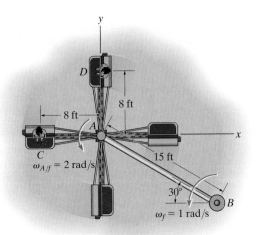

Probs. 16–144/145

16–146. If the slotted arm AB rotates about the pin A with a constant angular velocity of $\omega_{AB} = 10$ rad/s, determine the angular velocity of link CD at the instant shown.

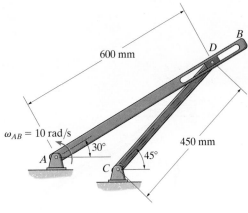

Prob. 16–146

16–147. At the instant shown, boat A travels with a speed of 15 m/s, which is decreasing at 3 m/s², while boat B travels with a speed of 10 m/s, which is increasing at 2 m/s². Determine the velocity and acceleration of boat A with respect to boat B at this instant.

*16–148. At the instant shown, boat A travels with a speed of 15 m/s, which is decreasing at 3 m/s², while boat B travels with a speed of 10 m/s, which is increasing at 2 m/s². Determine the velocity and acceleration of boat B with respect to boat A at this instant.

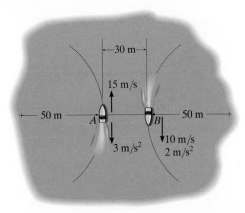

Probs. 16–147/148

16–149. If the piston is moving with a velocity of $v_A = 3$ m/s and acceleration of $a_A = 1.5$ m/s², determine the angular velocity and angular acceleration of the slotted link at the instant shown. Link AB slides freely along its slot on the fixed peg C.

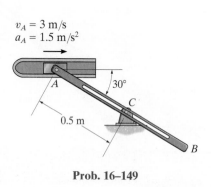

$v_A = 3$ m/s
$a_A = 1.5$ m/s²

30°

A

C

0.5 m

B

Prob. 16–149

16–150. The two-link mechanism serves to amplify angular motion. Link AB has a pin at B which is confined to move within the slot of link CD. If at the instant shown, AB (input) has an angular velocity of $\omega_{AB} = 2.5$ rad/s, determine the angular velocity of CD (output) at this instant.

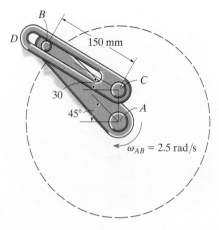

B

D

150 mm

C

30

45°

A

$\omega_{AB} = 2.5$ rad/s

Prob. 16–150

16–151. The gear has the angular motion shown. Determine the angular velocity and angular acceleration of the slotted link BC at this instant. The peg at A is fixed to the gear.

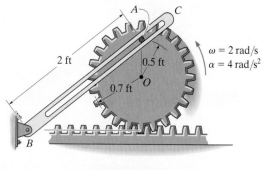

A

C

2 ft

0.5 ft

0.7 ft

O

$\omega = 2$ rad/s
$\alpha = 4$ rad/s²

B

Prob. 16–151

*16–152. The Geneva mechanism is used in a packaging system to convert constant angular motion into intermittent angular motion. The star wheel A makes one sixth of a revolution for each full revolution of the driving wheel B and the attached guide C. To do this, pin P, which is attached to B, slides into one of the radial slots of A, thereby turning wheel A, and then exits the slot. If B has a constant angular velocity of $\omega_B = 4$ rad/s, determine ω_A and α_A of wheel A at the instant shown.

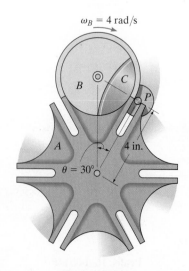

$\omega_B = 4$ rad/s

B

C

P

A

4 in.

$\theta = 30°$

Prob. 16–152

16

CONCEPTUAL PROBLEMS

P16–1. An electric motor turns the tire at *A* at a constant angular velocity, and friction then causes the tire to roll without slipping on the inside rim of the Ferris wheel. Using appropriate numerical values, determine the magnitude of the velocity and acceleration of passengers in one of the baskets. Do passengers in the other baskets experience this same motion? Explain.

P16–1

P16–2. The crank *AB* turns counterclockwise at a constant rate **ω** causing the connecting arm *CD* and rocking beam *DE* to move. Draw a sketch showing the location of the *IC* for the connecting arm when θ = 0°, 90°, 180°, and 270°. Also, how was the curvature of the head at *E* determined, and why is it curved in this way?

P16–2

P16–3. The bi-fold hangar door is opened by cables that move upward at a constant speed of 0.5 m/s. Determine the angular velocity of *BC* and the angular velocity of *AB* when θ = 45°. Panel *BC* is pinned at *C* and has a height which is the same as the height of *BA*. Use appropriate numerical values to explain your result.

P16–3

P16–4. If the tires do not slip on the pavement, determine the points on the tire that have a maximum and minimum speed and the points that have a maximum and minimum acceleration. Use appropriate numerical values for the car's speed and tire size to explain your result.

P16–4

16

CHAPTER REVIEW

Rigid-Body Planar Motion

A rigid body undergoes three types of planar motion: translation, rotation about a fixed axis, and general plane motion.

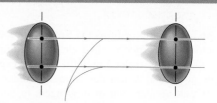

Path of rectilinear translation

Translation

When a body has rectilinear translation, all the particles of the body travel along parallel straight-line paths. If the paths have the same radius of curvature, then curvilinear translation occurs. Provided we know the motion of one of the particles, then the motion of all of the others is also known.

Path of curvilinear translation

Rotation about a Fixed Axis

For this type of motion, all of the particles move along circular paths. Here, all line segments in the body undergo the same angular displacement, angular velocity, and angular acceleration.

Once the angular motion of the body is known, then the velocity of any particle a distance r from the axis can be obtained.

The acceleration of any particle has two components. The tangential component accounts for the change in the magnitude of the velocity, and the normal component accounts for the change in the velocity's direction.

Rotation about a fixed axis

$$\omega = d\theta/dt \qquad\qquad \omega = \omega_0 + \alpha_c t$$

$$\alpha = d\omega/dt \quad \text{or} \quad \theta = \theta_0 + \omega_0 t + \tfrac{1}{2}\alpha_c t^2$$

$$\alpha\, d\theta = \omega\, d\omega \qquad\qquad \omega^2 = \omega_0^2 + 2\alpha_c(\theta - \theta_0)$$

$$\text{Constant } \alpha_c$$

$$v = \omega r \qquad\qquad a_t = \alpha r, \quad a_n = \omega^2 r$$

General Plane Motion

When a body undergoes general plane motion, it simultaneously translates and rotates. There are several methods for analyzing this motion.

Absolute Motion Analysis
If the motion of a point on a body or the angular motion of a line is known, then it may be possible to relate this motion to that of another point or line using an absolute motion analysis. To do so, linear position coordinates s or angular position coordinates θ are established (measured from a fixed point or line). These position coordinates are then related using the geometry of the body. The time derivative of this equation gives the relationship between the velocities and/or the angular velocities. A second time derivative relates the accelerations and/or the angular accelerations.

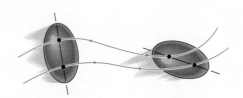

General plane motion

Relative-Motion using Translating Axes
General plane motion can also be analyzed using a relative-motion analysis between two points A and B located on the body. This method considers the motion in parts: first a translation of the selected base point A, then a relative "rotation" of the body about point A, which is measured from a translating axis. Since the relative motion is viewed as circular motion about the base point, point B will have a velocity $\mathbf{v}_{B/A}$ that is tangent to the circle. It also has two components of acceleration, $(\mathbf{a}_{B/A})_t$ and $(\mathbf{a}_{B/A})_n$. It is also important to realize that $\mathbf{a}_A$ and $\mathbf{a}_B$ will have tangential and normal components if these points move along curved paths.

$$\mathbf{v}_B = \mathbf{v}_A + \boldsymbol{\omega} \times \mathbf{r}_{B/A}$$

$$\mathbf{a}_B = \mathbf{a}_A + \boldsymbol{\alpha} \times \mathbf{r}_{B/A} - \omega^2 \mathbf{r}_{B/A}$$

Instantaneous Center of Zero Velocity
If the base point A is selected as having zero velocity, then the relative velocity equation becomes $\mathbf{v}_B = \boldsymbol{\omega} \times \mathbf{r}_{B/A}$. In this case, motion appears as if the body rotates about an instantaneous axis passing through A.

The instantaneous center of rotation (IC) can be established provided the directions of the velocities of any two points on the body are known, or the velocity of a point and the angular velocity are known. Since a radial line r will always be perpendicular to each velocity, then the IC is at the point of intersection of these two radial lines. Its measured location is determined from the geometry of the body. Once it is established, then the velocity of any point P on the body can be determined from $v = \omega r$, where r extends from the IC to point P.

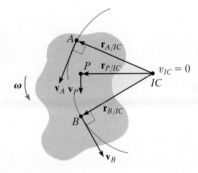

Relative Motion using Rotating Axes
Problems that involve connected members that slide relative to one another or points not located on the same body can be analyzed using a relative-motion analysis referenced from a rotating frame. This gives rise to the term $2\boldsymbol{\Omega} \times (\mathbf{v}_{B/A})_{xyz}$ that is called the Coriolis acceleration.

$$\mathbf{v}_B = \mathbf{v}_A + \boldsymbol{\Omega} \times \mathbf{r}_{B/A} + (\mathbf{v}_{B/A})_{xyz}$$

$$\mathbf{a}_B = \mathbf{a}_A + \dot{\boldsymbol{\Omega}} \times \mathbf{r}_{B/A} + \boldsymbol{\Omega} \times (\boldsymbol{\Omega} \times \mathbf{r}_{B/A}) + 2\boldsymbol{\Omega} \times (\mathbf{v}_{B/A})_{xyz} + (\mathbf{a}_{B/A})_{xyz}$$

16

Chapter 17

Tractors and other heavy equipment can be subjected to severe loadings due to dynamic loadings as they accelerate. In this chapter we will show how to determine these loadings for planar motion.

Planar Kinetics of a Rigid Body: Force and Acceleration

CHAPTER OBJECTIVES

■ To introduce the methods used to determine the mass moment of inertia of a body.

■ To develop the planar kinetic equations of motion for a symmetric rigid body.

■ To discuss applications of these equations to bodies undergoing translation, rotation about a fixed axis, and general plane motion.

17.1   Mass Moment of Inertia

Since a body has a definite size and shape, an applied nonconcurrent force system can cause the body to both translate and rotate. The translational aspects of the motion were studied in Chapter 13 and are governed by the equation $\mathbf{F} = m\mathbf{a}$. It will be shown in the next section that the rotational aspects, caused by a moment $\mathbf{M}$, are governed by an equation of the form $\mathbf{M} = I\boldsymbol{\alpha}$. The symbol I in this equation is termed the mass moment of inertia. By comparison, the *moment of inertia* is a measure of the resistance of a body to *angular acceleration* ($\mathbf{M} = I\boldsymbol{\alpha}$) in the same way that *mass* is a measure of the body's resistance to *acceleration* ($\mathbf{F} = m\mathbf{a}$).

The flywheel on the engine of this tractor has a large moment of inertia about its axis of rotation. Once it is set into motion, it will be difficult to stop, and this in turn will prevent the engine from stalling and instead will allow it to maintain a constant power.

Fig. 17–1

We define the *moment of inertia* as the integral of the "second moment" about an axis of all the elements of mass dm which compose the body.* For example, the body's moment of inertia about the z axis in Fig. 17–1 is

$$I = \int_m r^2 \, dm \qquad (17–1)$$

Here the "moment arm" r is the perpendicular distance from the z axis to the arbitrary element dm. Since the formulation involves r, the value of I is different for each axis about which it is computed. In the study of planar kinetics, the axis chosen for analysis generally passes through the body's mass center G and is always perpendicular to the plane of motion. The moment of inertia about this axis will be denoted as I_G. Since r is squared in Eq. 17–1, the mass moment of inertia is always a *positive* quantity. Common units used for its measurement are $kg \cdot m^2$ or $slug \cdot ft^2$.

If the body consists of material having a variable density, $\rho = \rho(x,y,z)$, the elemental mass dm of the body can be expressed in terms of its density and volume as $dm = \rho \, dV$. Substituting dm into Eq. 17–1, the body's moment of inertia is then computed using *volume elements* for integration; i.e.,

$$I = \int_V r^2 \rho \, dV \qquad (17–2)$$

---

*Another property of the body, which measures the symmetry of the body's mass with respect to a coordinate system, is the product of inertia. This property applies to the three-dimensional motion of a body and will be discussed in Chapter 21.

In the special case of ρ being a *constant*, this term may be factored out of the integral, and the integration is then purely a function of geometry,

$$I = \rho \int_V r^2 \, dV \qquad (17\text{--}3)$$

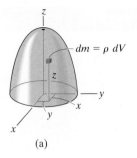

(a)

When the volume element chosen for integration has infinitesimal dimensions in all three directions, Fig. 17–2a, the moment of inertia of the body must be determined using "triple integration." The integration process can, however, be simplified to a *single integration* provided the chosen volume element has a differential size or thickness in only *one* direction. Shell or disk elements are often used for this purpose.

(b)

Procedure for Analysis

To obtain the moment of inertia by integration, we will consider only symmetric bodies having volumes which are generated by revolving a curve about an axis. An example of such a body is shown in Fig. 17–2a. Two types of differential elements can be chosen.

Shell Element.

- If a *shell element* having a height z, radius $r = y$, and thickness dy is chosen for integration, Fig. 17–2b, then the volume is $dV = (2\pi y)(z)dy$.

- This element may be used in Eq. 17–2 or 17–3 for determining the moment of inertia I_z of the body about the z axis, since the *entire element*, due to its "thinness," lies at the *same* perpendicular distance $r = y$ from the z axis (see Example 17.1).

Disk Element.

- If a disk element having a radius y and a thickness dz is chosen for integration, Fig. 17–2c, then the volume is $dV = (\pi y^2)dz$.

- This element is *finite* in the radial direction, and consequently its parts *do not* all lie at the *same radial distance* r from the z axis. As a result, Eq. 17–2 or 17–3 *cannot* be used to determine I_z directly. Instead, to perform the integration it is first necessary to determine the moment of inertia *of the element* about the z axis and then integrate this result (see Example 17.2).

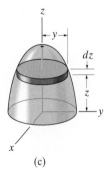

(c)

Fig. 17–2

17

EXAMPLE | 17.1

Determine the moment of inertia of the cylinder shown in Fig. 17–3a about the z axis. The density of the material, ρ, is constant.

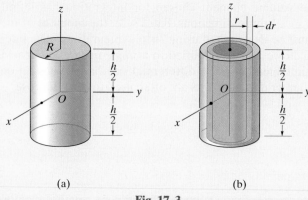

(a)   (b)

Fig. 17–3

SOLUTION

Shell Element.   This problem can be solved using the *shell element* in Fig. 17–3b and a single integration. The volume of the element is $dV = (2\pi r)(h)\, dr$, so that its mass is $dm = \rho dV = \rho(2\pi hr\, dr)$. Since the *entire element* lies at the same distance r from the z axis, the moment of inertia *of the element* is

$$dI_z = r^2 dm = \rho 2\pi hr^3\, dr$$

Integrating over the entire region of the cylinder yields

$$I_z = \int_m r^2\, dm = \rho 2\pi h \int_0^R r^3\, dr = \frac{\rho\pi}{2}R^4 h$$

The mass of the cylinder is

$$m = \int_m dm = \rho 2\pi h \int_0^R r\, dr = \rho\pi hR^2$$

so that

$$I_z = \frac{1}{2}mR^2 \qquad\qquad Ans.$$

EXAMPLE | 17.2

If the density of the material is 5 slug/ft³, determine the moment of inertia of the solid in Fig. 17–4a about the y axis.

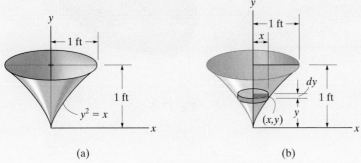

(a)   (b)

Fig. 17–4

SOLUTION

Disk Element.   The moment of inertia will be found using a *disk element*, as shown in Fig. 17–4b. Here the element intersects the curve at the arbitrary point (x, y) and has a mass

$$dm = \rho \, dV = \rho(\pi x^2) \, dy$$

Although all portions of the element are *not* located at the same distance from the y axis, it is still possible to determine the moment of inertia dI_y *of the element* about the y axis. In the preceding example it was shown that the moment of inertia of a cylinder about its longitudinal axis is $I = \frac{1}{2}mR^2$, where m and R are the mass and radius of the cylinder. Since the height is not involved in this formula, the disk itself can be thought of as a cylinder. Thus, for the disk element in Fig. 17–4b, we have

$$dI_y = \tfrac{1}{2}(dm)x^2 = \tfrac{1}{2}[\rho(\pi x^2) \, dy]x^2$$

Substituting $x = y^2$, $\rho = 5$ slug/ft³, and integrating with respect to y, from $y = 0$ to $y = 1$ ft, yields the moment of inertia for the entire solid.

$$I_y = \frac{\pi(5 \text{ slug/ft}^3)}{2} \int_0^{1 \text{ ft}} x^4 \, dy = \frac{\pi(5)}{2} \int_0^{1 \text{ ft}} y^8 \, dy = 0.873 \text{ slug} \cdot \text{ft}^2 \quad Ans.$$

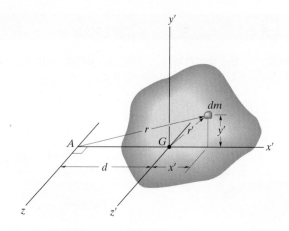

Fig. 17–5

Parallel-Axis Theorem.   If the moment of inertia of the body about an axis passing through the body's mass center is known, then the moment of inertia about any other *parallel axis* can be determined by using the *parallel-axis theorem*. This theorem can be derived by considering the body shown in Fig. 17–5. Here the z' axis passes through the mass center G, whereas the corresponding *parallel z axis* lies at a constant distance d away. Selecting the differential element of mass dm, which is located at point (x', y'), and using the Pythagorean theorem, $r^2 = (d + x')^2 + y'^2$, we can express the moment of inertia of the body about the z axis as

$$I = \int_m r^2 \, dm = \int_m [(d + x')^2 + y'^2] \, dm$$

$$= \int_m (x'^2 + y'^2) \, dm + 2d \int_m x' \, dm + d^2 \int_m dm$$

Since $r'^2 = x'^2 + y'^2$, the first integral represents I_G. The second integral equals *zero*, since the z' axis passes through the body's mass center, i.e., $\int x' dm = \bar{x}'m = 0$ since $\bar{x}' = 0$. Finally, the third integral represents the total mass m of the body. Hence, the moment of inertia about the z axis can be written as

$$I = I_G + md^2 \tag{17–4}$$

where

I_G = moment of inertia about the z' axis passing through the mass center G

m = mass of the body

d = perpendicular distance between the parallel z and z' axes

Radius of Gyration.   Occasionally, the moment of inertia of a body about a specified axis is reported in handbooks using the *radius of gyration, k*. This is a geometrical property which has units of length. When it and the body's mass m are known, the body's moment of inertia is determined from the equation

$$I = mk^2 \quad \text{or} \quad k = \sqrt{\frac{I}{m}} \tag{17–5}$$

Note the *similarity* between the definition of k in this formula and r in the equation $dI = r^2\, dm$, which defines the moment of inertia of an elemental mass dm of the body about an axis.

Composite Bodies.   If a body consists of a number of simple shapes such as disks, spheres, and rods, the moment of inertia of the body about any axis can be determined by adding algebraically the moments of inertia of all the composite shapes computed about the axis. Algebraic addition is necessary since a composite part must be considered as a negative quantity if it has already been counted as a piece of another part—for example, a "hole" subtracted from a solid plate. The parallel-axis theorem is needed for the calculations if the center of mass of each composite part does not lie on the axis. For the calculation, then, $I = \Sigma(I_G + md^2)$. Here I_G for each of the composite parts is determined by integration, or for simple shapes, such as rods and disks, it can be found from a table, such as the one given on the inside back cover of this book.

EXAMPLE | 17.3

If the plate shown in Fig. 17–6a has a density of 8000 kg/m³ and a thickness of 10 mm, determine its moment of inertia about an axis directed perpendicular to the page and passing through point O.

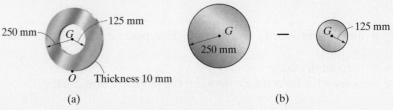

Thickness 10 mm

(a)   (b)

Fig. 17–6

SOLUTION

The plate consists of two composite parts, the 250-mm-radius disk *minus* a 125-mm-radius disk, Fig. 17–6b. The moment of inertia about O can be determined by computing the moment of inertia of each of these parts about O and then adding the results *algebraically*. The calculations are performed by using the parallel-axis theorem in conjunction with the data listed in the table on the inside back cover.

Disk. The moment of inertia of a disk about the centroidal axis perpendicular to the plane of the disk is $I_G = \frac{1}{2}mr^2$. The mass center of the disk is located at a distance of 0.25 m from point O. Thus,

$$m_d = \rho_d V_d = 8000 \text{ kg/m}^3 \, [\pi(0.25 \text{ m})^2(0.01 \text{ m})] = 15.71 \text{ kg}$$

$$(I_d)_O = \tfrac{1}{2}m_d r_d^2 + m_d d^2$$

$$= \frac{1}{2}(15.71 \text{ kg})(0.25 \text{ m})^2 + (15.71 \text{ kg})(0.25 \text{ m})^2$$

$$= 1.473 \text{ kg} \cdot \text{m}^2$$

Hole. For the 125-mm-radius disk (hole), we have

$$m_h = \rho_h V_h = 8000 \text{ kg/m}^3 \, [\pi(0.125 \text{ m})^2(0.01 \text{ m})] = 3.927 \text{ kg}$$

$$(I_h)_O = \tfrac{1}{2}m_h r_h^2 + m_h d^2$$

$$= \frac{1}{2}(3.927 \text{ kg})(0.125 \text{ m})^2 + (3.927 \text{ kg})(0.25 \text{ m})^2$$

$$= 0.276 \text{ kg} \cdot \text{m}^2$$

The moment of inertia of the plate about point O is therefore

$$I_O = (I_d)_O - (I_h)_O$$

$$= 1.473 \text{ kg} \cdot \text{m}^2 - 0.276 \text{ kg} \cdot \text{m}^2$$

$$= 1.20 \text{ kg} \cdot \text{m}^2 \qquad\qquad Ans.$$

EXAMPLE 17.4

The pendulum in Fig. 17–7 is suspended from the pin at O and consists of two thin rods. Rod OA weighs 10 lb, and BC weighs 8 lb. Determine the moment of inertia of the pendulum about an axis passing through (a) point O, and (b) the mass center G of the pendulum.

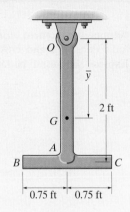

Fig. 17–7

SOLUTION

Part (a). Using the table on the inside back cover, the moment of inertia of rod OA about an axis perpendicular to the page and passing through point O of the rod is $I_O = \frac{1}{3}ml^2$. Hence,

$$(I_{OA})_O = \frac{1}{3}ml^2 = \frac{1}{3}\left(\frac{10 \text{ lb}}{32.2 \text{ ft/s}^2}\right)(2 \text{ ft})^2 = 0.414 \text{ slug} \cdot \text{ft}^2$$

This same value can be obtained using $I_G = \frac{1}{12}ml^2$ and the parallel-axis theorem.

$$(I_{OA})_O = \frac{1}{12}ml^2 + md^2 = \frac{1}{12}\left(\frac{10 \text{ lb}}{32.2 \text{ ft/s}^2}\right)(2 \text{ ft})^2 + \left(\frac{10 \text{ lb}}{32.2 \text{ ft/s}^2}\right)(1 \text{ ft})^2$$

$$= 0.414 \text{ slug} \cdot \text{ft}^2$$

For rod BC we have

$$(I_{BC})_O = \frac{1}{12}ml^2 + md^2 = \frac{1}{12}\left(\frac{8 \text{ lb}}{32.2 \text{ ft/s}^2}\right)(1.5 \text{ ft})^2 + \left(\frac{8 \text{ lb}}{32.2 \text{ ft/s}^2}\right)(2 \text{ ft})^2$$

$$= 1.040 \text{ slug} \cdot \text{ft}^2$$

The moment of inertia of the pendulum about O is therefore

$$I_O = 0.414 + 1.040 = 1.454 = 1.45 \text{ slug} \cdot \text{ft}^2 \qquad Ans.$$

Part (b). The mass center G will be located relative to point O. Assuming this distance to be $\bar{y}$, Fig. 17–7, and using the formula for determining the mass center, we have

$$\bar{y} = \frac{\Sigma \tilde{y}m}{\Sigma m} = \frac{1(10/32.2) + 2(8/32.2)}{(10/32.2) + (8/32.2)} = 1.444 \text{ ft}$$

The moment of inertia I_G may be found in the same manner as I_O, which requires successive applications of the parallel-axis theorem to transfer the moments of inertia of rods OA and BC to G. A more direct solution, however, involves using the result for I_O, i.e.,

$$I_O = I_G + md^2; \quad 1.454 \text{ slug} \cdot \text{ft}^2 = I_G + \left(\frac{18 \text{ lb}}{32.2 \text{ ft/s}^2}\right)(1.444 \text{ ft})^2$$

$$I_G = 0.288 \text{ slug} \cdot \text{ft}^2 \qquad Ans.$$

17

PROBLEMS

17–1. Determine the moment of inertia I_y for the slender rod. The rod's density ρ and cross-sectional area A are constant. Express the result in terms of the rod's total mass m.

17–3. Determine the moment of inertia of the thin ring about the z axis. The ring has a mass m.

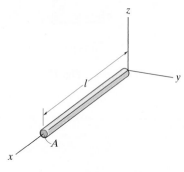

Prob. 17–1

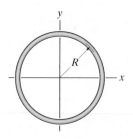

Prob. 17–3

17–2. The solid cylinder has an outer radius R, height h, and is made from a material having a density that varies from its center as $\rho = k + ar^2$, where k and a are constants. Determine the mass of the cylinder and its moment of inertia about the z axis.

17–4. Determine the moment of inertia of the semiellipsoid with respect to the x axis and express the result in terms of the mass m of the semiellipsoid. The material has a constant density ρ.

Prob. 17–2

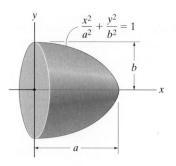

Prob. 17–4

17–5. The sphere is formed by revolving the shaded area around the x axis. Determine the moment of inertia I_x and express the result in terms of the total mass m of the sphere. The material has a constant density ρ.

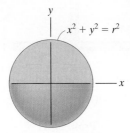

$$x^2 + y^2 = r^2$$

Prob. 17–5

17–6. Determine the mass moment of inertia I_z of the cone formed by revolving the shaded area around the z axis. The total density of the material is ρ. Express the result in terms of the mass m of the cone.

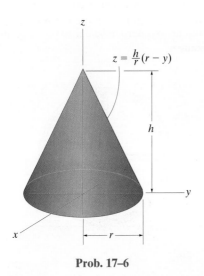

$$z = \frac{h}{r}(r - y)$$

Prob. 17–6

17–7. The solid is formed by revolving the shaded area around the y axis. Determine the radius of gyration k_y. The specific weight of the material is $\gamma = 380 \ \text{lb}/\text{ft}^3$.

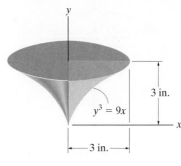

3 in.

$$y^3 = 9x$$

3 in.

Prob. 17–7

*17–8. The concrete shape is formed by rotating the shaded area about the y axis. Determine the moment of inertia I_y. The specific weight of concrete is $\gamma = 150 \ \text{lb}/\text{ft}^3$.

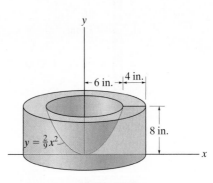

6 in. 4 in.

8 in.

$$y = \frac{2}{9}x^2$$

Prob. 17–8

17

17–9. Determine the moment of inertia I_z of the torus. The mass of the torus is m and the density ρ is constant. *Suggestion:* Use a shell element.

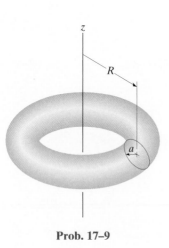

Prob. 17–9

17–10. Determine the mass moment of inertia of the pendulum about an axis perpendicular to the page and passing through point O. The slender rod has a mass of 10 kg and the sphere has a mass of 15 kg.

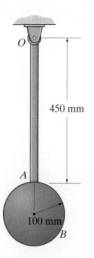

Prob. 17–10

17–11. The slender rods have a weight of 3 lb/ft. Determine the moment of inertia of the assembly about an axis perpendicular to the page and passing through the pin at A.

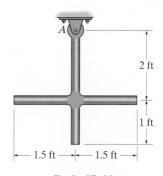

Prob. 17–11

*17–12. Determine the moment of inertia of the solid steel assembly about the x axis. Steel has a specific weight of $\gamma_{st} = 490$ lb/ft^3.

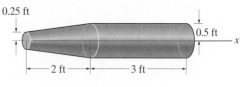

Prob. 17–12

17–13. The wheel consists of a thin ring having a mass of 10 kg and four spokes made from slender rods and each having a mass of 2 kg. Determine the wheel's moment of inertia about an axis perpendicular to the page and passing through point A.

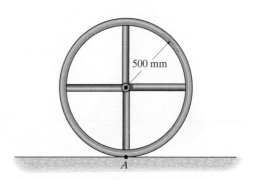

Prob. 17–13

17–14. If the large ring, small ring and each of the spokes weigh 100 lb, 15 lb, and 20 lb, respectively, determine the mass moment of inertia of the wheel about an axis perpendicular to the page and passing through point *A*.

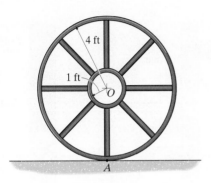

Prob. 17–14

***17–16.** Determine the mass moment of inertia of the thin plate about an axis perpendicular to the page and passing through point *O*. The material has a mass per unit area of 20 kg/m².

Prob. 17–16

17

17–15. Determine the moment of inertia about an axis perpendicular to the page and passing through the pin at *O*. The thin plate has a hole in its center. Its thickness is 50 mm, and the material has a density $\rho = 50 \text{ kg/m}^3$.

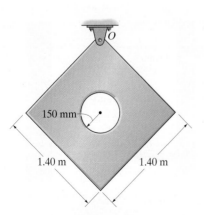

Prob. 17–15

17–17. The assembly consists of a disk having a mass of 6 kg and slender rods *AB* and *DC* which have a mass of 2 kg/m. Determine the length *L* of *DC* so that the center of mass is at the bearing *O*. What is the moment of inertia of the assembly about an axis perpendicular to the page and passing through *O*?

17–18. The assembly consists of a disk having a mass of 6 kg and slender rods *AB* and *DC* which have a mass of 2 kg/m. If *L* = 0.75 m, determine the moment of inertia of the assembly about an axis perpendicular to the page and passing through *O*.

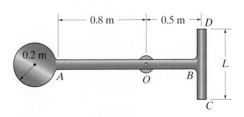

Probs. 17–17/18

17–19. The pendulum consists of two slender rods AB and OC which have a mass of 3 kg/m. The thin circular plate has a mass of 12 kg/m². Determine the location $\bar{y}$ of the center of mass G of the pendulum, then calculate the moment of inertia of the pendulum about an axis perpendicular to the page and passing through G.

*17–20. The pendulum consists of two slender rods AB and OC which have a mass of 3 kg/m. The thin circular plate has a mass of 12 kg/m². Determine the moment of inertia of the pendulum about an axis perpendicular to the page and passing through the pin at O.

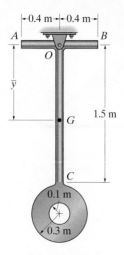

Probs. 17–19/20

17–21. The pendulum consists of the 3-kg slender rod and the 5-kg thin plate. Determine the location $\bar{y}$ of the center of mass G of the pendulum; then calculate the moment of inertia of the pendulum about an axis perpendicular to the page and passing through G.

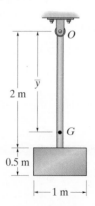

Prob. 17–21

*17–22. Determine the moment of inertia of the overhung crank about the x axis. The material is steel having a destiny of $\rho = 7.85$ Mg/m³.

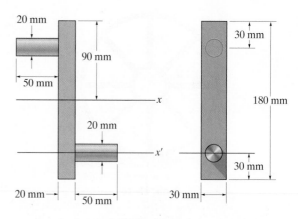

Prob. 17–22

17–23. Determine the moment of inertia of the overhung crank about the x' axis. The material is steel having a destiny of $\rho = 7.85$ Mg/m³.

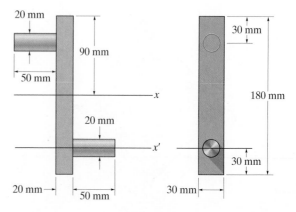

Prob. 17–23

17.2   Planar Kinetic Equations of Motion

In the following analysis we will limit our study of planar kinetics to rigid bodies which, along with their loadings, are considered to be *symmetrical* with respect to a fixed reference plane.* Since the motion of the body can be viewed within the reference plane, all the forces (and couple moments) acting on the body can then be projected onto the plane. An example of an arbitrary body of this type is shown in Fig. 17–8a. Here the *inertial frame of reference x, y, z* has its origin *coincident* with the arbitrary point *P* in the body. By definition, *these axes do not rotate and are either fixed or translate with constant velocity*

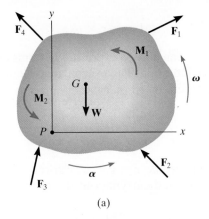

(a)

Fig. 17–8

Equation of Translational Motion.   The external forces acting on the body in Fig. 17–8a represent the effect of gravitational, electrical, magnetic, or contact forces between adjacent bodies. Since this force system has been considered previously in Sec. 13.3 for the analysis of a system of particles, the resulting Eq. 13–6 can be used here, in which case

$$\Sigma \mathbf{F} = m\mathbf{a}_G$$

This equation is referred to as the *translational equation of motion* for the mass center of a rigid body. It states that *the sum of all the external forces acting on the body is equal to the body's mass times the acceleration of its mass center G.*

For motion of the body in the *x–y* plane, the translational equation of motion may be written in the form of two independent scalar equations, namely,

$$\Sigma F_x = m(a_G)_x$$
$$\Sigma F_y = m(a_G)_y$$

*By doing this, the rotational equation of motion reduces to a rather simplified form. The more general case of body shape and loading is considered in Chapter 21.

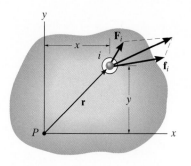

Particle free-body diagram

(b)

$\|$

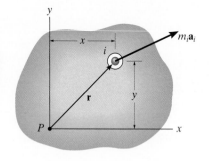

Particle kinetic diagram

(c)

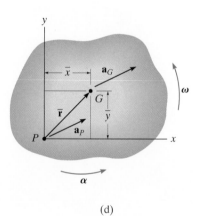

(d)

Fig. 17–8 (cont.)

Equation of Rotational Motion.   We will now determine the effects caused by the moments of the external force system computed about an axis perpendicular to the plane of motion (the z axis) and passing through point P. As shown on the free-body diagram of the ith particle, Fig. 17–8b, $\mathbf{F}_i$ represents the *resultant external force* acting on the particle, and $\mathbf{f}_i$ is the *resultant of the internal forces* caused by interactions with adjacent particles. If the particle has a mass m_i and its acceleration is $\mathbf{a}_i$, then its kinetic diagram is shown in Fig. 17–8c. Summing moments about point P, we require

$$\mathbf{r} \times \mathbf{F}_i + \mathbf{r} \times \mathbf{f}_i = \mathbf{r} \times m_i \mathbf{a}_i$$

or

$$(\mathbf{M}_P)_i = \mathbf{r} \times m_i \mathbf{a}_i$$

The moments about P can also be expressed in terms of the acceleration of point P, Fig. 17–8d. If the body has an angular acceleration $\boldsymbol{\alpha}$ and angular velocity $\boldsymbol{\omega}$, then using Eq. 16–18 we have

$$(\mathbf{M}_P)_i = m_i \mathbf{r} \times (\mathbf{a}_P + \boldsymbol{\alpha} \times \mathbf{r} - \omega^2 \mathbf{r})$$

$$= m_i[\mathbf{r} \times \mathbf{a}_P + \mathbf{r} \times (\boldsymbol{\alpha} \times \mathbf{r}) - \omega^2(\mathbf{r} \times \mathbf{r})]$$

The last term is zero, since $\mathbf{r} \times \mathbf{r} = \mathbf{0}$. Expressing the vectors with Cartesian components and carrying out the cross-product operations yields

$$(M_P)_i \mathbf{k} = m_i\{(x\mathbf{i} + y\mathbf{j}) \times [(a_P)_x \mathbf{i} + (a_P)_y \mathbf{j}]$$

$$+ (x\mathbf{i} + y\mathbf{j}) \times [\alpha \mathbf{k} \times (x\mathbf{i} + y\mathbf{j})]\}$$

$$(M_P)_i \mathbf{k} = m_i[-y(a_P)_x + x(a_P)_y + \alpha x^2 + \alpha y^2]\mathbf{k}$$

$$\zeta \, (M_P)_i = m_i[-y(a_P)_x + x(a_P)_y + \alpha r^2]$$

Letting $m_i \rightarrow dm$ and integrating with respect to the entire mass m of the body, we obtain the resultant moment equation

$$\zeta \, \Sigma M_P = -\left(\int_m y \, dm\right)(a_P)_x + \left(\int_m x \, dm\right)(a_P)_y + \left(\int_m r^2 dm\right)\alpha$$

Here ΣM_P represents only the moment of the *external forces* acting on the body about point P. The resultant moment of the internal forces is zero, since for the entire body these forces occur in equal and opposite collinear pairs and thus the moment of each pair of forces about P cancels. The integrals in the first and second terms on the right are used to locate the body's center of mass G with respect to P, since $\bar{y}m = \int y \, dm$ and $\bar{x}m = \int x \, dm$, Fig. 17–8d. Also, the last integral represents the body's moment of inertia about the z axis, i.e., $I_P = \int r^2 dm$. Thus,

$$\zeta \, \Sigma M_P = -\bar{y}m(a_P)_x + \bar{x}m(a_P)_y + I_P \alpha \qquad (17\text{–}6)$$

It is possible to reduce this equation to a simpler form if point P coincides with the mass center G for the body. If this is the case, then $\bar{x} = \bar{y} = 0$, and therefore*

$$\Sigma M_G = I_G \alpha \qquad (17\text{–}7)$$

This rotational equation of motion states that the sum of the moments of all the external forces about the body's mass center G is equal to the product of the moment of inertia of the body about an axis passing through G and the body's angular acceleration.

Equation 17–6 can also be rewritten in terms of the x and y components of $\mathbf{a}_G$ and the body's moment of inertia I_G. If point G is located at $(\bar{x}, \bar{y})$, Fig. 17–8d, then by the parallel-axis theorem, $I_P = I_G + m(\bar{x}^2 + \bar{y}^2)$. Substituting into Eq. 17–6 and rearranging terms, we get

$$\zeta \Sigma M_P = \bar{y}m[-(a_P)_x + \bar{y}\alpha] + \bar{x}m[(a_P)_y + \bar{x}\alpha] + I_G\alpha \qquad (17\text{–}8)$$

From the kinematic diagram of Fig. 17–8d, $\mathbf{a}_P$ can be expressed in terms of $\mathbf{a}_G$ as

$$\mathbf{a}_G = \mathbf{a}_P + \boldsymbol{\alpha} \times \bar{\mathbf{r}} - \omega^2 \bar{\mathbf{r}}$$
$$(a_G)_x\mathbf{i} + (a_G)_y\mathbf{j} = (a_P)_x\mathbf{i} + (a_P)_y\mathbf{j} + \alpha\mathbf{k} \times (\bar{x}\mathbf{i} + \bar{y}\mathbf{j}) - \omega^2(\bar{x}\mathbf{i} + \bar{y}\mathbf{j})$$

Carrying out the cross product and equating the respective $\mathbf{i}$ and $\mathbf{j}$ components yields the two scalar equations

$$(a_G)_x = (a_P)_x - \bar{y}\alpha - \bar{x}\omega^2$$
$$(a_G)_y = (a_P)_y + \bar{x}\alpha - \bar{y}\omega^2$$

From these equations, $[-(a_P)_x + \bar{y}\alpha] = [-(a_G)_x - \bar{x}\omega^2]$ and $[(a_P)_y + \bar{x}\alpha] = [(a_G)_y + \bar{y}\omega^2]$. Substituting these results into Eq. 17–8 and simplifying gives

$$\zeta \Sigma M_P = -\bar{y}m(a_G)_x + \bar{x}m(a_G)_y + I_G\alpha \qquad (17\text{–}9)$$

This important result indicates that when moments of the external forces shown on the free-body diagram are summed about point P, Fig. 17–8e, they are equivalent to the sum of the "kinetic moments" of the components of $m\mathbf{a}_G$ about P plus the "kinetic moment" of $I_G\boldsymbol{\alpha}$, Fig. 17–8f. In other words, when the "kinetic moments," $\Sigma(\mathcal{M}_k)_P$, are computed, Fig. 17–8f, the vectors $m(\mathbf{a}_G)_x$ and $m(\mathbf{a}_G)_y$ are treated as sliding vectors; that is, they can act at *any point* along their line of action. In a similar manner, $I_G\boldsymbol{\alpha}$ can be treated as a free vector and can therefore act at *any point*. It is important to keep in mind, however, that $m\mathbf{a}_G$ and $I_G\boldsymbol{\alpha}$ are not the same as a force or a couple moment. Instead, they are caused by the external effects of forces and couple moments acting on the body. With this in mind we can therefore write Eq. 17–9 in a more general form as

$$\Sigma M_P = \Sigma(\mathcal{M}_k)_P \qquad (17\text{–}10)$$

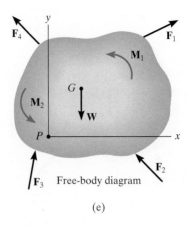

Free-body diagram

(e)

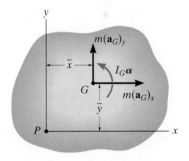

Kinetic diagram

(f)

Fig. 17–8 (cont.)

17

---

*It also reduces to this same simple form $\Sigma M_P = I_P\alpha$ if point P is a *fixed point* (see Eq. 17–16) or the acceleration of point P is directed along the line PG.

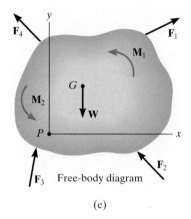

F₄ F₁ M₁ M₂ G W P x F₃ Free-body diagram F₂

(e)

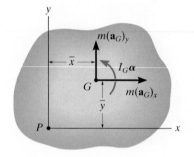

$m(\mathbf{a}_G)_y$ $\overline{x}$ $I_G\alpha$ G $m(\mathbf{a}_G)_x$ $\overline{y}$ P x

Kinetic diagram

(f)

Fig. 17–8 (cont.)

General Application of the Equations of Motion. To summarize this analysis, *three* independent scalar equations can be written to describe the general plane motion of a symmetrical rigid body.

$$\Sigma F_x = m(a_G)_x$$
$$\Sigma F_y = m(a_G)_y$$
$$\Sigma M_G = I_G\alpha$$

or

$$\Sigma M_P = \Sigma(\mathcal{M}_k)_P \qquad (17\text{–}11)$$

When applying these equations, one should *always* draw a free-body diagram, Fig. 17–8e, in order to account for the terms involved in ΣF_x, ΣF_y, ΣM_G, or ΣM_P. In some problems it may also be helpful to draw the *kinetic diagram* for the body, Fig. 17–8f. This diagram graphically accounts for the terms $m(\mathbf{a}_G)_x$, $m(\mathbf{a}_G)_y$, and $I_G\alpha$. It is especially convenient when used to determine the components of $m\mathbf{a}_G$ and the moment of these components in $\Sigma(\mathcal{M}_k)_P$.*

17.3   Equations of Motion: Translation

When the rigid body in Fig. 17–9a undergoes a *translation*, all the particles of the body have the *same acceleration*. Furthermore, $\alpha = \mathbf{0}$, in which case the rotational equation of motion applied at point G reduces to a simplified form, namely, $\Sigma M_G = 0$. Application of this and the force equations of motion will now be discussed for each of the two types of translation.

Rectilinear Translation. When a body is subjected to *rectilinear translation*, all the particles of the body (slab) travel along parallel straight-line paths. The free-body and kinetic diagrams are shown in Fig. 17–9b. Since $I_G\alpha = \mathbf{0}$, only $m\mathbf{a}_G$ is shown on the kinetic diagram. Hence, the equations of motion which apply in this case become

$$\boxed{\begin{aligned} \Sigma F_x &= m(a_G)_x \\ \Sigma F_y &= m(a_G)_y \\ \Sigma M_G &= 0 \end{aligned}} \qquad (17\text{–}12)$$

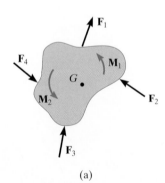

F₁ F₄ M₁ G M₂ F₂ F₃

(a)

Fig. 17–9

*For this reason, the kinetic diagram will be used in the solution of an example problem whenever $\Sigma M_P = \Sigma(\mathcal{M}_k)_P$ is applied.

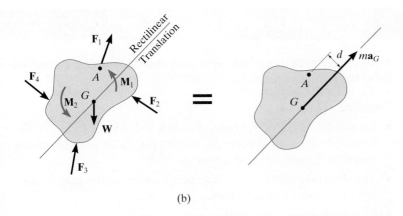

(b)

It is also possible to sum moments about other points on or off the body, in which case the moment of $m\mathbf{a}_G$ must be taken into account. For example, if point A is chosen, which lies at a perpendicular distance d from the line of action of $m\mathbf{a}_G$, the following moment equation applies:

$$\zeta + \Sigma M_A = \Sigma(\mathcal{M}_k)_A; \qquad \Sigma M_A = (ma_G)d$$

Here the sum of moments of the external forces and couple moments about A (ΣM_A, free-body diagram) equals the moment of $m\mathbf{a}_G$ about A ($\Sigma(\mathcal{M}_k)_A$, kinetic diagram).

Curvilinear Translation. When a rigid body is subjected to *curvilinear translation*, all the particles of the body have the same accelerations as they travel along *curved paths* as noted in Sec. 16–1. For analysis, it is often convenient to use an inertial coordinate system having an origin which coincides with the body's mass center at the instant considered, and axes which are oriented in the normal and tangential directions to the path of motion, Fig. 17–9c. The three scalar equations of motion are then

$$\boxed{\begin{aligned} \Sigma F_n &= m(a_G)_n \\ \Sigma F_t &= m(a_G)_t \\ \Sigma M_G &= 0 \end{aligned}} \qquad (17\text{–}13)$$

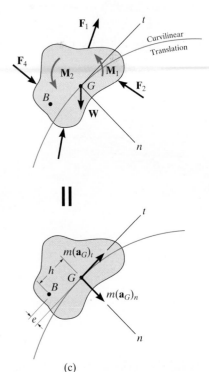

If moments are summed about the arbitrary point B, Fig. 17–9c, then it is necessary to account for the moments, $\Sigma(\mathcal{M}_k)_B$, of the two components $m(\mathbf{a}_G)_n$ and $m(\mathbf{a}_G)_t$ about this point. From the kinetic diagram, h and e represent the perpendicular distances (or "moment arms") from B to the lines of action of the components. The required moment equation therefore becomes

$$\zeta + \Sigma M_B = \Sigma(\mathcal{M}_k)_B; \qquad \Sigma M_B = e[m(a_G)_t] - h[m(a_G)_n]$$

Fig. 17–9

The free-body and kinetic diagrams for this boat and trailer are drawn first in order to apply the equations of motion. Here the forces on the free-body diagram cause the effect shown on the kinetic diagram. If moments are summed about the mass center, G, then $\Sigma M_G = 0$. However, if moments are summed about point B then $\zeta + \Sigma M_B = ma_G(d)$.

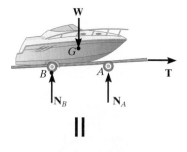

Procedure for Analysis

Kinetic problems involving rigid-body *translation* can be solved using the following procedure.

Free-Body Diagram.

- Establish the x, y or n, t inertial coordinate system and draw the free-body diagram in order to account for all the external forces and couple moments that act on the body.

- The direction and sense of the acceleration of the body's mass center $\mathbf{a}_G$ should be established.

- Identify the unknowns in the problem.

- If it is decided that the rotational equation of motion $\Sigma M_P = \Sigma(\mathcal{M}_k)_P$ is to be used in the solution, then consider drawing the kinetic diagram, since it graphically accounts for the components $m(\mathbf{a}_G)_x$, $m(\mathbf{a}_G)_y$ or $m(\mathbf{a}_G)_t$, $m(\mathbf{a}_G)_n$ and is therefore convenient for "visualizing" the terms needed in the moment sum $\Sigma(\mathcal{M}_k)_P$.

Equations of Motion.

- Apply the three equations of motion in accordance with the established sign convention.

- To simplify the analysis, the moment equation $\Sigma M_G = 0$ can be replaced by the more general equation $\Sigma M_P = \Sigma(\mathcal{M}_k)_P$, where point P is usually located at the intersection of the lines of action of as many unknown forces as possible.

- If the body is in contact with a *rough surface* and slipping occurs, use the friction equation $F = \mu_k N$. Remember, $\mathbf{F}$ always acts on the body so as to oppose the motion of the body relative to the surface it contacts.

Kinematics.

- Use kinematics to determine the velocity and position of the body.

- For rectilinear translation with *variable acceleration*

$$a_G = dv_G/dt \quad a_G ds_G = v_G dv_G$$

- For rectilinear translation with *constant acceleration*

$$v_G = (v_G)_0 + a_G t \quad v_G^2 = (v_G)_0^2 + 2a_G[s_G - (s_G)_0]$$

$$s_G = (s_G)_0 + (v_G)_0 t + \tfrac{1}{2}a_G t^2$$

- For curvilinear translation

$$(a_G)_n = v_G^2/\rho$$

$$(a_G)_t = dv_G/dt \quad (a_G)_t ds_G = v_G dv_G$$

EXAMPLE 17.5

The car shown in Fig. 17–10a has a mass of 2 Mg and a center of mass at G. Determine the acceleration if the rear "driving" wheels are always slipping, whereas the front wheels are free to rotate. Neglect the mass of the wheels. The coefficient of kinetic friction between the wheels and the road is $\mu_k = 0.25$.

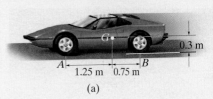

(a)

SOLUTION I

Free-Body Diagram. As shown in Fig. 17–10b, the rear-wheel frictional force $\mathbf{F}_B$ pushes the car forward, and since *slipping occurs*, $F_B = 0.25N_B$. The frictional forces acting on the *front wheels* are *zero*, since these wheels have negligible mass.* There are three unknowns in the problem, N_A, N_B, and a_G. Here we will sum moments about the mass center. The car (point G) accelerates to the left, i.e., in the negative x direction, Fig. 17–10b.

Equations of Motion.

$$\xrightarrow{+} \Sigma F_x = m(a_G)_x; \qquad -0.25N_B = -(2000 \text{ kg})a_G \qquad (1)$$

$$+\uparrow \Sigma F_y = m(a_G)_y; \qquad N_A + N_B - 2000(9.81) \text{ N} = 0 \qquad (2)$$

$$\zeta +\Sigma M_G = 0; \quad -N_A(1.25 \text{ m}) - 0.25N_B(0.3 \text{ m}) + N_B(0.75 \text{ m}) = 0 \quad (3)$$

Solving,

$$a_G = 1.59 \text{ m/s}^2 \leftarrow \qquad\qquad Ans.$$

$$N_A = 6.88 \text{ kN}$$

$$N_B = 12.7 \text{ kN}$$

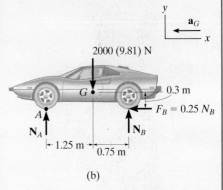

(b)

SOLUTION II

Free-Body and Kinetic Diagrams. If the "moment" equation is applied about point A, then the unknown N_A will be eliminated from the equation. To "visualize" the moment of ma_G about A, we will include the kinetic diagram as part of the analysis, Fig. 17–10c.

Equation of Motion.

$$\zeta +\Sigma M_A = \Sigma(\mathcal{M}_k)_A; \qquad N_B(2 \text{ m}) - [2000(9.81) \text{ N}](1.25 \text{ m}) =$$

$$(2000 \text{ kg})a_G(0.3 \text{ m})$$

Solving this and Eq. 1 for a_G leads to a simpler solution than that obtained from Eqs. 1 to 3.

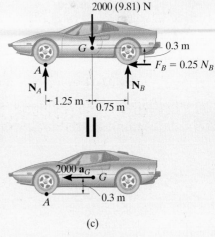

(c)

Fig. 17–10

---

* With negligible wheel mass, $I\alpha = 0$ and the frictional force at A required to turn the wheel is zero. If the wheels' mass were included, then the solution would be more involved, since a general-plane-motion analysis of the wheels would have to be considered (see Sec. 17.5).

EXAMPLE 17.6

The motorcycle shown in Fig. 17–11a has a mass of 125 kg and a center of mass at G_1, while the rider has a mass of 75 kg and a center of mass at G_2. Determine the minimum coefficient of static friction between the wheels and the pavement in order for the rider to do a "wheely," i.e., lift the front wheel off the ground as shown in the photo. What acceleration is necessary to do this? Neglect the mass of the wheels and assume that the front wheel is free to roll.

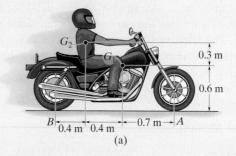

(a)

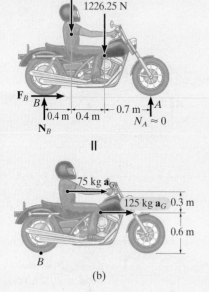

Fig. 17–11

SOLUTION

Free-Body and Kinetic Diagrams. In this problem we will consider both the motorcycle and the rider as a single *system*. It is possible first to determine the location of the center of mass for this "system" by using the equations $\bar{x} = \Sigma \tilde{x} m / \Sigma m$ and $\bar{y} = \Sigma \tilde{y} m / \Sigma m$. Here, however, we will consider the weight and mass of the motorcycle and rider seperate as shown on the free-body and kinetic diagrams, Fig. 17–11b. Both of these parts move with the *same* acceleration. We have assumed that the front wheel is *about* to leave the ground, so that the normal reaction $N_A \approx 0$. The three unknowns in the problem are N_B, F_B, and a_G.

Equations of Motion.

$$\xrightarrow{+} \Sigma F_x = m(a_G)_x; \qquad F_B = (75 \text{ kg} + 125 \text{ kg})a_G \qquad (1)$$

$$+\uparrow \Sigma F_y = m(a_G)_y; \qquad N_B - 735.75 \text{ N} - 1226.25 \text{ N} = 0$$

$$\zeta + \Sigma M_B = \Sigma(\mathcal{M}_k)_B; \quad -(735.75 \text{ N})(0.4 \text{ m}) - (1226.25 \text{ N})(0.8 \text{ m}) =$$
$$-(75 \text{ kg } a_G)(0.9 \text{ m}) - (125 \text{ kg } a_G)(0.6 \text{ m}) \qquad (2)$$

Solving,

$$a_G = 8.95 \text{ m/s}^2 \rightarrow \qquad \qquad \textit{Ans.}$$

$$N_B = 1962 \text{ N}$$

$$F_B = 1790 \text{ N}$$

Thus the minimum coefficient of static friction is

$$(\mu_s)_{\min} = \frac{F_B}{N_B} = \frac{1790 \text{ N}}{1962 \text{ N}} = 0.912 \qquad \textit{Ans.}$$

EXAMPLE 17.7

A uniform 50-kg crate rests on a horizontal surface for which the coefficient of kinetic friction is $\mu_k = 0.2$. Determine the acceleration if a force of $P = 600$ N is applied to the crate as shown in Fig. 17–12a.

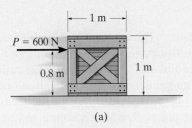

(a)

SOLUTION

Free-Body Diagram.   The force **P** can cause the crate either to slide or to tip over. As shown in Fig. 17–12b, it is assumed that the crate slides, so that $F = \mu_k N_C = 0.2 N_C$. Also, the resultant normal force $\mathbf{N}_C$ acts at O, a distance x (where $0 < x \le 0.5$ m) from the crate's center line.* The three unknowns are N_C, x, and a_G.

Equations of Motion.

$$\xrightarrow{+} \Sigma F_x = m(a_G)_x; \qquad 600\ \text{N} - 0.2N_C = (50\ \text{kg})a_G \qquad (1)$$

$$+\uparrow \Sigma F_y = m(a_G)_y; \qquad N_C - 490.5\ \text{N} = 0 \qquad (2)$$

$$\zeta + \Sigma M_G = 0; \quad -600\ \text{N}(0.3\ \text{m}) + N_C(x) - 0.2N_C(0.5\ \text{m}) = 0 \qquad (3)$$

Solving,

$$N_C = 490.5\ \text{N}$$
$$x = 0.467\ \text{m}$$
$$a_G = 10.0\ \text{m/s}^2 \rightarrow \qquad Ans.$$

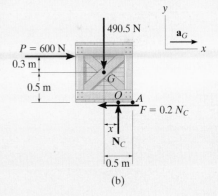

(b)

Fig. 17–12

Since $x = 0.467$ m < 0.5 m, indeed the crate slides as originally assumed.

NOTE: If the solution had given a value of $x > 0.5$ m, the problem would have to be reworked since tipping occurs. If this were the case, $\mathbf{N}_C$ would act at the *corner point A* and $F \le 0.2N_C$.

*The line of action of $\mathbf{N}_C$ does not necessarily pass through the mass center G ($x = 0$), since $\mathbf{N}_C$ must counteract the tendency for tipping caused by **P**. See Sec. 8.1 of *Engineering Mechanics: Statics*.

EXAMPLE | 17.8

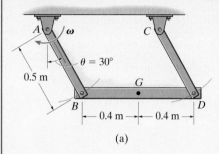

(a)

The 100-kg beam *BD* shown in Fig. 17–13*a* is supported by two rods having negligible mass. Determine the force developed in each rod if at the instant $\theta = 30°, \omega = 6$ rad/s.

SOLUTION

Free-Body Diagram.   The beam moves with *curvilinear translation* since all points on the beam move along circular paths, each path having the same radius of 0.5 m, but different centers of curvature. Using normal and tangential coordinates, the free-body diagram for the beam is shown in Fig. 17–13*b*. Because of the *translation*, *G* has the *same* motion as the pin at *B*, which is connected to both the rod and the beam. Note that the tangential component of acceleration acts downward to the left due to the clockwise direction of $\boldsymbol{\alpha}$, Fig. 17–13*c*. Furthermore, the normal component of acceleration is *always* directed toward the center of curvature (toward point *A* for rod *AB*). Since the angular velocity of *AB* is 6 rad/s when $\theta = 30°$, then

$$(a_G)_n = \omega^2 r = (6 \text{ rad/s})^2(0.5 \text{ m}) = 18 \text{ m/s}^2$$

The three unknowns are T_B, T_D, and $(a_G)_t$. The directions of $(\mathbf{a}_G)_n$ and $(\mathbf{a}_G)_t$ have been established, and are indicated on the coordinate axes.

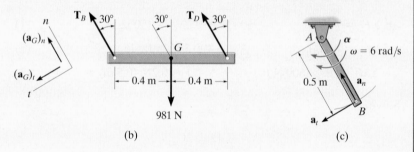

(b)   (c)

Fig. 17–13

Equations of Motion.

$$+\nwarrow\Sigma F_n = m(a_G)_n; \quad T_B + T_D - 981 \cos 30° \text{ N} = 100 \text{ kg}(18 \text{ m/s}^2) \quad (1)$$

$$+\swarrow\Sigma F_t = m(a_G)_t; \qquad 981 \sin 30° = 100 \text{ kg}(a_G)_t \qquad\qquad (2)$$

$$\zeta + \Sigma M_G = 0; \quad -(T_B \cos 30°)(0.4 \text{ m}) + (T_D \cos 30°)(0.4 \text{ m}) = 0 \quad (3)$$

Simultaneous solution of these three equations gives

$$T_B = T_D = 1.32 \text{ kN} \qquad\qquad\qquad Ans.$$

$$(a_G)_t = 4.905 \text{ m/s}^2$$

NOTE: It is also possible to apply the equations of motion along horizontal and vertical x, y axes, but the solution becomes more involved.

FUNDAMENTAL PROBLEMS

F17–1. The cart and its load have a total mass of 100 kg. Determine the acceleration of the cart and the normal reactions on the pair of wheels at A and B. Neglect the mass of the wheels.

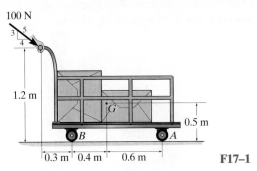

F17–1

F17–2. If the 80-kg cabinet is allowed to roll down the inclined plane, determine the acceleration of the cabinet and the normal reactions on the pair of rollers at A and B that have negligible mass.

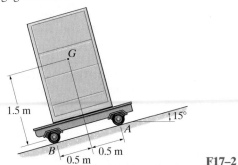

F17–2

F17–3. The 20-lb link AB is pinned to a moving frame at A and held in a vertical position by means of a string BC which can support a maximum tension of 10 lb. Determine the maximum acceleration of the frame without breaking the string. What are the corresponding components of reaction at the pin A?

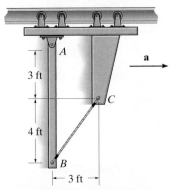

F17–3

F17–4. Determine the maximum acceleration of the truck without causing the assembly to move relative to the truck. Also what is the corresponding normal reaction on legs A and B? The 100-kg table has a mass center at G and the coefficient of static friction between the legs of the table and the bed of the truck is $\mu_s = 0.2$.

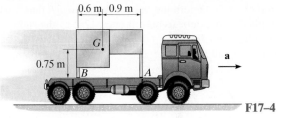

F17–4

F17–5. At the instant shown both rods of negligible mass swing with a counterclockwise angular velocity of $\omega = 5$ rad/s, while the 50-kg bar is subjected to the 100-N horizontal force. Determine the tension developed in the rods and the angular acceleration of the rods at this instant.

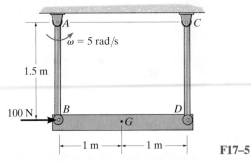

F17–5

F17–6. At the instant shown, link CD rotates with an angular velocity of $\omega = 6$ rad/s. If it is subjected to a couple moment $M = 450$ N·m, determine the force developed in link AB, the horizontal and vertical component of reaction on pin D, and the angular acceleration of link CD at this instant. The block has a mass of 50 kg and center of mass at G. Neglect the mass of links AB and CD.

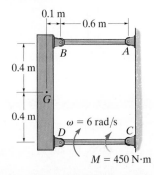

F17–6

PROBLEMS

*17–24. The door has a weight of 200 lb and a center of gravity at G. Determine how far the door moves in 2 s, starting from rest, if a man pushes on it at C with a horizontal force F = 30 lb. Also, find the vertical reactions at the rollers A and B.

17–25. The door has a weight of 200 lb and a center of gravity at G. Determine the constant force F that must be applied to the door to push it open 12 ft to the right in 5 s, starting from rest. Also, find the vertical reactions at the rollers A and B.

17–27. The drum truck supports the 600-lb drum that has a center of gravity at G. If the operator pushes it forward with a horizontal force of 20 lb, determine the acceleration of the truck and the normal reactions at each of the four wheels. Neglect the mass of the wheels.

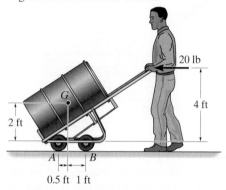

Prob. 17–27

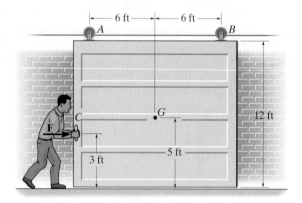

Probs. 17–24/25

17–26. The uniform pipe has a weight of 500 lb/ft and diameter of 2 ft. If it is hoisted as shown with an acceleration of 0.5 ft/s², determine the internal moment at the center A of the pipe due to the lift.

*17–28. If the cart is given a constant acceleration of a = 6 ft/s² up the inclined plane, determine the force developed in rod AC and the horizontal and vertical components of force at pin B. The crate has a weight of 150 lb with center of gravity at G, and it is secured on the platform, so that it does not slide. Neglect the platform's weight.

17–29. If the strut AC can withstand a maximum compression force of 150 lb before it fails, determine the cart's maximum permissible acceleration. The crate has a weight of 150 lb with center of gravity at G, and it is secured on the platform, so that it does not slide. Neglect the platform's weight.

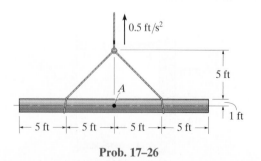

Prob. 17–26

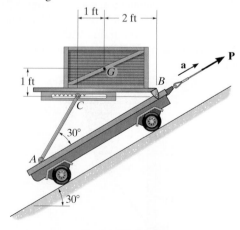

Probs. 17–28/29

17–30. The drop gate at the end of the trailer has a mass of 1.25 Mg and mass center at G. If it is supported by the cable AB and hinge at C, determine the tension in the cable when the truck begins to accelerate at 5 m/s². Also, what are the horizontal and vertical components of reaction at the hinge C?

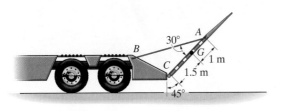

Prob. 17–30

17–31. The pipe has a length of 3 m and a mass of 500 kg. It is attached to the back of the truck using a 0.6-m-long chain AB. If the coefficient of kinetic friction at C is $\mu_k = 0.4$, determine the acceleration of the truck if the angle $\theta = 10°$ with the road as shown.

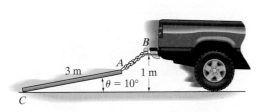

Prob. 17–31

*17–32. The mountain bike has a mass of 40 kg with center of mass at point G_1, while the rider has a mass of 60 kg with center of mass at point G_2. Determine the maximum deceleration when the brake is applied to the front wheel, without causing the rear wheel B to leave the road. Assume that the front wheel does not slip. Neglect the mass of all the wheels.

17–33. The mountain bike has a mass of 40 kg with center of mass at point G_1, while the rider has a mass of 60 kg with center of mass at point G_2. When the brake is applied to the front wheel, it causes the bike to decelerate at a constant rate of 3 m/s2. Determine the normal reaction the road exerts on the front and rear wheels. Assume that the rear wheel is free to roll. Neglect the mass of all the wheels.

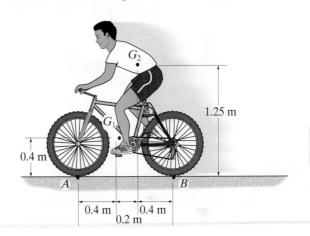

Probs. 17–32/33

17–34. The trailer with its load has a mass of 150 kg and a center of mass at G. If it is subjected to a horizontal force of $P = 600$ N, determine the trailer's acceleration and the normal force on the pair of wheels at A and at B. The wheels are free to roll and have negligible mass.

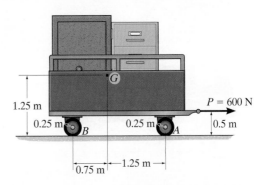

Prob. 17–34

17–35. At the start of a race, the rear drive wheels B of the 1550-lb car slip on the track. Determine the car's acceleration and the normal reaction the track exerts on the front pair of wheels A and rear pair of wheels B. The coefficient of kinetic friction is $\mu_k = 0.7$, and the mass center of the car is at G. The front wheels are free to roll. Neglect the mass of all the wheels.

*17–36. Determine the maximum acceleration that can be achieved by the car without having the front wheels A leave the track or the rear drive wheels B slip on the track. The coefficient of static friction is $\mu_s = 0.9$. The car's mass center is at G, and the front wheels are free to roll. Neglect the mass of all the wheels.

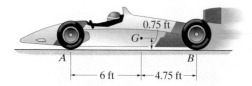

Probs. 17–35/36

17–37. If the 4500-lb van has front-wheel drive, and the coefficient of static friction between the front wheels A and the road is $\mu_s = 0.8$, determine the normal reactions on the pairs of front and rear wheels when the van has maximum acceleration. Also, find this maximum acceleration. The rear wheels are free to roll. Neglect the mass of the wheels.

17–38. If the 4500-lb van has rear-wheel drive, and the coefficient of static friction between the front wheels B and the road is $\mu_s = 0.8$, determine the normal reactions on the pairs of front and rear wheels when the van has maximum acceleration. The front wheels are free to roll. Neglect the mass of the wheels.

Probs. 17–37/38

17–39. The uniform bar of mass m is pin connected to the collar, which slides along the smooth horizontal rod. If the collar is given a constant acceleration of $\mathbf{a}$, determine the bar's inclination angle θ. Neglect the collar's mass.

Prob. 17–39

*17–40. The lift truck has a mass of 70 kg and mass center at G. If it lifts the 120-kg spool with an acceleration of 3 m/s^2, determine the reactions of each of the four wheels on the ground. The loading is symmetric. Neglect the mass of the movable arm CD.

17–41. The lift truck has a mass of 70 kg and mass center at G. Determine the largest upward acceleration of the 120-kg spool so that no reaction of the wheels on the ground exceeds 600 N.

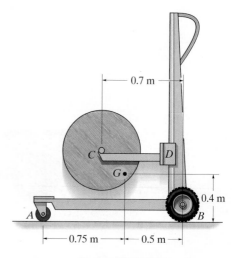

Probs. 17–40/41

17–42. The uniform crate has a mass of 50 kg and rests on the cart having an inclined surface. Determine the smallest acceleration that will cause the crate either to tip or slip relative to the cart. What is the magnitude of this acceleration? The coefficient of static friction between the crate and the cart is $\mu_s = 0.5$.

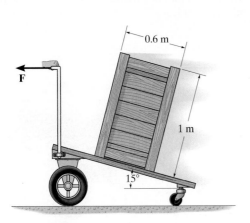

Prob. 17–42

17–43. Determine the acceleration of the 150-lb cabinet and the normal reaction under the legs A and B if $P = 35$ lb. The coefficients of static and kinetic friction between the cabinet and the plane are $\mu_s = 0.2$ and $\mu_k = 0.15$, respectively. The cabinet's center of gravity is located at G.

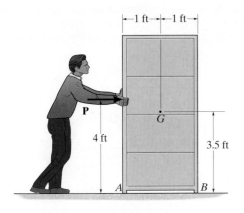

Prob. 17–43

*17–44. The assembly has a mass of 8 Mg and is hoisted using the boom and pulley system. If the winch at B draws in the cable with an acceleration of 2 m/s², determine the compressive force in the hydraulic cylinder needed to support the boom. The boom has a mass of 2 Mg and mass center at G.

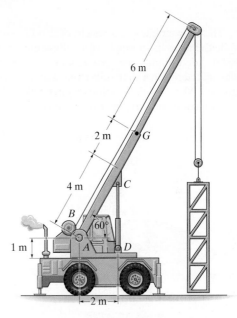

Prob. 17–44

17–45. The 2-Mg truck achieves a speed of 15 m/s with a constant acceleration after it has traveled a distance of 100 m, starting from rest. Determine the normal force exerted on each pair of front wheels B and rear driving wheels A. Also, find the traction force on the pair of wheels at A. The front wheels are free to roll. Neglect the mass of the wheels.

17–46. Determine the shortest time possible for the rear-wheel drive, 2-Mg truck to achieve a speed of 16 m/s with a constant acceleration starting from rest. The coefficient of static friction between the wheels and the road surface is $\mu_s = 0.8$. The front wheels are free to roll. Neglect the mass of the wheels.

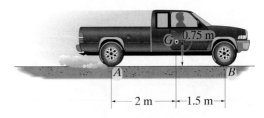

Probs. 17–45/46

17

17–47. The snowmobile has a weight of 250 lb, centered at G_1, while the rider has a weight of 150 lb, centered at G_2. If the acceleration is $a = 20$ ft/s^2, determine the maximum height h of G_2 of the rider so that the snowmobile's front skid does not lift off the ground. Also, what are the traction (horizontal) force and normal reaction under the rear tracks at A?

*17–48. The snowmobile has a weight of 250 lb, centered at G_1, while the rider has a weight of 150 lb, centered at G_2. If $h = 3$ ft, determine the snowmobile's maximum permissible acceleration **a** so that its front skid does not lift off the ground. Also, find the traction (horizontal) force and the normal reaction under the rear tracks at A.

17–51. The pipe has a mass of 800 kg and is being towed behind the truck. If the acceleration of the truck is $a_t = 0.5$ m/s^2, determine the angle θ and the tension in the cable. The coefficient of kinetic friction between the pipe and the ground is $\mu_k = 0.1$.

*17–52. The pipe has a mass of 800 kg and is being towed behind a truck. If the angle $\theta = 30°$, determine the acceleration of the truck and the tension in the cable. The coefficient of kinetic friction between the pipe and the ground is $\mu_k = 0.1$.

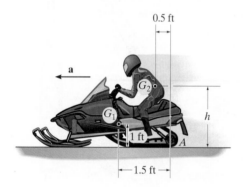

Probs. 17–47/48

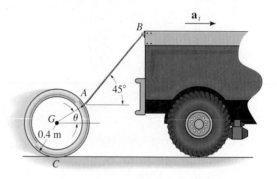

Probs. 17–51/52

17–49. If the cart's mass is 30 kg and it is subjected to a horizontal force of $P = 90$ N, determine the tension in cord AB and the horizontal and vertical components of reaction on end C of the uniform 15-kg rod BC.

17–50. If the cart's mass is 30 kg, determine the horizontal force P that should be applied to the cart so that the cord AB just becomes slack. The uniform rod BC has a mass of 15 kg.

17–53. The arched pipe has a mass of 80 kg and rests on the surface of the platform. As it is hoisted from one level to the next, $\alpha = 0.25$ rad/s^2 and $\omega = 0.5$ rad/s at the instant $\theta = 30°$. If it does not slip, determine the normal reactions of the arch on the platform at this instant.

17–54. The arched pipe has a mass of 80 kg and rests on the surface of the platform for which the coefficient of static friction is $\mu_s = 0.3$. Determine the greatest angular acceleration α of the platform, starting from rest when $\theta = 45°$, without causing the pipe to slip on the platform.

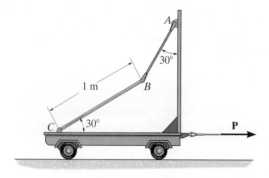

Probs. 17–49/50

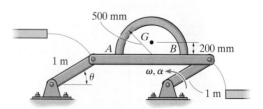

Probs. 17–53/54

17–55. At the instant shown, link *CD* rotates with an angular velocity of $\omega_{CD} = 8$ rad/s. If link *CD* is subjected to a couple moment of $M = 650$ lb · ft, determine the force developed in link *AB* and the angular acceleration of the links at this instant. Neglect the weight of the links and the platform. The crate weighs 150 lb and is fully secured on the platform.

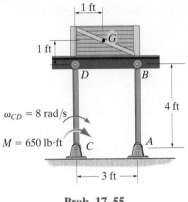

Prob. 17–55

*17–56. Determine the force developed in the links and the acceleration of the bar's mass center immediately after the cord fails. Neglect the mass of links *AB* and *CD*. The uniform bar has a mass of 20 kg.

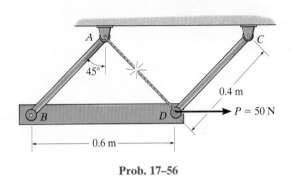

Prob. 17–56

17.4 Equations of Motion: Rotation about a Fixed Axis

Consider the rigid body (or slab) shown in Fig. 17–14*a*, which is constrained to rotate in the vertical plane about a fixed axis perpendicular to the page and passing through the pin at *O*. The angular velocity and angular acceleration are caused by the external force and couple moment system acting on the body. Because the body's center of mass *G* moves around a *circular path*, the acceleration of this point is best represented by its tangential and normal components. The *tangential component of acceleration* has a *magnitude* of $(a_G)_t = \alpha r_G$ and must act in a *direction* which is *consistent* with the body's angular acceleration α. The *magnitude* of the *normal component of acceleration* is $(a_G)_n = \omega^2 r_G$. This component is *always directed* from point *G* to *O*, regardless of the rotational sense of ω.

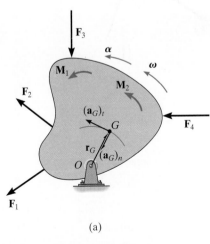

(a)

Fig. 17–14

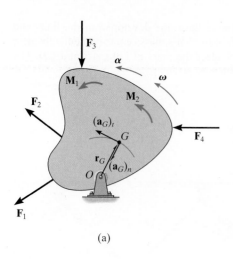

(a)

The free-body and kinetic diagrams for the body are shown in Fig. 17–14b. The two components $m(\mathbf{a}_G)_t$ and $m(\mathbf{a}_G)_n$, shown on the kinetic diagram, are associated with the tangential and normal components of acceleration of the body's mass center. The $I_G\boldsymbol{\alpha}$ vector acts in the same *direction* as $\boldsymbol{\alpha}$ and has a *magnitude* of $I_G\alpha$, where I_G is the body's moment of inertia calculated about an axis which is perpendicular to the page and passes through G. From the derivation given in Sec. 17.2, the equations of motion which apply to the body can be written in the form

$$\begin{aligned} \Sigma F_n &= m(a_G)_n = m\omega^2 r_G \\ \Sigma F_t &= m(a_G)_t = m\alpha r_G \\ \Sigma M_G &= I_G\alpha \end{aligned} \qquad (17\text{–}14)$$

The moment equation can be replaced by a moment summation about any arbitrary point P on or off the body provided one accounts for the moments $\Sigma(\mathcal{M}_k)_P$ produced by $I_G\boldsymbol{\alpha}$, $m(\mathbf{a}_G)_t$, and $m(\mathbf{a}_G)_n$ about the point.

Moment Equation About Point O.

Often it is convenient to sum moments about the pin at O in order to eliminate the *unknown* force $\mathbf{F}_O$. From the kinetic diagram, Fig. 17–14b, this requires

$$\zeta + \Sigma M_O = \Sigma(\mathcal{M}_k)_O; \qquad \Sigma M_O = r_G m(a_G)_t + I_G\alpha \qquad (17\text{–}15)$$

Note that the moment of $m(\mathbf{a}_G)_n$ is not included here since the line of action of this vector passes through O. Substituting $(a_G)_t = r_G\alpha$, we may rewrite the above equation as $\zeta + \Sigma M_O = (I_G + mr_G^2)\alpha$. From the parallel-axis theorem, $I_O = I_G + md^2$, and therefore the term in parentheses represents the *moment of inertia of the body about the fixed axis of rotation passing through O.** Consequently, we can write the three equations of motion for the body as

$$\begin{aligned} \Sigma F_n &= m(a_G)_n = m\omega^2 r_G \\ \Sigma F_t &= m(a_G)_t = m\alpha r_G \\ \Sigma M_O &= I_O\alpha \end{aligned} \qquad (17\text{–}16)$$

When using these equations, remember that "$I_O\alpha$" accounts for the "moment" of both $m(\mathbf{a}_G)_t$ and $I_G\boldsymbol{\alpha}$ about point O, Fig. 17–14b. In other words, $\Sigma M_O = \Sigma(\mathcal{M}_k)_O = I_O\alpha$, as indicated by Eqs. 17–15 and 17–16.

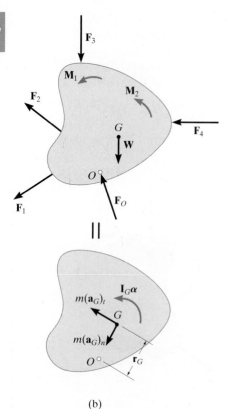

(b)

Fig. 17–14 (cont.)

*The result $\Sigma M_O = I_O\alpha$ can also be obtained *directly* from Eq. 17–6 by selecting point P to coincide with O, realizing that $(a_P)_x = (a_P)_y = 0$.

Procedure for Analysis

Kinetic problems which involve the rotation of a body about a fixed axis can be solved using the following procedure.

Free-Body Diagram.

- Establish the inertial n, t coordinate system and specify the direction and sense of the accelerations $(\mathbf{a}_G)_n$ and $(\mathbf{a}_G)_t$ and the angular acceleration $\boldsymbol{\alpha}$ of the body. Recall that $(\mathbf{a}_G)_t$ must act in a direction which is in accordance with the rotational sense of $\boldsymbol{\alpha}$, whereas $(\mathbf{a}_G)_n$ always acts toward the axis of rotation, point O.

- Draw the free-body diagram to account for all the external forces and couple moments that act on the body.

- Determine the moment of inertia I_G or I_O.

- Identify the unknowns in the problem.

- If it is decided that the rotational equation of motion $\Sigma M_P = \Sigma(\mathcal{M}_k)_P$ is to be used, i.e., P is a point other than G or O, then consider drawing the kinetic diagram in order to help "visualize" the "moments" developed by the components $m(\mathbf{a}_G)_n$, $m(\mathbf{a}_G)_t$, and $I_G\boldsymbol{\alpha}$ when writing the terms for the moment sum $\Sigma(\mathcal{M}_k)_P$.

Equations of Motion.

- Apply the three equations of motion in accordance with the established sign convention.

- If moments are summed about the body's mass center, G, then $\Sigma M_G = I_G\alpha$, since $(m\mathbf{a}_G)_t$ and $(m\mathbf{a}_G)_n$ create no moment about G.

- If moments are summed about the pin support O on the axis of rotation, then $(m\mathbf{a}_G)_n$ creates no moment about O, and it can be shown that $\Sigma M_O = I_O\alpha$.

Kinematics.

- Use kinematics if a complete solution cannot be obtained strictly from the equations of motion.

- If the angular acceleration is variable, use

$$\alpha = \frac{d\omega}{dt} \qquad \alpha \, d\theta = \omega \, d\omega \qquad \omega = \frac{d\theta}{dt}$$

- If the angular acceleration is constant, use

$$\omega = \omega_0 + \alpha_c t$$

$$\theta = \theta_0 + \omega_0 t + \tfrac{1}{2}\alpha_c t^2$$

$$\omega^2 = \omega_0^2 + 2\alpha_c(\theta - \theta_0)$$

17

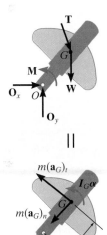

The crank on the oil-pumping rig undergoes rotation about a fixed axis which is caused by a driving torque $\mathbf{M}$ of the motor. The loadings shown on the free-body diagram cause the effects shown on the kinetic diagram. If moments are summed about the mass center, G, then $\Sigma M_G = I_G\alpha$. However, if moments are summed about point O, noting that $(a_G)_t = \alpha d$, then $\zeta + \Sigma M_O = I_G\alpha + m(a_G)_t d + m(a_G)_n(0) = (I_G + md^2)\alpha = I_O\alpha$.

EXAMPLE 17.9

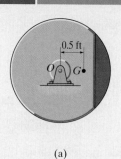

(a)

The unbalanced 50-lb flywheel shown in Fig. 17–15a has a radius of gyration of $k_G = 0.6$ ft about an axis passing through its mass center G. If it is released from rest, determine the horizontal and vertical components of reaction at the pin O.

SOLUTION

Free-Body and Kinetic Diagrams. Since G moves in a circular path, it will have both normal and tangential components of acceleration. Also, since α, which is caused by the flywheel's weight, acts clockwise, the tangential component of acceleration must act downward. Why? Since $\omega = 0$, only $m(a_G)_t = m\alpha r_G$ and $I_G\alpha$ are shown on the kinematic diagram in Fig. 17–15b. Here, the moment of inertia about G is

$$I_G = mk_G^2 = (50 \text{ lb}/32.2 \text{ ft/s}^2)(0.6 \text{ ft})^2 = 0.559 \text{ slug} \cdot \text{ft}^2$$

The three unknowns are O_n, O_t, and α.

Equations of Motion.

$$\xrightarrow{+} \Sigma F_n = m\omega^2 r_G; \qquad\qquad O_n = 0 \qquad\qquad Ans.$$

$$+\downarrow \Sigma F_t = m\alpha r_G; \qquad -O_t + 50 \text{ lb} = \left(\frac{50 \text{ lb}}{32.2 \text{ ft/s}^2}\right)(\alpha)(0.5 \text{ ft}) \qquad (1)$$

$$\zeta + \Sigma M_G = I_G\alpha; \qquad O_t(0.5 \text{ ft}) = (0.5590 \text{ slug} \cdot \text{ft}^2)\alpha$$

Solving,

$$\alpha = 26.4 \text{ rad/s}^2 \quad O_t = 29.5 \text{ lb} \qquad Ans.$$

Moments can also be summed about point O in order to eliminate $\mathbf{O}_n$ and $\mathbf{O}_t$ and thereby obtain a *direct solution* for $\boldsymbol{\alpha}$, Fig. 17–15b. This can be done in one of *two* ways.

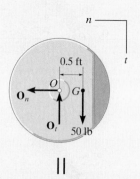

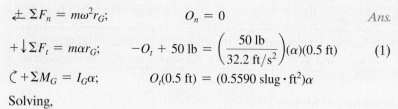

$$\zeta + \Sigma M_O = \Sigma(\mathcal{M}_k)_O;$$

$$(50 \text{ lb})(0.5 \text{ ft}) = (0.5590 \text{ slug} \cdot \text{ft}^2)\alpha + \left[\left(\frac{50 \text{ lb}}{32.2 \text{ ft/s}^2}\right)\alpha(0.5 \text{ ft})\right](0.5 \text{ ft})$$

$$50 \text{ lb}(0.5 \text{ ft}) = 0.9472\alpha \qquad (2)$$

If $\Sigma M_O = I_O\alpha$ is applied, then by the parallel-axis theorem the moment of inertia of the flywheel about O is

$$I_O = I_G + mr_G^2 = 0.559 + \left(\frac{50}{32.2}\right)(0.5)^2 = 0.9472 \text{ slug} \cdot \text{ft}^2$$

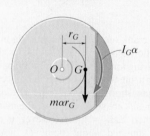

(b)

Fig. 17–15

Hence,

$$\zeta + \Sigma M_O = I_O\alpha; \quad (50 \text{ lb})(0.5 \text{ ft}) = (0.9472 \text{ slug} \cdot \text{ft}^2)\alpha$$

which is the same as Eq. 2. Solving for α and substituting into Eq. 1 yields the answer for O_t obtained previously.

EXAMPLE 17.10

At the instant shown in Fig. 17–16a, the 20-kg slender rod has an angular velocity of $\omega = 5$ rad/s. Determine the angular acceleration and the horizontal and vertical components of reaction of the pin on the rod at this instant.

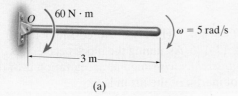

(a)

SOLUTION

Free-Body and Kinetic Diagrams. Fig. 17–16b. As shown on the kinetic diagram, point G moves around a circular path and so it has two components of acceleration. It is important that the tangential component $a_t = \alpha r_G$ act downward since it must be in accordance with the rotational sense of $\boldsymbol{\alpha}$. The three unknowns are O_n, O_t, and α.

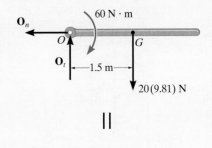

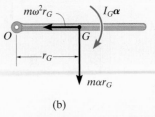

Equation of Motion.

$$\xrightarrow{+}\!\!\!\xleftarrow{} \Sigma F_n = m\omega^2 r_G; \qquad O_n = (20 \text{ kg})(5 \text{ rad/s})^2(1.5 \text{ m})$$

$$+\!\downarrow \Sigma F_t = m\alpha r_G; \qquad -O_t + 20(9.81)\text{N} = (20 \text{ kg})(\alpha)(1.5 \text{ m})$$

$$\zeta + \Sigma M_G = I_G \alpha; \qquad O_t(1.5 \text{ m}) + 60 \text{ N} \cdot \text{m} = \left[\tfrac{1}{12}(20 \text{ kg})(3 \text{ m})^2\right]\alpha$$

Solving

$$O_n = 750 \text{ N} \quad O_t = 19.05 \text{ N} \quad \alpha = 5.90 \text{ rad/s}^2 \qquad \textit{Ans.}$$

A more direct solution to this problem would be to sum moments about point O to eliminate $\mathbf{O}_n$ and $\mathbf{O}_t$ and obtain a *direct solution* for α. Here,

$$\zeta + \Sigma M_O = \Sigma(\mathcal{M}_k)_O; \quad 60 \text{ N} \cdot \text{m} + 20(9.81) \text{ N}(1.5 \text{ m}) =$$
$$\left[\tfrac{1}{12}(20 \text{ kg})(3 \text{ m})^2\right]\alpha + [20 \text{ kg}(\alpha)(1.5 \text{ m})](1.5 \text{ m})$$
$$\alpha = 5.90 \text{ rad/s}^2 \qquad \textit{Ans.}$$

Also, since $I_O = \tfrac{1}{3}ml^2$ for a slender rod, we can apply

$$\zeta + \Sigma M_O = I_O \alpha; \quad 60 \text{ N} \cdot \text{m} + 20(9.81) \text{ N}(1.5 \text{ m}) = \left[\tfrac{1}{3}(20 \text{ kg})(3 \text{ m})^2\right]\alpha$$
$$\alpha = 5.90 \text{ rad/s}^2 \qquad \textit{Ans.}$$

NOTE: By comparison, the last equation provides the simplest solution for α and *does not* require use of the kinetic diagram.

(b)

Fig. 17–16

17

EXAMPLE | 17.11

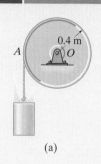

(a)

The drum shown in Fig. 17–17a has a mass of 60 kg and a radius of gyration $k_O = 0.25$ m. A cord of negligible mass is wrapped around the periphery of the drum and attached to a block having a mass of 20 kg. If the block is released, determine the drum's angular acceleration.

SOLUTION I

Free-Body Diagram. Here we will consider the drum and block separately, Fig. 17–17b. Assuming the block accelerates *downward* at **a**, it creates a *counterclockwise* angular acceleration α of the drum. The moment of inertia of the drum is

$$I_O = mk_O^2 = (60\text{ kg})(0.25\text{ m})^2 = 3.75\text{ kg}\cdot\text{m}^2$$

There are five unknowns, namely O_x, O_y, T, a, and α.

Equations of Motion. Applying the translational equations of motion $\Sigma F_x = m(a_G)_x$ and $\Sigma F_y = m(a_G)_y$ to the drum is of no consequence to the solution, since these equations involve the unknowns O_x and O_y. Thus, for the drum and block, respectively,

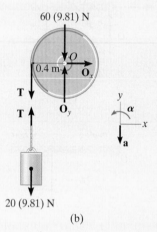

(b)

$$\zeta + \Sigma M_O = I_O\alpha; \qquad T(0.4\text{ m}) = (3.75\text{ kg}\cdot\text{m}^2)\alpha \qquad (1)$$

$$+\uparrow\Sigma F_y = m(a_G)_y; \qquad -20(9.81)\text{N} + T = -(20\text{ kg})a \qquad (2)$$

Kinematics. Since the point of contact A between the cord and drum has a tangential component of acceleration **a**, Fig. 17–17a, then

$$\zeta + a = \alpha r; \qquad\qquad a = \alpha(0.4\text{ m}) \qquad (3)$$

Solving the above equations,

$$T = 106\text{ N} \quad a = 4.52\text{ m/s}^2$$
$$\alpha = 11.3\text{ rad/s}^2 \circlearrowright \qquad\qquad Ans.$$

SOLUTION II

Free-Body and Kinetic Diagrams. The cable tension T can be eliminated from the analysis by considering the drum and block as a *single system*, Fig. 17–17c. The kinetic diagram is shown since moments will be summed about point O.

Equations of Motion. Using Eq. 3 and applying the moment equation about O to eliminate the unknowns O_x and O_y, we have

$$\zeta + \Sigma M_O = \Sigma(\mathcal{M}_k)_O; \qquad [20(9.81)\text{N}]\,(0.4\text{ m}) =$$
$$(3.75\text{ kg}\cdot\text{m}^2)\alpha + [20\text{ kg}(\alpha\,0.4\text{ m})](0.4\text{ m})$$
$$\alpha = 11.3\text{ rad/s}^2 \qquad\qquad Ans.$$

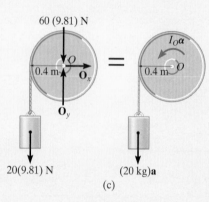

(c)

Fig. 17–17

NOTE: If the block were *removed* and a force of 20(9.81) N were applied to the cord, show that $\alpha = 20.9$ rad/s^2. This value is larger since the block has an inertia, or resistance to acceleration.

EXAMPLE 17.12

The slender rod shown in Fig. 17–18a has a mass m and length l and is released from rest when $\theta = 0°$. Determine the horizontal and vertical components of force which the pin at A exerts on the rod at the instant $\theta = 90°$.

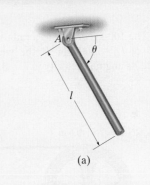

(a)

SOLUTION

Free-Body Diagram. The free-body diagram for the rod in the general position θ is shown in Fig. 17–18b. For convenience, the force components at A are shown acting in the n and t directions. Note that α acts clockwise and so $(\mathbf{a}_G)_t$ acts in the $+t$ direction.
The moment of inertia of the rod about point A is $I_A = \frac{1}{3}ml^2$.

Equations of Motion. Moments will be summed about A in order to eliminate A_n and A_t.

$$+\nwarrow\Sigma F_n = m\omega^2 r_G; \qquad A_n - mg \sin \theta = m\omega^2(l/2) \qquad (1)$$

$$+\swarrow\Sigma F_t = m\alpha r_G; \qquad A_t + mg \cos \theta = m\alpha(l/2) \qquad (2)$$

$$\zeta + \Sigma M_A = I_A\alpha; \qquad mg \cos \theta(l/2) = \left(\tfrac{1}{3}ml^2\right)\alpha \qquad (3)$$

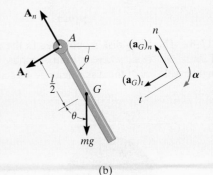

(b)

Fig. 17–18

Kinematics. For a given angle θ there are four unknowns in the above three equations: A_n, A_t, ω, and α. As shown by Eq. 3, α is *not constant*; rather, it depends on the position θ of the rod. The necessary fourth equation is obtained using kinematics, where α and ω can be related to θ by the equation

$$(\zeta +) \qquad\qquad \omega \, d\omega = \alpha \, d\theta \qquad (4)$$

Note that the positive clockwise direction for this equation *agrees* with that of Eq. 3. This is important since we are seeking a simultaneous solution.
In order to solve for ω at $\theta = 90°$, eliminate α from Eqs. 3 and 4, which yields

$$\omega \, d\omega = (1.5g/l) \cos \theta \, d\theta$$

Since $\omega = 0$ at $\theta = 0°$, we have

$$\int_0^\omega \omega \, d\omega = (1.5g/l) \int_{0°}^{90°} \cos \theta \, d\theta$$

$$\omega^2 = 3g/l$$

Substituting this value into Eq. 1 with $\theta = 90°$ and solving Eqs. 1 to 3 yields

$$\alpha = 0$$

$$A_t = 0 \quad A_n = 2.5mg \qquad\qquad Ans.$$

NOTE: If $\Sigma M_A = \Sigma(\mathcal{M}_k)_A$ is used, one must account for the moments of $I_G\alpha$ and $m(\mathbf{a}_G)_t$ about A.

FUNDAMENTAL PROBLEMS

F17–7. The 100-kg wheel has a radius of gyration about its center O of $k_O = 500$ mm. If the wheel starts from rest, determine its angular velocity in $t = 3$ s.

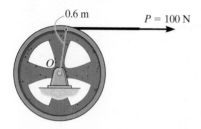

F17–7

F17–8. The 50-kg disk is subjected to the couple moment of $M = (9t)$ N·m, where t is in seconds. Determine the angular velocity of the disk when $t = 4$ s starting from rest.

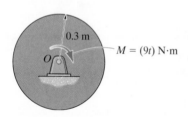

F17–8

F17–9. At the instant shown, the uniform 30-kg slender rod has a counterclockwise angular velocity of $\omega = 6$ rad/s. Determine the tangential and normal components of reaction of pin O on the rod and the angular acceleration of the rod at this instant.

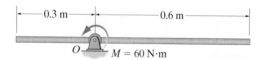

F17–9

F17–10. At the instant shown, the 30-kg disk has a counterclockwise angular velocity of $\omega = 10$ rad/s. Determine the tangential and normal components of reaction of the pin O on the disk and the angular acceleration of the disk at this instant.

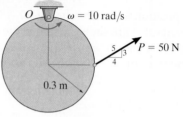

F17–10

F17–11. The uniform slender rod has a mass of 15 kg. Determine the horizontal and vertical components of reaction at the pin O, and the angular acceleration of the rod just after the cord is cut.

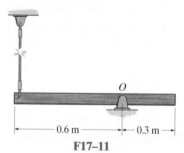

F17–11

F17–12. The uniform 30-kg slender rod is being pulled by the cord that passes over the small smooth peg at A. If the rod has a counterclockwise angular velocity of $\omega = 6$ rad/s at the instant shown, determine the tangential and normal components of reaction at the pin O and the angular acceleration of the rod.

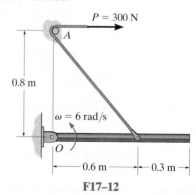

F17–12

PROBLEMS

17–57. The 10-kg wheel has a radius of gyration $k_A = 200$ mm. If the wheel is subjected to a moment $M = (5t)$ N · m, where t is in seconds, determine its angular velocity when $t = 3$ s starting from rest. Also, compute the reactions which the fixed pin A exerts on the wheel during the motion.

Prob. 17–57

17–58. The 80-kg disk is supported by a pin at A. If it is released from rest from the position shown, determine the initial horizontal and vertical components of reaction at the pin.

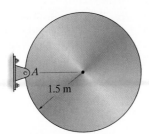

1.5 m

Prob. 17–58

17–59. The uniform slender rod has a mass m. If it is released from rest when $\theta = 0°$, determine the magnitude of the reactive force exerted on it by pin B when $\theta = 90°$.

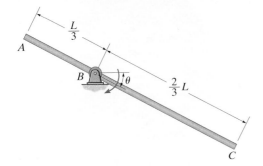

Prob. 17–59

*17–60.** The drum has a weight of 80 lb and a radius of gyration $k_O = 0.4$ ft. If the cable, which is wrapped around the drum, is subjected to a vertical force $P = 15$ lb, determine the time needed to increase the drum's angular velocity from $\omega_1 = 5$ rad/s to $\omega_2 = 25$ rad/s. Neglect the mass of the cable.

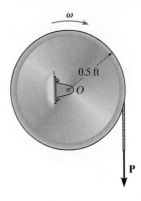

0.5 ft

Prob. 17–60

17–61. Cable is unwound from a spool supported on small rollers at A and B by exerting a force of $T = 300$ N on the cable in the direction shown. Compute the time needed to unravel 5 m of cable from the spool if the spool and cable have a total mass of 600 kg and a centroidal radius of gyration of $k_O = 1.2$ m. For the calculation, neglect the mass of the cable being unwound and the mass of the rollers at A and B. The rollers turn with no friction.

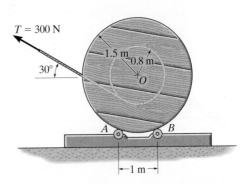

T = 300 N

1.5 m
0.8 m
30°
O

A   B

—1 m—

Prob. 17–61

17

17–62. The 10-lb bar is pinned at its center O and connected to a torsional spring. The spring has a stiffness $k = 5$ lb·ft/rad, so that the torque developed is $M = (5\theta)$ lb·ft, where θ is in radians. If the bar is released from rest when it is vertical at $\theta = 90°$, determine its angular velocity at the instant $\theta = 0°$.

17–63. The 10-lb bar is pinned at its center O and connected to a torsional spring. The spring has a stiffness $k = 5$ lb·ft/rad, so that the torque developed is $M = (5\theta)$ lb·ft, where θ is in radians. If the bar is released from rest when it is vertical at $\theta = 90°$, determine its angular velocity at the instant $\theta = 45°$.

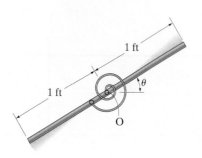

Probs. 17–62/63

***17–64.** If shaft BC is subjected to a torque of $M = (0.45t^{1/2})$ N·m, where t is in seconds, determine the angular velocity of the 3-kg rod AB when $t = 4$ s, starting from rest. Neglect the mass of shaft BC.

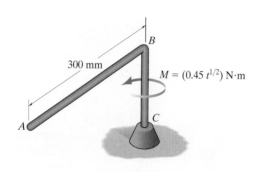

Prob. 17–64

17–65. Determine the vertical and horizontal components of reaction at the pin support A and the angular acceleration of the 12-kg rod at the instant shown, when the rod has an angular velocity of $\omega = 5$ rad/s.

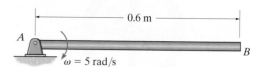

Prob. 17–65

17–66. The kinetic diagram representing the general rotational motion of a rigid body about a fixed axis passing through O is shown in the figure. Show that $I_G\alpha$ may be eliminated by moving the vectors $m(\mathbf{a}_G)_t$ and $m(\mathbf{a}_G)_n$ to point P, located a distance $r_{GP} = k_G^2/r_{OG}$ from the center of mass G of the body. Here k_G represents the radius of gyration of the body about an axis passing through G. The point P is called the *center of percussion* of the body.

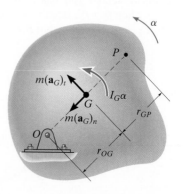

Prob. 17–66

17–67. Determine the position r_P of the center of percussion P of the 10-lb slender bar. (See Prob. 17–66.) What is the horizontal component of force that the pin at A exerts on the bar when it is struck at P with a force of $F = 20$ lb?

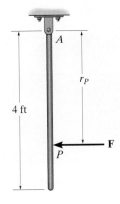

Prob. 17–67

*****17–68.** The disk has a mass M and a radius R. If a block of mass m is attached to the cord, determine the angular acceleration of the disk when the block is released from rest. Also, what is the velocity of the block after it falls a distance $2R$ starting from rest?

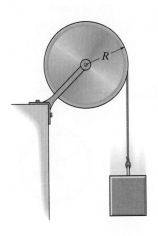

Prob. 17–68

17–69. The door will close automatically using torsional springs mounted on the hinges. Each spring has a stiffness $k = 50$ N·m/rad so that the torque on each hinge is $M = (50\theta)$ N·m, where θ is measured in radians. If the door is released from rest when it is open at $\theta = 90°$, determine its angular velocity at the instant $\theta = 0°$. For the calculation, treat the door as a thin plate having a mass of 70 kg.

17–70. The door will close automatically using torsional springs mounted on the hinges. If the torque on each hinge is $M = k\theta$, where θ is measured in radians, determine the required torsional stiffness k so that the door will close ($\theta = 0°$) with an angular velocity $\omega = 2$ rad/s when it is released from rest at $\theta = 90°$. For the calculation, treat the door as a thin plate having a mass of 70 kg.

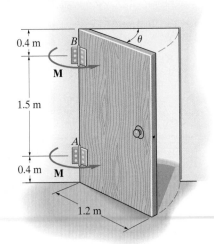

Probs. 17–69/70

17–71. The pendulum consists of a 10-kg uniform slender rod and a 15-kg sphere. If the pendulum is subjected to a torque of $M = 50$ N·m, and has an angular velocity of 3 rad/s when $\theta = 45°$, determine the magnitude of the reactive force pin O exerts on the pendulum at this instant.

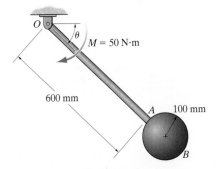

Prob. 17–71

17

*17–72. The disk has a mass of 20 kg and is originally spinning at the end of the strut with an angular velocity of $\omega = 60$ rad/s. If it is then placed against the wall, for which the coefficient of kinetic friction is $\mu_k = 0.3$ determine the time required for the motion to stop. What is the force in strut BC during this time?

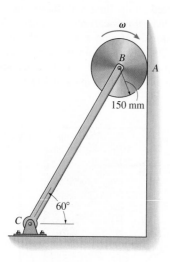

Prob. 17–72

17–73. The slender rod of length L and mass m is released from rest when $\theta = 0°$. Determine as a function of θ the normal and the frictional forces which are exerted by the ledge on the rod at A as it falls downward. At what angle θ does the rod begin to slip if the coefficient of static friction at A is μ?

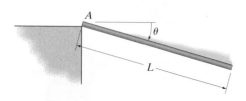

Prob. 17–73

17–74. The 5-kg cylinder is initially at rest when it is placed in contact with the wall B and the rotor at A. If the rotor always maintains a constant clockwise angular velocity $\omega = 6$ rad/s, determine the initial angular acceleration of the cylinder. The coefficient of kinetic friction at the contacting surfaces B and C is $\mu_k = 0.2$.

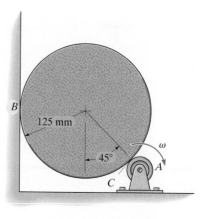

Prob. 17–74

17–75. The wheel has a mass of 25 kg and a radius of gyration $k_B = 0.15$ m. It is originally spinning at $\omega_1 = 40$ rad/s. If it is placed on the ground, for which the coefficient of kinetic friction is $\mu_C = 0.5$, determine the time required for the motion to stop. What are the horizontal and vertical components of reaction which the pin at A exerts on AB during this time? Neglect the mass of AB.

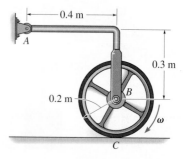

Prob. 17–75

*■**17–76.**   A 40-kg boy sits on top of the large wheel which has a mass of 400 kg and a radius of gyration $k_G = 5.5$ m. If the boy essentially starts from rest at $\theta = 0°$, and the wheel begins to rotate freely, determine the angle at which the boy begins to slip. The coefficient of static friction between the wheel and the boy is $\mu_s = 0.5$. Neglect the size of the boy in the calculation.

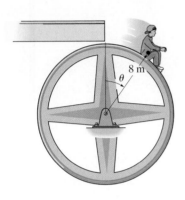

Prob. 17–76

17–77.   Gears A and B have a mass of 50 kg and 15 kg, respectively. Their radii of gyration about their respective centers of mass are $k_C = 250$ mm and $k_D = 150$ mm. If a torque of $M = 200(1 - e^{-0.2t})$ N·m, where t is in seconds, is applied to gear A, determine the angular velocity of both gears when $t = 3$ s, starting from rest.

17–78.   Block A has a mass m and rests on a surface having a coefficient of kinetic friction μ_k. The cord attached to A passes over a pulley at C and is attached to a block B having a mass $2m$. If B is released, determine the acceleration of A. Assume that the cord does not slip over the pulley. The pulley can be approximated as a thin disk of radius r and mass $\frac{1}{4}m$. Neglect the mass of the cord.

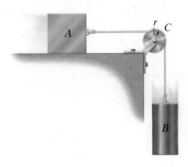

Prob. 17–78

17–79.   The two blocks A and B have a mass of 5 kg and 10 kg, respectively. If the pulley can be treated as a disk of mass 3 kg and radius 0.15 m, determine the acceleration of block A. Neglect the mass of the cord and any slipping on the pulley.

*■**17–80.**   The two blocks A and B have a mass m_A and m_B, respectively, where $m_B > m_A$. If the pulley can be treated as a disk of mass M, determine the acceleration of block A. Neglect the mass of the cord and any slipping on the pulley.

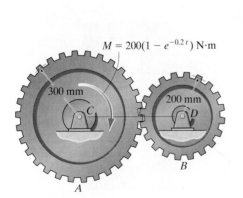

$M = 200(1 - e^{-0.2t})$ N·m

300 mm

200 mm

Prob. 17–77

Probs. 17–79/80

17–81. Determine the angular acceleration of the 25-kg diving board and the horizontal and vertical components of reaction at the pin A the instant the man jumps off. Assume that the board is uniform and rigid, and that at the instant he jumps off the spring is compressed a maximum amount of 200 mm, $\omega = 0$, and the board is horizontal. Take $k = 7 \text{ kN/m}$.

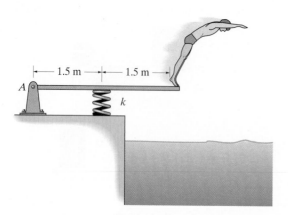

Prob. 17–81

17–82. The lightweight turbine consists of a rotor which is powered from a torque applied at its center. At the instant the rotor is horizontal it has an angular velocity of 15 rad/s and a clockwise angular acceleration of 8 rad/s². Determine the internal normal force, shear force, and moment at a section through A. Assume the rotor is a 50-m-long slender rod, having a mass of 3 kg/m.

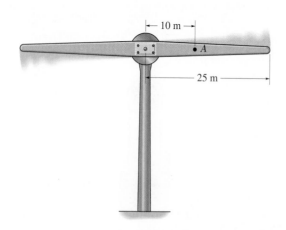

Prob. 17–82

17–83. The two-bar assembly is released from rest in the position shown. Determine the initial bending moment at the fixed joint B. Each bar has a mass m and length l.

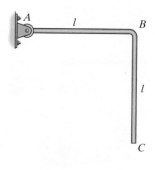

Prob. 17–83

*17–84. The armature (slender rod) AB has a mass of 0.2 kg and can pivot about the pin at A. Movement is controlled by the electromagnet E, which exerts a horizontal attractive force on the armature at B of $F_B = (0.2(10^{-3})l^{-2})$ N, where l in meters is the gap between the armature and the magnet at any instant. If the armature lies in the horizontal plane, and is originally at rest, determine the speed of the contact at B the instant $l = 0.01$ m. Originally $l = 0.02$ m.

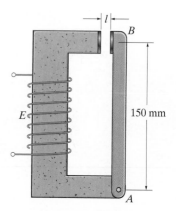

Prob. 17–84

17–85. The bar has a weight per length of w. If it is rotating in the vertical plane at a constant rate ω about point O, determine the internal normal force, shear force, and moment as a function of x and θ.

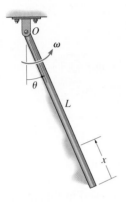

Prob. 17–85

17–86. A force $F = 2$ lb is applied perpendicular to the axis of the 5-lb rod and moves from O to A at a constant rate of 4 ft/s. If the rod is at rest when $\theta = 0°$ and $\mathbf{F}$ is at O when $t = 0$, determine the rod's angular velocity at the instant the force is at A. Through what angle has the rod rotated when this occurs? The rod rotates in the *horizontal plane*.

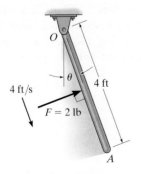

Prob. 17–86

17–87. The 15-kg block A and 20-kg cylinder B are connected by a light cord that passes over a 5-kg pulley (disk). If the system is released from rest, determine the cylinder's velocity after its has traveled downwards 2 m. Neglect friction between the plane and the block, and assume the cord does not slip over the pulley.

*17–88. The 15-kg block A and 20-kg cylinder B are connected by a light cord that passes over a 5-kg pulley (disk). If the system is released from rest, determine the cylinder's velocity after its has traveled downwards 2 m. The coefficient of kinetic friction between the block and the horizontal plane is $\mu_k = 0.3$. Assume the cord does not slip over the pulley.

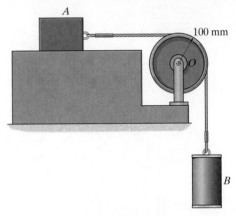

Probs. 17–87/88

17–89. The "Catherine wheel" is a firework that consists of a coiled tube of powder which is pinned at its center. If the powder burns at a constant rate of 20 g/s such as that the exhaust gases always exert a force having a constant magnitude of 0.3 N, directed tangent to the wheel, determine the angular velocity of the wheel when 75% of the mass is burned off. Initially, the wheel is at rest and has a mass of 100 g and a radius of $r = 75$ mm. For the calculation, consider the wheel to always be a thin disk.

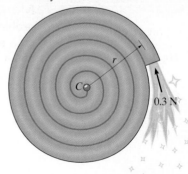

Prob. 17–89

17

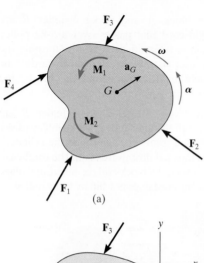

(a)

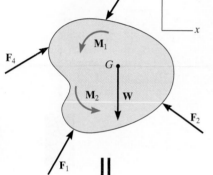

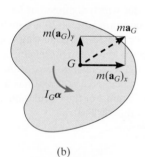

(b)

Fig. 17–19

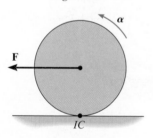

Fig. 17–20

17.5   Equations of Motion: General Plane Motion

The rigid body (or slab) shown in Fig. 17–19a is subjected to general plane motion caused by the externally applied force and couple-moment system. The free-body and kinetic diagrams for the body are shown in Fig. 17–19b. If an x and y inertial coordinate system is established as shown, the three equations of motion are

$$\Sigma F_x = m(a_G)_x$$
$$\Sigma F_y = m(a_G)_y \qquad (17\text{–}17)$$
$$\Sigma M_G = I_G \alpha$$

In some problems it may be convenient to sum moments about a point P other than G in order to eliminate as many unknown forces as possible from the moment summation. When used in this more general case, the three equations of motion are

$$\Sigma F_x = m(a_G)_x$$
$$\Sigma F_y = m(a_G)_y \qquad (17\text{–}18)$$
$$\Sigma M_P = \Sigma(\mathcal{M}_k)_P$$

Here $\Sigma(\mathcal{M}_k)_P$ represents the moment sum of $I_G\alpha$ and $m\mathbf{a}_G$ (or its components) about P as determined by the data on the kinetic diagram.

Moment Equation About the *IC*.

There is a particular type of problem that involves a uniform disk, or body of circular shape, that rolls on a rough surface *without slipping*, Fig 17–20. If we sum the moments about the instantaneous center of zero velocity, then $\Sigma(\mathcal{M}_k)_{IC}$ becomes $I_{IC}\alpha$, so that

$$\Sigma M_{IC} = I_{IC}\alpha \qquad (17\text{–}19)$$

This result compares with $\Sigma M_O = I_O\alpha$, which is used for a body pinned at point O, Eq. 17–16. See Prob. 17–90.

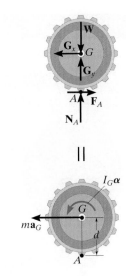

Procedure for Analysis

Kinetic problems involving general plane motion of a rigid body can be solved using the following procedure.

Free-Body Diagram.

- Establish the x, y inertial coordinate system and draw the free-body diagram for the body.
- Specify the direction and sense of the acceleration of the mass center, $\mathbf{a}_G$, and the angular acceleration $\boldsymbol{\alpha}$ of the body.
- Determine the moment of inertia I_G.
- Identify the unknowns in the problem.
- If it is decided that the rotational equation of motion $\Sigma M_P = \Sigma(\mathcal{M}_k)_P$ is to be used, then consider drawing the kinetic diagram in order to help "visualize" the "moments" developed by the components $m(\mathbf{a}_G)_x$, $m(\mathbf{a}_G)_y$, and $I_G\boldsymbol{\alpha}$ when writing the terms in the moment sum $\Sigma(\mathcal{M}_k)_P$.

Equations of Motion.

- Apply the three equations of motion in accordance with the established sign convention.
- When friction is present, there is the possibility for motion with no slipping or tipping. Each possibility for motion should be considered.

Kinematics.

- Use kinematics if a complete solution cannot be obtained strictly from the equations of motion.
- If the body's motion is *constrained* due to its supports, additional equations may be obtained by using $\mathbf{a}_B = \mathbf{a}_A + \mathbf{a}_{B/A}$, which relates the accelerations of any two points A and B on the body.
- When a wheel, disk, cylinder, or ball *rolls without slipping*, then $a_G = \alpha r$.

As the soil compactor, or "sheep's foot roller" moves forward, the roller has general plane motion. The forces shown on its free-body diagram cause the effects shown on the kinetic diagram. If moments are summed about the mass center, G, then $\Sigma M_G = I_G\alpha$. However, if moments are summed about point A (the IC) then $\zeta + \Sigma M_A = I_G\alpha + (ma_G)d = I_A\alpha$.

17

EXAMPLE | 17.13

(a)

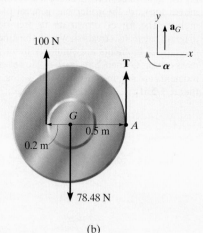

(b)

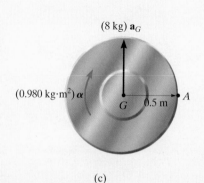

(c)

Fig. 17–21

Determine the angular acceleration of the spool in Fig. 17–21a. The spool has a mass of 8 kg and a radius of gyration of $k_G = 0.35$ m. The cords of negligible mass are wrapped around its inner hub and outer rim.

SOLUTION I

Free-Body Diagram. Fig. 17–21b. The 100-N force causes $\mathbf{a}_G$ to act upward. Also, $\boldsymbol{\alpha}$ acts clockwise, since the spool winds around the cord at A.

There are three unknowns T, a_G, and α. The moment of inertia of the spool about its mass center is

$$I_G = mk_G^2 = 8 \text{ kg}(0.35 \text{ m})^2 = 0.980 \text{ kg} \cdot \text{m}^2$$

Equations of Motion.

$$+\uparrow \Sigma F_y = m(a_G)_y; \qquad T + 100 \text{ N} - 78.48 \text{ N} = (8 \text{ kg})a_G \qquad (1)$$

$$\zeta + \Sigma M_G = I_G \alpha; \quad 100 \text{ N}(0.2 \text{ m}) - T(0.5 \text{ m}) = (0.980 \text{ kg} \cdot \text{m}^2)\alpha \qquad (2)$$

Kinematics. A complete solution is obtained if kinematics is used to relate a_G to α. In this case the spool "rolls without slipping" on the cord at A. Hence, we can use the results of Example 16.4 or 16.15 so that,

$$(\zeta+) \, a_G = \alpha r; \qquad\qquad a_G = \alpha \, (0.5 \text{ m}) \qquad (3)$$

Solving Eqs. 1 to 3, we have

$$\alpha = 10.3 \text{ rad/s}^2 \qquad\qquad Ans.$$
$$a_G = 5.16 \text{ m/s}^2$$
$$T = 19.8 \text{ N}$$

SOLUTION II

Equations of Motion. We can eliminate the unknown T by summing moments about point A. From the free-body and kinetic diagrams Figs. 17–21b and 17–21c, we have

$$\zeta + \Sigma M_A = \Sigma(\mathcal{M}_k)_A; \qquad 100 \text{ N}(0.7 \text{ m}) - 78.48 \text{ N}(0.5 \text{ m})$$
$$= (0.980 \text{ kg} \cdot \text{m}^2)\alpha + [(8 \text{ kg})a_G](0.5 \text{ m})$$

Using Eq. (3),

$$\alpha = 10.3 \text{ rad/s}^2 \qquad\qquad Ans.$$

SOLUTION III

Equations of Motion. The simplest way to solve this problem is to realize that point A is the *IC* for the spool. Then Eq. 17–19 applies.

$$\zeta + \Sigma M_A = I_A \alpha; \quad (100 \text{ N})(0.7 \text{ m}) - (78.48 \text{ N})(0.5 \text{ m})$$
$$= [0.980 \text{ kg} \cdot \text{m}^2 + (8 \text{ kg})(0.5 \text{ m})^2]\alpha$$
$$\alpha = 10.3 \text{ rad/s}^2$$

EXAMPLE 17.14

The 50-lb wheel shown in Fig. 17–22 has a radius of gyration $k_G = 0.70$ ft. If a 35-lb·ft couple moment is applied to the wheel, determine the acceleration of its mass center G. The coefficients of static and kinetic friction between the wheel and the plane at A are $\mu_s = 0.3$ and $\mu_k = 0.25$, respectively.

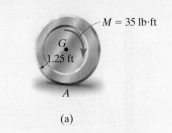

(a)

SOLUTION

Free-Body Diagram. By inspection of Fig. 17–22b, it is seen that the couple moment causes the wheel to have a clockwise angular acceleration of α. As a result, the acceleration of the mass center, $\mathbf{a}_G$, is directed to the right. The moment of inertia is

$$I_G = mk_G^2 = \frac{50 \text{ lb}}{32.2 \text{ ft/s}^2}(0.70 \text{ ft})^2 = 0.7609 \text{ slug} \cdot \text{ft}^2$$

The unknowns are N_A, F_A, a_G, and α.

Equations of Motion.

$$\xrightarrow{+} \Sigma F_x = m(a_G)_x; \qquad F_A = \left(\frac{50 \text{ lb}}{32.2 \text{ ft/s}^2}\right)a_G \qquad (1)$$

$$+\uparrow \Sigma F_y = m(a_G)_y; \qquad N_A - 50 \text{ lb} = 0 \qquad (2)$$

$$\zeta + \Sigma M_G = I_G\alpha; \qquad 35 \text{ lb} \cdot \text{ft} - 1.25 \text{ ft}(F_A) = (0.7609 \text{ slug} \cdot \text{ft}^2)\alpha \qquad (3)$$

A fourth equation is needed for a complete solution.

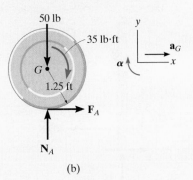

(b)

Fig. 17–22

Kinematics (No Slipping). If this assumption is made, then

$$(\zeta+) \qquad\qquad a_G = (1.25 \text{ ft})\alpha \qquad (4)$$

Solving Eqs. 1 to 4,

$$N_A = 50.0 \text{ lb} \qquad F_A = 21.3 \text{ lb}$$
$$\alpha = 11.0 \text{ rad/s}^2 \qquad a_G = 13.7 \text{ ft/s}^2$$

This solution requires that no slipping occurs, i.e., $F_A \leq \mu_s N_A$. However, since 21.3 lb $> 0.3(50$ lb$) = 15$ lb, the wheel slips as it rolls.

(Slipping). Equation 4 is not valid, and so $F_A = \mu_k N_A$, or

$$F_A = 0.25 N_A \qquad (5)$$

Solving Eqs. 1 to 3 and 5 yields

$$N_A = 50.0 \text{ lb} \qquad F_A = 12.5 \text{ lb}$$
$$\alpha = 25.5 \text{ rad/s}^2$$
$$a_G = 8.05 \text{ ft/s}^2 \rightarrow \qquad\qquad Ans.$$

17

EXAMPLE | 17.15

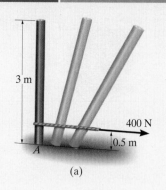

3 m

400 N

0.5 m

A

(a)

The uniform slender pole shown in Fig. 17–23a has a mass of 100 kg. If the coefficients of static and kinetic friction between the end of the pole and the surface are $\mu_s = 0.3$, and $\mu_k = 0.25$, respectively, determine the pole's angular acceleration at the instant the 400-N horizontal force is applied. The pole is originally at rest.

SOLUTION

Free-Body Diagram. Figure 17–23b. The path of motion of the mass center G will be along an unknown curved path having a radius of curvature ρ, which is initially on a vertical line. However, there is no normal or y component of acceleration since the pole is originally at rest, i.e., $\mathbf{v}_G = \mathbf{0}$, so that $(a_G)_y = v_G^2/\rho = 0$. We will assume the mass center accelerates to the right and that the pole has a clockwise angular acceleration of α. The unknowns are N_A, F_A, a_G, and α.

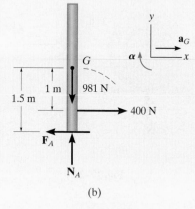

(b)

Fig. 17–23

Equation of Motion.

$$\xrightarrow{+} \Sigma F_x = m(a_G)_x; \qquad 400 \text{ N} - F_A = (100 \text{ kg})a_G \qquad (1)$$

$$+\uparrow \Sigma F_y = m(a_G)_y; \qquad N_A - 981 \text{ N} = 0 \qquad (2)$$

$$\zeta + \Sigma M_G = I_G\alpha; \; F_A(1.5 \text{ m}) - (400 \text{ N})(1 \text{ m}) = [\tfrac{1}{12}(100 \text{ kg})(3 \text{ m})^2]\alpha \quad (3)$$

A fourth equation is needed for a complete solution.

Kinematics (No Slipping). With this assumption, point A acts as a "pivot" so that α is clockwise, then a_G is directed to the right.

$$a_G = \alpha r_{AG}; \qquad\qquad a_G = (1.5 \text{ m})\,\alpha \qquad (4)$$

Solving Eqs. 1 to 4 yields

$$N_A = 981 \text{ N} \quad F_A = 300 \text{ N}$$

$$a_G = 1 \text{ m/s}^2 \quad \alpha = 0.667 \text{ rad/s}^2$$

The assumption of no slipping requires $F_A \le \mu_s N_A$. However, $300 \text{ N} > 0.3(981 \text{ N}) = 294 \text{ N}$ and so the pole slips at A.

(Slipping). For this case Eq. 4 does *not* apply. Instead the frictional equation $F_A = \mu_k N_A$ must be used. Hence,

$$F_A = 0.25N_A \qquad (5)$$

Solving Eqs. 1 to 3 and 5 simultaneously yields

$$N_A = 981 \text{ N} \quad F_A = 245 \text{ N} \quad a_G = 1.55 \text{ m/s}^2$$

$$\alpha = -0.428 \text{ rad/s}^2 = 0.428 \text{ rad/s}^2\,\zeta \qquad\qquad Ans.$$

17

EXAMPLE | 17.16

The uniform 50-kg bar in Fig. 17–24a is held in the equilibrium position by cords AC and BD. Determine the tension in BD and the angular acceleration of the bar immediately after AC is cut.

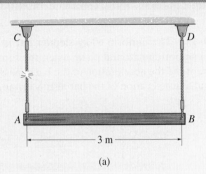

(a)

SOLUTION

Free-Body Diagram. Fig. 17–24b. There are four unknowns, T_B, $(a_G)_x$, $(a_G)_y$, and α.

Equations of Motion.

$$\xrightarrow{+} \Sigma F_x = m(a_G)_x; \qquad 0 = 50 \text{ kg } (a_G)_x$$

$$(a_G)_x = 0$$

$$+\uparrow \Sigma F_y = m(a_G)_y; \quad T_B - 50(9.81)\text{N} = -50 \text{ kg } (a_G)_y \qquad (1)$$

$$\zeta + \Sigma M_G = I_G\alpha; \qquad T_B(1.5 \text{ m}) = \left[\frac{1}{12}(50 \text{ kg})(3 \text{ m})^2\right]\alpha \qquad (2)$$

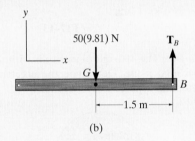

(b)

Kinematics. Since the bar is at rest just after the cable is cut, then its angular velocity and the velocity of point B at this instant are equal to zero. Thus $(a_B)_n = v_B^2/\rho_{BD} = 0$. Therefore, $\mathbf{a}_B$ only has a tangential component, which is directed along the x axis, Fig. 17–24c. Applying the relative acceleration equation to points G and B,

$$\mathbf{a}_G = \mathbf{a}_B + \boldsymbol{\alpha} \times \mathbf{r}_{G/B} - \omega^2\mathbf{r}_{G/B}$$

$$-(a_G)_y\mathbf{j} = a_B\mathbf{i} + (\alpha\mathbf{k}) \times (-1.5\mathbf{i}) - \mathbf{0}$$

$$-(a_G)_y\mathbf{j} = a_B\mathbf{i} - 1.5\alpha\mathbf{j}$$

Equating the $\mathbf{i}$ and $\mathbf{j}$ components of both sides of this equation,

$$0 = a_B$$

$$(a_G)_y = 1.5\alpha \qquad (3)$$

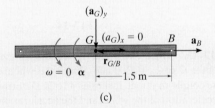

(c)

Fig. 17–24

Solving Eqs. (1) through (3) yields

$$\alpha = 4.905 \text{ rad/s}^2 \qquad \qquad Ans.$$

$$T_B = 123 \text{ N} \qquad \qquad Ans.$$

$$(a_G)_y = 7.36 \text{ m/s}^2$$

17

FUNDAMENTAL PROBLEMS

F17–13. The uniform 60-kg slender bar is initially at rest on a smooth horizontal plane when the forces are applied. Determine the acceleration of the bar's mass center and the angular acceleration of the bar at this instant.

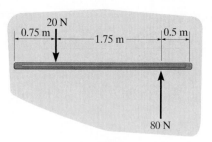

20 N
0.75 m
1.75 m
0.5 m
80 N

F17–13

F17–14. The 100-kg cylinder rolls without slipping on the horizontal plane. Determine the acceleration of its mass center and its angular acceleration.

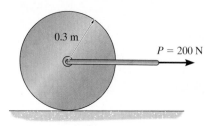

0.3 m
$P = 200$ N

F17–14

F17–15. The 20-kg wheel has a radius of gyration about its center O of $k_O = 300$ mm. When the wheel is subjected to the couple moment, it slips as it rolls. Determine the angular acceleration of the wheel and the acceleration of the wheel's center O. The coefficient of kinetic friction between the wheel and the plane is $\mu_k = 0.5$.

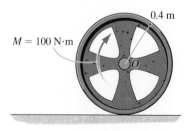

0.4 m
$M = 100$ N·m
O

F17–15

F17–16. The 20-kg sphere rolls down the inclined plane without slipping. Determine the angular acceleration of the sphere and the acceleration of its mass center.

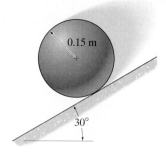

0.15 m
30°

F17–16

F17–17. The 200-kg spool has a radius of gyration about its mass center of $k_G = 300$ mm. If the couple moment is applied to the spool and the coefficient of kinetic friction between the spool and the ground is $\mu_k = 0.2$, determine the angular acceleration of the spool, the acceleration of G and the tension in the cable.

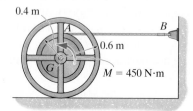

0.4 m
A
B
0.6 m
G
$M = 450$ N·m

F17–17

F17–18. The 12-kg slender rod is pinned to a small roller A that slides freely along the slot. If the rod is released from rest at $\theta = 0°$, determine the angular acceleration of the rod and the acceleration of the roller immediately after the release.

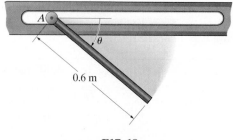

A
θ
0.6 m

F17–18

PROBLEMS

17–90. If the disk in Fig. 17–20 *rolls without slipping*, show that when moments are summed about the instantaneous center of zero velocity, *IC*, it is possible to use the moment equation $\Sigma M_{IC} = I_{IC}\alpha$, where I_{IC} represents the moment of inertia of the disk calculated about the instantaneous axis of zero velocity.

17–91. The 20-kg punching bag has a radius of gyration about its center of mass *G* of $k_G = 0.4$ m. If it is initially at rest and is subjected to a horizontal force $F = 30$ N, determine the initial angular acceleration of the bag and the tension in the supporting cable *AB*.

17–93. The rocket has a weight of 20 000 lb, mass center at *G*, and radius of gyration about the mass center of $k_G = 21$ ft when it is fired. Each of its two engines provides a thrust $T = 50 000$ lb. At a given instant engine *A* suddenly fails to operate. Determine the angular acceleration of the rocket and the acceleration of its nose *B*.

Prob. 17–93

17–94. The tire has a weight of 30 lb and a radius of gyration of $k_G = 0.6$ ft. If the coefficients of static and kinetic friction between the tire and the plane are $\mu_s = 0.2$ and $\mu_k = 0.15$, determine the tire's angular acceleration as it rolls down the incline. Set $\theta = 12°$.

17–95. The tire has a weight of 30 lb and a radius of gyration of $k_G = 0.6$ ft. If the coefficients of static and kinetic friction between the tire and the plane are $\mu_s = 0.2$ and $\mu_k = 0.15$, determine the maximum angle θ of the inclined plane so that the tire rolls without slipping.

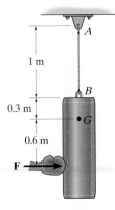

Prob. 17–91

*****17–92.** The uniform 150-lb beam is initially at rest when the forces are applied to the cables. Determine the magnitude of the acceleration of the mass center and the angular acceleration of the beam at this instant.

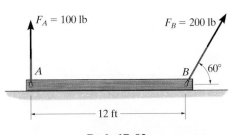

Prob. 17–92

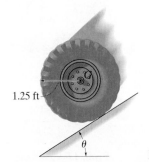

Probs. 17–94/95

*17–96. The spool has a mass of 100 kg and a radius of gyration of $k_G = 0.3$ m. If the coefficients of static and kinetic friction at A are $\mu_s = 0.2$ and $\mu_k = 0.15$, respectively, determine the angular acceleration of the spool if $P = 50$ N.

17–97. Solve Prob. 17–96 if the cord and force $P = 50$ N are directed vertically upwards.

17–98. The spool has a mass of 100 kg and a radius of gyration $k_G = 0.3$ m. If the coefficients of static and kinetic friction at A are $\mu_s = 0.2$ and $\mu_k = 0.15$, respectively, determine the angular acceleration of the spool if $P = 600$ N.

*17–100. A uniform rod having a weight of 10 lb is pin supported at A from a roller which rides on a horizontal track. If the rod is originally at rest, and a horizontal force of $F = 15$ lb is applied to the roller, determine the acceleration of the roller. Neglect the mass of the roller and its size d in the computations.

17–101. Solve Prob. 17–100 assuming that the roller at A is replaced by a slider block having a negligible mass. The coefficient of kinetic friction between the block and the track is $\mu_k = 0.2$. Neglect the dimension d and the size of the block in the computations.

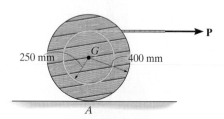

Probs. 17–96/97/98

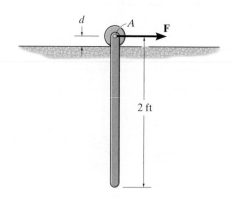

Probs. 17–100/101

17–99. The upper body of the crash dummy has a mass of 75 lb, a center of gravity at G, and a radius of gyration about G of $k_G = 0.7$ ft. By means of the seat belt this body segment is assumed to be pin-connected to the seat of the car at A. If a crash causes the car to decelerate at 50 ft/s², determine the angular velocity of the body when it has rotated to $\theta = 30°$.

17–102. The 2-kg slender bar is supported by cord BC and then released from rest at A. Determine the initial angular acceleration of the bar and the tension in the cord.

Prob. 17–99

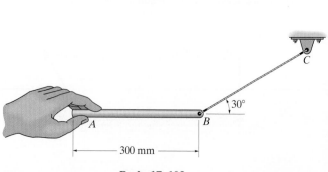

Prob. 17–102

17–103. If the truck accelerates at a constant rate of $6\,\text{m/s}^2$, starting from rest, determine the initial angular acceleration of the 20-kg ladder. The ladder can be considered as a uniform slender rod. The support at B is smooth.

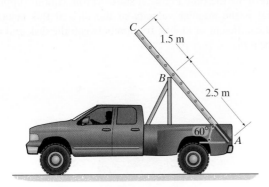

Prob. 17–103

*17–104. If $P = 30$ lb, determine the angular acceleration of the 50-lb roller. Assume the roller to be a uniform cylinder and that no slipping occurs.

17–105. If the coefficient of static friction between the 50-lb roller and the ground is $\mu_s = 0.25$, determine the maximum force P that can be applied to the handle, so that roller rolls on the ground without slipping. Also, find the angular acceleration of the roller. Assume the roller to be a uniform cylinder.

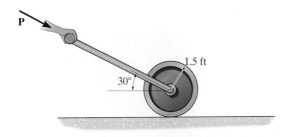

Probs. 17–104/105

17–106. The spool has a mass of 500 kg and a radius of gyration $k_G = 1.30$ m. It rests on the surface of a conveyor belt for which the coefficient of static friction is $\mu_s = 0.5$ and the coefficient of kinetic friction is $\mu_k = 0.4$. If the conveyor accelerates at $a_C = 1\,\text{m/s}^2$, determine the initial tension in the wire and the angular acceleration of the spool. The spool is originally at rest.

17–107. The spool has a mass of 500 kg and a radius of gyration $k_G = 1.30$ m. It rests on the surface of a conveyor belt for which the coefficient of static friction is $\mu_s = 0.5$. Determine the greatest acceleration a_C of the conveyor so that the spool will not slip. Also, what are the initial tension in the wire and the angular acceleration of the spool? The spool is originally at rest.

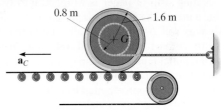

Probs. 17–106/107

*17–108. The semicircular disk having a mass of 10 kg is rotating at $\omega = 4\,\text{rad/s}$ at the instant $\theta = 60°$. If the coefficient of static friction at A is $\mu_s = 0.5$, determine if the disk slips at this instant.

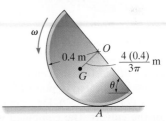

Prob. 17–108

17–109. The 500-kg concrete culvert has a mean radius of 0.5 m. If the truck has an acceleration of $3\,\text{m/s}^2$, determine the culvert's angular acceleration. Assume that the culvert does not slip on the truck bed, and neglect its thickness.

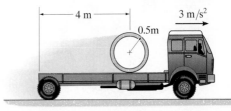

Prob. 17–109

17–110. The 10-lb hoop or thin ring is given an initial angular velocity of 6 rad/s when it is placed on the surface. If the coefficient of kinetic friction between the hoop and the surface is $\mu_k = 0.3$, determine the distance the hoop moves before it stops slipping.

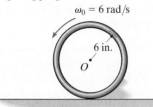

Prob. 17–110

17–111. A long strip of paper is wrapped into two rolls, each having a mass of 8 kg. Roll A is pin supported about its center whereas roll B is not centrally supported. If B is brought into contact with A and released from rest, determine the initial tension in the paper between the rolls and the angular acceleration of each roll. For the calculation, assume the rolls to be approximated by cylinders.

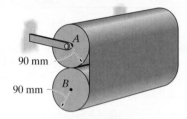

Prob. 17–111

*17–112. The circular concrete culvert rolls with an angular velocity of $\omega = 0.5$ rad/s when the man is at the position shown. At this instant the center of gravity of the culvert and the man is located at point G, and the radius of gyration about G is $k_G = 3.5$ ft. Determine the angular acceleration of the culvert. The combined weight of the culvert and the man is 500 lb. Assume that the culvert rolls without slipping, and the man does not move within the culvert.

Prob. 17–112

17–113. The uniform disk of mass m is rotating with an angular velocity of ω_0 when it is placed on the floor. Determine the initial angular acceleration of the disk and the acceleration of its mass center. The coefficient of kinetic friction between the disk and the floor is μ_k.

17–114. The uniform disk of mass m is rotating with an angular velocity of ω_0 when it is placed on the floor. Determine the time before it starts to roll without slipping. What is the angular velocity of the disk at this instant? The coefficient of kinetic friction between the disk and the floor is μ_k.

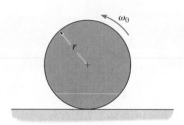

Probs. 17–113/114

17–115. The 16-lb bowling ball is cast horizontally onto a lane such that initially $\omega = 0$ and its mass center has a velocity $v = 8$ ft/s. If the coefficient of kinetic friction between the lane and the ball is $\mu_k = 0.12$, determine the distance the ball travels before it rolls without slipping. For the calculation, neglect the finger holes in the ball and assume the ball has a uniform density.

Prob. 17–115

*17–116. The uniform beam has a weight W. If it is originally at rest while being supported at A and B by cables, determine the tension in cable A if cable B suddenly fails. Assume the beam is a slender rod.

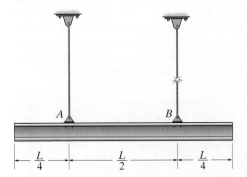

Prob. 17–116

17–117. A cord C is wrapped around each of the two 10-kg disks. If they are released from rest, determine the tension in the fixed cord D. Neglect the mass of the cord.

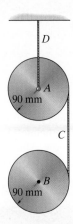

Prob. 17–117

17–118. The 500-lb beam is supported at A and B when it is subjected to a force of 1000 lb as shown. If the pin support at A suddenly fails, determine the beam's initial angular acceleration and the force of the roller support on the beam. For the calculation, assume that the beam is a slender rod so that its thickness can be neglected.

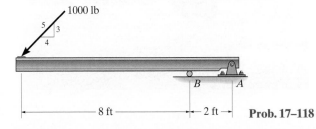

Prob. 17–118

17–119. The 30-kg uniform slender rod AB rests in the position shown when the couple moment of $M = 150$ N·m is applied. Determine the initial angular acceleration of the rod. Neglect the mass of the rollers.

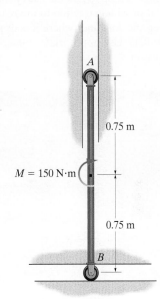

Prob. 17–119

*17–120. The 30-kg slender rod AB rests in the position shown when the horizontal force $P = 50$ N is applied. Determine the initial angular acceleration of the rod. Neglect the mass of the rollers.

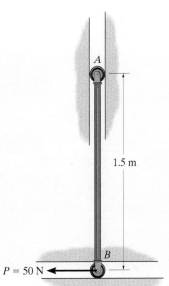

Prob. 17–120

CONCEPTUAL PROBLEMS

P17–1. The truck is used to pull the heavy container. To be most effective at providing traction to the rear wheels at *A*, is it best to keep the container where it is or place it at the front of the trailer? Use appropriate numerical values to explain your answer.

P17–1

P17–2. The tractor is about to tow the plane to the right. Is it possible for the driver to cause the front wheel of the plane to lift off the ground as he accelerates the tractor? Draw the free-body and kinetic diagrams and explain algebraically (letters) if and how this might be possible.

P17–2

P17–3. How can you tell the driver is accelerating this SUV? To explain your answer, draw the free-body and kinetic diagrams. Here power is supplied to the rear wheels. Would the photo look the same if power were supplied to the front wheels? Will the accelerations be the same? Use appropriate numerical values to explain your answers.

P17–3

P17–4. Here is something you should not try at home, at least not without wearing a helmet! Draw the free-body and kinetic diagrams and show what the rider must do to maintain this position. Use appropriate numerical values to explain your answer.

P17–4

CHAPTER REVIEW

Moment of Inertia

The moment of inertia is a measure of the resistance of a body to a change in its angular velocity. It is defined by $I = \int r^2 dm$ and will be different for each axis about which it is computed.

Many bodies are composed of simple shapes. If this is the case, then tabular values of I can be used, such as the ones given on the inside back cover of this book. To obtain the moment of inertia of a composite body about any specified axis, the moment of inertia of each part is determined about the axis and the results are added together. Doing this often requires use of the parallel-axis theorem.

$$I = I_G + md^2$$

Planar Equations of Motion

The equations of motion define the translational, and rotational motion of a rigid body. In order to account for all of the terms in these equations, a free-body diagram should always accompany their application, and for some problems, it may also be convenient to draw the kinetic diagram which shows $m\mathbf{a}_G$ and $I_G\boldsymbol{\alpha}$.

$$\Sigma F_x = m(a_G)_x$$

$$\Sigma F_y = m(a_G)_y$$

$$\Sigma M_G = 0$$

Rectilinear translation

$$\Sigma F_n = m(a_G)_n$$

$$\Sigma F_t = m(a_G)_t$$

$$\Sigma M_G = 0$$

Curvilinear translation

$$\Sigma F_n = m(a_G)_n = m\omega^2 r_G$$

$$\Sigma F_t = m(a_G)_t = m\alpha r_G$$

$$\Sigma M_G = I_G\alpha \text{ or } \Sigma M_O = I_O\alpha$$

Rotation About a Fixed Axis

$$\Sigma F_x = m(a_G)_x$$

$$\Sigma F_x = m(a_G)_x$$

$$\Sigma M_G = I_G\alpha \text{ or } \Sigma M_P = \Sigma(\mathcal{M}_k)_P$$

General Plane Motion

17

Chapter 18

Roller coasters must be able to coast over loops and through turns, and have enough energy to do so safely. Accurate calculation of this energy must account for the size of the car as it moves along the track.

Planar Kinetics of a Rigid Body: Work and Energy

CHAPTER OBJECTIVES

■ To develop formulations for the kinetic energy of a body, and define the various ways a force and couple do work.

■ To apply the principle of work and energy to solve rigid–body planar kinetic problems that involve force, velocity, and displacement.

■ To show how the conservation of energy can be used to solve rigid–body planar kinetic problems.

18.1 Kinetic Energy

In this chapter we will apply work and energy methods to solve planar motion problems involving force, velocity, and displacement. But first it will be necessary to develop a means of obtaining the body's kinetic energy when the body is subjected to translation, rotation about a fixed axis, or general plane motion.

To do this we will consider the rigid body shown in Fig. 18–1, which is represented here by a *slab* moving in the inertial x–y reference plane. An arbitrary ith particle of the body, having a mass dm, is located a distance r from the arbitrary point P. If at the *instant* shown the particle has a velocity $\mathbf{v}_i$, then the particle's kinetic energy is $T_i = \frac{1}{2} dm\, v_i^2$.

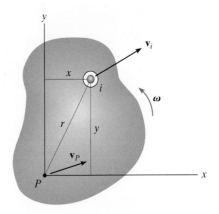

Fig. 18–1

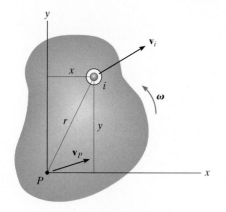

Fig. 18–1 (repeated)

The kinetic energy of the entire body is determined by writing similar expressions for each particle of the body and integrating the results, i.e.,

$$T = \frac{1}{2} \int_m dm \, v_i^2$$

This equation may also be expressed in terms of the velocity of point P. If the body has an angular velocity $\boldsymbol{\omega}$, then from Fig. 18–1 we have

$$\begin{aligned}
\mathbf{v}_i &= \mathbf{v}_P + \mathbf{v}_{i/P} \\
&= (v_P)_x \mathbf{i} + (v_P)_y \mathbf{j} + \omega \mathbf{k} \times (x\mathbf{i} + y\mathbf{j}) \\
&= [(v_P)_x - \omega y]\mathbf{i} + [(v_P)_y + \omega x]\mathbf{j}
\end{aligned}$$

The square of the magnitude of $\mathbf{v}_i$ is thus

$$\begin{aligned}
\mathbf{v}_i \cdot \mathbf{v}_i = v_i^2 &= [(v_P)_x - \omega y]^2 + [(v_P)_y + \omega x]^2 \\
&= (v_P)_x^2 - 2(v_P)_x \omega y + \omega^2 y^2 + (v_P)_y^2 + 2(v_P)_y \omega x + \omega^2 x^2 \\
&= v_P^2 - 2(v_P)_x \omega y + 2(v_P)_y \omega x + \omega^2 r^2
\end{aligned}$$

Substituting this into the equation of kinetic energy yields

$$T = \frac{1}{2}\left(\int_m dm\right)v_P^2 - (v_P)_x \omega \left(\int_m y \, dm\right) + (v_P)_y \omega \left(\int_m x \, dm\right) + \frac{1}{2}\omega^2\left(\int_m r^2 \, dm\right)$$

The first integral on the right represents the entire mass m of the body. Since $\bar{y}m = \int y \, dm$ and $\bar{x}m = \int x \, dm$, the second and third integrals locate the body's center of mass G with respect to P. The last integral represents the body's moment of inertia I_P, computed about the z axis passing through point P. Thus,

$$T = \tfrac{1}{2}mv_P^2 - (v_P)_x \omega \bar{y}m + (v_P)_y \omega \bar{x}m + \tfrac{1}{2}I_P\omega^2 \qquad (18\text{--}1)$$

As a special case, if point P coincides with the mass center G of the body, then $\bar{y} = \bar{x} = 0$, and therefore

$$T = \tfrac{1}{2}mv_G^2 + \tfrac{1}{2}I_G\omega^2 \qquad (18\text{--}2)$$

Both terms on the right side are *always positive*, since v_G and ω are squared. The first term represents the translational kinetic energy, referenced from the mass center, and the second term represents the body's rotational kinetic energy about the mass center.

Translation.   When a rigid body of mass m is subjected to either rectilinear or curvilinear *translation*, Fig. 18–2, the kinetic energy due to rotation is zero, since $\boldsymbol{\omega} = \mathbf{0}$. The kinetic energy of the body is therefore

$$T = \tfrac{1}{2}mv_G^2 \qquad (18\text{–}3)$$

Rotation About a Fixed Axis.   When a rigid body *rotates about a fixed axis* passing through point O, Fig. 18–3, the body has both *translational* and *rotational* kinetic energy so that

$$T = \tfrac{1}{2}mv_G^2 + \tfrac{1}{2}I_G\omega^2 \qquad (18\text{–}4)$$

The body's kinetic energy may also be formulated for this case by noting that $v_G = r_G\omega$, so that $T = \tfrac{1}{2}(I_G + mr_G^2)\omega^2$. By the parallel–axis theorem, the terms inside the parentheses represent the moment of inertia I_O of the body about an axis perpendicular to the plane of motion and passing through point O. Hence,*

$$T = \tfrac{1}{2}I_O\omega^2 \qquad (18\text{–}5)$$

From the derivation, this equation will give the same result as Eq. 18–4, since it accounts for *both* the translational and rotational kinetic energies of the body.

General Plane Motion.   When a rigid body is subjected to general plane motion, Fig. 18–4, it has an angular velocity $\boldsymbol{\omega}$ and its mass center has a velocity $\mathbf{v}_G$. Therefore, the kinetic energy is

$$T = \tfrac{1}{2}mv_G^2 + \tfrac{1}{2}I_G\omega^2 \qquad (18\text{–}6)$$

This equation can also be expressed in terms of the body's motion about its instantaneous center of zero velocity i.e.,

$$T = \tfrac{1}{2}I_{IC}\omega^2 \qquad (18\text{–}7)$$

where I_{IC} is the moment of inertia of the body about its instantaneous center. The proof is similar to that of Eq. 18–5. (See Prob. 18–1.)

*The similarity between this derivation and that of $\Sigma M_O = I_O\alpha$, should be noted. Also the same result can be obtained directly from Eq. 18–1 by selecting point P at O, realizing that $v_O = 0$.

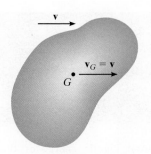
Translation
Fig. 18–2

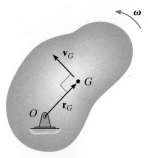
Rotation About a Fixed Axis
Fig. 18–3

General Plane Motion
Fig. 18–4

The total kinetic energy of this soil compactor consists of the kinetic energy of the body or frame of the machine due to its translation, and the translational and rotational kinetic energies of the roller and the wheels due to their general plane motion. Here we exclude the additional kinetic energy developed by the moving parts of the engine and drive train.

System of Bodies.   Because energy is a scalar quantity, the total kinetic energy for a system of *connected* rigid bodies is the sum of the kinetic energies of all its moving parts. Depending on the type of motion, the kinetic energy of *each body* is found by applying Eq. 18–2 or the alternative forms mentioned above.

18.2   The Work of a Force

Several types of forces are often encountered in planar kinetics problems involving a rigid body. The work of each of these forces has been presented in Sec. 14.1 and is listed below as a summary.

Work of a Variable Force.   If an external force $\mathbf{F}$ acts on a body, the work done by the force when the body moves along the path s, Fig. 18–5, is

$$U_F = \int \mathbf{F} \cdot d\mathbf{r} = \int_s F \cos \theta \, ds \qquad (18\text{–}8)$$

Here θ is the angle between the "tails" of the force and the differential displacement. The integration must account for the variation of the force's direction and magnitude.

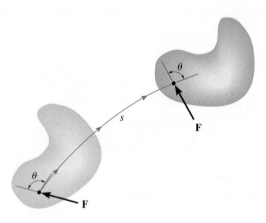

Fig. 18–5

Work of a Constant Force.   If an external force $\mathbf{F}_c$ acts on a body, Fig. 18–6, and maintains a constant magnitude F_c and constant direction θ, while the body undergoes a translation s, then the above equation can be integrated, so that the work becomes

$$U_{F_c} = (F_c \cos \theta)s \qquad (18\text{–}9)$$

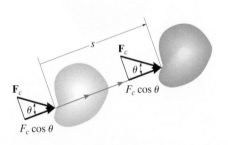

Fig. 18–6

Work of a Weight.   The weight of a body does work only when the body's center of mass G undergoes a *vertical displacement* Δy. If this displacement is *upward*, Fig. 18–7, the work is negative, since the weight is opposite to the displacement.

$$U_W = -W\,\Delta y \qquad (18\text{–}10)$$

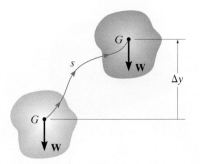

Fig. 18–7

Likewise, if the displacement is *downward* $(-\Delta y)$ the work becomes *positive*. In both cases the elevation change is considered to be small so that W, which is caused by gravitation, is constant.

Work of a Spring Force.   If a linear elastic spring is attached to a body, the spring force $F_s = ks$ *acting on the body* does work when the spring either stretches or compresses from s_1 to a *further* position s_2. In both cases the work will be *negative* since the *displacement of the body* is in the opposite direction to the force, Fig. 18–8. The work is

$$U_s = -\left(\tfrac{1}{2}ks_2^2 - \tfrac{1}{2}ks_1^2\right) \qquad (18\text{–}11)$$

where $|s_2| > |s_1|$.

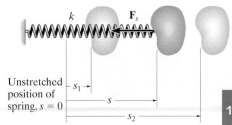

Fig. 18–8

Forces That Do No Work.   There are some external forces that do no work when the body is displaced. These forces act either at *fixed points* on the body, or they have a direction *perpendicular to their displacement*. Examples include the reactions at a pin support about which a body rotates, the normal reaction acting on a body that moves along a fixed surface, and the weight of a body when the center of gravity of the body moves in a *horizontal plane*, Fig. 18–9. A frictional force $\mathbf{F}_f$ acting on a round body as it *rolls without slipping* over a rough surface also does no work.* This is because, during any *instant of time* dt, $\mathbf{F}_f$ acts at a point on the body which has *zero velocity* (instantaneous center, IC) and so the work done by the force on the point is zero. In other words, the point is not displaced in the direction of the force during this instant. Since $\mathbf{F}_f$ contacts successive points for only an instant, the work of $\mathbf{F}_f$ will be zero.

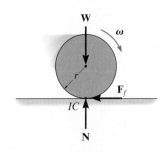

Fig. 18–9

*The work done by a frictional force *when the body slips* is discussed in Sec. 14.3.

18.3   The Work of a Couple Moment

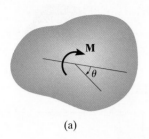

(a)

Consider the body in Fig. 18–10a, which is subjected to a couple moment $M = Fr$. If the body undergoes a differential displacement, then the work done by the couple forces can be found by considering the displacement as the sum of a separate translation plus rotation. When the body *translates*, the work of each force is produced only by the *component of displacement* along the line of action of the forces ds_t, Fig. 18–10b. Clearly the "positive" work of one force *cancels* the "negative" work of the other. When the body undergoes a differential rotation $d\theta$ about the arbitrary point O, Fig. 18–10c, then each force undergoes a displacement $ds_\theta = (r/2)\, d\theta$ in the direction of the force. Hence, the total work done is

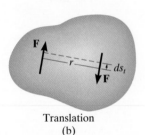

Translation
(b)

$$dU_M = F\left(\frac{r}{2}\, d\theta\right) + F\left(\frac{r}{2}\, d\theta\right) = (Fr)\, d\theta$$

$$= M\, d\theta$$

The work is *positive* when **M** and *dθ* have the *same sense of direction* and *negative* if these vectors are in the *opposite sense*.

When the body rotates in the plane through a finite angle θ measured in radians, from θ_1 to θ_2, the work of a couple moment is therefore

$$U_M = \int_{\theta_1}^{\theta_2} M\, d\theta \tag{18–12}$$

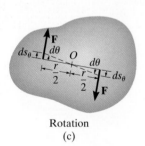

Rotation
(c)

Fig. 18–10

If the couple moment **M** has a *constant magnitude*, then

$$U_M = M(\theta_2 - \theta_1) \tag{18–13}$$

EXAMPLE | 18.1

The bar shown in Fig. 18–11a has a mass of 10 kg and is subjected to a couple moment of $M = 50$ N·m and a force of $P = 80$ N, which is always applied perpendicular to the end of the bar. Also, the spring has an unstretched length of 0.5 m and remains in the vertical position due to the roller guide at B. Determine the total work done by all the forces acting on the bar when it has rotated downward from $\theta = 0°$ to $\theta = 90°$.

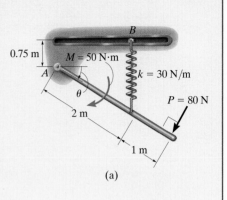

(a)

SOLUTION

First the free-body diagram of the bar is drawn in order to account for all the forces that act on it, Fig. 18–11b.

Weight W. Since the weight $10(9.81)$ N $= 98.1$ N is displaced downward 1.5 m, the work is

$$U_W = 98.1 \text{ N}(1.5 \text{ m}) = 147.2 \text{ J}$$

Why is the work positive?

Couple Moment M. The couple moment rotates through an angle of $\theta = \pi/2$ rad. Hence,

$$U_M = 50 \text{ N} \cdot \text{m}(\pi/2) = 78.5 \text{ J}$$

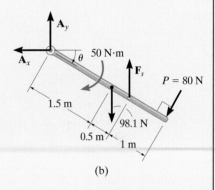

(b)

Fig. 18–11

Spring Force F_s. When $\theta = 0°$ the spring is stretched $(0.75 \text{ m} - 0.5 \text{ m}) = 0.25$ m, and when $\theta = 90°$, the stretch is $(2 \text{ m} + 0.75 \text{ m}) - 0.5$ m $= 2.25$ m. Thus,

$$U_s = -\left[\tfrac{1}{2}(30 \text{ N/m})(2.25 \text{ m})^2 - \tfrac{1}{2}(30 \text{ N/m})(0.25 \text{ m})^2\right] = -75.0 \text{ J}$$

By inspection the spring does negative work on the bar since $\mathbf{F}_s$ acts in the opposite direction to displacement. This checks with the result.

Force P. As the bar moves downward, the force is displaced through a distance of $(\pi/2)(3 \text{ m}) = 4.712$ m. The work is positive. Why?

$$U_P = 80 \text{ N}(4.712 \text{ m}) = 377.0 \text{ J}$$

Pin Reactions. Forces $\mathbf{A}_x$ and $\mathbf{A}_y$ do no work since they are not displaced.

Total Work. The work of all the forces when the bar is displaced is thus

$$U = 147.2 \text{ J} + 78.5 \text{ J} - 75.0 \text{ J} + 377.0 \text{ J} = 528 \text{ J} \qquad \textit{Ans.}$$

18

18.4   Principle of Work and Energy

By applying the principle of work and energy developed in Sec. 14.2 to each of the particles of a rigid body and adding the results algebraically, since energy is a scalar, the principle of work and energy for a rigid body becomes

$$T_1 + \Sigma U_{1-2} = T_2 \qquad\qquad (18\text{--}14)$$

This equation states that the body's initial translational *and* rotational kinetic energy, plus the work done by all the external forces and couple moments acting on the body as the body moves from its initial to its final position, is equal to the body's final translational *and* rotational kinetic energy. Note that the work of the body's *internal forces* does not have to be considered. These forces occur in equal but opposite collinear pairs, so that when the body moves, the work of one force cancels that of its counterpart. Furthermore, since the body is rigid, *no relative movement* between these forces occurs, so that no internal work is done.

When several rigid bodies are pin connected, connected by inextensible cables, or in mesh with one another, Eq. 18–14 can be applied to the *entire system* of connected bodies. In all these cases the internal forces, which hold the various members together, do no work and hence are eliminated from the analysis.

The work of the torque or moment developed by the driving gears on the motors is transformed into kinetic energy of rotation of the drum.

Procedure for Analysis

The principle of work and energy is used to solve kinetic problems that involve *velocity, force,* and *displacement,* since these terms are involved in the formulation. For application, it is suggested that the following procedure be used.

Kinetic Energy (Kinematic Diagrams).

- The kinetic energy of a body is made up of two parts. Kinetic energy of translation is referenced to the velocity of the mass center, $T = \frac{1}{2}mv_G^2$, and kinetic energy of rotation is determined using the moment of inertia of the body about the mass center, $T = \frac{1}{2}I_G\omega^2$. In the special case of rotation about a fixed axis (or rotation about the *IC*), these two kinetic energies are combined and can be expressed as $T = \frac{1}{2}I_O\omega^2$, where I_O is the moment of inertia about the axis of rotation.
- *Kinematic diagrams* for velocity may be useful for determining v_G and ω or for establishing a *relationship* between v_G and ω.*

Work (Free–Body Diagram).

- Draw a free–body diagram of the body when it is located at an intermediate point along the path in order to account for all the forces and couple moments which do work on the body as it moves along the path.
- A force does work when it moves through a displacement in the direction of the force.
- Forces that are functions of displacement must be integrated to obtain the work. Graphically, the work is equal to the area under the force–displacement curve.
- The work of a weight is the product of its magnitude and the vertical displacement, $U_W = Wy$. It is positive when the weight moves downwards.
- The work of a spring is of the form $U_s = \frac{1}{2}ks^2$, where k is the spring stiffness and s is the stretch or compression of the spring.
- The work of a couple is the product of the couple moment and the angle in radians through which it rotates, $U_M = M\theta$.
- Since *algebraic addition* of the work terms is required, it is important that the proper sign of each term be specified. Specifically, work is *positive* when the force (couple moment) is in the *same direction* as its displacement (rotation); otherwise, it is negative.

Principle of Work and Energy.

- Apply the principle of work and energy, $T_1 + \Sigma U_{1-2} = T_2$. Since this is a scalar equation, it can be used to solve for only one unknown when it is applied to a single rigid body.

*A brief review of Secs. 16.5 to 16.7 may prove helpful when solving problems, since computations for kinetic energy require a kinematic analysis of velocity.

18

EXAMPLE 18.2

The 30-kg disk shown in Fig. 18–12a is pin supported at its center. Determine the angle through which it must rotate to attain an angular velocity of 2 rad/s starting from rest. It is acted upon by a constant couple moment $M = 5$ N·m. The spring is orginally unstretched and its cord wraps around the rim of the disk.

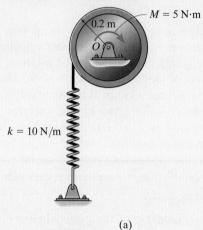

(a)

SOLUTION

Kinetic Energy.   Since the disk rotates about a fixed axis, and it is initially at rest, then

$$T_1 = 0$$

$$T_2 = \tfrac{1}{2}I_O\omega_2^2 = \tfrac{1}{2}\big[\tfrac{1}{2}(30 \text{ kg})(0.2 \text{ m})^2\big](2 \text{ rad/s})^2 = 1.2 \text{ J}$$

Work (Free–Body Diagram).   As shown in Fig. 18–12b, the pin reactions $\mathbf{O}_x$ and $\mathbf{O}_y$ and the weight (294.3 N) do no work, since they are not displaced. The *couple moment*, having a constant magnitude, does positive work $U_M = M\theta$ as the disk *rotates* through a clockwise angle of θ rad, and the spring does negative work $U_s = -\tfrac{1}{2}ks^2$.

Principle of Work and Energy.

$$\{T_1\} + \{\Sigma U_{1-2}\} = \{T_2\}$$

$$\{T_1\} + \Big\{M\theta - \tfrac{1}{2}ks^2\Big\} = \{T_2\}$$

$$\{0\} + \Big\{(5 \text{ N·m})\theta - \frac{1}{2}(10 \text{ N/m})[\theta(0.2 \text{ m})]^2\Big\} = \{1.2 \text{ J}\}$$

$$-0.2\theta^2 + 5\theta - 1.2 = 0$$

Solving this quadratic equation for the smallest positive root,

$$\theta = 0.2423 \text{ rad} = 0.2423 \text{ rad}\left(\frac{180°}{\pi \text{ rad}}\right) = 13.9° \qquad Ans.$$

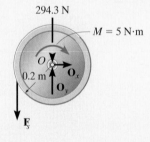

294.3 N

$M = 5$ N·m

0.2 m

$\mathbf{O}_x$

$\mathbf{O}_y$

$\mathbf{F}_s$

(b)

Fig. 18–12

EXAMPLE | 18.3

The wheel shown in Fig. 18–13a weighs 40 lb and has a radius of gyration $k_G = 0.6$ ft about its mass center G. If it is subjected to a clockwise couple moment of 15 lb·ft and rolls from rest without slipping, determine its angular velocity after its center G moves 0.5 ft. The spring has a stiffness $k = 10$ lb/ft and is initially unstretched when the couple moment is applied.

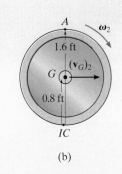

(a)

SOLUTION

Kinetic Energy (Kinematic Diagram). Since the wheel is initially at rest,

$$T_1 = 0$$

The kinematic diagram of the wheel when it is in the final position is shown in Fig. 18–13b. The final kinetic energy is determined from

$$T_2 = \tfrac{1}{2}I_{IC}\omega_2^2$$

$$= \frac{1}{2}\left[\frac{40 \text{ lb}}{32.2 \text{ ft/s}^2}(0.6 \text{ ft})^2 + \left(\frac{40 \text{ lb}}{32.2 \text{ ft/s}^2}\right)(0.8 \text{ ft})^2\right]\omega_2^2$$

$$T_2 = 0.6211\,\omega_2^2$$

(b)

Work (Free–Body Diagram). As shown in Fig. 18–13c, only the spring force $\mathbf{F}_s$ and the couple moment do work. The normal force does not move along its line of action and the frictional force does *no work*, since the wheel does not slip as it rolls.

The work of $\mathbf{F}_s$ is found using $U_s = -\tfrac{1}{2}ks^2$. Here the work is negative since $\mathbf{F}_s$ is in the opposite direction to displacement. Since the wheel does not slip when the center G moves 0.5 ft, then the wheel rotates $\theta = s_G/r_{G/IC} = 0.5 \text{ ft}/0.8 \text{ ft} = 0.625$ rad, Fig. 18–13b. Hence, the spring stretches $s = \theta r_{A/IC} = (0.625 \text{ rad})(1.6 \text{ ft}) = 1$ ft.

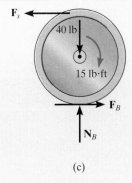

(c)

Fig. 18–13

Principle of Work and Energy.

$$\{T_1\} + \{\Sigma U_{1-2}\} = \{T_2\}$$

$$\{T_1\} + \{M\theta - \tfrac{1}{2}ks^2\} = \{T_2\}$$

$$\{0\} + \left\{15 \text{ lb} \cdot \text{ft}(0.625 \text{ rad}) - \frac{1}{2}(10 \text{ lb/ft})(1 \text{ ft})^2\right\} = \{0.6211\,\omega_2^2 \text{ ft} \cdot \text{lb}\}$$

$$\omega_2 = 2.65 \text{ rad/s} \;\circlearrowright \qquad\qquad Ans.$$

EXAMPLE 18.4

The 700-kg pipe is equally suspended from the two tines of the fork lift shown in the photo. It is undergoing a swinging motion such that when $\theta = 30°$ it is momentarily at rest. Determine the normal and frictional forces acting on each tine which are needed to support the pipe at the instant $\theta = 0°$. Measurements of the pipe and the suspender are shown in Fig. 18–14a. Neglect the mass of the suspender and the thickness of the pipe.

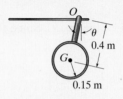

(a)

Fig. 18–14

SOLUTION

We must use the equations of motion to find the forces on the tines since these forces do no work. Before doing this, however, we will apply the principle of work and energy to determine the angular velocity of the pipe when $\theta = 0°$.

Kinetic Energy (Kinematic Diagram).   Since the pipe is originally at rest, then

$$T_1 = 0$$

The final kinetic energy may be computed with reference to either the fixed point O or the center of mass G. For the calculation we will consider the pipe to be a thin ring so that $I_G = mr^2$. If point G is considered, we have

$$T_2 = \tfrac{1}{2}m(v_G)_2^2 + \tfrac{1}{2}I_G\omega_2^2$$
$$= \tfrac{1}{2}(700 \text{ kg})[(0.4 \text{ m})\omega_2]^2 + \tfrac{1}{2}[700 \text{ kg}(0.15 \text{ m})^2]\omega_2^2$$
$$= 63.875\omega_2^2$$

If point O is considered then the parallel-axis theorem must be used to determine I_O. Hence,

$$T_2 = \tfrac{1}{2}I_O\omega_2^2 = \tfrac{1}{2}[700 \text{ kg}(0.15 \text{ m})^2 + 700 \text{ kg}(0.4 \text{ m})^2]\omega_2^2$$
$$= 63.875\omega_2^2$$

Work (Free-Body Diagram).   Fig. 18–14b. The normal and frictional forces on the tines do no work since they do not move as the pipe swings. The weight does positive work since the weight moves downward through a vertical distance $\Delta y = 0.4$ m $- 0.4 \cos 30°$ m $= 0.05359$ m.

Principle of Work and Energy.

$$\{T_1\} + \{\Sigma U_{1-2}\} = \{T_2\}$$

$$\{0\} + \{700(9.81) \text{ N}(0.05359 \text{ m})\} = \{63.875\omega_2^2\}$$

$$\omega_2 = 2.400 \text{ rad/s}$$

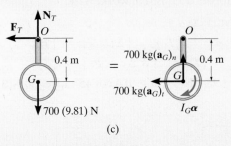

(b)

Equations of Motion.   Referring to the free-body and kinetic diagrams shown in Fig. 18–14c, and using the result for ω_2, we have

$$\xrightarrow{+} \Sigma F_t = m(a_G)_t; \quad F_T = (700 \text{ kg})(a_G)_t$$

$$+\uparrow \Sigma F_n = m(a_G)_n; \quad N_T - 700(9.81) \text{ N} = (700 \text{ kg})(2.400 \text{ rad/s})^2(0.4 \text{ m})$$

$$\zeta +\Sigma M_O = I_O\alpha; \quad 0 = [(700 \text{ kg})(0.15 \text{ m})^2 + (700 \text{ kg})(0.4 \text{ m})^2]\alpha$$

Since $(a_G)_t = (0.4 \text{ m})\alpha$, then

$$\alpha = 0, \quad (a_G)_t = 0$$

$$F_T = 0$$

$$N_T = 8.480 \text{ kN}$$

There are two tines used to support the load, therefore

$$F'_T = 0 \qquad\qquad Ans.$$

$$N'_T = \frac{8.480 \text{ kN}}{2} = 4.24 \text{ kN} \qquad\qquad Ans.$$

NOTE: Due to the swinging motion the tines are subjected to a *greater* normal force than would be the case if the load were static, in which case $N'_T = 700(9.81) \text{ N}/2 = 3.43$ kN.

(c)

Fig. 18–14

EXAMPLE 18.5

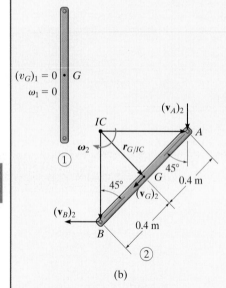

P = 50 N

B

(a)

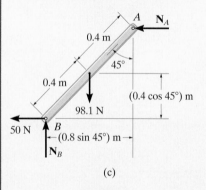

(b)

(c)

Fig. 18–15

The 10–kg rod shown in Fig. 18–15a is constrained so that its ends move along the grooved slots. The rod is initially at rest when $\theta = 0°$. If the slider block at B is acted upon by a horizontal force $P = 50$ N, determine the angular velocity of the rod at the instant $\theta = 45°$. Neglect friction and the mass of blocks A and B.

SOLUTION

Why can the principle of work and energy be used to solve this problem?

Kinetic Energy (Kinematic Diagrams). Two kinematic diagrams of the rod, when it is in the initial position 1 and final position 2, are shown in Fig. 18–15b. When the rod is in position 1, $T_1 = 0$ since $(\mathbf{v}_G)_1 = \boldsymbol{\omega}_1 = \mathbf{0}$. In position 2 the angular velocity is $\boldsymbol{\omega}_2$ and the velocity of the mass center is $(\mathbf{v}_G)_2$. Hence, the kinetic energy is

$$T_2 = \tfrac{1}{2}m(v_G)_2^2 + \tfrac{1}{2}I_G\omega_2^2$$
$$= \tfrac{1}{2}(10 \text{ kg})(v_G)_2^2 + \tfrac{1}{2}\big[\tfrac{1}{12}(10 \text{ kg})(0.8 \text{ m})^2\big]\omega_2^2$$
$$= 5(v_G)_2^2 + 0.2667(\omega_2)^2$$

The two unknowns $(v_G)_2$ and ω_2 can be related from the instantaneous center of zero velocity for the rod. Fig. 18–15b. It is seen that as A moves downward with a velocity $(\mathbf{v}_A)_2$, B moves horizontally to the left with a velocity $(\mathbf{v}_B)_2$, Knowing these directions, the IC is located as shown in the figure. Hence,

$$(v_G)_2 = r_{G/IC}\omega_2 = (0.4 \tan 45° \text{ m})\omega_2$$
$$= 0.4\omega_2$$

Therefore,

$$T_2 = 0.8\omega_2^2 + 0.2667\omega_2^2 = 1.0667\omega_2^2$$

Of course, we can also determine this result using $T_2 = \tfrac{1}{2}I_{IC}\omega_2^2$.

Work (Free–Body Diagram). Fig. 18–15c. The normal forces $\mathbf{N}_A$ and $\mathbf{N}_B$ do no work as the rod is displaced. Why? The 98.1-N weight is displaced a vertical distance of $\Delta y = (0.4 - 0.4 \cos 45°)$ m; whereas the 50-N force moves a horizontal distance of $s = (0.8 \sin 45°)$ m. Both of these forces do positive work. Why?

Principle of Work and Energy.

$$\{T_1\} + \{\Sigma U_{1-2}\} = \{T_2\}$$
$$\{T_1\} + \{W\,\Delta y + Ps\} = \{T_2\}$$
$$\{0\} + \{98.1 \text{ N}(0.4 \text{ m} - 0.4 \cos 45° \text{ m}) + 50 \text{ N}(0.8 \sin 45° \text{ m})\}$$
$$= \{1.0667\omega_2^2 \text{ J}\}$$

Solving for ω_2 gives

$$\omega_2 = 6.11 \text{ rad/s} \circlearrowright \qquad \qquad Ans.$$

FUNDAMENTAL PROBLEMS

F18–1. The 80-kg wheel has a radius of gyration about its mass center O of $k_O = 400$ mm. Determine its angular velocity after it has rotated 20 revolutions starting from rest.

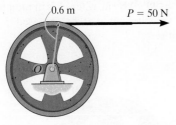

F18–1

F18–2. The uniform 50-lb slender rod is subjected to a couple moment of $M = 100$ lb · ft. If the rod is at rest when $\theta = 0°$, determine its angular velocity when $\theta = 90°$.

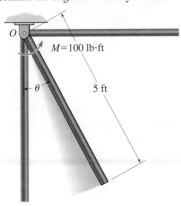

F18–2

F18–3. The uniform 50-kg slender rod is at rest in the position shown when $P = 600$ N is applied. Determine the angular velocity of the rod when the rod reaches the vertical position.

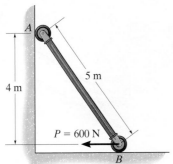

F18–3

F18–4. The 50-kg wheel is subjected to a force of 50 N. If the wheel starts from rest and rolls without slipping, determine   its angular velocity after it has rotated 10 revolutions. The radius of gyration of the wheel about its mass center O is $k_O = 0.3$ m.

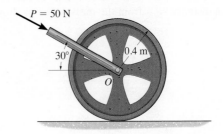

F18–4

F18–5. If the uniform 30-kg slender rod starts from rest at the position shown, determine its angular velocity after it has rotated 4 revolutions. The forces remain perpendicular to the rod.

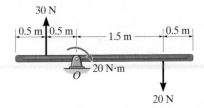

F18–5

F18–6. The 20-kg wheel has a radius of gyration about its center O of $k_O = 300$ mm. When it is subjected to a couple moment of $M = 50$ N · m, it rolls without slipping. Determine the angular velocity of the wheel after its center O has traveled through a distance of $s_O = 20$ m, starting from rest.

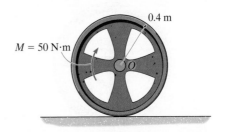

F18–6

18

PROBLEMS

18–1. At a given instant the body of mass m has an angular velocity $\boldsymbol{\omega}$ and its mass center has a velocity $\mathbf{v}_G$. Show that its kinetic energy can be represented as $T = \frac{1}{2}I_{IC}\omega^2$, where I_{IC} is the moment of inertia of the body determined about the instantaneous axis of zero velocity, located a distance $r_{G/IC}$ from the mass center as shown.

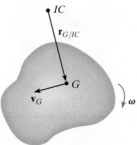

Prob. 18–1

18–2. The wheel is made from a 5-kg thin ring and two 2-kg slender rods. If the torsional spring attached to the wheel's center has a stiffness $k = 2$ N·m/rad, and the wheel is rotated until the torque $M = 25$ N·m is developed, determine the maximum angular velocity of the wheel if it is released from rest.

18–3. The wheel is made from a 5-kg thin ring and two 2-kg slender rods. If the torsional spring attached to the wheel's center has a stiffness $k = 2$ N·m/rad, so that the torque on the center of the wheel is $M = (2\theta)$ N·m, where θ is in radians, determine the maximum angular velocity of the wheel if it is rotated two revolutions and then released from rest.

Probs. 18–2/3

*18–4. The 50-kg flywheel has a radius of gyration of $k_0 = 200$ mm about its center of mass. If it is subjected to a torque of $M = (9\theta^{1/2})$ N·m, where θ is in radians, determine its angular velocity when it has rotated 5 revolutions, starting from rest.

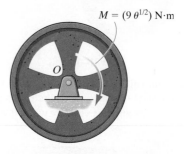

Prob. 18–4

18–5. The spool has a mass of 60 kg and a radius of gyration $k_G = 0.3$ m. If it is released from rest, determine how far its center descends down the smooth plane before it attains an angular velocity of $\omega = 6$ rad/s. Neglect friction and the mass of the cord which is wound around the central core.

18–6. Solve Prob. 18–5 if the coefficient of kinetic friction between the spool and plane at A is $\mu_k = 0.2$.

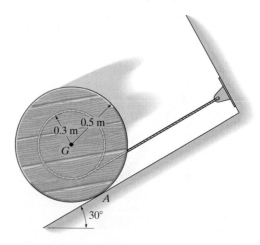

Probs. 18–5/6

18–7. The double pulley consists of two parts that are attached to one another. It has a weight of 50 lb and a centroidal radius of gyration of $k_O = 0.6$ ft and is turning with an angular velocity of 20 rad/s clockwise. Determine the kinetic energy of the system. Assume that neither cable slips on the pulley.

***18–8.** The double pulley consists of two parts that are attached to one another. It has a weight of 50 lb and a centroidal radius of gyration of $k_O = 0.6$ ft and is turning with an angular velocity of 20 rad/s clockwise. Determine the angular velocity of the pulley at the instant the 20-lb weight moves 2 ft downward.

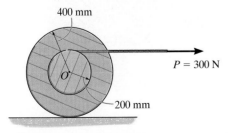

Probs. 18–7/8

18–9. If the cable is subjected to force of $P = 300$ N, and the spool starts from rest, determine its angular velocity after its center of mass O has moved 1.5 m. The mass of the spool is 100 kg and its radius of gyration about its center of mass is $k_O = 275$ mm. Assume that the spool rolls without slipping.

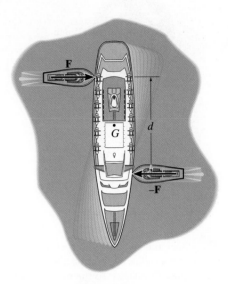

Prob. 18–9

18–10. The two tugboats each exert a constant force **F** on the ship. These forces are always directed perpendicular to the ship's centerline. If the ship has a mass m and a radius of gyration about its center of mass G of k_G, determine the angular velocity of the ship after it turns 90°. The ship is originally at rest.

Prob. 18–10

18–11. At the instant shown, link AB has an angular velocity $\omega_{AB} = 2$ rad/s. If each link is considered as a uniform slender bar with a weight of 0.5 lb/in., determine the total kinetic energy of the system.

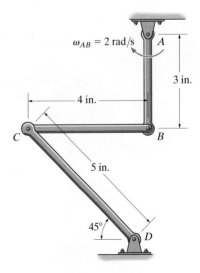

Prob. 18–11

18

*18–12. Determine the velocity of the 50-kg cylinder after it has descended a distance of 2 m. Initially, the system is at rest. The reel has a mass of 25 kg and a radius of gyration about its center of mass A of $k_A = 125$ mm.

75 mm

Prob. 18–12

18–13. The wheel and the attached reel have a combined weight of 50 lb and a radius of gyration about their center of $k_A = 6$ in. If pulley B attached to the motor is subjected to a torque of $M = 40(2 - e^{-0.1\theta})$ lb·ft, where θ is in radians, determine the velocity of the 200-lb crate after it has moved upwards a distance of 5 ft, starting from rest. Neglect the mass of pulley B.

18–14. The wheel and the attached reel have a combined weight of 50 lb and a radius of gyration about their center of $k_A = 6$ in. If pulley B that is attached to the motor is subjected to a torque of $M = 50$ lb·ft, determine the velocity of the 200-lb crate after the pulley has turned 5 revolutions. Neglect the mass of the pulley.

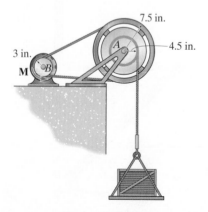

7.5 in.

3 in.

M

A

4.5 in.

B

Probs. 18–13/14

18–15. The 50-kg gear has a radius of gyration of 125 mm about its center of mass O. If gear rack B is stationary, while the 25-kg gear rack C is subjected to a horizontal force of $P = 150$ N, determine the speed of C after the gear's center O has moved to the right a distance of 0.3 m, starting from rest.

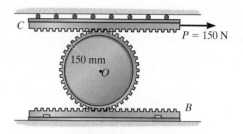

C

$P = 150$ N

150 mm

O

B

Prob. 18–15

*18–16. Gear B is rigidly attached to drum A and is supported by two small rollers at E and D. Gear B is in mesh with gear C and is subjected to a torque of $M = 50$ N·m. Determine the angular velocity of the drum after C has rotated 10 revolutions, starting from rest. Gear B and the drum have 100 kg and a radius of gyration about their rotating axis of 250 mm. Gear C has a mass of 30 kg and a radius of gyration about its rotating axis of 125 mm.

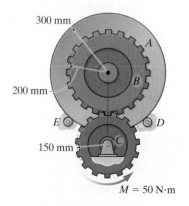

300 mm

A

200 mm

B

E

D

150 mm

C

$M = 50$ N·m

Prob. 18–16

18–17.   The center O of the thin ring of mass m is given an angular velocity of ω_0. If the ring rolls without slipping, determine its angular velocity after it has traveled a distance of s down the plane. Neglect its thickness.

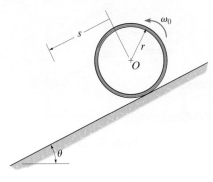

Prob. 18–17

18–19.   When $\theta = 0°$, the assembly is held at rest, and the torsional spring is untwisted. If the assembly is released and falls downward, determine its angular velocity at the instant $\theta = 90°$. Rod AB has a mass of 6 kg, and disk C has a mass of 9 kg.

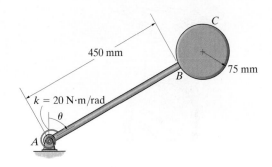

Prob. 18–19

18–18.   If the end of the cord is subjected to a force of $P = 75$ lb, determine the speed of the 100-lb block C after P has moved a distance of 4 ft, starting from rest. Pulleys A and B are identical, each of which has a weight of 10 lb and a radius of gyration of $k = 3$ in. about its center of mass.

***18–20.**   If $P = 200$ N and the 15-kg uniform slender rod starts from rest at $\theta = 0°$, determine the rod's angular velocity at the instant just before $\theta = 45°$.

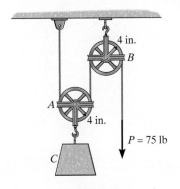

Prob. 18–18

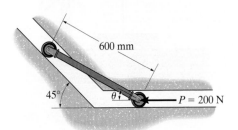

Prob. 18–20

18–21. A yo-yo has a weight of 0.3 lb and a radius of gyration $k_O = 0.06$ ft. If it is released from rest, determine how far it must descend in order to attain an angular velocity $\omega = 70$ rad/s. Neglect the mass of the string and assume that the string is wound around the central peg such that the mean radius at which it unravels is $r = 0.02$ ft.

18–23. The combined weight of the load and the platform is 200 lb, with the center of gravity located at G. If a couple moment of $M = 900$ lb · ft is applied to link AB, determine the angular velocity of links AB and CD at the instant $\theta = 60°$. The system is at rest when $\theta = 0°$. Neglect the weight of the links.

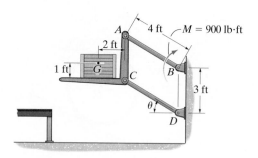

Prob. 18–23

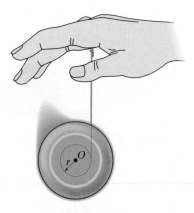

Prob. 18–21

18–22. If the 50-lb bucket is released from rest, determine its velocity after it has fallen a distance of 10 ft. The windlass A can be considered as a 30-lb cylinder, while the spokes are slender rods, each having a weight of 2 lb. Neglect the pulley's weight.

*18–24. The tub of the mixer has a weight of 70 lb and a radius of gyration $k_G = 1.3$ ft about its center of gravity. If a constant torque $M = 60$ lb · ft is applied to the dumping wheel, determine the angular velocity of the tub when it has rotated $\theta = 90°$. Originally the tub is at rest when $\theta = 0°$.

18–25. The tub of the mixer has a weight of 70 lb and a radius of gyration $k_G = 1.3$ ft about its center of gravity. If a constant torque $M = 60$ lb · ft is applied to the tub, determine its angular velocity when it has rotated $\theta = 45°$. Originally the tub is at rest when $\theta = 0°$.

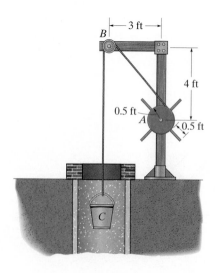

Prob. 18–22

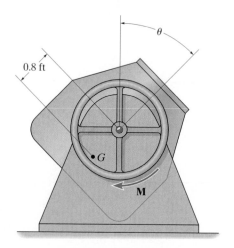

Probs. 18–24/25

18–26. Two wheels of negligible weight are mounted at corners A and B of the rectangular 75-lb plate. If the plate is released from rest at $\theta = 90°$, determine its angular velocity at the instant just before $\theta = 0°$.

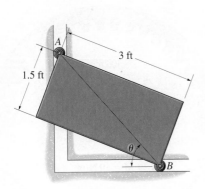

Prob. 18–26

18–27. The 100-lb block is transported a short distance by using two cylindrical rollers, each having a weight of 35 lb. If a horizontal force $P = 25$ lb is applied to the block, determine the block's speed after it has been displaced 2 ft to the left. Originally the block is at rest. No slipping occurs.

Prob. 18–27

*18–28. The hand winch is used to lift the 50-kg load. Determine the work required to rotate the handle five revolutions, starting and ending at rest. The gear at A has a radius of 20 mm.

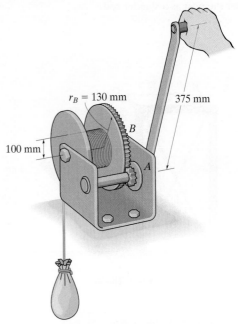

Prob. 18–28

18–29. A motor supplies a constant torque or twist of $M = 120$ lb · ft to the drum. If the drum has a weight of 30 lb and a radius of gyration of $k_O = 0.8$ ft, determine the speed of the 15-lb crate A after it rises $s = 4$ ft starting from rest. Neglect the mass of the cord.

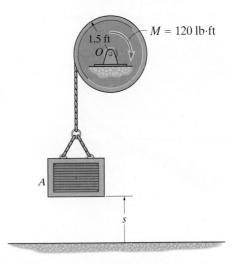

Prob. 18–29

18

18–30. Motor M exerts a constant force of $P = 750$ N on the rope. If the 100-kg post is at rest when $\theta = 0°$, determine the angular velocity of the post at the instant $\theta = 60°$. Neglect the mass of the pulley and its size, and consider the post as a slender rod.

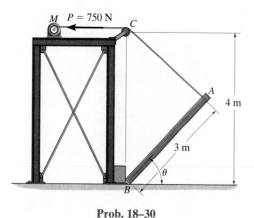

Prob. 18–30

18–31. The uniform bar has a mass m and length l. If it is released from rest when $\theta = 0°$, determine its angular velocity as a function of the angle θ before it slips.

*18–32. The uniform bar has a mass m and length l. If it is released from rest when $\theta = 0°$, determine the angle θ at which it first begins to slip. The coefficient of static friction at O is $\mu_s = 0.3$.

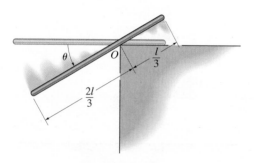

Probs. 18–31/32

18–33. The two 2-kg gears A and B are attached to the ends of a 3-kg slender bar. The gears roll within the fixed ring gear C, which lies in the horizontal plane. If a 10-N·m torque is applied to the center of the bar as shown, determine the number of revolutions the bar must rotate starting from rest in order for it to have an angular velocity of $\omega_{AB} = 20$ rad/s. For the calculation, assume the gears can be approximated by thin disks. What is the result if the gears lie in the vertical plane?

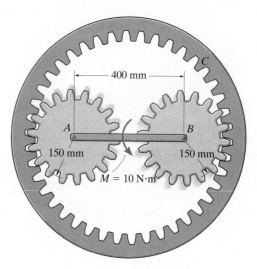

Prob. 18–33

18–34. A ball of mass m and radius r is cast onto the horizontal surface such that it rolls without slipping. Determine its angular velocity at the instant $\theta = 90°$, if it has an initial speed of v_G as shown.

18–35. A ball of mass m and radius r is cast onto the horizontal surface such that it rolls without slipping. Determine the minimum speed v_G of its mass center G so that it rolls completely around the loop of radius $R + r$ without leaving the track.

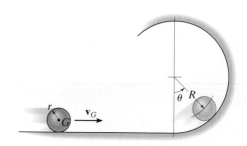

Probs. 18–34/35

18.5   Conservation of Energy

When a force system acting on a rigid body consists only of *conservative forces*, the conservation of energy theorem can be used to solve a problem that otherwise would be solved using the principle of work and energy. This theorem is often easier to apply since the work of a conservative force is *independent of the path* and depends only on the initial and final positions of the body. It was shown in Sec. 14.5 that the work of a conservative force can be expressed as the difference in the body's potential energy measured from an arbitrarily selected reference or datum.

Gravitational Potential Energy. Since the total weight of a body can be considered concentrated at its center of gravity, the *gravitational potential energy* of the body is determined by knowing the height of the body's center of gravity above or below a horizontal datum.

$$V_g = Wy_G \qquad (18\text{--}15)$$

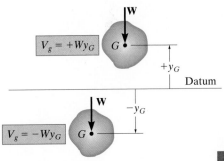

Gravitational potential energy

Fig. 18–16

Here the potential energy is *positive* when y_G is positive upward, since the weight has the ability to do *positive work* when the body moves back to the datum, Fig. 18–16. Likewise, if G is located *below* the datum ($-y_G$), the gravitational potential energy is *negative*, since the weight does *negative work* when the body returns to the datum.

Elastic Potential Energy. The force developed by an elastic spring is also a conservative force. The *elastic potential energy* which a spring imparts to an attached body when the spring is stretched or compressed from an initial undeformed position ($s = 0$) to a final position s, Fig. 18–17, is

$$V_e = +\tfrac{1}{2}ks^2 \qquad (18\text{--}16)$$

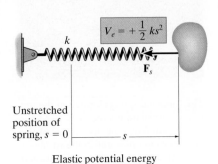

Elastic potential energy

Fig. 18–17

In the deformed position, the spring force acting *on the body* always has the ability for doing positive work when the spring returns back to its original undeformed position (see Sec. 14.5).

Conservation of Energy.

In general, if a body is subjected to both gravitational and elastic forces, the total *potential energy* can be expressed as a potential function represented as the algebraic sum

$$V = V_g + V_e \qquad (18\text{--}17)$$

Here measurement of V depends upon the location of the body with respect to the selected datum.

Realizing that the work of conservative forces can be written as a difference in their potential energies, i.e., $(\Sigma U_{1-2})_{cons} = V_1 - V_2$, Eq. 14–16, we can rewrite the principle of work and energy for a rigid body as

$$T_1 + V_1 + (\Sigma U_{1-2})_{noncons} = T_2 + V_2 \qquad (18\text{--}18)$$

Here $(\Sigma U_{1-2})_{noncons}$ represents the work of the nonconservative forces such as friction. If this term is zero, then

$$T_1 + V_1 = T_2 + V_2 \qquad (18\text{--}19)$$

This equation is referred to as the conservation of mechanical energy. It states that the *sum* of the potential and kinetic energies of the body remains *constant* when the body moves from one position to another. It also applies to a system of smooth, pin-connected rigid bodies, bodies connected by inextensible cords, and bodies in mesh with other bodies. In all these cases the forces acting at the points of contact are *eliminated* from the analysis, since they occur in equal but opposite collinear pairs and each pair of forces moves through an equal distance when the system undergoes a displacement.

It is important to remember that only problems involving conservative force systems can be solved by using Eq. 18–19. As stated in Sec. 14.5, friction or other drag-resistant forces, which depend on velocity or acceleration, are nonconservative. The work of such forces is transformed into thermal energy used to heat up the surfaces of contact, and consequently this energy is dissipated into the surroundings and may not be recovered. Therefore, problems involving frictional forces can be solved by using either the principle of work and energy written in the form of Eq. 18–18, if it applies, or the equations of motion.

The torsional springs located at the top of the garage door wind up as the door is lowered. When the door is raised, the potential energy stored in the springs is then transferred into gravitational potential energy of the door's weight, thereby making it easy to open.

Procedure for Analysis

The conservation of energy equation is used to solve problems involving *velocity, displacement*, and *conservative force systems*. For application it is suggested that the following procedure be used.

Potential Energy.

- Draw two diagrams showing the body located at its initial and final positions along the path.

- If the center of gravity, G, is subjected to a *vertical displacement*, establish a fixed horizontal datum from which to measure the body's gravitational potential energy V_g.

- Data pertaining to the elevation y_G of the body's center of gravity from the datum and the extension or compression of any connecting springs can be determined from the problem geometry and listed on the two diagrams.

- The potential energy is determined from $V = V_g + V_e$. Here $V_g = Wy_G$, which can be positive or negative, and $V_e = \frac{1}{2}ks^2$, which is always positive.

Kinetic Energy.

- The kinetic energy of the body consists of two parts, namely translational kinetic energy, $T = \frac{1}{2}mv_G^2$, and rotational kinetic energy, $T = \frac{1}{2}I_G\omega^2$.

- Kinematic diagrams for velocity may be useful for establishing a *relationship* between v_G and ω.

Conservation of Energy.

- Apply the conservation of energy equation $T_1 + V_1 = T_2 + V_2$.

18

EXAMPLE 18.6

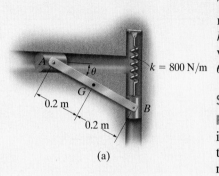

0.2 m

0.2 m

(a)

The 10-kg rod AB shown in Fig. 18–18a is confined so that its ends move in the horizontal and vertical slots. The spring has a stiffness of $k = 800$ N/m and is unstretched when $\theta = 0°$. Determine the angular velocity of AB when $\theta = 0°$, if the rod is released from rest when $\theta = 30°$. Neglect the mass of the slider blocks.

SOLUTION

Potential Energy. The two diagrams of the rod, when it is located at its initial and final positions, are shown in Fig. 18–18b. The datum, used to measure the gravitational potential energy, is placed in line with the rod when $\theta = 0°$.

When the rod is in position 1, the center of gravity G is located *below the datum* so its gravitational potential energy is *negative*. Furthermore, (positive) elastic potential energy is stored in the spring, since it is stretched a distance of $s_1 = (0.4 \sin 30°)$ m. Thus,

$$V_1 = -Wy_1 + \tfrac{1}{2}ks_1^2$$

$$= -(98.1 \text{ N})(0.2 \sin 30° \text{ m}) + \tfrac{1}{2}(800 \text{ N/m})(0.4 \sin 30° \text{ m})^2 = 6.19 \text{ J}$$

When the rod is in position 2, the potential energy of the rod is zero, since the center of gravity G is located at the datum, and the spring is unstretched, $s_2 = 0$. Thus,

$$V_2 = 0$$

Kinetic Energy. The rod is released from rest from position 1, thus $(\mathbf{v}_G)_1 = \boldsymbol{\omega}_1 = \mathbf{0}$, and so

$$T_1 = 0$$

In position 2, the angular velocity is $\boldsymbol{\omega}_2$ and the rod's mass center has a velocity of $(\mathbf{v}_G)_2$. Thus,

$$T_2 = \tfrac{1}{2}m(v_G)_2^2 + \tfrac{1}{2}I_G\omega_2^2$$

$$= \tfrac{1}{2}(10 \text{ kg})(v_G)_2^2 + \tfrac{1}{2}\left[\tfrac{1}{12}(10 \text{ kg})(0.4 \text{ m})^2\right]\omega_2^2$$

Using *kinematics*, $(\mathbf{v}_G)_2$ can be related to $\boldsymbol{\omega}_2$ as shown in Fig. 18–18c. At the instant considered, the instantaneous center of zero velocity (*IC*) for the rod is at point A; hence, $(v_G)_2 = (r_{G/IC})\omega_2 = (0.2 \text{ m})\omega_2$. Substituting into the above expression and simplifying (or using $\tfrac{1}{2}I_{IC}\omega_2^2$), we get

$$T_2 = 0.2667\omega_2^2$$

Conservation of Energy.

$$\{T_1\} + \{V_1\} = \{T_2\} + \{V_2\}$$

$$\{0\} + \{6.19 \text{ J}\} = \{0.2667\omega_2^2\} + \{0\}$$

$$\omega_2 = 4.82 \text{ rad/s} \curvearrowright \qquad Ans.$$

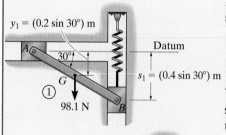

$y_1 = (0.2 \sin 30°)$ m

Datum

$s_1 = (0.4 \sin 30°)$ m

30°

98.1 N

①

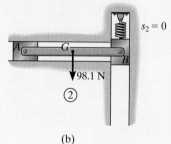

$s_2 = 0$

98.1 N

②

(b)

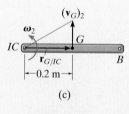

$(\mathbf{v}_G)_2$

$\boldsymbol{\omega}_2$

IC

$\mathbf{r}_{G/IC}$

0.2 m

(c)

Fig. 18–18

EXAMPLE 18.7

The wheel shown in Fig. 18–19a has a weight of 30 lb and a radius of gyration of $k_G = 0.6$ ft. It is attached to a spring which has a stiffness $k = 2$ lb/ft and an unstretched length of 1 ft. If the disk is released from rest in the position shown and rolls without slipping, determine its angular velocity at the instant G moves 3 ft to the left.

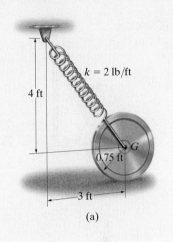

(a)

SOLUTION

Potential Energy. Two diagrams of the wheel, when it at the initial and final positions, are shown in Fig. 18–19b. A gravitational datum is not needed here since the weight is not displaced vertically. From the problem geometry the spring is stretched $s_1 = \left(\sqrt{3^2 + 4^2} - 1 \right) = 4$ ft in the initial position, and spring $s_2 = (4 - 1) = 3$ ft in the final position. Hence, the positive spring potential energy is

$$V_1 = \tfrac{1}{2}ks_1^2 = \tfrac{1}{2}(2 \text{ lb/ft})(4 \text{ ft})^2 = 16 \text{ ft} \cdot \text{lb}$$
$$V_2 = \tfrac{1}{2}ks_2^2 = \tfrac{1}{2}(2 \text{ lb/ft})(3 \text{ ft})^2 = 9 \text{ ft} \cdot \text{lb}$$

Kinetic Energy. The disk is released from rest and so $(v_G)_1 = \mathbf{0}$, $\boldsymbol{\omega}_1 = \mathbf{0}$. Therefore,

$$T_1 = 0$$

Since the instantaneous center of zero velocity is at the ground, Fig. 18–19c, we have

$$T_2 = \frac{1}{2}I_{IC}\omega_2^2$$
$$= \frac{1}{2}\left[\left(\frac{30 \text{ lb}}{32.2 \text{ ft/s}^2} \right)(0.6 \text{ ft})^2 + \left(\frac{30 \text{ lb}}{32.2 \text{ ft/s}^2} \right)(0.75 \text{ ft})^2 \right]\omega_2^2$$
$$= 0.4297\omega_2^2$$

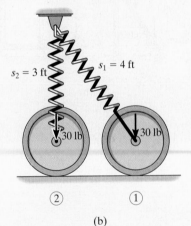

(b)

Conservation of Energy.

$$\{T_1\} + \{V_1\} = \{T_2\} + \{V_2\}$$
$$\{0\} + \{16 \text{ ft} \cdot \text{lb}\} = \{0.4297\omega_2^2\} + \{9 \text{ ft} \cdot \text{lb}\}$$
$$\omega_2 = 4.04 \text{ rad/s} \; \circlearrowright \qquad\qquad Ans.$$

(c)

Fig. 18–19

NOTE: If the principle of work and energy were used to solve this problem, then the work of the spring would have to be determined by considering both the change in magnitude and direction of the spring force.

EXAMPLE 18.8

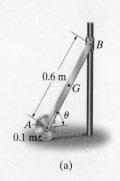

(a)

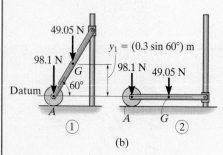

(b)

The 10-kg homogeneous disk shown in Fig. 18–20a is attached to a uniform 5-kg rod AB. If the assembly is released from rest when $\theta = 60°$, determine the angular velocity of the rod when $\theta = 0°$. Assume that the disk rolls without slipping. Neglect friction along the guide and the mass of the collar at B.

SOLUTION

Potential Energy. Two diagrams for the rod and disk, when they are located at their initial and final positions, are shown in Fig. 18–20b. For convenience the datum passes through point A.

When the system is in position 1, only the rod's weight has positive potential energy. Thus,

$$V_1 = W_r y_1 = (49.05 \text{ N})(0.3 \sin 60° \text{ m}) = 12.74 \text{ J}$$

When the system is in position 2, both the weight of the rod and the weight of the disk have zero potential energy. Why? Thus,

$$V_2 = 0$$

Kinetic Energy. Since the entire system is at rest at the initial position,

$$T_1 = 0$$

In the final position the rod has an angular velocity $(\omega_r)_2$ and its mass center has a velocity $(v_G)_2$, Fig. 18–20c. Since the rod is *fully extended* in this position, the disk is momentarily at rest, so $(\omega_d)_2 = 0$ and $(v_A)_2 = 0$. For the rod $(v_G)_2$ can be related to $(\omega_r)_2$ from the instantaneous center of zero velocity, which is located at point A, Fig. 18–20c. Hence, $(v_G)_2 = r_{G/IC}(\omega_r)_2$ or $(v_G)_2 = 0.3(\omega_r)_2$. Thus,

$$T_2 = \frac{1}{2}m_r(v_G)_2^2 + \frac{1}{2}I_G(\omega_r)_2^2 + \frac{1}{2}m_d(v_A)_2^2 + \frac{1}{2}I_A(\omega_d)_2^2$$

$$= \frac{1}{2}(5 \text{ kg})[(0.3 \text{ m})(\omega_r)_2]^2 + \frac{1}{2}\left[\frac{1}{12}(5 \text{ kg})(0.6 \text{ m})^2\right](\omega_r)_2^2 + 0 + 0$$

$$= 0.3(\omega_r)_2^2$$

Conservation of Energy.

$$\{T_1\} + \{V_1\} = \{T_2\} + \{V_2\}$$

$$\{0\} + \{12.74 \text{ J}\} = \{0.3(\omega_r)_2^2\} + \{0\}$$

$$(\omega_r)_2 = 6.52 \text{ rad/s} \circlearrowright \qquad Ans.$$

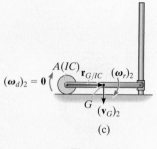

(c)

Fig. 18–20

NOTE: We can also determine the final kinetic energy of the rod using $T_2 = \frac{1}{2}I_{IC}\omega_2^2$.

FUNDAMENTAL PROBLEMS

F18–7. If the 30-kg disk is released from rest when $\theta = 0°$, determine its angular velocity when $\theta = 90°$.

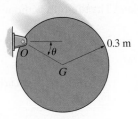

F18–7

F18–8. The 50-kg reel has a radius of gyration about its center O of $k_O = 300$ mm. If it is released from rest, determine its angular velocity when its center O has traveled 6 m down the smooth inclined plane.

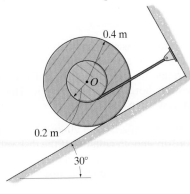

F18–8

F18–9. The 60-kg rod OA is released from rest when $\theta = 0°$. Determine its angular velocity when $\theta = 45°$. The spring remains vertical during the motion and is unstretched when $\theta = 0°$.

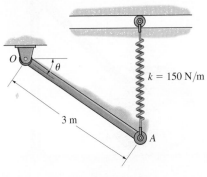

F18–9

F18–10. The 30-kg rod is released from rest when $\theta = 0°$. Determine the angular velocity of the rod when $\theta = 90°$. The spring is unstretched when $\theta = 0°$.

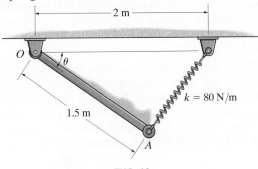

F18–10

F18–11. The 30-kg rod is released from rest when $\theta = 45°$. Determine the angular velocity of the rod when $\theta = 0°$. The spring is unstretched when $\theta = 45°$.

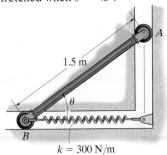

F18–11

F18–12. The 20-kg rod is released from rest when $\theta = 0°$. Determine its angular velocity when $\theta = 90°$. The spring has an unstretched length of 0.5 m.

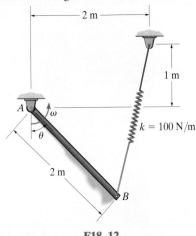

F18–12

18

PROBLEMS

*18–36. At the instant shown, the 50-lb bar rotates clockwise at 2 rad/s. The spring attached to its end always remains vertical due to the roller guide at C. If the spring has an unstretched length of 2 ft and a stiffness of $k = 6$ lb/ft, determine the angular velocity of the bar the instant it has rotated 30° clockwise.

18–37. At the instant shown, the 50-lb bar rotates clockwise at 2 rad/s. The spring attached to its end always remains vertical due to the roller guide at C. If the spring has an unstretched length of 2 ft and a stiffness of $k = 12$ lb/ft, determine the angle θ, measured from the horizontal, to which the bar rotates before it momentarily stops.

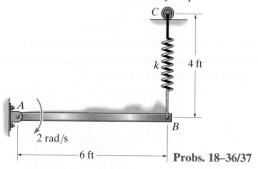

Probs. 18–36/37

18–38. The spool has a mass of 50 kg and a radius of gyration $k_O = 0.280$ m. If the 20-kg block A is released from rest, determine the distance the block must fall in order for the spool to have an angular velocity $\omega = 5$ rad/s. Also, what is the tension in the cord while the block is in motion? Neglect the mass of the cord.

18–39. The spool has a mass of 50 kg and a radius of gyration $k_O = 0.280$ m. If the 20-kg block A is released from rest, determine the velocity of the block when it descends 0.5 m.

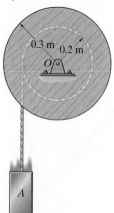

Probs. 18–38/39

*18–40. An automobile tire has a mass of 7 kg and radius of gyration $k_G = 0.3$ m. If it is released from rest at A on the incline, determine its angular velocity when it reaches the horizontal plane. The tire rolls without slipping.

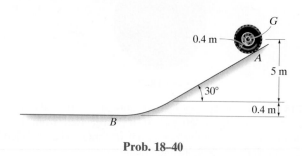

Prob. 18–40

18–41. The system consists of a 20-lb disk A, 4-lb slender rod BC, and a 1-lb smooth collar C. If the disk rolls without slipping, determine the velocity of the collar at the instant the rod becomes horizontal, i.e., $\theta = 0°$. The system is released from rest when $\theta = 45°$.

18–42. The system consists of a 20-lb disk A, 4-lb slender rod BC, and a 1-lb smooth collar C. If the disk rolls without slipping, determine the velocity of the collar at the instant $\theta = 30°$. The system is released from rest when $\theta = 45°$.

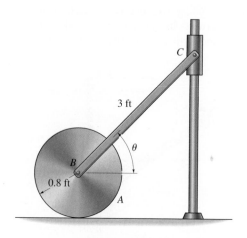

Probs. 18–41/42

18–43. The door is made from one piece, whose sides move along the horizontal and vertical tracks. If the door is in the open position, $\theta = 0°$, and then released, determine the speed at which its end A strikes the stop at C. Assume the door is a 180-lb thin plate having a width of 10 ft.

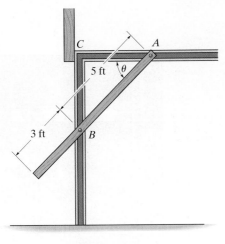

Prob. 18–43

*18–44. Determine the speed of the 50-kg cylinder after it has descended a distance of 2 m, starting from rest. Gear A has a mass of 10 kg and a radius of gyration of 125 mm about its center of mass. Gear B and drum C have a combined mass of 30 kg and a radius of gyration about their center of mass of 150 mm.

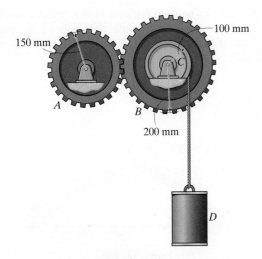

Prob. 18–44

18–45. The disk A is pinned at O and weighs 15 lb. A 1-ft rod weighing 2 lb and a 1-ft-diameter sphere weighing 10 lb are welded to the disk, as shown. If the spring is orginally stretched 1 ft and the sphere is released from the position shown, determine the angular velocity of the disk when it has rotated 90°.

18–46. The disk A is pinned at O and weighs 15 lb. A 1-ft rod weighing 2 lb and a 1-ft-diameter sphere weighing 10 lb are welded to the disk, as shown. If the spring is originally stretched 1 ft and the sphere is released from the position shown, determine the angular velocity of the disk when it has rotated 45°.

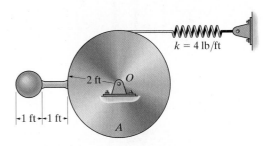

Probs. 18–45/46

18–47. At the instant the spring becomes undeformed, the center of the 40-kg disk has a speed of 4 m/s. From this point determine the distance d the disk moves down the plane before momentarily stopping. The disk rolls without slipping.

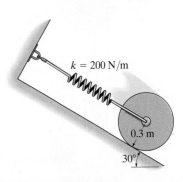

Prob. 18–47

18

*18–48. A chain that has a negligible mass is draped over the sprocket which has a mass of 2 kg and a radius of gyration of $k_O = 50$ mm. If the 4-kg block A is released from rest in the position $s = 1$ m, determine the angular velocity of the sprocket at the instant $s = 2$ m.

18–49. Solve Prob. 18–48 if the chain has a mass of 0.8 kg/m. For the calculation neglect the portion of the chain that wraps over the sprocket.

18–51. A spring having a stiffness of $k = 300$ N/m is attached to the end of the 15-kg rod, and it is unstretched when $\theta = 0°$. If the rod is released from rest when $\theta = 0°$, determine its angular velocity at the instant $\theta = 30°$. The motion is in the vertical plane.

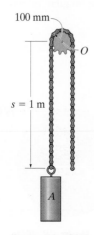

Probs. 18–48/49

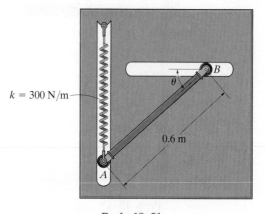

Prob. 18–51

18–50. The compound disk pulley consists of a hub and attached outer rim. If it has a mass of 3 kg and a radius of gyration $k_G = 45$ mm, determine the speed of block A after A descends 0.2 m from rest. Blocks A and B each have a mass of 2 kg. Neglect the mass of the cords.

*18–52. The two bars are released from rest at the position θ. Determine their angular velocities at the instant they become horizontal. Neglect the mass of the roller at C. Each bar has a mass m and length L.

18–53. The two bars are released from rest at the position $\theta = 90°$. Determine their angular velocities at the instant they become horizontal. Neglect the mass of the roller at C. Each bar has a mass m and length L.

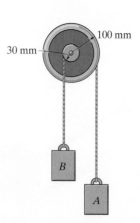

Prob. 18–50

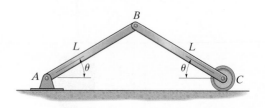

Probs. 18–52/53

18–54. If the 250-lb block is released from rest when the spring is unstretched, determine the velocity of the block after it has descended 5 ft. The drum has a weight of 50 lb and a radius of gyration of $k_O = 0.5$ ft about its center of mass O.

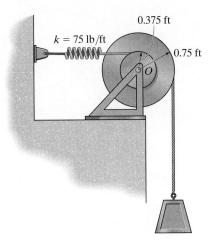

Prob. 18–54

18–55. The 6-kg rod ABC is connected to the 3-kg rod CD. If the system is released from rest when $\theta = 0°$, determine the angular velocity of rod ABC at the instant it becomes horizontal.

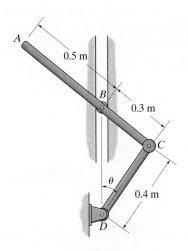

Prob. 18–55

*18–56. If the chain is released from rest from the position shown, determine the angular velocity of the pulley after the end B has risen 2 ft. The pulley has a weight of 50 lb and a radius of gyration of 0.375 ft about its axis. The chain weighs 6 lb/ft.

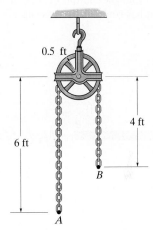

Prob. 18–56

18–57. If the gear is released from rest, determine its angular velocity after its center of gravity O has descended a distance of 4 ft. The gear has a weight of 100 lb and a radius of gyration about its center of gravity of $k = 0.75$ ft.

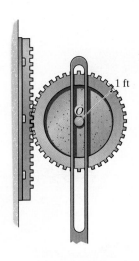

Prob. 18–57

18

18–58. When the slender 10-kg bar AB is horizontal it is at rest and the spring is unstretched. Determine the stiffness k of the spring so that the motion of the bar is momentarily stopped when it has rotated clockwise 90°.

18–59. When the slender 10-kg bar AB is horizontal it is at rest and the spring is unstretched. Determine the stiffness k of the spring so that the motion of the bar is momentarily stopped when it has rotated clockwise 45°.

18–61. A uniform ladder having a weight of 30 lb is released from rest when it is in the vertical position. If it is allowed to fall freely, determine the angle θ at which the bottom end A starts to slide to the right of A. For the calculation, assume the ladder to be a slender rod and neglect friction at A.

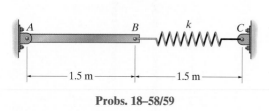

Probs. 18–58/59

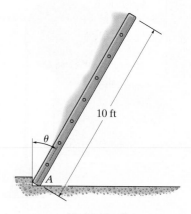

Prob. 18–61

*18–60.** If the 40-kg gear B is released from rest at $\theta = 0°$, determine the angular velocity of the 20-kg gear A at the instant $\theta = 90°$. The radii of gyration of gears A and B about their respective centers of mass are $k_A = 125$ mm and $k_B = 175$ mm. The outer gear ring P is fixed.

18–62. The 50-lb wheel has a radius of gyration about its center of gravity G of $k_G = 0.7$ ft. If it rolls without slipping, determine its angular velocity when it has rotated clockwise 90° from the position shown. The spring AB has a stiffness $k = 1.20$ lb/ft and an unstretched length of 0.5 ft. The wheel is released from rest.

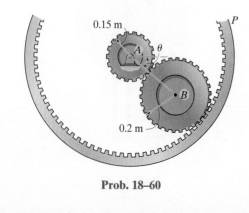

Prob. 18–60

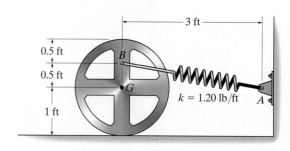

Prob. 18–62

18–63. The uniform window shade AB has a total weight of 0.4 lb. When it is released, it winds up around the spring-loaded core O. Motion is caused by a spring within the core, which is coiled so that it exerts a torque $M = 0.3(10^{-3})\theta$ lb·ft, where θ is in radians, on the core. If the shade is released from rest, determine the angular velocity of the core at the instant the shade is completely rolled up, i.e., after 12 revolutions. When this occurs, the spring becomes uncoiled and the radius of gyration of the shade about the axle at O is $k_O = 0.9$ in. *Note:* The elastic potential energy of the torsional spring is $V_e = \frac{1}{2}k\theta^2$, where $M = k\theta$ and $k = 0.3(10^{-3})$ lb·ft/rad.

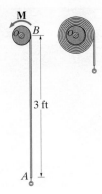

Prob. 18–63

*18–64.** The motion of the uniform 80-lb garage door is guided at its ends by the track. Determine the required initial stretch in the spring when the door is open, $\theta = 0°$, so that when it falls freely it comes to rest when it just reaches the fully closed position, $\theta = 90°$. Assume the door can be treated as a thin plate, and there is a spring and pulley system on each of the two sides of the door.

18–65. The motion of the uniform 80-lb garage door is guided at its ends by the track. If it is released from rest at $\theta = 0°$, determine the door's angular velocity at the instant $\theta = 30°$. The spring is originally stretched 1 ft when the door is held open, $\theta = 0°$. Assume the door can be treated as a thin plate, and there is a spring and pulley system on each of the two sides of the door.

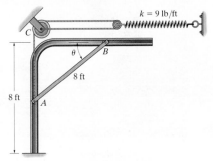

Probs. 18–64/65

18–66. The end A of the garage door AB travels along the horizontal track, and the end of member BC is attached to a spring at C. If the spring is originally unstretched, determine the stiffness k so that when the door falls downward from rest in the position shown, it will have zero angular velocity the moment it closes, i.e., when it and BC become vertical. Neglect the mass of member BC and assume the door is a thin plate having a weight of 200 lb and a width and height of 12 ft. There is a similar connection and spring on the other side of the door.

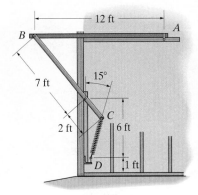

Prob. 18–66

18–67. Determine the stiffness k of the torsional spring at A, so that if the bars are released from rest when $\theta = 0°$, bar AB has an angular velocity of 0.5 rad/s at the closed position, $\theta = 90°$. The spring is uncoiled when $\theta = 0°$. The bars have a mass per unit length of 10 kg/m.

*18–68.** The torsional spring at A has a stiffness of $k = 900$ N·m/rad and is uncoiled when $\theta = 0°$. Determine the angular velocity of the bars, AB and BC, when $\theta = 0°$, if they are released from rest at the closed position, $\theta = 90°$. The bars have a mass per unit length of 10 kg/m.

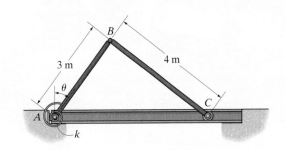

Probs. 18–67/68

CONCEPTUAL PROBLEMS

P18–1.   The bicycle and rider start from rest at the top of the hill. Show how to determine the speed of the rider when he freely coasts down the hill. Use appropriate dimensions of the wheels, and the mass of the rider, frame and wheels of the bicycle to explain your results.

P18–1

P18–2.   Two torsional springs, $M = k\theta$, are used to assist in opening and closing the hood of this truck. Assuming the springs are uncoiled ($\theta = 0°$) when the hood is opened, determine the stiffness k (N·m/rad) of each spring so that the hood can easily be lifted, i.e., practically no force applied to it, when it is closed in the unlocked position. Use appropriate numerical values to explain your result.

P18–2

P18–3.   The operation of this garage door is assisted using two springs AB and side members BCD, which are pinned at C. Assuming the springs are unstretched when the door is in the horizontal (open) position and $ABCD$ is vertical, determine each spring stiffness k so that when the door falls to the vertical (closed) position, it will slowly come to a stop. Use appropriate numerical values to explain your result.

P18–3

P18–4.   Determine the counterweight of A needed to balance the weight of the bridge deck when $\theta = 0°$. Show that this weight will maintain equilibrium of the deck by considering the potential energy of the system when the deck is in the arbitrary position θ. Both the deck and AB are horizontal when $\theta = 0°$. Neglect the weights of the other members. Use appropriate numerical values to explain this result.

P18–4

CHAPTER REVIEW

Kinetic Energy

The kinetic energy of a rigid body that undergoes planar motion can be referenced to its mass center. It includes a scalar sum of its translational and rotational kinetic energies.

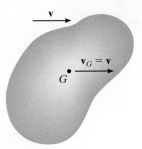

Translation

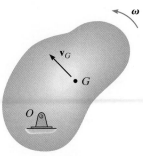

Rotation About a Fixed Axis

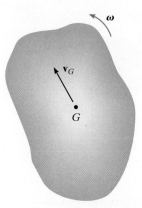

General Plane Motion

Translation

$$T = \tfrac{1}{2}mv_G^2$$

Rotation About a Fixed Axis

$$T = \tfrac{1}{2}mv_G^2 + \tfrac{1}{2}I_G\omega^2$$

or

$$T = \tfrac{1}{2}I_O\omega^2$$

General Plane Motion

$$T = \tfrac{1}{2}mv_G^2 + \tfrac{1}{2}I_G\omega^2$$

or

$$T = \tfrac{1}{2}I_{IC}\omega^2$$

18

Work of a Force and a Couple Moment

A force does work when it undergoes a displacement ds in the direction of the force. In particular, the frictional and normal forces that act on a cylinder or any circular body that rolls *without slipping* will do no work, since the normal force does not undergo a displacement and the frictional force acts on successive points on the surface of the body.

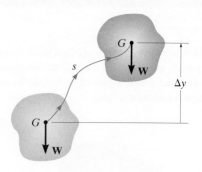

$$U_W = -W\Delta y$$

Weight

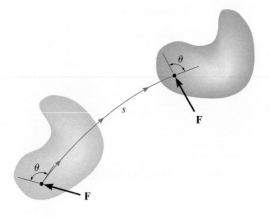

$$U_F = \int F \cos\theta \, ds$$

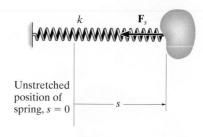

$$U = -\frac{1}{2}k\,s^2$$

Spring

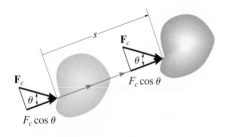

$$U_{F_c} = (F_c \cos\theta)s$$

Constant Force

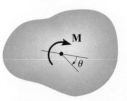

$$U_M = \int_{\theta_1}^{\theta_2} M \, d\theta$$

$$U_M = M(\theta_2 - \theta_1)$$

Constant magnitude

Principle of Work and Energy

Problems that involve velocity, force, and displacement can be solved using the principle of work and energy. The kinetic energy is the sum of both its rotational and translational parts. For application, a free-body diagram should be drawn in order to account for the work of all of the forces and couple moments that act on the body as it moves along the path.

$$T_1 + \Sigma U_{1-2} = T_2$$

Conservation of Energy

If a rigid body is subjected only to conservative forces, then the conservation-of-energy equation can be used to solve the problem. This equation requires that the sum of the potential and kinetic energies of the body remain the same at any two points along the path.

$$T_1 + V_1 = T_2 + V_2$$
$$\text{where } V = V_g + V_e$$

The potential energy is the sum of the body's gravitational and elastic potential energies. The gravitational potential energy will be positive if the body's center of gravity is located above a datum. If it is below the datum, then it will be negative. The elastic potential energy is always positive, regardless if the spring is stretched or compressed.

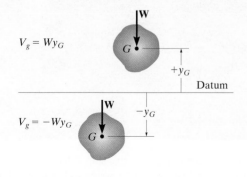

Gravitational potential energy

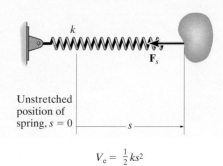

Elastic potential energy

Chapter 19

The impulse that this tugboat imparts to this ship will cause it to turn in a manner that can be predicted by applying the principles of impulse and momentum.

Planar Kinetics of a Rigid Body: Impulse and Momentum

CHAPTER OBJECTIVES

■ To develop formulations for the linear and angular momentum of a body.

■ To apply the principles of linear and angular impulse and momentum to solve rigid-body planar kinetic problems that involve force, velocity, and time.

■ To discuss application of the conservation of momentum.

■ To analyze the mechanics of eccentric impact.

19.1 Linear and Angular Momentum

In this chapter we will use the principles of linear and angular impulse and momentum to solve problems involving force, velocity, and time as related to the planar motion of a rigid body. Before doing this, we will first formalize the methods for obtaining a body's linear and angular momentum, assuming the body is symmetric with respect to an inertial x–y reference plane.

Linear Momentum. The linear momentum of a rigid body is determined by summing vectorially the linear momenta of all the particles of the body, i.e., $\mathbf{L} = \Sigma m_i \mathbf{v}_i$. Since $\Sigma m_i \mathbf{v}_i = m \mathbf{v}_G$ (see Sec. 15.2) we can also write

$$\mathbf{L} = m\mathbf{v}_G \qquad (19\text{–}1)$$

This equation states that the body's linear momentum is a vector quantity having a *magnitude* mv_G, which is commonly measured in units of kg · m/s or slug · ft/s and a *direction* defined by $\mathbf{v}_G$ the velocity of the body's mass center.

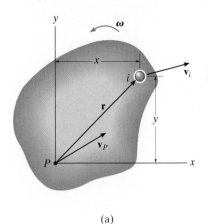

(a)

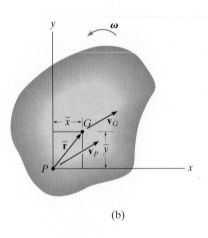

(b)

Fig. 19–1

Angular Momentum.   Consider the body in Fig. 19–1a, which is subjected to general plane motion. At the instant shown, the arbitrary point P has a known velocity $\mathbf{v}_P$, and the body has an angular velocity $\boldsymbol{\omega}$. Therefore the velocity of the ith particle of the body is

$$\mathbf{v}_i = \mathbf{v}_P + \mathbf{v}_{i/P} = \mathbf{v}_P + \boldsymbol{\omega} \times \mathbf{r}$$

The angular momentum of this particle about point P is equal to the "moment" of the particle's linear momentum about P, Fig. 19–1a. Thus,

$$(\mathbf{H}_P)_i = \mathbf{r} \times m_i \mathbf{v}_i$$

Expressing $\mathbf{v}_i$ in terms of $\mathbf{v}_P$ and using Cartesian vectors, we have

$$(H_P)_i \mathbf{k} = m_i(x\mathbf{i} + y\mathbf{j}) \times [(v_P)_x \mathbf{i} + (v_P)_y \mathbf{j} + \omega \mathbf{k} \times (x\mathbf{i} + y\mathbf{j})]$$
$$(H_P)_i = -m_i y(v_P)_x + m_i x(v_P)_y + m_i \omega r^2$$

Letting $m_i \rightarrow dm$ and integrating over the entire mass m of the body, we obtain

$$H_P = -\left(\int_m y\, dm \right)(v_P)_x + \left(\int_m x\, dm \right)(v_P)_y + \left(\int_m r^2\, dm \right)\omega$$

Here H_P represents the angular momentum of the body about an axis (the z axis) perpendicular to the plane of motion that passes through point P. Since $\bar{y}m = \int y\, dm$ and $\bar{x}m = \int x\, dm$, the integrals for the first and second terms on the right are used to locate the body's center of mass G with respect to P, Fig. 19–1b. Also, the last integral represents the body's moment of inertia about point P. Thus,

$$H_P = -\bar{y}m(v_P)_x + \bar{x}m(v_P)_y + I_P\omega \qquad (19\text{--}2)$$

This equation reduces to a simpler form if P coincides with the mass center G for the body,* in which case $\bar{x} = \bar{y} = 0$. Hence,

$$\boxed{H_G = I_G\omega} \qquad (19\text{--}3)$$

---

*It also reduces to the same simple form, $H_P = I_P\omega$, if point P is a *fixed point* (see Eq. 19–9) or the velocity of P is directed along the line PG.

Here the angular momentum of the body about G is equal to the product of the moment of inertia of the body about an axis passing through G and the body's angular velocity. Realize that $\mathbf{H}_G$ is a vector quantity having a *magnitude* $I_G\omega$, which is commonly measured in units of $kg \cdot m^2/s$ or $slug \cdot ft^2/s$, and a *direction* defined by $\boldsymbol{\omega}$, which is always perpendicular to the plane of motion.

Equation 19–2 can also be rewritten in terms of the x and y components of the velocity of the body's mass center, $(\mathbf{v}_G)_x$ and $(\mathbf{v}_G)_y$, and the body's moment of inertia I_G. Since G is located at coordinates $(\bar{x}, \bar{y})$, then by the parallel-axis theorem, $I_P = I_G + m(\bar{x}^2 + \bar{y}^2)$. Substituting into Eq. 19–2 and rearranging terms, we have

$$H_P = \bar{y}m[-(v_P)_x + \bar{y}\omega] + \bar{x}m[(v_P)_y + \bar{x}\omega] + I_G\omega \qquad (19\text{–}4)$$

From the kinematic diagram of Fig. 19–1*b*, $\mathbf{v}_G$ can be expressed in terms of $\mathbf{v}_P$ as

$$\mathbf{v}_G = \mathbf{v}_P + \boldsymbol{\omega} \times \bar{\mathbf{r}}$$

$$(v_G)_x\mathbf{i} + (v_G)_y\mathbf{j} = (v_P)_x\mathbf{i} + (v_P)_y\mathbf{j} + \omega\mathbf{k} \times (\bar{x}\mathbf{i} + \bar{y}\mathbf{j})$$

Carrying out the cross product and equating the respective $\mathbf{i}$ and $\mathbf{j}$ components yields the two scalar equations

$$(v_G)_x = (v_P)_x - \bar{y}\omega$$

$$(v_G)_y = (v_P)_y + \bar{x}\omega$$

Substituting these results into Eq. 19–4 yields

$$(\zeta +)H_P = -\bar{y}m(v_G)_x + \bar{x}m(v_G)_y + I_G\omega \qquad (19\text{–}5)$$

As shown in Fig. 19–1*c*, *this result indicates that when the angular momentum of the body is computed about point P, it is equivalent to the moment of the linear momentum* $m\mathbf{v}_G$, *or its components* $m(\mathbf{v}_G)_x$ *and* $m(\mathbf{v}_G)_y$, *about P plus the angular momentum* $I_G\boldsymbol{\omega}$. Using these results, we will now consider three types of motion.

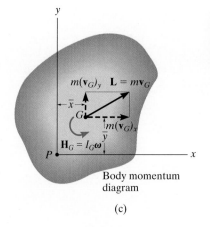

Body momentum diagram

(c)

Fig. 19–1

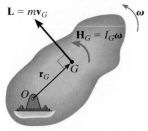

Translation

(a)

Translation. When a rigid body is subjected to either rectilinear or curvilinear *translation*, Fig. 19–2a, then $\boldsymbol{\omega} = \mathbf{0}$ and its mass center has a velocity of $\mathbf{v}_G = \mathbf{v}$. Hence, the linear momentum, and the angular momentum about G, become

$$\boxed{\begin{aligned} L &= mv_G \\ H_G &= 0 \end{aligned}} \tag{19–6}$$

If the angular momentum is computed about some other point A, the "moment" of the linear momentum $\mathbf{L}$ must be found about the point. Since d is the "moment arm" as shown in Fig. 19–2a, then in accordance with Eq. 19–5, $H_A = (d)(mv_G)\,\circlearrowright$.

Rotation About a Fixed Axis. When a rigid body is *rotating about a fixed axis*, Fig. 19–2b, the linear momentum, and the angular momentum about G, are

$$\boxed{\begin{aligned} L &= mv_G \\ H_G &= I_G\omega \end{aligned}} \tag{19–7}$$

It is sometimes convenient to compute the angular momentum about point O. Noting that $\mathbf{L}$ (or $\mathbf{v}_G$) is always *perpendicular* to $\mathbf{r}_G$, we have

$$(\circlearrowleft +)\ H_O = I_G\omega + r_G(mv_G) \tag{19–8}$$

Since $v_G = r_G\omega$, this equation can be written as $H_O = (I_G + mr_G^2)\omega$. Using the parallel-axis theorem,*

$$\boxed{H_O = I_O\omega} \tag{19–9}$$

Rotation about a fixed axis

(b)

Fig. 19–2

For the calculation, then, either Eq. 19–8 or 19–9 can be used.

---

*The similarity between this derivation and that of Eq. 17–16 ($\Sigma M_O = I_O\alpha$) and Eq. 18–5 $\left(T = \frac{1}{2}I_O\omega^2\right)$ should be noted. Also note that the same result can be obtained from Eq. 19–2 by selecting point P at O, realizing that $(v_O)_x = (v_O)_y = 0$.

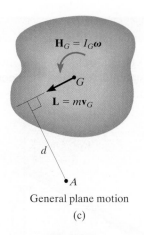

General plane motion

(c)

Fig. 19–2

General Plane Motion.

When a rigid body is subjected to general plane motion, Fig. 19–2c, the linear momentum, and the angular momentum about G, become

$$\boxed{\begin{aligned} L &= mv_G \\ H_G &= I_G\omega \end{aligned}} \qquad (19\text{--}10)$$

If the angular momentum is computed about point A, Fig. 19–2c, it is necessary to include the moment of $\mathbf{L}$ and $\mathbf{H}_G$ about this point. In this case,

$$(\zeta+) \quad H_A = I_G\omega + (d)(mv_G)$$

Here d is the moment arm, as shown in the figure.

As a special case, if point A is the instantaneous center of zero velocity then, like Eq. 19–9, we can write the above equation in simplified form as

$$\boxed{H_{IC} = I_{IC}\omega} \qquad (19\text{--}11)$$

where I_{IC} is the moment of inertia of the body about the IC. (See Prob. 19–2.)

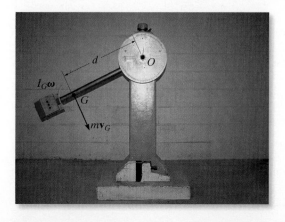

As the pendulum swings downward, its angular momentum about point O can be determined by computing the moment of $I_G\boldsymbol{\omega}$ and $m\mathbf{v}_G$ about O. This is $H_O = I_G\omega + (mv_G)d$. Since $v_G = \omega d$, then $H_O = I_G\omega + m(\omega d)d = (I_G + md^2)\omega = I_O\omega$.

EXAMPLE 19.1

At a given instant the 5-kg slender bar has the motion shown in Fig. 19–3a. Determine its angular momentum about point G and about the IC at this instant.

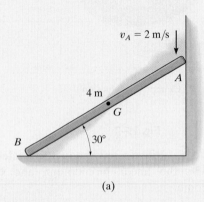

(a)

SOLUTION

Bar.   The bar undergoes *general plane motion*. The IC is established in Fig. 19–3b, so that

$$\omega = \frac{2 \text{ m/s}}{4 \text{ m cos } 30°} = 0.5774 \text{ rad/s}$$

$$v_G = (0.5774 \text{ rad/s})(2 \text{ m}) = 1.155 \text{ m/s}$$

Thus,

$$(\circlearrowleft+)H_G = I_G\omega = \left[\tfrac{1}{12}(5 \text{ kg})(4 \text{ m})^2\right](0.5774 \text{ rad/s}) = 3.85 \text{ kg} \cdot \text{m}^2/\text{s}\,\circlearrowright \quad \textit{Ans.}$$

Adding $I_G\omega$ and the moment of mv_G about the IC yields

$$(\circlearrowleft+) H_{IC} = I_G\omega + d(mv_G)$$

$$= \left[\tfrac{1}{12}(5 \text{ kg})(4 \text{ m})^2\right](0.5774 \text{ rad/s}) + (2 \text{ m})(5 \text{ kg})(1.155 \text{ m/s})$$

$$= 15.4 \text{ kg} \cdot \text{m}^2/\text{s}\,\circlearrowright \qquad \textit{Ans.}$$

We can also use

$$(\circlearrowleft+) H_{IC} = I_{IC}\omega$$

$$= \left[\tfrac{1}{12}\,(5 \text{ kg})(4 \text{ m})^2 + (5 \text{ kg})(2 \text{ m})^2\right](0.5774 \text{ rad/s})$$

$$= 15.4 \text{ kg} \cdot \text{m}^2/\text{s}\,\circlearrowright \qquad \textit{Ans.}$$

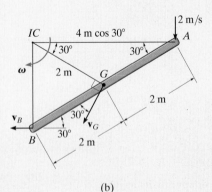

(b)

Fig. 19–3

19.2   Principle of Impulse and Momentum

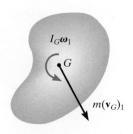

Like the case for particle motion, the principle of impulse and momentum for a rigid body can be developed by *combining* the equation of motion with kinematics. The resulting equation will yield a *direct solution to problems involving force, velocity, and time.*

Principle of Linear Impulse and Momentum.   The equation of translational motion for a rigid body can be written as $\Sigma \mathbf{F} = m\mathbf{a}_G = m(d\mathbf{v}_G/dt)$. Since the mass of the body is constant,

$$\Sigma \mathbf{F} = \frac{d}{dt}(m\mathbf{v}_G)$$

Multiplying both sides by dt and integrating from $t = t_1$, $\mathbf{v}_G = (\mathbf{v}_G)_1$ to $t = t_2$, $\mathbf{v}_G = (\mathbf{v}_G)_2$ yields

$$\Sigma \int_{t_1}^{t_2} \mathbf{F}\, dt = m(\mathbf{v}_G)_2 - m(\mathbf{v}_G)_1$$

This equation is referred to as the *principle of linear impulse and momentum*. It states that the sum of all the impulses created by the *external force system* which acts on the body during the time interval t_1 to t_2 is equal to the change in the linear momentum of the body during this time interval, Fig. 19–4.

Principle of Angular Impulse and Momentum.   If the body has *general plane motion* then $\Sigma M_G = I_G \alpha = I_G(d\omega/dt)$. Since the moment of inertia is constant,

$$\Sigma M_G = \frac{d}{dt}(I_G \omega)$$

Multiplying both sides by dt and integrating from $t = t_1$, $\omega = \omega_1$ to $t = t_2$, $\omega = \omega_2$ gives

$$\Sigma \int_{t_1}^{t_2} M_G\, dt = I_G \omega_2 - I_G \omega_1 \qquad (19\text{–}12)$$

In a similar manner, for *rotation about a fixed axis* passing through point O, Eq. 17–16 ($\Sigma M_O = I_O \alpha$) when integrated becomes

$$\Sigma \int_{t_1}^{t_2} M_O\, dt = I_O \omega_2 - I_O \omega_1 \qquad (19\text{–}13)$$

Equations 19–12 and 19–13 are referred to as the *principle of angular impulse and momentum*. Both equations state that the sum of the angular impulses acting on the body during the time interval t_1 to t_2 is equal to the change in the body's angular momentum during this time interval.

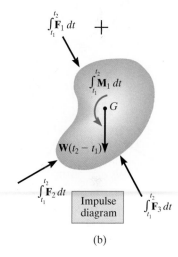

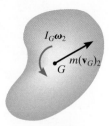

Fig. 19–4

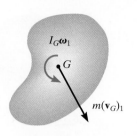

Initial
momentum
diagram

(a)

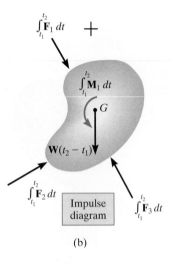

Impulse
diagram

(b)

$\parallel$

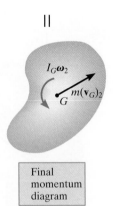

Final
momentum
diagram

(c)

Fig. 19–4 (repeated)

To summarize these concepts, if motion occurs in the x–y plane, the following *three scalar equations* can be written to describe the *planar motion* of the body.

$$m(v_{Gx})_1 + \Sigma \int_{t_1}^{t_2} F_x \, dt = m(v_{Gx})_2$$

$$m(v_{Gy})_1 + \Sigma \int_{t_1}^{t_2} F_y \, dt = m(v_{Gy})_2 \qquad (19\text{–}14)$$

$$I_G\omega_1 + \Sigma \int_{t_1}^{t_2} M_G \, dt = I_G\omega_2$$

The terms in these equations can be shown graphically by drawing a set of impulse and momentum diagrams for the body, Fig. 19–4. Note that the linear momentum $m\mathbf{v}_G$ is applied at the body's mass center, Figs. 19–4a and 19–4c; whereas the angular momentum $I_G\boldsymbol{\omega}$ is a free vector, and therefore, like a couple moment, it can be applied at any point on the body. When the impulse diagram is constructed, Fig. 19–4b, the forces $\mathbf{F}$ and moment $\mathbf{M}$ vary with time, and are indicated by the integrals. However, if $\mathbf{F}$ and $\mathbf{M}$ are *constant* integration of the impulses yields $\mathbf{F}(t_2 - t_1)$ and $\mathbf{M}(t_2 - t_1)$, respectively. Such is the case for the body's weight $\mathbf{W}$, Fig. 19–4b.

Equations 19–14 can also be applied to an entire system of connected bodies rather than to each body separately. This eliminates the need to include interaction impulses which occur at the connections since they are *internal* to the system. The resultant equations may be written in symbolic form as

$$\left(\Sigma \begin{matrix} \text{syst. linear} \\ \text{momentum} \end{matrix} \right)_{x1} + \left(\Sigma \begin{matrix} \text{syst. linear} \\ \text{impulse} \end{matrix} \right)_{x(1-2)} = \left(\Sigma \begin{matrix} \text{syst. linear} \\ \text{momentum} \end{matrix} \right)_{x2}$$

$$\left(\Sigma \begin{matrix} \text{syst. linear} \\ \text{momentum} \end{matrix} \right)_{y1} + \left(\Sigma \begin{matrix} \text{syst. linear} \\ \text{impulse} \end{matrix} \right)_{y(1-2)} = \left(\Sigma \begin{matrix} \text{syst. linear} \\ \text{momentum} \end{matrix} \right)_{y2}$$

$$\left(\Sigma \begin{matrix} \text{syst. angular} \\ \text{momentum} \end{matrix} \right)_{O1} + \left(\Sigma \begin{matrix} \text{syst. angular} \\ \text{impulse} \end{matrix} \right)_{O(1-2)} = \left(\Sigma \begin{matrix} \text{syst. angular} \\ \text{momentum} \end{matrix} \right)_{O2}$$

$$(19\text{–}15)$$

As indicated by the third equation, the system's angular momentum and angular impulse must be computed with respect to the *same reference point O* for all the bodies of the system.

Procedure For Analysis

Impulse and momentum principles are used to solve kinetic problems that involve *velocity, force*, and *time* since these terms are involved in the formulation.

Free-Body Diagram.

- Establish the *x, y, z* inertial frame of reference and draw the free-body diagram in order to account for all the forces and couple moments that produce impulses on the body.
- The direction and sense of the initial and final velocity of the body's mass center, $\mathbf{v}_G$, and the body's angular velocity $\boldsymbol{\omega}$ should be established. If any of these motions is unknown, assume that the sense of its components is in the direction of the positive inertial coordinates.
- Compute the moment of inertia I_G or I_O.
- As an alternative procedure, draw the impulse and momentum diagrams for the body or system of bodies. Each of these diagrams represents an outlined shape of the body which graphically accounts for the data required for each of the three terms in Eqs. 19–14 or 19–15, Fig. 19–4. These diagrams are particularly helpful in order to visualize the "moment" terms used in the principle of angular impulse and momentum, if application is about the *IC* or another point other than the body's mass center *G* or a fixed point *O*.

Principle of Impulse and Momentum.

- Apply the three scalar equations of impulse and momentum.
- The angular momentum of a rigid body rotating about a fixed axis is the moment of $m\mathbf{v}_G$ plus $I_G\boldsymbol{\omega}$ about the axis. This is equal to $H_O = I_O\omega$, where I_O is the moment of inertia of the body about the axis.
- All the forces acting on the body's free-body diagram will create an impulse; however, some of these forces will do no work.
- Forces that are functions of time must be integrated to obtain the impulse.
- The principle of angular impulse and momentum is often used to eliminate unknown impulsive forces that are parallel or pass through a common axis, since the moment of these forces is zero about this axis.

Kinematics.

- If more than three equations are needed for a complete solution, it may be possible to relate the velocity of the body's mass center to the body's angular velocity using *kinematics*. If the motion appears to be complicated, kinematic (velocity) diagrams may be helpful in obtaining the necessary relation.

19

EXAMPLE | 19.2

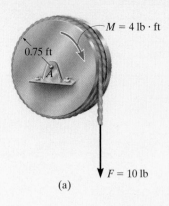

(a)

The 20-lb disk shown in Fig. 19–5a is acted upon by a constant couple moment of 4 lb·ft and a force of 10 lb which is applied to a cord wrapped around its periphery. Determine the angular velocity of the disk two seconds after starting from rest. Also, what are the force components of reaction at the pin?

SOLUTION

Since angular velocity, force, and time are involved in the problems, we will apply the principles of impulse and momentum to the solution.

Free-Body Diagram. Fig. 19–5b. The disk's mass center does not move; however, the loading causes the disk to rotate clockwise. The moment of inertia of the disk about its fixed axis of rotation is

$$I_A = \frac{1}{2}mr^2 = \frac{1}{2}\left(\frac{20 \text{ lb}}{32.2 \text{ ft/s}^2}\right)(0.75 \text{ ft})^2 = 0.1747 \text{ slug} \cdot \text{ft}^2$$

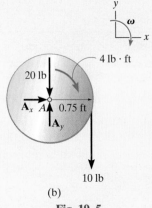

(b)

Fig. 19–5

Principle of Impulse and Momentum.

$$(\overset{+}{\rightarrow})\qquad m(v_{Ax})_1 + \Sigma \int_{t_1}^{t_2} F_x \, dt = m(v_{Ax})_2$$

$$0 + A_x(2 \text{ s}) = 0$$

$$(+\uparrow)\qquad m(v_{Ay})_1 + \Sigma \int_{t_1}^{t_2} F_y \, dt = m(v_{Ay})_2$$

$$0 + A_y(2 \text{ s}) - 20 \text{ lb}(2 \text{ s}) - 10 \text{ lb}(2 \text{ s}) = 0$$

$$(\zeta +)\qquad I_A\omega_1 + \Sigma \int_{t_1}^{t_2} M_A \, dt = I_A\omega_2$$

$$0 + 4 \text{ lb} \cdot \text{ft}(2 \text{ s}) + [10 \text{ lb}(2 \text{ s})](0.75 \text{ ft}) = 0.1747\omega_2$$

Solving these equations yields

$$A_x = 0 \qquad\qquad\qquad Ans.$$
$$A_y = 30 \text{ lb} \qquad\qquad Ans.$$
$$\omega_2 = 132 \text{ rad/s} \, \zeta \qquad Ans.$$

EXAMPLE 19.3

The 100-kg spool shown in Fig. 19–6a has a radius of gyration $k_G = 0.35$ m. A cable is wrapped around the central hub of the spool, and a horizontal force having a variable magnitude of $P = (t + 10)$ N is applied, where t is in seconds. If the spool is initially at rest, determine its angular velocity in 5 s. Assume that the spool rolls without slipping at A.

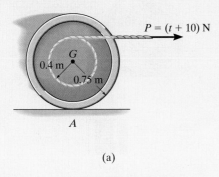

(a)

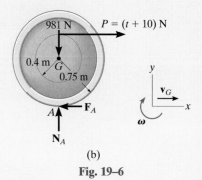

(b)

Fig. 19–6

SOLUTION

Free-Body Diagram.   From the free-body diagram, Fig. 19–6b, the *variable* force **P** will cause the friction force F_A to be variable, and thus the impulses created by both **P** and F_A must be determined by integration. Force **P** causes the mass center to have a velocity v_G to the right, and so the spool has a clockwise angular velocity ω.

Principle of Impulse and Momentum.   A direct solution for ω can be obtained by applying the principle of angular impulse and momentum about point A, the IC, in order to eliminate the unknown friction impulse.

$$(\zeta +) \qquad\qquad I_A\omega_1 + \Sigma \int M_A \, dt = I_A\omega_2$$

$$0 + \left[\int_0^{5\,s} (t + 10) \text{ N } dt \right](0.75 \text{ m} + 0.4 \text{ m}) = [100 \text{ kg } (0.35 \text{ m})^2 + (100 \text{ kg})(0.75 \text{ m})^2]\omega_2$$

$$62.5(1.15) = 68.5\omega_2$$

$$\omega_2 = 1.05 \text{ rad/s} \, \zeta \qquad\qquad\qquad Ans.$$

NOTE: Try solving this problem by applying the principle of impulse and momentum about G and using the principle of linear impulse and momentum in the x direction.

19

EXAMPLE | 19.4

The cylinder B, shown in Fig. 19–7a has a mass of 6 kg. It is attached to a cord which is wrapped around the periphery of a 20-kg disk that has a moment of inertia $I_A = 0.40$ kg $\cdot$ m^2. If the cylinder is initially moving downward with a speed of 2 m/s, determine its speed in 3 s. Neglect the mass of the cord in the calculation.

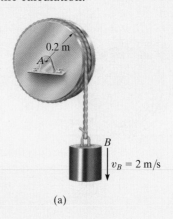

(a)

SOLUTION I

Free-Body Diagram.   The free-body diagrams of the cylinder and disk are shown in Fig. 19–7b. All the forces are *constant* since the weight of the cylinder causes the motion. The downward motion of the cylinder, $\mathbf{v}_B$, causes $\boldsymbol{\omega}$ of the disk to be clockwise.

Principle of Impulse and Momentum.   We can eliminate $\mathbf{A}_x$ and $\mathbf{A}_y$ from the analysis by applying the principle of angular impulse and momentum about point A. Hence

Disk

$(\zeta +)$
$$I_A \omega_1 + \Sigma \int M_A \, dt = I_A \omega_2$$

$$0.40 \text{ kg} \cdot \text{m}^2 (\omega_1) + T(3 \text{ s})(0.2 \text{ m}) = (0.40 \text{ kg} \cdot \text{m}^2)\omega_2$$

Cylinder

$(+\uparrow)$
$$m_B(v_B)_1 + \Sigma \int F_y \, dt = m_B(v_B)_2$$

$$-6 \text{ kg}(2 \text{ m/s}) + T(3 \text{ s}) - 58.86 \text{ N}(3 \text{ s}) = -6 \text{ kg}(v_B)_2$$

Kinematics.   Since $\omega = v_B/r$, then $\omega_1 = (2 \text{ m/s})/(0.2 \text{ m}) = 10$ rad/s and $\omega_2 = (v_B)_2/0.2 \text{ m} = 5(v_B)_2$. Substituting and solving the equations simultaneously for $(v_B)_2$ yields

$$(v_B)_2 = 13.0 \text{ m/s} \downarrow \qquad \qquad \textit{Ans.}$$

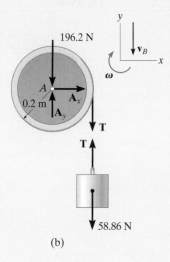

(b)

Fig. 19–7

19

SOLUTION II

Impulse and Momentum Diagrams. We can obtain $(v_B)_2$ *directly* by considering the *system* consisting of the cylinder, the cord, and the disk. The impulse and momentum diagrams have been drawn to clarify application of the principle of angular impulse and momentum about point A, Fig. 19–7c.

Principle of Angular Impulse and Momentum. Realizing that $\omega_1 = 10 \text{ rad/s}$ and $\omega_2 = 5(v_B)_2$, we have

$$(\zeta+)\left(\sum {}^{\text{syst. angular}}_{\text{momentum}}\right)_{A1} + \left(\sum {}^{\text{syst. angular}}_{\text{impulse}}\right)_{A(1-2)} = \left(\sum {}^{\text{syst. angular}}_{\text{momentum}}\right)_{A2}$$

$$(6 \text{ kg})(2 \text{ m/s})(0.2 \text{ m}) + (0.40 \text{ kg} \cdot \text{m}^2)(10 \text{ rad/s}) + (58.86 \text{ N})(3 \text{ s})(0.2 \text{ m})$$

$$= (6 \text{ kg})(v_B)_2(0.2 \text{ m}) + (0.40 \text{ kg} \cdot \text{m}^2)[5(v_B)_2]$$

$$(v_B)_2 = 13.0 \text{ m/s} \downarrow \qquad\qquad Ans.$$

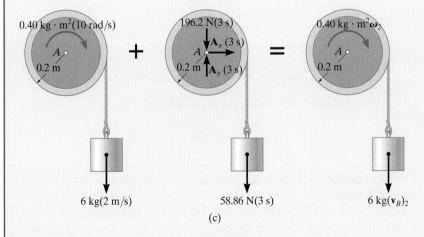

(c)

Fig. 19–7

EXAMPLE 19.5

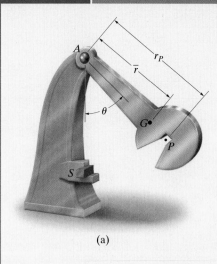

(a)

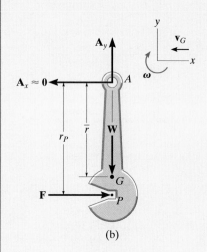

(b)

Fig. 19–8

The Charpy impact test is used in materials testing to determine the energy absorption characteristics of a material during impact. The test is performed using the pendulum shown in Fig. 19–8a, which has a mass m, mass center at G, and a radius of gyration k_G about G. Determine the distance r_P from the pin at A to the point P where the impact with the specimen S should occur so that the horizontal force at the pin A is essentially zero during the impact. For the calculation, assume the specimen absorbs all the pendulum's kinetic energy gained during the time it falls and thereby stops the pendulum from swinging when $\theta = 0°$.

SOLUTION

Free-Body Diagram. As shown on the free-body diagram, Fig. 19–8b, the conditions of the problem require the horizontal force at A to be zero. Just before impact, the pendulum has a clockwise angular velocity ω_1, and the mass center of the pendulum is moving to the left at $(v_G)_1 = \bar{r}\omega_1$.

Principle of Impulse and Momentum. We will apply the principle of angular impulse and momentum about point A. Thus,

$$I_A\omega_1 + \Sigma \int M_A\, dt = I_A\omega_2$$

$$(\zeta +) \qquad I_A\omega_1 - \left(\int F\, dt\right)r_P = 0$$

$$m(v_G)_1 + \Sigma \int F\, dt = m(v_G)_2$$

$$(\xrightarrow{+}) \qquad -m(\bar{r}\omega_1) + \int F\, dt = 0$$

Eliminating the impulse $\int F\, dt$ and substituting $I_A = mk_G^2 + m\bar{r}^2$ yields

$$[mk_G^2 + m\bar{r}^2]\omega_1 - m(\bar{r}\omega_1)r_P = 0$$

Factoring out $m\omega_1$ and solving for r_P, we obtain

$$r_P = \bar{r} + \frac{k_G^2}{\bar{r}} \qquad\qquad Ans.$$

NOTE: Point P, so defined, is called the *center of percussion*. By placing the striking point at P, the force developed at the pin will be minimized. Many sports rackets, clubs, etc. are designed so that collision with the object being struck occurs at the center of percussion. As a consequence, no "sting" or little sensation occurs in the hand of the player. (Also see Probs. 17–66 and 19–1.)

FUNDAMENTAL PROBLEMS

F19–1. The 60-kg wheel has a radius of gyration about its center O of $k_O = 300$ mm. If it is subjected to a couple moment of $M = (3t^2)$ N·m, where t is in seconds, determine the angular velocity of the wheel when $t = 4$ s, starting from rest.

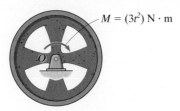

$M = (3t^2)$ N·m

F19–1

F19–2. The 300-kg wheel has a radius of gyration about its mass center O of $k_O = 400$ mm. If the wheel is subjected to a couple moment of $M = 300$ N·m, determine its angular velocity 6 s after it starts from rest and no slipping occurs. Also, determine the friction force that the ground applies to the wheel.

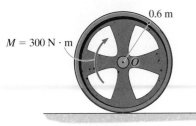

$M = 300$ N·m   0.6 m

F19–2

F19–3. If rod OA of negligible mass is subjected to the couple moment $M = 9$ N·m, determine the angular velocity of the 10-kg inner gear $t = 5$ s after it starts from rest. The gear has a radius of gyration about its mass center of $k_A = 100$ mm, and it rolls on the fixed outer gear. Motion occurs in the horizontal plane.

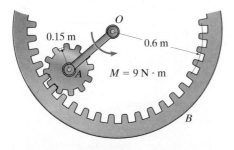

O
0.15 m   0.6 m
A
$M = 9$ N·m
B

F19–3

F19–4. Gears A and B of mass 10 kg and 50 kg have radii of gyration about their respective mass centers of $k_A = 80$ mm and $k_B = 150$ mm. If gear A is subjected to the couple moment $M = 10$ N·m when it is at rest, determine the angular velocity of gear B when $t = 5$ s.

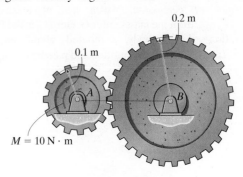

0.2 m
0.1 m
A B
$M = 10$ N·m

F19–4

F19–5. The 50-kg spool is subjected to a horizontal force of $P = 150$ N. If the spool rolls without slipping, determine its angular velocity 3 s after it starts from rest. The radius of gyration of the spool about its center of mass is $k_G = 175$ mm.

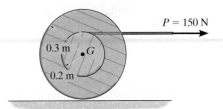

$P = 150$ N
0.3 m G
0.2 m

F19–5

F19–6. The reel has a weight of 150 lb and a radius of gyration about its center of gravity of $k_G = 1.25$ ft. If it is subjected to a torque of $M = 25$ lb·ft, and starts from rest when the torque is applied, determine its angular velocity in 3 seconds. The coefficient of kinetic friction between the reel and the horizontal plane is $\mu_k = 0.15$.

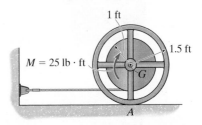

1 ft
1.5 ft
$M = 25$ lb·ft
G
A

F19–6

PROBLEMS

19–1. The rigid body (slab) has a mass m and rotates with an angular velocity ω about an axis passing through the fixed point O. Show that the momenta of all the particles composing the body can be represented by a single vector having a magnitude mv_G and acting through point P, called the *center of percussion*, which lies at a distance $r_{P/G} = k_G^2/r_{G/O}$ from the mass center G. Here k_G is the radius of gyration of the body, computed about an axis perpendicular to the plane of motion and passing through G.

19–3. Show that if a slab is rotating about a fixed axis perpendicular to the slab and passing through its mass center G, the angular momentum is the same when computed about any other point P.

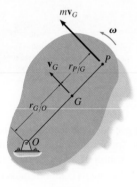

Prob. 19–1

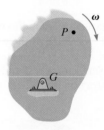

Prob. 19–3

19–2. At a given instant, the body has a linear momentum $\mathbf{L} = m\mathbf{v}_G$ and an angular momentum $\mathbf{H}_G = I_G\boldsymbol{\omega}$ computed about its mass center. Show that the angular momentum of the body computed about the instantaneous center of zero velocity IC can be expressed as $\mathbf{H}_{IC} = I_{IC}\boldsymbol{\omega}$, where I_{IC} represents the body's moment of inertia computed about the instantaneous axis of zero velocity. As shown, the IC is located at a distance $r_{G/IC}$ away from the mass center G.

*19–4.** The cable is subjected to a force of $P = (10t^2)$ lb, where t is in seconds. Determine the angular velocity of the spool 3 s after $\mathbf{P}$ is applied, starting from rest. The spool has a weight of 150 lb and a radius of gyration of 1.25 ft about its center of gravity.

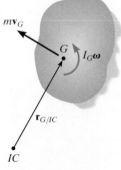

Prob. 19–2

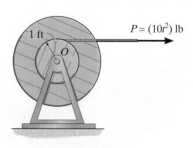

Prob. 19–4

19–5. The impact wrench consists of a slender 1-kg rod *AB* which is 580 mm long, and cylindrical end weights at *A* and *B* that each have a diameter of 20 mm and a mass of 1 kg. This assembly is free to turn about the handle and socket, which are attached to the lug nut on the wheel of a car. If the rod *AB* is given an angular velocity of 4 rad/s and it strikes the bracket *C* on the handle without rebounding, determine the angular impulse imparted to the lug nut.

19–7. The airplane is traveling in a straight line with a speed of 300 km/h, when the engines *A* and *B* produce a thrust of $T_A = 40$ kN and $T_B = 20$ kN, respectively. Determine the angular velocity of the airplane in $t = 5$ s. The plane has a mass of 200 Mg, its center of mass is located at *G*, and its radius of gyration about *G* is $k_G = 15$ m.

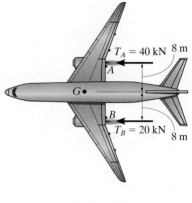

Prob. 19–7

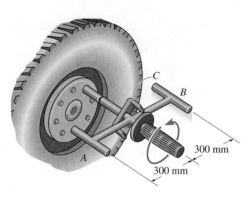

Prob. 19–5

19–6. The space capsule has a mass of 1200 kg and a moment of inertia $I_G = 900$ kg·m² about an axis passing through *G* and directed perpendicular to the page. If it is traveling forward with a speed $v_G = 800$ m/s and executes a turn by means of two jets, which provide a constant thrust of 400 N for 0.3 s, determine the capsule's angular velocity just after the jets are turned off.

*19–8. The assembly weighs 10 lb and has a radius of gyration $k_G = 0.6$ ft about its center of mass *G*. The kinetic energy of the assembly is 31 ft·lb when it is in the position shown. If it rolls counterclockwise on the surface without slipping, determine its linear momentum at this instant.

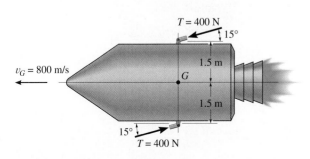

Prob. 19–6

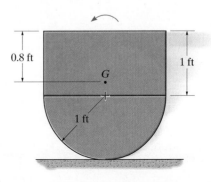

Prob. 19–8

19–9. The wheel having a mass of 100 kg and a radius of gyration about the z axis of $k_z = 300$ mm, rests on the smooth horizontal plane. If the belt is subjected to a force of $P = 200$ N, determine the angular velocity of the wheel and the speed of its center of mass O, three seconds after the force is applied.

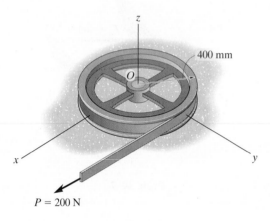

Prob. 19–9

19–11. The 30-kg reel is mounted on the 20-kg cart. If the cable wrapped around the inner hub of the reel is subjected to a force of $P = 50$ N, determine the velocity of the cart and the angular velocity of the reel when $t = 4$ s. The radius of gyration of the reel about its center of mass O is $k_O = 250$ mm. Neglect the size of the small wheels.

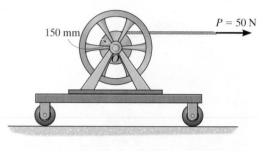

Prob. 19–11

19–10. The 30-kg gear A has a radius of gyration about its center of mass O of $k_O = 125$ mm. If the 20-kg gear rack B is subjected to a force of $P = 200$ N, determine the time required for the gear to obtain an angular velocity of 20 rad/s, starting from rest. The contact surface between the gear rack and the horizontal plane is smooth.

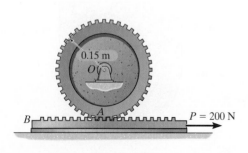

Prob. 19–10

*19–12. The spool has a weight of 75 lb and a radius of gyration $k_O = 1.20$ ft. If the block B weighs 60 lb, and a force $P = 25$ lb is applied to the cord, determine the speed of the block in 5 s starting from rest. Neglect the mass of the cord.

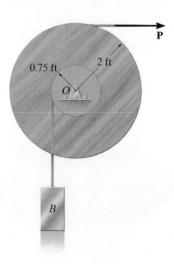

Prob. 19–12

19–13. The slender rod has a mass m and is suspended at its end A by a cord. If the rod receives a horizontal blow giving it an impulse **I** at its bottom B, determine the location y of the point P about which the rod appears to rotate during the impact.

19–15. The assembly shown consists of a 10-kg rod AB and a 20-kg circular disk C. If it is subjected to a torque of $M = (20t^{3/2})$ N·m, where t is it in seconds, determine its angular velocity when $t = 3$ s. When $t = 0$ the assembly is rotating at $\boldsymbol{\omega}_1 = \{-6\mathbf{k}\}$ rad/s.

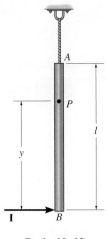

Prob. 19–13

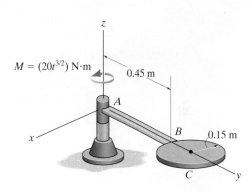

Prob. 19–15

19–14. If the ball has a weight W and radius r and is thrown onto a *rough surface* with a velocity $\mathbf{v}_0$ parallel to the surface, determine the amount of backspin, $\boldsymbol{\omega}_0$, it must be given so that it stops spinning at the same instant that its forward velocity is zero. It is not necessary to know the coefficient of friction at A for the calculation.

*****19–16.** The frame of a tandem drum roller has a weight of 4000 lb excluding the two rollers. Each roller has a weight of 1500 lb and a radius of gyration about its axle of 1.25 ft. If a torque of $M = 300$ lb·ft is supplied to the rear roller A, determine the speed of the drum roller 10 s later, starting from rest.

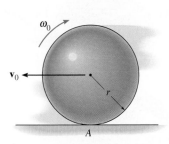

Prob. 19–14

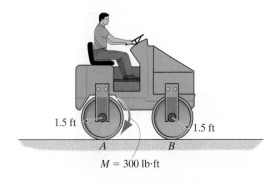

Prob. 19–16

19–17. A motor transmits a torque of $M = 0.05$ N·m to the center of gear A. Determine the angular velocity of each of the three (equal) smaller gears in 2 s starting from rest. The smaller gears (B) are pinned at their centers, and the masses and centroidal radii of gyration of the gears are given in the figure.

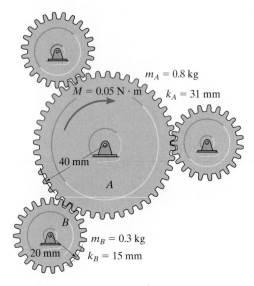

$m_A = 0.8$ kg
$M = 0.05$ N · m $k_A = 31$ mm

40 mm

A

B
20 mm $m_B = 0.3$ kg
$k_B = 15$ mm

Prob. 19–17

19–18. The man pulls the rope off the reel with a constant force of 8 lb in the direction shown. If the reel has a weight of 250 lb and radius of gyration $k_G = 0.8$ ft about the trunnion (pin) at A, determine the angular velocity of the reel in 3 s starting from rest. Neglect friction and the weight of rope that is removed.

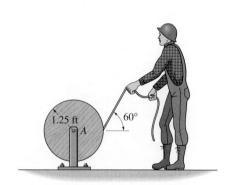

1.25 ft
A 60°

Prob. 19–18

19–19. The double pulley consists of two wheels which are attached to one another and turn at the same rate. The pulley has a mass of 15 kg and a radius of gyration $k_O = 110$ mm. If the block at A has a mass of 40 kg, determine the speed of the block in 3 s after a constant force $F = 2$ kN is applied to the rope wrapped around the inner hub of the pulley. The block is originally at rest. Neglect the mass of the rope.

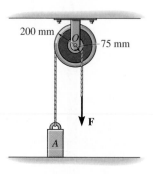

200 mm O 75 mm

F

A

Prob. 19–19

*19–20.** The cable is subjected to a force of $P = 20$ lb, and the spool rolls up the rail without slipping. Determine the angular velocity of the spool in 5 s, starting from rest. The spool has a weight of 100 lb and a radius of gyration about its center of gravity O of $k_O = 0.75$ ft.

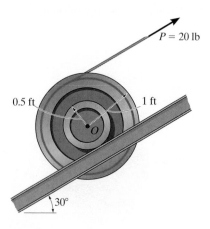

$P = 20$ lb

0.5 ft   1 ft

O

30°

Prob. 19–20

19–21. The inner hub of the wheel rests on the horizontal track. If it does not slip at A, determine the speed of the 10-lb block in 2 s after the block is released from rest. The wheel has a weight of 30 lb and a radius of gyration $k_G = 1.30$ ft. Neglect the mass of the pulley and cord.

19–23. The 100-kg reel has a radius of gyration about its center of mass G of $k_G = 200$ mm. If the cable B is subjected to a force of $P = 300$ N, determine the time required for the reel to obtain an angular velocity of 20 rad/s. The coefficient of kinetic friction between the reel and the plane is $\mu_k = 0.15$.

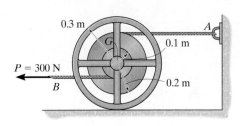

Prob. 19–23

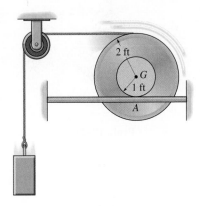

Prob. 19–21

19–22. The 1.25-lb tennis racket has a center of gravity at G and a radius of gyration about G of $k_G = 0.625$ ft. Determine the position P where the ball must be hit so that 'no sting' is felt by the hand holding the racket, i.e., the horizontal force exerted by the racket on the hand is zero.

*****19–24.** The 30-kg gear is subjected to a force of $P = (20t)$ N, where t is in seconds. Determine the angular velocity of the gear at $t = 4$ s, starting from rest. Gear rack B is fixed to the horizontal plane, and the gear's radius of gyration about its mass center O is $k_O = 125$ mm.

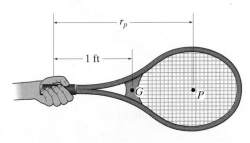

Prob. 19–22

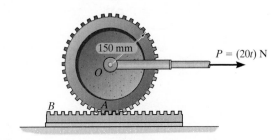

Prob. 19–24

19–25. The double pulley consists of two wheels which are attached to one another and turn at the same rate. The pulley has a mass of 30 kg and a radius of gyration $k_O = 250$ mm. If two men A and B grab the suspended ropes and step off the ledges at the same time, determine their speeds in 4 s starting from rest. The men A and B have a mass of 60 kg and 70 kg, respectively. Assume they do not move relative to the rope during the motion. Neglect the mass of the rope.

Prob. 19–25

19–26. If the shaft is subjected to a torque of $M = (15t^2)$ N·m, where t is in seconds, determine the angular velocity of the assembly when $t = 3$ s, starting from rest. Rods AB and BC each have a mass of 9 kg.

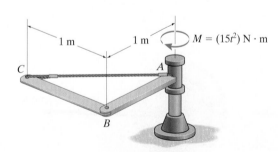

Prob. 19–26

19–27. The square plate has a mass m and is suspended at its corner A by a cord. If it receives a horizontal impulse **I** at corner B, determine the location y of the point P about which the plate appears to rotate during the impact.

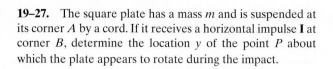

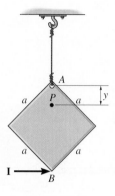

Prob. 19–27

***19–28.** The crate has a mass m_c. Determine the constant speed v_0 it acquires as it moves down the conveyor. The rollers each have a radius of r, mass m, and are spaced d apart. Note that friction causes each roller to rotate when the crate comes in contact with it.

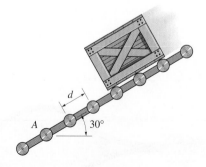

Prob. 19–28

19.3   Conservation of Momentum

Conservation of Linear Momentum.   If the sum of all the *linear impulses* acting on a system of connected rigid bodies is *zero* in a specific direction, then the linear momentum of the system is constant, or conserved in this direction, that is,

$$\left(\Sigma\,\frac{\text{syst. linear}}{\text{momentum}}\right)_1 = \left(\Sigma\,\frac{\text{syst. linear}}{\text{momentum}}\right)_2 \qquad (19\text{--}16)$$

This equation is referred to as the *conservation of linear momentum.*

   Without inducing appreciable errors in the calculations, it may be possible to apply Eq. 19–16 in a specified direction for which the linear impulses are small or *nonimpulsive.* Specifically, nonimpulsive forces occur when small forces act over very short periods of time. Typical examples include the force of a slightly deformed spring, the initial contact force with soft ground, and in some cases the weight of the body.

Conservation of Angular Momentum.   The angular momentum of a system of connected rigid bodies is conserved about the system's center of mass G, or a fixed point O, when the sum of all the angular impulses about these points is zero or appreciably small (nonimpulsive). The third of Eqs. 19–15 then becomes

$$\left(\Sigma\,\frac{\text{syst. angular}}{\text{momentum}}\right)_{O1} = \left(\Sigma\,\frac{\text{syst. angular}}{\text{momentum}}\right)_{O2} \qquad (19\text{--}17)$$

This equation is referred to as the *conservation of angular momentum.* In the case of a single rigid body, Eq. 19–17 applied to point G becomes $(I_G\omega)_1 = (I_G\omega)_2$. For example, consider a swimmer who executes a somersault after jumping off a diving board. By tucking his arms and legs in close to his chest, he *decreases* his body's moment of inertia and thus *increases* his angular velocity ($I_G\omega$ must be constant). If he straightens out just before entering the water, his body's moment of inertia is *increased,* and so his angular velocity *decreases.* Since the weight of his body creates a linear impulse during the time of motion, this example also illustrates how the angular momentum of a body can be conserved and yet the linear momentum is *not.* Such cases occur whenever the external forces creating the linear impulse pass through either the center of mass of the body or a fixed axis of rotation.

19

Procedure for Analysis

The conservation of linear or angular momentum should be applied using the following procedure.

Free-Body Diagram.

- Establish the *x, y* inertial frame of reference and draw the free-body diagram for the body or system of bodies during the time of impact. From this diagram classify each of the applied forces as being either "impulsive" or "nonimpulsive."

- By inspection of the free-body diagram, the *conservation of linear momentum* applies in a given direction when *no* external impulsive forces act on the body or system in that direction; whereas the *conservation of angular momentum* applies about a fixed point *O* or at the mass center *G* of a body or system of bodies when all the external impulsive forces acting on the body or system create zero moment (or zero angular impulse) about *O* or *G*.

- As an alternative procedure, draw the impulse and momentum diagrams for the body or system of bodies. These diagrams are particularly helpful in order to visualize the "moment" terms used in the conservation of angular momentum equation, when it has been decided that angular momenta are to be computed about a point other than the body's mass center *G*.

Conservation of Momentum.

- Apply the conservation of linear or angular momentum in the appropriate directions.

Kinematics.

- If the motion appears to be complicated, kinematic (velocity) diagrams may be helpful in obtaining the necessary kinematic relations.

19

EXAMPLE | 19.6

The 10-kg wheel shown in Fig. 19–9a has a moment of inertia $I_G = 0.156 \text{ kg} \cdot \text{m}^2$. Assuming that the wheel does not slip or rebound, determine the minimum velocity $\mathbf{v}_G$ it must have to just roll over the obstruction at A.

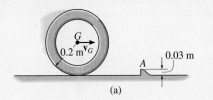

(a)

SOLUTION

Impulse and Momentum Diagrams. Since no slipping or rebounding occurs, the wheel essentially *pivots* about point A during contact. This condition is shown in Fig. 19–9b, which indicates, respectively, the momentum of the wheel *just before impact*, the impulses given to the wheel *during impact*, and the momentum of the wheel *just after impact*. Only two impulses (forces) act on the wheel. By comparison, the force at A is much greater than that of the weight, and since the time of impact is very short, the weight can be considered nonimpulsive. The impulsive force $\mathbf{F}$ at A has both an unknown magnitude and an unknown direction θ. To eliminate this force from the analysis, note that angular momentum about A is essentially *conserved* since $(98.1\Delta t)d \approx 0$.

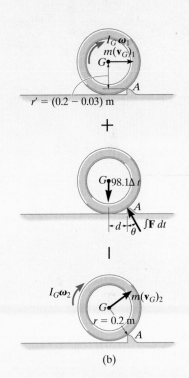

Conservation of Angular Momentum. With reference to Fig. 19–9b,

$$(\zeta +) \qquad (H_A)_1 = (H_A)_2$$

$$r'm(v_G)_1 + I_G\omega_1 = rm(v_G)_2 + I_G\omega_2$$

$$(0.2 \text{ m} - 0.03 \text{ m})(10 \text{ kg})(v_G)_1 + (0.156 \text{ kg} \cdot \text{m}^2)(\omega_1) =$$

$$(0.2 \text{ m})(10 \text{ kg})(v_G)_2 + (0.156 \text{ kg} \cdot \text{m}^2)(\omega_2)$$

Kinematics. Since no slipping occurs, in general $\omega = v_G/r = v_G/0.2 \text{ m} = 5v_G$. Substituting this into the above equation and simplifying yields

$$(v_G)_2 = 0.8921(v_G)_1 \qquad (1)$$

Conservation of Energy.* In order to roll over the obstruction, the wheel must pass position 3 shown in Fig. 19–9c. Hence, if $(v_G)_2$ [or $(v_G)_1$] is to be a minimum, it is necessary that the kinetic energy of the wheel at position 2 be equal to the potential energy at position 3. Placing the datum through the center of gravity, as shown in the figure, and applying the conservation of energy equation, we have

$$\{T_2\} + \{V_2\} = \{T_3\} + \{V_3\}$$

$$\left\{\tfrac{1}{2}(10 \text{ kg})(v_G)_2^2 + \tfrac{1}{2}(0.156 \text{ kg} \cdot \text{m}^2)\omega_2^2\right\} + \{0\} =$$

$$\{0\} + \{(98.1 \text{ N})(0.03 \text{ m})\}$$

Substituting $\omega_2 = 5(v_G)_2$ and Eq. 1 into this equation, and solving,

$$(v_G)_1 = 0.729 \text{ m/s} \rightarrow \qquad\qquad Ans.$$

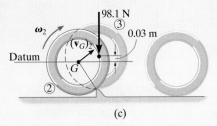

(c)

Fig. 19–9

*This principle *does not apply during impact*, since energy is *lost* during the collision. However, just after impact, as in Fig. 19–9c, it can be used.

19

EXAMPLE | 19.7

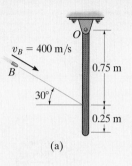

(a)

The 5-kg slender rod shown in Fig. 19–10a is pinned at O and is initially at rest. If a 4-g bullet is fired into the rod with a velocity of 400 m/s, as shown in the figure, determine the angular velocity of the rod just after the bullet becomes embedded in it.

SOLUTION

Impulse and Momentum Diagrams.   The impulse which the bullet exerts on the rod can be eliminated from the analysis, and the angular velocity of the rod just after impact can be determined by considering the bullet and rod as a single system. To clarify the principles involved, the impulse and momentum diagrams are shown in Fig. 19–10b. The momentum diagrams are drawn *just before and just after impact.* During impact, the bullet and rod exert equal but *opposite internal impulses* at A. As shown on the impulse diagram, the impulses that are external to the system are due to the reactions at O and the weights of the bullet and rod. Since the time of impact, Δt, is very short, the rod moves only a slight amount, and so the "moments" of the weight impulses about point O are essentially zero. Therefore angular momentum is conserved about this point.

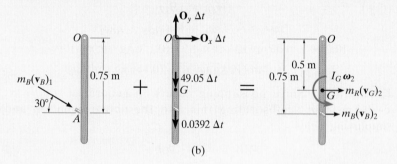

(b)

Conservation of Angular Momentum.   From Fig. 19–10b, we have

$(\zeta +)$ $\qquad\qquad\qquad \Sigma(H_O)_1 = \Sigma(H_O)_2$

$m_B(v_B)_1 \cos 30°(0.75 \text{ m}) = m_B(v_B)_2(0.75 \text{ m}) + m_R(v_G)_2(0.5 \text{ m}) + I_G\omega_2$

$(0.004 \text{ kg})(400 \cos 30° \text{ m/s})(0.75 \text{ m}) =$

$(0.004 \text{ kg})(v_B)_2(0.75 \text{ m}) + (5 \text{ kg})(v_G)_2(0.5 \text{ m}) + \left[\frac{1}{12}(5 \text{ kg})(1 \text{ m})^2\right]\omega_2$ (1)

or

$\qquad\qquad 1.039 = 0.003(v_B)_2 + 2.50(v_G)_2 + 0.4167\omega_2$

Kinematics.   Since the rod is pinned at O, from Fig. 19–10c we have

$\qquad\qquad (v_G)_2 = (0.5 \text{ m})\omega_2 \quad (v_B)_2 = (0.75 \text{ m})\omega_2$

Substituting into Eq. 1 and solving yields

$\qquad\qquad\qquad \omega_2 = 0.623 \text{ rad/s} \, \zeta$ $\qquad\qquad\qquad$ *Ans.*

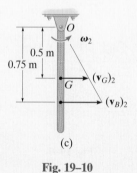

(c)

Fig. 19–10

19

*19.4   Eccentric Impact

The concepts involving central and oblique impact of particles were presented in Sec. 15.4. We will now expand this treatment and discuss the eccentric impact of two bodies. *Eccentric impact* occurs when the line connecting the *mass centers* of the two bodies *does not* coincide with the line of impact.* This type of impact often occurs when one or both of the bodies are constrained to rotate about a fixed axis. Consider, for example, the collision at C between the two bodies A and B, shown in Fig. 19–11a. It is assumed that just before collision B is rotating counterclockwise with an angular velocity $(\boldsymbol{\omega}_B)_1$, and the velocity of the contact point C located on A is $(\mathbf{u}_A)_1$. Kinematic diagrams for both bodies just before collision are shown in Fig. 19–11b. Provided the bodies are smooth, the *impulsive forces* they exert on each other *are directed along the line of impact*. Hence, the component of velocity of point C on body B, which is directed along the line of impact, is $(v_B)_1 = (\omega_B)_1 r$, Fig. 19–11b. Likewise, on body A the component of velocity $(\mathbf{u}_A)_1$ along the line of impact is $(v_A)_1$. In order for a collision to occur, $(v_A)_1 > (v_B)_1$.

During the impact an equal but opposite impulsive force $\mathbf{P}$ is exerted between the bodies which *deforms* their shapes at the point of contact. The resulting impulse is shown on the impulse diagrams for both bodies, Fig. 19–11c. Note that the impulsive force at point C on the rotating body creates impulsive pin reactions at O. On these diagrams it is assumed that the impact creates forces which are much larger than the nonimpulsive weights of the bodies, which are not shown. When the deformation at point C is a maximum, C on both the bodies moves with a common velocity $\mathbf{v}$ along the line of impact, Fig. 19–11d. A period of *restitution* then occurs in which the bodies tend to regain their original shapes. The restitution phase creates an equal but opposite impulsive force $\mathbf{R}$ acting between the bodies as shown on the impulse diagram, Fig. 19–11e. After restitution the bodies move apart such that point C on body B has a velocity $(\mathbf{v}_B)_2$ and point C on body A has a velocity $(\mathbf{u}_A)_2$, Fig. 19–11f, where $(v_B)_2 > (v_A)_2$.

In general, a problem involving the impact of two bodies requires determining the *two unknowns* $(v_A)_2$ and $(v_B)_2$, assuming $(v_A)_1$ and $(v_B)_1$ are known (or can be determined using kinematics, energy methods, the equations of motion, etc.). To solve such problems, two equations must be written. The *first equation* generally involves application of *the conservation of angular momentum to the two bodies*. In the case of both bodies A and B, we can state that angular momentum is conserved about point O since the impulses at C are internal to the system and the impulses at O create zero moment (or zero angular impulse) about O. The *second equation* can be obtained using the definition of the *coefficient of restitution, e*, which is a ratio of the restitution impulse to the deformation impulse.

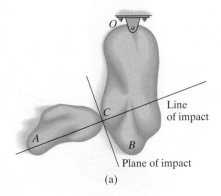

Line
of impact

Plane of impact

(a)

Fig. 19–11

19

Here is an example of eccentric impact occurring between this bowling ball and pin.

---

*When these lines coincide, central impact occurs and the problem can be analyzed as discussed in Sec. 15.4.

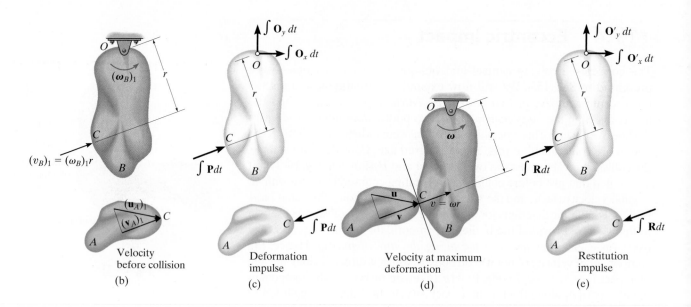

Velocity
before collision
(b)

Deformation
impulse
(c)

Velocity at maximum
deformation
(d)

Restitution
impulse
(e)

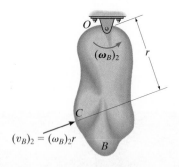

Velocity
after collision
(f)

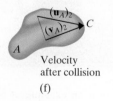

Fig. 19–11 (cont.)

Is is important to realize, however, that *this analysis has only a very limited application in engineering, because values of e for this case have been found to be highly sensitive to the material, geometry, and the velocity of each of the colliding bodies.* To establish a useful form of the coefficient of restitution equation we must first apply the principle of angular impulse and momentum about point O to bodies B and A separately. Combining the results, we then obtain the necessary equation. Proceeding in this manner, the principle of impulse and momentum applied to body B from the time just before the collision to the instant of maximum deformation, Figs. 19–11b, 19–11c, and 19–11d, becomes

$$(\zeta+) \qquad I_O(\omega_B)_1 + r\int P \, dt = I_O\omega \qquad (19\text{--}18)$$

Here I_O is the moment of inertia of body B about point O. Similarly, applying the principle of angular impulse and momentum from the instant of maximum deformation to the time just after the impact, Figs. 19–11d, 19–11e, and 19–11f, yields

$$(\zeta+) \qquad I_O\omega + r\int R \, dt = I_O(\omega_B)_2 \qquad (19\text{--}19)$$

Solving Eqs. 19–18 and 19–19 for $\int P \, dt$ and $\int R \, dt$, respectively, and formulating e, we have

$$e = \frac{\displaystyle\int R \, dt}{\displaystyle\int P \, dt} = \frac{r(\omega_B)_2 - r\omega}{r\omega - r(\omega_B)_1} = \frac{(v_B)_2 - v}{v - (v_B)_1}$$

In the same manner, we can write an equation which relates the magnitudes of velocity $(v_A)_1$ and $(v_A)_2$ of body A. The result is

$$e = \frac{v - (v_A)_2}{(v_A)_1 - v}$$

Combining the above two equations by eliminating the common velocity v yields the desired result, i.e.,

$(+\nearrow)$

$$e = \frac{(v_B)_2 - (v_A)_2}{(v_A)_1 - (v_B)_1}$$

(19–20)

This equation is identical to Eq. 15–11, which was derived for the central impact between two particles. It states that the coefficient of restitution is equal to the ratio of the relative velocity of *separation* of the points of contact (C) *just after impact* to the relative velocity at which the points *approach* one another *just* before impact. In deriving this equation, we assumed that the points of contact for both bodies move up and to the right *both* before and after impact. If motion of any one of the contacting points occurs down and to the left, the velocity of this point should be considered a negative quantity in Eq. 19–20.

During impact the columns of many highway signs are intended to break out of their supports and easily collapse at their joints. This is shown by the slotted connections at their base and the breaks at the column's midsection.

EXAMPLE 19.8

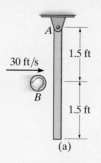

(a)

The 10-lb slender rod is suspended from the pin at A, Fig. 19–12a. If a 2-lb ball B is thrown at the rod and strikes its center with a velocity of 30 ft/s, determine the angular velocity of the rod just after impact. The coefficient of restitution is $e = 0.4$.

SOLUTION

Conservation of Angular Momentum. Consider the ball and rod as a system, Fig. 19–12b. Angular momentum is conserved about point A since the impulsive force between the rod and ball is *internal*. Also, the *weights* of the ball and rod are *nonimpulsive*. Noting the directions of the velocities of the ball and rod just after impact as shown on the kinematic diagram, Fig. 19–12c, we require

$$(\zeta +) \qquad\qquad\qquad (H_A)_1 = (H_A)_2$$

$$m_B(v_B)_1(1.5\text{ ft}) = m_B(v_B)_2(1.5\text{ ft}) + m_R(v_G)_2(1.5\text{ ft}) + I_G\omega_2$$

$$\left(\frac{2\text{ lb}}{32.2\text{ ft/s}^2}\right)(30\text{ ft/s})(1.5\text{ ft}) = \left(\frac{2\text{ lb}}{32.2\text{ ft/s}^2}\right)(v_B)_2(1.5\text{ ft}) +$$

$$\left(\frac{10\text{ lb}}{32.2\text{ ft/s}^2}\right)(v_G)_2(1.5\text{ ft}) + \left[\frac{1}{12}\left(\frac{10\text{ lb}}{32.2\text{ ft/s}^2}\right)(3\text{ ft})^2\right]\omega_2$$

Since $(v_G)_2 = 1.5\omega_2$ then

$$2.795 = 0.09317(v_B)_2 + 0.9317\omega_2 \qquad (1)$$

(b)

Coefficient of Restitution. With reference to Fig. 19–12c, we have

$$(\overset{+}{\rightarrow}) \qquad e = \frac{(v_G)_2 - (v_B)_2}{(v_B)_1 - (v_G)_1} \qquad 0.4 = \frac{(1.5\text{ ft})\omega_2 - (v_B)_2}{30\text{ ft/s} - 0}$$

$$12.0 = 1.5\omega_2 - (v_B)_2 \qquad (2)$$

Solving Eqs. 1 and 2, yields

$$(v_B)_2 = -6.52\text{ ft/s} = 6.52\text{ ft/s} \leftarrow$$

$$\omega_2 = 3.65\text{ rad/s} \,\circlearrowright \qquad\qquad\qquad Ans.$$

(c)

Fig. 19–12

PROBLEMS

19–29. A man has a moment of inertia I_z about the z axis. He is originally at rest and standing on a small platform which can turn freely. If he is handed a wheel which is rotating at $\boldsymbol{\omega}$ and has a moment of inertia I about its spinning axis, determine his angular velocity if (a) he holds the wheel upright as shown, (b) turns the wheel out, $\theta = 90°$, and (c) turns the wheel downward, $\theta = 180°$. Neglect the effect of holding the wheel a distance d away from the z axis.

Prob. 19–29

19–30. Two wheels A and B have masses m_A and m_B, and radii of gyration about their central vertical axes of k_A and k_B, respectively. If they are freely rotating in the same direction at $\boldsymbol{\omega}_A$ and $\boldsymbol{\omega}_B$ about the same vertical axis, determine their common angular velocity after they are brought into contact and slipping between them stops.

19–31. A 150-lb man leaps off the circular platform with a velocity of $v_{m/p} = 5$ ft/s, relative to the platform. Determine the angular velocity of the platform afterwards. Initially the man and platform are at rest. The platform weighs 300 lb and can be treated as a uniform circular disk.

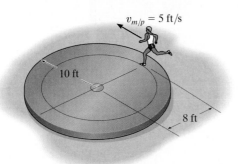

Prob. 19–31

*19–32. The space satellite has a mass of 125 kg and a moment of inertia $I_z = 0.940$ kg·m², excluding the four solar panels A, B, C, and D. Each solar panel has a mass of 20 kg and can be approximated as a thin plate. If the satellite is originally spinning about the z axis at a constant rate $\omega_z = 0.5$ rad/s when $\theta = 90°$, determine the rate of spin if all the panels are raised and reach the upward position, $\theta = 0°$, at the same instant.

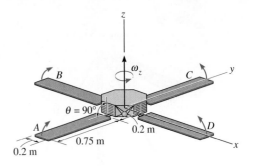

Prob. 19–32

19–33. The 80-kg man is holding two dumbbells while standing on a turntable of negligible mass, which turns freely about a vertical axis. When his arms are fully extended, the turntable is rotating with an angular velocity of 0.5 rev/s. Determine the angular velocity of the man when he retracts his arms to the position shown. When his arms are fully extended, approximate each arm as a uniform 6-kg rod having a length of 650 mm, and his body as a 68-kg solid cylinder of 400-mm diameter. With his arms in the retracted position, assume the man is an 80-kg solid cylinder of 450-mm diameter. Each dumbbell consists of two 5-kg spheres of negligible size.

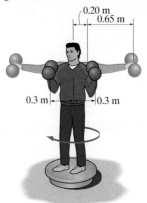

Prob. 19–33

19–34. The 75-kg gymnast lets go of the horizontal bar in a fully stretched position A, rotating with an angular velocity of $\omega_A = 3$ rad/s. Estimate his angular velocity when he assumes a tucked position B. Assume the gymnast at positions A and B as a uniform slender rod and a uniform circular disk, respectively.

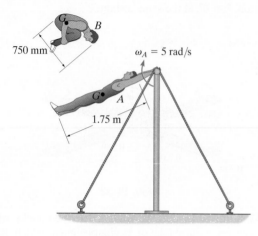

Prob. 19–34

19–35. The 2-kg rod ACB supports the two 4-kg disks at its ends. If both disks are given a clockwise angular velocity $(\omega_A)_1 = (\omega_B)_1 = 5$ rad/s while the rod is held stationary and then released, determine the angular velocity of the rod after both disks have stopped spinning relative to the rod due to frictional resistance at the pins A and B. Motion is in the *horizontal plane*. Neglect friction at pin C.

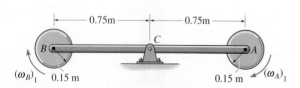

Prob. 19–35

*19–36. The 5-lb rod AB supports the 3-lb disk at its end. If the disk is given an angular velocity $\omega_D = 8$ rad/s while the rod is held stationary and then released, determine the angular velocity of the rod after the disk has stopped spinning relative to the rod due to frictional resistance at the bearing A. Motion is in the *horizontal plane*. Neglect friction at the fixed bearing B.

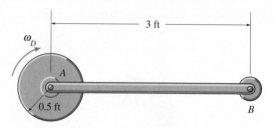

Prob. 19–36

19–37. The pendulum consists of a 5-lb slender rod AB and a 10-lb wooden block. A projectile weighing 0.2 lb is fired into the center of the block with a velocity of 1000 ft/s. If the pendulum is initially at rest, and the projectile embeds itself into the block, determine the angular velocity of the pendulum just after the impact.

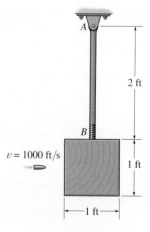

Prob. 19–37

19–38. The 20-kg cylinder *A* is free to slide along rod *BC*. When the cylinder is at *x* = 0, the 50-kg circular disk *D* is rotating with an angular velocity of 5 rad/s. If the cylinder is given a slight push, determine the angular velocity of the disk when the cylinder strikes *B* at *x* = 600 mm. Neglect the mass of the brackets and the smooth rod.

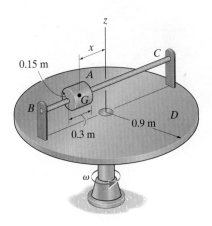

Prob. 19–38

19–39. The slender bar of mass *m* pivots at support *A* when it is released from rest in the vertical position. When it falls and rotates 90°, pin *C* will strike support *B*, and the pin at *A* will leave its support. Determine the angular velocity of the bar immediately after the impact. Assume the pin at *B* will not rebound.

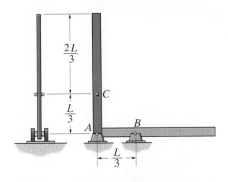

Prob. 19–39

***19–40.** The uniform rod assembly rotates with an angular velocity of ω_0 on the smooth horizontal plane just before the hook strikes the peg *P* without rebound. Determine the angular velocity of the assembly immediately after the impact. Each rod has a mass of *m*.

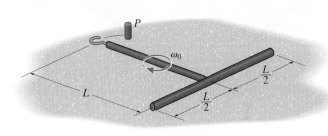

Prob. 19–40

19–41. A thin disk of mass *m* has an angular velocity ω_1 while rotating on a smooth surface. Determine its new angular velocity just after the hook at its edge strikes the peg *P* and the disk starts to rotate about *P* without rebounding.

Prob. 19–41

19–42. The vertical shaft is rotating with an angular velocity of 3 rad/s when $\theta = 0°$. If a force **F** is applied to the collar so that $\theta = 90°$, determine the angular velocity of the shaft. Also, find the work done by force **F**. Neglect the mass of rods GH and EF and the collars I and J. The rods AB and CD each have a mass of 10 kg.

*19–44. A 7-g bullet having a velocity of 800 m/s is fired into the edge of the 5-kg disk as shown. Determine the angular velocity of the disk just after the bullet becomes embedded in it. Also, calculate how far θ the disk will swing until it stops. The disk is originally at rest.

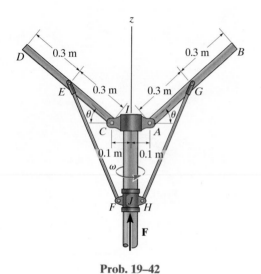

Prob. 19–42

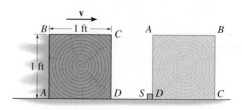

Prob. 19–44

19–43. The mass center of the 3-lb ball has a velocity of $(v_G)_1 = 6$ ft/s when it strikes the end of the smooth 5-lb slender bar which is at rest. Determine the angular velocity of the bar about the z axis just after impact if $e = 0.8$.

19–45. The 10-lb block is sliding on the smooth surface when the corner D hits a stop block S. Determine the minimum velocity **v** the block should have which would allow it to tip over on its side and land in the position shown. Neglect the size of S. *Hint:* During impact consider the weight of the block to be nonimpulsive.

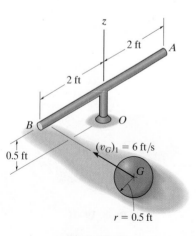

Prob. 19–43

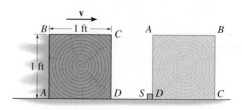

Prob. 19–45

19–46. The two disks each weigh 10 lb. If they are released from rest when $\theta = 30°$, determine θ after they collide and rebound from each other. The coefficient of restitution is $e = 0.75$. When $\theta = 0°$, the disks hang so that they just touch one another.

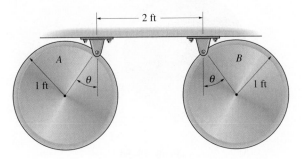

Prob. 19–46

19–47. The pendulum consists of a 10-lb solid ball and 4-lb rod. If it is released from rest when $\theta_1 = 0°$, determine the angle θ_2 after the ball strikes the wall, rebounds, and the pendulum swings up to the point of momentary rest. Take $e = 0.6$.

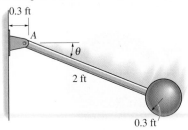

Prob. 19–47

*****19–48.** The 4-lb rod AB is hanging in the vertical position. A 2-lb block, sliding on a smooth horizontal surface with a velocity of 12 ft/s, strikes the rod at its end B. Determine the velocity of the block immediately after the collision. The coefficient of restitution between the block and the rod at B is $e = 0.8$.

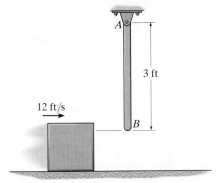

Prob. 19–48

19–49. The hammer consists of a 10-kg solid cylinder C and 6-kg uniform slender rod AB. If the hammer is released from rest when $\theta = 90°$ and strikes the 30-kg block D when $\theta = 0°$, determine the velocity of block D and the angular velocity of the hammer immediately after the impact. The coefficient of restitution between the hammer and the block is $e = 0.6$.

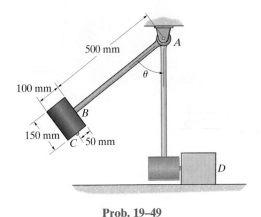

Prob. 19–49

19–50. The 6-lb slender rod AB is originally at rest, suspended in the vertical position. A 1-lb ball is thrown at the rod with a velocity $v = 50$ ft/s and strikes the rod at C. Determine the angular velocity of the rod just after the impact. Take $e = 0.7$ and $d = 2$ ft.

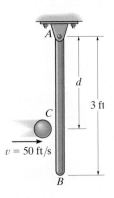

Prob. 19–50

19

19–51. The solid ball of mass m is dropped with a velocity $\mathbf{v}_1$ onto the edge of the rough step. If it rebounds horizontally off the step with a velocity $\mathbf{v}_2$, determine the angle θ at which contact occurs. Assume no slipping when the ball strikes the step. The coefficient of restitution is e.

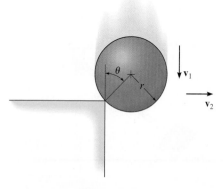

Prob. 19–51

*19–52. The wheel has a mass of 50 kg and a radius of gyration of 125 mm about its center of mass G. Determine the minimum value of the angular velocity ω_1 of the wheel, so that it strikes the step at A without rebounding and then rolls over it without slipping.

19–53. The wheel has a mass of 50 kg and a radius of gyration of 125 mm about its center of mass G. If it rolls without slipping with an angular velocity of $\omega_1 = 5$ rad/s before it strikes the step at A, determine its angular velocity after it rolls over the step. The wheel does not lose contact with the step when it strikes it.

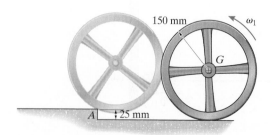

Probs. 19–52/53

19–54. The disk has a mass m and radius r. If it strikes the step without rebounding, determine the largest angular velocity ω_1 the disk can have and not lose contact with the step.

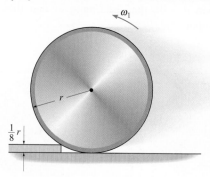

Prob. 19–54

19–55. A solid ball with a mass m is thrown on the ground such that at the instant of contact it has an angular velocity ω_1 and velocity components $(\mathbf{v}_G)_{x1}$ and $(\mathbf{v}_G)_{y1}$ as shown. If the ground is rough so no slipping occurs, determine the components of the velocity of its mass center just after impact. The coefficient of restitution is e.

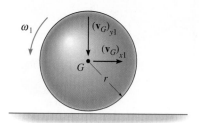

Prob. 19–55

*19–56. The pendulum consists of a 10-lb sphere and 4-lb rod. If it is released from rest when $\theta = 90°$, determine the angle θ of rebound after the sphere strikes the floor. Take $e = 0.8$.

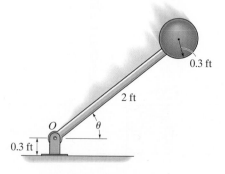

Prob. 19–56

CONCEPTUAL PROBLEMS

P19–1. The soil compactor moves forward at constant velocity by supplying power to the rear wheels. Use appropriate numerical data for the wheel, roller, and body and calculate the angular momentum of this system about point *A* at the ground, point *B* on the rear axle, and point *G*, the center of gravity for the system.

P19–1

P19–2. The swing bridge opens and closes by turning 90° using a motor located under the center of the deck at *A* that applies a torque **M** to the bridge. If the bridge was supported at its end *B*, would the same torque open the bridge at the same time, or would it open slower or faster? Explain your answer using numerical values and an impulse and momentum analysis. Also, what are the benefits of making the bridge have the variable depth as shown?

P19–2

P19–3. Why is it necessary to have the tail blade *B* on the helicopter that spins perpendicular to the spin of the main blade *A*? Explain your answer using numerical values and an impulse and momentum analysis.

P19–3

P19–4. The amusement park ride consists of two gondolas *A* and *B*, and counterweights *C* and *D* that swing in opposite directions. Using realistic dimensions and mass, calculate the angular momentum of this system for any angular position of the gondolas. Explain through analysis why it is a good idea to design this system to have counterweights with each gondola.

P19–4

CHAPTER REVIEW

Linear and Angular Momentum

The linear and angular momentum of a rigid body can be referenced to its mass center G.

If the angular momentum is to be determined about an axis other than the one passing through the mass center, then the angular momentum is determined by summing vector $\mathbf{H}_G$ and the moment of vector $\mathbf{L}$ about this axis.

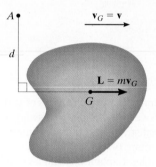

Translation

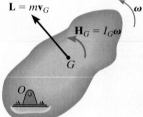

Rotation about a fixed axis

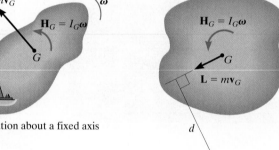

General plane motion

$$L = mv_G$$

$$H_G = 0$$

$$H_A = (mv_G)d$$

$$L = mv_G$$

$$H_G = I_G\omega$$

$$H_O = I_O\omega$$

$$L = mv_G$$

$$H_G = I_G\omega$$

$$H_A = I_G\omega + (mv_G)d$$

Principle of Impulse and Momentum

The principles of linear and angular impulse and momentum are used to solve problems that involve force, velocity, and time. Before applying these equations, it is important to establish the x, y, z inertial coordinate system. The free-body diagram for the body should also be drawn in order to account for all of the forces and couple moments that produce impulses on the body.

$$m(v_{Gx})_1 + \Sigma \int_{t_1}^{t_2} F_x \, dt = m(v_{Gx})_2$$

$$m(v_{Gy})_1 + \Sigma \int_{t_1}^{t_2} F_y \, dt = m(v_{Gy})_2$$

$$I_G\omega_1 + \Sigma \int_{t_1}^{t_2} M_G \, dt = I_G\omega_2$$

Conservation of Momentum

Provided the sum of the linear impulses acting on a system of connected rigid bodies is zero in a particular direction, then the linear momentum for the system is conserved in this direction. Conservation of angular momentum occurs if the impulses pass through an axis or are parallel to it. Momentum is also conserved if the external forces are small and thereby create nonimpulsive forces on the system. A free-body diagram should accompany any application in order to classify the forces as impulsive or nonimpulsive and to determine an axis about which the angular momentum may be conserved.

$$\left(\Sigma \begin{matrix} \text{syst. linear} \\ \text{momentum} \end{matrix} \right)_1 = \left(\Sigma \begin{matrix} \text{syst. linear} \\ \text{momentum} \end{matrix} \right)_2$$

$$\left(\Sigma \begin{matrix} \text{syst. angular} \\ \text{momentum} \end{matrix} \right)_{O1} = \left(\Sigma \begin{matrix} \text{syst. angular} \\ \text{momentum} \end{matrix} \right)_{O2}$$

Eccentric Impact

If the line of impact does not coincide with the line connecting the mass centers of two colliding bodies, then eccentric impact will occur. If the motion of the bodies just after the impact is to be determined, then it is necessary to consider a conservation of momentum equation for the system and use the coefficient of restitution equation.

$$e = \frac{(v_B)_2 - (v_A)_2}{(v_A)_1 - (v_B)_1}$$

19

Review

2

Planar Kinematics and Kinetics of a Rigid Body

Having presented the various topics in planar kinematics and kinetics in Chapters 16 through 19, we will now summarize these principles and provide an opportunity for applying them to the solution of various types of problems.

Kinematics. Here we are interested in studying the geometry of motion, without concern for the forces which cause the motion. Before solving a planar kinematics problem, it is *first* necessary to *classify the motion* as being either rectilinear or curvilinear translation, rotation about a fixed axis, or general plane motion. In particular, problems involving general plane motion can be solved either with reference to a fixed axis (absolute motion analysis) or using translating or rotating frames of reference (relative motion analysis). The choice generally depends upon the type of constraints and the problem's geometry. In all cases, application of the necessary equations can be clarified by drawing a kinematic diagram. Remember that the *velocity* of a point is always *tangent* to its path of motion, and the *acceleration* of a point can have *components* in the $n-t$ directions when the path is *curved*.

Translation. When the body moves with rectilinear or curvilinear translation, *all* the points on the body have the *same motion*.

$$\mathbf{v}_B = \mathbf{v}_A \qquad \mathbf{a}_B = \mathbf{a}_A$$

Rotation About a Fixed Axis. Angular Motion.
Variable Angular Acceleration. Provided a mathematical relationship is given between *any two* of the *four* variables θ, ω, α, and t, then a *third* variable can be determined by solving one of the following equations which relate all three variables.

$$\omega = \frac{d\theta}{dt} \qquad \alpha = \frac{d\omega}{dt} \qquad \alpha \, d\theta = \omega \, d\omega$$

Constant Angular Acceleration. The following equations apply when it is *absolutely certain* that the angular acceleration is constant.

$$\theta = \theta_0 + \omega_0 t + \tfrac{1}{2}\alpha_c t^2 \qquad \omega = \omega_0 + \alpha_c t \qquad \omega^2 = \omega_0^2 + 2\alpha_c(\theta - \theta_0)$$

Motion of Point P.

Once $\boldsymbol{\omega}$ and $\boldsymbol{\alpha}$ have been determined, then the circular motion of point P can be specified using the following scalar or vector equations.

$$v = \omega r \qquad\qquad \mathbf{v} = \boldsymbol{\omega} \times \mathbf{r}$$

$$a_t = \alpha r \quad a_n = \omega^2 r \qquad \mathbf{a} = \boldsymbol{\alpha} \times \mathbf{r} - \omega^2 \mathbf{r}$$

General Plane Motion—Relative-Motion Analysis.

Recall that when *translating axes* are placed at the "base point" A, the *relative motion* of point B with respect to A is simply *circular motion of B about A*. The following equations apply to two points A and B located on the *same* rigid body.

$$\mathbf{v}_B = \mathbf{v}_A + \mathbf{v}_{B/A} = \mathbf{v}_A + \boldsymbol{\omega} \times \mathbf{r}_{B/A}$$

$$\mathbf{a}_B = \mathbf{a}_A + \mathbf{a}_{B/A} = \mathbf{a}_A + \boldsymbol{\alpha} \times \mathbf{r}_{B/A} - \omega^2 \mathbf{r}_{B/A}$$

Rotating and translating axes are often used to analyze the motion of rigid bodies which are connected together by collars or slider blocks.

$$\mathbf{v}_B = \mathbf{v}_A + \boldsymbol{\Omega} \times \mathbf{r}_{B/A} + (\mathbf{v}_{B/A})_{xyz}$$

$$\mathbf{a}_B = \mathbf{a}_A + \dot{\boldsymbol{\Omega}} \times \mathbf{r}_{B/A} + \boldsymbol{\Omega} \times (\boldsymbol{\Omega} \times \mathbf{r}_{B/A}) + 2\boldsymbol{\Omega} \times (\mathbf{v}_{B/A})_{xyz} + (\mathbf{a}_{B/A})_{xyz}$$

Kinetics.

To analyze the forces which cause the motion we must use the principles of kinetics. When applying the necessary equations, it is important to first establish the inertial coordinate system and define the positive directions of the axes. The *directions* should be the *same* as those selected when writing any equations of kinematics if *simultaneous solution* of equations becomes necessary.

Equations of Motion.

These equations are used to determine accelerated motions or forces causing the motion. If used to determine position, velocity, or time of motion, then kinematics will have to be considered to complete the solution. Before applying the equations of motion, *always draw a free-body diagram* in order to identify all the forces acting on the body. Also, establish the directions of the acceleration of the mass center and the angular acceleration of the body. (A kinetic diagram may also be drawn in order to represent $m\mathbf{a}_G$ and $I_G\boldsymbol{\alpha}$ graphically. This diagram is particularly convenient for resolving $m\mathbf{a}_G$ into components and for identifying the terms in the moment sum $\Sigma(\mathcal{M}_k)_P$.)

The three equations of motion are

$$\Sigma F_x = m(a_G)_x$$

$$\Sigma F_y = m(a_G)_y$$

$$\Sigma M_G = I_G\alpha \quad \text{or} \quad \Sigma M_P = \Sigma(\mathcal{M}_k)_P$$

In particular, if the body is *rotating about a fixed axis,* moments may also be summed about point O on the axis, in which case

$$\Sigma M_O = \Sigma(\mathcal{M}_k)_O = I_O\alpha$$

Work and Energy.   *The equation of work and energy is used to solve problems involving force, velocity, and displacement.* Before applying this equation, *always draw a free-body diagram of the body in order to identify the forces which do work.* Recall that the kinetic energy of the body is due to translational motion of the mass center, $\mathbf{v}_G$, *and* rotational motion of the body, $\boldsymbol{\omega}$.

$$T_1 + \Sigma U_{1-2} = T_2$$

where

$$T = \tfrac{1}{2}mv_G^2 + \tfrac{1}{2}I_G\omega^2$$

$$U_F = \int F\cos\theta\, ds \qquad \text{(variable force)}$$

$$U_{F_c} = F_c\cos\theta(s_2 - s_1) \qquad \text{(constant force)}$$

$$U_W = -W\,\Delta y \qquad \text{(weight)}$$

$$U_s = -\left(\tfrac{1}{2}ks_2^2 - \tfrac{1}{2}ks_1^2\right) \qquad \text{(spring)}$$

$$U_M = M\theta \qquad \text{(constant couple moment)}$$

If the forces acting on the body are *conservative forces,* then apply the *conservation of energy equation.* This equation is easier to use than the equation of work and energy, since it applies only at *two points* on the path and *does not* require calculation of the work done by a force as the body moves along the path.

$$T_1 + V_1 = T_2 + V_2$$

where $V = V_g + V_e$ and

$$V_g = Wy \quad \text{(gravitational potential energy)}$$

$$V_e = \tfrac{1}{2}ks^2 \quad \text{(elastic potential energy)}$$

Impulse and Momentum.   *The principles of linear and angular impulse and momentum are used to solve problems involving force, velocity, and time.* Before applying the equations, *draw a free-body diagram* in order to identify all the forces which cause linear and angular impulses on the body. Also, establish the directions of the velocity of the mass center and the angular velocity of the body just before and just after the impulses are applied. (As an alternative procedure, the impulse and momentum diagrams may accompany the solution in order to graphically account for the terms in the equations. These diagrams are particularly advantageous when computing the angular impulses and angular momenta about a point other than the body's mass center.)

$$m(\mathbf{v}_G)_1 + \Sigma \int \mathbf{F}\, dt = m(\mathbf{v}_G)_2$$

$$(\mathbf{H}_G)_1 + \Sigma \int \mathbf{M}_G\, dt = (\mathbf{H}_G)_2$$

or

$$(\mathbf{H}_O)_1 + \Sigma \int \mathbf{M}_O\, dt = (\mathbf{H}_O)_2$$

Conservation of Momentum.   If nonimpulsive forces or no impulsive forces act on the body in a particular direction, or if the motions of several bodies are involved in the problem, then consider applying the conservation of linear or angular momentum for the solution. Investigation of the free-body diagram (or the impulse diagram) will aid in determining the directions along which the impulsive forces are zero, or axes about which the impulsive forces create zero angular impulse. For these cases,

$$m(\mathbf{v}_G)_1 = m(\mathbf{v}_G)_2$$

$$(\mathbf{H}_O)_1 = (\mathbf{H}_O)_2$$

The problems that follow involve application of all the above concepts. They are presented in *random order* so that practice may be gained at identifying the various types of problems and developing the skills necessary for their solution.

R2

REVIEW PROBLEMS

R2–1. Blocks A and B weigh 50 and 10 lb, respectively. If $P = 100$ lb, determine the normal force exerted by block A on block B. Neglect friction and the weights of the pulleys, cord, and bars of the triangular frame.

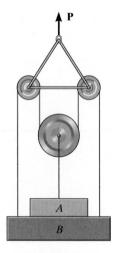

Prob. R2–1

R2–2. The handcart has a mass of 200 kg and center of mass at G. Determine the normal reactions at *each* of the wheels at A and B if a force $P = 50$ N is applied to the handle. Neglect the mass and rolling resistance of the wheels.

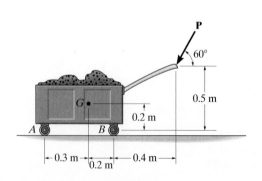

Prob. R2–2

R2–3. The truck carries the 800-lb crate which has a center of gravity at G_c. Determine the largest acceleration of the truck so that the crate will not slip or tip on the truck bed. The coefficient of static friction between the crate and the truck is $\mu_s = 0.6$.

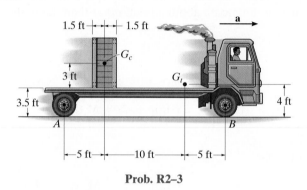

Prob. R2–3

*R2–4. The spool has a weight of 30 lb and a radius of gyration $k_O = 0.65$ ft. If a force of 40 lb is applied to the cord at A, determine the angular velocity of the spool in $t = 3$ s starting from rest. Neglect the mass of the pulley and cord.

R2–5. Solve Prob. R2–4 if a 40-lb block is suspended from the cord at A, rather than applying the 40-lb force.

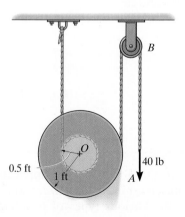

Probs. R2–4/5

R2–6. The uniform plate weighs 40 lb and is supported by a roller at A. If a horizontal force $F = 70$ lb is suddenly applied to the roller, determine the acceleration of the center of the roller at the instant the force is applied. The plate has a moment of inertia about its center of mass of $I_G = 0.414$ slug $\cdot$ ft^2. Neglect the weight of the roller.

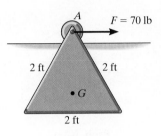

Prob. R2–6

R2–7. The center of the pulley is being lifted vertically with an acceleration of 4 m/s^2 at the instant it has a velocity of 2 m/s. If the cable does not slip on the pulley's surface, determine the accelerations of the cylinder B and point C on the pulley.

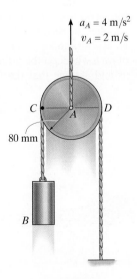

Prob. R2–7

*R2–8. The double pendulum consists of two rods. Rod AB has a constant angular velocity of 3 rad/s, and rod BC has a constant angular velocity of 2 rad/s. Both of these absolute motions are measured counterclockwise. Determine the velocity and acceleration of point C at the instant shown.

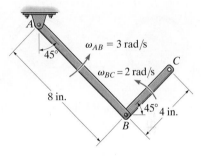

Prob. R2–8

R2–9. The link OA is pinned at O and rotates because of the sliding action of rod R along the horizontal groove. If R starts from rest when $\theta = 0°$ and has a constant acceleration $a_R = 60$ mm/s^2 to the right, determine the angular velocity and angular acceleration of OA when $t = 2$ s.

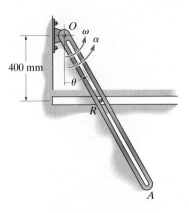

Prob. R2–9

R2–10. The drive wheel A has a constant angular velocity of ω_A. At a particular instant, the radius of rope wound on each wheel is as shown. If the rope has a thickness T, determine the angular acceleration of wheel B.

*R2–12. If the ball has a weight of 15 lb and is thrown onto a *rough surface* so that its center has a velocity of 6 ft/s parallel to the surface, determine the amount of backspin, ω, the ball must be given so that it stops spinning at the same instant that its forward velocity is zero. It is not necessary to know the coefficient of kinetic friction at A for the calculation.

Prob. R2–10

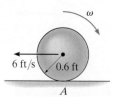

Prob. R2–12

R2–11. The dresser has a weight of 80 lb and is pushed along the floor. If the coefficient of static friction at A and B is $\mu_s = 0.3$ and the coefficient of kinetic friction is $\mu_k = 0.2$, determine the smallest horizontal force P needed to cause motion. If this force is increased slightly, determine the acceleration of the dresser. Also, what are the normal reactions at A and B when it begins to move?

R2–13. The dragster has a mass of 1500 kg and a center of mass at G. If the coefficient of kinetic friction between the rear wheels and the pavement is $\mu_k = 0.6$, determine if it is possible for the driver to lift the front wheels, A, off the ground while the rear wheels are slipping. If so, what acceleration is necessary to do this? Neglect the mass of the wheels and assume that the front wheels are free to roll.

R2–14. The dragster has a mass of 1500 kg and a center of mass at G. If no slipping occurs, determine the friction force F_B which must be applied to *each* of the rear wheels B in order to develop an acceleration $a = 6$ m/s^2. What are the normal reactions of *each* wheel on the ground? Neglect the mass of the wheels and assume that the front wheels are free to roll.

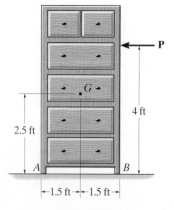

Prob. R2–11

Probs. R2–13/14

R2–15. If the operator initially drives the pedals at 20 rev/min, and then begins an angular acceleration of 30 rev/min², determine the angular velocity of the flywheel F when $t = 3$ s. Note that the pedal arm is fixed connected to the chain wheel A, which in turn drives the sheave B using the fixed connected clutch gear D. The belt wraps around the sheave then drives the pulley E and fixed-connected flywheel.

*R2–16. If the operator initially drives the pedals at 12 rev/min, and then begins an angular acceleration of 8 rev/min², determine the angular velocity of the flywheel F after the pedal arm has rotated 2 revolutions. Note that the pedal arm is fixed connected to the chain wheel A, which in turn drives the sheave B using the fixed-connected clutch gear D. The belt wraps around the sheave then drives the pulley E and fixed-connected flywheel.

$r_A = 125$ mm $\quad r_B = 175$ mm
$r_D = 20$ mm $\quad r_E = 30$ mm

Probs. R2–15/16

R2–17. The drum has a mass of 50 kg and a radius of gyration about the pin at O of $k_O = 0.23$ m. Starting from rest, the suspended 15-kg block B is allowed to fall 3 m without applying the brake ACD. Determine the speed of the block at this instant. If the coefficient of kinetic friction at the brake pad C is $\mu_k = 0.5$, determine the force $\mathbf{P}$ that must be applied at the brake handle which will then stop the block after it descends *another* 3 m. Neglect the thickness of the handle.

R2–18. The drum has a mass of 50 kg and a radius of gyration about the pin at O of $k_O = 0.23$ m. If the 15-kg block is moving downward at 3 m/s, and a force of $P = 100$ N is applied to the brake arm, determine how far the block descends from the instant the brake is applied until it stops. Neglect the thickness of the handle. The coefficient of kinetic friction at the brake pad is $\mu_k = 0.5$.

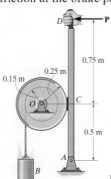

Probs. R2–17/18

R2–19. The 1.6-Mg car shown has been "raked" by increasing the height $h = 0.2$ m of its center of mass. This was done by raising the springs on the rear axle. If the coefficient of static friction between the rear wheels and the ground is $\mu_s = 0.3$, show that the car can accelerate slightly faster than its counterpart for which $h = 0$. Neglect the mass of the wheels and driver and assume the front wheels at B are free to roll while the rear wheels slip.

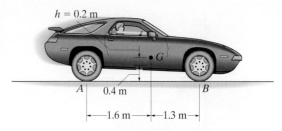

$h = 0.2$ m

0.4 m

A $\qquad$ B

1.6 m $\quad$ 1.3 m

Prob. R2–19

*R2–20. The disk is rotating at a constant rate $\omega = 4$ rad/s, and as it falls freely, its center has an acceleration of 32.2 ft/s². Determine the acceleration of point A on the rim of the disk at the instant shown.

R2–21. The disk is rotating at a constant rate $\omega = 4$ rad/s, and as it falls freely, its center has an acceleration of 32.2 ft/s². Determine the acceleration of point B on the rim of the disk at the instant shown.

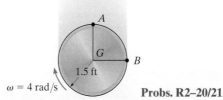

A

G

B

1.5 ft

$\omega = 4$ rad/s

Probs. R2–20/21

R2–22. The board rests on the surface of two drums. At the instant shown, it has an acceleration of 0.5 m/s² to the right, while at the same instant points on the outer rim of each drum have an acceleration with a magnitude of 3 m/s². If the board does not slip on the drums, determine its speed due to the motion.

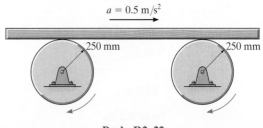

$a = 0.5$ m/s²

250 mm $\qquad$ 250 mm

Prob. R2–22

R2–23. A 20-kg roll of paper, originally at rest, is pin-supported at its ends to bracket AB. The roll rests against a wall for which the coefficient of kinetic friction at C is $\mu_C = 0.3$. If a force of 40 N is applied uniformly to the end of the sheet, determine the initial angular acceleration of the roll and the tension in the bracket as the paper unwraps. For the calculation, treat the roll as a cylinder.

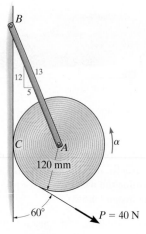

Prob. R2–23

*R2–24. At the instant shown, link AB has an angular velocity $\omega_{AB} = 2 \text{ rad/s}$ and an angular acceleration $\alpha_{AB} = 6 \text{ rad/s}^2$. Determine the acceleration of the pin at C and the angular acceleration of link CB at this instant, when $\theta = 60°$.

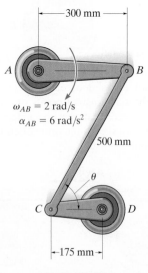

Prob. R2–24

R2–25. The truck has a weight of 8000 lb and center of gravity at G_t. It carries the 800-lb crate, which has a center of gravity at G_c. Determine the normal reaction at *each* of its four tires if it accelerates at $a = 0.5 \text{ ft/s}^2$. Also, what is the frictional force acting between the crate and the truck, and between *each* of the rear tires and the road? Assume that power is delivered only to the rear tires. The front tires are free to roll. Neglect the mass of the tires. The crate does not slip or tip on the truck.

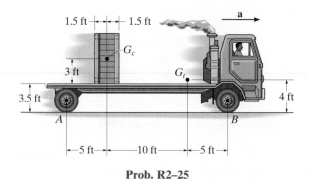

Prob. R2–25

R2–26. The 15-lb cylinder is initially at rest on a 5-lb plate. If a couple moment $M = 40 \text{ lb} \cdot \text{ft}$ is applied to the cylinder, determine the angular acceleration of the cylinder and the time needed for the end B of the plate to travel 3 ft and strike the wall. Assume the cylinder does not slip on the plate, and neglect the mass of the rollers under the plate.

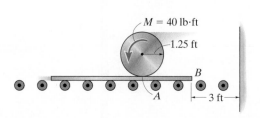

Prob. R2–26

R2–27. At the instant shown, two forces act on the 30-lb slender rod which is pinned at O. Determine the magnitude of force **F** and the initial angular acceleration of the rod so that the horizontal reaction which the *pin exerts on the rod* is 5 lb directed to the right.

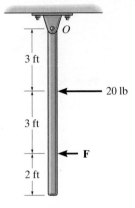

3 ft

20 lb

3 ft

F

2 ft

Prob. R2–27

R2–30. The wheelbarrow and its contents have a mass of 40 kg and a mass center at G, excluding the wheel. The wheel has a mass of 4 kg and a radius of gyration $k_O = 0.120$ m. If the wheelbarrow is released from rest from the position shown, determine its speed after it travels 4 m down the incline. The coefficient of kinetic friction between the incline and A is $\mu_A = 0.3$. The wheels roll without slipping at B.

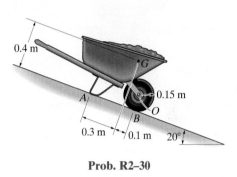

0.4 m

G

A

0.15 m

B

O

0.3 m   0.1 m   20°

Prob. R2–30

*R2–28. The 20-lb solid ball is cast on the floor such that it has a backspin $\omega = 15$ rad/s and its center has an initial horizontal velocity $v_G = 20$ ft/s. If the coefficient of kinetic friction between the floor and the ball is $\mu_A = 0.3$, determine the distance it travels before it stops spinning.

R2–29. Determine the backspin ω which should be given to the 20-lb ball so that when its center is given an initial horizontal velocity $v_G = 20$ ft/s it stops spinning and translating at the same instant. The coefficient of kinetic friction is $\mu_A = 0.3$.

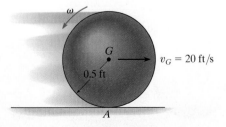

ω

G

$v_G = 20$ ft/s

0.5 ft

A

Probs. R2–28/29

R2–31. At the given instant member AB has the angular motions shown. Determine the velocity and acceleration of the slider block C at this instant.

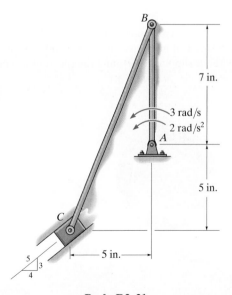

B

7 in.

3 rad/s

2 rad/s²

A

5 in.

C

5

3

4

5 in.

Prob. R2–31

*R2–32. The spool and wire wrapped around its core have a mass of 20 kg and a centroidal radius of gyration $k_G = 250$ mm. If the coefficient of kinetic friction at the ground is $\mu_B = 0.1$, determine the angular acceleration of the spool when the 30-N · m couple moment is applied.

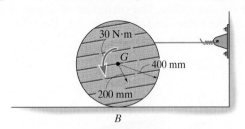

Prob. R2–32

R2–33. The car has a mass of 1.50 Mg and a mass center at G. Determine the maximum acceleration it can have if (a) power is supplied only to the rear wheels, (b) power is supplied only to the front wheels. Neglect the mass of the wheels in the calculation, and assume that the wheels that do not receive power are free to roll. Also, assume that slipping of the powered wheels occurs, where the coefficient of kinetic friction is $\mu_k = 0.3$.

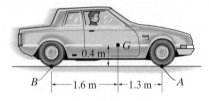

Prob. R2–33

R2–34. The tire has a mass of 9 kg and a radius of gyration $k_O = 225$ mm. If it is released from rest and rolls down the plane without slipping, determine the speed of its center O when $t = 3$ s.

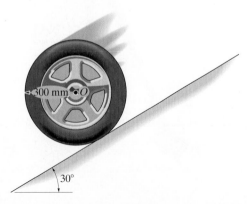

Prob. R2–34

R2–35. The bar has a mass m and length l. If it is released from rest from the position $\theta = 30°$, determine its angular acceleration and the horizontal and vertical components of reaction at the pin O.

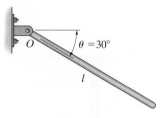

Prob. R2–35

*R2–36. The pendulum consists of a 30-lb sphere and a 10-lb slender rod. Compute the reaction at the pin O just after the cord AB is cut.

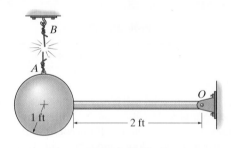

Prob. R2–36

R2–37. Spool B is at rest and spool A is rotating at 6 rad/s when the slack in the cord connecting them is taken up. If the cord does not stretch, determine the angular velocity of each spool immediately after the cord is jerked tight. The spools A and B have weights and radii of gyration $W_A = 30$ lb, $k_A = 0.8$ ft and $W_B = 15$ lb, $k_B = 0.6$ ft, respectively.

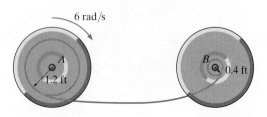

Prob. R2–37

R2–38. The rod is bent into the shape of a sine curve and is forced to rotate about the y axis by connecting the spindle S to a motor. If the rod starts from rest in the position shown and a motor drives it for a short time with an angular acceleration $\alpha = (1.5e^t)\,\text{rad/s}^2$, where t is in seconds, determine the magnitudes of the angular velocity and angular displacement of the rod when $t = 3$ s. Locate the point on the rod which has the greatest velocity and acceleration, and compute the magnitudes of the velocity and acceleration of this point when $t = 3$ s. The curve defining the rod is $z = 0.25 \sin(\pi y)$, where the argument for the sine is given in radians when y is in meters.

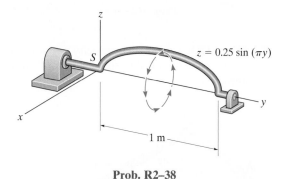

Prob. R2–38

R2–39. The scaffold S is raised by moving the roller at A toward the pin at B. If A is approaching B with a speed of 1.5 ft/s, determine the speed at which the platform rises as a function of θ. The 4-ft links are pin connected at their midpoint.

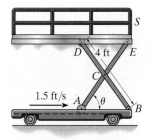

Prob. R2–39

***R2–40.** The pendulum of the Charpy impact machine has a mass of 50 kg and a radius of gyration of $k_A = 1.75$ m. If it is released from rest when $\theta = 0°$, determine its angular velocity just before it strikes the specimen S, $\theta = 90°$.

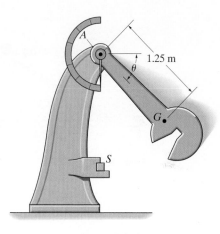

Prob. R2–40

R2–41. The gear rack has a mass of 6 kg, and the gears each have a mass of 4 kg and a radius of gyration $k = 30$ mm at their centers. If the rack is originally moving downward at 2 m/s, when $s = 0$. determine the speed of the rack when $s = 600$ mm. The gears are free to turn about their centers, A and B.

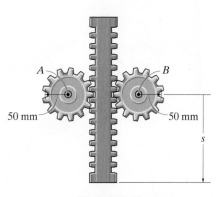

Prob. R2–41

R2–42.   A 7-kg automobile tire is released from rest at A on the incline and rolls without slipping to point B, where it then travels in free flight. Determine the maximum height h the tire attains. The radius of gyration of the tire about its mass center is $k_G = 0.3$ m.

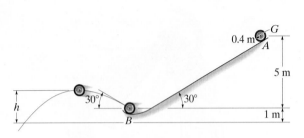

Prob. R2–42

R2–43.   The two 3-lb rods EF and HI are fixed (welded) to the link AC at E. Determine the internal axial force E_x, shear force E_y, and moment M_E, which the bar AC exerts on FE at E if at the instant $\theta = 30°$ link AB has an angular velocity $\omega = 5$ rad/s and an angular acceleration $\alpha = 8$ rad/s^2 as shown.

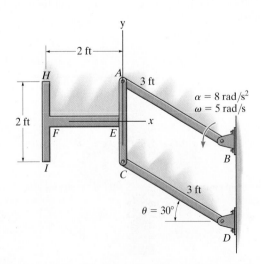

Prob. R2–43

*R2–44.   The uniform connecting rod BC has a mass of 3 kg and is pin-connected at its end points. Determine the vertical forces which the pins exert on the ends B and C of the rod at the instant (a) $\theta = 0°$, and (b) $\theta = 90°$. The crank AB is turning with a constant angular velocity $\omega_{AB} = 5$ rad/s.

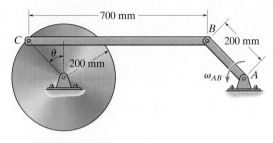

Prob. R2–44

R2–45.   If bar AB has an angular velocity $\omega_{AB} = 6$ rad/s, determine the velocity of the slider block C at the instant shown.

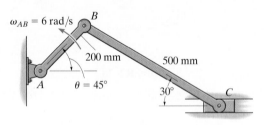

Prob. R2–45

R2–46.   The drum of mass m, radius r, and radius of gyration k_O rolls along an inclined plane for which the coefficient of static friction is μ. If the drum is released from rest, determine the maximum angle θ for the incline so that it rolls without slipping.

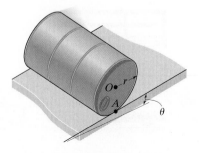

Prob. R2–46

R2–47. Determine the velocity and acceleration of rod R for any angle θ of cam C if the cam rotates with a constant angular velocity ω. The pin connection at O does not cause an interference with the motion of A on C.

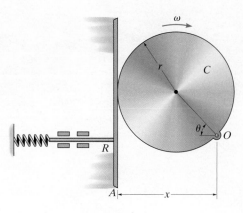

Prob. R2–47

*R2–48. When the crank on the Chinese windlass is turning, the rope on shaft A unwinds while that on shaft B winds up. Determine the speed at which the block lowers if the crank is turning with an angular velocity $\omega = 4$ rad/s. What is the angular velocity of the pulley at C? The rope segments on each side of the pulley are both parallel and vertical, and the rope does not slip on the pulley.

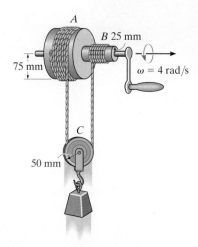

Prob. R2–48

R2–49. The semicircular disk has a mass of 50 kg and is released from rest from the position shown. The coefficients of static and kinetic friction between the disk and the beam are $\mu_s = 0.5$ and $\mu_k = 0.3$, respectively. Determine the initial reactions at the pin A and roller B, used to support the beam. Neglect the mass of the beam for the calculation.

R2–50. The semicircular disk has a mass of 50 kg and is released from rest from the position shown. The coefficients of static and kinetic friction between the disk and the beam are $\mu_s = 0.2$ and $\mu_k = 0.1$, respectively. Determine the initial reactions at the pin A and roller B used to support the beam. Neglect the mass of the beam for the calculation.

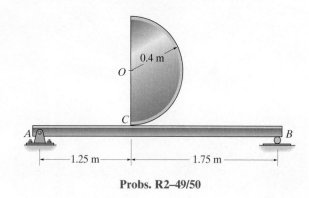

Probs. R2–49/50

R2–51. The hoisting gear A has an initial angular velocity of 60 rad/s and a constant deceleration of 1 rad/s². Determine the velocity and deceleration of the block which is being hoisted by the hub on gear B when $t = 3$ s.

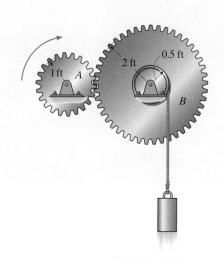

Prob. R2–51

Chapter 20

Design of industrial robots requires knowing the kinematics of their three-dimensional motions.

Three-Dimensional Kinematics of a Rigid Body

CHAPTER OBJECTIVES

- To analyze the kinematics of a body subjected to rotation about a fixed point and general plane motion.

- To provide a relative-motion analysis of a rigid body using translating and rotating axes.

20.1   Rotation About a Fixed Point

When a rigid body rotates about a fixed point, the distance r from the point to a particle located on the body is the *same* for *any position* of the body. Thus, the path of motion for the particle lies on the *surface of a sphere* having a radius r and centered at the fixed point. Since motion along this path occurs only from a series of rotations made during a finite time interval, we will first develop a familiarity with some of the properties of rotational displacements.

The boom can rotate up and down, and because it is hinged at a point on the vertical axis about which it turns, it is subjected to rotation about a fixed point.

Euler's Theorem.

Euler's theorem states that two "component" rotations about different axes passing through a point are equivalent to a single resultant rotation about an axis passing through the point. If more than two rotations are applied, they can be combined into pairs, and each pair can be further reduced and combined into one rotation.

Finite Rotations.

If component rotations used in Euler's theorem are *finite*, it is important that the *order* in which they are applied be maintained. To show this, consider the two finite rotations $\boldsymbol{\theta}_1 + \boldsymbol{\theta}_2$ applied to the block in Fig. 20–1a. Each rotation has a magnitude of 90° and a direction defined by the right-hand rule, as indicated by the arrow. The final position of the block is shown at the right. When these two rotations are applied in the order $\boldsymbol{\theta}_2 + \boldsymbol{\theta}_1$, as shown in Fig. 20–1b, the final position of the block is *not* the same as it is in Fig. 20–1a. Because *finite rotations* do not obey the commutative law of addition $(\boldsymbol{\theta}_1 + \boldsymbol{\theta}_2 \neq \boldsymbol{\theta}_2 + \boldsymbol{\theta}_1)$, *they cannot be classified as vectors*. If smaller, yet finite, rotations had been used to illustrate this point, e.g., 10° instead of 90°, the *final position* of the block after each combination of rotations would also be different; however, in this case, the difference is only a small amount.

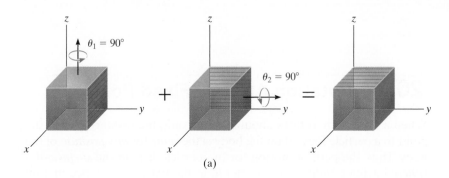

(a)

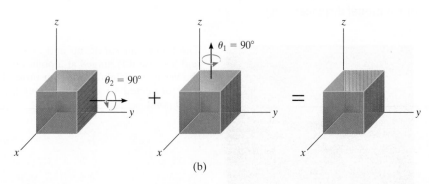

(b)

Fig. 20–1

Infinitesimal Rotations.

When defining the angular motions of a body subjected to three-dimensional motion, only rotations which are *infinitesimally small* will be considered. *Such rotations can be classified as vectors, since they can be added vectorially in any manner.* To show this, for purposes of simplicity let us consider the rigid body itself to be a sphere which is allowed to rotate about its central fixed point O, Fig. 20–2a. If we impose two infinitesimal rotations $d\boldsymbol{\theta}_1 + d\boldsymbol{\theta}_2$ on the body, it is seen that point P moves along the path $d\boldsymbol{\theta}_1 \times \mathbf{r} + d\boldsymbol{\theta}_2 \times \mathbf{r}$ and ends up at P'. Had the two successive rotations occurred in the order $d\boldsymbol{\theta}_2 + d\boldsymbol{\theta}_1$, then the resultant displacements of P would have been $d\boldsymbol{\theta}_2 \times \mathbf{r} + d\boldsymbol{\theta}_1 \times \mathbf{r}$. Since the vector cross product obeys the distributive law, by comparison $(d\boldsymbol{\theta}_1 + d\boldsymbol{\theta}_2) \times \mathbf{r} = (d\boldsymbol{\theta}_2 + d\boldsymbol{\theta}_1) \times \mathbf{r}$. Here infinitesimal rotations $d\boldsymbol{\theta}$ are vectors, since these quantities have both a magnitude and direction for which the order of (vector) addition is not important, i.e., $d\boldsymbol{\theta}_1 + d\boldsymbol{\theta}_2 = d\boldsymbol{\theta}_2 + d\boldsymbol{\theta}_1$. As a result, as shown in Fig. 20–2a, the two "component" rotations $d\boldsymbol{\theta}_1$ and $d\boldsymbol{\theta}_2$ are equivalent to a single resultant rotation $d\boldsymbol{\theta} = d\boldsymbol{\theta}_1 + d\boldsymbol{\theta}_2$, a consequence of Euler's theorem.

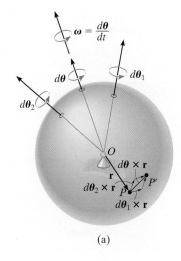

(a)

Angular Velocity.

If the body is subjected to an angular rotation $d\boldsymbol{\theta}$ about a fixed point, the angular velocity of the body is defined by the time derivative,

$$\boldsymbol{\omega} = \dot{\boldsymbol{\theta}} \qquad (20\text{–}1)$$

The line specifying the direction of $\boldsymbol{\omega}$, which is collinear with $d\boldsymbol{\theta}$, is referred to as the *instantaneous axis of rotation*, Fig. 20–2b. In general, this axis changes direction during each instant of time. Since $d\boldsymbol{\theta}$ is a vector quantity, so too is $\boldsymbol{\omega}$, and it follows from vector addition that if the body is subjected to two component angular motions, $\boldsymbol{\omega}_1 = \dot{\boldsymbol{\theta}}_1$ and $\boldsymbol{\omega}_2 = \dot{\boldsymbol{\theta}}_2$, the resultant angular velocity is $\boldsymbol{\omega} = \boldsymbol{\omega}_1 + \boldsymbol{\omega}_2$.

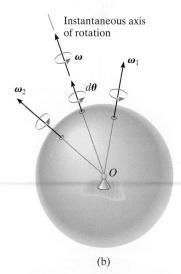

(b)

Fig. 20–2

Angular Acceleration.

The body's angular acceleration is determined from the time derivative of its angular velocity, i.e.,

$$\boldsymbol{\alpha} = \dot{\boldsymbol{\omega}} \qquad (20\text{–}2)$$

For motion about a fixed point, $\boldsymbol{\alpha}$ must account for a change in *both* the magnitude and direction of $\boldsymbol{\omega}$, so that, in general, $\boldsymbol{\alpha}$ is not directed along the instantaneous axis of rotation, Fig. 20–3.

As the direction of the instantaneous axis of rotation (or the line of action of $\boldsymbol{\omega}$) changes in space, the locus of the axis generates a fixed *space cone*, Fig. 20–4. If the change in the direction of this axis is viewed with respect to the rotating body, the locus of the axis generates a *body cone*.

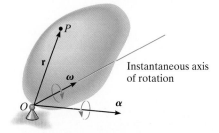

Fig. 20–3

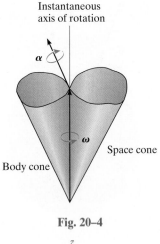

Fig. 20–4

At any given instant, these cones meet along the instantaneous axis of rotation, and when the body is in motion, the body cone appears to roll either on the inside or the outside surface of the fixed space cone. Provided the paths defined by the open ends of the cones are described by the head of the $\boldsymbol{\omega}$ vector, then $\boldsymbol{\alpha}$ must act tangent to these paths at any given instant, since the time rate of change of $\boldsymbol{\omega}$ is equal to $\boldsymbol{\alpha}$. Fig. 20–4.

To illustrate this concept, consider the disk in Fig. 20–5a that spins about the rod at $\boldsymbol{\omega}_s$, while the rod and disk precess about the vertical axis at $\boldsymbol{\omega}_p$. The resultant angular velocity of the disk is therefore $\boldsymbol{\omega} = \boldsymbol{\omega}_s + \boldsymbol{\omega}_p$. Since both point O and the contact point P have zero velocity, then both $\boldsymbol{\omega}$ and the instantaneous axis of rotation are along OP. Therefore, as the disk rotates, this axis appears to move along the surface of the fixed space cone shown in Fig. 20–5b. If the axis is observed from the rotating disk, the axis then appears to move on the surface of the body cone. At any instant, though, these two cones meet each other along the axis OP. If $\boldsymbol{\omega}$ has a constant magnitude, then $\boldsymbol{\alpha}$ indicates only the change in the direction of $\boldsymbol{\omega}$, which is tangent to the cones at the tip of $\boldsymbol{\omega}$ as shown in Fig. 20–5b.

Velocity.   Once $\boldsymbol{\omega}$ is specified, the velocity of any point on a body rotating about a fixed point can be determined using the same methods as for a body rotating about a fixed axis. Hence, by the cross product,

$$\mathbf{v} = \boldsymbol{\omega} \times \mathbf{r} \tag{20–3}$$

Here $\mathbf{r}$ defines the position of the point measured from the fixed point O, Fig. 20–3.

Acceleration.   If $\boldsymbol{\omega}$ and $\boldsymbol{\alpha}$ are known at a given instant, the acceleration of a point can be obtained from the time derivative of Eq. 20–3, which yields

$$\mathbf{a} = \boldsymbol{\alpha} \times \mathbf{r} + \boldsymbol{\omega} \times (\boldsymbol{\omega} \times \mathbf{r}) \tag{20–4}$$

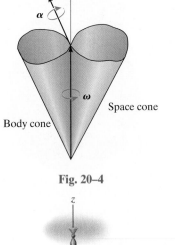

(a)

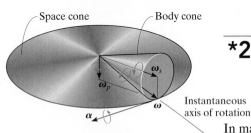

(b)

Fig. 20–5

*20.2   The Time Derivative of a Vector Measured from Either a Fixed or Translating-Rotating System

In many types of problems involving the motion of a body about a fixed point, the angular velocity $\boldsymbol{\omega}$ is specified in terms of its components. Then, if the angular acceleration $\boldsymbol{\alpha}$ of such a body is to be determined, it is often easier to compute the time derivative of $\boldsymbol{\omega}$ using a coordinate system that has a *rotation* defined by one or more of the components of $\boldsymbol{\omega}$. For example, in the case of the disk in Fig. 20–5a, where $\boldsymbol{\omega} = \boldsymbol{\omega}_s + \boldsymbol{\omega}_p$, the x, y, z axes can be given an angular velocity of $\boldsymbol{\omega}_p$. For this reason, and for other uses later, an equation will now be derived, which relates the time derivative of any vector $\mathbf{A}$ defined from a translating-rotating reference to its time derivative defined from a fixed reference.

Consider the x, y, z axes of the moving frame of reference to be rotating with an angular velocity $\boldsymbol{\Omega}$, which is measured from the fixed X, Y, Z axes, Fig. 20–6a. In the following discussion, it will be convenient to express vector $\mathbf{A}$ in terms of its $\mathbf{i}, \mathbf{j}, \mathbf{k}$ components, which define the directions of the moving axes. Hence,

$$\mathbf{A} = A_x \mathbf{i} + A_y \mathbf{j} + A_z \mathbf{k}$$

In general, the time derivative of $\mathbf{A}$ must account for the change in both its magnitude and direction. However, if this derivative is taken *with respect to the moving frame of reference*, only the change in the magnitudes of the components of $\mathbf{A}$ must be accounted for, since the directions of the components do not change with respect to the moving reference. Hence,

$$(\dot{\mathbf{A}})_{xyz} = \dot{A}_x \mathbf{i} + \dot{A}_y \mathbf{j} + \dot{A}_z \mathbf{k} \qquad (20\text{–}5)$$

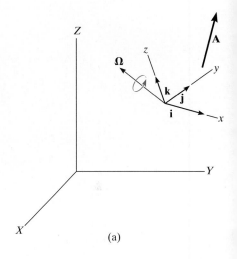

When the time derivative of $\mathbf{A}$ is taken *with respect to the fixed frame of reference*, the *directions* of $\mathbf{i}, \mathbf{j}$, and $\mathbf{k}$ change only on account of the *rotation* $\boldsymbol{\Omega}$ of the axes and not their translation. Hence, in general,

$$\dot{\mathbf{A}} = \dot{A}_x \mathbf{i} + \dot{A}_y \mathbf{j} + \dot{A}_z \mathbf{k} + A_x \dot{\mathbf{i}} + A_y \dot{\mathbf{j}} + A_z \dot{\mathbf{k}}$$

The time derivatives of the unit vectors will now be considered. For example, $\dot{\mathbf{i}} = d\mathbf{i}/dt$ represents only the change in the *direction* of $\mathbf{i}$ with respect to time, since $\mathbf{i}$ always has a magnitude of 1 unit. As shown in Fig. 20–6b, the change, $d\mathbf{i}$, is *tangent to the path* described by the arrowhead of $\mathbf{i}$ as $\mathbf{i}$ swings due to the rotation $\boldsymbol{\Omega}$. Accounting for both the magnitude and direction of $d\mathbf{i}$, we can therefore define $\dot{\mathbf{i}}$ using the cross product, $\dot{\mathbf{i}} = \boldsymbol{\Omega} \times \mathbf{i}$. In general, then

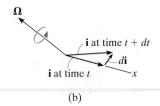

$$\dot{\mathbf{i}} = \boldsymbol{\Omega} \times \mathbf{i} \qquad \dot{\mathbf{j}} = \boldsymbol{\Omega} \times \mathbf{j} \qquad \dot{\mathbf{k}} = \boldsymbol{\Omega} \times \mathbf{k}$$

These formulations were also developed in Sec. 16.8, regarding planar motion of the axes. Substituting these results into the above equation and using Eq. 20–5 yields

$$\boxed{\dot{\mathbf{A}} = (\dot{\mathbf{A}})_{xyz} + \boldsymbol{\Omega} \times \mathbf{A}} \qquad (20\text{–}6)$$

This result is important, and will be used throughout Sec. 20.4 and Chapter 21. It states that the time derivative of *any vector* $\mathbf{A}$ as observed from the fixed X, Y, Z frame of reference is equal to the time rate of change of $\mathbf{A}$ as observed from the x, y, z translating-rotating frame of reference, Eq. 20–5, plus $\boldsymbol{\Omega} \times \mathbf{A}$, the change of $\mathbf{A}$ caused by the rotation of the x, y, z frame. As a result, Eq. 20–6 should always be used whenever $\boldsymbol{\Omega}$ produces a change in the direction of $\mathbf{A}$ as seen from the X, Y, Z reference. If this change does not occur, i.e., $\boldsymbol{\Omega} = \mathbf{0}$, then $\dot{\mathbf{A}} = (\dot{\mathbf{A}})_{xyz}$, and so the time rate of change of $\mathbf{A}$ as observed from both coordinate systems will be the *same*.

Fig. 20–6

20

EXAMPLE   20.1

The disk shown in Fig. 20–7 spins about its axle with a constant angular velocity $\omega_s = 3$ rad/s, while the horizontal platform on which the disk is mounted rotates about the vertical axis at a constant rate $\omega_p = 1$ rad/s. Determine the angular acceleration of the disk and the velocity and acceleration of point A on the disk when it is in the position shown.

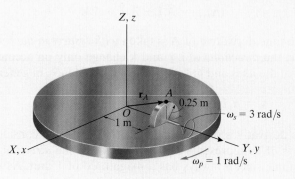

Fig. 20–7

SOLUTION

Point O represents a fixed point of rotation for the disk if one considers a hypothetical extension of the disk to this point. To determine the velocity and acceleration of point A, it is first necessary to determine the angular velocity $\boldsymbol{\omega}$ and angular acceleration $\boldsymbol{\alpha}$ of the disk, since these vectors are used in Eqs. 20–3 and 20–4.

Angular Velocity.   The angular velocity, which is measured from X, Y, Z, is simply the vector addition of its two component motions. Thus,

$$\boldsymbol{\omega} = \boldsymbol{\omega}_s + \boldsymbol{\omega}_p = \{3\mathbf{j} - 1\mathbf{k}\} \text{ rad/s}$$

Angular Acceleration.   Since the magnitude of $\boldsymbol{\omega}$ is constant, only a change in its direction, as seen from the fixed reference, creates the angular acceleration $\boldsymbol{\alpha}$ of the disk. One way to obtain $\boldsymbol{\alpha}$ is to compute the time derivative of *each of the two components* of $\boldsymbol{\omega}$ using Eq. 20–6. At the instant shown in Fig. 20–7, imagine the fixed X, Y, Z and a rotating x, y, z frame to be coincident. If the rotating x, y, z frame is chosen to have an angular velocity of $\boldsymbol{\Omega} = \boldsymbol{\omega}_p = \{-1\mathbf{k}\}$ rad/s, then $\boldsymbol{\omega}_s$ will *always* be directed along the y (not Y) axis, and the time rate of change of $\boldsymbol{\omega}_s$ as seen from x, y, z is zero; i.e., $(\dot{\boldsymbol{\omega}}_s)_{xyz} = \mathbf{0}$ (the magnitude and direction of $\boldsymbol{\omega}_s$ is constant). Thus,

$$\dot{\boldsymbol{\omega}}_s = (\dot{\boldsymbol{\omega}}_s)_{xyz} + \boldsymbol{\omega}_p \times \boldsymbol{\omega}_s = \mathbf{0} + (-1\mathbf{k}) \times (3\mathbf{j}) = \{3\mathbf{i}\} \ \text{rad/s}^2$$

By the same choice of axes rotation, $\boldsymbol{\Omega} = \boldsymbol{\omega}_p$, or even with $\boldsymbol{\Omega} = \mathbf{0}$, the time derivative $(\dot{\boldsymbol{\omega}}_p)_{xyz} = \mathbf{0}$, since $\boldsymbol{\omega}_p$ has a constant magnitude and direction with respect to x, y, z. Hence,

$$\dot{\boldsymbol{\omega}}_p = (\dot{\boldsymbol{\omega}}_p)_{xyz} + \boldsymbol{\omega}_p \times \boldsymbol{\omega}_p = \mathbf{0} + \mathbf{0} = \mathbf{0}$$

The angular acceleration of the disk is therefore

$$\boldsymbol{\alpha} = \dot{\boldsymbol{\omega}} = \dot{\boldsymbol{\omega}}_s + \dot{\boldsymbol{\omega}}_p = \{3\mathbf{i}\} \ \text{rad/s}^2 \qquad \textit{Ans.}$$

Velocity and Acceleration.   Since $\boldsymbol{\omega}$ and $\boldsymbol{\alpha}$ have now been determined, the velocity and acceleration of point A can be found using Eqs. 20–3 and 20–4. Realizing that $\mathbf{r}_A = \{1\mathbf{j} + 0.25\mathbf{k}\}$ m, Fig. 20–7, we have

$$\mathbf{v}_A = \boldsymbol{\omega} \times \mathbf{r}_A = (3\mathbf{j} - 1\mathbf{k}) \times (1\mathbf{j} + 0.25\mathbf{k}) = \{1.75\mathbf{i}\} \ \text{m/s} \qquad \textit{Ans.}$$

$$\mathbf{a}_A = \boldsymbol{\alpha} \times \mathbf{r}_A + \boldsymbol{\omega} \times (\boldsymbol{\omega} \times \mathbf{r}_A)$$
$$= (3\mathbf{i}) \times (1\mathbf{j} + 0.25\mathbf{k}) + (3\mathbf{j} - 1\mathbf{k}) \times [(3\mathbf{j} - 1\mathbf{k}) \times (1\mathbf{j} + 0.25\mathbf{k})]$$
$$= \{-2.50\mathbf{j} - 2.25\mathbf{k}\} \ \text{m/s}^2 \qquad \textit{Ans.}$$

EXAMPLE | **20.2**

At the instant $\theta = 60°$, the gyrotop in Fig. 20–8 has three components of angular motion directed as shown and having magnitudes defined as:

Spin: $\omega_s = 10$ rad/s, increasing at the rate of 6 rad/s²

Nutation: $\omega_n = 3$ rad/s, increasing at the rate of 2 rad/s²

Precession: $\omega_p = 5$ rad/s, increasing at the rate of 4 rad/s²

Determine the angular velocity and angular acceleration of the top.

SOLUTION

Angular Velocity.   The top rotates about the fixed point O. If the fixed and rotating frames are coincident at the instant shown, then the angular velocity can be expressed in terms of $\mathbf{i}, \mathbf{j}, \mathbf{k}$ components, with reference to the x, y, z frame; i.e.,

$$\boldsymbol{\omega} = -\omega_n \mathbf{i} + \omega_s \sin\theta \mathbf{j} + (\omega_p + \omega_s \cos\theta)\mathbf{k}$$
$$= -3\mathbf{i} + 10\sin 60°\mathbf{j} + (5 + 10\cos 60°)\mathbf{k}$$
$$= \{-3\mathbf{i} + 8.66\mathbf{j} + 10\mathbf{k}\} \text{ rad/s} \qquad \textit{Ans.}$$

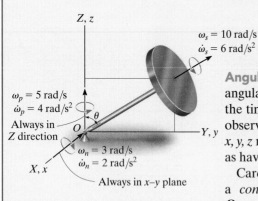

$\omega_s = 10$ rad/s
$\dot{\omega}_s = 6$ rad/s²

$\omega_p = 5$ rad/s
$\dot{\omega}_p = 4$ rad/s²
Always in
Z direction

$\omega_n = 3$ rad/s
$\dot{\omega}_n = 2$ rad/s²

Always in x–y plane

Fig. 20–8

Angular Acceleration.   As in the solution of Example 20.1, the angular acceleration $\boldsymbol{\alpha}$ will be determined by investigating separately the time rate of change of *each of the angular velocity components* as observed from the fixed X, Y, Z reference. We will choose an $\boldsymbol{\Omega}$ for the x, y, z reference so that the component of $\boldsymbol{\omega}$ being considered is viewed as having a *constant direction* when observed from x, y, z.

Careful examination of the motion of the top reveals that $\boldsymbol{\omega}_s$ has a *constant direction* relative to x, y, z if these axes rotate at $\boldsymbol{\Omega} = \boldsymbol{\omega}_n + \boldsymbol{\omega}_p$. Thus,

$$\dot{\boldsymbol{\omega}}_s = (\dot{\boldsymbol{\omega}}_s)_{xyz} + (\boldsymbol{\omega}_n + \boldsymbol{\omega}_p) \times \boldsymbol{\omega}_s$$
$$= (6\sin 60°\mathbf{j} + 6\cos 60°\mathbf{k}) + (-3\mathbf{i} + 5\mathbf{k}) \times (10\sin 60°\mathbf{j} + 10\cos 60°\mathbf{k})$$
$$= \{-43.30\mathbf{i} + 20.20\mathbf{j} - 22.98\mathbf{k}\} \text{ rad/s}^2$$

Since $\boldsymbol{\omega}_n$ *always* lies in the fixed X–Y plane, this vector has a *constant direction* if the motion is viewed from axes x, y, z having a rotation of $\boldsymbol{\Omega} = \boldsymbol{\omega}_p$ (not $\boldsymbol{\Omega} = \boldsymbol{\omega}_s + \boldsymbol{\omega}_p$). Thus,

$$\dot{\boldsymbol{\omega}}_n = (\dot{\boldsymbol{\omega}}_n)_{xyz} + \boldsymbol{\omega}_p \times \boldsymbol{\omega}_n = -2\mathbf{i} + (5\mathbf{k}) \times (-3\mathbf{i}) = \{-2\mathbf{i} - 15\mathbf{j}\} \text{ rad/s}^2$$

Finally, the component $\boldsymbol{\omega}_p$ is *always directed* along the Z axis so that here it is not necessary to think of x, y, z as rotating, i.e., $\boldsymbol{\Omega} = \mathbf{0}$. Expressing the data in terms of the $\mathbf{i}, \mathbf{j}, \mathbf{k}$ components, we therefore have

$$\dot{\boldsymbol{\omega}}_p = (\dot{\boldsymbol{\omega}}_p)_{xyz} + \mathbf{0} \times \boldsymbol{\omega}_p = \{4\mathbf{k}\} \text{ rad/s}^2$$

Thus, the angular acceleration of the top is

$$\boldsymbol{\alpha} = \dot{\boldsymbol{\omega}}_s + \dot{\boldsymbol{\omega}}_n + \dot{\boldsymbol{\omega}}_p = \{-45.3\mathbf{i} + 5.20\mathbf{j} - 19.0\mathbf{k}\} \text{ rad/s}^2 \qquad \textit{Ans.}$$

20.3 General Motion

Shown in Fig. 20–9 is a rigid body subjected to general motion in three dimensions for which the angular velocity is $\boldsymbol{\omega}$ and the angular acceleration is $\boldsymbol{\alpha}$. If point A has a known motion of $\mathbf{v}_A$ and $\mathbf{a}_A$, the motion of any other point B can be determined by using a relative-motion analysis. In this section a *translating coordinate system* will be used to define the relative motion, and in the next section a reference that is both rotating and translating will be considered.

If the origin of the translating coordinate system x, y, z ($\boldsymbol{\Omega} = \mathbf{0}$) is located at the "base point" A, then, at the instant shown, the motion of the body can be regarded as the sum of an instantaneous translation of the body having a motion of $\mathbf{v}_A$, and $\mathbf{a}_A$, and a rotation of the body about an instantaneous axis passing through point A. Since the body is rigid, the motion of point B measured by an observer located at A is therefore the same as *the rotation of the body about a fixed point*. This relative motion occurs about the instantaneous axis of rotation and is defined by $\mathbf{v}_{B/A} = \boldsymbol{\omega} \times \mathbf{r}_{B/A}$, Eq. 20–3, and $\mathbf{a}_{B/A} = \boldsymbol{\alpha} \times \mathbf{r}_{B/A} + \boldsymbol{\omega} \times (\boldsymbol{\omega} \times \mathbf{r}_{B/A})$, Eq. 20–4. For translating axes, the relative motions are related to absolute motions by $\mathbf{v}_B = \mathbf{v}_A + \mathbf{v}_{B/A}$ and $\mathbf{a}_B = \mathbf{a}_A + \mathbf{a}_{B/A}$, Eqs. 16–15 and 16–17, so that the absolute velocity and acceleration of point B can be determined from the equations

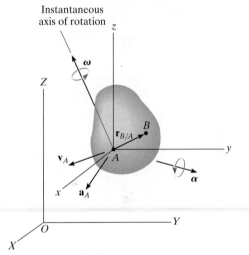

Instantaneous axis of rotation

Fig. 20–9

$$\mathbf{v}_B = \mathbf{v}_A + \boldsymbol{\omega} \times \mathbf{r}_{B/A} \tag{20–7}$$

and

$$\mathbf{a}_B = \mathbf{a}_A + \boldsymbol{\alpha} \times \mathbf{r}_{B/A} + \boldsymbol{\omega} \times (\boldsymbol{\omega} \times \mathbf{r}_{B/A}) \tag{20–8}$$

These two equations are essentially the same as to those describing the general plane motion of a rigid body, Eqs. 16–16 and 16–18. However, difficulty in application arises for three-dimensional motion, because $\boldsymbol{\alpha}$ now measures the change in *both* the magnitude and direction of $\boldsymbol{\omega}$.

20

EXAMPLE 20.3

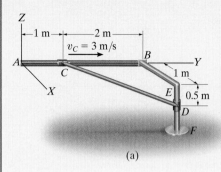

(a)

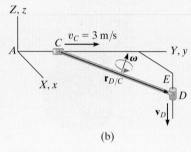

(b)

Fig. 20–10

If the collar at C in Fig. 20–10a moves towards B with a speed of 3 m/s, determine the velocity of the collar at D and the angular velocity of the bar at the instant shown. The bar is connected to the collars at its end points by ball-and-socket joints.

SOLUTION

Bar CD is subjected to general motion. Why? The velocity of point D on the bar can be related to the velocity of point C by the equation

$$\mathbf{v}_D = \mathbf{v}_C + \boldsymbol{\omega} \times \mathbf{r}_{D/C}$$

The fixed and translating frames of reference are assumed to coincide at the instant considered, Fig. 20–10b. We have

$$\mathbf{v}_D = -v_D\mathbf{k} \qquad \mathbf{v}_C = \{3\mathbf{j}\} \text{ m/s}$$
$$\mathbf{r}_{D/C} = \{1\mathbf{i} + 2\mathbf{j} - 0.5\mathbf{k}\} \text{ m} \qquad \boldsymbol{\omega} = \omega_x\mathbf{i} + \omega_y\mathbf{j} + \omega_z\mathbf{k}$$

Substituting into the above equation we get

$$-v_D\mathbf{k} = 3\mathbf{j} + \begin{vmatrix} \mathbf{i} & \mathbf{j} & \mathbf{k} \\ \omega_x & \omega_y & \omega_z \\ 1 & 2 & -0.5 \end{vmatrix}$$

Expanding and equating the respective $\mathbf{i}, \mathbf{j}, \mathbf{k}$ components yields

$$-0.5\omega_y - 2\omega_z = 0 \tag{1}$$
$$0.5\omega_x + 1\omega_z + 3 = 0 \tag{2}$$
$$2\omega_x - 1\omega_y + v_D = 0 \tag{3}$$

These equations contain four unknowns.[*] A fourth equation can be written if the direction of $\boldsymbol{\omega}$ is specified. In particular, any component of $\boldsymbol{\omega}$ acting along the bar's axis has no effect on moving the collars. This is because the bar is *free to rotate* about its axis. Therefore, if $\boldsymbol{\omega}$ is specified as acting *perpendicular* to the axis of the bar, then $\boldsymbol{\omega}$ must have a unique magnitude to satisfy the above equations. Perpendicularity is guaranteed provided the dot product of $\boldsymbol{\omega}$ and $\mathbf{r}_{D/C}$ is zero (see Eq. C–14 of Appendix C). Hence,

$$\boldsymbol{\omega} \cdot \mathbf{r}_{D/C} = (\omega_x\mathbf{i} + \omega_y\mathbf{j} + \omega_z\mathbf{k}) \cdot (1\mathbf{i} + 2\mathbf{j} - 0.5\mathbf{k}) = 0$$
$$1\omega_x + 2\omega_y - 0.5\omega_z = 0 \tag{4}$$

Solving Eqs. 1 through 4 simultaneously yields

$$\omega_x = -4.86 \text{ rad/s} \quad \omega_y = 2.29 \text{ rad/s} \quad \omega_z = -0.571 \text{ rad/s} \qquad Ans.$$
$$v_D = 12.0 \text{ m/s} \downarrow \qquad Ans.$$

---

[*]Although this is the case, the magnitude of $\mathbf{v}_D$ can be obtained. For example, solve Eqs. 1 and 2 for ω_y and ω_x in terms of ω_z and substitute into Eq. 3. It will be noted that ω_z will cancel out, which will allow a solution for v_D.

PROBLEMS

20–1. At a given instant, the satellite dish has an angular motion $\omega_1 = 6$ rad/s and $\dot{\omega}_1 = 3$ rad/s² about the z axis. At this same instant $\theta = 25°$, the angular motion about the x axis is $\omega_2 = 2$ rad/s, and $\dot{\omega}_2 = 1.5$ rad/s². Determine the velocity and acceleration of the signal horn A at this instant.

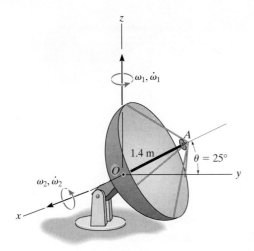

Prob. 20–1

20–2. Gears A and B are fixed, while gears C and D are free to rotate about the shaft S. If the shaft turns about the z axis at a constant rate of $\omega_1 = 4$ rad/s, determine the angular velocity and angular acceleration of gear C.

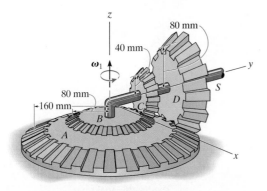

Prob. 20–2

20–3. The ladder of the fire truck rotates around the z axis with an angular velocity $\omega_1 = 0.15$ rad/s, which is increasing at 0.8 rad/s². At the same instant it is rotating upward at a constant rate $\omega_2 = 0.6$ rad/s. Determine the velocity and acceleration of point A located at the top of the ladder at this instant.

*20–4. The ladder of the fire truck rotates around the z axis with an angular velocity of $\omega_1 = 0.15$ rad/s, which is increasing at 0.2 rad/s². At the same instant it is rotating upwards at $\omega_2 = 0.6$ rad/s while increasing at 0.4 rad/s². Determine the velocity and acceleration of point A located at the top of the ladder at this instant.

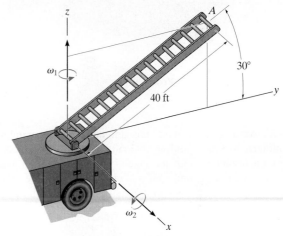

Probs. 20–3/4

20–5. Gear B is connected to the rotating shaft, while the plate gear A is fixed. If the shaft is turning at a constant rate of $\omega_z = 10$ rad/s about the z axis, determine the magnitudes of the angular velocity and the angular acceleration of gear B. Also, determine the magnitudes of the velocity and acceleration of point P.

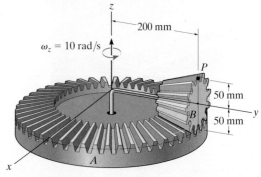

Prob. 20–5

20

20–6. Gear A is fixed while gear B is free to rotate on the shaft S. If the shaft is turning about the z axis at $\omega_z = 5$ rad/s, while increasing at 2 rad/s², determine the velocity and acceleration of point P at the instant shown. The face of gear B lies in a vertical plane.

*20–8. The cone rolls without slipping such that at the instant shown $\omega_z = 4$ rad/s and $\dot{\omega}_z = 3$ rad/s². Determine the velocity and acceleration of point A at this instant.

20–9. The cone rolls without slipping such that at the instant shown $\omega_z = 4$ rad/s and $\dot{\omega}_z = 3$ rad/s². Determine the velocity and acceleration of point B at this instant.

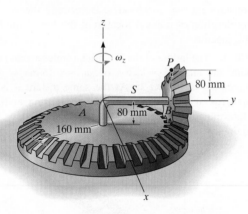

Prob. 20–6

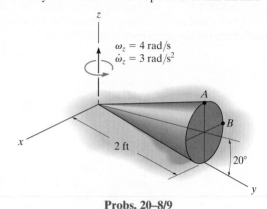

Probs. 20–8/9

20–7. At a given instant, the antenna has an angular motion $\omega_1 = 3$ rad/s and $\dot{\omega}_1 = 2$ rad/s² about the z axis. At this same instant $\theta = 30°$, the angular motion about the x axis is $\omega_2 = 1.5$ rad/s, and $\dot{\omega}_2 = 4$ rad/s². Determine the velocity and acceleration of the signal horn A at this instant. The distance from O to A is $d = 3$ ft.

20–10. At the instant when $\theta = 90°$, the satellite's body is rotating with an angular velocity of $\omega_1 = 15$ rad/s and angular acceleration of $\dot{\omega}_1 = 3$ rad/s². Simultaneously, the solar panels rotate with an angular velocity of $\omega_2 = 6$ rad/s and angular acceleration of $\dot{\omega}_2 = 1.5$ rad/s². Determine the velocity and acceleration of point B on the solar panel at this instant.

20–11. At the instant when $\theta = 90°$, the satellite's body travels in the x direction with a velocity of $\mathbf{v}_O = \{500\mathbf{i}\}$ m/s and acceleration of $\mathbf{a}_O = \{50\mathbf{i}\}$ m/s². Simultaneously, the body also rotates with an angular velocity of $\omega_1 = 15$ rad/s and angular acceleration of $\dot{\omega}_1 = 3$ rad/s². At the same time, the solar panels rotate with an angular velocity of $\omega_2 = 6$ rad/s and angular acceleration of $\dot{\omega}_2 = 1.5$ rad/s² Determine the velocity and acceleration of point B on the solar panel.

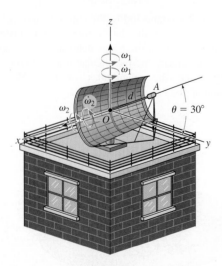

Prob. 20–7

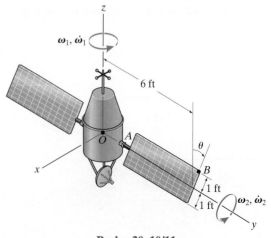

Probs. 20–10/11

*20–12. The disk is free to rotate on the shaft S. If the shaft is turning about the z axis at $\omega_z = 2 \text{ rad/s}$, while increasing at 8 rad/s^2, determine the velocity and acceleration of point A at the instant shown.

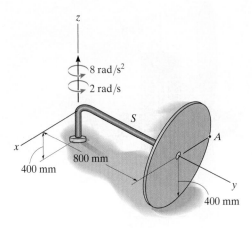

Prob. 20–12

20–13. The disk spins about the arm with an angular velocity of $\omega_s = 8 \text{ rad/s}$, which is increasing at a constant rate of $\dot{\omega}_s = 3 \text{ rad/s}^2$ at the instant shown. If the shaft rotates with a constant angular velocity of $\omega_p = 6 \text{ rad/s}$, determine the velocity and acceleration of point A located on the rim of the disk at this instant.

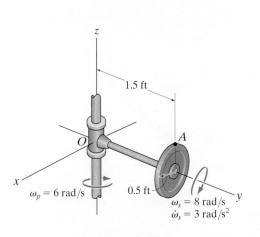

Prob. 20–13

20–14. The wheel is spinning about shaft AB with an angular velocity of $\omega_s = 10 \text{ rad/s}$, which is increasing at a constant rate of $\dot{\omega}_2 = 6 \text{ rad/s}^2$, while the frame precesses about the z axis with an angular velocity of $\omega_p = 12 \text{ rad/s}$, which is increasing at a constant rate of $\dot{\omega}_p = 3 \text{ rad/s}^2$. Determine the velocity and acceleration of point C located on the rim of the wheel at this instant.

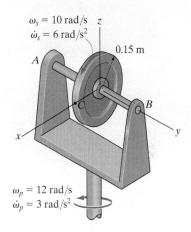

Prob. 20–14

20–15. At the instant shown, the tower crane rotates about the z axis with an angular velocity $\omega_1 = 0.25 \text{ rad/s}$, which is increasing at 0.6 rad/s^2. The boom OA rotates downward with an angular velocity $\omega_2 = 0.4 \text{ rad/s}$, which is increasing at 0.8 rad/s^2. Determine the velocity and acceleration of point A located at the end of the boom at this instant.

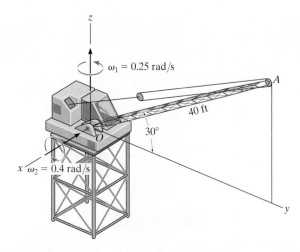

Prob. 20–15

*20–16. If the top gear B rotates at a constant rate of ω, determine the angular velocity of gear A, which is free to rotate about the shaft and rolls on the bottom fixed gear C.

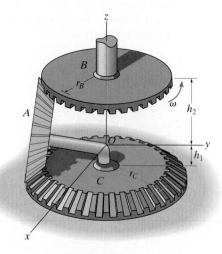

Prob. 20–16

20–17. When $\theta = 0°$, the radar disk rotates about the y axis with an angular velocity of $\dot{\theta} = 2$ rad/s, increasing at a constant rate of $\ddot{\theta} = 1.5$ rad/s². Simultaneously, the disk also precesses about the z axis with an angular velocity of $\omega_p = 5$ rad/s, increasing at a constant rate of $\dot{\omega}_p = 3$ rad/s². Determine the velocity and acceleration of the receiver A at this instant.

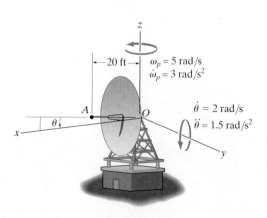

Prob. 20–17

20–18. Gear A is fixed to the crankshaft S, while gear C is fixed. Gear B and the propeller are free to rotate. The crankshaft is turning at 80 rad/s about its axis. Determine the magnitudes of the angular velocity of the propeller and the angular acceleration of gear B.

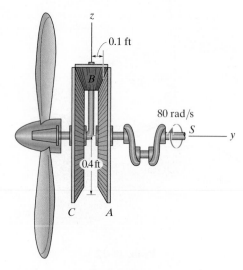

Prob. 20–18

20–19. Shaft BD is connected to a ball-and-socket joint at B, and a beveled gear A is attached to its other end. The gear is in mesh with a fixed gear C. If the shaft and gear A are *spinning* with a constant angular velocity $\omega_1 = 8$ rad/s, determine the angular velocity and angular acceleration of gear A.

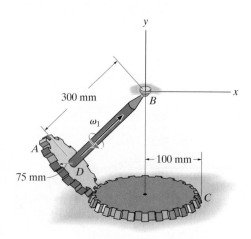

Prob. 20–19

*20–20. Gear B is driven by a motor mounted on turntable C. If gear A is held fixed, and the motor shaft rotates with a constant angular velocity of $\omega_y = 30$ rad/s, determine the angular velocity and angular acceleration of gear B.

20–21. Gear B is driven by a motor mounted on turntable C. If gear A and the motor shaft rotate with constant angular speeds of $\omega_A = \{10\mathbf{k}\}$ rad/s and $\omega_y = \{30\mathbf{j}\}$ rad/s, respectively, determine the angular velocity and angular acceleration of gear B.

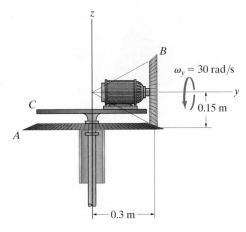

Probs. 20–20/21

20–22. The crane boom OA rotates about the z axis with a constant angular velocity of $\omega_1 = 0.15$ rad/s, while it is rotating downward with a constant angular velocity of $\omega_2 = 0.2$ rad/s. Determine the velocity and acceleration of point A located at the end of the boom at the instant shown.

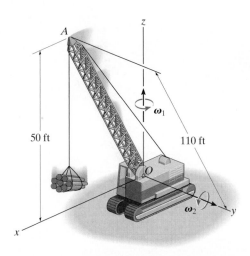

Prob. 20–22

20–23. The differential of an automobile allows the two rear wheels to rotate at different speeds when the automobile travels along a curve. For operation, the rear axles are attached to the wheels at one end and have beveled gears A and B on their other ends. The differential case D is placed over the left axle but can rotate about C independent of the axle. The case supports a pinion gear E on a shaft, which meshes with gears A and B. Finally, a ring gear G is *fixed* to the differential case so that the case rotates with the ring gear when the latter is driven by the drive pinion H. This gear, like the differential case, is free to rotate about the left wheel axle. If the drive pinion is turning at $\omega_H = 100$ rad/s and the pinion gear E is spinning about its shaft at $\omega_E = 30$ rad/s, determine the angular velocity, ω_A and ω_B, of each axle.

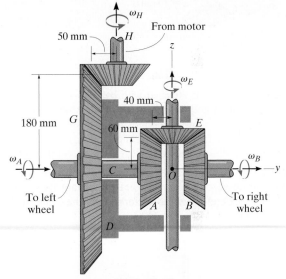

Prob. 20–23

*20–24. The truncated cone rotates about the z axis at a constant rate $\omega_z = 0.4$ rad/s without slipping on the horizontal plane. Determine the velocity and acceleration of point A on the cone.

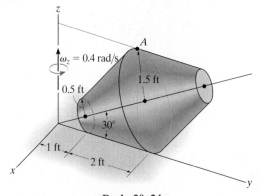

Prob. 20–24

20

20–25. Disk A rotates at a constant angular velocity of 10 rad/s. If rod BC is joined to the disk and a collar by ball-and-socket joints, determine the velocity of collar B at the instant shown. Also, what is the rod's angular velocity ω_{BC} if it is directed perpendicular to the axis of the rod?

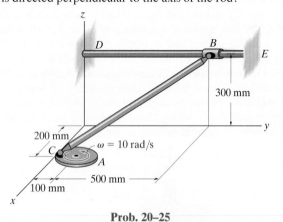

Prob. 20–25

20–26. If the rod is attached with ball-and-socket joints to smooth collars A and B at its end points, determine the speed of B at the instant shown if A is moving downward at a constant speed of $v_A = 8$ ft/s. Also, determine the angular velocity of the rod if it is directed perpendicular to the axis of the rod.

20–27. If the collar at A is moving downward with an acceleration $\mathbf{a}_A = \{-5\mathbf{k}\}$ ft/s², at the instant its speed is $v_A = 8$ ft/s, determine the acceleration of the collar at B at this instant.

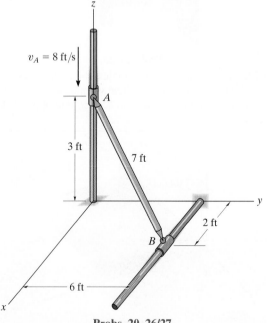

Probs. 20–26/27

*20–28.** If wheel C rotates with a constant angular velocity of $\omega_C = 10$ rad/s, determine the velocity of the collar at B when rod AB is in the position shown.

20–29. At the instant rod AB is in the position shown wheel C rotates with an angular velocity of $\omega_C = 10$ rad/s and has an angular acceleration of $\alpha_C = 1.5$ rad/s². Determine the acceleration of collar B at this instant.

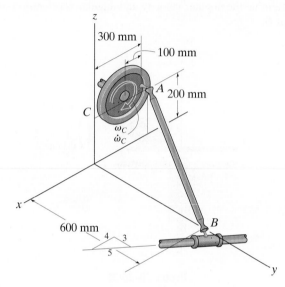

Probs. 20–28/29

20–30. If wheel D rotates with an angular velocity of $\omega_D = 6$ rad/s, determine the angular velocity of the follower link BC at the instant shown. The link rotates about the z axis at $z = 2$ ft.

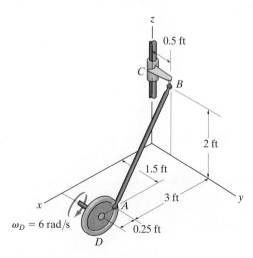

Prob. 20–30

20–31. Rod AB is attached to the rotating arm using ball-and-socket joints. If AC is rotating with a constant angular velocity of 8 rad/s about the pin at C, determine the angular velocity of link BD at the instant shown.

*20–32. Rod AB is attached to the rotating arm using ball-and-socket joints. If AC is rotating about point C with an angular velocity of 8 rad/s and has an angular acceleration of $\alpha_{AC} = \{6\mathbf{k}\}$ rad/s² at the instant shown, determine the angular velocity and angular acceleration of link BD at this instant.

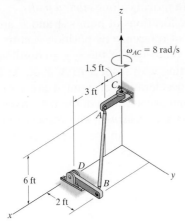

Probs. 20–31/32

20–33. Rod AB is attached to collars at its ends by ball-and-socket joints. If collar A moves upward with a velocity of $\mathbf{v}_A = \{8\mathbf{k}\}$ ft/s, determine the angular velocity of the rod and the speed of collar B at the instant shown. Assume that the rod's angular velocity is directed perpendicular to the rod.

20–34. Rod AB is attached to collars at its ends by ball-and-socket joints. If collar A moves upward with an acceleration of $\mathbf{a}_A = \{4\mathbf{k}\}$ ft/s², determine the angular acceleration of rod AB and the magnitude of acceleration of collar B. Assume that the rod's angular acceleration is directed perpendicular to the rod, and use the result of Prob. 20–33 for $\boldsymbol{\omega}_{AB}$.

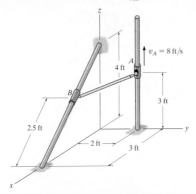

Probs. 20–33/34

20–35. Solve Prob. 20–25 if the connection at B consists of a pin as shown in the figure below, rather than a ball-and-socket joint. *Hint:* The constraint allows rotation of the rod both about bar DE (**j** direction) and about the axis of the pin (**n** direction). Since there is no rotational component in the **u** direction, i.e., perpendicular to **n** and **j** where $\mathbf{u} = \mathbf{j} \times \mathbf{n}$, an additional equation for solution can be obtained from $\boldsymbol{\omega} \cdot \mathbf{u} = 0$. The vector **n** is in the same direction as $\mathbf{r}_{B/C} \times \mathbf{r}_{D/C}$.

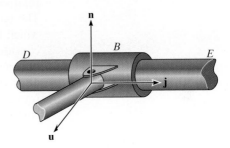

Prob. 20–35

*20–36. The rod assembly is supported at B by a ball-and-socket joint and at A by a clevis. If the collar at B moves in the x–z plane with a speed $v_B = 5$ ft/s, determine the velocity of points A and C on the rod assembly at the instant shown. *Hint:* See Prob. 20–35.

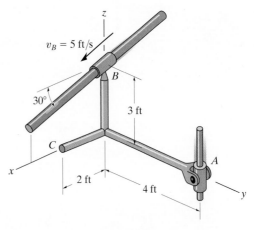

Prob. 20–36

20

*20.4   Relative-Motion Analysis Using Translating and Rotating Axes

The most general way to analyze the three-dimensional motion of a rigid body requires the use of x, y, z axes that both translate and rotate relative to a second frame X, Y, Z. This analysis also provides a means to determine the motions of two points A and B located on separate members of a mechanism, and the relative motion of one particle with respect to another when one or both particles are moving along *curved paths*.

As shown in Fig. 20–11, the locations of points A and B are specified relative to the X, Y, Z frame of reference by position vectors $\mathbf{r}_A$ and $\mathbf{r}_B$. The base point A represents the origin of the x, y, z coordinate system, which is translating and rotating with respect to X, Y, Z. At the instant considered, the velocity and acceleration of point A are $\mathbf{v}_A$ and $\mathbf{a}_A$, and the angular velocity and angular acceleration of the x, y, z axes are $\mathbf{\Omega}$ and $\dot{\mathbf{\Omega}} = d\mathbf{\Omega}/dt$. All these vectors are *measured* with respect to the X, Y, Z frame of reference, although they can be expressed in Cartesian component form along either set of axes.

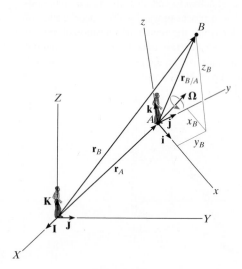

Fig. 20–11

Position.   If the position of "B with respect to A" is specified by the *relative-position vector* $\mathbf{r}_{B/A}$, Fig. 20–11, then, by vector addition,

$$\boxed{\mathbf{r}_B = \mathbf{r}_A + \mathbf{r}_{B/A}} \qquad (20\text{–}9)$$

where

$$\mathbf{r}_B = \text{position of } B$$
$$\mathbf{r}_A = \text{position of the origin } A$$
$$\mathbf{r}_{B/A} = \text{position of "} B \text{ with respect to } A\text{"}$$

Velocity.   The velocity of point B measured from X, Y, Z can be determined by taking the time derivative of Eq. 20–9,

$$\dot{\mathbf{r}}_B = \dot{\mathbf{r}}_A + \dot{\mathbf{r}}_{B/A}$$

The first two terms represent $\mathbf{v}_B$ and $\mathbf{v}_A$. The last term must be evaluated by applying Eq. 20–6, since $\mathbf{r}_{B/A}$ is measured with respect to a rotating reference. Hence,

$$\dot{\mathbf{r}}_{B/A} = (\dot{\mathbf{r}}_{B/A})_{xyz} + \boldsymbol{\Omega} \times \mathbf{r}_{B/A} = (\mathbf{v}_{B/A})_{xyz} + \boldsymbol{\Omega} \times \mathbf{r}_{B/A} \qquad (20\text{–}10)$$

Therefore,

$$\boxed{\mathbf{v}_B = \mathbf{v}_A + \boldsymbol{\Omega} \times \mathbf{r}_{B/A} + (\mathbf{v}_{B/A})_{xyz}} \qquad (20\text{–}11)$$

where

$$\mathbf{v}_B = \text{velocity of } B$$
$$\mathbf{v}_A = \text{velocity of the origin } A \text{ of the } x, y, z \text{ frame of reference}$$
$$(\mathbf{v}_{B/A})_{xyz} = \text{velocity of "} B \text{ with respect to } A\text{" as measured by an}$$
$$\text{observer attached to the rotating } x, y, z \text{ frame of reference}$$
$$\boldsymbol{\Omega} = \text{angular velocity of the } x, y, z \text{ frame of reference}$$
$$\mathbf{r}_{B/A} = \text{position of "} B \text{ with respect to } A\text{"}$$

20

Acceleration.   The acceleration of point B measured from X, Y, Z is determined by taking the time derivative of Eq. 20–11.

$$\dot{\mathbf{v}}_B = \dot{\mathbf{v}}_A + \dot{\boldsymbol{\Omega}} \times \mathbf{r}_{B/A} + \boldsymbol{\Omega} \times \dot{\mathbf{r}}_{B/A} + \frac{d}{dt}(\mathbf{v}_{B/A})_{xyz}$$

The time derivatives defined in the first and second terms represent $\mathbf{a}_B$ and $\mathbf{a}_A$, respectively. The fourth term can be evaluated using Eq. 20–10, and the last term is evaluated by applying Eq. 20–6, which yields

$$\frac{d}{dt}(\mathbf{v}_{B/A})_{xyz} = (\dot{\mathbf{v}}_{B/A})_{xyz} + \boldsymbol{\Omega} \times (\mathbf{v}_{B/A})_{xyz} = (\mathbf{a}_{B/A})_{xyz} + \boldsymbol{\Omega} \times (\mathbf{v}_{B/A})_{xyz}$$

Here $(\mathbf{a}_{B/A})_{xyz}$ is the acceleration of B with respect to A measured from x, y, z. Substituting this result and Eq. 20–10 into the above equation and simplifying, we have

$$\mathbf{a}_B = \mathbf{a}_A + \dot{\boldsymbol{\Omega}} \times \mathbf{r}_{B/A} + \boldsymbol{\Omega} \times (\boldsymbol{\Omega} \times \mathbf{r}_{B/A}) + 2\boldsymbol{\Omega} \times (\mathbf{v}_{B/A})_{xyz} + (\mathbf{a}_{B/A})_{xyz}$$

$$(20\text{–}12)$$

where

$$\mathbf{a}_B = \text{acceleration of } B$$

$$\mathbf{a}_A = \text{acceleration of the origin } A \text{ of the } x, y, z \text{ frame of reference}$$

$$(\mathbf{a}_{B/A})_{xyz}, (\mathbf{v}_{B/A})_{xyz} = \text{relative acceleration and relative velocity of ``}B \text{ with respect to } A\text{'' as measured by an observer attached to the rotating } x, y, z \text{ frame of reference}$$

$$\dot{\boldsymbol{\Omega}}, \boldsymbol{\Omega} = \text{angular acceleration and angular velocity of the } x, y, z \text{ frame of reference}$$

$$\mathbf{r}_{B/A} = \text{position of ``}B \text{ with respect to } A\text{''}$$

Equations 20–11 and 20–12 are identical to those used in Sec. 16.8 for analyzing relative plane motion.* In that case, however, application is simplified since $\boldsymbol{\Omega}$ and $\dot{\boldsymbol{\Omega}}$ have a *constant direction* which is always perpendicular to the plane of motion. For three-dimensional motion, $\dot{\boldsymbol{\Omega}}$ must be computed by using Eq. 20–6, since $\dot{\boldsymbol{\Omega}}$ depends on the change in *both* the magnitude and direction of $\boldsymbol{\Omega}$.

*Refer to Sec. 16.8 for an interpretation of the terms.

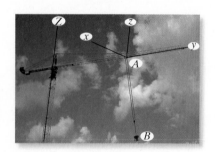

Complicated spatial motion of the concrete bucket B occurs due to the rotation of the boom about the Z axis, motion of the carriage A along the boom, and extension and swinging of the cable AB. A translating-rotating x, y, z coordinate system can be established on the carriage, and a relative-motion analysis can then be applied to study this motion.

20

Procedure for Analysis

Three-dimensional motion of particles or rigid bodies can be analyzed with Eqs. 20–11 and 20–12 by using the following procedure.

Coordinate Axes.

- Select the location and orientation of the X, Y, Z and x, y, z coordinate axes. Most often solutions can be easily obtained if at the instant considered:

  (1) the origins are *coincident*

  (2) the axes are collinear

  (3) the axes are parallel

- If several components of angular velocity are involved in a problem, the calculations will be reduced if the x, y, z axes are selected such that only one component of angular velocity is observed with respect to this frame ($\mathbf{\Omega}_{xyz}$) and the frame rotates with $\mathbf{\Omega}$ defined by the other components of angular velocity.

Kinematic Equations.

- After the origin of the moving reference, A, is defined and the moving point B is specified, Eqs. 20–11 and 20–12 should then be written in symbolic form as

$$\mathbf{v}_B = \mathbf{v}_A + \mathbf{\Omega} \times \mathbf{r}_{B/A} + (\mathbf{v}_{B/A})_{xyz}$$

$$\mathbf{a}_B = \mathbf{a}_A + \dot{\mathbf{\Omega}} \times \mathbf{r}_{B/A} + \mathbf{\Omega} \times (\mathbf{\Omega} \times \mathbf{r}_{B/A}) + 2\mathbf{\Omega} \times (\mathbf{v}_{B/A})_{xyz} + (\mathbf{a}_{B/A})_{xyz}$$

- If $\mathbf{r}_A$ and $\mathbf{\Omega}$ appear to *change direction* when observed from the fixed X, Y, Z reference then use a set of primed reference axes, x', y', z' having a rotation $\mathbf{\Omega}' = \mathbf{\Omega}$. Equation 20–6 is then used to determine $\dot{\mathbf{\Omega}}$ and the motion $\mathbf{v}_A$ and $\mathbf{a}_A$ of the origin of the moving x, y, z axes.

- If $\mathbf{r}_{B/A}$ and $\mathbf{\Omega}_{xyz}$ appear to change direction as observed from x, y, z, then use a set of double-primed reference axes x'', y'', z'' having $\mathbf{\Omega}'' = \mathbf{\Omega}_{xyz}$ and apply Eq. 20–6 to determine $\dot{\mathbf{\Omega}}_{xyz}$ and the relative motion $(\mathbf{v}_{B/A})_{xyz}$ and $(\mathbf{a}_{B/A})_{xyz}$.

- After the final forms of $\dot{\mathbf{\Omega}}, \mathbf{v}_A, \mathbf{a}_A, \dot{\mathbf{\Omega}}_{xyz}, (\mathbf{v}_{B/A})_{xyz}$, and $(\mathbf{a}_{B/A})_{xyz}$ are obtained, numerical problem data can be substituted and the kinematic terms evaluated. The components of all these vectors can be selected either along the X, Y, Z or along the x, y, z axes. The choice is arbitrary, provided a consistent set of unit vectors is used.

20

EXAMPLE 20.4

A motor and attached rod AB have the angular motions shown in Fig. 20–12. A collar C on the rod is located 0.25 m from A and is moving downward along the rod with a velocity of 3 m/s and an acceleration of 2 m/s². Determine the velocity and acceleration of C at this instant.

SOLUTION

Coordinate Axes.
The origin of the fixed X, Y, Z reference is chosen at the center of the platform, and the origin of the moving x, y, z frame at point A, Fig. 20–12. Since the collar is subjected to two components of angular motion, $\boldsymbol{\omega}_p$ and $\boldsymbol{\omega}_M$, it will be viewed as having an angular velocity of $\boldsymbol{\Omega}_{xyz} = \boldsymbol{\omega}_M$ in x, y, z. Therefore, the x, y, z axes will be attached to the platform so that $\boldsymbol{\Omega} = \boldsymbol{\omega}_p$.

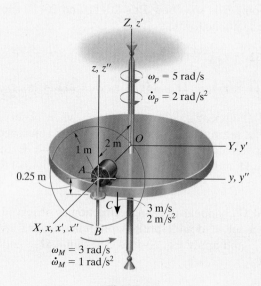

Fig. 20–12

Kinematic Equations.   Equations 20–11 and 20–12, applied to points C and A, become

$$\mathbf{v}_C = \mathbf{v}_A + \mathbf{\Omega} \times \mathbf{r}_{C/A} + (\mathbf{v}_{C/A})_{xyz}$$

$$\mathbf{a}_C = \mathbf{a}_A + \dot{\mathbf{\Omega}} \times \mathbf{r}_{C/A} + \mathbf{\Omega} \times (\mathbf{\Omega} \times \mathbf{r}_{C/A}) + 2\mathbf{\Omega} \times (\mathbf{v}_{C/A})_{xyz} + (\mathbf{a}_{C/A})_{xyz}$$

Motion of A.   Here $\mathbf{r}_A$ changes direction relative to X, Y, Z. To find the time derivatives of $\mathbf{r}_A$ we will use a set of x', y', z' axes coincident with the X, Y, Z axes that rotate at $\mathbf{\Omega}' = \boldsymbol{\omega}_p$. Thus,

$$\mathbf{\Omega} = \boldsymbol{\omega}_p = \{5\mathbf{k}\} \text{ rad/s } (\mathbf{\Omega} \text{ does not change direction relative to } X, Y, Z.)$$

$$\dot{\mathbf{\Omega}} = \dot{\boldsymbol{\omega}}_p = \{2\mathbf{k}\} \text{ rad/s}^2$$

$$\mathbf{r}_A = \{2\mathbf{i}\} \text{ m}$$

$$\mathbf{v}_A = \dot{\mathbf{r}}_A = (\dot{\mathbf{r}}_A)_{x'y'z'} + \boldsymbol{\omega}_p \times \mathbf{r}_A = 0 + 5\mathbf{k} \times 2\mathbf{i} = \{10\mathbf{j}\} \text{ m/s}$$

$$\mathbf{a}_A = \ddot{\mathbf{r}}_A = [(\ddot{\mathbf{r}}_A)_{x'y'z'} + \boldsymbol{\omega}_p \times (\dot{\mathbf{r}}_A)_{x'y'z'}] + \dot{\boldsymbol{\omega}}_p \times \mathbf{r}_A + \boldsymbol{\omega}_p \times \dot{\mathbf{r}}_A$$

$$= [0 + 0] + 2\mathbf{k} \times 2\mathbf{i} + 5\mathbf{k} \times 10\mathbf{j} = \{-50\mathbf{i} + 4\mathbf{j}\} \text{ m/s}^2$$

Motion of C with Respect to A.   Here $\mathbf{r}_{C/A}$ changes direction relative to x, y, z, and so to find its time derivatives use a set of x'', y'', z'' axes that rotate at $\mathbf{\Omega}'' = \mathbf{\Omega}_{xyz} = \boldsymbol{\omega}_M$. Thus,

$$\mathbf{\Omega}_{xyz} = \boldsymbol{\omega}_M = \{3\mathbf{i}\} \text{ rad/s } (\mathbf{\Omega}_{xyz} \text{ does not change direction relative to } x, y, z.)$$

$$\dot{\mathbf{\Omega}}_{xyz} = \dot{\boldsymbol{\omega}}_M = \{1\mathbf{i}\} \text{ rad/s}^2$$

$$\mathbf{r}_{C/A} = \{-0.25\mathbf{k}\} \text{ m}$$

$$(\mathbf{v}_{C/A})_{xyz} = (\dot{\mathbf{r}}_{C/A})_{xyz} = (\dot{\mathbf{r}}_{C/A})_{x''y''z''} + \boldsymbol{\omega}_M \times \mathbf{r}_{C/A}$$

$$= -3\mathbf{k} + [3\mathbf{i} \times (-0.25\mathbf{k})] = \{0.75\mathbf{j} - 3\mathbf{k}\} \text{ m/s}$$

$$(\mathbf{a}_{C/A})_{xyz} = (\ddot{\mathbf{r}}_{C/A})_{xyz} = [(\ddot{\mathbf{r}}_{C/A})_{x''y''z''} + \boldsymbol{\omega}_M \times (\dot{\mathbf{r}}_{C/A})_{x''y''z''}] + \dot{\boldsymbol{\omega}}_M \times \mathbf{r}_{C/A} + \boldsymbol{\omega}_M \times (\dot{\mathbf{r}}_{C/A})_{xyz}$$

$$= [-2\mathbf{k} + 3\mathbf{i} \times (-3\mathbf{k})] + (1\mathbf{i}) \times (-0.25\mathbf{k}) + (3\mathbf{i}) \times (0.75\mathbf{j} - 3\mathbf{k})$$

$$= \{18.25\mathbf{j} + 0.25\mathbf{k}\} \text{ m/s}^2$$

Motion of C.

$$\mathbf{v}_C = \mathbf{v}_A + \mathbf{\Omega} \times \mathbf{r}_{C/A} + (\mathbf{v}_{C/A})_{xyz}$$

$$= 10\mathbf{j} + [5\mathbf{k} \times (-0.25\mathbf{k})] + (0.75\mathbf{j} - 3\mathbf{k})$$

$$= \{10.75\mathbf{j} - 3\mathbf{k}\} \text{ m/s} \qquad\qquad\qquad Ans.$$

$$\mathbf{a}_C = \mathbf{a}_A + \dot{\mathbf{\Omega}} \times \mathbf{r}_{C/A} + \mathbf{\Omega} \times (\mathbf{\Omega} \times \mathbf{r}_{C/A}) + 2\mathbf{\Omega} \times (\mathbf{v}_{C/A})_{xyz} + (\mathbf{a}_{C/A})_{xyz}$$

$$= (-50\mathbf{i} + 4\mathbf{j}) + [2\mathbf{k} \times (-0.25\mathbf{k})] + 5\mathbf{k} \times [5\mathbf{k} \times (-0.25\mathbf{k})]$$

$$+ 2[5\mathbf{k} \times (0.75\mathbf{j} - 3\mathbf{k})] + (18.25\mathbf{j} + 0.25\mathbf{k})$$

$$= \{-57.5\mathbf{i} + 22.25\mathbf{j} + 0.25\mathbf{k}\} \text{ m/s}^2 \qquad\qquad Ans.$$

20

EXAMPLE | 20.5

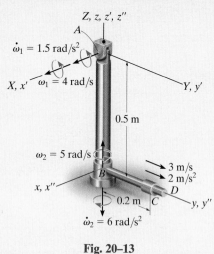

$\omega_1 = 4$ rad/s

$\omega_2 = 5$ rad/s

3 m/s
2 m/s^2

0.5 m

0.2 m | C

$\dot{\omega}_2 = 6$ rad/s^2

Fig. 20–13

The pendulum shown in Fig. 20–13 consists of two rods; AB is pin supported at A and swings only in the Y–Z plane, whereas a bearing at B allows the attached rod BD to spin about rod AB. At a given instant, the rods have the angular motions shown. Also, a collar C, located 0.2 m from B, has a velocity of 3 m/s and an acceleration of 2 m/s^2 along the rod. Determine the velocity and acceleration of the collar at this instant.

SOLUTION I

Coordinate Axes.   The origin of the fixed X, Y, Z frame will be placed at A. Motion of the collar is conveniently observed from B, so the origin of the x, y, z frame is located at this point. We will choose $\mathbf{\Omega} = \boldsymbol{\omega}_1$ and $\mathbf{\Omega}_{xyz} = \boldsymbol{\omega}_2$.

Kinematic Equations.

$$\mathbf{v}_C = \mathbf{v}_B + \mathbf{\Omega} \times \mathbf{r}_{C/B} + (\mathbf{v}_{C/B})_{xyz}$$

$$\mathbf{a}_C = \mathbf{a}_B + \dot{\mathbf{\Omega}} \times \mathbf{r}_{C/B} + \mathbf{\Omega} \times (\mathbf{\Omega} \times \mathbf{r}_{C/B}) + 2\mathbf{\Omega} \times (\mathbf{v}_{C/B})_{xyz} + (\mathbf{a}_{C/B})_{xyz}$$

Motion of B.   To find the time derivatives of $\mathbf{r}_B$ let the x', y', z' axes rotate with $\mathbf{\Omega}' = \boldsymbol{\omega}_1$. Then

$$\mathbf{\Omega}' = \boldsymbol{\omega}_1 = \{4\mathbf{i}\} \text{ rad/s} \quad \dot{\mathbf{\Omega}}' = \dot{\boldsymbol{\omega}}_1 = \{1.5\mathbf{i}\} \text{ rad/s}^2$$

$$\mathbf{r}_B = \{-0.5\mathbf{k}\} \text{ m}$$

$$\mathbf{v}_B = \dot{\mathbf{r}}_B = (\dot{\mathbf{r}}_B)_{x'y'z'} + \boldsymbol{\omega}_1 \times \mathbf{r}_B = \mathbf{0} + 4\mathbf{i} \times (-0.5\mathbf{k}) = \{2\mathbf{j}\} \text{ m/s}$$

$$\mathbf{a}_B = \ddot{\mathbf{r}}_B = [(\ddot{\mathbf{r}}_B)_{x'y'z'} + \boldsymbol{\omega}_1 \times (\dot{\mathbf{r}}_B)_{x'y'z'}] + \dot{\boldsymbol{\omega}}_1 \times \mathbf{r}_B + \boldsymbol{\omega}_1 \times \dot{\mathbf{r}}_B$$

$$= [0 + 0] + 1.5\mathbf{i} \times (-0.5\mathbf{k}) + 4\mathbf{i} \times 2\mathbf{j} = \{0.75\mathbf{j} + 8\mathbf{k}\} \text{ m/s}^2$$

Motion of C with Respect to B.   To find the time derivatives of $\mathbf{r}_{C/B}$ relative to x, y, z, let the x'', y'', z'' axes rotate with $\mathbf{\Omega}_{xyz} = \boldsymbol{\omega}_2$. Then

$$\mathbf{\Omega}_{xyz} = \boldsymbol{\omega}_2 = \{5\mathbf{k}\} \text{ rad/s} \quad \dot{\mathbf{\Omega}}_{xyz} = \dot{\boldsymbol{\omega}}_2 = \{-6\mathbf{k}\} \text{ rad/s}^2$$

$$\mathbf{r}_{C/B} = \{0.2\mathbf{j}\} \text{ m}$$

$$(\mathbf{v}_{C/B})_{xyz} = (\dot{\mathbf{r}}_{C/B})_{xyz} = (\dot{\mathbf{r}}_{C/B})_{x''y''z''} + \boldsymbol{\omega}_2 \times \mathbf{r}_{C/B} = 3\mathbf{j} + 5\mathbf{k} \times 0.2\mathbf{j} = \{-1\mathbf{i} + 3\mathbf{j}\} \text{ m/s}$$

$$(\mathbf{a}_{C/B})_{xyz} = (\ddot{\mathbf{r}}_{C/B})_{xyz} = [(\ddot{\mathbf{r}}_{C/B})_{x''y''z''} + \boldsymbol{\omega}_2 \times (\dot{\mathbf{r}}_{C/B})_{x''y''z''}] + \dot{\boldsymbol{\omega}}_2 \times \mathbf{r}_{C/B} + \boldsymbol{\omega}_2 \times (\dot{\mathbf{r}}_{C/B})_{xyz}$$

$$= (2\mathbf{j} + 5\mathbf{k} \times 3\mathbf{j}) + (-6\mathbf{k} \times 0.2\mathbf{j}) + [5\mathbf{k} \times (-1\mathbf{i} + 3\mathbf{j})]$$

$$= \{-28.8\mathbf{i} - 3\mathbf{j}\} \text{ m/s}^2$$

Motion of C.

$$\mathbf{v}_C = \mathbf{v}_B + \mathbf{\Omega} \times \mathbf{r}_{C/B} + (\mathbf{v}_{C/B})_{xyz} = 2\mathbf{j} + 4\mathbf{i} \times 0.2\mathbf{j} + (-1\mathbf{i} + 3\mathbf{j})$$

$$= \{-1\mathbf{i} + 5\mathbf{j} + 0.8\mathbf{k}\} \text{ m/s} \qquad \textit{Ans.}$$

$$\mathbf{a}_C = \mathbf{a}_B + \dot{\mathbf{\Omega}} \times \mathbf{r}_{C/B} + \mathbf{\Omega} \times (\mathbf{\Omega} \times \mathbf{r}_{C/B}) + 2\mathbf{\Omega} \times (\mathbf{v}_{C/B})_{xyz} + (\mathbf{a}_{C/B})_{xyz}$$

$$= (0.75\mathbf{j} + 8\mathbf{k}) + (1.5\mathbf{i} \times 0.2\mathbf{j}) + [4\mathbf{i} \times (4\mathbf{i} \times 0.2\mathbf{j})]$$

$$+ 2[4\mathbf{i} \times (-1\mathbf{i} + 3\mathbf{j})] + (-28.8\mathbf{i} - 3\mathbf{j})$$

$$= \{-28.8\mathbf{i} - 5.45\mathbf{j} + 32.3\mathbf{k}\} \text{ m/s}^2 \qquad \textit{Ans.}$$

20

SOLUTION II

Coordinate Axes. Here we will let the x, y, z axes rotate at

$$\Omega = \omega_1 + \omega_2 = \{4\mathbf{i} + 5\mathbf{k}\} \text{ rad/s}$$

Then $\Omega_{xyz} = 0$.

Motion of B. From the constraints of the problem ω_1 does not change direction relative to X, Y, Z; however, the direction of ω_2 is changed by ω_1. Thus, to obtain $\dot{\Omega}$ consider x', y', z' axes coincident with the X, Y, Z axes at A, so that $\Omega' = \omega_1$. Then taking the derivative of the components of Ω,

$$\dot{\Omega} = \dot{\omega}_1 + \dot{\omega}_2 = [(\dot{\omega}_1)_{x'y'z'} + \omega_1 \times \omega_1] + [(\dot{\omega}_2)_{x'y'z'} + \omega_1 \times \omega_2]$$
$$= [1.5\mathbf{i} + 0] + [-6\mathbf{k} + 4\mathbf{i} \times 5\mathbf{k}] = \{1.5\mathbf{i} - 20\mathbf{j} - 6\mathbf{k}\} \text{ rad/s}^2$$

Also, ω_1 changes the direction of $\mathbf{r}_B$ so that the time derivatives of $\mathbf{r}_B$ can be found using the primed axes defined above. Hence,

$$\mathbf{v}_B = \dot{\mathbf{r}}_B = (\dot{\mathbf{r}}_B)_{x'y'z'} + \omega_1 \times \mathbf{r}_B$$
$$= 0 + 4\mathbf{i} \times (-0.5\mathbf{k}) = \{2\mathbf{j}\} \text{ m/s}$$
$$\mathbf{a}_B = \ddot{\mathbf{r}}_B = [(\ddot{\mathbf{r}}_B)_{x'y'z'} + \omega_1 \times (\dot{\mathbf{r}}_B)_{x'y'z'}] + \dot{\omega}_1 \times \mathbf{r}_B + \omega_1 \times \dot{\mathbf{r}}_B$$
$$= [0 + 0] + 1.5\mathbf{i} \times (-0.5\mathbf{k}) + 4\mathbf{i} \times 2\mathbf{j} = \{0.75\mathbf{j} + 8\mathbf{k}\} \text{ m/s}^2$$

Motion of C with Respect to B.

$$\Omega_{xyz} = 0$$
$$\dot{\Omega}_{xyz} = 0$$
$$\mathbf{r}_{C/B} = \{0.2\mathbf{j}\} \text{ m}$$
$$(\mathbf{v}_{C/B})_{xyz} = \{3\mathbf{j}\} \text{ m/s}$$
$$(\mathbf{a}_{C/B})_{xyz} = \{2\mathbf{j}\} \text{ m/s}^2$$

Motion of C.

$$\mathbf{v}_C = \mathbf{v}_B + \Omega \times \mathbf{r}_{C/B} + (\mathbf{v}_{C/B})_{xyz}$$
$$= 2\mathbf{j} + [(4\mathbf{i} + 5\mathbf{k}) \times (0.2\mathbf{j})] + 3\mathbf{j}$$
$$= \{-1\mathbf{i} + 5\mathbf{j} + 0.8\mathbf{k}\} \text{ m/s} \qquad \text{Ans.}$$

$$\mathbf{a}_C = \mathbf{a}_B + \dot{\Omega} \times \mathbf{r}_{C/B} + \Omega \times (\Omega \times \mathbf{r}_{C/B}) + 2\Omega \times (\mathbf{v}_{C/B})_{xyz} + (\mathbf{a}_{C/B})_{xyz}$$
$$= (0.75\mathbf{j} + 8\mathbf{k}) + [(1.5\mathbf{i} - 20\mathbf{j} - 6\mathbf{k}) \times (0.2\mathbf{j})]$$
$$+ (4\mathbf{i} + 5\mathbf{k}) \times [(4\mathbf{i} + 5\mathbf{k}) \times 0.2\mathbf{j}] + 2[(4\mathbf{i} + 5\mathbf{k}) \times 3\mathbf{j}] + 2\mathbf{j}$$
$$= \{-28.8\mathbf{i} - 5.45\mathbf{j} + 32.3\mathbf{k}\} \text{ m/s}^2 \qquad \text{Ans.}$$

20

PROBLEMS

20–37. Solve Example 20.5 such that the x, y, z axes move with curvilinear translation, $\Omega = 0$ in which case the collar appears to have both an angular velocity $\Omega_{xyz} = \omega_1 + \omega_2$ and radial motion.

20–38. Solve Example 20.5 by fixing x, y, z axes to rod BD so that $\Omega = \omega_1 + \omega_2$. In this case the collar appears only to move radially outward along BD; hence $\Omega_{xyz} = 0$.

20–39. At the instant $\theta = 60°$, the telescopic boom AB of the construction lift is rotating with a constant angular velocity about the z axis of $\omega_1 = 0.5$ rad/s and about the pin at A with a constant angular speed of $\omega_2 = 0.25$ rad/s. Simultaneously, the boom is extending with a velocity of 1.5 ft/s, and it has an acceleration of 0.5 ft/s², both measured relative to the construction lift. Determine the velocity and acceleration of point B located at the end of the boom at this instant.

***20–40.** At the instant $\theta = 60°$, the construction lift is rotating about the z axis with an angular velocity of $\omega_1 = 0.5$ rad/s and an angular acceleration of $\dot\omega_1 = 0.25$ rad/s² while the telescopic boom AB rotates about the pin at A with an angular velocity of $\omega_2 = 0.25$ rad/s and angular acceleration of $\dot\omega_2 = 0.1$ rad/s². Simultaneously, the boom is extending with a velocity of 1.5 ft/s, and it has an acceleration of 0.5 ft/s², both measured relative to the frame. Determine the velocity and acceleration of point B located at the end of the boom at this instant.

20–41. At a given instant, rod BD is rotating about the y axis with an angular velocity $\omega_{BD} = 2$ rad/s and an angular acceleration $\dot\omega_{BD} = 5$ rad/s². Also, when $\theta = 60°$ link AC is rotating downward such that $\dot\theta = 2$ rad/s and $\ddot\theta = 8$ rad/s². Determine the velocity and acceleration of point A on the link at this instant.

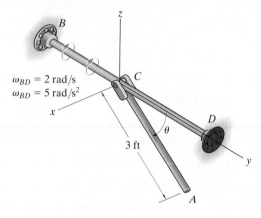

Prob. 20–41

20–42. At the instant $\theta = 30°$, the frame of the crane and the boom AB rotate with a constant angular velocity of $\omega_1 = 1.5$ rad/s and $\omega_2 = 0.5$ rad/s, respectively. Determine the velocity and acceleration of point B at this instant.

20–43. At the instant $\theta = 30°$, the frame of the crane is rotating with an angular velocity of $\omega_1 = 1.5$ rad/s and angular acceleration of $\dot\omega_1 = 0.5$ rad/s², while the boom AB rotates with an angular velocity of $\omega_2 = 0.5$ rad/s and angular acceleration of $\dot\omega_2 = 0.25$ rad/s². Determine the velocity and acceleration of point B at this instant.

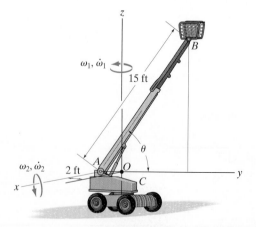

Probs. 20–39/40

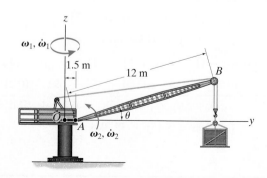

Probs. 20–42/43

*20–44. At the instant shown, the boom is rotating about the z axis with an angular velocity $\omega_1 = 2$ rad/s and angular acceleration $\dot{\omega}_1 = 0.8$ rad/s². At this same instant the swivel is rotating at $\omega_2 = 3$ rad/s when $\dot{\omega}_2 = 2$ rad/s², both measured relative to the boom. Determine the velocity and acceleration of point P on the pipe at this instant.

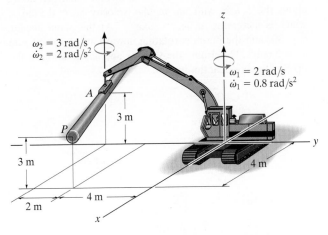

Prob. 20–44

20–45. During the instant shown the frame of the X-ray camera is rotating about the vertical axis at $\omega_z = 5$ rad/s and $\dot{\omega}_z = 2$ rad/s². Relative to the frame the arm is rotating at $\omega_{rel} = 2$ rad/s and $\dot{\omega}_{rel} = 1$ rad/s². Determine the velocity and acceleration of the center of the camera C at this instant.

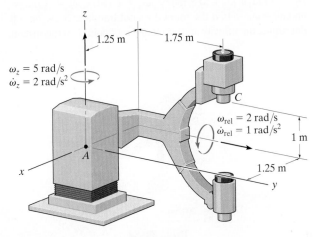

Prob. 20–45

20–46. The boom AB of the crane is rotating about the z axis with an angular velocity $\omega_z = 0.75$ rad/s, which is increasing at $\dot{\omega}_z = 2$ rad/s². At the same instant, $\theta = 60°$ and the boom is rotating upward at a constant rate $\dot{\theta} = 0.5$ rad/s². Determine the velocity and acceleration of the tip B of the boom at this instant.

*20–47. The boom AB of the crane is rotating about the z axis with an angular velocity of $\omega_z = 0.75$ rad/s, which is increasing at $\dot{\omega}_z = 2$ rad/s². At the same instant, $\theta = 60°$ and the boom is rotating upward at $\dot{\theta} = 0.5$ rad/s², which is increasing at $\ddot{\theta} = 0.75$ rad/s². Determine the velocity and acceleration of the tip B of the boom at this instant.

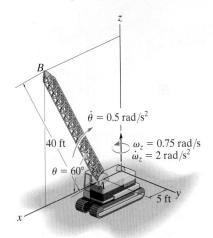

Probs. 20–46/47

20-48. At the instant shown, the motor rotates about the z axis with an angular velocity of $\omega_1 = 3$ rad/s and angular acceleration of $\dot{\omega}_1 = 1.5$ rad/s². Simultaneously, shaft OA rotates with an angular velocity of $\omega_2 = 6$ rad/s, and angular acceleration of $\dot{\omega}_2 = 3$ rad/s², and collar C slides along rod AB with a velocity and acceleration of 6 m/s and 3 m/s². Determine the velocity and acceleration of collar C at this instant.

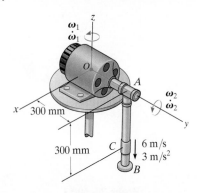

Prob. 20–48

20–49. The motor rotates about the z axis with a constant angular velocity of $\omega_1 = 3$ rad/s. Simultaneously, shaft OA rotates with a constant angular velocity of $\omega_2 = 6$ rad/s. Also, collar C slides along rod AB with a velocity and acceleration of 6 m/s and 3 m/s². Determine the velocity and acceleration of collar C at the instant shown.

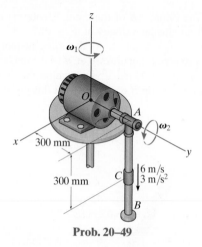

Prob. 20–49

20–50. At the instant shown, the arm OA of the conveyor belt is rotating about the z axis with a constant angular velocity $\omega_1 = 6$ rad/s, while at the same instant the arm is rotating upward at a constant rate $\omega_2 = 4$ rad/s. If the conveyor is running at a constant rate $\dot{r} = 5$ ft/s, determine the velocity and acceleration of the package P at the instant shown. Neglect the size of the package.

20–51. At the instant shown, the arm OA of the conveyor belt is rotating about the z axis with a constant angular velocity $\omega_1 = 6$ rad/s, while at the same instant the arm is rotating upward at a constant rate $\omega_2 = 4$ rad/s. If the conveyor is running at a rate $\dot{r} = 5$ ft/s, which is increasing at $\ddot{r} = 8$ ft/s², determine the velocity and acceleration of the package P at the instant shown. Neglect the size of the package.

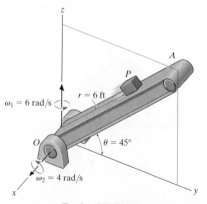

Probs. 20–50/51

*20–52.** The boom AB of the locomotive crane is rotating about the z axis with an angular velocity $\omega_1 = 0.5$ rad/s, which is increasing at $\dot{\omega}_1 = 3$ rad/s². At this same instant, $\theta = 30°$ and the boom is rotating upward at a constant rate of $\dot{\theta} = 3$ rad/s. Determine the velocity and acceleration of the tip B of the boom at this instant.

20–53. The locomotive crane is traveling to the right at 2 m/s and has an acceleration of 1.5 m/s², while the boom is rotating about the z axis with an angular velocity $\omega_1 = 0.5$ rad/s, which is increasing at $\dot{\omega}_1 = 3$ rad/s². At this same instant, $\theta = 30°$ and the boom is rotating upward at a constant rate $\dot{\theta} = 3$ rad/s. Determine the velocity and acceleration of the tip B of the boom at this instant.

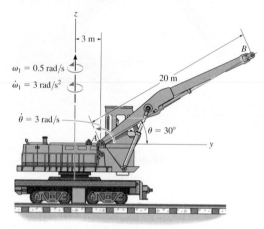

Probs. 20–52/53

20–54. The robot shown has four degrees of rotational freedom, namely, arm OA rotates about the x and z axes, arm AB rotates about the x axis, and CB rotates about the y axis. At the instant shown, $\omega_2 = 1.5$ rad/s, $\dot{\omega}_2 = 1$ rad/s², $\omega_3 = 3$ rad/s, $\dot{\omega}_3 = 0.5$ rad/s², $\omega_4 = 6$ rad/s, $\dot{\omega}_4 = 3$ rad/s², and $\omega_1 = \dot{\omega}_1 = 0$. If the robot does not translate, i.e., $\mathbf{v} = \mathbf{a} = \mathbf{0}$, determine the velocity and acceleration of point C at this instant.

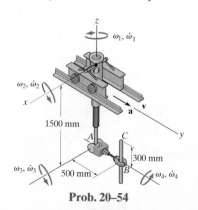

Prob. 20–54

CHAPTER REVIEW

Rotation About a Fixed Point

When a body rotates about a fixed point O, then points on the body follow a path that lies on the surface of a sphere centered at O.

Since the angular acceleration is a time rate of change in the angular velocity, then it is necessary to account for both the magnitude and directional changes of $\boldsymbol{\omega}$ when finding its time derivative. To do this, the angular velocity is often specified in terms of its component motions, such that the direction of some of these components will remain constant relative to rotating x, y, z axes. If this is the case, then the time derivative relative to the fixed axis can be determined using $\dot{\mathbf{A}} = (\dot{\mathbf{A}})_{xyz} + \boldsymbol{\Omega} \times \mathbf{A}$.

Once $\boldsymbol{\omega}$ and $\boldsymbol{\alpha}$ are known, the velocity and acceleration of any point P in the body can then be determined.

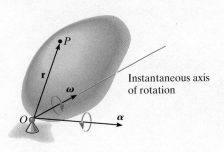

Instantaneous axis of rotation

$$\mathbf{v}_P = \boldsymbol{\omega} \times \mathbf{r}$$
$$\mathbf{a}_P = \boldsymbol{\alpha} \times \mathbf{r} + \boldsymbol{\omega} \times (\boldsymbol{\omega} \times \mathbf{r})$$

General Motion

If the body undergoes general motion, then the motion of a point B on the body can be related to the motion of another point A using a relative motion analysis, with translating axes attached to A.

$$\mathbf{v}_B = \mathbf{v}_A + \boldsymbol{\omega} \times \mathbf{r}_{B/A}$$
$$\mathbf{a}_B = \mathbf{a}_A + \boldsymbol{\alpha} \times \mathbf{r}_{B/A} + \boldsymbol{\omega} \times (\boldsymbol{\omega} \times \mathbf{r}_{B/A})$$

Relative Motion Analysis Using Translating and Rotating Axes

The motion of two points A and B on a body, a series of connected bodies, or each point located on two different paths, can be related using a relative motion analysis with rotating and translating axes at A.

When applying the equations, to find $\mathbf{v}_B$ and $\mathbf{a}_B$, it is important to account for both the magnitude and directional changes of $\mathbf{r}_A$, $\mathbf{r}_{B/A}$, $\boldsymbol{\Omega}$, and $\boldsymbol{\Omega}_{xyz}$ when taking their time derivatives to find $\mathbf{v}_A$, $\mathbf{a}_A$, $(\mathbf{v}_{B/A})_{xyz}$, $(\mathbf{a}_{B/A})_{xyz}$, $\dot{\boldsymbol{\Omega}}$, and $\dot{\boldsymbol{\Omega}}_{xyz}$. To do this properly, one must use Eq. 20–6.

$$\mathbf{v}_B = \mathbf{v}_A + \boldsymbol{\Omega} \times \mathbf{r}_{B/A} + (\mathbf{v}_{B/A})_{xyz}$$
$$\mathbf{a}_B = \mathbf{a}_A + \dot{\boldsymbol{\Omega}} \times \mathbf{r}_{B/A} + \boldsymbol{\Omega} \times (\boldsymbol{\Omega} \times \mathbf{r}_{B/A}) + 2\boldsymbol{\Omega} \times (\mathbf{v}_{B/A})_{xyz} + (\mathbf{a}_{B/A})_{xyz}$$

20

The forces acting on each of these motorcycles can be determined using the equations of motion as discussed in this chapter.

Three-Dimensional Kinetics of a Rigid Body

CHAPTER OBJECTIVES

- To introduce the methods for finding the moments of inertia and products of inertia of a body about various axes.

- To show how to apply the principles of work and energy and linear and angular momentum to a rigid body having three-dimensional motion.

- To develop and apply the equations of motion in three dimensions.

- To study gyroscopic and torque-free motion.

*21.1 Moments and Products of Inertia

When studying the planar kinetics of a body, it was necessary to introduce the moment of inertia I_G, which was computed about an axis perpendicular to the plane of motion and passing through the body's mass center G. For the kinetic analysis of three-dimensional motion it will sometimes be necessary to calculate six inertial quantities. These terms, called the moments and products of inertia, describe in a particular way the distribution of mass for a body relative to a given coordinate system that has a specified orientation and point of origin.

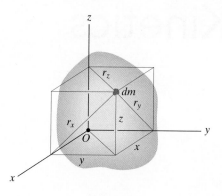

Fig. 21–1

Moment of Inertia.

Moment of Inertia. Consider the rigid body shown in Fig. 21–1. The *moment of inertia* for a differential element dm of the body about any one of the three coordinate axes is defined as the product of the mass of the element and the square of the shortest distance from the axis to the element. For example, as noted in the figure, $r_x = \sqrt{y^2 + z^2}$, so that the mass moment of inertia of the element about the x axis is

$$dI_{xx} = r_x^2 \, dm = (y^2 + z^2) \, dm$$

The moment of inertia I_{xx} for the body can be determined by integrating this expression over the entire mass of the body. Hence, for each of the axes, we can write

$$
\begin{aligned}
I_{xx} &= \int_m r_x^2 \, dm = \int_m (y^2 + z^2) \, dm \\
I_{yy} &= \int_m r_y^2 \, dm = \int_m (x^2 + z^2) \, dm \\
I_{zz} &= \int_m r_z^2 \, dm = \int_m (x^2 + y^2) \, dm
\end{aligned}
\tag{21–1}
$$

Here it is seen that the moment of inertia is *always a positive quantity*, since it is the summation of the product of the mass dm, which is always positive, and the distances squared.

Product of Inertia.

Product of Inertia. The *product of inertia* for a differential element dm with respect to a set of *two orthogonal planes* is defined as the product of the mass of the element and the perpendicular (or shortest) distances from the planes to the element. For example, this distance is x to the y–z plane and it is y to the x–z plane, Fig. 21–1. The product of inertia dI_{xy} for the element is therefore

$$dI_{xy} = xy \, dm$$

Note also that $dI_{yx} = dI_{xy}$. By integrating over the entire mass, the products of inertia of the body with respect to each combination of planes can be expressed as

$$
\begin{aligned}
I_{xy} &= I_{yx} = \int_m xy \, dm \\
I_{yz} &= I_{zy} = \int_m yz \, dm \\
I_{xz} &= I_{zx} = \int_m xz \, dm
\end{aligned}
\tag{21–2}
$$

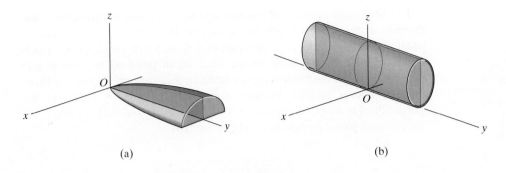

(a)   (b)

Fig. 21–2

Unlike the moment of inertia, which is always positive, the product of inertia may be positive, negative, or zero. The result depends on the algebraic signs of the two defining coordinates, which vary independently from one another. In particular, if either one or both of the orthogonal planes are *planes of symmetry* for the mass, the *product of inertia* with respect to these planes will be *zero*. In such cases, elements of mass will occur in *pairs* located on each side of the plane of symmetry. On one side of the plane the product of inertia for the element will be positive, while on the other side the product of inertia of the corresponding element will be negative, the sum therefore yielding zero. Examples of this are shown in Fig. 21–2. In the first case, Fig. 21–2a, the y–z plane is a plane of symmetry, and hence $I_{xy} = I_{xz} = 0$. Calculation of I_{yz} will yield a *positive* result, since all elements of mass are located using only positive y and z coordinates. For the cylinder, with the coordinate axes located as shown in Fig. 21–2b, the x–z and y–z planes are both planes of symmetry. Thus, $I_{xy} = I_{yz} = I_{zx} = 0$.

Parallel-Axis and Parallel-Plane Theorems.
The techniques of integration used to determine the moment of inertia of a body were described in Sec. 17.1. Also discussed were methods to determine the moment of inertia of a composite body, i.e., a body that is composed of simpler segments, as tabulated on the inside back cover. In both of these cases the *parallel-axis theorem* is often used for the calculations. This theorem, which was developed in Sec. 17.1, allows us to transfer the moment of inertia of a body from an axis passing through its mass center G to a parallel axis passing through some other point. If G has coordinates x_G, y_G, z_G defined with respect to the x, y, z axes, Fig. 21–3, then the parallel-axis equations used to calculate the moments of inertia about the x, y, z axes are

$$\begin{aligned}
I_{xx} &= (I_{x'x'})_G + m(y_G^2 + z_G^2) \\
I_{yy} &= (I_{y'y'})_G + m(x_G^2 + z_G^2) \\
I_{zz} &= (I_{z'z'})_G + m(x_G^2 + y_G^2)
\end{aligned}$$

(21–3)

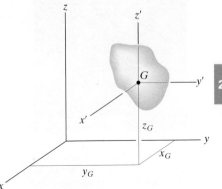

Fig. 21–3

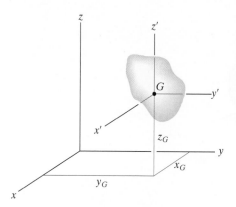

Fig. 21–3 (repeated)

The products of inertia of a composite body are computed in the same manner as the body's moments of inertia. Here, however, the *parallel-plane theorem* is important. This theorem is used to transfer the products of inertia of the body with respect to a set of three orthogonal planes passing through the body's mass center to a corresponding set of three parallel planes passing through some other point O. Defining the perpendicular distances between the planes as x_G, y_G and z_G, Fig. 21–3, the parallel-plane equations can be written as

$$\begin{aligned}
I_{xy} &= (I_{x'y'})_G + mx_G y_G \\
I_{yz} &= (I_{y'z'})_G + my_G z_G \\
I_{zx} &= (I_{z'x'})_G + mz_G x_G
\end{aligned} \tag{21–4}$$

The derivation of these formulas is similar to that given for the parallel-axis equation, Sec. 17.1.

Inertia Tensor. The inertial properties of a body are therefore completely characterized by nine terms, six of which are independent of one another. This set of terms is defined using Eqs. 21–1 and 21–2 and can be written as

$$\begin{pmatrix}
I_{xx} & -I_{xy} & -I_{xz} \\
-I_{yx} & I_{yy} & -I_{yz} \\
-I_{zx} & -I_{zy} & I_{zz}
\end{pmatrix}$$

This array is called an *inertia tensor*.* It has a unique set of values for a body when it is determined for each location of the origin O and orientation of the coordinate axes.

In general, for point O we can specify a unique axes inclination for which the products of inertia for the body are zero when computed with respect to these axes. When this is done, the inertia tensor is said to be "diagonalized" and may be written in the simplified form

$$\begin{pmatrix}
I_x & 0 & 0 \\
0 & I_y & 0 \\
0 & 0 & I_z
\end{pmatrix}$$

Here $I_x = I_{xx}$, $I_y = I_{yy}$, and $I_z = I_{zz}$ are termed the *principal moments of inertia* for the body, which are computed with respect to the *principal axes of inertia*. Of these three principal moments of inertia, one will be a maximum and another a minimum of the body's moment of inertia.

The dynamics of the space shuttle while it orbits the earth can be predicted only if its moments and products of inertia are known relative to its mass center.

*The negative signs are here as a consequence of the development of angular momentum, Eqs. 21–10.

The mathematical determination of the directions of principal axes of inertia will not be discussed here (see Prob. 21–22). However, there are many cases in which the principal axes can be determined by inspection. From the previous discussion it was noted that if the coordinate axes are oriented such that *two* of the three orthogonal planes containing the axes are planes of *symmetry* for the body, then all the products of inertia for the body are zero with respect to these coordinate planes, and hence these coordinate axes are principal axes of inertia. For example, the x, y, z axes shown in Fig. 21–2b represent the principal axes of inertia for the cylinder at point O.

Moment of Inertia About an Arbitrary Axis.

Consider the body shown in Fig. 21–4, where the nine elements of the inertia tensor have been determined with respect to the x, y, z axes having an origin at O. Here we wish to determine the moment of inertia of the body about the Oa axis, which has a direction defined by the unit vector $\mathbf{u}_a$. By definition $I_{Oa} = \int b^2\, dm$, where b is the *perpendicular distance* from dm to Oa. If the position of dm is located using $\mathbf{r}$, then $b = r\sin\theta$, which represents the *magnitude* of the cross product $\mathbf{u}_a \times \mathbf{r}$. Hence, the moment of inertia can be expressed as

$$I_{Oa} = \int_m |(\mathbf{u}_a \times \mathbf{r})|^2\, dm = \int_m (\mathbf{u}_a \times \mathbf{r}) \cdot (\mathbf{u}_a \times \mathbf{r})\, dm$$

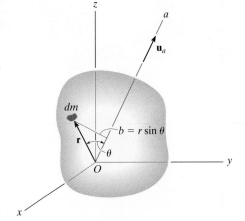

Fig. 21–4

Provided $\mathbf{u}_a = u_x\mathbf{i} + u_y\mathbf{j} + u_z\mathbf{k}$ and $\mathbf{r} = x\mathbf{i} + y\mathbf{j} + z\mathbf{k}$, then $\mathbf{u}_a \times \mathbf{r} = (u_y z - u_z y)\mathbf{i} + (u_z x - u_x z)\mathbf{j} + (u_x y - u_y x)\mathbf{k}$. After substituting and performing the dot-product operation, the moment of inertia is

$$I_{Oa} = \int_m [(u_y z - u_z y)^2 + (u_z x - u_x z)^2 + (u_x y - u_y x)^2]\, dm$$

$$= u_x^2 \int_m (y^2 + z^2)\, dm + u_y^2 \int_m (z^2 + x^2)\, dm + u_z^2 \int_m (x^2 + y^2)\, dm$$

$$- 2u_x u_y \int_m xy\, dm - 2u_y u_z \int_m yz\, dm - 2u_z u_x \int_m zx\, dm$$

Recognizing the integrals to be the moments and products of inertia of the body, Eqs. 21–1 and 21–2, we have

$$I_{Oa} = I_{xx}u_x^2 + I_{yy}u_y^2 + I_{zz}u_z^2 - 2I_{xy}u_x u_y - 2I_{yz}u_y u_z - 2I_{zx}u_z u_x \quad (21\text{–}5)$$

Thus, if the inertia tensor is specified for the x, y, z axes, the moment of inertia of the body about the inclined Oa axis can be found. For the calculation, the direction cosines u_x, u_y, u_z of the axes must be determined. These terms specify the cosines of the coordinate direction angles α, β, γ made between the positive Oa axis and the positive x, y, z axes, respectively (see Appendix B).

21

EXAMPLE | 21.1

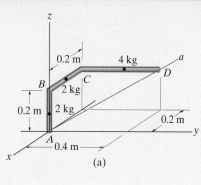

(a)

Determine the moment of inertia of the bent rod shown in Fig. 21–5a about the Aa axis. The mass of each of the three segments is given in the figure.

SOLUTION

Before applying Eq. 21–5, it is first necessary to determine the moments and products of inertia of the rod with respect to the x, y, z axes. This is done using the formula for the moment of inertia of a slender rod, $I = \frac{1}{12}ml^2$, and the parallel-axis and parallel-plane theorems, Eqs. 21–3 and 21–4. Dividing the rod into three parts and locating the mass center of each segment, Fig. 21–5b, we have

$$I_{xx} = \left[\frac{1}{12}(2)(0.2)^2 + 2(0.1)^2 \right] + [0 + 2(0.2)^2]$$
$$+ \left[\frac{1}{12}(4)(0.4)^2 + 4((0.2)^2 + (0.2)^2) \right] = 0.480 \text{ kg} \cdot \text{m}^2$$

$$I_{yy} = \left[\frac{1}{12}(2)(0.2)^2 + 2(0.1)^2 \right] + \left[\frac{1}{12}(2)(0.2)^2 + 2((-0.1)^2 + (0.2)^2) \right]$$
$$+ [0 + 4((-0.2)^2 + (0.2)^2)] = 0.453 \text{ kg} \cdot \text{m}^2$$

$$I_{zz} = \left[0 + 0 \right] + \left[\frac{1}{12}(2)(0.2)^2 + 2(-0.1)^2 \right] + \left[\frac{1}{12}(4)(0.4)^2 + \right.$$
$$\left. 4((-0.2)^2 + (0.2)^2) \right] = 0.400 \text{ kg} \cdot \text{m}^2$$

$$I_{xy} = [0 + 0] + [0 + 0] + [0 + 4(-0.2)(0.2)] = -0.160 \text{ kg} \cdot \text{m}^2$$

$$I_{yz} = [0 + 0] + [0 + 0] + [0 + 4(0.2)(0.2)] = 0.160 \text{ kg} \cdot \text{m}^2$$

$$I_{zx} = [0 + 0] + [0 + 2(0.2)(-0.1)] +$$
$$[0 + 4(0.2)(-0.2)] = -0.200 \text{ kg} \cdot \text{m}^2$$

The Aa axis is defined by the unit vector

$$\mathbf{u}_{Aa} = \frac{\mathbf{r}_D}{r_D} = \frac{-0.2\mathbf{i} + 0.4\mathbf{j} + 0.2\mathbf{k}}{\sqrt{(-0.2)^2 + (0.4)^2 + (0.2)^2}} = -0.408\mathbf{i} + 0.816\mathbf{j} + 0.408\mathbf{k}$$

Thus,

$$u_x = -0.408 \quad u_y = 0.816 \quad u_z = 0.408$$

Substituting these results into Eq. 21–5 yields

$$I_{Aa} = I_{xx}u_x^2 + I_{yy}u_y^2 + I_{zz}u_z^2 - 2I_{xy}u_xu_y - 2I_{yz}u_yu_z - 2I_{zx}u_zu_x$$

$$= 0.480(-0.408)^2 + (0.453)(0.816)^2 + 0.400(0.408)^2$$

$$- 2(-0.160)(-0.408)(0.816) - 2(0.160)(0.816)(0.408)$$

$$- 2(-0.200)(0.408)(-0.408)$$

$$= 0.169 \text{ kg} \cdot \text{m}^2 \qquad \qquad Ans.$$

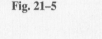

(b)

Fig. 21–5

21

PROBLEMS

21–1. Show that the sum of the moments of inertia of a body, $I_{xx} + I_{yy} + I_{zz}$, is independent of the orientation of the x, y, z axes and thus depends only on the location of the origin.

21–2. Determine the moment of inertia of the cone with respect to a vertical $\bar{y}$ axis passing through the cone's center of mass. What is the moment of inertia about a parallel axis y' that passing through the diameter of the base of the cone? The cone has a mass m.

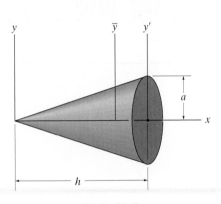

Prob. 21–2

21–3. Determine the moments of inertia I_x and I_y of the paraboloid of revolution. The mass of the paraboloid is m.

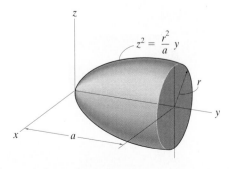

Prob. 21–3

*21–4. Determine the radii of gyration k_x and k_y for the solid formed by revolving the shaded area about the y axis. The density of the material is ρ.

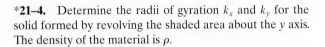
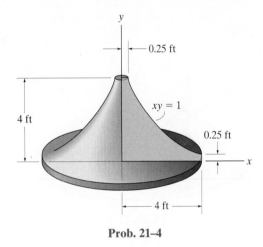

Prob. 21–4

21–5. Determine by direct integration the product of inertia I_{yz} for the homogeneous prism. The density of the material is ρ. Express the result in terms of the total mass m of the prism.

21–6. Determine by direct integration the product of inertia I_{xy} for the homogeneous prism. The density of the material is ρ. Express the result in terms of the total mass m of the prism.

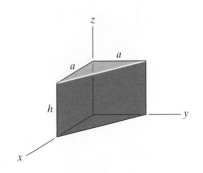

Probs. 21–5/6

21

21–7. Determine the product of inertia I_{xy} of the object formed by revolving the shaded area about the line $x = 5$ ft. Express the result in terms of the density of the material, ρ.

*21–8. Determine the moment of inertia I_y of the object formed by revolving the shaded area about the line $x = 5$ ft. Express the result in terms of the density of the material, ρ.

21–10. Determine the mass moment of inertia of the homogeneous block with respect to its centroidal x' axis. The mass of the block is m.

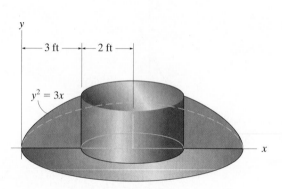

Probs. 21–7/8

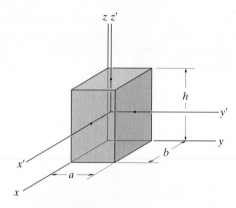

Prob. 21–10

21–9. Determine the elements of the inertia tensor for the cube with respect to the x, y, z coordinate system. The mass of the cube is m.

21–11. Determine the moment of inertia of the cylinder with respect to the a–a axis of the cylinder. The cylinder has a mass m.

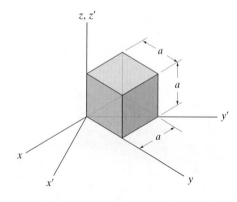

Prob. 21–9

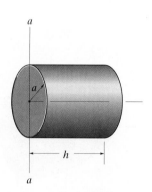

Prob. 21–11

*21–12. Determine the moment of inertia I_{xx} of the composite plate assembly. The plates have a specific weight of 6 lb/ft².

21–13. Determine the product of inertia I_{yz} of the composite plate assembly. The plates have a weight of 6 lb/ft².

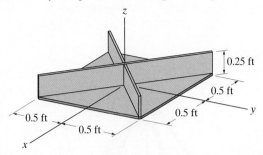

0.25 ft
0.5 ft
0.5 ft
0.5 ft
0.5 ft
0.5 ft

Probs. 21–12/13

21–14. Determine the products of inertia I_{xy}, I_{yz}, and I_{xz}, of the thin plate. The material has a density per unit area of 50 kg/m².

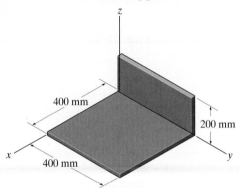

400 mm
200 mm
400 mm

Prob. 21–14

21–15. Determine the products of inertia I_{xy}, I_{yz} and I_{xz} of the solid. The material is steel, which has a specific weight of 490 lb/ft³.

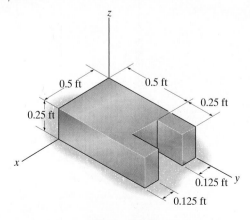

0.5 ft
0.5 ft
0.25 ft
0.25 ft
0.125 ft
0.125 ft

Prob. 21–15

*21–16. The bent rod has a mass of 4 kg/m. Determine the moment of inertia of the rod about the Oa axis.

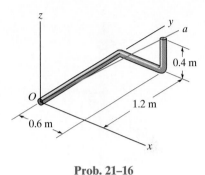

0.4 m
O
0.6 m
1.2 m

Prob. 21–16

21–17. The bent rod has a weight of 1.5 lb/ft. Locate the center of gravity $G(\bar{x}, \bar{y})$ and determine the principal moments of inertia $I_{x'}$, $I_{y'}$, and $I_{z'}$ of the rod with respect to the x', y', z' axes.

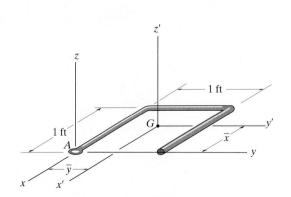

1 ft
1 ft
G
A
Prob. 21–17

21–18. Determine the moments of inertia about the x, y, z axes of the rod assembly. The rods have a mass of 0.75 kg/m.

*21–20. The assembly consists of a 15-lb plate A, 40-lb plate B, and four 7-lb rods. Determine the moments of inertia of the assembly with respect to the principal x, y, z axes.

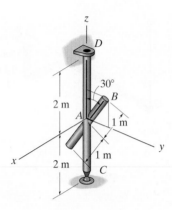

Prob. 21–18

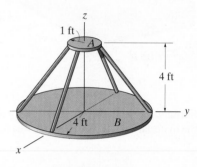

Prob. 21–20

21–21. Determine the moment of inertia of the rod-and-thin-ring assembly about the z axis. The rods and ring have a mass per unit length of 2 kg/m.

21–19. Determine the moment of inertia of the composite body about the aa axis. The cylinder weighs 20 lb, and each hemisphere weighs 10 lb.

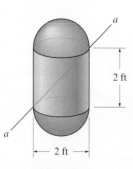

Prob. 21–19

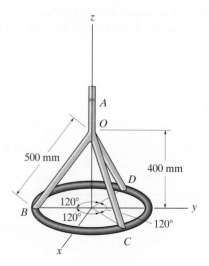

Prob. 21–21

21.2   Angular Momentum

In this section we will develop the necessary equations used to determine the angular momentum of a rigid body about an arbitrary point. These equations will provide a means for developing both the principle of impulse and momentum and the equations of rotational motion for a rigid body.

Consider the rigid body in Fig. 21–6, which has a mass m and center of mass at G. The X, Y, Z coordinate system represents an inertial frame of reference, and hence, its axes are fixed or translate with a constant velocity. The angular momentum as measured from this reference will be determined relative to the arbitrary point A. The position vectors $\mathbf{r}_A$ and $\boldsymbol{\rho}_A$ are drawn from the origin of coordinates to point A and from A to the ith particle of the body. If the particle's mass is m_i, the angular momentum about point A is

$$(\mathbf{H}_A)_i = \boldsymbol{\rho}_A \times m_i \mathbf{v}_i$$

where $\mathbf{v}_i$ represents the particle's velocity measured from the X, Y, Z coordinate system. If the body has an angular velocity $\boldsymbol{\omega}$ at the instant considered, $\mathbf{v}_i$ may be related to the velocity of A by applying Eq. 20–7, i.e.,

$$\mathbf{v}_i = \mathbf{v}_A + \boldsymbol{\omega} \times \boldsymbol{\rho}_A$$

Thus,

$$(\mathbf{H}_A)_i = \boldsymbol{\rho}_A \times m_i(\mathbf{v}_A + \boldsymbol{\omega} \times \boldsymbol{\rho}_A)$$

$$= (\boldsymbol{\rho}_A m_i) \times \mathbf{v}_A + \boldsymbol{\rho}_A \times (\boldsymbol{\omega} \times \boldsymbol{\rho}_A)m_i$$

Summing the moments of all the particles of the body requires an integration. Since $m_i \rightarrow dm$, we have

$$\mathbf{H}_A = \left(\int_m \boldsymbol{\rho}_A \, dm \right) \times \mathbf{v}_A + \int_m \boldsymbol{\rho}_A \times (\boldsymbol{\omega} \times \boldsymbol{\rho}_A)\,dm \qquad (21\text{–}6)$$

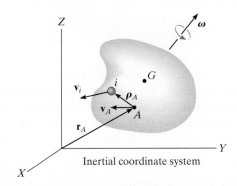

Inertial coordinate system

Fig. 21–6

21

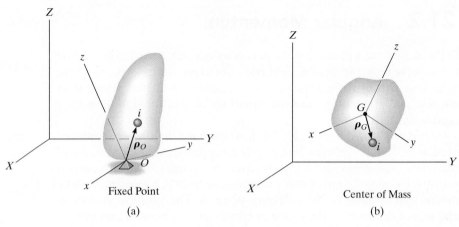

Fig. 21–7

Fixed Point O.

If A becomes a *fixed point O* in the body, Fig. 21–7a, then $\mathbf{v}_A = \mathbf{0}$ and Eq. 21–6 reduces to

$$\mathbf{H}_O = \int_m \boldsymbol{\rho}_O \times (\boldsymbol{\omega} \times \boldsymbol{\rho}_O)\,dm \qquad (21\text{–}7)$$

Center of Mass G.

If A is located at the *center of mass G* of the body, Fig. 21–7b, then $\int_m \boldsymbol{\rho}_A\,dm = \mathbf{0}$ and

$$\mathbf{H}_G = \int_m \boldsymbol{\rho}_G \times (\boldsymbol{\omega} \times \boldsymbol{\rho}_G)\,dm \qquad (21\text{–}8)$$

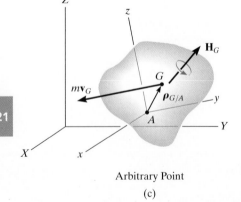

Arbitrary Point

(c)

Arbitrary Point A.

In general, A can be a point other than O or G, Fig. 21–7c, in which case Eq. 21–6 may nevertheless be simplified to the following form (see Prob. 21–23).

$$\mathbf{H}_A = \boldsymbol{\rho}_{G/A} \times m\mathbf{v}_G + \mathbf{H}_G \qquad (21\text{–}9)$$

Here the angular momentum consists of two parts—the moment of the linear momentum $m\mathbf{v}_G$ of the body about point A added (vectorially) to the angular momentum $\mathbf{H}_G$. Equation 21–9 can also be used to determine the angular momentum of the body about a fixed point O. The results, of course, will be the same as those found using the more convenient Eq. 21–7.

Rectangular Components of H.

To make practical use of Eqs. 21–7 through 21–9, the angular momentum must be expressed in terms of its scalar components. For this purpose, it is convenient to choose a second set of x, y, z axes having an arbitrary orientation relative to the X, Y, Z axes, Fig. 21–7, and for a general formulation, note that Eqs. 21–7 and 21–8 are both of the form

$$\mathbf{H} = \int_m \boldsymbol{\rho} \times (\boldsymbol{\omega} \times \boldsymbol{\rho}) dm$$

Expressing $\mathbf{H}, \boldsymbol{\rho}$, and $\boldsymbol{\omega}$ in terms of x, y, z components, we have

$$H_x \mathbf{i} + H_y \mathbf{j} + H_z \mathbf{k} = \int_m (x\mathbf{i} + y\mathbf{j} + z\mathbf{k}) \times [(\omega_x \mathbf{i} + \omega_y \mathbf{j} + \omega_z \mathbf{k})$$
$$\times (x\mathbf{i} + y\mathbf{j} + z\mathbf{k})] dm$$

Expanding the cross products and combining terms yields

$$H_x \mathbf{i} + H_y \mathbf{j} + H_z \mathbf{k} = \left[\omega_x \int_m (y^2 + z^2) dm - \omega_y \int_m xy\, dm - \omega_z \int_m xz\, dm \right] \mathbf{i}$$
$$+ \left[-\omega_x \int_m xy\, dm + \omega_y \int_m (x^2 + z^2) dm - \omega_z \int_m yz\, dm \right] \mathbf{j}$$
$$+ \left[-\omega_x \int_m zx\, dm - \omega_y \int_m yz\, dm + \omega_z \int_m (x^2 + y^2) dm \right] \mathbf{k}$$

Equating the respective $\mathbf{i}, \mathbf{j}, \mathbf{k}$ components and recognizing that the integrals represent the moments and products of inertia, we obtain

$$\begin{aligned} H_x &= I_{xx}\omega_x - I_{xy}\omega_y - I_{xz}\omega_z \\ H_y &= -I_{yx}\omega_x + I_{yy}\omega_y - I_{yz}\omega_z \\ H_z &= -I_{zx}\omega_x - I_{zy}\omega_y + I_{zz}\omega_z \end{aligned} \qquad (21\text{–}10)$$

These equations can be simplified further if the x, y, z coordinate axes are oriented such that they become *principal axes of inertia* for the body at the point. When these axes are used, the products of inertia $I_{xy} = I_{yz} = I_{zx} = 0$, and if the principal moments of inertia about the x, y, z axes are represented as $I_x = I_{xx}$, $I_y = I_{yy}$, and $I_z = I_{zz}$, the three components of angular momentum become

$$H_x = I_x \omega_x \quad H_y = I_y \omega_y \quad H_z = I_z \omega_z \qquad (21\text{–}11)$$

The motion of the astronaut is controlled by use of small directional jets attached to his or her space suit. The impulses these jets provide must be carefully specified in order to prevent tumbling and loss of orientation.

Principle of Impulse and Momentum. Now that the formulation of the angular momentum for a body has been developed, the *principle of impulse and momentum*, as discussed in Sec. 19.2, can be used to solve kinetic problems which involve *force, velocity, and time*. For this case, the following two vector equations are available:

$$m(\mathbf{v}_G)_1 + \Sigma \int_{t_1}^{t_2} \mathbf{F} \, dt = m(\mathbf{v}_G)_2 \tag{21–12}$$

$$(\mathbf{H}_O)_1 + \Sigma \int_{t_1}^{t_2} \mathbf{M}_O \, dt = (\mathbf{H}_O)_2 \tag{21–13}$$

In three dimensions each vector term can be represented by three scalar components, and therefore a total of *six scalar equations* can be written. Three equations relate the linear impulse and momentum in the x, y, z directions, and the other three equations relate the body's angular impulse and momentum about the x, y, z axes. Before applying Eqs. 21–12 and 21–13 to the solution of problems, the material in Secs. 19.2 and 19.3 should be reviewed.

21.3 Kinetic Energy

In order to apply the principle of work and energy to solve problems involving general rigid body motion, it is first necessary to formulate expressions for the kinetic energy of the body. To do this, consider the rigid body shown in Fig. 21–8, which has a mass m and center of mass at G. The kinetic energy of the ith particle of the body having a mass m_i and velocity $\mathbf{v}_i$, measured relative to the inertial X, Y, Z frame of reference, is

$$T_i = \tfrac{1}{2} m_i v_i^2 = \tfrac{1}{2} m_i (\mathbf{v}_i \cdot \mathbf{v}_i)$$

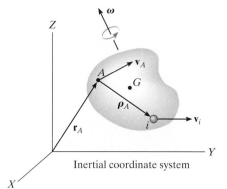

Inertial coordinate system

Fig. 21–8

Provided the velocity of an arbitrary point A in the body is known, $\mathbf{v}_i$ can be related to $\mathbf{v}_A$ by the equation $\mathbf{v}_i = \mathbf{v}_A + \boldsymbol{\omega} \times \boldsymbol{\rho}_A$, where $\boldsymbol{\omega}$ is the angular velocity of the body, measured from the X, Y, Z coordinate system, and $\boldsymbol{\rho}_A$ is a position vector extending from A to i. Using this expression, the kinetic energy for the particle can be written as

$$T_i = \tfrac{1}{2} m_i (\mathbf{v}_A + \boldsymbol{\omega} \times \boldsymbol{\rho}_A) \cdot (\mathbf{v}_A + \boldsymbol{\omega} \times \boldsymbol{\rho}_A)$$

$$= \tfrac{1}{2} (\mathbf{v}_A \cdot \mathbf{v}_A) m_i + \mathbf{v}_A \cdot (\boldsymbol{\omega} \times \boldsymbol{\rho}_A) m_i + \tfrac{1}{2} (\boldsymbol{\omega} \times \boldsymbol{\rho}_A) \cdot (\boldsymbol{\omega} \times \boldsymbol{\rho}_A) m_i$$

The kinetic energy for the entire body is obtained by summing the kinetic energies of all the particles of the body. This requires an integration. Since $m_i \to dm$, we get

$$T = \tfrac{1}{2} m (\mathbf{v}_A \cdot \mathbf{v}_A) + \mathbf{v}_A \cdot \left(\boldsymbol{\omega} \times \int_m \boldsymbol{\rho}_A \, dm \right) + \tfrac{1}{2} \int_m (\boldsymbol{\omega} \times \boldsymbol{\rho}_A) \cdot (\boldsymbol{\omega} \times \boldsymbol{\rho}_A) \, dm$$

21

The last term on the right can be rewritten using the vector identity $\mathbf{a} \times \mathbf{b} \cdot \mathbf{c} = \mathbf{a} \cdot \mathbf{b} \times \mathbf{c}$, where $\mathbf{a} = \boldsymbol{\omega}$, $\mathbf{b} = \boldsymbol{\rho}_A$, and $\mathbf{c} = \boldsymbol{\omega} \times \boldsymbol{\rho}_A$. The final result is

$$T = \tfrac{1}{2}m(\mathbf{v}_A \cdot \mathbf{v}_A) + \mathbf{v}_A \cdot \left(\boldsymbol{\omega} \times \int_m \boldsymbol{\rho}_A dm\right)$$

$$+ \tfrac{1}{2}\boldsymbol{\omega} \cdot \int_m \boldsymbol{\rho}_A \times (\boldsymbol{\omega} \times \boldsymbol{\rho}_A)dm \qquad (21\text{--}14)$$

This equation is rarely used because of the computations involving the integrals. Simplification occurs, however, if the reference point A is either a fixed point or the center of mass.

Fixed Point O.   If A is a *fixed point O* in the body, Fig. 21–7a, then $\mathbf{v}_A = \mathbf{0}$, and using Eq. 21–7, we can express Eq. 21–14 as

$$T = \tfrac{1}{2}\boldsymbol{\omega} \cdot \mathbf{H}_O$$

If the x, y, z axes represent the principal axes of inertia for the body, then $\boldsymbol{\omega} = \omega_x \mathbf{i} + \omega_y \mathbf{j} + \omega_z \mathbf{k}$ and $\mathbf{H}_O = I_x \omega_x \mathbf{i} + I_y \omega_y \mathbf{j} + I_z \omega_z \mathbf{k}$. Substituting into the above equation and performing the dot-product operations yields

$$\boxed{T = \tfrac{1}{2}I_x \omega_x^2 + \tfrac{1}{2}I_y \omega_y^2 + \tfrac{1}{2}I_z \omega_z^2} \qquad (21\text{--}15)$$

Center of Mass G.   If A is located at the *center of mass G* of the body, Fig. 21–7b, then $\int \boldsymbol{\rho}_A \, dm = \mathbf{0}$ and, using Eq. 21–8, we can write Eq. 21–14 as

$$T = \tfrac{1}{2}mv_G^2 + \tfrac{1}{2}\boldsymbol{\omega} \cdot \mathbf{H}_G$$

In a manner similar to that for a fixed point, the last term on the right side may be represented in scalar form, in which case

$$\boxed{T = \tfrac{1}{2}mv_G^2 + \tfrac{1}{2}I_x \omega_x^2 + \tfrac{1}{2}I_y \omega_y^2 + \tfrac{1}{2}I_z \omega_z^2} \qquad (21\text{--}16)$$

Here it is seen that the kinetic energy consists of two parts; namely, the translational kinetic energy of the mass center, $\tfrac{1}{2}mv_G^2$, and the body's rotational kinetic energy.

Principle of Work and Energy.   Having formulated the kinetic energy for a body, the *principle of work and energy* can be applied to solve kinetics problems which involve *force, velocity, and displacement*. For this case only one scalar equation can be written for each body, namely,

$$\boxed{T_1 + \Sigma U_{1\text{--}2} = T_2} \qquad (21\text{--}17)$$

Before applying this equation, the material in Chapter 18 should be reviewed.

21

EXAMPLE | 21.2

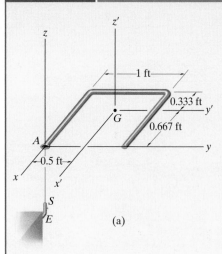

(a)

The rod in Fig. 21–9a has a weight per unit length of 1.5 lb/ft. Determine its angular velocity just after the end A falls onto the hook at E. The hook provides a permanent connection for the rod due to the spring-lock mechanism S. Just before striking the hook the rod is falling downward with a speed $(v_G)_1 = 10$ ft/s.

SOLUTION

The principle of impulse and momentum will be used since impact occurs.

Impulse and Momentum Diagrams.   Fig. 21–9b. During the short time Δt, the impulsive force $\mathbf{F}$ acting at A changes the momentum of the rod. (The impulse created by the rod's weight $\mathbf{W}$ during this time is small compared to $\int \mathbf{F}\, dt$, so that it can be neglected, i.e., the weight is a nonimpulsive force.) Hence, the angular momentum of the rod is *conserved* about point A since the moment of $\int \mathbf{F}\, dt$ about A is zero.

Conservation of Angular Momentum.   Equation 21–9 must be used to find the angular momentum of the rod, since A does not become a *fixed point* until *after* the impulsive interaction with the hook. Thus, with reference to Fig. 21–9b, $(\mathbf{H}_A)_1 = (\mathbf{H}_A)_2$, or

$$\mathbf{r}_{G/A} \times m(\mathbf{v}_G)_1 = \mathbf{r}_{G/A} \times m(\mathbf{v}_G)_2 + (\mathbf{H}_G)_2 \qquad (1)$$

From Fig. 21–9a, $\mathbf{r}_{G/A} = \{-0.667\mathbf{i} + 0.5\mathbf{j}\}$ ft. Furthermore, the primed axes are principal axes of inertia for the rod because $I_{x'y'} = I_{x'z'} = I_{z'y'} = 0$. Hence, from Eqs. 21–11, $(\mathbf{H}_G)_2 = I_{x'}\omega_x\mathbf{i} + I_{y'}\omega_y\mathbf{j} + I_{z'}\omega_z\mathbf{k}$. The principal moments of inertia are $I_{x'} = 0.0272$ slug $\cdot$ ft^2, $I_{y'} = 0.0155$ slug $\cdot$ ft^2, $I_{z'} = 0.0427$ slug $\cdot$ ft^2 (see Prob. 21–17). Substituting into Eq. 1, we have

$$(-0.667\mathbf{i} + 0.5\mathbf{j}) \times \left[\left(\frac{4.5}{32.2}\right)(-10\mathbf{k})\right] = (-0.667\mathbf{i} + 0.5\mathbf{j}) \times \left[\left(\frac{4.5}{32.2}\right)(-v_G)_2\mathbf{k}\right]$$
$$+ 0.0272\omega_x\mathbf{i} + 0.0155\omega_y\mathbf{j} + 0.0427\omega_z\mathbf{k}$$

Expanding and equating the respective $\mathbf{i}, \mathbf{j}, \mathbf{k}$ components yields

$$-0.699 = -0.0699(v_G)_2 + 0.0272\omega_x \qquad (2)$$
$$-0.932 = -0.0932(v_G)_2 + 0.0155\omega_y \qquad (3)$$
$$0 = 0.0427\omega_z \qquad (4)$$

Kinematics.   There are four unknowns in the above equations; however, another equation may be obtained by relating $\boldsymbol{\omega}$ to $(\mathbf{v}_G)_2$ using *kinematics*. Since $\omega_z = 0$ (Eq. 4) and after impact the rod rotates about the fixed point A, Eq. 20–3 can be applied, in which case $(\mathbf{v}_G)_2 = \boldsymbol{\omega} \times \mathbf{r}_{G/A}$, or

$$-(v_G)_2\mathbf{k} = (\omega_x\mathbf{i} + \omega_y\mathbf{j}) \times (-0.667\mathbf{i} + 0.5\mathbf{j})$$
$$-(v_G)_2 = 0.5\omega_x + 0.667\omega_y \qquad (5)$$

Solving Eqs. 2, 3 and 5 simultaneously yields

$$(\mathbf{v}_G)_2 = \{-8.41\mathbf{k}\} \text{ ft/s} \quad \boldsymbol{\omega} = \{-4.09\mathbf{i} - 9.55\mathbf{j}\} \text{ rad/s} \quad \textit{Ans.}$$

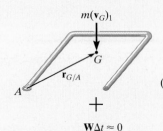

(b)

Fig. 21–9

21

EXAMPLE | 21.3

A 5-N · m torque is applied to the vertical shaft CD shown in Fig. 21–10a, which allows the 10-kg gear A to turn freely about CE. Assuming that gear A starts from rest, determine the angular velocity of CD after it has turned two revolutions. Neglect the mass of shaft CD and axle CE and assume that gear A can be approximated by a thin disk. Gear B is fixed.

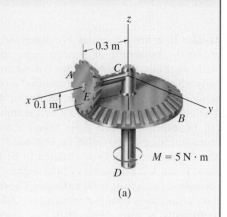

(a)

SOLUTION

The principle of work and energy may be used for the solution. Why?

Work. If shaft CD, the axle CE, and gear A are considered as a system of connected bodies, only the applied torque $\mathbf{M}$ does work. For two revolutions of CD, this work is $\Sigma U_{1-2} = (5\text{ N} \cdot \text{m})(4\pi \text{ rad}) = 62.83 \text{ J}$.

Kinetic Energy. Since the gear is initially at rest, its initial kinetic energy is zero. A kinematic diagram for the gear is shown in Fig. 21–10b. If the angular velocity of CD is taken as $\boldsymbol{\omega}_{CD}$, then the angular velocity of gear A is $\boldsymbol{\omega}_A = \boldsymbol{\omega}_{CD} + \boldsymbol{\omega}_{CE}$. The gear may be imagined as a portion of a massless extended body which is rotating about the *fixed point C*. The instantaneous axis of rotation for this body is along line CH, because both points C and H on the body (gear) have zero velocity and must therefore lie on this axis. This requires that the components $\boldsymbol{\omega}_{CD}$ and $\boldsymbol{\omega}_{CE}$ be related by the equation $\omega_{CD}/0.1 \text{ m} = \omega_{CE}/0.3 \text{ m}$ or $\omega_{CE} = 3\omega_{CD}$. Thus,

$$\boldsymbol{\omega}_A = -\omega_{CE}\mathbf{i} + \omega_{CD}\mathbf{k} = -3\omega_{CD}\mathbf{i} + \omega_{CD}\mathbf{k} \qquad (1)$$

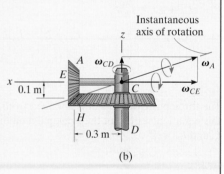

(b)

Fig. 21–10

The x, y, z axes in Fig. 21–10a represent *principal axes of inertia* at C for the gear. Since point C is a fixed point of rotation, Eq. 21–15 may be applied to determine the kinetic energy, i.e.,

$$T = \tfrac{1}{2}I_x\omega_x^2 + \tfrac{1}{2}I_y\omega_y^2 + \tfrac{1}{2}I_z\omega_z^2 \qquad (2)$$

Using the parallel-axis theorem, the moments of inertia of the gear about point C are as follows:

$$I_x = \tfrac{1}{2}(10 \text{ kg})(0.1 \text{ m})^2 = 0.05 \text{ kg} \cdot \text{m}^2$$

$$I_y = I_z = \tfrac{1}{4}(10 \text{ kg})(0.1 \text{ m})^2 + 10 \text{ kg}(0.3 \text{ m})^2 = 0.925 \text{ kg} \cdot \text{m}^2$$

Since $\omega_x = -3\omega_{CD}$, $\omega_y = 0$, $\omega_z = \omega_{CD}$, Eq. 2 becomes

$$T_A = \tfrac{1}{2}(0.05)(-3\omega_{CD})^2 + 0 + \tfrac{1}{2}(0.925)(\omega_{CD})^2 = 0.6875\omega_{CD}^2$$

Principle of Work and Energy. Applying the principle of work and energy, we obtain

$$T_1 + \Sigma U_{1-2} = T_2$$

$$0 + 62.83 = 0.6875\omega_{CD}^2$$

$$\omega_{CD} = 9.56 \text{ rad/s} \qquad \qquad Ans.$$

21

PROBLEMS

21–22. If a body contains *no planes of symmetry*, the principal moments of inertia can be determined mathematically. To show how this is done, consider the rigid body which is spinning with an angular velocity $\boldsymbol{\omega}$, directed along one of its principal axes of inertia. If the principal moment of inertia about this axis is I, the angular momentum can be expressed as $\mathbf{H} = I\boldsymbol{\omega} = I\omega_x\mathbf{i} + I\omega_y\mathbf{j} + I\omega_z\mathbf{k}$. The components of $\mathbf{H}$ may also be expressed by Eqs. 21–10, where the inertia tensor is assumed to be known. Equate the $\mathbf{i}$, $\mathbf{j}$, and $\mathbf{k}$ components of both expressions for $\mathbf{H}$ and consider ω_x, ω_y, and ω_z to be unknown. The solution of these three equations is obtained provided the determinant of the coefficients is zero. Show that this determinant, when expanded, yields the cubic equation

$$I^3 - (I_{xx} + I_{yy} + I_{zz})I^2$$
$$+ (I_{xx}I_{yy} + I_{yy}I_{zz} + I_{zz}I_{xx} - I_{xy}^2 - I_{yz}^2 - I_{zx}^2)I$$
$$- (I_{xx}I_{yy}I_{zz} - 2I_{xy}I_{yz}I_{zx} - I_{xx}I_{yz}^2$$
$$- I_{yy}I_{zx}^2 - I_{zz}I_{xy}^2) = 0$$

The three positive roots of I, obtained from the solution of this equation, represent the principal moments of inertia I_x, I_y, and I_z.

21–23. Show that if the angular momentum of a body is determined with respect to an arbitrary point A, then $\mathbf{H}_A$ can be expressed by Eq. 21–9. This requires substituting $\boldsymbol{\rho}_A = \boldsymbol{\rho}_G + \boldsymbol{\rho}_{G/A}$ into Eq. 21–6 and expanding, noting that $\int \boldsymbol{\rho}_G \, dm = \mathbf{0}$ by definition of the mass center and $\mathbf{v}_G = \mathbf{v}_A + \boldsymbol{\omega} \times \boldsymbol{\rho}_{G/A}$.

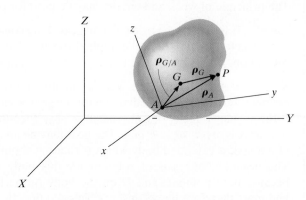

Prob. 21–23

*21–24. The 15-kg circular disk spins about its axle with a constant angular velocity of $\omega_1 = 10$ rad/s. Simultaneously, the yoke is rotating with a constant angular velocity of $\omega_2 = 5$ rad/s. Determine the angular momentum of the disk about its center of mass O, and its kinetic energy.

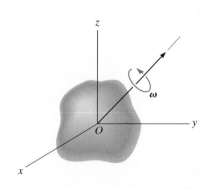

Prob. 21–22

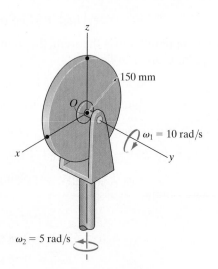

Prob. 21–24

21–25. The cone has a mass m and rolls without slipping on the conical surface so that it has an angular velocity about the vertical axis of $\boldsymbol{\omega}$. Determine the kinetic energy of the cone due to this motion.

*21–28. The space capsule has a mass of 5 Mg and the radii of gyration are $k_x = k_z = 1.30$ m and $k_y = 0.45$ m. If it travels with a velocity $\mathbf{v}_G = \{400\mathbf{j} + 200\mathbf{k}\}$ m/s, compute its angular velocity just after it is struck by a meteoroid having a mass of 0.80 kg and a velocity $\mathbf{v}_m = \{-300\mathbf{i} + 200\mathbf{j} - 150\mathbf{k}\}$ m/s. Assume that the meteoroid embeds itself into the capsule at point A and that the capsule initially has no angular velocity.

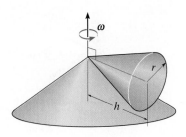

Prob. 21–25

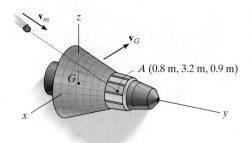

Prob. 21–28

21–26. The circular disk has a weight of 15 lb and is mounted on the shaft AB at an angle of 45° with the horizontal. Determine the angular velocity of the shaft when $t = 3$ s if a constant torque $M = 2$ lb·ft is applied to the shaft. The shaft is originally spinning at $\omega_1 = 8$ rad/s when the torque is applied.

21–27. The circular disk has a weight of 15 lb and is mounted on the shaft AB at an angle of 45° with the horizontal. Determine the angular velocity of the shaft when $t = 2$ s if a torque $M = (4e^{0.1t})$ lb·ft, where t is in seconds, is applied to the shaft. The shaft is originally spinning at $\omega_1 = 8$ rad/s when the torque is applied.

21–29. The 2-kg gear A rolls on the fixed plate gear C. Determine the angular velocity of rod OB about the z axis after it rotates one revolution about the z axis, starting from rest. The rod is acted upon by the constant moment $M = 5$ N·m. Neglect the mass of rod OB. Assume that gear A is a uniform disk having a radius of 100 mm.

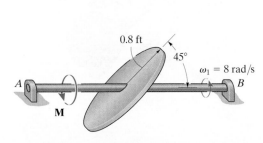

Probs. 21–26/27

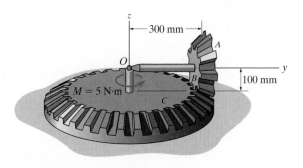

Prob. 21–29

21–30. The rod weighs 3 lb/ft and is suspended from parallel cords at A and B. If the rod has an angular velocity of 2 rad/s about the z axis at the instant shown, determine how high the center of the rod rises at the instant the rod momentarily stops swinging.

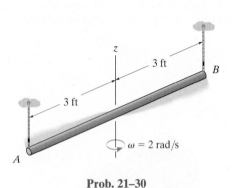

Prob. 21–30

21–31. Rod AB has a weight of 6 lb and is attached to two smooth collars at its ends by ball-and-socket joints. If collar A is moving downward with a speed of 8 ft/s when $z = 3$ ft, determine the speed of A at the instant $z = 0$. The spring has an unstretched length of 2 ft. Neglect the mass of the collars. Assume the angular velocity of rod AB is perpendicular to its axis.

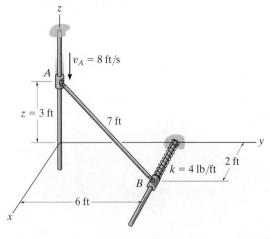

Prob. 21–31

*21–32. The 5-kg circular disk spins about AB with a constant angular velocity of $\omega_1 = 15$ rad/s. Simultaneously, the shaft to which arm OAB is rigidly attached, rotates with a constant angular velocity of $\omega_2 = 6$ rad/s. Determine the angular momentum of the disk about point O, and its kinetic energy.

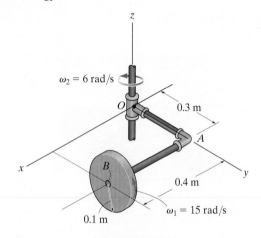

Prob. 21–32

21–33. The 20-kg sphere rotates about the axle with a constant angular velocity of $\omega_s = 60$ rad/s. If shaft AB is subjected to a torque of $M = 50$ N·m, causing it to rotate, determine the value of ω_p after the shaft has turned 90° from the position shown. Initially, $\omega_p = 0$. Neglect the mass of arm CDE.

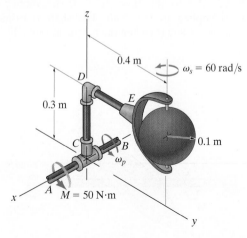

Prob. 21–33

21–34. The 200-kg satellite has its center of mass at point G. Its radii of gyration about the z', x', y' axes are $k_{z'} = 300$ mm, $k_{x'} = k_{y'} = 500$ mm, respectively. At the instant shown, the satellite rotates about the x', y', and z' axes with the angular velocity shown, and its center of mass G has a velocity of $\mathbf{v}_G = \{-250\mathbf{i} + 200\mathbf{j} + 120\mathbf{k}\}$ m/s. Determine the angular momentum of the satellite about point A at this instant.

21–35. The 200-kg satellite has its center of mass at point G. Its radii of gyration about the z', x', y' axes are $k_{z'} = 300$ mm, $k_{x'} = k_{y'} = 500$ mm, respectively. At the instant shown, the satellite rotates about the x', y', and z' axes with the angular velocity shown, and its center of mass G has a velocity of $\mathbf{v}_G = \{-250\mathbf{i} + 200\mathbf{j} + 120\mathbf{k}\}$ m/s. Determine the kinetic energy of the satellite at this instant.

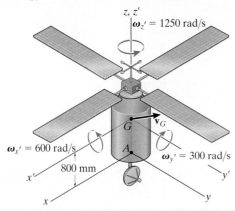

Probs. 21–34/35

***21–36.** The 15-kg rectangular plate is free to rotate about the y axis because of the bearing supports at A and B. When the plate is balanced in the vertical plane, a 3-g bullet is fired into it, perpendicular to its surface, with a velocity $\mathbf{v} = \{-2000\mathbf{i}\}$ m/s. Compute the angular velocity of the plate at the instant it has rotated 180°. If the bullet strikes corner D with the same velocity $\mathbf{v}$, instead of at C, does the angular velocity remain the same? Why or why not?

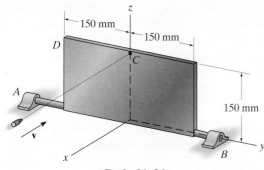

Prob. 21–36

21–37. The circular plate has a weight of 19 lb and a diameter of 1.5 ft. If it is released from rest and falls horizontally 2.5 ft onto the hook at S, which provides a permanent connection, determine the velocity of the mass center of the plate just after the connection with the hook is made.

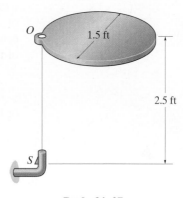

Prob. 21–37

21–38. The 10-kg disk rolls on the horizontal plane without slipping. Determine the magnitude of its angular momentum when it is spinning about the y axis at 2 rad/s.

21–39. If arm OA is subjected to a torque of $M = 5$ N·m, determine the spin angular velocity of the 10-kg disk after the arm has turned 2 rev, starting from rest. The disk rolls on the horizontal plane without slipping. Neglect the mass of the arm.

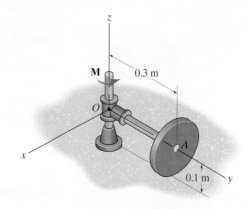

Probs. 21–38/39

21

*21.4   Equations of Motion

Having become familiar with the techniques used to describe both the inertial properties and the angular momentum of a body, we can now write the equations which describe the motion of the body in their most useful forms.

Equations of Translational Motion.

The *translational motion* of a body is defined in terms of the acceleration of the body's mass center, which is measured from an inertial X, Y, Z reference. The equation of translational motion for the body can be written in vector form as

$$\Sigma \mathbf{F} = m\mathbf{a}_G \qquad (21\text{-}18)$$

or by the three scalar equations

$$\boxed{\begin{aligned} \Sigma F_x &= m(a_G)_x \\ \Sigma F_y &= m(a_G)_y \\ \Sigma F_z &= m(a_G)_z \end{aligned}} \qquad (21\text{-}19)$$

Here, $\Sigma \mathbf{F} = \Sigma F_x \mathbf{i} + \Sigma F_y \mathbf{j} + \Sigma F_z \mathbf{k}$ represents the sum of all the external forces acting on the body.

Equations of Rotational Motion.

In Sec. 15.6, we developed Eq. 15–17, namely,

$$\Sigma \mathbf{M}_O = \dot{\mathbf{H}}_O \qquad (21\text{-}20)$$

which states that the sum of the moments of all the external forces acting on a system of particles (contained in a rigid body) about a fixed point O is equal to the time rate of change of the total angular momentum of the body about point O. When moments of the external forces acting on the particles are summed about the system's *mass center* G, one again obtains the same simple form of Eq. 21–20, relating the moment summation $\Sigma \mathbf{M}_G$ to the angular momentum $\mathbf{H}_G$. To show this, consider the system of particles in Fig. 21–11, where X, Y, Z represents an inertial frame of reference and the x, y, z axes, with origin at G, *translate* with respect to this frame. In general, G is *accelerating*, so by definition the translating frame is *not* an inertial reference. The angular momentum of the ith particle with respect to this frame is, however,

$$(\mathbf{H}_i)_G = \mathbf{r}_{i/G} \times m_i \mathbf{v}_{i/G}$$

where $\mathbf{r}_{i/G}$ and $\mathbf{v}_{i/G}$ represent the position and velocity of the ith particle with respect to G. Taking the time derivative we have

$$(\dot{\mathbf{H}}_i)_G = \dot{\mathbf{r}}_{i/G} \times m_i \mathbf{v}_{i/G} + \mathbf{r}_{i/G} \times m_i \dot{\mathbf{v}}_{i/G}$$

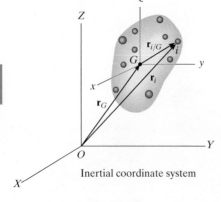

Inertial coordinate system

Fig. 21–11

By definition, $\mathbf{v}_{i/G} = \dot{\mathbf{r}}_{i/G}$. Thus, the first term on the right side is zero since the cross product of the same vectors is zero. Also, $\mathbf{a}_{i/G} = \dot{\mathbf{v}}_{i/G}$, so that

$$(\dot{\mathbf{H}}_i)_G = (\mathbf{r}_{i/G} \times m_i \mathbf{a}_{i/G})$$

Similar expressions can be written for the other particles of the body. When the results are summed, we get

$$\dot{\mathbf{H}}_G = \Sigma(\mathbf{r}_{i/G} \times m_i \mathbf{a}_{i/G})$$

Here $\dot{\mathbf{H}}_G$ is the time rate of change of the total angular momentum of the body computed about point G.

The relative acceleration for the ith particle is defined by the equation $\mathbf{a}_{i/G} = \mathbf{a}_i - \mathbf{a}_G$, where $\mathbf{a}_i$ and $\mathbf{a}_G$ represent, respectively, the accelerations of the ith particle and point G measured with respect to the *inertial frame of reference*. Substituting and expanding, using the distributive property of the vector cross product, yields

$$\dot{\mathbf{H}}_G = \Sigma(\mathbf{r}_{i/G} \times m_i \mathbf{a}_i) - (\Sigma m_i \mathbf{r}_{i/G}) \times \mathbf{a}_G$$

By definition of the mass center, the sum $(\Sigma m_i \mathbf{r}_{i/G}) = (\Sigma m_i)\bar{\mathbf{r}}$ is equal to zero, since the position vector $\bar{\mathbf{r}}$ relative to G is zero. Hence, the last term in the above equation is zero. Using the equation of motion, the product $m_i \mathbf{a}_i$ can be replaced by the resultant *external force* $\mathbf{F}_i$ acting on the ith particle. Denoting $\Sigma \mathbf{M}_G = \Sigma(\mathbf{r}_{i/G} \times \mathbf{F}_i)$, the final result can be written as

$$\Sigma \mathbf{M}_G = \dot{\mathbf{H}}_G \qquad (21\text{–}21)$$

The rotational equation of motion for the body will now be developed from either Eq. 21–20 or 21–21. In this regard, the scalar components of the angular momentum $\mathbf{H}_O$ or $\mathbf{H}_G$ are defined by Eqs. 21–10 or, if principal axes of inertia are used either at point O or G, by Eqs. 21–11. If these components are computed about x, y, z axes that are *rotating* with an angular velocity $\boldsymbol{\Omega}$ that is *different* from the body's angular velocity $\boldsymbol{\omega}$, then the time derivative $\dot{\mathbf{H}} = d\mathbf{H}/dt$, as used in Eqs. 21–20 and 21–21, must account for the rotation of the x, y, z axes as measured from the inertial X, Y, Z axes. This requires application of Eq. 20–6, in which case Eqs. 21–20 and 21–21 become

$$\Sigma \mathbf{M}_O = (\dot{\mathbf{H}}_O)_{xyz} + \boldsymbol{\Omega} \times \mathbf{H}_O$$
$$\Sigma \mathbf{M}_G = (\dot{\mathbf{H}}_G)_{xyz} + \boldsymbol{\Omega} \times \mathbf{H}_G \qquad (21\text{–}22)$$

Here $(\dot{\mathbf{H}})_{xyz}$ is the time rate of change of $\mathbf{H}$ measured from the x, y, z reference.

There are three ways in which one can define the motion of the x, y, z axes. Obviously, motion of this reference should be chosen so that it will yield the simplest set of moment equations for the solution of a particular problem.

21

x, y, z Axes Having Motion $\mathbf{\Omega} = 0$.

If the body has general motion, the x, y, z axes can be chosen with origin at G, such that the axes only *translate* relative to the inertial X, Y, Z frame of reference. Doing this simplifies Eq. 21–22, since $\mathbf{\Omega} = \mathbf{0}$. However, the body may have a rotation $\boldsymbol{\omega}$ about these axes, and therefore the moments and products of inertia of the body would have to be expressed as *functions of time*. In most cases this would be a difficult task, so that such a choice of axes has restricted application.

x, y, z Axes Having Motion $\mathbf{\Omega} = \boldsymbol{\omega}$.

The x, y, z axes can be chosen such that they are *fixed in and move with the body*. The moments and products of inertia of the body relative to these axes will then be *constant* during the motion. Since $\mathbf{\Omega} = \boldsymbol{\omega}$, Eqs. 21–22 become

$$\Sigma \mathbf{M}_O = (\dot{\mathbf{H}}_O)_{xyz} + \boldsymbol{\omega} \times \mathbf{H}_O$$
$$\Sigma \mathbf{M}_G = (\dot{\mathbf{H}}_G)_{xyz} + \boldsymbol{\omega} \times \mathbf{H}_G$$

$$(21\text{–}23)$$

We can express each of these vector equations as three scalar equations using Eqs. 21–10. Neglecting the subscripts O and G yields

$$\Sigma M_x = I_{xx}\dot{\omega}_x - (I_{yy} - I_{zz})\omega_y\omega_z - I_{xy}(\dot{\omega}_y - \omega_z\omega_x)$$
$$- I_{yz}(\omega_y^2 - \omega_z^2) - I_{zx}(\dot{\omega}_z + \omega_x\omega_y)$$
$$\Sigma M_y = I_{yy}\dot{\omega}_y - (I_{zz} - I_{xx})\omega_z\omega_x - I_{yz}(\dot{\omega}_z - \omega_x\omega_y) \qquad (21\text{–}24)$$
$$- I_{zx}(\omega_z^2 - \omega_x^2) - I_{xy}(\dot{\omega}_x + \omega_y\omega_z)$$
$$\Sigma M_z = I_{zz}\dot{\omega}_z - (I_{xx} - I_{yy})\omega_x\omega_y - I_{zx}(\dot{\omega}_x - \omega_y\omega_z)$$
$$- I_{xy}(\omega_x^2 - \omega_y^2) - I_{yz}(\dot{\omega}_y + \omega_z\omega_x)$$

If the x, y, z axes are chosen as *principal axes of inertia*, the products of inertia are zero, $I_{xx} = I_x$, etc., and the above equations become

$$\boxed{\begin{aligned}
\Sigma M_x &= I_x\dot{\omega}_x - (I_y - I_z)\omega_y\omega_z \\
\Sigma M_y &= I_y\dot{\omega}_y - (I_z - I_x)\omega_z\omega_x \\
\Sigma M_z &= I_z\dot{\omega}_z - (I_x - I_y)\omega_x\omega_y
\end{aligned}} \qquad (21\text{–}25)$$

This set of equations is known historically as the *Euler equations of motion*, named after the Swiss mathematician Leonhard Euler, who first developed them. They apply *only* for moments summed about either point O or G.

When applying these equations it should be realized that $\dot{\omega}_x$, $\dot{\omega}_y$, $\dot{\omega}_z$ represent the time derivatives of the magnitudes of the x, y, z components of $\boldsymbol{\omega}$ as observed from x, y, z. To determine these components, it is first necessary to find ω_x, ω_y, ω_z when the x, y, z axes are oriented in a *general position* and *then* take the time derivative of the magnitude of these components, i.e., $(\dot{\boldsymbol{\omega}})_{xyz}$. However, since the x, y, z axes are rotating at $\boldsymbol{\Omega} = \boldsymbol{\omega}$, then from Eq. 20–6, it should be noted that $\dot{\boldsymbol{\omega}} = (\dot{\boldsymbol{\omega}})_{xyz} + \boldsymbol{\omega} \times \boldsymbol{\omega}$. Since $\boldsymbol{\omega} \times \boldsymbol{\omega} = \mathbf{0}$, then $\dot{\boldsymbol{\omega}} = (\dot{\boldsymbol{\omega}})_{xyz}$. This important result indicates that the time derivative of $\boldsymbol{\omega}$ with respect to the fixed X, Y, Z axes, that is $\dot{\boldsymbol{\omega}}$, can also be used to obtain $(\dot{\boldsymbol{\omega}})_{xyz}$. Generally this is the easiest way to determine the result. See Example 21.5.

x, y, z Axes Having Motion $\Omega \neq \omega$.

To simplify the calculations for the time derivative of $\boldsymbol{\omega}$, it is often convenient to choose the x, y, z axes having an angular velocity $\boldsymbol{\Omega}$ which is different from the angular velocity $\boldsymbol{\omega}$ of the body. This is particularly suitable for the analysis of spinning tops and gyroscopes which are *symmetrical* about their spinning axes.* When this is the case, the moments and products of inertia remain constant about the axis of spin.

Equations 21–22 are applicable for such a set of axes. Each of these two vector equations can be reduced to a set of three scalar equations which are derived in a manner similar to Eqs. 21–25,† i.e.,

$$\Sigma M_x = I_x \dot{\omega}_x - I_y \Omega_z \omega_y + I_z \Omega_y \omega_z$$
$$\Sigma M_y = I_y \dot{\omega}_y - I_z \Omega_x \omega_z + I_x \Omega_z \omega_x \qquad (21\text{--}26)$$
$$\Sigma M_z = I_z \dot{\omega}_z - I_x \Omega_y \omega_x + I_y \Omega_x \omega_y$$

Here Ω_x, Ω_y, Ω_z represent the x, y, z components of $\boldsymbol{\Omega}$, measured from the inertial frame of reference, and $\dot{\omega}_x$, $\dot{\omega}_y$, $\dot{\omega}_z$ must be determined relative to the x, y, z axes that have the rotation $\boldsymbol{\Omega}$. See Example 21.6.

Any one of these sets of moment equations, Eqs. 21–24, 21–25, or 21–26, represents a series of three first-order nonlinear differential equations. These equations are "coupled," since the angular-velocity components are present in all the terms. Success in determining the solution for a particular problem therefore depends upon what is unknown in these equations. Difficulty certainly arises when one attempts to solve for the unknown components of $\boldsymbol{\omega}$ when the external moments are functions of time. Further complications can arise if the moment equations are coupled to the three scalar equations of translational motion, Eqs. 21–19. This can happen because of the existence of kinematic constraints which relate the rotation of the body to the translation of its mass center, as in the case of a hoop which rolls

21

---

*A detailed discussion of such devices is given in Sec. 21.5.
†See Prob. 21–42.

without slipping. Problems that require the simultaneous solution of differential equations are generally solved using numerical methods with the aid of a computer. In many engineering problems, however, we are given information about the motion of the body and are required to determine the applied moments acting on the body. Most of these problems have direct solutions, so that there is no need to resort to computer techniques.

Procedure for Analysis

Problems involving the three-dimensional motion of a rigid body can be solved using the following procedure.

Free-Body Diagram.

- Draw a *free-body diagram* of the body at the instant considered and specify the x, y, z coordinate system. The origin of this reference must be located either at the body's mass center G, or at point O, considered fixed in an inertial reference frame and located either in the body or on a massless extension of the body.

- Unknown reactive force components can be shown having a positive sense of direction.

- Depending on the nature of the problem, decide what type of rotational motion $\mathbf{\Omega}$ the x, y, z coordinate system should have, i.e., $\mathbf{\Omega} = \mathbf{0}$, $\mathbf{\Omega} = \boldsymbol{\omega}$, or $\mathbf{\Omega} \neq \boldsymbol{\omega}$. When choosing, keep in mind that the moment equations are simplified when the axes move in such a manner that they represent principal axes of inertia for the body at all times.

- Compute the necessary moments and products of inertia for the body relative to the x, y, z axes.

Kinematics.

- Determine the x, y, z components of the body's angular velocity and find the time derivatives of $\boldsymbol{\omega}$.

- Note that if $\mathbf{\Omega} = \boldsymbol{\omega}$, then $\dot{\boldsymbol{\omega}} = (\dot{\boldsymbol{\omega}})_{xyz}$. Therefore we can either find the time derivative of $\boldsymbol{\omega}$ with respect to the X, Y, Z axes, $\dot{\boldsymbol{\omega}}$, and then determine its components $\dot{\omega}_x$, $\dot{\omega}_y$, $\dot{\omega}_z$, or we can find the components of $\boldsymbol{\omega}$ along the x, y, z axes, when the axes are oriented in a general position, and then take the time derivative of the magnitudes of these components, $(\dot{\boldsymbol{\omega}})_{xyz}$.

Equations of Motion.

- Apply either the two vector equations 21–18 and 21–22 or the six scalar component equations appropriate for the x, y, z coordinate axes chosen for the problem.

21

EXAMPLE | 21.4

The gear shown in Fig. 21–12a has a mass of 10 kg and is mounted at an angle of 10° with the rotating shaft having negligible mass. If $I_z = 0.1 \text{ kg} \cdot \text{m}^2$, $I_x = I_y = 0.05 \text{ kg} \cdot \text{m}^2$, and the shaft is rotating with a constant angular velocity of $\omega = 30 \text{ rad/s}$, determine the components of reaction that the thrust bearing A and journal bearing B exert on the shaft at the instant shown.

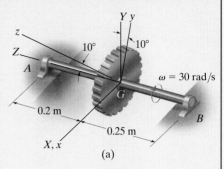

(a)

SOLUTION

Free-Body Diagram. Fig. 21–12b. The origin of the x, y, z coordinate system is located at the gear's center of mass G, which is also a fixed point. The axes are fixed in and rotate with the gear so that these axes will then always represent the principal axes of inertia for the gear. Hence $\boldsymbol{\Omega} = \boldsymbol{\omega}$.

Kinematics. As shown in Fig. 21–12c, the angular velocity $\boldsymbol{\omega}$ of the gear is constant in magnitude and is always directed along the axis of the shaft AB. Since this vector is measured from the X, Y, Z inertial frame of reference, for any position of the x, y, z axes,

$$\omega_x = 0 \quad \omega_y = -30 \sin 10° \quad \omega_z = 30 \cos 10°$$

These components remain constant for any general orientation of the x, y, z axes, and so $\dot{\omega}_x = \dot{\omega}_y = \dot{\omega}_z = 0$. Also note that since $\boldsymbol{\Omega} = \boldsymbol{\omega}$, then $\dot{\boldsymbol{\omega}} = (\dot{\boldsymbol{\omega}})_{xyz}$. Therefore, we can find these time derivatives relative to the X, Y, Z axes. In this regard $\boldsymbol{\omega}$ has a constant magnitude and direction (+Z) since $\dot{\boldsymbol{\omega}} = \mathbf{0}$, and so $\dot{\omega}_x = \dot{\omega}_y = \dot{\omega}_z = 0$. Furthermore, since G is a fixed point, $(a_G)_x = (a_G)_y = (a_G)_z = 0$.

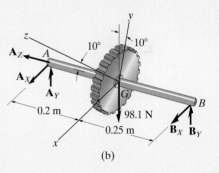

(b)

Equations of Motion. Applying Eqs. 21–25 ($\boldsymbol{\Omega} = \boldsymbol{\omega}$) yields

$$\Sigma M_x = I_x \dot{\omega}_x - (I_y - I_z)\omega_y\omega_z$$
$$-(A_Y)(0.2) + (B_Y)(0.25) = 0 - (0.05 - 0.1)(-30 \sin 10°)(30 \cos 10°)$$
$$-0.2A_Y + 0.25B_Y = -7.70 \tag{1}$$
$$\Sigma M_y = I_y \dot{\omega}_y - (I_z - I_x)\omega_z\omega_x$$
$$A_X(0.2) \cos 10° - B_X(0.25) \cos 10° = 0 - 0$$
$$A_X = 1.25B_X \tag{2}$$
$$\Sigma M_z = I_z \dot{\omega}_z - (I_x - I_y)\omega_x\omega_y$$
$$A_X(0.2) \sin 10° - B_X(0.25) \sin 10° = 0 - 0$$
$$A_X = 1.25B_X \text{ (check)}$$

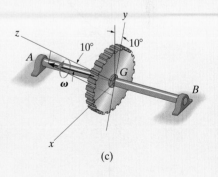

(c)

Fig. 21–12

Applying Eqs. 21–19, we have
$$\Sigma F_X = m(a_G)_X; \qquad\qquad A_X + B_X = 0 \tag{3}$$
$$\Sigma F_Y = m(a_G)_Y; \qquad\quad A_Y + B_Y - 98.1 = 0 \tag{4}$$
$$\Sigma F_Z = m(a_G)_Z; \qquad\qquad\qquad A_Z = 0 \qquad\qquad Ans.$$
Solving Eqs. 1 through 4 simultaneously gives
$$A_X = B_X = 0 \quad A_Y = 71.6 \text{ N} \quad B_Y = 26.5 \text{ N} \qquad Ans.$$

21

EXAMPLE 21.5

The airplane shown in Fig. 21–13a is in the process of making a steady *horizontal* turn at the rate of ω_p. During this motion, the propeller is spinning at the rate of ω_s. If the propeller has two blades, determine the moments which the propeller shaft exerts on the propeller at the instant the blades are in the vertical position. For simplicity, assume the blades to be a uniform slender bar having a moment of inertia I about an axis perpendicular to the blades passing through the center of the bar, and having zero moment of inertia about a longitudinal axis.

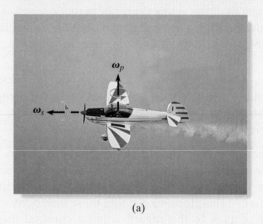

(a)

SOLUTION

Free-Body Diagram.   Fig. 21–13b. The reactions of the connecting shaft on the propeller are indicated by the resultants $\mathbf{F}_R$ and $\mathbf{M}_R$. (The propeller's weight is assumed to be negligible.) The x, y, z axes will be taken fixed to the propeller, since these axes always represent the principal axes of inertia for the propeller. Thus, $\boldsymbol{\Omega} = \boldsymbol{\omega}$. The moments of inertia I_x and I_y are equal ($I_x = I_y = I$) and $I_z = 0$.

Kinematics.   The angular velocity of the propeller observed from the X, Y, Z axes, coincident with the x, y, z axes, Fig. 21–13c, is $\boldsymbol{\omega} = \boldsymbol{\omega}_s + \boldsymbol{\omega}_p = \omega_s\mathbf{i} + \omega_p\mathbf{k}$, so that the x, y, z components of $\boldsymbol{\omega}$ are

$$\omega_x = \omega_s \qquad \omega_y = 0 \qquad \omega_z = \omega_p$$

Since $\boldsymbol{\Omega} = \boldsymbol{\omega}$, then $\dot{\boldsymbol{\omega}} = (\dot{\boldsymbol{\omega}})_{xyz}$. To find $\dot{\boldsymbol{\omega}}$, which is the time derivative with respect to the fixed X, Y, Z axes, we can use Eq. 20–6 since $\boldsymbol{\omega}$ changes direction relative to X, Y, Z. The time rate of change of each of these components $\dot{\boldsymbol{\omega}} = \dot{\boldsymbol{\omega}}_s + \dot{\boldsymbol{\omega}}_p$ relative to the X, Y, Z axes can be obtained by introducing a third coordinate system x', y', z', which has an angular velocity $\boldsymbol{\Omega}' = \boldsymbol{\omega}_p$ and is coincident with the X, Y, Z axes at the instant shown. Thus

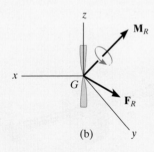

(b)

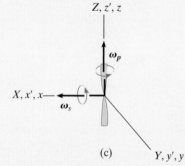

(c)

Fig. 21–13

$$\dot{\omega} = (\dot{\omega})_{x'y'z'} + \omega_p \times \omega$$

$$= (\dot{\omega}_s)_{x'y'z'} + (\dot{\omega}_p)_{x'y'z'} + \omega_p \times (\omega_s + \omega_p)$$

$$= 0 + 0 + \omega_p \times \omega_s + \omega_p \times \omega_p$$

$$= 0 + 0 + \omega_p\mathbf{k} \times \omega_s\mathbf{i} + 0 = \omega_p\omega_s\mathbf{j}$$

Since the X, Y, Z axes are coincident with the x, y, z axes at the instant shown, the components of $\dot{\omega}$ along x, y, z are therefore

$$\dot{\omega}_x = 0 \quad \dot{\omega}_y = \omega_p\omega_s \quad \dot{\omega}_z = 0$$

These same results can also be determined by direct calculation of $(\dot{\omega})_{xyz}$; however, this will involve a bit more work. To do this, it will be necessary to view the propeller (or the x, y, z axes) in some *general position* such as shown in Fig. 21–13d. Here the plane has turned through an angle ϕ (phi) and the propeller has turned through an angle ψ (psi) relative to the plane. Notice that ω_p is always directed along the fixed Z axis and ω_s follows the x axis. Thus the general components of ω are

$$\omega_x = \omega_s \quad \omega_y = \omega_p \sin \psi \quad \omega_z = \omega_p \cos \psi$$

Since ω_s and ω_p are constant, the time derivatives of these components become

$$\dot{\omega}_x = 0 \quad \dot{\omega}_y = \omega_p \cos \psi \, \dot{\psi} \quad \dot{\omega}_z = -\omega_p \sin \psi \, \dot{\psi}$$

But $\phi = \psi = 0°$ and $\dot{\psi} = \omega_s$ at the instant considered. Thus,

$$\omega_x = \omega_s \quad \omega_y = 0 \quad \omega_z = \omega_p$$

$$\dot{\omega}_x = 0 \quad \dot{\omega}_y = \omega_p\omega_s \quad \dot{\omega}_z = 0$$

which are the same results as those obtained previously.

Equations of Motion.   Using Eqs. 21–25, we have

$$\Sigma M_x = I_x\dot{\omega}_x - (I_y - I_z)\omega_y\omega_z = I(0) - (I - 0)(0)\omega_p$$

$$M_x = 0 \qquad\qquad Ans.$$

$$\Sigma M_y = I_y\dot{\omega}_y - (I_z - I_x)\omega_z\omega_x = I(\omega_p\omega_s) - (0 - I)\omega_p\omega_s$$

$$M_y = 2I\omega_p\omega_s \qquad\qquad Ans.$$

$$\Sigma M_z = I_z\dot{\omega}_z - (I_x - I_y)\omega_x\omega_y = 0(0) - (I - I)\omega_s(0)$$

$$M_z = 0 \qquad\qquad Ans.$$

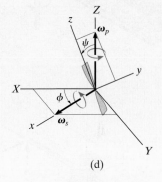

(d)

Fig. 21–13

EXAMPLE | 21.6

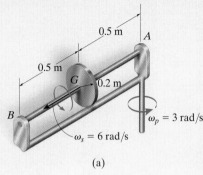

(a)

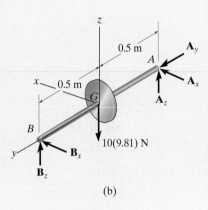

(b)

Fig. 21–14

The 10-kg flywheel (or thin disk) shown in Fig. 21–14a rotates (spins) about the shaft at a constant angular velocity of $\omega_s = 6$ rad/s. At the same time, the shaft rotates (precessing) about the bearing at A with an angular velocity of $\omega_p = 3$ rad/s. If A is a thrust bearing and B is a journal bearing, determine the components of force reaction at each of these supports due to the motion.

SOLUTION I

Free-Body Diagram. Fig. 21–14b. The origin of the x, y, z coordinate system is located at the center of mass G of the flywheel. Here we will let these coordinates have an angular velocity of $\boldsymbol{\Omega} = \boldsymbol{\omega}_p = \{3\mathbf{k}\}$ rad/s. Although the wheel spins relative to these axes, the moments of inertia *remain constant*,* i.e.,

$$I_x = I_z = \tfrac{1}{4}(10 \text{ kg})(0.2 \text{ m})^2 = 0.1 \text{ kg} \cdot \text{m}^2$$
$$I_y = \tfrac{1}{2}(10 \text{ kg})(0.2 \text{ m})^2 = 0.2 \text{ kg} \cdot \text{m}^2$$

Kinematics. From the coincident inertial X, Y, Z frame of reference, Fig. 21–14c, the flywheel has an angular velocity of $\boldsymbol{\omega} = \{6\mathbf{j} + 3\mathbf{k}\}$ rad/s, so that

$$\omega_x = 0 \quad \omega_y = 6 \text{ rad/s} \quad \omega_z = 3 \text{ rad/s}$$

The time derivative of $\boldsymbol{\omega}$ must be determined relative to the x, y, z axes. In this case both $\boldsymbol{\omega}_p$ and $\boldsymbol{\omega}_s$ do not change their magnitude or direction, and so

$$\dot{\omega}_x = 0 \quad \dot{\omega}_y = 0 \quad \dot{\omega}_z = 0$$

Equations of Motion. Applying Eqs. 21–26 ($\boldsymbol{\Omega} \neq \boldsymbol{\omega}$) yields

$$\Sigma M_x = I_x \dot{\omega}_x - I_y \Omega_z \omega_y + I_z \Omega_y \omega_z$$

$$-A_z(0.5) + B_z(0.5) = 0 - (0.2)(3)(6) + 0 = -3.6$$

$$\Sigma M_y = I_y \dot{\omega}_y - I_z \Omega_x \omega_z + I_x \Omega_z \omega_x$$

$$0 = 0 - 0 + 0$$

$$\Sigma M_z = I_z \dot{\omega}_z - I_x \Omega_y \omega_x + I_y \Omega_x \omega_y$$

$$A_x(0.5) - B_x(0.5) = 0 - 0 + 0$$

*This would not be true for the propeller in Example 21.5.

Applying Eqs. 21–19, we have

$$\Sigma F_X = m(a_G)_X; \qquad A_x + B_x = 0$$

$$\Sigma F_Y = m(a_G)_Y; \qquad A_y = -10(0.5)(3)^2$$

$$\Sigma F_Z = m(a_G)_Z; \qquad A_z + B_z - 10(9.81) = 0$$

Solving these equations, we obtain

$$A_x = 0 \quad A_y = -45.0 \text{ N} \quad A_z = 52.6 \text{ N} \qquad \textit{Ans.}$$
$$B_x = 0 \qquad\qquad\qquad\quad B_z = 45.4 \text{ N} \qquad \textit{Ans.}$$

NOTE: If the precession ω_p had not occurred, the z component of force at A and B would be equal to 49.05 N. In this case, however, the difference in these components is caused by the "gyroscopic moment" created whenever a spinning body precesses about another axis. We will study this effect in detail in the next section.

SOLUTION II

This example can also be solved using Euler's equations of motion, Eqs. 21–25. In this case $\boldsymbol{\Omega} = \boldsymbol{\omega} = \{6\mathbf{j} + 3\mathbf{k}\}$ rad/s, and the time derivative $(\dot{\boldsymbol{\omega}})_{xyz}$ can be conveniently obtained with reference to the fixed X, Y, Z axes since $\dot{\boldsymbol{\omega}} = (\dot{\boldsymbol{\omega}})_{xyz}$. This calculation can be performed by choosing x', y', z' axes to have an angular velocity of $\boldsymbol{\Omega}' = \boldsymbol{\omega}_p$, Fig. 21–14c, so that

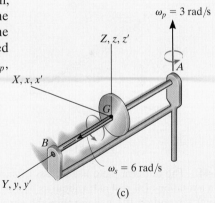

$\omega_p = 3$ rad/s

$\omega_s = 6$ rad/s

(c)

Fig. 21–14

$$\dot{\boldsymbol{\omega}} = (\dot{\boldsymbol{\omega}})_{x'y'z'} + \boldsymbol{\omega}_p \times \boldsymbol{\omega} = 0 + 3\mathbf{k} \times (6\mathbf{j} + 3\mathbf{k}) = \{-18\mathbf{i}\} \text{ rad/s}^2$$

$$\dot{\omega}_x = -18 \text{ rad/s} \quad \dot{\omega}_y = 0 \quad \dot{\omega}_z = 0$$

The moment equations then become

$$\Sigma M_x = I_x \dot{\omega}_x - (I_y - I_z)\omega_y\omega_z$$

$$-A_z(0.5) + B_z(0.5) = 0.1(-18) - (0.2 - 0.1)(6)(3) = -3.6$$

$$\Sigma M_y = I_y \dot{\omega}_y - (I_z - I_x)\omega_z\omega_x$$

$$0 = 0 - 0$$

$$\Sigma M_z = I_z \dot{\omega}_z - (I_x - I_y)\omega_x\omega_y$$

$$A_x(0.5) - B_x(0.5) = 0 - 0$$

The solution then proceeds as before.

PROBLEMS

*21–40. Derive the scalar form of the rotational equation of motion about the x axis if $\boldsymbol{\Omega} \neq \boldsymbol{\omega}$ and the moments and products of inertia of the body are *not constant* with respect to time.

21–41. Derive the scalar form of the rotational equation of motion about the x axis if $\boldsymbol{\Omega} \neq \boldsymbol{\omega}$ and the moments and products of inertia of the body are *constant* with respect to time.

21–42. Derive the Euler equations of motion for $\boldsymbol{\Omega} \neq \boldsymbol{\omega}$, i.e., Eqs. 21–26.

21–43. The 4-lb bar rests along the smooth corners of an open box. At the instant shown, the box has a velocity $\mathbf{v} = \{3\mathbf{j}\}$ ft/s and an acceleration $\mathbf{a} = \{-6\mathbf{j}\}$ ft/s². Determine the x, y, z components of force which the corners exert on the bar.

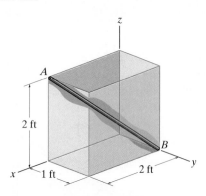

Prob. 21–43

*21–44. The uniform plate has a mass of $m = 2$ kg and is given a rotation of $\omega = 4$ rad/s about its bearings at A and B. If $a = 0.2$ m and $c = 0.3$ m, determine the vertical reactions at the instant shown. Use the x, y, z axes shown and note that $I_{zx} = -\left(\dfrac{mac}{12}\right)\left(\dfrac{c^2 - a^2}{c^2 + a^2}\right)$.

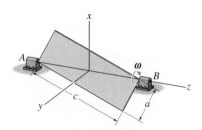

Prob. 21–44

21–45. If the shaft AB is rotating with a constant angular velocity of $\omega = 30$ rad/s, determine the X, Y, Z components of reaction at the thrust bearing A and journal bearing B at the instant shown. The disk has a weight of 15 lb. Neglect the weight of the shaft AB.

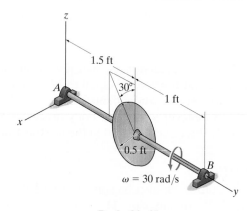

Prob. 21–45

21–46. The 40-kg flywheel (disk) is mounted 20 mm off its true center at G. If the shaft is rotating at a constant speed $\omega = 8$ rad/s, determine the maximum reactions exerted on the journal bearings at A and B.

21–47. The 40-kg flywheel (disk) is mounted 20 mm off its true center at G. If the shaft is rotating at a constant speed $\omega = 8$ rad/s, determine the minimum reactions exerted on the journal bearings at A and B during the motion.

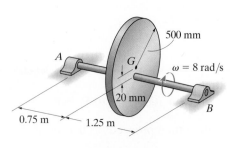

Probs. 21–46/47

*21–48. The man sits on a swivel chair which is rotating with a constant angular velocity of 3 rad/s. He holds the uniform 5-lb rod *AB* horizontal. He suddenly gives it an angular acceleration of 2 rad/s², measured relative to him, as shown. Determine the required force and moment components at the grip, *A*, necessary to do this. Establish axes at the rod's center of mass *G*, with +*z* upward, and +*y* directed along the axis of the rod towards *A*.

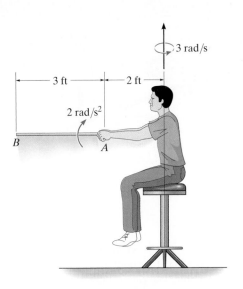

Prob. 21–48

21–49. The 5-kg rod *AB* is supported by a rotating arm. The support at *A* is a journal bearing, which develops reactions normal to the rod. The support at *B* is a thrust bearing, which develops reactions both normal to the rod and along the axis of the rod. Neglecting friction, determine the *x, y, z* components of reaction at these supports when the frame rotates with a constant angular velocity of $\omega = 10$ rad/s.

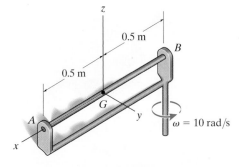

Prob. 21–49

21–50. The rod assembly is supported by a ball-and-socket joint at *C* and a journal bearing at *D*, which develops only *x* and *y* force reactions. The rods have a mass of 0.75 kg/m. Determine the angular acceleration of the rods and the components of reaction at the supports at the instant $\omega = 8$ rad/s as shown.

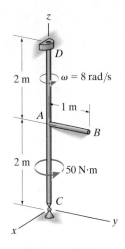

Prob. 21–50

21–51. The uniform hatch door, having a mass of 15 kg and a mass center at *G*, is supported in the horizontal plane by bearings at *A* and *B*. If a vertical force $F = 300$ N is applied to the door as shown, determine the components of reaction at the bearings and the angular acceleration of the door. The bearing at *A* will resist a component of force in the *y* direction, whereas the bearing at *B* will not. For the calculation, assume the door to be a thin plate and neglect the size of each bearing. The door is originally at rest.

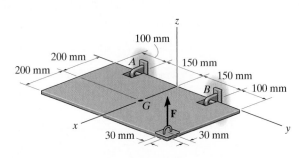

Prob. 21–51

21

*21–52. The conical pendulum consists of a bar of mass m and length L that is supported by the pin at its end A. If the pin is subjected to a rotation $\boldsymbol{\omega}$, determine the angle θ that the bar makes with the vertical as it rotates.

21–54. The rod assembly is supported by journal bearings at A and B, which develop only x and z force reactions on the shaft. If the shaft AB is rotating in the direction shown at $\boldsymbol{\omega} = \{-5\mathbf{j}\}$ rad/s, determine the reactions at the bearings when the assembly is in the position shown. Also, what is the shaft's angular acceleration? The mass of each rod is 1.5 kg/m.

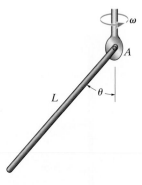

Prob. 21–52

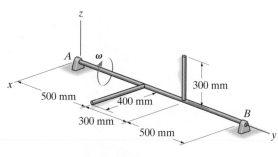

Prob. 21–54

21–55. The 20-kg sphere is rotating with a constant angular speed of $\omega_1 = 150$ rad/s about axle CD, which is mounted on the circular ring. The ring rotates about shaft AB with a constant angular speed of $\omega_2 = 50$ rad/s. If shaft AB is supported by a thrust bearing at A and a journal bearing at B, determine the X, Y, Z components of reaction at these bearings at the instant shown. Neglect the mass of the ring and shaft.

21–53. The car travels around the curved road of radius ρ such that its mass center has a constant speed v_G. Write the equations of rotational motion with respect to the x, y, z axes. Assume that the car's six moments and products of inertia with respect to these axes are known.

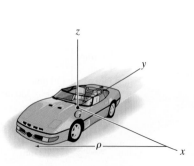

Prob. 21–53

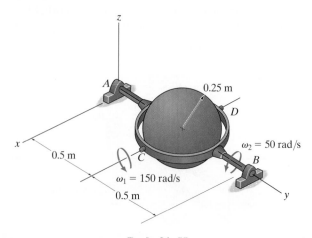

Prob. 21–55

*21–56. The rod assembly has a weight of 5 lb/ft. It is supported at B by a smooth journal bearing, which develops x and y force reactions, and at A by a smooth thrust bearing, which develops x, y, and z force reactions. If a 50-lb · ft torque is applied along rod AB, determine the components of reaction at the bearings when the assembly has an angular velocity $\omega = 10$ rad/s at the instant shown.

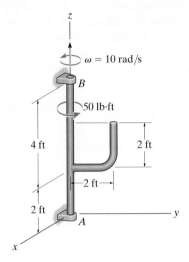

Prob. 21–56

21–57. The blades of a wind turbine spin about the shaft S with a constant angular speed of ω_s, while the frame precesses about the vertical axis with a constant angular speed of ω_p. Determine the x, y, and z components of moment that the shaft exerts on the blades as a function of θ. Consider each blade as a slender rod of mass m and length l.

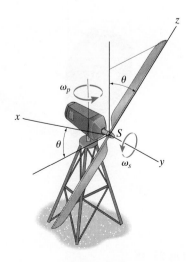

Prob. 21–57

21–58. The cylinder has a mass of 30 kg and is mounted on an axle that is supported by bearings at A and B. If the axle is turning at $\omega = \{-40\mathbf{j}\}$ rad/s, determine the vertical components of force acting at the bearings at this instant.

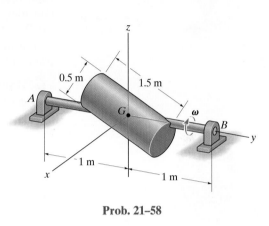

Prob. 21–58

21–59. The *thin rod* has a mass of 0.8 kg and a total length of 150 mm. It is rotating about its midpoint at a constant rate $\dot{\theta} = 6$ rad/s, while the table to which its axle A is fastened is rotating at 2 rad/s. Determine the x, y, z moment components which the axle exerts on the rod when the rod is in any position θ.

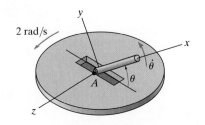

Prob. 21–59

21

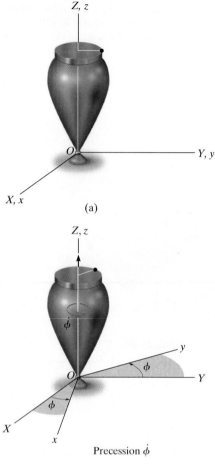

(a)

Precession $\dot{\phi}$

(b)

*21.5 Gyroscopic Motion

In this section we will develop the equations defining the motion of a body (top) which is symmetrical with respect to an axis and rotating about a fixed point. These equations also apply to the motion of a particularly interesting device, the gyroscope.

The body's motion will be analyzed using *Euler angles* ϕ, θ, ψ (phi, theta, psi). To illustrate how they define the position of a body, consider the top shown in Fig. 21–15a. To define its final position, Fig. 21–15d, a second set of x, y, z axes is fixed in the top. Starting with the X, Y, Z and x, y, z axes in coincidence, Fig. 21–15a, the final position of the top can be determined using the following three steps:

1. Rotate the top about the Z (or z) axis through an angle ϕ ($0 \le \phi < 2\pi$), Fig. 21–15b.

2. Rotate the top about the x axis through an angle θ ($0 \le \theta \le \pi$), Fig. 21–15c.

3. Rotate the top about the z axis through an angle ψ ($0 \le \psi < 2\pi$) to obtain the final position, Fig. 21–15d.

The sequence of these three angles, ϕ, θ, then ψ, must be maintained, since finite rotations are *not vectors* (see Fig. 20–1). Although this is the case, the differential rotations $d\phi$, $d\theta$, and $d\psi$ are vectors, and thus the angular velocity $\boldsymbol{\omega}$ of the top can be expressed in terms of the time derivatives of the Euler angles. The angular-velocity components $\dot{\phi}$, $\dot{\theta}$, and $\dot{\psi}$ are known as the *precession, nutation,* and *spin,* respectively.

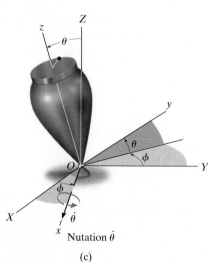

Nutation $\dot{\theta}$

(c)

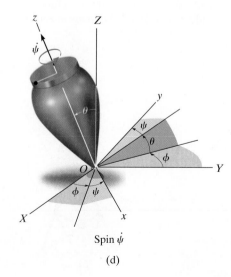

Spin $\dot{\psi}$

(d)

Fig. 21–15

Their positive directions are shown in Fig. 21–16. It is seen that these vectors are not all perpendicular to one another; however, $\boldsymbol{\omega}$ of the top can still be expressed in terms of these three components.

Since the body (top) is symmetric with respect to the z or spin axis, there is no need to attach the x, y, z axes to the top since the inertial properties of the top will remain constant with respect to this frame during the motion. Therefore $\boldsymbol{\Omega} = \boldsymbol{\omega}_p + \boldsymbol{\omega}_n$, Fig. 21–16. Hence, the angular velocity of the body is

$$\boldsymbol{\omega} = \omega_x \mathbf{i} + \omega_y \mathbf{j} + \omega_z \mathbf{k}$$

$$= \dot{\theta}\mathbf{i} + (\dot{\phi}\sin\theta)\mathbf{j} + (\dot{\phi}\cos\theta + \dot{\psi})\mathbf{k} \qquad (21\text{–}27)$$

And the angular velocity of the axes is

$$\boldsymbol{\Omega} = \Omega_x \mathbf{i} + \Omega_y \mathbf{j} + \Omega_z \mathbf{k}$$

$$= \dot{\theta}\mathbf{i} + (\dot{\phi}\sin\theta)\mathbf{j} + (\dot{\phi}\cos\theta)\mathbf{k} \qquad (21\text{–}28)$$

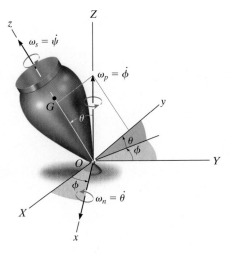

Fig. 21–16

Have the x, y, z axes represent principal axes of inertia for the top, and so the moments of inertia will be represented as $I_{xx} = I_{yy} = I$ and $I_{zz} = I_z$. Since $\boldsymbol{\Omega} \neq \boldsymbol{\omega}$, Eqs. 21–26 are used to establish the rotational equations of motion. Substituting into these equations the respective angular-velocity components defined by Eqs. 21–27 and 21–28, their corresponding time derivatives, and the moment of inertia components, yields

$$\Sigma M_x = I(\ddot{\theta} - \dot{\phi}^2 \sin\theta\cos\theta) + I_z\dot{\phi}\sin\theta(\dot{\phi}\cos\theta + \dot{\psi})$$

$$\Sigma M_y = I(\ddot{\phi}\sin\theta + 2\dot{\phi}\dot{\theta}\cos\theta) - I_z\dot{\theta}(\dot{\phi}\cos\theta + \dot{\psi}) \qquad (21\text{–}29)$$

$$\Sigma M_z = I_z(\ddot{\psi} + \ddot{\phi}\cos\theta - \dot{\phi}\dot{\theta}\sin\theta)$$

Each moment summation applies only at the fixed point O or the center of mass G of the body. Since the equations represent a coupled set of nonlinear second-order differential equations, in general a closed-form solution may not be obtained. Instead, the Euler angles ϕ, θ, and ψ may be obtained graphically as functions of time using numerical analysis and computer techniques.

A special case, however, does exist for which simplification of Eqs. 21–29 is possible. Commonly referred to as *steady precession*, it occurs when the nutation angle θ, precession $\dot{\phi}$, and spin $\dot{\psi}$ all remain *constant*. Equations 21–29 then reduce to the form

$$\boxed{\Sigma M_x = -I\dot{\phi}^2 \sin\theta\cos\theta + I_z\dot{\phi}\sin\theta(\dot{\phi}\cos\theta + \dot{\psi})} \qquad (21\text{–}30)$$

$$\Sigma M_y = 0$$

$$\Sigma M_z = 0$$

21

Equation 21–30 can be further simplified by noting that, from Eq. 21–27, $\omega_z = \dot{\phi} \cos \theta + \dot{\psi}$, so that

$$\Sigma M_x = -I\dot{\phi}^2 \sin \theta \cos \theta + I_z \dot{\phi}(\sin \theta)\omega_z$$

or

$$\boxed{\Sigma M_x = \dot{\phi} \sin \theta (I_z \omega_z - I\dot{\phi} \cos \theta)} \qquad (21\text{–}31)$$

It is interesting to note what effects the spin $\dot{\psi}$ has on the moment about the x axis. To show this, consider the spinning rotor in Fig. 21–17. Here $\theta = 90°$, in which case Eq. 21–30 reduces to the form

$$\Sigma M_x = I_z \dot{\phi}\dot{\psi}$$

or

$$\boxed{\Sigma M_x = I_z \Omega_y \omega_z} \qquad (21\text{–}32)$$

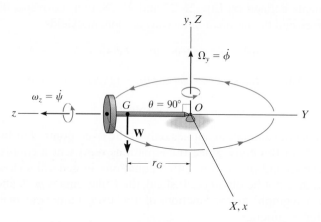

Fig. 21–17

From the figure it can be seen that $\mathbf{\Omega}_y$ and $\boldsymbol{\omega}_z$ act along their respective *positive axes* and therefore are mutually perpendicular. Instinctively, one would expect the rotor to fall down under the influence of gravity! However, this is not the case at all, provided the product $I_z \Omega_y \omega_z$ is correctly chosen to counterbalance the moment $\Sigma M_x = Wr_G$ of the rotor's weight about O. This unusual phenomenon of rigid-body motion is often referred to as the *gyroscopic effect*.

Perhaps a more intriguing demonstration of the gyroscopic effect comes from studying the action of a *gyroscope*, frequently referred to as a *gyro*. A gyro is a rotor which spins at a very high rate about its axis of symmetry. This rate of spin is considerably greater than its precessional rate of rotation about the vertical axis. Hence, for all practical purposes, the angular momentum of the gyro can be assumed directed along its axis of spin. Thus, for the gyro rotor shown in Fig. 21–18, $\omega_z \gg \Omega_y$, and the magnitude of the angular momentum about point O, as determined from Eqs. 21–11, reduces to the form $H_O = I_z\omega_z$. Since both the magnitude and direction of $\mathbf{H}_O$ are constant as observed from x, y, z, direct application of Eq. 21–22 yields

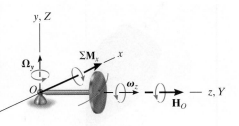

$$\boxed{\Sigma \mathbf{M}_x = \mathbf{\Omega}_y \times \mathbf{H}_O} \qquad (21\text{–}33)$$

Fig. 21–18

Using the right-hand rule applied to the cross product, it can be seen that $\mathbf{\Omega}_y$ always swings $\mathbf{H}_O$ (or $\boldsymbol{\omega}_z$) toward the sense of $\Sigma\mathbf{M}_x$. In effect, the *change in direction* of the gyro's angular momentum, $d\mathbf{H}_O$, is equivalent to the angular impulse caused by the gyro's weight about O, i.e., $d\mathbf{H}_O = \Sigma\mathbf{M}_x \, dt$, Eq. 21–20. Also, since $H_O = I_z\omega_z$ and $\Sigma\mathbf{M}_x$, $\mathbf{\Omega}_y$, and $\mathbf{H}_O$ are mutually perpendicular, Eq. 21–33 reduces to Eq. 21–32.

When a gyro is mounted in gimbal rings, Fig. 21–19, it becomes *free* of external moments applied to its base. Thus, in theory, its angular momentum $\mathbf{H}$ will never precess but, instead, maintain its same fixed orientation along the axis of spin when the base is rotated. This type of gyroscope is called a *free gyro* and is useful as a gyrocompass when the spin axis of the gyro is directed north. In reality, the gimbal mechanism is never completely free of friction, so such a device is useful only for the local navigation of ships and aircraft. The gyroscopic effect is also useful as a means of stabilizing both the rolling motion of ships at sea and the trajectories of missiles and projectiles. Furthermore, this effect is of significant importance in the design of shafts and bearings for rotors which are subjected to forced precessions.

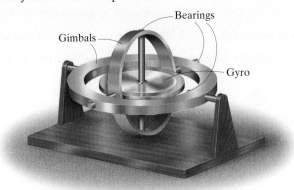

Fig. 21–19

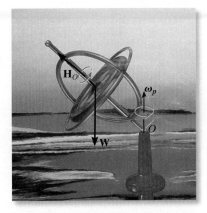

The spinning of the gyro within the frame of this toy gyroscope produces angular momentum $\mathbf{H}_O$, which is changing direction as the frame precesses $\boldsymbol{\omega}_p$ about the vertical axis. The gyroscope will not fall down since the moment of its weight $\mathbf{W}$ about the support is balanced by the change in the direction of $\mathbf{H}_O$.

21

EXAMPLE | 21.7

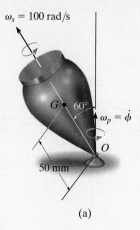

$\omega_s = 100$ rad/s

G 60°

$\omega_p = \dot{\phi}$

O

50 mm

(a)

Fig. 21–20

The top shown in Fig. 21–20a has a mass of 0.5 kg and is precessing about the vertical axis at a constant angle of $\theta = 60°$. If it spins with an angular velocity $\omega_s = 100$ rad/s, determine the precession ω_p. Assume that the axial and transverse moments of inertia of the top are $0.45(10^{-3})$ kg·m² and $1.20(10^{-3})$ kg·m², respectively, measured with respect to the fixed point O.

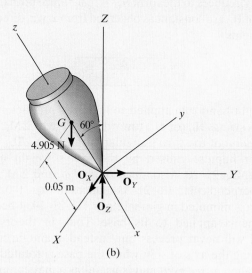

(b)

SOLUTION

Equation 21–30 will be used for the solution since the motion is *steady precession*. As shown on the free-body diagram, Fig. 21–20b, the coordinate axes are established in the usual manner, that is, with the positive z axis in the direction of spin, the positive Z axis in the direction of precession, and the positive x axis in the direction of the moment ΣM_x (refer to Fig. 21–16). Thus,

$$\Sigma M_x = -I\dot{\phi}^2 \sin\theta\cos\theta + I_z\dot{\phi}\sin\theta(\dot{\phi}\cos\theta + \dot{\psi})$$

$$4.905 \text{ N}(0.05 \text{ m})\sin 60° = -[1.20(10^{-3})\text{ kg}\cdot\text{m}^2\ \dot{\phi}^2]\sin 60°\cos 60°$$
$$+ [0.45(10^{-3})\text{ kg}\cdot\text{m}^2]\dot{\phi}\sin 60°(\dot{\phi}\cos 60° + 100 \text{ rad/s})$$

or

$$\dot{\phi}^2 - 120.0\dot{\phi} + 654.0 = 0 \qquad (1)$$

Solving this quadratic equation for the precession gives

$$\dot{\phi} = 114 \text{ rad/s} \quad \text{(high precession)} \qquad \textit{Ans.}$$

and

$$\dot{\phi} = 5.72 \text{ rad/s} \quad \text{(low precession)} \qquad \textit{Ans.}$$

NOTE: In reality, low precession of the top would generally be observed, since high precession would require a larger kinetic energy.

21

EXAMPLE | 21.8

The 1-kg disk shown in Fig. 21–21a spins about its axis with a constant angular velocity $\omega_D = 70$ rad/s. The block at B has a mass of 2 kg, and by adjusting its position s one can change the precession of the disk about its supporting pivot at O while the shaft remains horizontal. Determine the position s that will enable the disk to have a constant precession $\omega_p = 0.5$ rad/s about the pivot. Neglect the weight of the shaft.

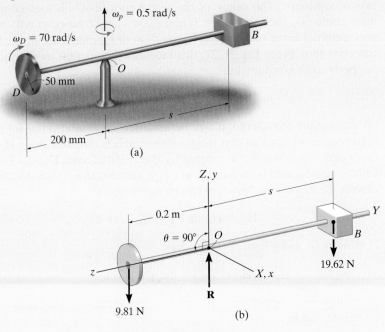

(a)

(b)

Fig. 21–21

SOLUTION
The free-body diagram of the assembly is shown in Fig. 21–21b. The origin for both the x, y, z and X, Y, Z coordinate systems is located at the fixed point O. In the conventional sense, the Z axis is chosen along the axis of precession, and the z axis is along the axis of spin, so that $\theta = 90°$. Since the precession is *steady*, Eq. 21–32 can be used for the solution.

$$\Sigma M_x = I_z \Omega_y \omega_z$$

Substituting the required data gives

$$(9.81 \text{ N})(0.2 \text{ m}) - (19.62 \text{ N})s = \left[\tfrac{1}{2}(1 \text{ kg})(0.05 \text{ m})^2\right]0.5 \text{ rad/s}(-70 \text{ rad/s})$$

$$s = 0.102 \text{ m} = 102 \text{ mm} \qquad \textit{Ans.}$$

21

21.6   Torque-Free Motion

When the only external force acting on a body is caused by gravity, the general motion of the body is referred to as *torque-free motion*. This type of motion is characteristic of planets, artificial satellites, and projectiles—provided air friction is neglected.

In order to describe the characteristics of this motion, the distribution of the body's mass will be assumed *axisymmetric*. The satellite shown in Fig. 21–22 is an example of such a body, where the z axis represents an axis of symmetry. The origin of the x, y, z coordinates is located at the mass center G, such that $I_{zz} = I_z$ and $I_{xx} = I_{yy} = I$. Since gravity is the only external force present, the summation of moments about the mass center is zero. From Eq. 21–21, this requires the angular momentum of the body to be constant, i.e.,

$$\mathbf{H}_G = \text{constant}$$

At the instant considered, it will be assumed that the inertial frame of reference is oriented so that the positive Z axis is directed along $\mathbf{H}_G$ and the y axis lies in the plane formed by the z and Z axes, Fig. 21–22. The Euler angle formed between Z and z is θ, and therefore, with this choice of axes the angular momentum can be expressed as

$$\mathbf{H}_G = H_G \sin \theta \, \mathbf{j} + H_G \cos \theta \, \mathbf{k}$$

Furthermore, using Eqs. 21–11, we have

$$\mathbf{H}_G = I\omega_x \mathbf{i} + I\omega_y \mathbf{j} + I_z\omega_z \mathbf{k}$$

Equating the respective $\mathbf{i}$, $\mathbf{j}$, and $\mathbf{k}$ components of the above two equations yields

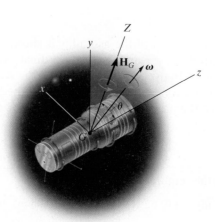

Fig. 21–22

$$\omega_x = 0 \quad \omega_y = \frac{H_G \sin \theta}{I} \quad \omega_z = \frac{H_G \cos \theta}{I_z} \qquad (21\text{–}34)$$

or

$$\boldsymbol{\omega} = \frac{H_G \sin \theta}{I}\mathbf{j} + \frac{H_G \cos \theta}{I_z}\mathbf{k} \qquad (21\text{–}35)$$

In a similar manner, equating the respective **i**, **j**, **k** components of Eq. 21–27 to those of Eq. 21–34, we obtain

$$\dot{\theta} = 0$$

$$\dot{\phi} \sin \theta = \frac{H_G \sin \theta}{I}$$

$$\dot{\phi} \cos \theta + \dot{\psi} = \frac{H_G \cos \theta}{I_z}$$

Solving, we get

$$\theta = \text{constant}$$

$$\dot{\phi} = \frac{H_G}{I} \qquad (21\text{–}36)$$

$$\dot{\psi} = \frac{I - I_z}{I I_z} H_G \cos \theta$$

Thus, for torque-free motion of an axisymmetrical body, the angle θ formed between the angular-momentum vector and the spin of the body remains constant. Furthermore, the angular momentum $\mathbf{H}_G$, precession $\dot{\phi}$, and spin $\dot{\psi}$ for the body remain constant at all times during the motion.

Eliminating H_G from the second and third of Eqs. 21–36 yields the following relation between the spin and precession:

$$\dot{\psi} = \frac{I - I_z}{I_z} \dot{\phi} \cos \theta \qquad (21\text{–}37)$$

21

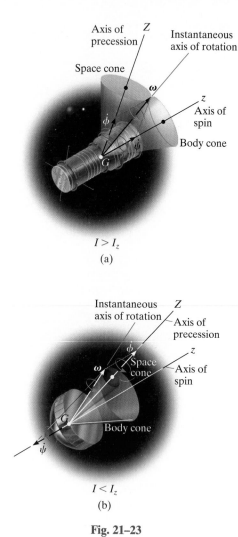

$I > I_z$

(a)

$I < I_z$

(b)

Fig. 21–23

These two components of angular motion can be studied by using the body and space cone models introduced in Sec. 20.1. The *space cone* defining the precession is fixed from rotating, since the precession has a fixed direction, while the outer surface of the *body cone* rolls on the space cone's outer surface. Try to imagine this motion in Fig. 21–23a. The interior angle of each cone is chosen such that the resultant angular velocity of the body is directed along the line of contact of the two cones. This line of contact represents the instantaneous axis of rotation for the body cone, and hence the angular velocity of both the body cone and the body must be directed along this line. Since the spin is a function of the moments of inertia I and I_z of the body, Eq. 21–36, the cone model in Fig. 21–23a is satisfactory for describing the motion, provided $I > I_z$. Torque-free motion which meets these requirements is called *regular precession*. If $I < I_z$, the spin is negative and the precession positive. This motion is represented by the satellite motion shown in Fig. 21–23b ($I < I_z$). The cone model can again be used to represent the motion; however, to preserve the correct vector addition of spin and precession to obtain the angular velocity $\boldsymbol{\omega}$, the inside surface of the body cone must roll on the outside surface of the (fixed) space cone. This motion is referred to as *retrograde precession*.

Satellites are often given a spin before they are launched. If their angular momentum is not collinear with the axis of spin, they will exhibit precession. In the photo on the left, regular precession will occur since $I > I_z$, and in the photo on the right, retrograde precession will occur since $I < I_z$.

EXAMPLE | 21.9

The motion of a football is observed using a slow-motion projector. From the film, the spin of the football is seen to be directed 30° from the horizontal, as shown in Fig. 21–24a. Also, the football is precessing about the vertical axis at a rate $\dot{\phi} = 3$ rad/s. If the ratio of the axial to transverse moments of inertia of the football is $\frac{1}{3}$, measured with respect to the center of mass, determine the magnitude of the football's spin and its angular velocity. Neglect the effect of air resistance.

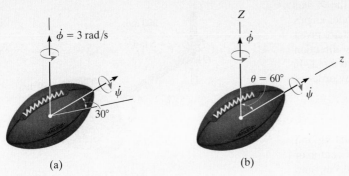

Fig. 21–24

SOLUTION

Since the weight of the football is the only force acting, the motion is torque-free. In the conventional sense, if the z axis is established along the axis of spin and the Z axis along the precession axis, as shown in Fig. 21–24b, then the angle $\theta = 60°$. Applying Eq. 21–37, the spin is

$$\dot{\psi} = \frac{I - I_z}{I_z}\dot{\phi}\cos\theta = \frac{I - \frac{1}{3}I}{\frac{1}{3}I}(3)\cos 60°$$

$$= 3 \text{ rad/s} \qquad \qquad Ans.$$

Using Eqs. 21–34, where $H_G = \dot{\phi}I$ (Eq. 21–36), we have

$$\omega_x = 0$$

$$\omega_y = \frac{H_G \sin\theta}{I} = \frac{3I \sin 60°}{I} = 2.60 \text{ rad/s}$$

$$\omega_z = \frac{H_G \cos\theta}{I_z} = \frac{3I \cos 60°}{\frac{1}{3}I} = 4.50 \text{ rad/s}$$

Thus,

$$\omega = \sqrt{(\omega_x)^2 + (\omega_y)^2 + (\omega_z)^2}$$

$$= \sqrt{(0)^2 + (2.60)^2 + (4.50)^2}$$

$$= 5.20 \text{ rad/s} \qquad \qquad Ans.$$

21

PROBLEMS

*21–60. Show that the angular velocity of a body, in terms of Euler angles ϕ, θ, and ψ, can be expressed as $\omega = (\dot\phi \sin\theta \sin\psi + \dot\theta \cos\psi)\mathbf{i} + (\dot\phi \sin\theta \cos\psi - \dot\theta \sin\psi)\mathbf{j} + (\dot\phi \cos\theta + \dot\psi)\mathbf{k}$, where $\mathbf{i}$, $\mathbf{j}$, and $\mathbf{k}$ are directed along the x, y, z axes as shown in Fig. 21–15d.

21–61. A thin rod is initially coincident with the Z axis when it is given three rotations defined by the Euler angles $\phi = 30°$, $\theta = 45°$, and $\psi = 60°$. If these rotations are given in the order stated, determine the coordinate direction angles α, β, γ of the axis of the rod with respect to the X, Y, and Z axes. Are these directions the same for any order of the rotations? Why?

21–62. The turbine on a ship has a mass of 400 kg and is mounted on bearings A and B as shown. Its center of mass is at G, its radius of gyration is $k_z = 0.3$ m, and $k_x = k_y = 0.5$ m. If it is spinning at 200 rad/s, determine the vertical reactions at the bearings when the ship undergoes each of the following motions: (a) rolling, $\omega_1 = 0.2$ rad/s, (b) turning, $\omega_2 = 0.8$ rad/s, (c) pitching, $\omega_3 = 1.4$ rad/s.

21–63. The 10-kg disk spins about axle AB at a constant rate of $\omega_s = 100$ rad/s. If the supporting arm precesses about the vertical axis at a constant rate of $\omega_p = 5$ rad/s, determine the internal moment at O caused only by the gyroscopic action.

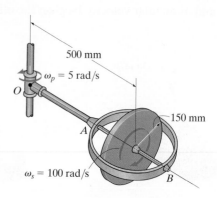

Prob. 21–63

*21–64. The 10-kg disk spins about axle AB at a constant rate of $\omega_s = 250$ rad/s, and $\theta = 30°$. Determine the rate of precession of arm OA. Neglect the mass of arm OA, axle AB, and the circular ring D.

21–65. When OA precesses at a constant rate of $\omega_p = 5$ rad/s, when $\theta = 90°$, determine the required spin of the 10-kg disk C. Neglect the mass of arm OA, axle AB, and the circular ring D.

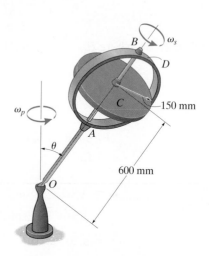

Probs. 21–64/65

Prob. 21–62

21–66. The car travels at a constant speed of $v_C = 100$ km/h around the horizontal curve having a radius of 80 m. If each wheel has a mass of 16 kg, a radius of gyration $k_G = 300$ mm about its spinning axis, and a radius of 400 mm, determine the difference between the normal forces of the rear wheels, caused by the gyroscopic effect. The distance between the wheels is 1.30 m.

*21–68. The top consists of a thin disk that has a weight of 8 lb and a radius of 0.3 ft. The rod has a negligible mass and a length of 0.5 ft. If the top is spinning with an angular velocity $\omega_s = 300$ rad/s, determine the steady-state precessional angular velocity ω_p of the rod when $\theta = 40°$.

21–69. Solve Prob. 21–68 when $\theta = 90°$.

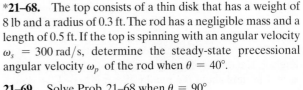

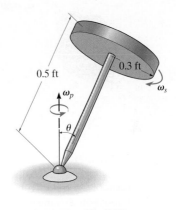

Probs. 21–68/69

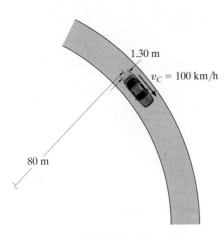

Prob. 21–66

21–70. The top has a mass of 90 g, a center of mass at G, and a radius of gyration $k = 18$ mm about its axis of symmetry. About any transverse axis acting through point O the radius of gyration is $k_t = 35$ mm. If the top is connected to a ball-and-socket joint at O and the precession is $\omega_p = 0.5$ rad/s, determine the spin ω_s.

21–67. A wheel of mass m and radius r rolls with constant spin ω about a circular path having a radius a. If the angle of inclination is θ, determine the rate of precession. Treat the wheel as a thin ring. No slipping occurs.

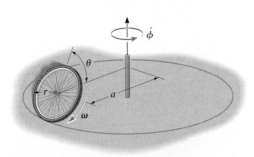

Prob. 21–67

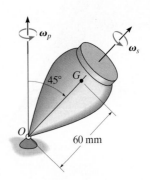

Prob. 21–70

21

21–71. The 1-lb top has a center of gravity at point G. If it spins about its axis of symmetry and precesses about the vertical axis at constant rates of $\omega_s = 60$ rad/s and $\omega_p = 10$ rad/s, respectively, determine the steady state angle θ. The radius of gyration of the top about the z axis is $k_z = 1$ in., and about the x and y axes it is $k_x = k_y = 4$ in.

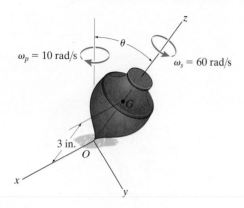

Prob. 21–71

*21–72. While the rocket is in free flight, it has a spin of 3 rad/s and precesses about an axis measured 10° from the axis of spin. If the ratio of the axial to transverse moments of inertia of the rocket is 1/15, computed about axes which pass through the mass center G, determine the angle which the resultant angular velocity makes with the spin axis. Construct the body and space cones used to describe the motion. Is the precession regular or retrograde?

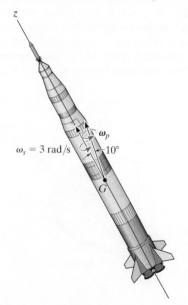

Prob. 21–72

21–73. The 0.2-kg football is thrown with a spin $\omega_z = 35$ rad/s. If the angle θ is measured as 60°, determine the precession about the Z axis. The radius of gyration about the spin axis is $k_z = 0.05$ m, and about a transverse axis it is $k_t = 0.1$ m.

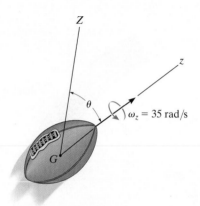

Prob. 21–73

21–74. The projectile shown is subjected to torque-free motion. The transverse and axial moments of inertia are I and I_z, respectively. If θ represents the angle between the precessional axis Z and the axis of symmetry z, and β is the angle between the angular velocity ω and the z axis, show that β and θ are related by the equation $\tan \theta = (I/I_z) \tan \beta$.

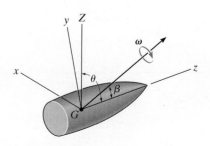

Prob. 21–74

21–75. The 4-kg disk is thrown with a spin $\omega_z = 6$ rad/s. If the angle θ is measured as 160°, determine the precession about the Z axis.

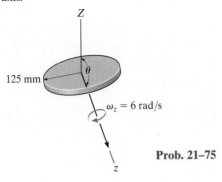

Prob. 21–75

*21–76. The rocket has a mass of 4 Mg and radii of gyration $k_z = 0.85$ m and $k_y = 2.3$ m. It is initially spinning about the z axis at $\omega_z = 0.05$ rad/s when a meteoroid M strikes it at A and creates an impulse $\mathbf{I} = \{300\mathbf{i}\}$ N·s. Determine the axis of precession after the impact.

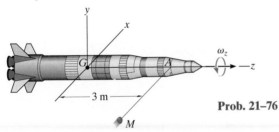

Prob. 21–76

21–77. The football has a mass of 450 g and radii of gyration about its axis of symmetry (z axis) and its transverse axes (x or y axis) of $k_z = 30$ mm and $k_x = k_y = 50$ mm, respectively. If the football has an angular momentum of $H_G = 0.02$ kg·m²/s, determine its precession $\dot{\phi}$ and spin $\dot{\psi}$. Also, find the angle β that the angular velocity vector makes with the z axis.

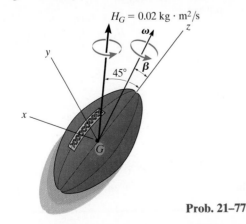

Prob. 21–77

21–78. The projectile precesses about the Z axis at a constant rate of $\dot{\phi} = 15$ rad/s when it leaves the barrel of a gun. Determine its spin $\dot{\psi}$ and the magnitude of its angular momentum $\mathbf{H}_G$. The projectile has a mass of 1.5 kg and radii of gyration about its axis of symmetry (z axis) and about its transverse axes (x and y axes) of $k_z = 65$ mm and $k_x = k_y = 125$ mm, respectively.

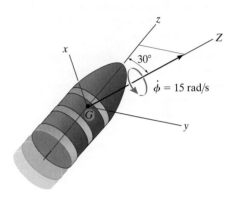

Prob. 21–78

21–79. The space capsule has a mass of 3.2 Mg, and about axes passing through the mass center G the axial and transverse radii of gyration are $k_z = 0.90$ m and $k_t = 1.85$ m, respectively. If it is spinning at $\omega_s = 0.8$ rev/s, determine its angular momentum. Precession occurs about the Z axis.

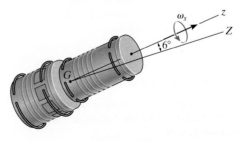

Prob. 21–79

21

CHAPTER REVIEW

Moments and Products of Inertia

A body has six components of inertia for any specified x, y, z axes. Three of these are moments of inertia about each of the axes, I_{xx}, I_{yy}, I_{zz}, and three are products of inertia, each defined from two orthogonal planes, I_{xy}, I_{yz}, I_{xz}. If either one or both of these planes are planes of symmetry, then the product of inertia with respect to these planes will be zero.

The moments and products of inertia can be determined by direct integration or by using tabulated values. If these quantities are to be determined with respect to axes or planes that do not pass through the mass center, then parallel-axis and parallel-plane theorems must be used.

Provided the six components of inertia are known, then the moment of inertia about any axis can be determined using the inertia transformation equation.

$$I_{xx} = \int_m r_x^2 \, dm = \int_m (y^2 + z^2) \, dm \qquad I_{xy} = I_{yx} = \int_m xy \, dm$$

$$I_{yy} = \int_m r_y^2 \, dm = \int_m (x^2 + z^2) \, dm \qquad I_{yz} = I_{zy} = \int_m yz \, dm$$

$$I_{zz} = \int_m r_z^2 \, dm = \int_m (x^2 + y^2) \, dm \qquad I_{xz} = I_{zx} = \int_m xz \, dm$$

$$I_{Oa} = I_{xx}u_x^2 + I_{yy}u_y^2 + I_{zz}u_z^2 - 2I_{xy}u_xu_y - 2I_{yz}u_yu_z - 2I_{zx}u_zu_x$$

Principal Moments of Inertia

At any point on or off the body, the x, y, z axes can be oriented so that the products of inertia will be zero. The resulting moments of inertia are called the principal moments of inertia, one of which will be a maximum and the other a minimum.

$$\begin{pmatrix} I_x & 0 & 0 \\ 0 & I_y & 0 \\ 0 & 0 & I_z \end{pmatrix}$$

Principle of Impulse and Momentum

The angular momentum for a body can be determined about any arbitrary point A.

Once the linear and angular momentum for the body have been formulated, then the principle of impulse and momentum can be used to solve problems that involve force, velocity, and time.

$$m(\mathbf{v}_G)_1 + \Sigma \int_{t_1}^{t_2} \mathbf{F} \, dt = m(\mathbf{v}_G)_2$$

$$\mathbf{H}_O = \int_m \boldsymbol{\rho}_O \times (\boldsymbol{\omega} \times \boldsymbol{\rho}_O) \, dm$$
Fixed Point O

$$\mathbf{H}_G = \int_m \boldsymbol{\rho}_G \times (\boldsymbol{\omega} \times \boldsymbol{\rho}_G) \, dm$$
Center of Mass

$$\mathbf{H}_A = \boldsymbol{\rho}_{G/A} \times m\mathbf{v}_G + \mathbf{H}_G$$
Arbitrary Point

$$(\mathbf{H}_O)_1 + \Sigma \int_{t_1}^{t_2} \mathbf{M}_O \, dt = (\mathbf{H}_O)_2$$
where
$$H_x = I_{xx}\omega_x - I_{xy}\omega_y - I_{xz}\omega_z$$
$$H_y = -I_{yx}\omega_x + I_{yy}\omega_y - I_{yz}\omega_z$$
$$H_z = -I_{zx}\omega_x - I_{zy}\omega_y + I_{zz}\omega_z$$

Principle of Work and Energy

The kinetic energy for a body is usually determined relative to a fixed point or the body's mass center.

$$T = \tfrac{1}{2}I_x\omega_x^2 + \tfrac{1}{2}I_y\omega_y^2 + \tfrac{1}{2}I_z\omega_z^2 \qquad T = \tfrac{1}{2}mv_G^2 + \tfrac{1}{2}I_x\omega_x^2 + \tfrac{1}{2}I_y\omega_y^2 + \tfrac{1}{2}I_z\omega_z^2$$
Fixed Point   Center of Mass

These formulations can be used with the principle of work and energy to solve problems that involve force, velocity, and displacement.	$$T_1 + \Sigma U_{1-2} = T_2$$

Equations of Motion

There are three scalar equations of translational motion for a rigid body that moves in three dimensions.

The three scalar equations of rotational motion depend upon the motion of the x, y, z reference. Most often, these axes are oriented so that they are principal axes of inertia. If the axes are fixed in and move with the body so that $\mathbf{\Omega} = \mathbf{\omega}$, then the equations are referred to as the Euler equations of motion.

A free-body diagram should always accompany the application of the equations of motion.

$$\Sigma F_x = m(a_G)_x$$
$$\Sigma F_y = m(a_G)_y$$
$$\Sigma F_z = m(a_G)_z$$

$$\Sigma M_x = I_x \dot{\omega}_x - (I_y - I_z)\omega_y\omega_z$$
$$\Sigma M_y = I_y \dot{\omega}_y - (I_z - I_x)\omega_z\omega_x$$
$$\Sigma M_z = I_z \dot{\omega}_z - (I_x - I_y)\omega_x\omega_y$$

$$\mathbf{\Omega} = \mathbf{\omega}$$

$$\Sigma M_x = I_x \dot{\omega}_x - I_y\Omega_z\omega_y + I_z\Omega_y\omega_z$$
$$\Sigma M_y = I_y \dot{\omega}_y - I_z\Omega_x\omega_z + I_x\Omega_z\omega_x$$
$$\Sigma M_z = I_z \dot{\omega}_z - I_x\Omega_y\omega_x + I_y\Omega_x\omega_y$$

$$\mathbf{\Omega} \neq \mathbf{\omega}$$

Gyroscopic Motion

The angular motion of a gyroscope is best described using the three Euler angles ϕ, θ, and ψ. The angular velocity components are called the precession $\dot{\phi}$, the nutation $\dot{\theta}$, and the spin $\dot{\psi}$.

If $\dot{\theta} = 0$ and $\dot{\phi}$ and $\dot{\psi}$ are constant, then the motion is referred to as steady precession.

It is the spin of a gyro rotor that is responsible for holding a rotor from falling downward, and instead causing it to precess about a vertical axis. This phenomenon is called the gyroscopic effect.

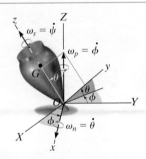

$$\Sigma M_x = -I\dot{\phi}^2 \sin\theta\cos\theta + I_z\dot{\phi}\sin\theta(\dot{\phi}\cos\theta + \dot{\psi})$$

$$\Sigma M_y = 0, \ \Sigma M_z = 0$$

Torque-Free Motion

A body that is only subjected to a gravitational force will have no moments on it about its mass center, and so the motion is described as torque-free motion. The angular momentum for the body about its mass center will remain constant. This causes the body to have both a spin and a precession. The motion depends upon the magnitude of the moment of inertia of a symmetric body about the spin axis, I_z, versus that about a perpendicular axis, I.

$$\theta = \text{constant}$$

$$\dot{\phi} = \frac{H_G}{I}$$

$$\dot{\psi} = \frac{I - I_z}{I I_z}H_G \cos\theta$$

21

Chapter 22

The analysis of vibrations plays an important role in the study of the behavior of structures subjected to earthquakes.

Vibrations

CHAPTER OBJECTIVES

■ To discuss undamped one-degree-of-freedom vibration of a rigid body using the equation of motion and energy methods.

■ To study the analysis of undamped forced vibration and viscous damped forced vibration.

*22.1 Undamped Free Vibration

A *vibration* is the oscillating motion of a body or system of connected bodies displaced from a position of equilibrium. In general, there are two types of vibration, free and forced. *Free vibration* occurs when the motion is maintained by gravitational or elastic restoring forces, such as the swinging motion of a pendulum or the vibration of an elastic rod. *Forced vibration* is caused by an external periodic or intermittent force applied to the system. Both of these types of vibration can either be damped or undamped. *Undamped* vibrations exclude frictional effects in the analysis. Since in reality both internal and external frictional forces are present, the motion of all vibrating bodies is actually *damped*.

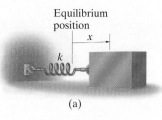

Equilibrium
position

(a)

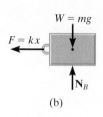

(b)

Fig. 22–1

The simplest type of vibrating motion is undamped free vibration, represented by the block and spring model shown in Fig. 22–1a. Vibrating motion occurs when the block is released from a displaced position x so that the spring pulls on the block. The block will attain a velocity such that it will proceed to move out of equilibrium when x = 0, and provided the supporting surface is smooth, the block will oscillate back and forth.

The time-dependent path of motion of the block can be determined by applying the equation of motion to the block when it is in the displaced position x. The free-body diagram is shown in Fig. 22–1b. The elastic restoring force $F = kx$ is always directed toward the equilibrium position, whereas the acceleration **a** is assumed to act in the direction of *positive displacement*. Since $a = d^2x/dt^2 = \ddot{x}$, we have

$$\xrightarrow{+} \Sigma F_x = ma_x; \qquad -kx = m\ddot{x}$$

Note that the acceleration is proportional to the block's displacement. Motion described in this manner is called *simple harmonic motion*. Rearranging the terms into a "standard form" gives

$$\ddot{x} + \omega_n^2 x = 0 \qquad (22\text{--}1)$$

The constant ω_n is called the *natural frequency*, and in this case

$$\boxed{\omega_n = \sqrt{\frac{k}{m}}} \qquad (22\text{--}2)$$

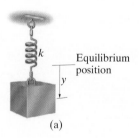

Equilibrium
position

(a)

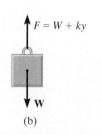

(b)

Fig. 22–2

Equation 22–1 can also be obtained by considering the block to be suspended so that the displacement y is measured from the block's *equilibrium position*, Fig. 22–2a. When the block is in equilibrium, the spring exerts an upward force of $F = W = mg$ on the block. Hence, when the block is displaced a distance y downward from this position, the magnitude of the spring force is $F = W + ky$, Fig. 22–2b. Applying the equation of motion gives

$$+\downarrow \Sigma F_y = ma_y; \qquad -W - ky + W = m\ddot{y}$$

or

$$\ddot{y} + \omega_n^2 y = 0$$

which is the same form as Eq. 22–1 and ω_n is defined by Eq. 22–2.

Equation 22–1 is a homogeneous, second-order, linear, differential equation with constant coefficients. It can be shown, using the methods of differential equations, that the general solution is

$$x = A \sin \omega_n t + B \cos \omega_n t \qquad (22\text{–}3)$$

Here A and B represent two constants of integration. The block's velocity and acceleration are determined by taking successive time derivatives, which yields

$$v = \dot{x} = A\omega_n \cos \omega_n t - B\omega_n \sin \omega_n t \qquad (22\text{–}4)$$

$$a = \ddot{x} = -A\omega_n^2 \sin \omega_n t - B\omega_n^2 \cos \omega_n t \qquad (22\text{–}5)$$

When Eqs. 22–3 and 22–5 are substituted into Eq. 22–1, the differential equation will be satisfied, showing that Eq. 22–3 is indeed the solution to Eq. 22–1.

The integration constants in Eq. 22–3 are generally determined from the initial conditions of the problem. For example, suppose that the block in Fig. 22–1a has been displaced a distance x_1 to the right from its equilibrium position and given an initial (positive) velocity $\mathbf{v}_1$ directed to the right. Substituting $x = x_1$ when $t = 0$ into Eq. 22–3 yields $B = x_1$. And since $v = v_1$ when $t = 0$, using Eq. 22–4 we obtain $A = v_1/\omega_n$. If these values are substituted into Eq. 22–3, the equation describing the motion becomes

$$x = \frac{v_1}{\omega_n} \sin \omega_n t + x_1 \cos \omega_n t \qquad (22\text{–}6)$$

Equation 22–3 may also be expressed in terms of simple sinusoidal motion. To show this, let

$$A = C \cos \phi \qquad (22\text{–}7)$$

and

$$B = C \sin \phi \qquad (22\text{–}8)$$

where C and ϕ are new constants to be determined in place of A and B. Substituting into Eq. 22–3 yields

$$x = C \cos \phi \sin \omega_n t + C \sin \phi \cos \omega_n t$$

And since $\sin(\theta + \phi) = \sin \theta \cos \phi + \cos \theta \sin \phi$, then

$$x = C \sin(\omega_n t + \phi) \qquad (22\text{–}9)$$

If this equation is plotted on an x versus $\omega_n t$ axis, the graph shown in Fig. 22–3 is obtained. The maximum displacement of the block from its equilibrium position is defined as the *amplitude* of vibration. From either the figure or Eq. 22–9 the amplitude is C. The angle ϕ is called the *phase angle* since it represents the amount by which the curve is displaced from the origin when $t = 0$. We can relate these two constants to A and B using Eqs. 22–7 and 22–8. Squaring and adding these two equations, the amplitude becomes

$$C = \sqrt{A^2 + B^2} \qquad (22\text{–}10)$$

If Eq. 22–8 is divided by Eq. 22–7, the phase angle is then

$$\phi = \tan^{-1}\frac{B}{A} \qquad (22\text{–}11)$$

Note that the sine curve, Eq. 22–9, completes one *cycle* in time $t = \tau$ (tau) when $\omega_n\tau = 2\pi$, or

$$\tau = \frac{2\pi}{\omega_n} \qquad (22\text{–}12)$$

This time interval is called a *period*, Fig. 22–3. Using Eq. 22–2, the period can also be represented as

$$\tau = 2\pi\sqrt{\frac{m}{k}} \qquad (22\text{–}13)$$

Finally, the *frequency f* is defined as the number of cycles completed per unit of time, which is the reciprocal of the period; that is,

$$f = \frac{1}{\tau} = \frac{\omega_n}{2\pi} \qquad (22\text{–}14)$$

or

$$f = \frac{1}{2\pi}\sqrt{\frac{k}{m}} \qquad (22\text{–}15)$$

The frequency is expressed in cycles/s. This ratio of units is called a *hertz* (Hz), where $1\ \text{Hz} = 1\ \text{cycle/s} = 2\pi\ \text{rad/s}$.

When a body or system of connected bodies is given an initial displacement from its equilibrium position and released, it will vibrate with the *natural frequency*, ω_n. Provided the system has a single degree of freedom, that is, it requires only one coordinate to specify completely the position of the system at any time, then the vibrating motion will have the same characteristics as the simple harmonic motion of the block and spring just presented. Consequently, the motion is described by a differential equation of the same "standard form" as Eq. 22–1, i.e.,

$$\ddot{x} + \omega_n^2 x = 0 \qquad (22\text{–}16)$$

Hence, if the natural frequency ω_n is known, the period of vibration τ, frequency f, and other vibrating characteristics can be established using Eqs. 22–3 through 22–15.

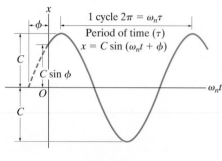

Fig. 22–3

Important Points

- Free vibration occurs when the motion is maintained by gravitational or elastic restoring forces.
- The amplitude is the maximum displacement of the body.
- The period is the time required to complete one cycle.
- The frequency is the number of cycles completed per unit of time, where 1 Hz = 1 cycle/s.
- Only one position coordinate is needed to describe the location of a one-degree-of-freedom system.

Procedure for Analysis

As in the case of the block and spring, the natural frequency ω_n of a body or system of connected bodies having a single degree of freedom can be determined using the following procedure:

Free-Body Diagram.

- Draw the free-body diagram of the body when the body is displaced a *small amount* from its equilibrium position.
- Locate the body with respect to its equilibrium position by using an appropriate *inertial coordinate q*. The acceleration of the body's mass center $\mathbf{a}_G$ or the body's angular acceleration $\boldsymbol{\alpha}$ should have an assumed sense of direction which is in the *positive direction* of the position coordinate.
- If the rotational equation of motion $\Sigma M_P = \Sigma(\mathcal{M}_k)_P$ is to be used, then it may be beneficial to also draw the kinetic diagram since it graphically accounts for the components $m(\mathbf{a}_G)_x$, $m(\mathbf{a}_G)_y$, and $I_G\boldsymbol{\alpha}$, and thereby makes it convenient for visualizing the terms needed in the moment sum $\Sigma(\mathcal{M}_k)_P$.

Equation of Motion.

- Apply the equation of motion to relate the elastic or gravitational *restoring* forces and couple moments acting on the body to the body's accelerated motion.

Kinematics.

- Using kinematics, express the body's accelerated motion in terms of the second time derivative of the position coordinate, $\ddot{q}$.
- Substitute the result into the equation of motion and determine ω_n by rearranging the terms so that the resulting equation is in the "standard form," $\ddot{q} + \omega_n^2 q = 0$.

22

EXAMPLE | 22.1

(a)

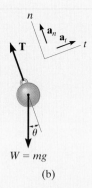

(b)

Fig. 22–4

Determine the period of oscillation for the simple pendulum shown in Fig. 22–4a. The bob has a mass m and is attached to a cord of length l. Neglect the size of the bob.

SOLUTION

Free-Body Diagram. Motion of the system will be related to the position coordinate $(q =) \theta$, Fig. 22–4b. When the bob is displaced by a small angle θ, the *restoring force* acting on the bob is created by the tangential component of its weight, $mg \sin \theta$. Furthermore, $\mathbf{a}_t$ acts in the direction of *increasing s* (or θ).

Equation of Motion. Applying the equation of motion in the *tangential direction*, since it involves the restoring force, yields

$$+\nearrow \Sigma F_t = ma_t; \qquad -mg \sin \theta = ma_t \qquad (1)$$

Kinematics. $a_t = d^2s/dt^2 = \ddot{s}$. Furthermore, s can be related to θ by the equation $s = l\theta$, so that $a_t = l\ddot{\theta}$. Hence, Eq. 1 reduces to

$$\ddot{\theta} + \frac{g}{l} \sin \theta = 0 \qquad (2)$$

The solution of this equation involves the use of an elliptic integral. For *small displacements*, however, $\sin \theta \approx \theta$, in which case

$$\ddot{\theta} + \frac{g}{l} \theta = 0 \qquad (3)$$

Comparing this equation with Eq. 22–16 ($\ddot{x} + \omega_n^2 x = 0$), it is seen that $\omega_n = \sqrt{g/l}$. From Eq. 22–12, the period of time required for the bob to make one complete swing is therefore

$$\tau = \frac{2\pi}{\omega_n} = 2\pi \sqrt{\frac{l}{g}} \qquad \textit{Ans.}$$

This interesting result, originally discovered by Galileo Galilei through experiment, indicates that the period depends only on the length of the cord and not on the mass of the pendulum bob or the angle θ.

NOTE: The solution of Eq. 3 is given by Eq. 22–3, where $\omega_n = \sqrt{g/l}$ and θ is substituted for x. Like the block and spring, the constants A and B in this problem can be determined if, for example, one knows the displacement and velocity of the bob at a given instant.

22

EXAMPLE | 22.2

The 10-kg rectangular plate shown in Fig. 22–5a is suspended at its center from a rod having a torsional stiffness $k = 1.5\ \text{N} \cdot \text{m/rad}$. Determine the natural period of vibration of the plate when it is given a small angular displacement θ in the plane of the plate.

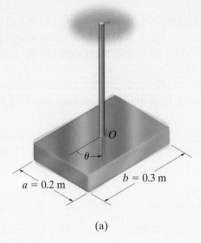

$a = 0.2\ \text{m}$ $b = 0.3\ \text{m}$

(a)

SOLUTION

Free-Body Diagram. Fig. 22–5b. Since the plate is displaced in its own plane, the torsional *restoring* moment created by the rod is $M = k\theta$. This moment acts in the direction opposite to the angular displacement θ. The angular acceleration $\ddot{\theta}$ acts in the direction of *positive* θ.

Equation of Motion.

$$\Sigma M_O = I_O \alpha; \qquad -k\theta = I_O \ddot{\theta}$$

or

$$\ddot{\theta} + \frac{k}{I_O}\theta = 0$$

Since this equation is in the "standard form," the natural frequency is $\omega_n = \sqrt{k/I_O}$.

From the table on the inside back cover, the moment of inertia of the plate about an axis coincident with the rod is $I_O = \frac{1}{12}m(a^2 + b^2)$. Hence,

$$I_O = \frac{1}{12}(10\ \text{kg})\left[(0.2\ \text{m})^2 + (0.3\ \text{m})^2\right] = 0.1083\ \text{kg} \cdot \text{m}^2$$

The natural period of vibration is therefore,

$$\tau = \frac{2\pi}{\omega_n} = 2\pi\sqrt{\frac{I_O}{k}} = 2\pi\sqrt{\frac{0.1083}{1.5}} = 1.69\ \text{s} \qquad Ans.$$

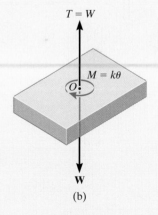

$T = W$

$M = k\theta$

O

W

(b)

Fig. 22–5

22

EXAMPLE 22.3

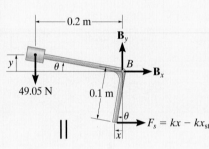

(a)

The bent rod shown in Fig. 22–6a has a negligible mass and supports a 5-kg collar at its end. If the rod is in the equilibrium position shown, determine the natural period of vibration for the system.

SOLUTION

Free-Body and Kinetic Diagrams. Fig. 22–6b. Here the rod is displaced by a small angle θ from the equilibrium position. Since the spring is subjected to an initial compression of x_{st} for equilibrium, then when the displacement $x > x_{st}$ the spring exerts a force of $F_s = kx - kx_{st}$ on the rod. To obtain the "standard form," Eq. 22–16, $5\mathbf{a}_y$ must act *upward*, which is in accordance with positive θ displacement.

Equation of Motion. Moments will be summed about point B to eliminate the unknown reaction at this point. Since θ is small,

$$\zeta + \Sigma M_B = \Sigma(\mathcal{M}_k)_B;$$

$$kx(0.1 \text{ m}) - kx_{st}(0.1 \text{ m}) + 49.05 \text{ N}(0.2 \text{ m}) = -(5 \text{ kg})a_y(0.2 \text{ m})$$

The second term on the left side, $-kx_{st}(0.1 \text{ m})$, represents the moment created by the spring force which is necessary to hold the collar in *equilibrium*, i.e., at $x = 0$. Since this moment is equal and opposite to the moment 49.05 N(0.2 m) created by the weight of the collar, these two terms cancel in the above equation, so that

$$kx(0.1) = -5a_y(0.2) \tag{1}$$

Kinematics. The deformation of the spring and the position of the collar can be related to the angle θ, Fig. 22–6c. Since θ is small, $x = (0.1 \text{ m})\theta$ and $y = (0.2 \text{ m})\theta$. Therefore, $a_y = \ddot{y} = 0.2\ddot{\theta}$. Substituting into Eq. 1 yields

$$400(0.1\theta)\,0.1 = -5(0.2\ddot{\theta})0.2$$

Rewriting this equation in the "standard form" gives

$$\ddot{\theta} + 20\theta = 0$$

Compared with $\ddot{x} + \omega_n^2 x = 0$ (Eq. 22–16), we have

$$\omega_n^2 = 20 \quad \omega_n = 4.47 \text{ rad/s}$$

The natural period of vibration is therefore

$$\tau = \frac{2\pi}{\omega_n} = \frac{2\pi}{4.47} = 1.40 \text{ s} \qquad \textit{Ans.}$$

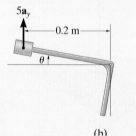

(b)

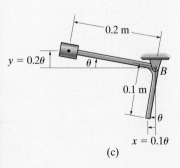

(c)

Fig. 22–6

22

EXAMPLE | 22.4

A 10-lb block is suspended from a cord that passes over a 15-lb disk, as shown in Fig. 22–7a. The spring has a stiffness $k = 200$ lb/ft. Determine the natural period of vibration for the system.

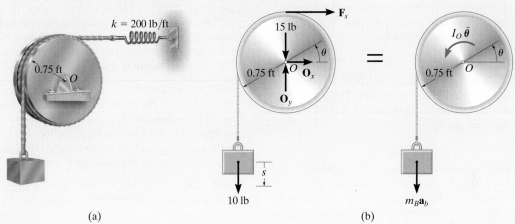

(a)   (b)

SOLUTION

Free-Body and Kinetic Diagrams. Fig. 22–7b. The *system* consists of the disk, which undergoes a rotation defined by the angle θ, and the block, which translates by an amount s. The vector $I_O\ddot{\theta}$ acts in the direction of *positive* θ, and consequently $m_B\mathbf{a}_b$ acts downward in the direction of *positive* s.

Equation of Motion. Summing moments about point O to eliminate the reactions $\mathbf{O}_x$ and $\mathbf{O}_y$, realizing that $I_O = \frac{1}{2}mr^2$, yields

$$\zeta + \Sigma M_O = \Sigma(\mathcal{M}_k)_O;$$

$$10 \text{ lb}(0.75 \text{ ft}) - F_s(0.75 \text{ ft})$$

$$= \frac{1}{2}\left(\frac{15 \text{ lb}}{32.2 \text{ ft/s}^2}\right)(0.75 \text{ ft})^2\,\ddot{\theta} + \left(\frac{10 \text{ lb}}{32.2 \text{ ft/s}^2}\right)a_b(0.75 \text{ ft}) \quad (1)$$

Kinematics. As shown on the kinematic diagram in Fig. 22–7c, a small positive displacement θ of the disk causes the block to lower by an amount $s = 0.75\theta$; hence, $a_b = \ddot{s} = 0.75\ddot{\theta}$. When $\theta = 0°$, the spring force required for *equilibrium* of the disk is 10 lb, acting to the right. For position θ, the spring force is $F_s = (200 \text{ lb/ft})(0.75\theta \text{ ft}) + 10 \text{ lb}$. Substituting these results into Eq. 1 and simplifying yields

$$\ddot{\theta} + 368\theta = 0$$

Hence,

$$\omega_n^2 = 368 \qquad \omega_n = 19.18 \text{ rad/s}$$

Therefore, the natural period of vibration is

$$\tau = \frac{2\pi}{\omega_n} = \frac{2\pi}{19.18} = 0.328 \text{ s} \qquad \textit{Ans.}$$

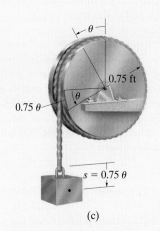

(c)

Fig. 22–7

22

PROBLEMS

22–1. A spring has a stiffness of 600 N/m. If a 4-kg block is attached to the spring, pushed 50 mm above its equilibrium position, and released from rest, determine the equation which describes the block's motion. Assume that positive displacement is measured downward.

22–2. When a 2-kg block is suspended from a spring, the spring is stretched a distance of 40 mm. Determine the frequency and the period of vibration for a 0.5-kg block attached to the same spring.

22–3. A spring is stretched 200 mm by a 15-kg block. If the block is displaced 100 mm downward from its equilibrium position and given a downward velocity of 0.75 m/s, determine the equation which describes the motion. What is the phase angle? Assume that positive displacement is downward.

*22–4. When a 20-lb weight is suspended from a spring, the spring is stretched a distance of 4 in. Determine the natural frequency and the period of vibration for a 10-lb weight attached to the same spring.

22–5. When a 3-kg block is suspended from a spring, the spring is stretched a distance of 60 mm. Determine the natural frequency and the period of vibration for a 0.2-kg block attached to the same spring.

22–6. An 8-kg block is suspended from a spring having a stiffness $k = 80$ N/m. If the block is given an upward velocity of 0.4 m/s when it is 90 mm above its equilibrium position, determine the equation which describes the motion and the maximum upward displacement of the block measured from the equilibrium position. Assume that positive displacement is measured downward.

22–7. A 2-lb weight is suspended from a spring having a stiffness $k = 2$ lb/in. If the weight is pushed 1 in. upward from its equilibrium position and then released from rest, determine the equation which describes the motion. What is the amplitude and the natural frequency of the vibration?

*22–8. A 6-lb weight is suspended from a spring having a stiffness $k = 3$ lb/in. If the weight is given an upward velocity of 20 ft/s when it is 2 in. above its equilibrium position, determine the equation which describes the motion and the maximum upward displacement of the weight, measured from the equilibrium position. Assume positive displacement is downward.

22–9. A 3-kg block is suspended from a spring having a stiffness of $k = 200$ N/m. If the block is pushed 50 mm upward from its equilibrium position and then released from rest, determine the equation that describes the motion. What are the amplitude and the natural frequency of the vibration? Assume that positive displacement is downward.

22–10. Determine the frequency of vibration for the block. The springs are originally compressed Δ.

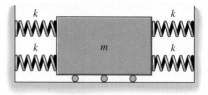

Prob. 22–10

22–11. The semicircular disk weighs 20 lb. Determine the natural period of vibration if it is displaced a small amount and released.

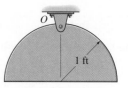

Prob. 22–11

*22–12. The uniform beam is supported at its ends by two springs A and B, each having the same stiffness k. When nothing is supported on the beam, it has a period of vertical vibration of 0.83 s. If a 50-kg mass is placed at its center, the period of vertical vibration is 1.52 s. Compute the stiffness of each spring and the mass of the beam.

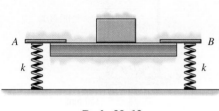

Prob. 22–12

22–13. The body of arbitrary shape has a mass m, mass center at G, and a radius of gyration about G of k_G. If it is displaced a slight amount θ from its equilibrium position and released, determine the natural period of vibration.

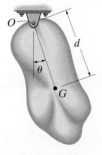

Prob. 22–13

22–14. The connecting rod is supported by a knife edge at A and the period of vibration is measured as $\tau_A = 3.38$ s. It is then removed and rotated $180°$ so that it is supported by the knife edge at B. In this case the period of vibration is measured as $\tau_B = 3.96$ s. Determine the location d of the center of gravity G, and compute the radius of gyration k_G.

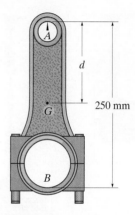

Prob. 22–14

22–15. The thin hoop of mass m is supported by a knife-edge. Determine the natural period of vibration for small amplitudes of swing.

Prob. 22–15

22

*22–16. A block of mass m is suspended from two springs having a stiffness of k_1 and k_2, arranged a) parallel to each other, and b) as a series. Determine the equivalent stiffness of a single spring with the same oscillation characteristics and the period of oscillation for each case.

22–17. The 15-kg block is suspended from two springs having a different stiffness and arranged a) parallel to each other, and b) as a series. If the natural periods of oscillation of the parallel system and series system are observed to be 0.5 s and 1.5 s, respectively, determine the spring stiffnesses k_1 and k_2.

22–19. The 50-kg block is suspended from the 10-kg pulley that has a radius of gyration about its center of mass of 125 mm. If the block is given a small vertical displacement and then released, determine the natural frequency of oscillation.

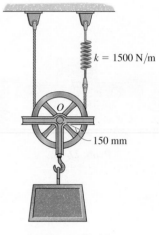

Prob. 22–19

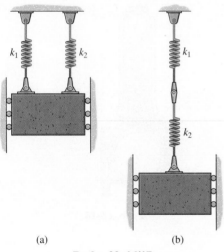

(a)   (b)

Probs. 22–16/17

22–18. The pointer on a metronome supports a 0.4-lb slider A, which is positioned at a fixed distance from the pivot O of the pointer. When the pointer is displaced, a torsional spring at O exerts a restoring torque on the pointer having a magnitude $M = (1.2\theta)$ lb · ft, where θ represents the angle of displacement from the vertical, measured in radians. Determine the natural period of vibration when the pointer is displaced a small amount θ and released. Neglect the mass of the pointer.

*22–20. A uniform board is supported on two wheels which rotate in opposite directions at a constant angular speed. If the coefficient of kinetic friction between the wheels and board is μ, determine the frequency of vibration of the board if it is displaced slightly, a distance x from the midpoint between the wheels, and released.

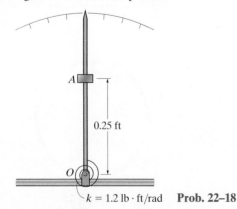

$k = 1.2$ lb · ft/rad   **Prob. 22–18**

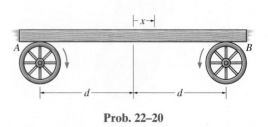

Prob. 22–20

22–21. If the 20-kg block is given a downward velocity of 6 m/s at its equilibrium position, determine the equation that describes the amplitude of the block's oscillation.

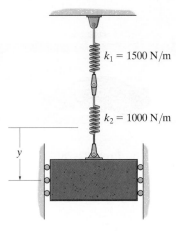

Prob. 22–21

22–23. The 50-lb spool is attached to two springs. If the spool is displaced a small amount and released, determine the natural period of vibration. The radius of gyration of the spool is $k_G = 1.5$ ft. The spool rolls without slipping.

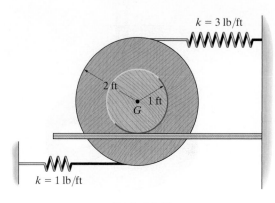

$k = 3$ lb/ft

$k = 1$ lb/ft

Prob. 22–23

22–22. The bar has a length l and mass m. It is supported at its ends by rollers of negligible mass. If it is given a small displacement and released, determine the natural frequency of vibration.

*22–24.** The cart has a mass of m and is attached to two springs, each having a stiffness of $k_1 = k_2 = k$, unstretched length of l_0, and a stretched length of l when the cart is in the equilibrium position. If the cart is displaced a distance of $x = x_0$ such that both springs remain in tension $(x_0 < l - l_0)$, determine the natural frequency of oscillation.

22–25. The cart has a mass of m and is attached to two springs, each having a stiffness of k_1 and k_2, respectively. If both springs are unstretched when the cart is in the equilibrium position shown, determine the natural frequency of oscillation.

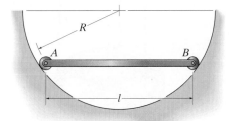

Prob. 22–22

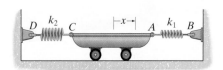

Probs. 22–24/25

22–26. A flywheel of mass m, which has a radius of gyration about its center of mass of k_O, is suspended from a circular shaft that has a torsional resistance of $M = C\theta$. If the flywheel is given a small angular displacement of θ and released, determine the natural period of oscillation.

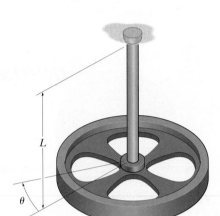

Prob. 22–26

22–27. If a block D of negligible size and of mass m is attached at C, and the bell crank of mass M is given a small angular displacement of θ, the natural period of oscillation is τ_1. When D is removed, the natural period of oscillation is τ_2. Determine the bell crank's radius of gyration about its center of mass, pin B, and the spring's stiffness k. The spring is unstretched at $\theta = 0°$, and the motion occurs in the *horizontal plane*.

Prob. 22–27

22–28. The platform AB when empty has a mass of 400 kg, center of mass at G_1, and natural period of oscillation $\tau_1 = 2.38$ s. If a car, having a mass of 1.2 Mg and center of mass at G_2, is placed on the platform, the natural period of oscillation becomes $\tau_2 = 3.16$ s. Determine the moment of inertia of the car about an axis passing through G_2.

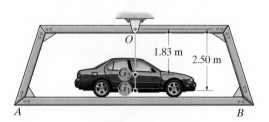

Prob. 22–28

22–29. A wheel of mass m is suspended from three equal-length cords. When it is given a small angular displacement of θ about the z axis and released, it is observed that the period of oscillation is τ. Determine the radius of gyration of the wheel about the z axis.

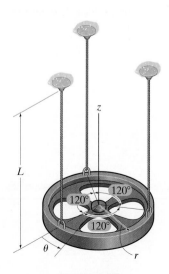

Prob. 22–29

*22.2   Energy Methods

The simple harmonic motion of a body, discussed in the previous section, is due only to gravitational and elastic restoring forces acting on the body. Since these forces are *conservative*, it is also possible to use the conservation of energy equation to obtain the body's natural frequency or period of vibration. To show how to do this, consider again the block and spring model in Fig. 22–8. When the block is displaced x from the equilibrium position, the kinetic energy is $T = \frac{1}{2}mv^2 = \frac{1}{2}m\dot{x}^2$ and the potential energy is $V = \frac{1}{2}kx^2$. Since energy is conserved, it is necessary that

$$T + V = \text{constant}$$

$$\tfrac{1}{2}m\dot{x}^2 + \tfrac{1}{2}kx^2 = \text{constant} \qquad (22\text{--}17)$$

The differential equation describing the *accelerated motion* of the block can be obtained by *differentiating* this equation with respect to time; i.e.,

$$m\dot{x}\ddot{x} + kx\dot{x} = 0$$

$$\dot{x}(m\ddot{x} + kx) = 0$$

Since the velocity $\dot{x}$ is not *always* zero in a vibrating system,

$$\ddot{x} + \omega_n^2 x = 0 \qquad \omega_n = \sqrt{k/m}$$

which is the same as Eq. 22–1.

If the conservation of energy equation is written for a *system of connected bodies*, the natural frequency or the equation of motion can also be determined by time differentiation. It is *not necessary* to dismember the system to account for the internal forces because they do no work.

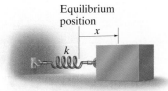

Equilibrium position

Fig. 22–8

22

The suspension of a railroad car consists of a set of springs which are mounted between the frame of the car and the wheel truck. This will give the car a natural frequency of vibration which can be determined.

Procedure for Analysis

The natural frequency ω_n of a body or system of connected bodies can be determined by applying the conservation of energy equation using the following procedure.

Energy Equation.

- Draw the body when it is displaced by a *small amount* from its equilibrium position and define the location of the body from its equilibrium position by an appropriate position coordinate q.

- Formulate the conservation of energy for the body, $T + V =$ constant, in terms of the position coordinate.

- In general, the kinetic energy must account for both the body's translational and rotational motion, $T = \frac{1}{2}mv_G^2 + \frac{1}{2}I_G\omega^2$, Eq. 18–2.

- The potential energy is the sum of the gravitational and elastic potential energies of the body, $V = V_g + V_e$, Eq. 18–17. In particular, V_g should be measured from a datum for which $q = 0$ (equilibrium position).

Time Derivative.

- Take the time derivative of the energy equation using the chain rule of calculus and factor out the common terms. The resulting differential equation represents the equation of motion for the system. The natural frequency of ω_n is obtained after rearranging the terms in the "standard form," $\ddot{q} + \omega_n^2 q = 0$.

EXAMPLE | 22.5

The thin hoop shown in Fig. 22–9a is supported by the peg at O. Determine the natural period of oscillation for small amplitudes of swing. The hoop has a mass m.

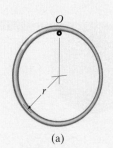

(a)

SOLUTION

Energy Equation. A diagram of the hoop when it is displaced a small amount $(q =) \theta$ from the equilibrium position is shown in Fig. 22–9b. Using the table on the inside back cover and the parallel-axis theorem to determine I_O, the kinetic energy is

$$T = \tfrac{1}{2}I_O \omega_n^2 = \tfrac{1}{2}[mr^2 + mr^2]\dot{\theta}^2 = mr^2\dot{\theta}^2$$

If a horizontal datum is placed through point O, then in the displaced position, the potential energy is

$$V = -mg(r \cos \theta)$$

The total energy in the system is

$$T + V = mr^2\dot{\theta}^2 - mgr \cos \theta$$

Time Derivative.

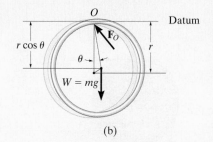

(b)

Fig. 22–9

$$mr^2(2\dot{\theta})\ddot{\theta} + mgr(\sin \theta)\dot{\theta} = 0$$

$$mr\dot{\theta}(2r\ddot{\theta} + g \sin \theta) = 0$$

Since $\dot{\theta}$ is not always equal to zero, from the terms in parentheses,

$$\ddot{\theta} + \frac{g}{2r}\sin \theta = 0$$

For small angle θ, $\sin \theta \approx \theta$.

$$\ddot{\theta} + \frac{g}{2r}\theta = 0$$

$$\omega_n = \sqrt{\frac{g}{2r}}$$

so that

$$\tau = \frac{2\pi}{\omega_n} = 2\pi\sqrt{\frac{2r}{g}} \qquad \qquad Ans.$$

22

EXAMPLE 22.6

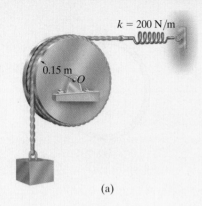

$k = 200$ N/m

0.15 m

O

(a)

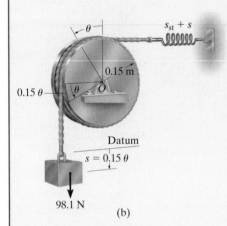

θ

$s_{st} + s$

0.15 m

$0.15\,\theta$ θ O

Datum

$s = 0.15\,\theta$

98.1 N

(b)

Fig. 22–10

A 10-kg block is suspended from a cord wrapped around a 5-kg disk, as shown in Fig. 22–10a. If the spring has a stiffness $k = 200$ N/m, determine the natural period of vibration for the system.

SOLUTION

Energy Equation.   A diagram of the block and disk when they are displaced by respective amounts s and θ from the equilibrium position is shown in Fig. 22–10b. Since $s = (0.15\text{ m})\theta$, then $v_b \approx \dot{s} = (0.15\text{ m})\dot{\theta}$. Thus, the kinetic energy of the system is

$$T = \tfrac{1}{2}m_b v_b^2 + \tfrac{1}{2}I_O \omega_d^2$$
$$= \tfrac{1}{2}(10\text{ kg})[(0.15\text{ m})\dot{\theta}]^2 + \tfrac{1}{2}\Big[\tfrac{1}{2}(5\text{ kg})(0.15\text{ m})^2\Big](\dot{\theta})^2$$
$$= 0.1406(\dot{\theta})^2$$

Establishing the datum at the equilibrium position of the block and realizing that the spring stretches s_{st} for equilibrium, the potential energy is

$$V = \tfrac{1}{2}k(s_{st} + s)^2 - Ws$$
$$= \tfrac{1}{2}(200\text{ N/m})[s_{st} + (0.15\text{ m})\theta]^2 - 98.1\text{ N}[(0.15\text{ m})\theta]$$

The total energy for the system is therefore,

$$T + V = 0.1406(\dot{\theta})^2 + 100(s_{st} + 0.15\theta)^2 - 14.715\theta$$

Time Derivative.

$$0.28125(\dot{\theta})\ddot{\theta} + 200(s_{st} + 0.15\theta)0.15\dot{\theta} - 14.72\dot{\theta} = 0$$

Since $s_{st} = 98.1/200 = 0.4905$ m, the above equation reduces to the "standard form"

$$\ddot{\theta} + 16\theta = 0$$

so that

$$\omega_n = \sqrt{16} = 4\text{ rad/s}$$

Thus,

$$\tau = \frac{2\pi}{\omega_n} = \frac{2\pi}{4} = 1.57\text{ s} \qquad\qquad Ans.$$

22

PROBLEMS

22–30. Determine the differential equation of motion of the 3-kg block when it is displaced slightly and released. The surface is smooth and the springs are originally unstretched.

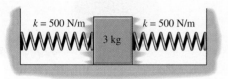

Prob. 22–30

22–31. Determine the natural period of vibration of the pendulum. Consider the two rods to be slender, each having a weight of 8 lb/ft.

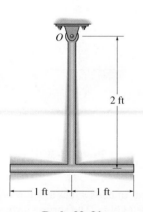

Prob. 22–31

*22–32. The uniform rod of mass m is supported by a pin at A and a spring at B. If the end B is given a small downward displacement and released, determine the natural period of vibration.

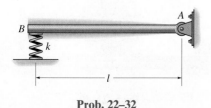

Prob. 22–32

22–33. The 7-kg disk is pin connected at its midpoint. Determine the natural period of vibration of the disk if the springs have sufficient tension in them to prevent the cord from slipping on the disk as it oscillates. *Hint:* Assume that the initial stretch in each spring is δ_O. This term will cancel out after taking the time derivative of the energy equation.

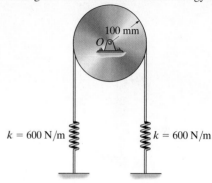

Prob. 22–33

22–34. The machine has a mass m and is uniformly supported by *four* springs, each having a stiffness k. Determine the natural period of vertical vibration.

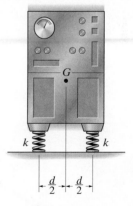

Prob. 22–34

22–35. Determine the natural period of vibration of the 3-kg sphere. Neglect the mass of the rod and the size of the sphere.

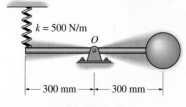

Prob. 22–35

*22–36. The slender rod has a mass m and is pinned at its end O. When it is vertical, the springs are unstretched. Determine the natural period of vibration.

Prob. 22–36

22–37. Determine the natural frequency of vibration of the 20-lb disk. Assume the disk does not slip on the inclined surface.

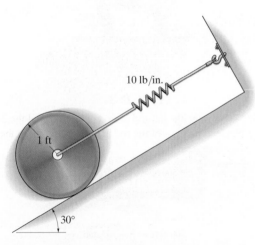

Prob. 22–37

22–38. If the disk has a mass of 8 kg, determine the natural frequency of vibration. The springs are originally unstretched.

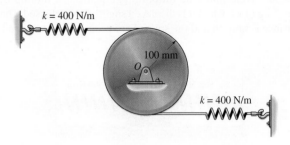

Prob. 22–38

22–39. The semicircular disk has a mass m and radius r, and it rolls without slipping in the semicircular trough. Determine the natural period of vibration of the disk if it is displaced slightly and released. *Hint:* $I_O = \frac{1}{2} mr^2$.

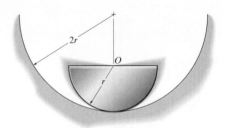

Prob. 22–39

*22–40. The gear of mass m has a radius of gyration about its center of mass O of k_O. The springs have stiffnesses of k_1 and k_2, respectively, and both springs are unstretched when the gear is in an equilibrium position. If the gear is given a small angular displacement of θ and released, determine its natural period of oscillation.

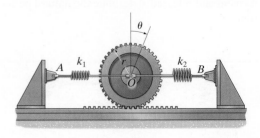

Prob. 22–40

*22.3   Undamped Forced Vibration

Undamped forced vibration is considered to be one of the most important types of vibrating motion in engineering. Its principles can be used to describe the motion of many types of machines and structures.

Periodic Force.   The block and spring shown in Fig. 22–11a provide a convenient model which represents the vibrational characteristics of a system subjected to a periodic force $F = F_0 \sin \omega_0 t$. This force has an amplitude of F_0 and a *forcing frequency* ω_0. The free-body diagram for the block when it is displaced a distance x is shown in Fig. 22–11b. Applying the equation of motion, we have

$$\xrightarrow{+} \Sigma F_x = ma_x; \qquad F_0 \sin \omega_0 t - kx = m\ddot{x}$$

or

$$\ddot{x} + \frac{k}{m}x = \frac{F_0}{m}\sin \omega_0 t \qquad (22\text{–}18)$$

This equation is a nonhomogeneous second-order differential equation. The general solution consists of a complementary solution, x_c, *plus* a particular solution, x_p.

The *complementary solution* is determined by setting the term on the right side of Eq. 22–18 equal to zero and solving the resulting homogeneous equation. The solution is defined by Eq. 22–9, i.e.,

$$x_c = C \sin(\omega_n t + \phi) \qquad (22\text{–}19)$$

where ω_n is the natural frequency, $\omega_n = \sqrt{k/m}$, Eq. 22–2.

Since the motion is periodic, the *particular solution* of Eq. 22–18 can be determined by assuming a solution of the form

$$x_p = X \sin \omega_0 t \qquad (22\text{–}20)$$

where X is a constant. Taking the second time derivative and substituting into Eq. 22–18 yields

$$-X\omega_0^2 \sin \omega_0 t + \frac{k}{m}(X \sin \omega_0 t) = \frac{F_0}{m}\sin \omega_0 t$$

Factoring out $\sin \omega_0 t$ and solving for X gives

$$X = \frac{F_0/m}{(k/m) - \omega_0^2} = \frac{F_0/k}{1 - (\omega_0/\omega_n)^2} \qquad (22\text{–}21)$$

Substituting into Eq. 22–20, we obtain the particular solution

$$x_p = \frac{F_0/k}{1 - (\omega_0/\omega_n)^2}\sin \omega_0 t \qquad (22\text{–}22)$$

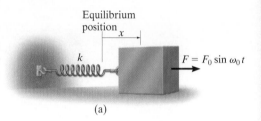

Equilibrium position

(a)

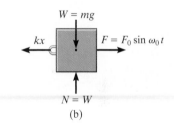

$W = mg$

(b)

Fig. 22–11

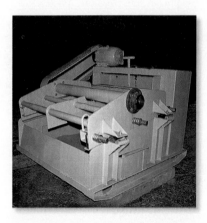

Shaker tables provide forced vibration and are used to separate out granular materials.

The *general solution* is therefore the sum of two sine functions having different frequencies.

$$x = x_c + x_p = C \sin(\omega_n t + \phi) + \frac{F_0/k}{1 - (\omega_0/\omega_n)^2} \sin \omega_0 t \quad (22\text{–}23)$$

The *complementary solution* x_c defines the *free vibration*, which depends on the natural frequency $\omega_n = \sqrt{k/m}$ and the constants C and ϕ. The *particular solution* x_p describes the *forced vibration* of the block caused by the applied force $F = F_0 \sin \omega_0 t$. Since all vibrating systems are subject to *friction*, the free vibration, x_c, will in time dampen out. For this reason the free vibration is referred to as *transient*, and the forced vibration is called *steady-state*, since it is the only vibration that remains.

From Eq. 22–21 it is seen that the *amplitude* of forced or steady-state vibration depends on the *frequency ratio* ω_0/ω_n. If the *magnification factor* MF is defined as the ratio of the amplitude of steady-state vibration, X, to the static deflection, F_0/k, which would be produced by the amplitude of the periodic force F_0, then, from Eq. 22–21,

The soil compactor operates by forced vibration developed by an internal motor. It is important that the forcing frequency not be close to the natural frequency of vibration of the compactor, which can be determined when the motor is turned off; otherwise resonance will occur and the machine will become uncontrollable.

$$\text{MF} = \frac{X}{F_0/k} = \frac{1}{1 - (\omega_0/\omega_n)^2} \qquad (22\text{--}24)$$

This equation is graphed in Fig. 22–12. Note that if the force or displacement is applied with a frequency close to the natural frequency of the system, i.e., $\omega_0/\omega_n \approx 1$, the amplitude of vibration of the block becomes extremely large. This occurs because the force **F** is applied to the block so that it always follows the motion of the block. This condition is called *resonance*, and in practice, resonating vibrations can cause tremendous stress and rapid failure of parts.*

Periodic Support Displacement.

Forced vibrations can also arise from the periodic excitation of the support of a system. The model shown in Fig. 22–13a represents the periodic vibration of a block which is caused by harmonic movement $\delta = \delta_0 \sin \omega_0 t$ of the support. The free-body diagram for the block in this case is shown in Fig. 22–13b. The displacement δ of the support is measured from the point of zero displacement, i.e., when the radial line OA coincides with OB. Therefore, general deformation of the spring is $(x - \delta_0 \sin \omega_0 t)$. Applying the equation of motion yields

$$\xrightarrow{+} F_x = ma_x; \qquad -k(x - \delta_0 \sin \omega_0 t) = m\ddot{x}$$

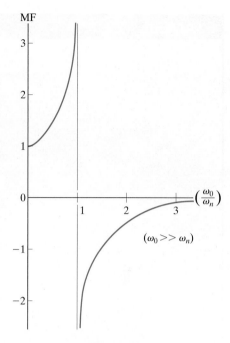

Fig. 22–12

or

$$\ddot{x} + \frac{k}{m}x = \frac{k\delta_0}{m}\sin \omega_0 t \qquad (22\text{--}25)$$

By comparison, this equation is identical to the form of Eq. 22–18, *provided F_0 is replaced by $k\delta_0$*. If this substitution is made into the solutions defined by Eqs. 22–21 to 22–23, the results are appropriate for describing the motion of the block when subjected to the support displacement $\delta = \delta_0 \sin \omega_0 t$.

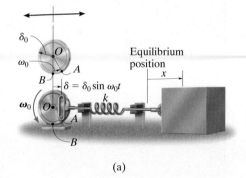

(a)

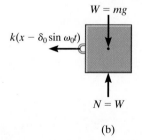

(b)

*A swing has a natural period of vibration, as determined in Example 22.1. If someone pushes on the swing only when it reaches its highest point, neglecting drag or wind resistance, resonance will occur since the natural and forcing frequencies are the same.

Fig. 22–13

EXAMPLE | 22.7

The instrument shown in Fig. 22–14 is rigidly attached to a platform P, which in turn is supported by *four* springs, each having a stiffness $k = 800$ N/m. If the floor is subjected to a vertical displacement $\delta = 10 \sin(8t)$ mm, where t is in seconds, determine the amplitude of steady-state vibration. What is the frequency of the floor vibration required to cause resonance? The instrument and platform have a total mass of 20 kg.

Fig. 22–14

SOLUTION
The natural frequency is

$$\omega_n = \sqrt{\frac{k}{m}} = \sqrt{\frac{4(800 \text{ N/m})}{20 \text{ kg}}} = 12.65 \text{ rad/s}$$

The amplitude of steady-state vibration is found using Eq. 22–21, with $k\delta_0$ replacing F_0.

$$X = \frac{\delta_0}{1 - (\omega_0/\omega_n)^2} = \frac{10}{1 - [(8 \text{ rad/s})/(12.65 \text{ rad/s})]^2} = 16.7 \text{ mm} \quad \textit{Ans.}$$

Resonance will occur when the amplitude of vibration X caused by the floor displacement approaches infinity. This requires

$$\omega_0 = \omega_n = 12.6 \text{ rad/s} \qquad \textit{Ans.}$$

*22.4   Viscous Damped Free Vibration

The vibration analysis considered thus far has not included the effects of friction or damping in the system, and as a result, the solutions obtained are only in close agreement with the actual motion. Since all vibrations die out in time, the presence of damping forces should be included in the analysis.

In many cases damping is attributed to the resistance created by the substance, such as water, oil, or air, in which the system vibrates. Provided the body moves slowly through this substance, the resistance to motion is directly proportional to the body's speed. The type of force developed under these conditions is called a *viscous damping force*. The magnitude of this force is expressed by an equation of the form

$$F = c\dot{x} \qquad (22\text{–}26)$$

where the constant c is called the *coefficient of viscous damping* and has units of $N \cdot s/m$ or $lb \cdot s/ft$.

The vibrating motion of a body or system having viscous damping can be characterized by the block and spring shown in Fig. 22–15a. The effect of damping is provided by the *dashpot* connected to the block on the right side. Damping occurs when the piston P moves to the right or left within the enclosed cylinder. The cylinder contains a fluid, and the motion of the piston is retarded since the fluid must flow around or through a small hole in the piston. The dashpot is assumed to have a coefficient of viscous damping c.

If the block is displaced a distance x from its equilibrium position, the resulting free-body diagram is shown in Fig. 22–15b. Both the spring and damping force oppose the forward motion of the block, so that applying the equation of motion yields

$$\xrightarrow{+} \Sigma F_x = ma_x; \qquad -kx - c\dot{x} = m\ddot{x}$$

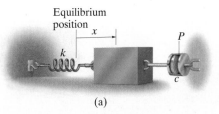

(a)

or

$$m\ddot{x} + c\dot{x} + kx = 0 \qquad (22\text{–}27)$$

This linear, second-order, homogeneous, differential equation has a solution of the form

$$x = e^{\lambda t}$$

where e is the base of the natural logarithm and λ (lambda) is a constant. The value of λ can be obtained by substituting this solution and its time derivatives into Eq. 22–27, which yields

$$m\lambda^2 e^{\lambda t} + c\lambda e^{\lambda t} + k e^{\lambda t} = 0$$

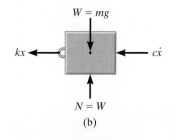

(b)

or

$$e^{\lambda t}(m\lambda^2 + c\lambda + k) = 0$$

Fig. 22–15

Since $e^{\lambda t}$ can never be zero, a solution is possible provided

$$m\lambda^2 + c\lambda + k = 0$$

Hence, by the quadratic formula, the two values of λ are

$$\lambda_1 = -\frac{c}{2m} + \sqrt{\left(\frac{c}{2m}\right)^2 - \frac{k}{m}}$$

$$\lambda_2 = -\frac{c}{2m} - \sqrt{\left(\frac{c}{2m}\right)^2 - \frac{k}{m}}$$

(22–28)

The general solution of Eq. 22–27 is therefore a combination of exponentials which involves both of these roots. There are three possible combinations of λ_1 and λ_2 which must be considered. Before discussing these combinations, however, we will first define the critical damping coefficient c_c as the value of c which makes the radical in Eqs. 22–28 equal to zero; i.e.,

$$\left(\frac{c_c}{2m}\right)^2 - \frac{k}{m} = 0$$

or

$$c_c = 2m\sqrt{\frac{k}{m}} = 2m\omega_n$$

(22–29)

Overdamped System.   When $c > c_c$, the roots λ_1 and λ_2 are both real. The general solution of Eq. 22–27 can then be written as

$$x = Ae^{\lambda_1 t} + Be^{\lambda_2 t}$$

(22–30)

Motion corresponding to this solution is *nonvibrating*. The effect of damping is so strong that when the block is displaced and released, it simply creeps back to its original position without oscillating. The system is said to be *overdamped*.

Critically Damped System.   If $c = c_c$, then $\lambda_1 = \lambda_2 = -c_c/2m = -\omega_n$. This situation is known as *critical damping*, since it represents a condition where c has the smallest value necessary to cause the system to be nonvibrating. Using the methods of differential equations, it can be shown that the solution to Eq. 22–27 for critical damping is

$$x = (A + Bt)e^{-\omega_n t}$$

(22–31)

Underdamped System.

Most often $c < c_c$, in which case the system is referred to as *underdamped*. In this case the roots λ_1 and λ_2 are complex numbers, and it can be shown that the general solution of Eq. 22–27 can be written as

$$x = D[e^{-(c/2m)t} \sin(\omega_d t + \phi)] \tag{22–32}$$

where D and ϕ are constants generally determined from the initial conditions of the problem. The constant ω_d is called the *damped natural frequency* of the system. It has a value of

$$\omega_d = \sqrt{\frac{k}{m} - \left(\frac{c}{2m}\right)^2} = \omega_n \sqrt{1 - \left(\frac{c}{c_c}\right)^2} \tag{22–33}$$

where the ratio c/c_c is called the *damping factor*.

The graph of Eq. 22–32 is shown in Fig. 22–16. The initial limit of motion, D, diminishes with each cycle of vibration, since motion is confined within the bounds of the exponential curve. Using the damped natural frequency ω_d, the period of damped vibration can be written as

$$\tau_d = \frac{2\pi}{\omega_d} \tag{22–34}$$

Since $\omega_d < \omega_n$, Eq. 22–33, the period of damped vibration, τ_d, will be greater than that of free vibration, $\tau = 2\pi/\omega_n$.

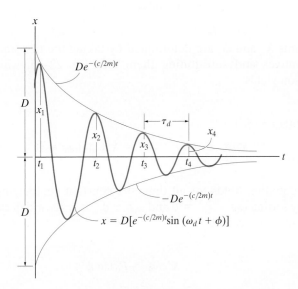

Fig. 22–16

*22.5   Viscous Damped Forced Vibration

The most general case of single-degree-of-freedom vibrating motion occurs when the system includes the effects of forced motion and induced damping. The analysis of this particular type of vibration is of practical value when applied to systems having significant damping characteristics.

If a dashpot is attached to the block and spring shown in Fig. 22–11a, the differential equation which describes the motion becomes

$$m\ddot{x} + c\dot{x} + kx = F_0 \sin \omega_0 t \tag{22–35}$$

A similar equation can be written for a block and spring having a periodic support displacement, Fig. 22–13a, which includes the effects of damping. In that case, however, F_0 is replaced by $k\delta_0$. Since Eq. 22–35 is nonhomogeneous, the general solution is the sum of a complementary solution, x_c, and a particular solution, x_p. The complementary solution is determined by setting the right side of Eq. 22–35 equal to zero and solving the homogeneous equation, which is equivalent to Eq. 22–27. The solution is therefore given by Eq. 22–30, 22–31, or 22–32, depending on the values of λ_1 and λ_2. Because all systems are subjected to friction, then this solution will dampen out with time. Only the particular solution, which describes the *steady-state vibration* of the system, will remain. Since the applied forcing function is harmonic, the steady-state motion will also be harmonic. Consequently, the particular solution will be of the form

$$X_P = X' \sin(\omega_0 t - \phi') \tag{22–36}$$

The constants X' and ϕ' are determined by taking the first and second time derivatives and substituting them into Eq. 22–35, which after simplification yields

$$-X'm\omega_0^2 \sin(\omega_0 t - \phi') +$$
$$X'c\omega_0 \cos(\omega_0 t - \phi') + X'k \sin(\omega_0 t - \phi') = F_0 \sin \omega_0 t$$

Since this equation holds for all time, the constant coefficients can be obtained by setting $\omega_0 t - \phi' = 0$ and $\omega_0 t - \phi' = \pi/2$, which causes the above equation to become

$$X'c\omega_0 = F_0 \sin \phi'$$
$$-X'm\omega_0^2 + X'k = F_0 \cos \phi'$$

22

The amplitude is obtained by squaring these equations, adding the results, and using the identity $\sin^2\phi' + \cos^2\phi' = 1$, which gives

$$X' = \frac{F_0}{\sqrt{(k - m\omega_0^2)^2 + c^2\omega_0^2}} \qquad (22\text{–}37)$$

Dividing the first equation by the second gives

$$\phi' = \tan^{-1}\left[\frac{c\omega_0}{k - m\omega_0^2}\right] \qquad (22\text{–}38)$$

Since $\omega_n = \sqrt{k/m}$ and $c_c = 2m\omega_n$, then the above equations can also be written as

$$X' = \frac{F_0/k}{\sqrt{[1 - (\omega_0/\omega_n)^2]^2 + [2(c/c_c)(\omega_0/\omega_n)]^2}}$$

$$\phi' = \tan^{-1}\left[\frac{2(c/c_c)(\omega_0/\omega_n)}{1 - (\omega_0/\omega_n)^2}\right] \qquad (22\text{–}39)$$

The angle ϕ' represents the phase difference between the applied force and the resulting steady-state vibration of the damped system.

The *magnification factor* MF has been defined in Sec. 22.3 as the ratio of the amplitude of deflection caused by the forced vibration to the deflection caused by a static force F_0. Thus,

$$\text{MF} = \frac{X'}{F_0/k} = \frac{1}{\sqrt{[1 - (\omega_0/\omega_n)^2]^2 + [2(c/c_c)(\omega_0/\omega_n)]^2}} \qquad (22\text{–}40)$$

The MF is plotted in Fig. 22–17 versus the frequency ratio ω_0/ω_n for various values of the damping factor c/c_c. It can be seen from this graph that the magnification of the amplitude increases as the damping factor decreases. Resonance obviously occurs only when the damping factor is zero and the frequency ratio equals 1.

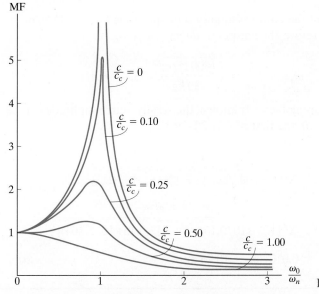

Fig. 22–17

EXAMPLE | 22.8

The 30-kg electric motor shown in Fig. 22–18 is confined to move vertically, and is supported by *four* springs, each spring having a stiffness of 200 N/m. If the rotor is unbalanced such that its effect is equivalent to a 4-kg mass located 60 mm from the axis of rotation, determine the amplitude of vibration when the rotor is turning at $\omega_0 = 10$ rad/s. The damping factor is $c/c_c = 0.15$.

Fig. 22–18

SOLUTION
The periodic force which causes the motor to vibrate is the centrifugal force due to the unbalanced rotor. This force has a constant magnitude of

$$F_0 = ma_n = mr\omega_0^2 = 4 \text{ kg}(0.06 \text{ m})(10 \text{ rad/s})^2 = 24 \text{ N}$$

The stiffness of the entire system of four springs is $k = 4(200 \text{ N/m}) = 800$ N/m. Therefore, the natural frequency of vibration is

$$\omega_n = \sqrt{\frac{k}{m}} = \sqrt{\frac{800 \text{ N/m}}{30 \text{ kg}}} = 5.164 \text{ rad/s}$$

Since the damping factor is known, the steady-state amplitude can be determined from the first of Eqs. 22–39, i.e.,

$$X' = \frac{F_0/k}{\sqrt{[1 - (\omega_0/\omega_n)^2]^2 + [2(c/c_c)(\omega_0/\omega_n)]^2}}$$

$$= \frac{24/800}{\sqrt{[1 - (10/5.164)^2]^2 + [2(0.15)(10/5.164)]^2}}$$

$$= 0.0107 \text{ m} = 10.7 \text{ mm} \qquad\qquad Ans.$$

22

*22.6 Electrical Circuit Analogs

The characteristics of a vibrating mechanical system can be represented by an electric circuit. Consider the circuit shown in Fig. 22–19a, which consists of an inductor L, a resistor R, and a capacitor C. When a voltage $E(t)$ is applied, it causes a current of magnitude i to flow through the circuit. As the current flows past the inductor the voltage drop is $L(di/dt)$, when it flows across the resistor the drop is Ri, and when it arrives at the capacitor the drop is $(1/C) \int i\, dt$. Since current cannot flow past a capacitor, it is only possible to measure the charge q acting on the capacitor. The charge can, however, be related to the current by the equation $i = dq/dt$. Thus, the voltage drops which occur across the inductor, resistor, and capacitor become $L\, d^2q/dt^2$, $R\, dq/dt$, and q/C, respectively. According to Kirchhoff's voltage law, the applied voltage balances the sum of the voltage drops around the circuit. Therefore,

$$L\frac{d^2q}{dt^2} + R\frac{dq}{dt} + \frac{1}{C}q = E(t) \qquad (22\text{--}41)$$

Consider now the model of a single-degree-of-freedom mechanical system, Fig. 22–19b, which is subjected to both a general forcing function $F(t)$ and damping. The equation of motion for this system was established in the previous section and can be written as

$$m\frac{d^2x}{dt^2} + c\frac{dx}{dt} + kx = F(t) \qquad (22\text{--}42)$$

By comparison, it is seen that Eqs. 22–41 and 22–42 have the same form, and hence mathematically the procedure of analyzing an electric circuit is the same as that of analyzing a vibrating mechanical system. The analogs between the two equations are given in Table 22–1.

This analogy has important application to experimental work, for it is much easier to simulate the vibration of a complex mechanical system using an electric circuit, which can be constructed on an analog computer, than to make an equivalent mechanical spring-and-dashpot model.

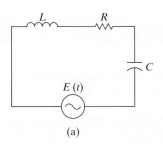

(a)

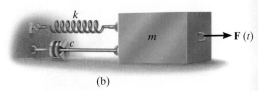

(b)

Fig. 22–19

TABLE 22–1
Electrical–Mechanical Analogs

Electrical		Mechanical	
Electric charge	q	Displacement	x
Electric current	i	Velocity	dx/dt
Voltage	$E(t)$	Applied force	$F(t)$
Inductance	L	Mass	m
Resistance	R	Viscous damping coefficient	c
Reciprocal of capacitance	$1/C$	Spring stiffness	k

22

PROBLEMS

22–41. If the block is subjected to the periodic force $F = F_0 \cos \omega t$, show that the differential equation of motion is $\ddot{y} + (k/m)y = (F_0/m) \cos \omega t$, where y is measured from the equilibrium position of the block. What is the general solution of this equation?

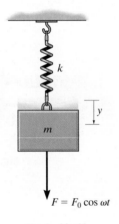

$F = F_0 \cos \omega t$

Prob. 22–41

22–42. The block shown in Fig. 22–15 has a mass of 20 kg, and the spring has a stiffness $k = 600$ N/m. When the block is displaced and released, two successive amplitudes are measured as $x_1 = 150$ mm and $x_2 = 87$ mm. Determine the coefficient of viscous damping, c.

22–43. A 4-lb weight is attached to a spring having a stiffness $k = 10$ lb/ft. The weight is drawn downward a distance of 4 in. and released from rest. If the support moves with a vertical displacement $\delta = (0.5 \sin 4t)$ in., where t is in seconds, determine the equation which describes the position of the weight as a function of time.

*22–44. A 4-kg block is suspended from a spring that has a stiffness of $k = 600$ N/m. The block is drawn downward 50 mm from the equilibrium position and released from rest when $t = 0$. If the support moves with an impressed displacement of $\delta = (10 \sin 4t)$ mm, where t is in seconds, determine the equation that describes the vertical motion of the block. Assume positive displacement is downward.

22–45. Use a block-and-spring model like that shown in Fig. 22–13a, but suspended from a vertical position and subjected to a periodic support displacement $\delta = \delta_0 \sin \omega_0 t$, determine the equation of motion for the system, and obtain its general solution. Define the displacement y measured from the static equilibrium position of the block when $t = 0$.

22–46. A 5-kg block is suspended from a spring having a stiffness of 300 N/m. If the block is acted upon by a vertical force $F = (7 \sin 8t)$ N, where t is in seconds, determine the equation which describes the motion of the block when it is pulled down 100 mm from the equilibrium position and released from rest at $t = 0$. Assume that positive displacement is downward.

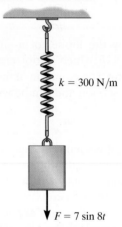

$k = 300$ N/m

$F = 7 \sin 8t$

Prob. 22–46

22–47. The electric motor has a mass of 50 kg and is supported by *four springs*, each spring having a stiffness of 100 N/m. If the motor turns a disk D which is mounted eccentrically, 20 mm from the disk's center, determine the angular velocity ω at which resonance occurs. Assume that the motor only vibrates in the vertical direction.

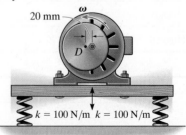

20 mm

D

$k = 100$ N/m $k = 100$ N/m

Prob. 22–47

*22–48.   The 20-lb block is attached to a spring having a stiffness of 20 lb/ft. A force $F = (6 \cos 2t)$ lb, where t is in seconds, is applied to the block. Determine the maximum speed of the block after frictional forces cause the free vibrations to dampen out.

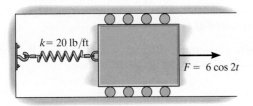

$k = 20$ lb/ft

$F = 6 \cos 2t$

Prob. 22–48

22–49.   The light elastic rod supports a 4-kg sphere. When an 18-N vertical force is applied to the sphere, the rod deflects 14 mm. If the wall oscillates with harmonic frequency of 2 Hz and has an amplitude of 15 mm, determine the amplitude of vibration for the sphere.

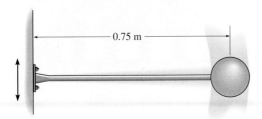

0.75 m

Prob. 22–49

22–50.   The instrument is centered uniformly on a platform P, which in turn is supported by *four* springs, each spring having a stiffness $k = 130$ N/m. If the floor is subjected to a vibration $\omega = 7$ Hz, having a vertical displacement amplitude $\delta_0 = 0.17$ ft, determine the vertical displacement amplitude of the platform and instrument. The instrument and the platform have a total weight of 18 lb.

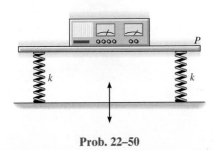

P

k k

Prob. 22–50

22–51.   The uniform rod has a mass of m. If it is acted upon by a periodic force of $F = F_0 \sin \omega t$, determine the amplitude of the steady-state vibration.

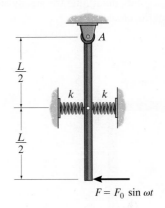

A

$\dfrac{L}{2}$

k k

$\dfrac{L}{2}$

$F = F_0 \sin \omega t$

Prob. 22–51

*22–52.   Using a block-and-spring model, like that shown in Fig. 22–13a, but suspended from a vertical position and subjected to a periodic support displacement of $\delta = \delta_0 \cos \omega_0 t$, determine the equation of motion for the system, and obtain its general solution. Define the displacement y measured from the static equilibrium position of the block when $t = 0$.

22–53.   The fan has a mass of 25 kg and is fixed to the end of a horizontal beam that has a negligible mass. The fan blade is mounted eccentrically on the shaft such that it is equivalent to an unbalanced 3.5-kg mass located 100 mm from the axis of rotation. If the static deflection of the beam is 50 mm as a result of the weight of the fan, determine the angular velocity of the fan blade at which resonance will occur. *Hint:* See the first part of Example 22.8.

22–54.   In Prob. 22–53, determine the amplitude of steady-state vibration of the fan if its angular velocity is 10 rad/s.

22–55.   What will be the amplitude of steady-state vibration of the fan in Prob. 22–53 if the angular velocity of the fan blade is 18 rad/s? *Hint:* See the first part of Example 22.8.

ω

Probs. 22–53/54/55

22

*22–56. The small block at A has a mass of 4 kg and is mounted on the bent rod having negligible mass. If the rotor at B causes a harmonic movement $\delta_B = (0.1 \cos 15t)$ m, where t is in seconds, determine the steady-state amplitude of vibration of the block.

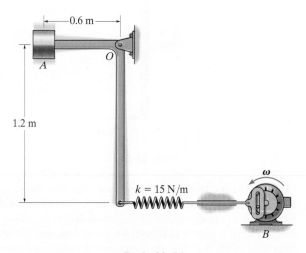

Prob. 22–56

22–57. The electric motor turns an eccentric flywheel which is equivalent to an unbalanced 0.25-lb weight located 10 in. from the axis of rotation. If the static deflection of the beam is 1 in. because of the weight of the motor, determine the angular velocity of the flywheel at which resonance will occur. The motor weighs 150 lb. Neglect the mass of the beam.

22–58. What will be the amplitude of steady-state vibration of the motor in Prob. 22–57 if the angular velocity of the flywheel is 20 rad/s?

22–59. Determine the angular velocity of the flywheel in Prob. 22–57 which will produce an amplitude of vibration of 0.25 in.

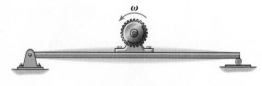

Probs. 22–57/58/59

*22–60. The engine is mounted on a foundation block which is spring supported. Describe the steady-state vibration of the system if the block and engine have a total weight of 1500 lb and the engine, when running, creates an impressed force $F = (50 \sin 2t)$ lb, where t is in seconds. Assume that the system vibrates only in the vertical direction, with the positive displacement measured downward, and that the total stiffness of the springs can be represented as $k = 2000$ lb/ft.

22–61. Determine the rotational speed ω of the engine in Prob. 22–60 which will cause resonance.

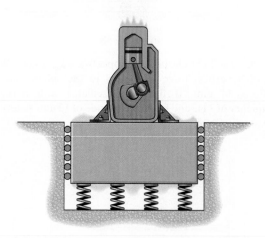

Probs. 22–60/61

22–62. The motor of mass M is supported by a simply supported beam of negligible mass. If block A of mass m is clipped onto the rotor, which is turning at constant angular velocity of ω, determine the amplitude of the steady-state vibration. *Hint:* When the beam is subjected to a concentrated force of P at its mid-span, it deflects $\delta = PL^3/48EI$ at this point. Here E is Young's modulus of elasticity, a property of the material, and I is the moment of inertia of the beam's cross-sectional area.

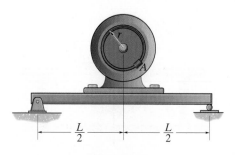

Prob. 22–62

22–63. A block having a mass of 0.8 kg is suspended from a spring having a stiffness of 120 N/m. If a dashpot provides a damping force of 2.5 N when the speed of the block is 0.2 m/s, determine the period of free vibration.

*22–64. The block, having a weight of 15 lb, is immersed in a liquid such that the damping force acting on the block has a magnitude of $F = (0.8|v|)$ lb, where v is the velocity of the block in ft/s. If the block is pulled down 0.8 ft and released from rest, determine the position of the block as a function of time. The spring has a stiffness of $k = 40$ lb/ft. Consider positive displacement to be downward.

*22–68. The 4-kg circular disk is attached to three springs, each spring having a stiffness $k = 180$ N/m. If the disk is immersed in a fluid and given a downward velocity of 0.3 m/s at the equilibrium position, determine the equation which describes the motion. Consider positive displacement to be measured downward, and that fluid resistance acting on the disk furnishes a damping force having a magnitude $F = (60|v|)$ N, where v is the velocity of the block in m/s.

Prob. 22–68

Prob. 22–64

22–69. If the 12-kg rod is subjected to a periodic force of $F = (30 \sin 6t)$ N, where t is in seconds, determine the steady-state vibration amplitude θ_{max} of the rod about the pin B. Assume θ is small.

22–65. A 7-lb block is suspended from a spring having a stiffness of $k = 75$ lb/ft. The support to which the spring is attached is given simple harmonic motion which may be expressed as $\delta = (0.15 \sin 2t)$ ft, where t is in seconds. If the damping factor is $c/c_c = 0.8$, determine the phase angle ϕ of forced vibration.

22–66. Determine the magnification factor of the block, spring, and dashpot combination in Prob. 22–65.

22–67. A block having a mass of 7 kg is suspended from a spring that has a stiffness $k = 600$ N/m. If the block is given an upward velocity of 0.6 m/s from its equilibrium position at $t = 0$, determine its position as a function of time. Assume that positive displacement of the block is downward and that motion takes place in a medium which furnishes a damping force $F = (50|v|)$ N, where v is in m/s.

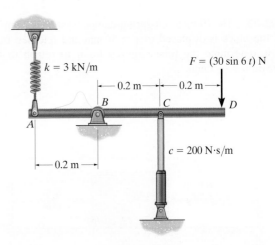

Prob. 22–69

22

22–70. The damping factor, c/c_c, may be determined experimentally by measuring the successive amplitudes of vibrating motion of a system. If two of these maximum displacements can be approximated by x_1 and x_2, as shown in Fig. 22–16, show that the ratio $\ln(x_1/x_2) = 2\pi(c/c_c)/\sqrt{1-(c/c_c)^2}$. The quantity $\ln(x_1/x_2)$ is called the *logarithmic decrement*.

22–71. If the amplitude of the 50-lb cylinder's steady-vibration is 6 in., determine the wheel's angular velocity ω.

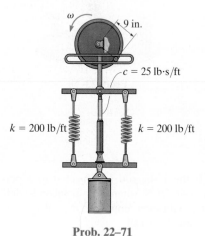

Prob. 22–71

*22–72.** The 10-kg block-spring-damper system is damped. If the block is displaced to $x = 50$ mm and released from rest, determine the time required for it to return to the position $x = 2$ mm.

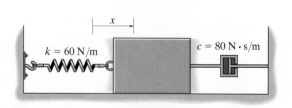

Prob. 22–72

22–73. The 20-kg block is subjected to the action of the harmonic force $F = (90 \cos 6t)$ N, where t is in seconds. Write the equation which describes the steady-state motion.

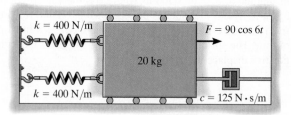

Prob. 22–73

22–74. A bullet of mass m has a velocity of v_0 just before it strikes the target of mass M. If the bullet embeds in the target, and the vibration is to be critically damped, determine the dashpot's critical damping coefficient, and the springs' maximum compression. The target is free to move along the two horizontal guides that are "nested" in the springs.

22–75. A bullet of mass m has a velocity v_0 just before it strikes the target of mass M. If the bullet embeds in the target, and the dashpot's damping coefficient is $0 < c \ll c_c$, determine the springs' maximum compression. The target is free to move along the two horizontal guides that are "nested" in the springs.

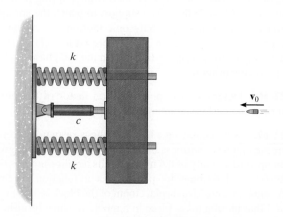

Probs. 22–74/75

*22–76. Determine the differential equation of motion for the damped vibratory system shown. What type of motion occurs? Take $k = 100$ N/m, $c = 200$ N·s/m, $m = 25$ kg.

22–78. Draw the electrical circuit that is equivalent to the mechanical system shown. What is the differential equation which describes the charge q in the circuit?

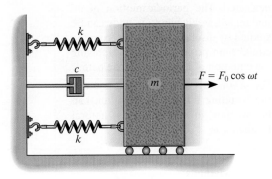

Prob. 22–78

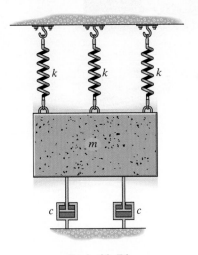

Prob. 22–76

22–79. Draw the electrical circuit that is equivalent to the mechanical system shown. Determine the differential equation which describes the charge q in the circuit.

22–77. Draw the electrical circuit that is equivalent to the mechanical system shown. Determine the differential equation which describes the charge q in the circuit.

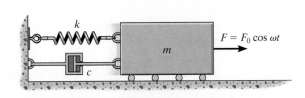

Prob. 22–77

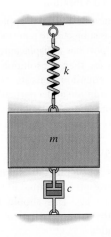

Prob. 22–79

22

CHAPTER REVIEW

Undamped Free Vibration

A body has free vibration when gravitational or elastic restoring forces cause the motion. This motion is undamped when friction forces are neglected. The periodic motion of an undamped, freely vibrating body can be studied by displacing the body from the equilibrium position and then applying the equation of motion along the path.

For a one-degree-of-freedom system, the resulting differential equation can be written in terms of its natural frequency ω_n.

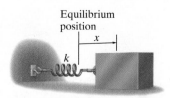

Equilibrium position

x

k

$$\ddot{x} + \omega_n^2 x = 0 \qquad \tau = \frac{2\pi}{\omega_n} \qquad f = \frac{1}{\tau} = \frac{\omega_n}{2\pi}$$

Energy Methods

Provided the restoring forces acting on the body are gravitational and elastic, then conservation of energy can also be used to determine its simple harmonic motion. To do this, the body is displaced a small amount from its equilibrium position, and an expression for its kinetic and potential energy is written. The time derivative of this equation can then be rearranged in the standard form $\ddot{x} + \omega_n^2 x = 0$.

Undamped Forced Vibration

When the equation of motion is applied to a body, which is subjected to a periodic force, or the support has a displacement with a frequency ω_0, then the solution of the differential equation consists of a complementary solution and a particular solution. The complementary solution is caused by the free vibration and can be neglected. The particular solution is caused by the forced vibration.

Resonance will occur if the natural frequency of vibration ω_n is equal to the forcing frequency ω_0. This should be avoided, since the motion will tend to become unbounded.

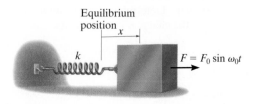

Equilibrium position

x

k

$F = F_0 \sin \omega_0 t$

$$x_p = \frac{F_0/k}{1 - (\omega_0/\omega_n)^2} \sin \omega_0 t$$

22

Viscous Damped Free Vibration

A viscous damping force is caused by fluid drag on the system as it vibrates. If the motion is slow, this drag force will be proportional to the velocity, that is, $F = c\dot{x}$. Here c is the coefficient of viscous damping. By comparing its value to the critical damping coefficient $c_c = 2m\omega_n$, we can specify the type of vibration that occurs. If $c > c_c$, it is an overdamped system; if $c = c_c$, it is a critically damped system; if $c < c_c$, it is an underdamped system.

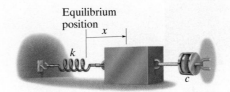

Equilibrium position
x
k
c

Viscous Damped Forced Vibration

The most general type of vibration for a one-degree-of-freedom system occurs when the system is damped and subjected to periodic forced motion. The solution provides insight as to how the damping factor, c/c_c, and the frequency ratio, ω_0/ω_n, influence the vibration.

Resonance is avoided provided $c/c_c \neq 0$ and $\omega_0/\omega_n \neq 1$.

Electrical Circuit Analogs

The vibrating motion of a complex mechanical system can be studied by modeling it as an electrical circuit. This is possible since the differential equations that govern the behavior of each system are the same.

22

APPENDIX A

Mathematical Expressions

Quadratic Formula

If $ax^2 + bx + c = 0$, then $x = \dfrac{-b \pm \sqrt{b^2 - 4ac}}{2a}$

Hyperbolic Functions

$\sinh x = \dfrac{e^x - e^{-x}}{2}$, $\cosh x = \dfrac{e^x + e^{-x}}{2}$, $\tanh x = \dfrac{\sinh x}{\cosh x}$

Trigonometric Identities

$\sin \theta = \dfrac{A}{C}$, $\csc \theta = \dfrac{C}{A}$

$\cos \theta = \dfrac{B}{C}$, $\sec \theta = \dfrac{C}{B}$

$\tan \theta = \dfrac{A}{B}$, $\cot \theta = \dfrac{B}{A}$

$\sin^2 \theta + \cos^2 \theta = 1$

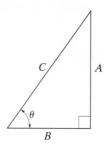

$\sin(\theta \pm \phi) = \sin \theta \cos \phi \pm \cos \theta \sin \phi$

$\sin 2\theta = 2 \sin \theta \cos \theta$

$\cos(\theta \pm \phi) = \cos \theta \cos \phi \mp \sin \theta \sin \phi$

$\cos 2\theta = \cos^2 \theta - \sin^2 \theta$

$\cos \theta = \pm\sqrt{\dfrac{1 + \cos 2\theta}{2}}$, $\sin \theta = \pm\sqrt{\dfrac{1 - \cos 2\theta}{2}}$

$\tan \theta = \dfrac{\sin \theta}{\cos \theta}$

$1 + \tan^2 \theta = \sec^2 \theta \quad 1 + \cot^2 \theta = \csc^2 \theta$

Power-Series Expansions

$\sin x = x - \dfrac{x^3}{3!} + \cdots \qquad \sinh x = x + \dfrac{x^3}{3!} + \cdots$

$\cos x = 1 - \dfrac{x^2}{2!} + \cdots \qquad \cosh x = 1 + \dfrac{x^2}{2!} + \cdots$

Derivatives

$\dfrac{d}{dx}(u^n) = nu^{n-1}\dfrac{du}{dx}$

$\dfrac{d}{dx}(uv) = u\dfrac{dv}{dx} + v\dfrac{du}{dx}$

$\dfrac{d}{dx}\left(\dfrac{u}{v}\right) = \dfrac{v\dfrac{du}{dx} - u\dfrac{dv}{dx}}{v^2}$

$\dfrac{d}{dx}(\cot u) = -\csc^2 u\dfrac{du}{dx}$

$\dfrac{d}{dx}(\sec u) = \tan u \sec u\dfrac{du}{dx}$

$\dfrac{d}{dx}(\csc u) = -\csc u \cot u\dfrac{du}{dx}$

$\dfrac{d}{dx}(\sin u) = \cos u\dfrac{du}{dx}$

$\dfrac{d}{dx}(\cos u) = -\sin u\dfrac{du}{dx}$

$\dfrac{d}{dx}(\tan u) = \sec^2 u\dfrac{du}{dx}$

$\dfrac{d}{dx}(\sinh u) = \cosh u\dfrac{du}{dx}$

$\dfrac{d}{dx}(\cosh u) = \sinh u\dfrac{du}{dx}$

670

Integrals

$$\int x^n \, dx = \frac{x^{n+1}}{n+1} + C, n \neq -1$$

$$\int \frac{dx}{a + bx} = \frac{1}{b} \ln(a + bx) + C$$

$$\int \frac{dx}{a + bx^2} = \frac{1}{2\sqrt{-ba}} \ln\left[\frac{a + x\sqrt{-ab}}{a - x\sqrt{-ab}}\right] + C, \, ab < 0$$

$$\int \frac{x \, dx}{a + bx^2} = \frac{1}{2b} \ln(bx^2 + a) + C$$

$$\int \frac{x^2 \, dx}{a + bx^2} = \frac{x}{b} - \frac{a}{b\sqrt{ab}} \tan^{-1} \frac{x\sqrt{ab}}{a} + C, \, ab > 0$$

$$\int \frac{dx}{a^2 - x^2} = \frac{1}{2a} \ln\left[\frac{a + x}{a - x}\right] + C, \, a^2 > x^2$$

$$\int \sqrt{a + bx} \, dx = \frac{2}{3b} \sqrt{(a + bx)^3} + C$$

$$\int x\sqrt{a + bx} \, dx = \frac{-2(2a - 3bx)\sqrt{(a + bx)^3}}{15b^2} + C$$

$$\int x^2\sqrt{a + bx} \, dx = \frac{2(8a^2 - 12abx + 15b^2x^2)\sqrt{(a + bx)^3}}{105b^3} + C$$

$$\int \sqrt{a^2 - x^2} \, dx = \frac{1}{2}\left[x\sqrt{a^2 - x^2} + a^2 \sin^{-1}\frac{x}{a}\right] + C, \, a > 0$$

$$\int x\sqrt{x^2 \pm a^2} \, dx = \frac{1}{3}\sqrt{(x^2 \pm a^2)^3} + C$$

$$\int x^2\sqrt{a^2 - x^2} \, dx = -\frac{x}{4}\sqrt{(a^2 - x^2)^3}$$

$$+ \frac{a^2}{8}\left(x\sqrt{a^2 - x^2} + a^2 \sin^{-1}\frac{x}{a}\right) + C, \, a > 0$$

$$\int \sqrt{x^2 \pm a^2} \, dx = \frac{1}{2}\left[x\sqrt{x^2 \pm a^2} \pm a^2 \ln\left(x + \sqrt{x^2 \pm a^2}\right)\right] + C$$

$$\int x\sqrt{a^2 - x^2} \, dx = -\frac{1}{3}\sqrt{(a^2 - x^2)^3} + C$$

$$\int x^2\sqrt{x^2 \pm a^2} \, dx = \frac{x}{4}\sqrt{(x^2 \pm a^2)^3} \mp \frac{a^2}{8}x\sqrt{x^2 \pm a^2}$$

$$- \frac{a^4}{8} \ln\left(x + \sqrt{x^2 \pm a^2}\right) + C$$

$$\int \frac{dx}{\sqrt{a + bx}} = \frac{2\sqrt{a + bx}}{b} + C$$

$$\int \frac{x \, dx}{\sqrt{x^2 \pm a^2}} = \sqrt{x^2 \pm a^2} + C$$

$$\int \frac{dx}{\sqrt{a + bx + cx^2}} = \frac{1}{\sqrt{c}} \ln\left[\sqrt{a + bx + cx^2}\right.$$

$$\left. + x\sqrt{c} + \frac{b}{2\sqrt{c}}\right] + C, \, c > 0$$

$$= \frac{1}{\sqrt{-c}} \sin^{-1}\left(\frac{-2cx - b}{\sqrt{b^2 - 4ac}}\right) + C, \, c < 0$$

$$\int \sin x \, dx = -\cos x + C$$

$$\int \cos x \, dx = \sin x + C$$

$$\int x \cos(ax) \, dx = \frac{1}{a^2}\cos(ax) + \frac{x}{a}\sin(ax) + C$$

$$\int x^2 \cos(ax) \, dx = \frac{2x}{a^2}\cos(ax)$$

$$+ \frac{a^2x^2 - 2}{a^3}\sin(ax) + C$$

$$\int e^{ax} \, dx = \frac{1}{a}e^{ax} + C$$

$$\int xe^{ax} \, dx = \frac{e^{ax}}{a^2}(ax - 1) + C$$

$$\int \sinh x \, dx = \cosh x + C$$

$$\int \cosh x \, dx = \sinh x + C$$

A

APPENDIX

B

Vector Analysis

The following discussion provides a brief review of vector analysis. A more detailed treatment of these topics is given in *Engineering Mechanics: Statics.*

Vector.   A vector, **A**, is a quantity which has magnitude and direction, and adds according to the parallelogram law. As shown in Fig. B–1, **A** = **B** + **C**, where **A** is the *resultant vector* and **B** and **C** are *component vectors.*

Unit Vector.   A unit vector, $\mathbf{u}_A$, has a magnitude of one "dimensionless" unit and acts in the same direction as **A**. It is determined by dividing **A** by its magnitude A, i.e,

$$\mathbf{u}_A = \frac{\mathbf{A}}{A} \qquad \text{(B–1)}$$

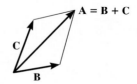

Fig. B–1

Cartesian Vector Notation.

The directions of the positive x, y, z axes are defined by the Cartesian unit vectors $\mathbf{i}$, $\mathbf{j}$, $\mathbf{k}$, respectively.

As shown in Fig. B–2, vector $\mathbf{A}$ is formulated by the addition of its x, y, z components as

$$\mathbf{A} = A_x\mathbf{i} + A_y\mathbf{j} + A_z\mathbf{k} \qquad \text{(B–2)}$$

The *magnitude* of $\mathbf{A}$ is determined from

$$A = \sqrt{A_x^2 + A_y^2 + A_z^2} \qquad \text{(B–3)}$$

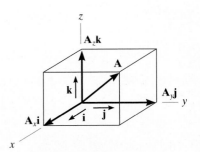

Fig. B–2

The *direction* of $\mathbf{A}$ is defined in terms of its *coordinate direction angles*, α, β, γ, measured from the *tail* of $\mathbf{A}$ to the *positive x, y, z axes*, Fig. B–3. These angles are determined from the *direction cosines* which represent the $\mathbf{i}$, $\mathbf{j}$, $\mathbf{k}$ components of the unit vector $\mathbf{u}_A$; i.e., from Eqs. B–1 and B–2

$$\mathbf{u}_A = \frac{A_x}{A}\mathbf{i} + \frac{A_y}{A}\mathbf{j} + \frac{A_z}{A}\mathbf{k} \qquad \text{(B–4)}$$

so that the direction cosines are

$$\cos\alpha = \frac{A_x}{A} \qquad \cos\beta = \frac{A_y}{A} \qquad \cos\gamma = \frac{A_z}{A} \qquad \text{(B–5)}$$

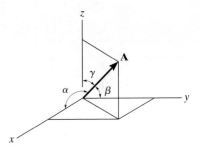

Fig. B–3

Hence, $\mathbf{u}_A = \cos\alpha\,\mathbf{i} + \cos\beta\,\mathbf{j} + \cos\gamma\,\mathbf{k}$, and using Eq. B–3, it is seen that

$$\cos^2\alpha + \cos^2\beta + \cos^2\gamma = 1 \qquad \text{(B–6)}$$

The Cross Product.

The cross product of two vectors $\mathbf{A}$ and $\mathbf{B}$, which yields the resultant vector $\mathbf{C}$, is written as

$$\mathbf{C} = \mathbf{A} \times \mathbf{B} \qquad \text{(B–7)}$$

and reads $\mathbf{C}$ equals $\mathbf{A}$ "cross" $\mathbf{B}$. The *magnitude* of $\mathbf{C}$ is

$$C = AB\sin\theta \qquad \text{(B–8)}$$

where θ is the angle made between the *tails* of $\mathbf{A}$ and $\mathbf{B}$ ($0° \leq \theta \leq 180°$). The *direction* of $\mathbf{C}$ is determined by the right-hand rule, whereby the fingers of the right hand are curled *from $\mathbf{A}$ to $\mathbf{B}$* and the thumb points in the direction of $\mathbf{C}$, Fig. B–4. This vector is perpendicular to the plane containing vectors $\mathbf{A}$ and $\mathbf{B}$.

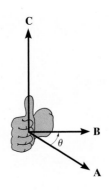

Fig. B–4

B

The vector cross product is *not* commutative, i.e., $\mathbf{A} \times \mathbf{B} \neq \mathbf{B} \times \mathbf{A}$. Rather,

$$\mathbf{A} \times \mathbf{B} = -\mathbf{B} \times \mathbf{A} \qquad \text{(B–9)}$$

The distributive law is valid; i.e.,

$$\mathbf{A} \times (\mathbf{B} + \mathbf{D}) = \mathbf{A} \times \mathbf{B} + \mathbf{A} \times \mathbf{D} \qquad \text{(B–10)}$$

And the cross product may be multiplied by a scalar m in any manner; i.e.,

$$m(\mathbf{A} \times \mathbf{B}) = (m\mathbf{A}) \times \mathbf{B} = \mathbf{A} \times (m\mathbf{B}) = (\mathbf{A} \times \mathbf{B})m \qquad \text{(B–11)}$$

Fig. B–5

Equation B–7 can be used to find the cross product of any pair of Cartesian unit vectors. For example, to find $\mathbf{i} \times \mathbf{j}$, the magnitude is $(i)(j) \sin 90° = (1)(1)(1) = 1$, and its direction $+\mathbf{k}$ is determined from the right-hand rule, applied to $\mathbf{i} \times \mathbf{j}$, Fig. B–2. A simple scheme shown in Fig. B–5 may be helpful in obtaining this and other results when the need arises. If the circle is constructed as shown, then "crossing" two of the unit vectors in a *counterclockwise* fashion around the circle yields a *positive* third unit vector, e.g., $\mathbf{k} \times \mathbf{i} = \mathbf{j}$. Moving *clockwise*, a *negative* unit vector is obtained, e.g., $\mathbf{i} \times \mathbf{k} = -\mathbf{j}$.

If $\mathbf{A}$ and $\mathbf{B}$ are expressed in Cartesian component form, then the cross product, Eq. B–7, may be evaluated by expanding the determinant

$$\mathbf{C} = \mathbf{A} \times \mathbf{B} = \begin{vmatrix} \mathbf{i} & \mathbf{j} & \mathbf{k} \\ A_x & A_y & A_z \\ B_x & B_y & B_z \end{vmatrix} \qquad \text{(B–12)}$$

which yields

$$\mathbf{C} = (A_y B_z - A_z B_y)\mathbf{i} - (A_x B_z - A_z B_x)\mathbf{j} + (A_x B_y - A_y B_x)\mathbf{k}$$

Recall that the cross product is used in statics to define the moment of a force $\mathbf{F}$ about point O, in which case

$$\mathbf{M}_O = \mathbf{r} \times \mathbf{F} \qquad \text{(B–13)}$$

where $\mathbf{r}$ is a position vector directed from point O to *any point* on the line of action of $\mathbf{F}$.

B

The Dot Product.

The dot product of two vectors **A** and **B**, which yields a scalar, is defined as

$$\mathbf{A} \cdot \mathbf{B} = AB \cos \theta \tag{B–14}$$

and reads **A** "dot" **B**. The angle θ is formed between the *tails* of **A** and **B** ($0° \leq \theta \leq 180°$).

The dot product is commutative; i.e.,

$$\mathbf{A} \cdot \mathbf{B} = \mathbf{B} \cdot \mathbf{A} \tag{B–15}$$

The distributive law is valid; i.e.,

$$\mathbf{A} \cdot (\mathbf{B} + \mathbf{D}) = \mathbf{A} \cdot \mathbf{B} + \mathbf{A} \cdot \mathbf{D} \tag{B–16}$$

And scalar multiplication can be performed in any manner, i.e.,

$$m(\mathbf{A} \cdot \mathbf{B}) = (m\mathbf{A}) \cdot \mathbf{B} = \mathbf{A} \cdot (m\mathbf{B}) = (\mathbf{A} \cdot \mathbf{B})m \tag{B–17}$$

Using Eq. B–14, the dot product between any two Cartesian vectors can be determined. For example, $\mathbf{i} \cdot \mathbf{i} = (1)(1) \cos 0° = 1$ and $\mathbf{i} \cdot \mathbf{j} = (1)(1) \cos 90° = 0$.

If **A** and **B** are expressed in Cartesian component form, then the dot product, Eq. C–14, can be determined from

$$\mathbf{A} \cdot \mathbf{B} = A_x B_x + A_y B_y + A_z B_z \tag{B–18}$$

The dot product may be used to determine the *angle θ formed between two vectors*. From Eq. B–14,

$$\theta = \cos^{-1}\left(\frac{\mathbf{A} \cdot \mathbf{B}}{AB}\right) \tag{B–19}$$

B

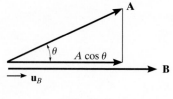

Fig. B–6

It is also possible to find the *component of a vector in a given direction* using the dot product. For example, the magnitude of the component (or projection) of vector **A** in the direction of **B**, Fig. B–6, is defined by $A \cos \theta$. From Eq. B–14, this magnitude is

$$A \cos \theta = \mathbf{A} \cdot \frac{\mathbf{B}}{B} = \mathbf{A} \cdot \mathbf{u}_B \qquad (B–20)$$

where $\mathbf{u}_B$ represents a unit vector acting in the direction of **B**, Fig. B–6.

Differentiation and Integration of Vector Functions.
The rules for differentiation and integration of the sums and products of scalar functions also apply to vector functions. Consider, for example, the two vector functions $\mathbf{A}(s)$ and $\mathbf{B}(s)$. Provided these functions are smooth and continuous for all s, then

$$\frac{d}{ds}(\mathbf{A} + \mathbf{B}) = \frac{d\mathbf{A}}{ds} + \frac{d\mathbf{B}}{ds} \qquad (B–21)$$

$$\int (\mathbf{A} + \mathbf{B}) \, ds = \int \mathbf{A} \, ds + \int \mathbf{B} \, ds \qquad (B–22)$$

For the cross product,

$$\frac{d}{ds}(\mathbf{A} \times \mathbf{B}) = \left(\frac{d\mathbf{A}}{ds} \times \mathbf{B} \right) + \left(\mathbf{A} \times \frac{d\mathbf{B}}{ds} \right) \qquad (B–23)$$

Similarly, for the dot product,

$$\frac{d}{ds}(\mathbf{A} \cdot \mathbf{B}) = \frac{d\mathbf{A}}{ds} \cdot \mathbf{B} + \mathbf{A} \cdot \frac{d\mathbf{B}}{ds} \qquad (B–24)$$

B

The Chain Rule

The chain rule of calculus can be used to determine the time derivative of a composite function. For example, if y is a function of x and x is a function of t, then we can find the derivative of y with respect to t as follows

$$\dot{y} = \frac{dy}{dt} = \frac{dy}{dx}\frac{dx}{dt} \tag{C-1}$$

In other words, to find $\dot{y}$ we take the ordinary derivative (dy/dx) and multiply it by the time derivative (dx/dt).

If several variables are functions of time and they are multiplied together, then the product rule $d(uv) = du\,v + u\,dv$ must be used along with the chain rule when taking the time derivatives. Here are some examples.

EXAMPLE C–1

If $y = x^3$ and $x = t^4$, find $\ddot{y}$, the second derivative of y with respect to time.

SOLUTION
Using the chain rule, Eq. C–1,

$$\dot{y} = 3x^2\dot{x}$$

To obtain the second time derivative we must use the product rule since x and $\dot{x}$ are both functions of time, and also, for $3x^2$ the chain rule must be applied. Thus, with $u = 3x^2$ and $v = \dot{x}$, we have

$$\ddot{y} = [6x\dot{x}]\dot{x} + 3x^2[\ddot{x}]$$

$$= 3x[2\dot{x}^2 + x\ddot{x}]$$

Since $x = t^4$, then $\dot{x} = 4t^3$ and $\ddot{x} = 12t^2$ so that

$$\ddot{y} = 3(t^4)[2(4t^3)^2 + t^4(12t^2)]$$

$$= 132t^{10}$$

Note that this result can also be obtained by combining the functions, then taking the time derivatives, that is,

$$y = x^3 = (t^4)^3 = t^{12}$$

$$\dot{y} = 12t^{11}$$

$$\ddot{y} = 132t^{10}$$

EXAMPLE C–2

If $y = xe^x$, find $\ddot{y}$.

SOLUTION
Since x and e^x are both functions of time the product and chain rules must be applied. Have $u = x$ and $v = e^x$.

$$\dot{y} = [\dot{x}]e^x + x[e^x\dot{x}]$$

The second time derivative also requires application of the product and chain rules. Note that the product rule applies to the three time variables in the last term, i.e., x, e^x, and $\dot{x}$.

$$\ddot{y} = \{[\ddot{x}]e^x + \dot{x}[e^x\dot{x}]\} + \{[\dot{x}]e^x\dot{x} + x[e^x\dot{x}]\dot{x} + xe^x[\ddot{x}]\}$$

$$= e^x[\ddot{x}(1 + x) + \dot{x}^2(2 + x)]$$

If $x = t^2$ then $\dot{x} = 2t$, $\ddot{x} = 2$ so that in terms in t, we have

$$\ddot{y} = e^{t^2}[2(1 + t^2) + 4t^2(2 + t^2)]$$

EXAMPLE | C–3

If the path in radial coordinates is given as $r = 5\theta^2$, where θ is a known function of time, find $\ddot{r}$.

SOLUTION
First, using the chain rule then the chain and product rules where $u = 10\theta$ and $v = \dot{\theta}$, we have

$$r = 5\theta^2$$

$$\dot{r} = 10\theta\dot{\theta}$$

$$\ddot{r} = 10[(\dot{\theta})\dot{\theta} + \theta(\ddot{\theta})]$$

$$= 10\dot{\theta}^2 + 10\theta\ddot{\theta}$$

EXAMPLE | C–4

If $r^2 = 6\theta^3$, find $\ddot{r}$.

SOLUTION
Here the chain and product rules are applied as follows.

$$r^2 = 6\theta^3$$

$$2r\dot{r} = 18\theta^2\dot{\theta}$$

$$2[(\dot{r})\dot{r} + r(\ddot{r})] = 18[(2\theta\dot{\theta})\dot{\theta} + \theta^2(\ddot{\theta})]$$

$$\dot{r}^2 + r\ddot{r} = 9(2\theta\dot{\theta}^2 + \theta^2\ddot{\theta})$$

To find $\ddot{r}$ at a specified value of θ which is a known function of time, we can first find $\dot{\theta}$ and $\ddot{\theta}$. Then using these values, evaluate r from the first equation, $\dot{r}$ from the second equation and $\ddot{r}$ using the last equation.

C

Fundamental Problems
Partial Solutions And Answers

Chapter 12

F12–1. $v = v_0 + a_c t$
$10 = 35 + a_c(15)$
$a_c = -1.67 \text{ m/s}^2 = 1.67 \text{ m/s}^2 \leftarrow$ *Ans.*

F12–2. $s = s_0 + v_0 t + \frac{1}{2} a_c t^2$
$0 = 0 + 15t + \frac{1}{2}(-9.81)t^2$
$t = 3.06 \text{ s}$ *Ans.*

F12–3. $ds = v\, dt$
$$\int_0^s ds = \int_0^t \left(4t - 3t^2\right) dt$$
$s = \left(2t^2 - t^3\right)\text{m}$
$s = 2\left(4^2\right) - 4^3$
$= -32 \text{ m} = 32 \text{ m} \leftarrow$ *Ans.*

F12–4. $a = \frac{dv}{dt} = \frac{d}{dt}\left(0.5t^3 - 8t\right)$
$a = \left(1.5t^2 - 8\right) \text{m/s}^2$
When $t = 2$ s,
$a = 1.5\left(2^2\right) - 8 = -2 \text{ m/s}^2 = 2 \text{ m/s}^2 \leftarrow$ *Ans.*

F12–5. $v = \frac{ds}{dt} = \frac{d}{dt}(2t^2 - 8t + 6) = (4t - 8) \text{ m/s}$
$v = 0 = (4t - 8)$
$t = 2 \text{ s}$ *Ans.*
$s|_{t=0} = 2\left(0^2\right) - 8(0) + 6 = 6 \text{ m}$
$s|_{t=2} = 2\left(2^2\right) - 8(2) + 6 = -2 \text{ m}$
$s|_{t=3} = 2\left(3^2\right) - 8(3) + 6 = 0 \text{ m}$
$(\Delta s)_{\text{Tot}} = 8 \text{ m} + 2 \text{ m} = 10 \text{ m}$ *Ans.*

F12–6. $\int v\, dv = \int a\, ds$
$$\int_{5\text{ m/s}}^v v\, dv = \int_0^s (10 - 0.2s)ds$$
$v = \left(\sqrt{20s - 0.2s^2 + 25}\right) \text{m/s}$
At $s = 10$ m,
$v = \sqrt{20(10) - 0.2(10^2) + 25}$
$= 14.3 \text{ m/s} \rightarrow$ *Ans.*

F12–7. $v = \int (4t^2 - 2)\, dt$
$v = \frac{4}{3}t^3 - 2t + C_1$
$s = \int \left(\frac{4}{3}t^3 - 2t + C_1\right) dt$
$s = \frac{1}{3}t^4 - t^2 + C_1 t + C_2$
$t = 0, s = -2, C_2 = -2$
$t = 2, s = -20, C_1 = -9.67$
$t = 4, s = 28.7 \text{ m}$ *Ans.*

F12–8. $a = v\frac{dv}{ds}$
$= \left(20 - 0.05s^2\right)\left(-0.1s\right)$
At $s = 15$ m,
$a = -13.1 \text{ m/s}^2 = 13.1 \text{ m/s}^2 \leftarrow$ *Ans.*

F12–9. $v = \frac{ds}{dt} = \frac{d}{dt}\left(0.5t^3\right) = 1.5t^2$
$v = \frac{ds}{dt} = \frac{d}{dt}(108) = 0$ *Ans.*

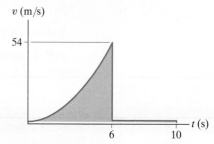

F12–10. $ds = v\, dt$
$$\int_0^s ds = \int_0^t (-4t + 80)\, dt$$
$s = -2t^2 + 80t$
$a = \frac{dv}{dt} = \frac{d}{dt}(-4t + 80) = -4 \text{ ft/s}^2 = 4 \text{ ft/s}^2 \leftarrow$
Also,
$a = \frac{\Delta v}{\Delta t} = \frac{0 - 80 \text{ ft/s}}{20 \text{ s} - 0} = -4 \text{ ft/s}^2$

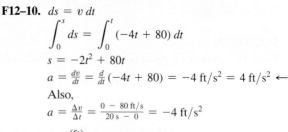

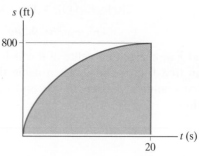

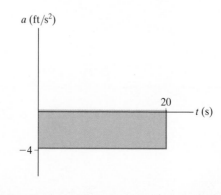

F12–11. $a\,ds = v\,dv$

$$a = v\frac{dv}{ds} = 0.25s\frac{d}{ds}(0.25s) = 0.0625s$$

$$a|_{s=40\,m} = 0.0625(40\ \text{m}) = 2.5\ \text{m/s}^2 \rightarrow$$

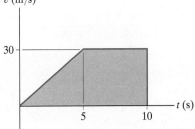

F12–12. $0 \le t < 5$ s,

$$v = \frac{ds}{dt} = \frac{d}{dt}(3t^2) = (6t)\ \text{m/s}$$

5 s $< t \le 10$ s,

$$v = \frac{ds}{dt} = \frac{d}{dt}(30t - 75) = 30\ \text{m/s}$$

$$v = \frac{\Delta s}{\Delta t} = \frac{225\ \text{m} - 75\ \text{m}}{10\ \text{m} - 5\ \text{m}} = 30\ \text{m/s}$$

$0 \le t < 5$ s,

$$a = \frac{dv}{dt} = \frac{d}{dt}(6t) = 6\ \text{m/s}^2$$

5 s $< t \le 10$ s,

$$a = \frac{dv}{dt} = \frac{d}{dt}(30) = 0$$

$0 \le t < 5$ s, $a = \Delta v/\Delta t = 6\ \text{m/s}^2$

5 s $< t \le 10$ s, $a = \Delta v/\Delta t = 0$

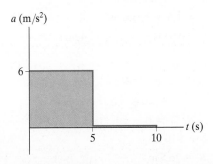

F12–13. $0 \le t < 5$ s,

$$dv = a\,dt \qquad \int_0^v dv = \int_0^t 20\,dt$$

$$v = (20t)\ \text{m/s}$$

5 s $< t \le t'$,

$$(\overset{+}{\rightarrow}) \quad dv = a\,dt \qquad \int_{100\,\text{m/s}}^v dv = \int_{5\,\text{s}}^t -10\,dt$$

$$v - 100 = (50 - 10t)\ \text{m/s},$$

$$0 = 150 - 10t'$$

$$t' = 15\ \text{s}$$

Also,

$\Delta v = 0 = $ Area under the a–t graph

$$0 = (20\ \text{m/s}^2)(5\ \text{s}) + [-(10\ \text{m/s})(t' - 5)\ \text{s}]$$

$$t' = 15\ \text{s}$$

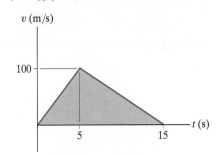

F12–14. $0 \le t \le 5$ s,

$$ds = v\,dt \qquad \int_0^s ds = \int_0^t 30t\,dt$$

$$s|_0^s = 15t^2|_0^t$$

$$s = (15t^2)\ \text{m}$$

5 s $< t \le 15$ s,

$$(\overset{+}{\rightarrow})\,ds = v\,dt; \qquad \int_{375\,\text{m}}^s ds = \int_{5\,\text{s}}^t (-15t + 225)\,dt$$

$$s = (-7.5t^2 + 225t - 562.5)\ \text{m}$$

$$s = (-7.5)(15)^2 + 225(15) - 562.5\ \text{m}$$

$$= 1125\ \text{m} \qquad \qquad Ans.$$

Also,

$\Delta s = $ Area under the v–t graph

$$= \tfrac{1}{2}(150\ \text{m/s})(15\ \text{s})$$

$$= 1125\ \text{m} \qquad \qquad Ans.$$

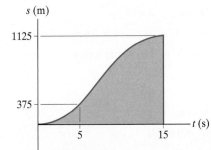

F12–15. $\int_0^x dx = \int_0^t 32t \, dt$

$x = \left(16t^2\right)$ m   (1)

$\int_0^y dy = \int_0^t 8 \, dt$

$t = \dfrac{y}{8}$ (2)

Substituting Eq. (2) into Eq. (1), get

$y = 2\sqrt{x}$ *Ans.*

F12–16. $y = 0.75(8t) = 6t$

$v_x = \dot{x} = \frac{dx}{dt} = \frac{d}{dt}(8t) = 8$ m/s $\rightarrow$

$v_y = \dot{y} = \frac{dy}{dt} = \frac{d}{dt}(6t) = 6$ m/s $\uparrow$

The magnitude of the particle's velocity is

$v = \sqrt{v_x^2 + v_y^2} = \sqrt{(8 \text{ m/s})^2 + (6 \text{ m/s})^2}$

$= 10$ m/s   *Ans.*

F12–17. $y = \left(4t^2\right)$ m

$v_x = \dot{x} = \frac{d}{dt}\left(4t^4\right) = \left(16t^3\right)$ m/s $\rightarrow$

$v_y = \dot{y} = \frac{d}{dt}\left(4t^2\right) = (8t)$ m/s $\uparrow$

When $t = 0.5$ s,

$v = \sqrt{v_x^2 + v_y^2} = \sqrt{(2 \text{ m/s})^2 + (4 \text{ m/s})^2}$

$= 4.47$ m/s   *Ans.*

$a_x = \dot{v}_x = \frac{d}{dt}\left(16t^3\right) = \left(48t^2\right)$ m/s^2

$a_y = \dot{v}_y = \frac{d}{dt}(8t) = 8$ m/s^2

When $t = 0.5$ s,

$a = \sqrt{a_x^2 + a_y^2} = \sqrt{(12 \text{ m/s}^2)^2 + (8 \text{ m/s}^2)^2}$

$= 14.4$ m/s^2 *Ans.*

F12–18. $y = 0.5x$

$\dot{y} = 0.5\dot{x}$

$v_y = t^2$

When $t = 4$ s,

$v_x = 32$ m/s $v_y = 16$ m/s

$v = \sqrt{v_x^2 + v_y^2} = 35.8$ m/s   *Ans.*

$a_x = \dot{v}_x = 4t$

$a_y = \dot{v}_y = 2t$

When $t = 4$ s,

$a_x = 16$ m/s^2 $a_y = 8$ m/s^2

$a = \sqrt{a_x^2 + a_y^2} = \sqrt{16^2 + 8^2} = 17.9$ m/s^2 *Ans.*

F12–19. $y = \left(t^4\right)$ m

$v_x = \dot{x} = (4t)$ m/s $v_y = \dot{y} = \left(4t^3\right)$ m/s

When $t = 2$ s,

$v_x = 8$ m/s $v_y = 32$ m/s

$v = \sqrt{v_x^2 + v_y^2} = 33.0$ m/s   *Ans.*

$a_x = \dot{v}_x = 4$ m/s^2

$a_y = \dot{v}_y = \left(12t^2\right)$ m/s^2

When $t = 2$ s,

$a_x = 4$ m/s^2 $a_y = 48$ m/s^2

$a = \sqrt{a_x^2 + a_y^2} = \sqrt{4^2 + 48^2} = 48.2$ m/s^2 *Ans.*

F12–20. $\dot{y} = 0.1x\dot{x}$

$v_y = 0.1(5)(-3) = -1.5$ m/s $= 1.5$ m/s $\downarrow$ *Ans.*

$\ddot{y} = 0.1\left[\dot{x}\dot{x} + x\ddot{x}\right]$

$a_y = 0.1\left[(-3)^2 + 5(-1.5)\right] = 0.15$ m/s^2 $\uparrow$ *Ans.*

F12–21. $(v_B)_y^2 = (v_A)_y^2 + 2a_y(y_B - y_A)$

$0^2 = (5 \text{ m/s})^2 + 2(-9.81 \text{ m/s}^2)(h - 0)$

$h = 1.27$ m   *Ans.*

F12–22. $y_C = y_A + (v_A)_y t_{AC} + \frac{1}{2} a_y t_{AC}^2$

$0 = 0 + (5 \text{ m/s})t_{AC} + \frac{1}{2}(-9.81 \text{ m/s}^2)t_{AC}^2$

$t_{AC} = 1.0194$ s

$(v_C)_y = (v_A)_y + a_y t_{AC}$

$(v_C)_y = 5 \text{ m/s} + (-9.81 \text{ m/s}^2)(1.0194 \text{ s})$

$= -5 \text{ m/s} = 5$ m/s $\downarrow$

$v_C = \sqrt{(v_C)_x^2 + (v_C)_y^2}$

$= \sqrt{(8.660 \text{ m/s})^2 + (5 \text{ m/s})^2} = 10$ m/s   *Ans.*

$R = x_A + (v_A)_x t_{AC} = 0 + (8.660 \text{ m/s})(1.0194 \text{ s})$

$= 8.83$ m   *Ans.*

F12–23. $s = s_0 + v_0 t$

$10 = 0 + v_A \cos 30° t$

$s = s_0 + v_0 t + \frac{1}{2} a_c t^2$

$3 = 1.5 + v_A \sin 30° t + \frac{1}{2}(-9.81)t^2$

$t = 0.9334$ s, $v_A = 12.4$ m/s   *Ans.*

F12–24. $s = s_0 + v_0 t$

$R\left(\frac{4}{5}\right) = 0 + 20\left(\frac{3}{5}\right)t$

$s = s_0 + v_0 t + \frac{1}{2} a_c t^2$

$-R\left(\frac{3}{5}\right) = 0 + 20\left(\frac{4}{5}\right)t + \frac{1}{2}(-9.81)t^2$

$t = 5.10$ s

$R = 76.5$ m   *Ans.*

F12–25. $x_B = x_A + (v_A)_x t_{AB}$

$\qquad 12 \text{ ft} = 0 + (0.8660 \, v_A) t_{AB}$

$\qquad v_A t_{AB} = 13.856 \qquad\qquad\qquad\qquad (1)$

$\qquad y_B = y_A + (v_A)_y t_{AB} + \frac{1}{2} a_y t_{AB}^2$

$\qquad (8 - 3) \text{ ft} = 0 + 0.5 v_A t_{AB} + \frac{1}{2}(-32.2 \text{ ft/s}^2) t_{AB}^2$

$\qquad$ Using Eq. (1),

$\qquad 5 = 0.5(13.856) - 16.1 \, t_{AB}^2$

$\qquad t_{AB} = 0.3461 \text{ s}$

$\qquad v_A = 40.0 \text{ ft/s} \qquad\qquad\qquad\qquad Ans.$

F12–26. $y_B = y_A + (v_A)_y t_{AB} + \frac{1}{2} a_y t_{AB}^2$

$\qquad -150 \text{ m} = 0 + (90 \text{ m/s}) t_{AB} + \frac{1}{2}(-9.81 \text{ m/s}^2) t_{AB}^2$

$\qquad t_{AB} = 19.89 \text{ s}$

$\qquad x_B = x_A + (v_A)_x t_{AB}$

$\qquad R = 0 + 120 \text{ m/s}(19.89 \text{ s}) = 2386.37 \text{ m}$

$\qquad = 2.39 \text{ km} \qquad\qquad\qquad\qquad Ans.$

F12–27. $a_t = \dot{v} = \frac{dv}{dt} = \frac{d}{dt}(0.0625t^2) = (0.125t) \text{ m/s}^2 \big|_{t=10 \text{ s}}$

$\qquad\qquad = 1.25 \text{ m/s}^2$

$\qquad a_n = \frac{v^2}{\rho} = \frac{(0.0625t^2)^2}{40 \text{ m}} = \left[97.656(10^{-6})t^4\right] \text{ m/s}^2 \big|_{t=10 \text{ s}}$

$\qquad\qquad = 0.9766 \text{ m/s}^2$

$\qquad a = \sqrt{a_t^2 + a_n^2} = \sqrt{(1.25 \text{ m/s}^2)^2 + (0.9766 \text{ m/s}^2)^2}$

$\qquad\qquad = 1.59 \text{ m/s}^2 \qquad\qquad\qquad\qquad Ans.$

F12–28. $v = 2s \big|_{s=10} = 20 \text{ m/s}$

$\qquad a_n = \frac{v^2}{\rho} = \frac{(20 \text{ m/s})^2}{50 \text{ m}} = 8 \text{ m/s}^2$

$\qquad a_t = v \frac{dv}{ds} = 4s \big|_{s=10} = 40 \text{ m/s}^2$

$\qquad a = \sqrt{a_t^2 + a_n^2} = \sqrt{(40 \text{ m/s}^2)^2 + (8 \text{ m/s}^2)^2}$

$\qquad\qquad = 40.8 \text{ m/s}^2 \qquad\qquad\qquad\qquad Ans.$

F12–29. $v_C^2 = v_A^2 + 2a_t(s_C - s_A)$

$\qquad (15 \text{ m/s})^2 = (25 \text{ m/s})^2 + 2a_t(300 \text{ m} - 0)$

$\qquad a_t = -0.6667 \text{ m/s}^2$

$\qquad v_B^2 = v_A^2 + 2a_t(s_B - s_A)$

$\qquad v_B^2 = (25 \text{ m/s})^2 + 2(-0.6667 \text{ m/s}^2)(250 \text{ m} - 0)$

$\qquad v_B = 17.08 \text{ m/s}$

$\qquad (a_B)_n = \frac{v_B^2}{\rho} = \frac{(17.08 \text{ m/s})^2}{300 \text{ m}} = 0.9722 \text{ m/s}^2$

$\qquad a_B = \sqrt{(a_B)_t^2 + (a_B)_n^2}$

$\qquad\quad = \sqrt{(-0.6667 \text{ m/s}^2)^2 + (0.9722 \text{ m/s}^2)^2}$

$\qquad\quad = 1.18 \text{ m/s}^2 \qquad\qquad\qquad\qquad Ans.$

F12–30. $\tan \theta = \frac{dy}{dx} = \frac{d}{dx}\left(\frac{1}{24}x^2\right) = \frac{1}{12}x$

$\qquad \theta = \tan^{-1}\left(\frac{1}{12}x\right)\bigg|_{x=10 \text{ ft}}$

$\qquad\quad = \tan^{-1}\left(\frac{10}{12}\right) = 39.81° = 39.8° \, \angle \qquad Ans.$

$\qquad \rho = \frac{\left[1 + (dy/dx)^2\right]^{3/2}}{|d^2y/dx^2|} = \frac{\left[1 + \left(\frac{1}{12}x\right)^2\right]^{3/2}}{\left|\frac{1}{12}\right|}\bigg|_{x=10 \text{ ft}}$

$\qquad\quad = 26.468 \text{ ft}$

$\qquad a_n = \frac{v^2}{\rho} = \frac{(20 \text{ ft/s})^2}{26.468 \text{ ft}} = 15.11 \text{ ft/s}^2$

$\qquad a = \sqrt{(a_t)^2 + (a_n)^2} = \sqrt{(6 \text{ ft/s}^2)^2 + (15.11 \text{ ft/s}^2)^2}$

$\qquad\quad = 16.3 \text{ ft/s}^2 \qquad\qquad\qquad\qquad Ans.$

F12–31. $(a_B)_t = -0.001s = (-0.001)(300 \text{ m})\left(\frac{\pi}{2}\text{ rad}\right) \text{ m/s}^2$

$\qquad\qquad = -0.4712 \text{ m/s}^2$

$\qquad v \, dv = a_t \, ds$

$\qquad \int_{25 \text{ m/s}}^{v_B} v \, dv = \int_0^{150\pi \text{ m}} -0.001s \, ds$

$\qquad v_B = 20.07 \text{ m/s}$

$\qquad (a_B)_n = \frac{v_B^2}{\rho} = \frac{(20.07 \text{ m/s})^2}{300 \text{ m}} = 1.343 \text{ m/s}^2$

$\qquad a_B = \sqrt{(a_B)_t^2 + (a_B)_n^2}$

$\qquad\quad = \sqrt{(-0.4712 \text{ m/s}^2)^2 + (1.343 \text{ m/s}^2)^2}$

$\qquad\quad = 1.42 \text{ m/s}^2 \qquad\qquad\qquad\qquad Ans.$

F12–32. $a_t\,ds = v\,dv$

$$a_t = v\frac{dv}{ds} = (0.2s)(0.2) = (0.04s)\text{ m/s}^2$$

$$a_t = 0.04(50\text{ m}) = 2\text{ m/s}^2$$

$$v = 0.2(50\text{ m}) = 10\text{ m/s}$$

$$a_n = \frac{v^2}{\rho} = \frac{(10\text{ m/s})^2}{500\text{ m}} = 0.2\text{ m/s}^2$$

$$a = \sqrt{a_t^2 + a_n^2} = \sqrt{(2\text{ m/s}^2)^2 + (0.2\text{ m/s}^2)^2}$$

$$= 2.01\text{ m/s}^2 \qquad \textit{Ans.}$$

F12–33. $v_r = \dot{r} = 0$

$$v_\theta = r\dot{\theta} = (400\dot{\theta})\text{ ft/s}$$

$$v = \sqrt{v_r^2 + v_\theta^2}$$

$$55\text{ ft/s} = \sqrt{0^2 + [(400\dot{\theta})\text{ ft/s}]^2}$$

$$\dot{\theta} = 0.1375\text{ rad/s} \qquad \textit{Ans.}$$

F12–34. $r = 0.1t^3\big|_{t=1.5\text{ s}} = 0.3375\text{ m}$

$$\dot{r} = 0.3t^2\big|_{t=1.5\text{ s}} = 0.675\text{ m/s}$$

$$\ddot{r} = 0.6t\big|_{t=1.5\text{ s}} = 0.900\text{ m/s}^2$$

$$\theta = 4t^{3/2}\big|_{t=1.5\text{ s}} = 7.348\text{ rad}$$

$$\dot{\theta} = 6t^{1/2}\big|_{t=1.5\text{ s}} = 7.348\text{ rad/s}$$

$$\ddot{\theta} = 3t^{-1/2}\big|_{t=1.5\text{ s}} = 2.449\text{ rad/s}^2$$

$$v_r = \dot{r} = 0.675\text{ m/s}$$

$$v_\theta = r\dot{\theta} = (0.3375\text{ m})(7.348\text{ rad/s}) = 2.480\text{ m/s}$$

$$a_r = \ddot{r} - r\dot{\theta}^2$$

$$= (0.900\text{ m/s}^2) - (0.3375\text{ m})(7.348\text{ rad/s})^2$$

$$= -17.325\text{ m/s}^2$$

$$a_\theta = r\ddot{\theta} + 2\dot{r}\dot{\theta} = (0.3375\text{ m})(2.449\text{ rad/s}^2)$$

$$+\ 2(0.675\text{ m/s})(7.348\text{ rad/s}) = 10.747\text{ m/s}^2$$

$$v = \sqrt{v_r^2 + v_\theta^2}$$

$$= \sqrt{(0.675\text{ m/s})^2 + (2.480\text{ m/s})^2}$$

$$= 2.57\text{ m/s} \qquad \textit{Ans.}$$

$$a = \sqrt{a_r^2 + a_\theta^2}$$

$$= \sqrt{(-17.325\text{ m/s}^2)^2 + (10.747\text{ m/s}^2)^2}$$

$$= 20.4\text{ m/s}^2 \qquad \textit{Ans.}$$

F12–35. $r = 2\theta$

$$\dot{r} = 2\dot{\theta}$$

$$\ddot{r} = 2\ddot{\theta}$$

At $\theta = \pi/4$ rad,

$$r = 2\left(\tfrac{\pi}{4}\right) = \tfrac{\pi}{2}\text{ ft}$$

$$\dot{r} = 2(3\text{ rad/s}) = 6\text{ ft/s}$$

$$\ddot{r} = 2(1\text{ rad/s}) = 2\text{ ft/s}^2$$

$$a_r = \ddot{r} - r\dot{\theta}^2 = 2\text{ ft/s}^2 - \left(\tfrac{\pi}{2}\text{ ft}\right)(3\text{ rad/s})^2$$

$$= -12.14\text{ ft/s}^2$$

$$a_\theta = r\ddot{\theta} + 2\dot{r}\dot{\theta}$$

$$= \left(\tfrac{\pi}{2}\text{ ft}\right)(1\text{ rad/s}^2) + 2(6\text{ ft/s})(3\text{ rad/s})$$

$$= 37.57\text{ ft/s}^2$$

$$a = \sqrt{a_r^2 + a_\theta^2}$$

$$= \sqrt{(-12.14\text{ ft/s}^2)^2 + (37.57\text{ ft/s}^2)^2}$$

$$= 39.5\text{ ft/s}^2 \qquad \textit{Ans.}$$

F12–36. $r = e^\theta$

$$\dot{r} = e^\theta\dot{\theta}$$

$$\ddot{r} = e^\theta\ddot{\theta} + e^\theta\dot{\theta}^2$$

$$a_r = \ddot{r} - r\dot{\theta}^2 = (e^\theta\ddot{\theta} + e^\theta\dot{\theta}^2) - e^\theta\dot{\theta}^2 = e^{\pi/4}(4)$$

$$= 8.77\text{ m/s}^2 \qquad \textit{Ans.}$$

$$a_\theta = r\ddot{\theta} + 2\dot{r}\dot{\theta} = (e^\theta\ddot{\theta}) + (2(e^\theta\dot{\theta})\dot{\theta}) = e^\theta(\ddot{\theta} + 2\dot{\theta}^2)$$

$$= e^{\pi/4}(4 + 2(2)^2)$$

$$= 26.3\text{ m/s}^2 \qquad \textit{Ans.}$$

F12–37. $r = [0.2(1 + \cos\theta)]\text{ m}\big|_{\theta=30°} = 0.3732\text{ m}$

$$\dot{r} = \left[-0.2(\sin\theta)\dot{\theta}\right]\text{ m/s}\big|_{\theta=30°}$$

$$= -0.2\sin 30°(3\text{ rad/s})$$

$$= -0.3\text{ m/s}$$

$$v_r = \dot{r} = -0.3\text{ m/s}$$

$$v_\theta = r\dot{\theta} = (0.3732\text{ m})(3\text{ rad/s}) = 1.120\text{ m/s}$$

$$v = \sqrt{v_r^2 + v_\theta^2} = \sqrt{(-0.3\text{ m/s})^2 + (1.120\text{ m/s})^2}$$

$$= 1.16\text{ m/s} \qquad \textit{Ans.}$$

F12–38. $30\text{ m} = r\sin\theta$

$$r = \left(\tfrac{30\text{ m}}{\sin\theta}\right) = (30\csc\theta)\text{ m}$$

$$r = (30\csc\theta)\big|_{\theta=45°} = 42.426\text{ m}$$

$$\dot{r} = -30\csc\theta\,\text{ctn}\,\theta\,\dot{\theta}\big|_{\theta=45°} = -\left(42.426\dot{\theta}\right)\text{ m/s}$$

$$v_r = \dot{r} = -(42.426\dot{\theta})\text{ m/s}$$

$$v_\theta = r\dot{\theta} = (42.426\dot{\theta})\text{ m/s}$$

$$v = \sqrt{v_r^2 + v_\theta^2}$$

$$2 = \sqrt{(-42.426\dot{\theta})^2 + (42.426\dot{\theta})^2}$$

$$\dot{\theta} = 0.0333\text{ rad/s} \qquad \textit{Ans.}$$

F12–39. $l_T = 3s_D + s_A$

$$0 = 3v_D + v_A$$

$$0 = 3v_D + 3\text{ m/s}$$

$$v_D = -1\text{ m/s} = 1\text{ m/s}\uparrow \qquad \textit{Ans.}$$

F12–40. $s_B + 2s_A + 2h = l$

$v_B + 2v_A = 0$

$6 + 2v_A = 0$ $v_A = -3$ m/s $= 3$ m/s ↑   *Ans.*

F12–41. $3s_A + s_B = l$

$3v_A + v_B = 0$

$3v_A + 1.5 = 0$ $v_A = -0.5$ m/s $= 0.5$ m/s ↑ *Ans.*

F12–42. $l_T = 4\,s_A + s_F$

$0 = 4\,v_A + v_F$

$0 = 4\,v_A + 3$ m/s

$v_A = -0.75$ m/s $= 0.75$ m/s ↑   *Ans.*

F12–43. $s_A + 2(s_A - a) + (s_A - s_P) = l$

$4s_A - s_P = l + 2a$

$4v_A - v_P = 0$

$4v_A - (-4) = 0$

$4v_A + 4 = 0$ $v_A = -1$ m/s $= 1$ m/s ↗   *Ans.*

F12–44. $s_C + s_B = l_{CED}$ (1)

$(s_A - s_C) + (s_B - s_C) + s_B = l_{ACDF}$

$s_A + 2s_B - 2s_C = l_{ACDF}$ (2)

Thus

$v_C + v_B = 0$

$v_A + 2v_B - 2v_C = 0$

Eliminating v_C,

$v_A + 4v_B = 0$

Thus,

4 ft/s $+ 4v_B = 0$

$v_B = -1$ ft/s $= 1$ ft/s ↑   *Ans.*

F12–45. $\mathbf{v}_B = \mathbf{v}_A + \mathbf{v}_{B/A}$

$100\mathbf{i} = 80\mathbf{j} + \mathbf{v}_{B/A}$

$\mathbf{v}_{B/A} = 100\mathbf{i} - 80\mathbf{j}$

$v_{B/A} = \sqrt{(v_{B/A})_x^2 + (v_{B/A})_y^2}$

$= \sqrt{(100 \text{ km/h})^2 + (-80 \text{ km/h})^2}$

$= 128$ km/h   *Ans.*

$\theta = \tan^{-1}\left[\dfrac{(v_{B/A})_y}{(v_{B/A})_x}\right] = \tan^{-1}\left(\dfrac{80 \text{ km/h}}{100 \text{ km/h}}\right) = 38.7° \,↘$ *Ans.*

F12–46. $\mathbf{v}_B = \mathbf{v}_A + \mathbf{v}_{B/A}$

$(-400\mathbf{i} - 692.82\mathbf{j}) = (650\mathbf{i}) + \mathbf{v}_{B/A}$

$\mathbf{v}_{B/A} = [-1050\mathbf{i} - 692.82\mathbf{j}]$ km/h

$v_{B/A} = \sqrt{(v_{B/A})_x^2 + (v_{B/A})_y^2}$

$= \sqrt{(1050 \text{ km/h})^2 + (692.82 \text{ km/h})^2}$

$= 1258$ km/h   *Ans.*

$\theta = \tan^{-1}\left[\dfrac{(v_{B/A})_y}{(v_{B/A})_x}\right] = \tan^{-1}\left(\dfrac{692.82 \text{ km/h}}{1050 \text{ km/h}}\right) = 33.4° \,↗$ *Ans.*

F12–47. $\mathbf{v}_B = \mathbf{v}_A + \mathbf{v}_{B/A}$

$(5\mathbf{i} + 8.660\mathbf{j}) = (12.99\mathbf{i} + 7.5\mathbf{j}) + \mathbf{v}_{B/A}$

$\mathbf{v}_{B/A} = [-7.990\mathbf{i} + 1.160\mathbf{j}]$ m/s

$v_{B/A} = \sqrt{(-7.990 \text{ m/s})^2 + (1.160 \text{ m/s})^2}$

$= 8.074$ m/s

$d_{AB} = v_{B/A}t = (8.074 \text{ m/s})(4 \text{ s}) = 32.3$ m   *Ans.*

F12–48. $\mathbf{v}_A = \mathbf{v}_B + \mathbf{v}_{A/B}$

$-20 \cos 45°\mathbf{i} + 20 \sin 45°\mathbf{j} = 65\mathbf{i} + \mathbf{v}_{A/B}$

$\mathbf{v}_{A/B} = -79.14\mathbf{i} + 14.14\mathbf{j}$

$v_{A/B} = \sqrt{(-79.14)^2 + (14.14)^2}$

$= 80.4$ km/h   *Ans.*

$\mathbf{a}_A = \mathbf{a}_B + \mathbf{a}_{A/B}$

$\dfrac{(20)^2}{0.1} \cos 45°\mathbf{i} + \dfrac{(20)^2}{0.1} \sin 45°\mathbf{j} = 1200\mathbf{i} + \mathbf{a}_{A/B}$

$\mathbf{a}_{A/B} = 1628\mathbf{i} + 2828\mathbf{j}$

$a_{A/B} = \sqrt{(1628)^2 + (2828)^2}$

$= 3.26(10^3)$ km/h²   *Ans.*

Chapter 13

F13–1. $s = s_0 + v_0 t + \frac{1}{2} a_c t^2$

6 m $= 0 + 0 + \frac{1}{2} a(3 \text{ s})^2$

$a = 1.333$ m/s²

$\Sigma F_y = ma_y;$ $N_A - 20(9.81)$ N $\cos 30° = 0$

$N_A = 169.91$ N

$\Sigma F_x = ma_x;$ $T - 20(9.81)$ N $\sin 30°$

$- 0.3(169.91 \text{ N}) = (20 \text{ kg})(1.333 \text{ m/s}^2)$

$T = 176$ N   *Ans.*

F13–2. $(F_f)_{max} = \mu_s N_A = 0.3(245.25 \text{ N}) = 73.575$ N.

Since $F = 100$ N $> (F_f)_{max}$ when $t = 0$, the crate will start to move immediately after $\mathbf{F}$ is applied.

$+\uparrow \Sigma F_y = ma_y;$ $N_A - 25(9.81)$ N $= 0$

$N_A = 245.25$ N

$\xrightarrow{+} \Sigma F_x = ma_x;$

$10t^2 + 100 - 0.25(245.25 \text{ N}) = (25 \text{ kg})a$

$a = (0.4t^2 + 1.5475)$ m/s²

$dv = a\,dt$

$\displaystyle\int_0^v dv = \int_0^{4 \text{ s}} (0.4t^2 + 1.5475)dt$

$v = 14.7$ m/s →   *Ans.*

F13–3. $\xrightarrow{+} \Sigma F_x = ma_x;$

$\left(\frac{4}{5}\right)500 \text{ N} - (500s)\text{N} = (10 \text{ kg})a$

$a = (40 - 50s) \text{ m/s}^2$

$v \, dv = a \, ds$

$\int_0^v v \, dv = \int_0^{0.5 \text{ m}} (40 - 50s) \, ds$

$\left.\frac{v^2}{2}\right|_0^v = \left(40s - 25s^2\right)\Big|_0^{0.5 \text{ m}}$

$v = 5.24 \text{ m/s}$ *Ans.*

F13–4. $\xrightarrow{+} \Sigma F_x = ma_x$ $100(s + 1) \text{ N} = (2000 \text{ kg})a$

$a = (0.05(s + 1)) \text{ m/s}^2$

$v \, dv = a \, ds$

$\int_0^v v \, dv = \int_0^{10 \text{ m}} 0.05(s + 1) \, ds$

$v = 2.45 \text{ m/s}$

F13–5. $F_{sp} = k(l - l_0) = (200 \text{ N/m})(0.5 \text{ m} - 0.3 \text{ m})$

$= 40 \text{ N}$

$\theta = \tan^{-1}\left(\frac{0.3 \text{ m}}{0.4 \text{ m}}\right) = 36.86°$

$\xrightarrow{+} \Sigma F_x = ma_x;$

$100 \text{ N} - (40 \text{ N})\cos 36.86° = (25 \text{ kg})a$

$a = 2.72 \text{ m/s}^2$

F13–6. Blocks A and B:

$\xrightarrow{+} \Sigma F_x = ma_x; \ 6 = \frac{70}{32.2}a; \ a = 2.76 \text{ ft/s}^2$

Check if slipping occurs between A and B.

$\xrightarrow{+} \Sigma F_x = ma_x; \ 6 - F = \frac{20}{32.2}(2.76);$

$F = 4.29 \text{ lb} < 0.4(20) = 8 \text{ lb}$

$a_A = a_B = 2.76 \text{ ft/s}^2$ *Ans.*

F13–7. $\Sigma F_n = m\frac{v^2}{\rho}; \ (0.3)m(9.81) = m\frac{v^2}{2}$

$v = 2.43 \text{ m/s}$ *Ans.*

F13–8. $+\downarrow \Sigma F_n = ma_n; \ m(32.2) = m\left(\frac{v^2}{250}\right)$

$v = 89.7 \text{ ft/s}$ *Ans.*

F13–9. $+\downarrow \Sigma F_n = ma_n; \ 150 + N_p = \frac{150}{32.2}\left(\frac{(120)^2}{400}\right)$

$N_p = 17.7 \text{ lb}$ *Ans.*

F13–10. $\xleftarrow{+} \Sigma F_n = ma_n;$

$N_c \sin 30° + 0.2 N_c \cos 30° = m\frac{v^2}{500}$

$+\uparrow \Sigma F_b = 0;$

$N_c \cos 30° - 0.2N_c \sin 30° - m(32.2) = 0$

$v = 119 \text{ ft/s}$ *Ans.*

F13–11. $\Sigma F_t = ma_t; \quad 10(9.81) \text{ N} \cos 45° = (10 \text{ kg})a_t$

$a_t = 6.94 \text{ m/s}^2$ *Ans.*

$\Sigma F_n = ma_n;$

$T - 10(9.81) \text{ N} \sin 45° = (10 \text{ kg})\frac{(3 \text{ m/s})^2}{2 \text{ m}}$

$T = 114 \text{ N}$ *Ans.*

F13–12. $\Sigma F_n = ma_n;$

$F_n = (500 \text{ kg})\frac{(15 \text{ m/s})^2}{200 \text{ m}} = 562.5 \text{ N}$

$\Sigma F_t = ma_t;$

$F_t = (500 \text{ kg})(1.5 \text{ m/s}^2) = 750 \text{ N}$

$F = \sqrt{F_n^2 + F_t^2} = \sqrt{(562.5 \text{ N})^2 + (750 \text{ N})^2}$

$= 938 \text{ N}$ *Ans.*

F13–13. $a_r = \ddot{r} - r\dot{\theta}^2 = 0 - (1.5 \text{ m} + (8 \text{ m})\sin 45°)\dot{\theta}^2$

$= (-7.157 \, \dot{\theta}^2) \text{ m/s}^2$

$\Sigma F_z = ma_z;$

$T \cos 45° - m(9.81) = m(0) \quad T = 13.87 \, m$

$\Sigma F_r = ma_r;$

$-(13.87m) \sin 45° = m(-7.157 \, \dot{\theta}^2)$

$\dot{\theta} = 1.17 \text{ rad/s}$ *Ans.*

F13–14. $\theta = \pi t^2\big|_{t=0.5 \text{ s}} = (\pi/4) \text{ rad}$

$\dot{\theta} = 2\pi t\big|_{t=0.5 \text{ s}} = \pi \text{ rad/s}$

$\ddot{\theta} = 2\pi \text{ rad/s}^2$

$r = 0.6 \sin \theta\big|_{\theta = \pi/4 \text{ rad}} = 0.4243 \text{ m}$

$\dot{r} = 0.6 (\cos \theta)\dot{\theta}\big|_{\theta = \pi/4 \text{ rad}} = 1.3329 \text{ m/s}$

$\ddot{r} = 0.6 [(\cos \theta)\ddot{\theta} - (\sin\theta)\dot{\theta}^2]\big|_{\theta = \pi/4 \text{ rad}} = -1.5216 \text{ m/s}^2$

$a_r = \ddot{r} - r\dot{\theta}^2 = -1.5216 \text{ m/s}^2 - (0.4243 \text{ m})(\pi \text{ rad/s})^2$

$= -5.7089 \text{ m/s}^2$

$a_\theta = r\ddot{\theta} + 2\dot{r}\,\dot{\theta} = 0.4243 \text{ m}(2\pi \text{ rad/s}^2)$

$+ 2(1.3329 \text{ m/s})(\pi \text{ rad/s})$

$= 11.0404 \text{ m/s}^2$

$\Sigma F_r = ma_r;$

$F\cos 45° - N \cos 45° - 0.2(9.81)\cos 45°$

$= 0.2(-5.7089)$

$\Sigma F_\theta = ma_\theta;$

$F \sin 45° + N \sin 45° - 0.2(9.81)\sin 45°$

$= 0.2(11.0404)$

$N = 2.37 \text{ N}$ $F = 2.72 \text{ N}$ *Ans.*

F13–15. $r = 50e^{2\theta}\big|_{\theta=\pi/6 \text{ rad}} = \left[50e^{2(\pi/6)}\right]$ m $= 142.48$ m

$\dot{r} = 50\left(2e^{2\theta}\,\dot{\theta}\right) = 100e^{2\theta}\,\dot{\theta}\big|_{\theta=\pi/6 \text{ rad}}$

$\quad = \left[100e^{2(\pi/6)}(0.05)\right] = 14.248$ m/s

$\ddot{r} = 100\left((2e^{2\theta}\dot{\theta})\dot{\theta} + e^{2\theta}(\ddot{\theta})\right)\big|_{\theta=\pi/6 \text{ rad}}$

$\quad = 100\left[2e^{2(\pi/6)}(0.05^2) + e^{2(\pi/6)}(0.01)\right]$

$\quad = 4.274$ m/s^2

$a_r = \ddot{r} - r\dot{\theta}^2 = 4.274$ m/s$^2 - 142.48$ m$(0.05 \text{ rad/s})^2$

$\quad = 3.918$ m/s^2

$a_\theta = r\ddot{\theta} + 2\dot{r}\dot{\theta} = 142.48$ m(0.01 rad/s^2)

$\qquad + 2(14.248 \text{ m/s})(0.05 \text{ rad/s})$

$\qquad = 2.850$ m/s^2

$\Sigma F_r = ma_r;$

$\quad F_r = (2000 \text{ kg})(3.918 \text{ m/s}^2) = 7836.55$ N

$\Sigma F_\theta = ma_\theta;$

$\quad F_\theta = (2000 \text{ kg})(2.850 \text{ m/s}^2) = 5699.31$ N

$\quad F = \sqrt{F_r^2 + F_\theta^2}$

$\quad = \sqrt{(7836.55 \text{ N})^2 + (5699.31 \text{ N})^2}$

$\quad = 9689.87$ N $= 9.69$ kN

F13–16. $r = (0.6 \cos 2\theta)$ m $\big|_{\theta=0°} = [0.6 \cos 2(0°)]$ m $= 0.6$ m

$\dot{r} = (-1.2 \sin 2\theta\dot{\theta})$ m/s $\big|_{\theta=0°}$

$\quad = \left[-1.2 \sin 2(0°)(-3)\right]$ m/s $= 0$

$\ddot{r} = -1.2\left(\sin 2\theta\ddot{\theta} + 2\cos 2\theta\dot{\theta}^2\right)$ m/s$^2\big|_{\theta=0°}$

$\quad = -21.6$ m/s^2

Thus,

$a_r = \ddot{r} - r\dot{\theta}^2 = -21.6$ m/s$^2 - 0.6$ m$(-3 \text{ rad/s})^2$

$\quad = -27$ m/s^2

$a_\theta = r\ddot{\theta} + 2\dot{r}\dot{\theta} = 0.6$ m$(0) + 2(0)(-3 \text{ rad/s}) = 0$

$\Sigma F_\theta = ma_\theta; \quad F - 0.2(9.81)$ N $= 0.2$ kg(0)

$\qquad\qquad F = 1.96$ N $\uparrow$ *Ans.*

Chapter 14

F14–1. $T_1 + \Sigma U_{1-2} = T_2$

$0 + \left(\frac{4}{5}\right)(500 \text{ N})(0.5 \text{ m}) - \frac{1}{2}(500 \text{ N/m})(0.5 \text{ m})^2$

$\qquad = \frac{1}{2}(10 \text{ kg})v^2$

$v = 5.24$ m/s *Ans.*

F14–2. $\Sigma F_y = ma_y; \quad N_A - 20(9.81)$ N $\cos 30° = 0$

$N_A = 169.91$ N

$T_1 + \Sigma U_{1-2} = T_2$

$0 + 300$ N$(10$ m$) - 0.3(169.91 \text{ N})(10 \text{ m})$

$\qquad - 20(9.81)$N $(10$ m$) \sin 30°$

$\quad = \frac{1}{2}(20 \text{ kg})v^2$

$v = 12.3$ m/s *Ans.*

F14–3. $T_1 + \Sigma U_{1-2} = T_2$

$0 + 2\left[\int_0^{15 \text{ m}} (600 + 2s^2) \text{ N } ds\right] - 100(9.81)$ N$(15$ m$)$

$\qquad = \frac{1}{2}(100 \text{ kg})v^2$

$v = 12.5$ m/s *Ans.*

F14–4. $T_1 + \Sigma U_{1-2} = T_2$

$\frac{1}{2}(1800 \text{ kg})(125 \text{ m/s})^2 - \left[\frac{(50\,000 \text{ N } + 20\,000 \text{ N})}{2}(400 \text{ m})\right]$

$\qquad\qquad = \frac{1}{2}(1800 \text{ kg})v^2$

$v = 8.33$ m/s *Ans.*

F14–5. $T_1 + \Sigma U_{1-2} = T_2$

$\frac{1}{2}(10 \text{ kg})(5 \text{ m/s})^2 + 100 \text{ N}s' + [10(9.81) \text{ N}]\, s' \sin 30°$

$\qquad -\frac{1}{2}(200 \text{ N/m})(s')^2 = 0$

$s' = 2.09$ m

$s = 0.6$ m $+ 2.09$ m $= 2.69$ m *Ans.*

F14–6. $T_A + \Sigma U_{A-B} = T_B$

Consider difference in cord length $AC - BC$, which is distance F moves.

$0 + 10 \text{ lb}(\sqrt{(3 \text{ ft})^2 + (4 \text{ ft})^2} - 3 \text{ ft})$

$\qquad = \frac{1}{2}\left(\frac{5}{32.2} \text{ slug}\right)v_B^2$

$v_B = 16.0$ ft/s *Ans.*

F14–7. $\xrightarrow{+} \Sigma F_x = ma_x;$

$30\left(\frac{4}{5}\right) = 20a \quad a = 1.2 \text{ m/s}^2 \rightarrow$

$v = v_0 + a_ct$

$v = 0 + 1.2(4) = 4.8$ m/s

$P = \mathbf{F} \cdot \mathbf{v} = F(\cos\theta)v$

$\quad = 30\left(\frac{4}{5}\right)(4.8)$

$\quad = 115$ W *Ans.*

F14–8. $\xrightarrow{+} \Sigma F_x = ma_x;$

$10s = 20a \quad a = 0.5s \text{ m/s}^2 \rightarrow$

$v\,dv = a\,ds$

$\int_1^v v\,dv = \int_0^{5 \text{ m}} 0.5\, s\, ds$

$v = 3.674$ m/s

$P = \mathbf{F} \cdot \mathbf{v} = [10(5)](3.674) = 184$ W *Ans.*

F14–9. $(+\uparrow)\Sigma F_y = 0$;

$T_1 - 100\text{ lb} = 0 \qquad T_1 = 100\text{ lb}$

$(+\uparrow)\Sigma F_y = 0$;

$100\text{ lb} + 100\text{ lb} - T_2 = 0 \quad T_2 = 200\text{ lb}$

$P_{\text{out}} = \mathbf{T}_B \cdot \mathbf{v}_B = (200\text{ lb})(3\text{ ft/s}) = 1.091\text{ hp}$

$P_{\text{in}} = \dfrac{P_{\text{out}}}{\varepsilon} = \dfrac{1.091\text{ hp}}{0.8} = 1.36\text{ hp}$ *Ans.*

F14–10. $\Sigma F_{y'} = ma_{y'};\quad N - 20(9.81)\cos 30° = 20(0)$

$N = 169.91\text{ N}$

$\Sigma F_{x'} = ma_{x'};$

$F - 20(9.81)\sin 30° - 0.2(169.91) = 0$

$F = 132.08\text{ N}$

$P = \mathbf{F} \cdot \mathbf{v} = 132.08(5) = 660\text{ W}$ *Ans.*

F14–11. $+\uparrow\Sigma F_y = ma_y;$

$T - 50(9.81) = 50(0) \quad T = 490.5\text{ N}$

$P_{\text{out}} = \mathbf{T} \cdot \mathbf{v} = 490.5(1.5) = 735.75\text{ W}$

Also, for a point on the other cable

$P_{\text{out}} = \left(\dfrac{490.5}{2}\right)(1.5)(2) = 735.75\text{ W}$

$P_{\text{in}} = \dfrac{P_{\text{out}}}{\varepsilon} = \dfrac{735.75}{0.8} = 920\text{ W}$ *Ans.*

F14–12. $2s_A + s_P = l$

$2a_A + a_P = 0$

$2a_A + 6 = 0$

$a_A = -3\text{ m/s}^2 = 3\text{ m/s}^2 \uparrow$

$\Sigma F_y = ma_y;\quad T_A - 490.5\text{ N} = (50\text{ kg})(3\text{ m/s}^2)$

$T_A = 640.5\text{ N}$

$P_{\text{out}} = \mathbf{T} \cdot \mathbf{v} = (640.5\text{ N}/2)(12) = 3843\text{ W}$

$P_{\text{in}} = \dfrac{P_{\text{out}}}{\varepsilon} = \dfrac{3843}{0.8} = 4803.75\text{ W} = 4.80\text{ kW}$ *Ans.*

F14–13. $T_A + V_A = T_B + V_B$

$0 + 2(9.81)(1.5) = \frac{1}{2}(2)(v_B)^2 + 0$

$v_B = 5.42\text{ m/s}$ *Ans.*

$+\uparrow\Sigma F_n = ma_n;\ T - 2(9.81) = 2\left(\dfrac{(5.42)^2}{1.5}\right)$

$T = 58.9\text{ N}$ *Ans.*

F14–14. $T_A + V_A = T_B + V_B$

$\frac{1}{2}m_A v_A^2 + mgh_A = \frac{1}{2}m_B v_B^2 + mgh_B$

$\left[\frac{1}{2}(2\text{ kg})(1\text{ m/s})^2\right] + [2\,(9.81)\text{ N}(4\text{ m})]$

$= \left[\frac{1}{2}(2\text{ kg})v_B^2\right] + [0]$

$v_B = 8.915\text{ m/s} = 8.92\text{ m/s}$ *Ans.*

$+\uparrow\Sigma F_n = ma_n;\quad N_B - 2(9.81)\text{ N}$

$= (2\text{ kg})\left(\dfrac{(8.915\text{ m/s})^2}{2\text{ m}}\right)$

$N_B = 99.1\text{ N}$ *Ans.*

F14–15. $T_1 + V_1 = T_2 + V_2$

$\frac{1}{2}(2)(4)^2 + \frac{1}{2}(30)(2 - 1)^2$

$= \frac{1}{2}(2)(v)^2 - 2(9.81)(1) + \frac{1}{2}(30)\left(\sqrt{5} - 1\right)^2$

$v = 5.26\text{ m/s}$ *Ans.*

F14–16. $T_A + V_A = T_B + V_B$

$0 + \frac{1}{2}(4)(2.5 - 0.5)^2 + 5(2.5)$

$= \frac{1}{2}\left(\dfrac{5}{32.2}\right)v_B^2 + \frac{1}{2}(4)(1 - 0.5)^2$

$v_B = 16.0\text{ ft/s}$ *Ans.*

F14–17. $T_1 + V_1 = T_2 + V_2$

$\frac{1}{2}mv_1^2 + mgy_1 + \frac{1}{2}ks_1^2$

$\qquad = \frac{1}{2}mv_2^2 + mgy_2 + \frac{1}{2}ks_2^2$

$[0] + [0] + [0] = [0] +$

$\left[-75\text{ lb}(5\text{ ft} + s)\right] + \left[2\left(\frac{1}{2}(1000\text{ lb/ft})s^2\right)\right.$

$\left. + \frac{1}{2}(1500\text{ lb/ft})(s - 0.25\text{ ft})^2\right]$

$s = s_A = s_C = 0.580\text{ ft}$ *Ans.*

Also,

$s_B = 0.5803\text{ ft} - 0.25\text{ ft} = 0.330\text{ ft}$ *Ans.*

F14–18. $T_A + V_A = T_B + V_B$

$\frac{1}{2}mv_A^2 + \left(\frac{1}{2}ks_A^2 + mgy_A\right)$

$\qquad = \frac{1}{2}mv_B^2 + \left(\frac{1}{2}ks_B^2 + mgy_B\right)$

$\frac{1}{2}(4\text{ kg})(2\text{ m/s})^2 + \frac{1}{2}(400\text{ N/m})(0.1\text{ m} - 0.2\text{ m})^2 + 0$

$= \frac{1}{2}(4\text{ kg})v_B^2 + \frac{1}{2}(400\text{ N/m})(\sqrt{(0.4\text{ m})^2 + (0.3\text{ m})^2}$

$\qquad - 0.2\text{ m})^2 + [4(9.81)\text{ N}](-(0.1\text{ m} + 0.3\text{ m}))$

$v_B = 1.962\text{ m/s} = 1.96\text{ m/s}$ *Ans.*

Chapter 15

F15–1. $(\xrightarrow{+})\quad m(v_1)_x + \Sigma\displaystyle\int_{t1}^{t2} F_x\,dt = m(v_2)_x$

$(0.5\text{ kg})(25\text{ m/s})\cos 45° - \displaystyle\int F_x\,dt$

$\qquad = (0.5\text{ kg})(10\text{ m/s})\cos 30°$

$I_x = \displaystyle\int F_x\,dt = 4.509\text{ N} \cdot \text{s}$

$(+\uparrow)\quad m(v_1)_y + \Sigma\displaystyle\int_{t1}^{t2} F_y\,dt = m(v_2)_y$

$-(0.5\text{ kg})(25\text{ m/s})\sin 45° + \displaystyle\int F_y\,dt$

$\qquad = (0.5\text{ kg})(10\text{ m/s})\sin 30°$

$I_y = \displaystyle\int F_y\,dt = 11.339\text{ N} \cdot \text{s}$

$I = \displaystyle\int F\,dt = \sqrt{(4.509\text{ N} \cdot \text{s})^2 + (11.339\text{ N} \cdot \text{s})^2}$

$\qquad = 12.2\text{ N} \cdot \text{s}$ *Ans.*

F15–2. $(+\uparrow)$ $m(v_1)_y + \Sigma \int_{t1}^{t2} F_y\,dt = m(v_2)_y$

$0 + N(4\text{ s}) + (100\text{ lb})(4\text{ s})\sin 30°$
$\qquad - (150\text{ lb})(4\text{ s}) = 0$

$N = 100\text{ lb}$

$(\xrightarrow{+})$ $m(v_1)_x + \Sigma \int_{t1}^{t2} F_x\,dt = m(v_2)_x$

$0 + (100\text{ lb})(4\text{ s})\cos 30° - 0.2(100\text{ lb})(4\text{ s})$
$\qquad = \left(\frac{150}{32.2}\text{ slug}\right)v$

$v = 57.2\text{ ft/s}$ *Ans.*

F15–3. Time to start motion,

$+\uparrow \Sigma F_y = 0;\quad N - 25(9.81)\text{ N} = 0\quad N = 245.25\text{ N}$

$\xrightarrow{+} \Sigma F_x = 0;\quad 20t^2 - 0.3(245.25\text{ N}) = 0\quad t = 1.918\text{ s}$

$(\xrightarrow{+})$ $m(v_1)_x + \Sigma \int_{t1}^{t2} F_x\,dt = m(v_2)_x$

$0 + \int_{1.918\text{ s}}^{4\text{ s}} 20t^2\,dt - (0.25(245.25\text{ N}))(4\text{ s} - 1.918\text{ s})$
$\qquad = (25\text{ kg})v$

$v = 10.1\text{ m/s}$ *Ans.*

F15–4. $(\xrightarrow{+})$ $m(v_1)_x + \Sigma \int_{t1}^{t2} F_x\,dt = m(v_2)_x$

$(1500\text{ kg})(0) + \left[\tfrac{1}{2}(6000\text{ N})(2\text{ s}) + (6000\text{ N})(6\text{ s} - 2\text{ s})\right]$
$\qquad = (1500\text{ kg})\,v$

$v = 20\text{ m/s}$ *Ans.*

F15–5. SUV and trailer,

$m(v_1)_x + \Sigma \int_{t1}^{t2} F_x\,dt = m(v_2)_x$

$0 + (9000\text{ N})(20\text{ s}) = (1500\text{ kg} + 2500\text{ kg})v$

$v = 45.0\text{ m/s}$ *Ans.*

Trailer,

$m(v_1)_x + \Sigma \int_{t1}^{t2} F_x\,dt = m(v_2)_x$

$0 + T(20\text{ s}) = (1500\text{ kg})(45.0\text{ m/s})$

$T = 3375\text{ N} = 3.375\text{ kN}$ *Ans.*

F15–6. Block B:

$(+\downarrow)\ mv_1 + \int F\,dt = mv_2$

$0 + 8(5) - T(5) = \tfrac{8}{32.2}(1)$

$T = 7.95\text{ lb}$ *Ans.*

Block A:

$(\xrightarrow{+})\ mv_1 + \int F\,dt = mv_2$

$0 + 7.95(5) - \mu_k(10)(5) = \tfrac{10}{32.2}(1)$

$\mu_k = 0.789$ *Ans.*

F15–7. $(\xrightarrow{+})$ $m_A(v_A)_1 + m_B(v_B)_1 = m_A(v_A)_2 + m_B(v_B)_2$

$(20(10^3)\text{ kg})(3\text{ m/s}) + (15(10^3)\text{ kg})(-1.5\text{ m/s})$
$\qquad = (20(10^3)\text{ kg})(v_A)_2 + (15(10^3)\text{ kg})(2\text{ m/s})$

$(v_A)_2 = 0.375\text{ m/s} \rightarrow$ *Ans.*

$(\xrightarrow{+})$ $m(v_B)_1 + \Sigma \int_{t1}^{t2} F\,dt = m(v_B)_2$

$(15(10^3)\text{ kg})(-1.5\text{ m/s}) + F_{avg}\,(0.5\text{ s})$
$\qquad = (15(10^3)\text{ kg})(2\text{ m/s})$

$F_{avg} = 105(10^3)\text{ N} = 105\text{ kN}$ *Ans.*

F15–8. $(\xrightarrow{+})$ $m_p[(v_p)_1]_x + m_c[(v)_1]_x = (m_p + m_c)v_2$

$5\left[10\left(\tfrac{4}{5}\right)\right] + 0 = (5 + 20)v_2$

$v_2 = 1.6\text{ m/s}$ *Ans.*

F15–9. $T_1 + V_1 = T_2 + V_2$

$\tfrac{1}{2}m_A(v_A)_1^2 + (V_g)_1 = \tfrac{1}{2}m_A(v_A)_2^2 + (V_g)_2$

$\tfrac{1}{2}(5)(5)^2 + 5(9.81)(1.5) = \tfrac{1}{2}(5)(v_A)_2^2$

$(v_A)_2 = 7.378\text{ m/s}$

$(\xrightarrow{+})$ $m_A(v_A)_2 + m_B(v_B)_2 = (m_A + m_B)v$

$5(7.378) + 0 = (5 + 8)v$

$v = 2.84\text{ m/s}$ *Ans.*

F15–10. $(\xrightarrow{+})$ $m_A(v_A)_1 + m_B(v_B)_1 = m_A(v_A)_2 + m_B(v_B)_2$

$0 + 0 = 10(v_A)_2 + 15(v_B)_2$ (1)

$T_1 + V_1 = T_2 + V_2$

$\tfrac{1}{2}m_A(v_A)_1^2 + \tfrac{1}{2}m_B(v_B)_1^2 + (V_e)_1$
$\qquad = \tfrac{1}{2}m_A(v_A)_2^2 + \tfrac{1}{2}m_B(v_B)_2^2 + (V_e)_2$

$0 + 0 + \tfrac{1}{2}\left[5(10^3)\right](0.2^2)$
$\qquad = \tfrac{1}{2}(10)(v_A)_2^2 + \tfrac{1}{2}(15)(v_B)_2^2 + 0$

$5(v_A)_2^2 + 7.5\,(v_B)_2^2 = 100$ (2)

Solving Eqs. (1) and (2),

$(v_B)_2 = 2.31\text{ m/s} \rightarrow$ *Ans.*

$(v_A)_2 = -3.464\text{ m/s} = 3.46\text{ m/s} \leftarrow$ *Ans.*

F15–11. $(\xrightarrow{+})$ $m_A(v_A)_1 + m_B(v_B)_1 = (m_A + m_B)v_2$

$0 + 10(15) = (15 + 10)v_2$

$v_2 = 6\text{ m/s}$

$T_1 + V_1 = T_2 + V_2$

$\tfrac{1}{2}(m_A + m_B)v_2^2 + (V_e)_2 = \tfrac{1}{2}(m_A + m_B)v_3^2 + (V_e)_3$

$\tfrac{1}{2}(15 + 10)(6^2) + 0 = 0 + \tfrac{1}{2}\left[10(10^3)\right]s_{max}^2$

$s_{max} = 0.3\text{ m} = 300\text{ mm}$ *Ans.*

F15–12. $(\xrightarrow{+})$ $0 + 0 = m_p(v_p)_x - m_c v_c$

$0 = (20 \text{ kg})(v_p)_x - (250 \text{ kg})v_c$

$(v_p)_x = 12.5 \, v_c$ (1)

$\mathbf{v}_p = \mathbf{v}_c + \mathbf{v}_{p/c}$

$(v_p)_x\mathbf{i} + (v_p)_y\mathbf{j} = -v_c\mathbf{i} + [(400 \text{ m/s}) \cos 30°\mathbf{i}$
$\qquad\qquad\qquad\qquad + (400 \text{ m/s}) \sin 30°\mathbf{j}]$

$(v_p)_x\mathbf{i} + (v_p)_y\mathbf{j} = (346.41 - v_c)\mathbf{i} + 200\mathbf{j}$

$(v_p)_x = 346.41 - v_c$

$(v_p)_y = 200 \text{ m/s}$

$(v_p)_x = 320.75 \text{ m/s} \quad v_c = 25.66 \text{ m/s}$

$v_p = \sqrt{(v_p)_x^2 + (v_p)_y^2}$

$\quad = \sqrt{(320.75 \text{ m/s})^2 + (200 \text{ m/s})^2}$

$\quad = 378 \text{ m/s}$ *Ans.*

F15–13. $(\xrightarrow{+})$ $e = \dfrac{(v_B)_2 - (v_A)_2}{(v_A)_1 - (v_B)_1}$

$\qquad = \dfrac{(9 \text{ m/s}) - (1 \text{ m/s})}{(8 \text{ m/s}) - (-2 \text{ m/s})} = 0.8$

F15–14. $(\xrightarrow{+})$ $m_A(v_A)_1 + m_B(v_B)_1 = m_A(v_A)_2 + m_B(v_B)_2$

$[15(10^3) \text{ kg}](5 \text{ m/s}) + [25(10^3)](-7 \text{ m/s})$
$\qquad = [15(10^3) \text{ kg}](v_A)_2 + [25(10^3)](v_B)_2$

$15(v_A)_2 + 25(v_B)_2 = -100$ (1)

Using the coefficient of restitution equation,

$(\xrightarrow{+})$ $e = \dfrac{(v_B)_2 - (v_A)_2}{(v_A)_1 - (v_B)_1}$

$0.6 = \dfrac{(v_B)_2 - (v_A)_2}{5 \text{ m/s} - (-7 \text{ m/s})}$

$(v_B)_2 - (v_A)_2 = 7.2$ (2)

Solving,

$(v_B)_2 = 0.2 \text{ m/s} \rightarrow$ *Ans.*

$(v_A)_2 = -7 \text{ m/s} = 7 \text{ m/s} \leftarrow$ *Ans.*

F15–15. $T_1 + V_1 = T_2 + V_2$

$\frac{1}{2} m(v_A)_1^2 + mg(h_A)_1 = \frac{1}{2} m(v_A)_2^2 + mg(h_A)_2$

$\frac{1}{2}\left(\frac{30}{32.2} \text{ slug}\right)(5 \text{ ft/s})^2 + (30 \text{ lb})(10\text{ft})$
$\qquad = \frac{1}{2}\left(\frac{30}{32.2} \text{ slug}\right)(v_A)_2^2 + 0$

$(v_A)_2 = 25.87 \text{ ft/s} \leftarrow$

$(\xleftarrow{+})$ $m_A(v_A)_2 + m_B(v_B)_2 = m_A(v_A)_3 + m_B(v_B)_3$

$\left(\frac{30}{32.2} \text{ slug}\right)(25.87 \text{ ft/s}) + 0$
$\qquad = \left(\frac{30}{32.2} \text{ slug}\right)(v_A)_3 + \left(\frac{80}{32.2} \text{ slug}\right)(v_B)_3$

$30(v_A)_3 + 80(v_B)_3 = 775.95$ (1)

$(\xleftarrow{+})$ $e = \dfrac{(v_B)_3 - (v_A)_3}{(v_A)_2 - (v_B)_2}$

$0.6 = \dfrac{(v_B)_3 - (v_A)_3}{25.87 \text{ ft/s} - 0}$

$(v_B)_3 - (v_A)_3 = 15.52$ (2)

Solving Eqs. (1) and (2), yields

$(v_B)_3 = 11.3 \text{ ft/s} \leftarrow$

$(v_A)_3 = -4.23 \text{ ft/s} = 4.23 \text{ ft/s} \rightarrow$ *Ans.*

F15–16. $(+\uparrow)$ $m[(v_b)_1]_y = m[(v_b)_2]_y$

$[(v_b)_2]_y = [(v_b)_1]_y = (20 \text{ m/s}) \sin 30° = 10 \text{ m/s} \uparrow$

$(\xrightarrow{+})$ $e = \dfrac{(v_w)_2 - [(v_b)_2]_x}{[(v_b)_1]_x - (v_w)_1}$

$0.75 = \dfrac{0 - [(v_b)_2]_x}{(20 \text{ m/s})\cos 30° - 0}$

$[(v_b)_2]_x = -12.99 \text{ m/s} = 12.99 \text{ m/s} \leftarrow$

$(v_b)_2 = \sqrt{[(v_b)_2]_x^2 + [(v_b)_2]_y^2}$

$\quad = \sqrt{(12.99 \text{ m/s})^2 + (10 \text{ m/s})^2}$

$\quad = 16.4 \text{ m/s}$ *Ans.*

$\theta = \tan^{-1}\left(\dfrac{[(v_b)_2]_y}{[(v_b)_2]_x}\right) = \tan^{-1}\left(\dfrac{10 \text{ m/s}}{12.99 \text{ m/s}}\right)$

$\quad = 37.6°$ *Ans.*

F15–17. $\Sigma m(v_x)_1 = \Sigma m(v_x)_2$

$0 + 0 = \frac{2}{32.2}(1) + \frac{11}{32.2}(v_{Bx})_2$

$(v_{Bx})_2 = -0.1818 \text{ ft/s}$

$\Sigma m(v_y)_1 = \Sigma m(v_y)_2$

$\frac{2}{32.2}(3) + 0 = 0 + \frac{11}{32.2}(v_{By})_2$

$(v_{By})_2 = 0.545 \text{ ft/s}$

$(v_B)_2 = \sqrt{(-0.1818)^2 + (0.545)^2}$

$\quad = 0.575 \text{ ft/s}$ *Ans.*

F15–18. $+\nearrow$ $\frac{2}{32.2}(3)\left(\frac{3}{5}\right) - \frac{2}{32.2}(4)\left(\frac{4}{5}\right)$

$\qquad\qquad = \frac{2}{32.2}(v_B)_{2x} + \frac{2}{32.2}(v_A)_{2x}$

$+\nearrow$ $0.5 = [(v_A)_{2x} - (v_B)_{2x}]/[(3)\left(\frac{3}{5}\right) - (-4)\left(\frac{4}{5}\right)]$

Solving,

$(v_A)_{2x} = 0.550 \text{ ft/s}, (v_B)_{2x} = -1.95 \text{ ft/s}$

Disc A,

$+\nwarrow$ $-\frac{2}{32.2}(4)\left(\frac{3}{5}\right) = \frac{2}{32.2}(v_A)_{2y}$

$(v_A)_{2y} = -2.40 \text{ ft/s}$

Disc B,

$$- \tfrac{2}{32.2}(3)\left(\tfrac{4}{5}\right) = \tfrac{2}{32.2}(v_B)_{2y}$$

$$(v_B)_{2y} = -2.40 \text{ ft/s}$$

$$(v_A)_2 = \sqrt{(0.550)^2 + (2.40)^2} = 2.46 \text{ ft/s} \qquad \textit{Ans.}$$

$$(v_B)_2 = \sqrt{(1.95)^2 + (2.40)^2} = 3.09 \text{ ft/s} \qquad \textit{Ans.}$$

F15–19. $H_O = \Sigma mvd;$

$$H_O = \left[2(10)\left(\tfrac{4}{5}\right)\right](4) - \left[2(10)\left(\tfrac{3}{5}\right)\right](3)$$

$$= 28 \text{ kg} \cdot \text{m}^2/\text{s}$$

F15–20. $H_P = \Sigma mvd;$

$$H_P = [2(15) \sin 30°](2) - [2(15) \cos 30°](5)$$

$$= -99.9 \text{ kg} \cdot \text{m}^2/\text{s} = 99.9 \text{ kg} \cdot \text{m}^2/\text{s} \; \circlearrowleft$$

F15–21. $(H_z)_1 + \Sigma \displaystyle\int M_z \, dt = (H_z)_2$

$$5(2)(1.5) + 5(1.5)(3) = 5v(1.5)$$

$$v = 5 \text{ m/s} \qquad \textit{Ans.}$$

F15–22. $(H_z)_1 + \Sigma \displaystyle\int M_z \, dt = (H_z)_2$

$$0 + \int_0^{4 \text{ s}} (10t)\left(\tfrac{4}{5}\right)(1.5)dt = 5v(1.5)$$

$$v = 12.8 \text{ m/s} \qquad \textit{Ans.}$$

F15–23. $(H_z)_1 + \Sigma \displaystyle\int M_z \, dt = (H_z)_2$

$$0 + \int_0^{5 \text{ s}} 0.9t^2 \, dt = 2v(0.6)$$

$$v = 31.2 \text{ m/s} \qquad \textit{Ans.}$$

F15–24. $(H_z)_1 + \Sigma \displaystyle\int M_z \, dt = (H_z)_2$

$$0 + \int_0^{4 \text{ s}} 8t \, dt + 2(10)(0.5)(4) = 2[10v(0.5)]$$

$$v = 10.4 \text{ m/s} \qquad \textit{Ans.}$$

Chapter 16

F16–1. $\theta = (20 \text{ rev})\left(\tfrac{2\pi \text{ rad}}{1 \text{ rev}}\right) = 40\pi \text{ rad}$

$$\omega^2 = \omega_0^2 + 2\alpha_c (\theta - \theta_0)$$

$$(30 \text{ rad/s})^2 = 0^2 + 2\alpha_c [(40\pi \text{ rad}) - 0]$$

$$\alpha_c = 3.581 \text{ rad/s}^2 = 3.58 \text{ rad/s}^2 \qquad \textit{Ans.}$$

$$\omega = \omega_0 + \alpha_c t$$

$$30 \text{ rad/s} = 0 + (3.581 \text{ rad/s}^2)t$$

$$t = 8.38 \text{ s} \qquad \textit{Ans.}$$

F16–2. $\tfrac{d\omega}{d\theta} = 2(0.005\theta) = (0.01\theta)$

$$\alpha = \omega\tfrac{d\omega}{d\theta} = \left(0.005\,\theta^2\right)(0.01\theta) = 50\left(10^{-6}\right)\theta^3 \text{ rad/s}^2$$

When $\theta = 20 \text{ rev}(2\pi \text{ rad}/1 \text{ rev}) = 40\pi \text{ rad}$,

$$\alpha = \left[50\left(10^{-6}\right)(40\pi)^3\right] \text{ rad/s}^2$$

$$= 99.22 \text{ rad/s}^2 = 99.2 \text{ rad/s}^2 \qquad \textit{Ans.}$$

F16–3. $\omega = 4\theta^{1/2}$

$$150 \text{ rad/s} = 4\,\theta^{1/2}$$

$$\theta = 1406.25 \text{ rad}$$

$$dt = \tfrac{d\theta}{\omega}$$

$$\int_0^t dt = \int_{1 \text{ rad}}^\theta \frac{d\theta}{4\theta^{1/2}}$$

$$t\big|_0^t = \tfrac{1}{2}\,\theta^{1/2}\big|_{1 \text{ rad}}^\theta$$

$$t = \tfrac{1}{2}\,\theta^{1/2} - \tfrac{1}{2}$$

$$t = \tfrac{1}{2}(1406.25)^{1/2} - \tfrac{1}{2} = 18.25 \text{ s} \qquad \textit{Ans.}$$

F16–4. $\omega = \tfrac{d\theta}{dt} = (1.5t^2 + 15) \text{ rad/s}$

$$\alpha = \tfrac{d\omega}{dt} = (3t) \text{ rad/s}$$

$$\omega = [1.5(3^2) + 15] \text{ rad/s} = 28.5 \text{ rad/s}$$

$$\alpha = 3(3) \text{ rad/s}^2 = 9 \text{ rad/s}^2$$

$$v = \omega r = (28.5 \text{ rad/s})(0.75 \text{ ft}) = 21.4 \text{ ft/s} \qquad \textit{Ans.}$$

$$a = \alpha r = (9 \text{ rad/s}^2)(0.75 \text{ ft}) = 6.75 \text{ ft/s}^2 \qquad \textit{Ans.}$$

F16–5. $\omega \, d\omega = \alpha \, d\theta$

$$\int_{2 \text{ rad/s}}^\omega \omega \, d\omega = \int_0^\theta 0.5\theta \, d\theta$$

$$\tfrac{\omega^2}{2}\Big|_{2 \text{ rad/s}}^\omega = 0.25\theta^2\big|_0^\theta$$

$$\omega = (0.5\,\theta^2 + 4)^{1/2} \text{ rad/s}$$

When $\theta = 2 \text{ rev} = 4\pi \text{ rad}$,

$$\omega = [0.5(4\pi)^2 + 4]^{1/2} \text{ rad/s} = 9.108 \text{ rad/s}$$

$$v_P = \omega r = (9.108 \text{ rad/s})(0.2 \text{ m}) = 1.82 \text{ m/s} \quad \textit{Ans.}$$

$$(a_P)_t = \alpha r = (0.5\theta \text{ rad/s}^2)(0.2 \text{ m})\big|_{\theta = 4\pi \text{ rad}}$$

$$= 1.257 \text{ m/s}^2$$

$$(a_P)_n = \omega^2 r = (9.108 \text{ rad/s})^2(0.2 \text{ m}) = 16.59 \text{ m/s}^2$$

$$a_P = \sqrt{(a_P)_t^2 + (a_P)_n^2}$$

$$= \sqrt{(1.257 \text{ m/s}^2)^2 + (16.59 \text{ m/s}^2)^2}$$

$$= 16.6 \text{ m/s}^2 \qquad \textit{Ans.}$$

F16–6. $\alpha_B = \alpha_A\left(\dfrac{r_A}{r_B}\right)$

$\qquad\quad = (4.5 \text{ rad/s}^2)\left(\dfrac{0.075 \text{ m}}{0.225 \text{ m}}\right) = 1.5 \text{ rad/s}^2$

$\qquad \omega_B = (\omega_B)_0 + \alpha_B t$

$\qquad \omega_B = 0 + (1.5 \text{ rad/s}^2)(3 \text{ s}) = 4.5 \text{ rad/s}$

$\qquad \theta_B = (\theta_B)_0 + (\omega_B)_0 t + \frac{1}{2}\alpha_B t^2$

$\qquad \theta_B = 0 + 0 + \frac{1}{2}(1.5 \text{ rad/s}^2)(3 \text{ s})^2$

$\qquad \theta_B = 6.75 \text{ rad}$

$\qquad v_C = \omega_B r_D = (4.5 \text{ rad/s})(0.125 \text{ m})$

$\qquad\quad = 0.5625 \text{ m/s}$ $\hfill$ *Ans.*

$\qquad s_C = \theta_B r_D = (6.75 \text{ rad})(0.125 \text{ m}) = 0.84375 \text{ m}$

$\qquad\quad = 844 \text{ mm}$ $\hfill$ *Ans.*

F16–7. Vector Analysis

$\qquad \mathbf{v}_B = \mathbf{v}_A + \boldsymbol{\omega} \times \mathbf{r}_{B/A}$

$\qquad -v_B\mathbf{j} = (3\mathbf{i}) \text{ m/s}$

$\qquad\qquad + (\omega\mathbf{k}) \times (-1.5\cos 30°\mathbf{i} + 1.5\sin 30°\mathbf{j})$

$\qquad -v_B\mathbf{j} = [3 - \omega_{AB}(1.5\sin 30°)]\mathbf{i} - \omega(1.5\cos 30°)\mathbf{j}$

$\qquad\quad 0 = 3 - \omega(1.5\sin 30°)$ $\hfill$ (1)

$\qquad -v_B = 0 - \omega(1.5\cos 30°)$ $\hfill$ (2)

$\qquad \omega = 4 \text{ rad/s} \qquad v_B = 5.20 \text{ m/s}$ $\hfill$ *Ans.*

Scalar Solution

$\qquad \mathbf{v}_B = \mathbf{v}_A + \mathbf{v}_{B/A}$

$\qquad \begin{bmatrix} \downarrow v_B \end{bmatrix} = \begin{bmatrix} 3 \\ \rightarrow \end{bmatrix} + \begin{bmatrix} \omega(1.5) \measuredangle 30° \end{bmatrix}$

This yields Eqs. (1) and (2).

F16–8. Vector Analysis

$\qquad \mathbf{v}_B = \mathbf{v}_A + \boldsymbol{\omega} \times \mathbf{r}_{B/A}$

$\qquad (v_B)_x\mathbf{i} + (v_B)_y\mathbf{j} = \mathbf{0} + (-10\mathbf{k}) \times (-0.6\mathbf{i} + 0.6\mathbf{j})$

$\qquad (v_B)_x\mathbf{i} + (v_B)_y\mathbf{j} = 6\mathbf{i} + 6\mathbf{j}$

$\qquad (v_B)_x = 6 \text{ m/s} \text{ and } (v_B)_y = 6 \text{ m/s}$

$\qquad v_B = \sqrt{(v_B)_x^2 + (v_B)_y^2}$

$\qquad\quad = \sqrt{(6 \text{ m/s})^2 + (6 \text{ m/s})^2}$

$\qquad\quad = 8.49 \text{ m/s}$ $\hfill$ *Ans.*

Scalar Solution

$\qquad \mathbf{v}_B = \mathbf{v}_A + \mathbf{v}_{B/A}$

$\qquad \begin{bmatrix} (v_B)_x \\ \rightarrow \end{bmatrix} + \begin{bmatrix} (v_B)_y \uparrow \end{bmatrix} = \begin{bmatrix} 0 \end{bmatrix} + \begin{bmatrix} \measuredangle 45° \quad 10\left(\dfrac{0.6}{\cos 45°}\right) \end{bmatrix}$

$\qquad \xrightarrow{+} (v_B)_x = 0 + 10(0.6/\cos 45°)\cos 45° = 6 \text{ m/s} \rightarrow$

$\qquad +\uparrow (v_B)_y = 0 + 10(0.6/\cos 45°)\sin 45° = 6 \text{ m/s}\uparrow$

F16–9. Vector Analysis

$\qquad \mathbf{v}_B = \mathbf{v}_A + \boldsymbol{\omega} \times \mathbf{r}_{B/A}$

$\qquad (4 \text{ ft/s})\mathbf{i} = (-2 \text{ ft/s})\mathbf{i} + (-\omega\mathbf{k}) \times (3 \text{ ft})\mathbf{j}$

$\qquad 4\mathbf{i} = (-2 + 3\omega)\mathbf{i}$

$\qquad \omega = 2 \text{ rad/s}$ $\hfill$ *Ans.*

Scalar Solution

$\qquad \mathbf{v}_B = \mathbf{v}_A + \mathbf{v}_{B/A}$

$\qquad \begin{bmatrix} 4 \\ \rightarrow \end{bmatrix} = \begin{bmatrix} 2 \\ \leftarrow \end{bmatrix} + \begin{bmatrix} \omega(3) \\ \rightarrow \end{bmatrix}$

$\qquad \xrightarrow{+} \quad 4 = -2 + \omega(3); \qquad \omega = 2 \text{ rad/s}$

F16–10. Vector Analysis

$\qquad \mathbf{v}_A = \boldsymbol{\omega}_{OA} \times \mathbf{r}_A$

$\qquad\quad = (12 \text{ rad/s})\mathbf{k} \times (0.3 \text{ m})\mathbf{j}$

$\qquad\quad = [-3.6\mathbf{i}] \text{ m/s}$

$\qquad \mathbf{v}_B = \mathbf{v}_A + \boldsymbol{\omega}_{AB} \times \mathbf{r}_{B/A}$

$\qquad v_B\mathbf{j} = (-3.6 \text{ m/s})\mathbf{i}$

$\qquad\qquad + (\omega_{AB}\mathbf{k}) \times (0.6\cos 30°\mathbf{i} - 0.6\sin 30°\mathbf{j}) \text{ m}$

$\qquad v_B\mathbf{j} = [\omega_{AB}(0.6\sin 30°) - 3.6]\mathbf{i} + \omega_{AB}(0.6\cos 30°)\mathbf{j}$

$\qquad\quad 0 = \omega_{AB}(0.6\sin 30°) - 3.6$ $\hfill$ (1)

$\qquad v_B = \omega_{AB}(0.6\cos 30°)$ $\hfill$ (2)

$\qquad \omega_{AB} = 12 \text{ rad/s} \quad v_B = 6.24 \text{ m/s} \uparrow$

Scalar Solution

$\qquad \mathbf{v}_B = \mathbf{v}_A + \mathbf{v}_{B/A}$

$\qquad \begin{bmatrix} v_B \uparrow \end{bmatrix} = \begin{bmatrix} \leftarrow \\ 12(0.3) \end{bmatrix} + \begin{bmatrix} \nearrow 30° \omega(0.6) \end{bmatrix}$

This yields Eqs. (1) and (2).

F16–11. Vector Analysis

$\qquad \mathbf{v}_C = \mathbf{v}_B + \boldsymbol{\omega}_{BC} \times \mathbf{r}_{C/B}$

$\qquad v_C\mathbf{j} = (-60\mathbf{i}) \text{ ft/s}$

$\qquad\qquad + (-\omega_{BC}\mathbf{k}) \times (-2.5\cos 30°\mathbf{i} + 2.5\sin 30°\mathbf{j}) \text{ ft}$

$\qquad v_C\mathbf{j} = (-60)\mathbf{i} + 2.165\omega_{BC}\mathbf{j} + 1.25\omega_{BC}\mathbf{i}$

$\qquad\quad 0 = -60 + 1.25\omega_{BC}$ $\hfill$ (1)

$\qquad v_C = 2.165 \omega_{BC}$ $\hfill$ (2)

$\qquad \omega_{BC} = 48 \text{ rad/s}$ $\hfill$ *Ans.*

$\qquad v_C = 104 \text{ ft/s}$

Scalar Solution

$\qquad \mathbf{v}_C = \mathbf{v}_B + \mathbf{v}_{C/B}$

$\qquad \begin{bmatrix} v_C \uparrow \end{bmatrix} = \begin{bmatrix} v_B \\ \leftarrow \end{bmatrix} + \begin{bmatrix} \nearrow 30° \omega (2.5) \end{bmatrix}$

This yields Eqs. (1) and (2).

F16–12. Vector Analysis

$$\mathbf{v}_B = \mathbf{v}_A + \boldsymbol{\omega} \times \mathbf{r}_{B/A}$$

$$-v_B \cos 30° \,\mathbf{i} + v_B \sin 30° \,\mathbf{j} = (-3 \text{ m/s})\mathbf{j} +$$

$$(-\omega\mathbf{k}) \times (-2 \sin 45°\mathbf{i} - 2 \cos 45°\mathbf{j}) \text{ m}$$

$$-0.8660 v_B \,\mathbf{i} + 0.5 v_B \,\mathbf{j}$$

$$= -1.4142\omega\mathbf{i} + (1.4142\omega - 3)\mathbf{j}$$

$$-0.8660 v_B = -1.4142\omega \qquad (1)$$

$$0.5 v_B = 1.4142\omega - 3 \qquad (2)$$

$$\omega = 5.02 \text{ rad/s} \quad v_B = 8.20 \text{ m/s} \qquad Ans.$$

Scalar Solution

$$\mathbf{v}_B = \mathbf{v}_A + \mathbf{v}_{B/A}$$

$$\left[\begin{smallmatrix} \\ \end{smallmatrix} 30° \ v_B \right] = \left[\downarrow 3 \right] + \left[\begin{smallmatrix} \\ \end{smallmatrix} 45° \ \omega(2) \right]$$

This yields Eqs. (1) and (2).

F16–13. $\omega_{AB} = \dfrac{v_A}{r_{A/IC}} = \dfrac{6}{3} = 2 \text{ rad/s} \qquad Ans.$

$$\phi = \tan^{-1}\left(\tfrac{2}{1.5}\right) = 53.13°$$

$$r_{C/IC} = \sqrt{(3)^2 + (2.5)^2 - 2(3)(2.5)\cos 53.13°} = 2.5 \text{ m}$$

$$v_C = \omega_{AB}\, r_{C/IC} = 2(2.5) = 5 \text{ m/s} \qquad Ans.$$

$$\theta = 90° - \phi = 90° - 53.13° = 36.9° \ \begin{smallmatrix}\\\end{smallmatrix} \qquad Ans.$$

F16–14. $v_B = \omega_{AB}\, r_{B/A} = 12(0.6) = 7.2 \text{ m/s} \downarrow$

$$v_C = 0 \qquad Ans.$$

$$\omega_{BC} = \dfrac{v_B}{r_{B/IC}} = \dfrac{7.2}{1.2} = 6 \text{ rad/s} \qquad Ans.$$

F16–15. $\omega = \dfrac{v_O}{r_{O/IC}} = \dfrac{6}{0.3} = 20 \text{ rad/s} \qquad Ans.$

$$r_{A/IC} = \sqrt{0.3^2 + 0.6^2} = 0.6708 \text{ m}$$

$$\phi = \tan^{-1}\left(\tfrac{0.3}{0.6}\right) = 26.57°$$

$$v_A = \omega r_{A/IC} = 20(0.6708) = 13.4 \text{ m/s} \qquad Ans.$$

$$\theta = 90° - \phi = 90° - 26.57° = 63.4° \ \angle \qquad Ans.$$

F16–16. The location of *IC* can be determined using similar triangles.

$$\dfrac{0.5 - r_{C/IC}}{3} = \dfrac{r_{C/IC}}{1.5} \qquad r_{C/IC} = 0.1667 \text{ m}$$

$$\omega = \dfrac{v_C}{r_{C/IC}} = \dfrac{1.5}{0.1667} = 9 \text{ rad/s} \qquad Ans.$$

Also, $r_{O/IC} = 0.3 - r_{C/IC} = 0.3 - 0.1667$

$$= 0.1333 \text{ m}.$$

$$v_O = \omega r_{O/IC} = 9(0.1333) = 1.20 \text{ m/s} \qquad Ans.$$

F16–17. $v_B = \omega r_{B/A} = 6(0.2) = 1.2 \text{ m/s}$

$$r_{B/IC} = 0.8 \tan 60° = 1.3856 \text{ m}$$

$$r_{C/IC} = \dfrac{0.8}{\cos 60°} = 1.6 \text{ m}$$

$$\omega_{BC} = \dfrac{v_B}{r_{B/IC}} = \dfrac{1.2}{1.3856} = 0.8660 \text{ rad/s}$$

$$= 0.866 \text{ rad/s} \qquad Ans.$$

Then,

$$v_C = \omega_{BC}\, r_{C/IC} = 0.8660(1.6) = 1.39 \text{ m/s} \qquad Ans.$$

F16–18. $v_B = \omega_{AB}\, r_{B/A} = 10(0.2) = 2 \text{ m/s}$

$$v_C = \omega_{CD}\, r_{C/D} = \omega_{CD}\,(0.2) \rightarrow$$

$$r_{B/IC} = \dfrac{0.4}{\cos 30°} = 0.4619 \text{ m}$$

$$r_{C/IC} = 0.4 \tan 30° = 0.2309 \text{ m}$$

$$\omega_{BC} = \dfrac{v_B}{r_{B/IC}} = \dfrac{2}{0.4619} = 4.330 \text{ rad/s}$$

$$= 4.33 \text{ rad/s} \qquad Ans.$$

$$v_C = \omega_{BC}\, r_{C/IC}$$

$$\omega_{CD}\,(0.2) = 4.330(0.2309)$$

$$\omega_{CD} = 5 \text{ rad/s} \qquad Ans.$$

F16–19. $\omega = \dfrac{v_A}{r_{A/IC}} = \dfrac{6}{3} = 2 \text{ rad/s}$

Vector Analysis

$$\mathbf{a}_B = \mathbf{a}_A + \boldsymbol{\alpha} \times \mathbf{r}_{B/A} - \omega^2\, \mathbf{r}_{B/A}$$

$$a_B\mathbf{i} = -5\mathbf{j} + (\alpha\mathbf{k}) \times (3\mathbf{i} - 4\mathbf{j}) - 2^2(3\mathbf{i} - 4\mathbf{j})$$

$$a_B\mathbf{i} = (4\alpha - 12)\mathbf{i} + (3\alpha + 11)\mathbf{j}$$

$$a_B = 4\alpha - 12 \qquad (1)$$

$$0 = 3\alpha + 11 \qquad (2)$$

$$\alpha = -3.67 \text{ rad/s}^2 \qquad Ans.$$

$$a_B = -26.7 \text{ m/s}^2 \qquad Ans.$$

Scalar Solution

$$\mathbf{a}_B = \mathbf{a}_A + \mathbf{a}_{B/A}$$

$$\left[\begin{smallmatrix} a_B \\ \rightarrow \end{smallmatrix} \right] = \left[\downarrow 5 \right] + \left[\alpha\,(5)\ \tfrac{5}{4}3 \right] + \left[4\ \tfrac{5}{3}\,(2)^2(5) \right]$$

This yields Eqs. (1) and (2).

F16–20. Vector Analysis

$$\mathbf{a}_A = \mathbf{a}_O + \boldsymbol{\alpha} \times \mathbf{r}_{A/O} - \omega^2 \mathbf{r}_{A/O}$$
$$= 1.8\mathbf{i} + (-6\mathbf{k}) \times (0.3\mathbf{j}) - 12^2(0.3\mathbf{j})$$
$$= \{3.6\mathbf{i} - 43.2\mathbf{j}\}\,\text{m/s}^2 \qquad Ans.$$

Scalar Analysis

$$\mathbf{a}_A = \mathbf{a}_O + \mathbf{a}_{A/O}$$

$$\begin{bmatrix}(a_A)_x \\ \rightarrow\end{bmatrix} + \begin{bmatrix}(a_A)_y\uparrow\end{bmatrix} = \begin{bmatrix}(6)(0.3) \\ \rightarrow\end{bmatrix} + \begin{bmatrix}(6)(0.3) \\ \rightarrow\end{bmatrix}$$
$$+ \begin{bmatrix}\downarrow(12)^2(0.3)\end{bmatrix}$$

$$\overset{+}{\rightarrow} \quad (a_A)_x = 1.8 + 1.8 = 3.6\ \text{m/s}^2 \rightarrow$$
$$+\uparrow \quad (a_A)_y = -43.2\ \text{m/s}^2$$

F16–21. Using

$$v_O = \omega r; \qquad 6 = \omega(0.3)$$
$$\omega = 20\ \text{rad/s}$$
$$a_O = \alpha r; \qquad 3 = \alpha(0.3)$$
$$\alpha = 10\ \text{rad/s}^2 \qquad Ans.$$

Vector Analysis

$$\mathbf{a}_A = \mathbf{a}_O + \boldsymbol{\alpha} \times \mathbf{r}_{A/O} - \omega^2 \mathbf{r}_{A/O}$$
$$= 3\mathbf{i} + (-10\mathbf{k}) \times (-0.6\mathbf{i}) - 20^2(-0.6\mathbf{i})$$
$$= \{243\mathbf{i} + 6\mathbf{j}\}\ \text{m/s}^2 \qquad Ans.$$

Scalar Analysis

$$\mathbf{a}_A = \mathbf{a}_O + \mathbf{a}_{A/O}$$

$$\begin{bmatrix}(a_A)_x \\ \rightarrow\end{bmatrix} + \begin{bmatrix}(a_A)_y \\ \uparrow\end{bmatrix} = \begin{bmatrix}3 \\ \rightarrow\end{bmatrix} + \begin{bmatrix}10(0.6) \\ \uparrow\end{bmatrix} + \begin{bmatrix}(20)^2(0.6) \\ \rightarrow\end{bmatrix}$$

$$\overset{+}{\rightarrow} \quad (a_A)_x = 3 + 240 = 243\ \text{m/s}^2$$
$$+\uparrow \quad (a_A)_y = 10(0.6) = 6\ \text{m/s}^2 \uparrow$$

F16–22. $\dfrac{r_{A/IC}}{3} = \dfrac{0.5 - r_{A/IC}}{1.5}; \qquad r_{A/IC} = 0.3333\ \text{m}$

$$\omega = \frac{v_A}{r_{A/IC}} = \frac{3}{0.3333} = 9\ \text{rad/s}$$

Vector Analysis

$$\mathbf{a}_A = \mathbf{a}_C + \boldsymbol{\alpha} \times \mathbf{r}_{A/C} - \omega^2\,\mathbf{r}_{A/C}$$
$$1.5\mathbf{i} - (a_A)_n\mathbf{j} = -0.75\mathbf{i} + (a_C)_n\mathbf{j}$$
$$+ (-\alpha\mathbf{k}) \times 0.5\mathbf{j} - 9^2(0.5\mathbf{j})$$
$$1.5\mathbf{i} - (a_A)_n\mathbf{j} = (0.5\alpha - 0.75)\mathbf{i} + \big[(a_C)_n - 40.5\big]\mathbf{j}$$
$$1.5 = 0.5\alpha - 0.75$$
$$\alpha = 4.5\ \text{rad/s}^2 \qquad Ans.$$

Scalar Analysis

$$\mathbf{a}_A = \mathbf{a}_C + \mathbf{a}_{A/C}$$

$$\begin{bmatrix}1.5 \\ \rightarrow\end{bmatrix} + \begin{bmatrix}(a_A)_n \\ \downarrow\end{bmatrix} = \begin{bmatrix}0.75 \\ \leftarrow\end{bmatrix} + \begin{bmatrix}(a_C)_n \\ \uparrow\end{bmatrix} + \begin{bmatrix}\alpha(0.5) \\ \rightarrow\end{bmatrix}$$
$$+ \begin{bmatrix}(9)^2(0.5) \\ \downarrow\end{bmatrix}$$

$$\overset{+}{\rightarrow} \quad 1.5 = -0.75 + \alpha(0.5)$$
$$\alpha = 4.5\ \text{rad/s}^2$$

F16–23. $v_B = \omega r_{B/A} = 12(0.3) = 3.6\ \text{m/s}$

$$\omega_{BC} = \frac{v_B}{r_{B/IC}} = \frac{3.6}{1.2} = 3\ \text{rad/s}$$

Vector Analysis

$$\mathbf{a}_B = \boldsymbol{\alpha} \times \mathbf{r}_{B/A} - \omega^2 \mathbf{r}_{B/A}$$
$$= (-6\mathbf{k}) \times (0.3\mathbf{i}) - 12^2(0.3\mathbf{i})$$
$$= \{-43.2\mathbf{i} - 1.8\mathbf{j}\}\ \text{m/s}^2$$
$$\mathbf{a}_C = \mathbf{a}_B + \boldsymbol{\alpha}_{BC} \times \mathbf{r}_{C/B} - \omega_{BC}^2\mathbf{r}_{C/B}$$
$$a_C\mathbf{i} = (-43.2\mathbf{i} - 1.8\mathbf{j})$$
$$+ (\alpha_{BC}\mathbf{k}) \times (1.2\mathbf{i}) - 3^2(1.2\mathbf{i})$$
$$a_C\mathbf{i} = -54\mathbf{i} + (1.2\alpha_{BC} - 1.8)\mathbf{j}$$
$$a_C = -54\ \text{m/s}^2 = 54\ \text{m/s}^2 \leftarrow \qquad Ans.$$
$$0 = 1.2\alpha_{BC} - 1.8 \quad \alpha_{BC} = 1.5\ \text{rad/s}^2 \qquad Ans.$$

Scalar Analysis

$$\mathbf{a}_C = \mathbf{a}_B + \mathbf{a}_{C/B}$$

$$\begin{bmatrix}a_C \\ \leftarrow\end{bmatrix} = \begin{bmatrix}6(0.3) \\ \downarrow\end{bmatrix} + \begin{bmatrix}(12)^2(0.3) \\ \leftarrow\end{bmatrix} + \begin{bmatrix}\alpha_{BC}(1.2) \\ \uparrow\end{bmatrix} + \begin{bmatrix}(3)^2(1.2) \\ \leftarrow\end{bmatrix}$$

$$\overset{+}{\leftarrow} \quad a_C = 43.2 + 10.8 = 54\ \text{m/s}^2 \leftarrow$$
$$+\uparrow \quad 0 = -6(0.3) + 1.2\alpha_{BC}$$
$$\alpha_{BC} = 1.5\ \text{rad/s}^2$$

F16–24. $v_B = \omega\, r_{B/A} = 6(0.2) = 1.2\ \text{m/s} \rightarrow$

$$r_{B/IC} = 0.8 \tan 60° = 1.3856\ \text{m}$$

$$\omega_{BC} = \frac{v_B}{r_{B/IC}} = \frac{1.2}{1.3856} = 0.8660\ \text{rad/s}$$

Vector Analysis

$$\mathbf{a}_B = \boldsymbol{\alpha} \times \mathbf{r}_{B/A} - \omega^2 \mathbf{r}_{B/A}$$
$$= (-3\mathbf{k}) \times (0.2\mathbf{j}) - 6^2(0.2\mathbf{j})$$
$$= [0.6\mathbf{i} - 7.2\mathbf{j}]\ \text{m/s}$$
$$\mathbf{a}_C = \mathbf{a}_B + \boldsymbol{\alpha}_{BC} \times \mathbf{r}_{C/B} - \omega^2\mathbf{r}_{C/B}$$
$$a_C \cos 30°\mathbf{i} + a_C \sin 30°\mathbf{j}$$
$$= (0.6\mathbf{i} - 7.2\mathbf{j}) + (\alpha_{BC}\mathbf{k} \times 0.8\mathbf{i}) - 0.8660^2(0.8\mathbf{i})$$

$0.8660a_C \mathbf{i} + 0.5a_C \mathbf{j} = (0.8\alpha_{BC} - 7.2)\mathbf{j}$

$0.8660a_C = 0$ (1)

$0.5a_C = 0.8\alpha_{BC} - 7.2$ (2)

$a_C = 0$ $\alpha_{BC} = 9 \text{ rad/s}^2$ *Ans.*

Scalar Analysis

$\mathbf{a}_C = \mathbf{a}_B + \mathbf{a}_{C/B}$

$$\begin{bmatrix} a_C \\ \measuredangle 30° \end{bmatrix} = \begin{bmatrix} 3(0.2) \\ \rightarrow \end{bmatrix} + \begin{bmatrix} (6)^2(0.2) \\ \downarrow \end{bmatrix} + \begin{bmatrix} \alpha_{BC}(0.8) \\ \uparrow \end{bmatrix}$$
$$+ \begin{bmatrix} (0.8660)^2(0.8) \\ \leftarrow \end{bmatrix}$$

This yields Eqs. (1) and (2).

Chapter 17

F17–1. $\xrightarrow{+} \Sigma F_x = m(a_G)_x;$ $100\left(\frac{4}{5}\right) = 100a$

$a = 0.8 \text{ m/s}^2 \rightarrow$ *Ans.*

$+\uparrow \Sigma F_y = m(a_G)_y;$

$N_A + N_B - 100\left(\frac{3}{5}\right) - 100(9.81) = 0$ (1)

$\zeta + \Sigma M_G = 0;$

$N_A(0.6) + 100\left(\frac{3}{5}\right)(0.7)$

      $- N_B(0.4) - 100\left(\frac{4}{5}\right)(0.7) = 0$ (2)

$N_A = 430.4 \text{ N} = 430 \text{ N}$ *Ans.*

$N_B = 610.6 \text{ N} = 611 \text{ N}$ *Ans.*

F17–2. $\Sigma F_{x'} = m(a_G)_{x'};$ $80(9.81) \sin 15° = 80a$

$a = 2.54 \text{ m/s}^2$ *Ans.*

$\Sigma F_{y'} = m(a_G)_{y'};$

$N_A + N_B - 80(9.81) \cos 15° = 0$ (1)

$\zeta + \Sigma M_G = 0;$

$N_A(0.5) - N_B(0.5) = 0$ (2)

$N_A = N_B = 379 \text{ N}$ *Ans.*

F17–3. $\zeta + \Sigma M_A = \Sigma(M_k)_A;$ $10\left(\frac{3}{5}\right)(7) = \frac{20}{32.2} a(3.5)$

$a = 19.3 \text{ ft/s}^2$ *Ans.*

$\xrightarrow{+} \Sigma F_x = m(a_G)_x;$ $A_x + 10\left(\frac{3}{5}\right) = \frac{20}{32.2}(19.32)$

$A_x = 6 \text{ lb}$ *Ans.*

$+\uparrow \Sigma F_y = m(a_G)_y;$ $A_y - 20 + 10\left(\frac{4}{5}\right) = 0$

$A_y = 12 \text{ lb}$ *Ans.*

F17–4. $F_A = \mu_s N_A = 0.2N_A$ $F_B = \mu_s N_B = 0.2N_B$

$\xrightarrow{+} \Sigma F_x = m(a_G)_x;$

$0.2N_A + 0.2N_B = 100a$ (1)

$+\uparrow \Sigma F_y = m(a_G)_y;$

$N_A + N_B - 100(9.81) = 0$ (2)

$\zeta + \Sigma M_G = 0;$

$0.2N_A(0.75) + N_A(0.9) + 0.2N_B(0.75)$

      $- N_B(0.6) = 0$ (3)

Solving Eqs. (1), (2), and (3),

$N_A = 294.3 \text{ N} = 294 \text{ N}$

$N_B = 686.7 \text{ N} = 687 \text{ N}$

$a = 1.96 \text{ m/s}^2$ *Ans.*

Since N_A is positive, the table will indeed slide before it tips.

F17–5. $(a_G)_t = \alpha r = \alpha(1.5 \text{ m})$

$(a_G)_n = \omega^2 r = (5 \text{ rad/s})^2(1.5 \text{ m}) = 37.5 \text{ m/s}^2$

$\Sigma F_t = m(a_G)_t;$ $100 \text{ N} = 50 \text{ kg}[\alpha(1.5 \text{ m})]$

                    $\alpha = 1.33 \text{ rad/s}^2$ *Ans.*

$\Sigma F_n = m(a_G)_n;$ $T_{AB} + T_{CD} - 50(9.81) \text{ N}$

                    $= 50 \text{ kg}(37.5 \text{ m/s}^2)$

$T_{AB} + T_{CD} = 2365.5$

$\zeta + \Sigma M_G = 0;$ $T_{CD}(1 \text{ m}) - T_{AB}(1 \text{ m}) = 0$

$T_{AB} = T_{CD} = 1182.75 \text{ N} = 1.18 \text{ kN}$ *Ans.*

F17–6. $\zeta + \Sigma M_C = 0;$

$\mathbf{a}_G = \mathbf{a}_D = \mathbf{a}_B$

$D_y(0.6) - 450 = 0$ $D_y = 750 \text{ N}$ *Ans.*

$(a_G)_n = \omega^2 r = 6^2(0.6) = 21.6 \text{ m/s}^2$

$(a_G)_t = \alpha r = \alpha(0.6)$

$+\uparrow \Sigma F_t = m(a_G)_t;$

$750 - 50(9.81) = 50[\alpha(0.6)]$

$\alpha = 8.65 \text{ rad/s}^2$ *Ans.*

$\xrightarrow{+} \Sigma F_n = m(a_G)_n;$

$F_{AB} + D_x = 50(21.6)$ (1)

$\zeta + \Sigma M_G = 0;$

$D_x(0.4) + 750(0.1) - F_{AB}(0.4) = 0$ (2)

$D_x = 446.25 \text{ N} = 446 \text{ N}$ *Ans.*

$F_{AB} = 633.75 \text{ N} = 634 \text{ N}$ *Ans.*

F17–7. $I_O = mk_O^2 = 100(0.5^2) = 25 \text{ kg} \cdot \text{m}^2$

$\zeta + \Sigma M_O = I_O\alpha; \qquad -100(0.6) = -25\alpha$

$\alpha = 2.4 \text{ rad/s}^2$

$\omega = \omega_0 + \alpha_c t$

$\omega = 0 + 2.4(3) = 7.2 \text{ rad/s} \qquad\qquad Ans.$

F17–8. $I_O = \frac{1}{2}mr^2 = \frac{1}{2}(50)(0.3^2) = 2.25 \text{ kg} \cdot \text{m}^2$

$\zeta + \Sigma M_O = I_O\alpha;$

$\qquad -9t = -2.25\alpha \qquad \alpha = (4t) \text{ rad/s}^2$

$d\omega = \alpha \, dt$

$\int_0^\omega d\omega = \int_0^t 4t \, dt$

$\omega = (2t^2) \text{ rad/s}$

$\omega = 2(4^2) = 32 \text{ rad/s} \qquad\qquad Ans.$

F17–9. $(a_G)_t = \alpha r_G = \alpha(0.15)$

$(a_G)_n = \omega^2 r_G = 6^2(0.15) = 5.4 \text{ m/s}^2$

$I_O = I_G + md^2 = \frac{1}{12}(30)(0.9^2) + 30(0.15^2)$

$\qquad = 2.7 \text{ kg} \cdot \text{m}^2$

$\zeta + \Sigma M_O = I_O\alpha; \quad 60 - 30(9.81)(0.15) = 2.7\alpha$

$\alpha = 5.872 \text{ rad/s}^2 = 5.87 \text{ rad/s}^2 \qquad Ans.$

$\overset{+}{\leftarrow} \Sigma F_n = m(a_G)_n; \quad O_n = 30(5.4) = 162 \text{ N} \quad Ans.$

$+\uparrow \Sigma F_t = m(a_G)_t;$

$O_t - 30(9.81) = 30[5.872(0.15)]$

$O_t = 320.725 \text{ N} = 321 \text{ N} \qquad\qquad Ans.$

F17–10. $(a_G)_t = \alpha r_G = \alpha(0.3)$

$(a_G)_n = \omega^2 r_G = 10^2(0.3) = 30 \text{ m/s}^2$

$I_O = I_G + md^2 = \frac{1}{2}(30)(0.3^2) + 30(0.3^2)$

$\qquad = 4.05 \text{ kg} \cdot \text{m}^2$

$\zeta + \Sigma M_O = I_O\alpha;$

$50\left(\frac{3}{5}\right)(0.3) + 50\left(\frac{4}{5}\right)(0.3) = 4.05\alpha$

$\alpha = 5.185 \text{ rad/s}^2 = 5.19 \text{ rad/s}^2 \qquad Ans.$

$+\uparrow \Sigma F_n = m(a_G)_n;$

$O_n + 50\left(\frac{3}{5}\right) - 30(9.81) = 30(30)$

$O_n = 1164.3 \text{ N} = 1.16\text{kN} \qquad\qquad Ans.$

$\overset{+}{\to} \Sigma F_t = m(a_G)_t;$

$O_t + 50\left(\frac{4}{5}\right) = 30[5.185(0.3)]$

$O_t = 6.67 \text{ N} \qquad\qquad Ans.$

F17–11. $I_G = \frac{1}{12}ml^2 = \frac{1}{12}(15 \text{ kg})(0.9 \text{ m})^2 = 1.0125 \text{ kg} \cdot \text{m}^2$

$(a_G)_n = \omega^2 r_G = 0$

$(a_G)_t = \alpha(0.15)$

$I_O = I_G + md_{OG}^2$

$\qquad = 1.0125 \text{ kg} \cdot \text{m}^2 + 15 \text{ kg}(0.15 \text{ m})^2$

$\qquad = 1.35 \text{ kg} \cdot \text{m}^2$

$\zeta + \Sigma M_O = I_O\alpha;$

$[15(9.81) \text{ N}](0.15 \text{ m}) = (1.35 \text{ kg} \cdot \text{m}^2)\alpha$

$\alpha = 16.35 \text{ rad/s}^2 \qquad\qquad Ans.$

$+\downarrow \Sigma F_t = m(a_G)_t; \quad -O_t + 15(9.81)\text{N}$

$\qquad = (15 \text{ kg})[16.35 \text{ rad/s}^2(0.15 \text{ m})]$

$O_t = 110.36 \text{ N} = 110 \text{ N} \qquad\qquad Ans.$

$\overset{+}{\to} \Sigma F_n = m(a_G)_n; \qquad O_n = 0 \qquad\qquad Ans.$

F17–12. $(a_G)_t = \alpha r_G = \alpha(0.45)$

$(a_G)_n = \omega^2 r_G = 6^2(0.45) = 16.2 \text{ m/s}^2$

$I_O = \frac{1}{3}ml^2 = \frac{1}{3}(30)(0.9^2) = 8.1 \text{ kg} \cdot \text{m}^2$

$\zeta + \Sigma M_O = I_O\alpha;$

$300\left(\frac{4}{5}\right)(0.6) - 30(9.81)(0.45) = 8.1\alpha$

$\alpha = 1.428 \text{ rad/s}^2 = 1.43 \text{ rad/s}^2 \qquad Ans.$

$\overset{+}{\leftarrow} \Sigma F_n = m(a_G)_n; \quad O_n + 300\left(\frac{3}{5}\right) = 30(16.2)$

$\qquad\qquad\qquad O_n = 306 \text{ N} \qquad\qquad Ans.$

$+\uparrow \Sigma F_t = m(a_G)_t; \quad O_t + 300\left(\frac{4}{5}\right) - 30(9.81)$

$\qquad\qquad\qquad = 30[1.428(0.45)]$

$O_t = 73.58 \text{ N} = 73.6 \text{ N} \qquad\qquad Ans.$

F17–13. $I_G = \frac{1}{12}ml^2 = \frac{1}{12}(60)(3^2) = 45 \text{ kg} \cdot \text{m}^2$

$+\uparrow \Sigma F_y = m(a_G)_y;$

$80 - 20 = 60a_G \quad a_G = 1 \text{ m/s}^2\uparrow$

$\zeta + \Sigma M_G = I_G\alpha; \qquad 80(1) + 20(0.75) = 45\alpha$

$\alpha = 2.11 \text{ rad/s}^2 \qquad\qquad Ans.$

F17–14. $\zeta + \Sigma M_A = (\mathcal{M}_k)_A;$

$-200(0.3) = -100a_G(0.3) - 4.5\alpha$

$30a_G + 4.5\alpha = 60 \qquad (1)$

$a_G = \alpha r = \alpha(0.3) \qquad (2)$

$\alpha = 4.44 \text{ rad/s}^2 \quad a_G = 1.33 \text{ m/s}^2 \to \qquad Ans.$

F17–15. $+\uparrow \Sigma F_y = m(a_G)_y;$

$N - 20(9.81) = 0 \quad N = 196.2 \text{ N}$

$\overset{+}{\to} \Sigma F_x = m(a_G)_x; \quad 0.5(196.2) = 20a_O$

$a_O = 4.905 \text{ m/s}^2 \rightarrow$ *Ans.*

$\zeta + \Sigma M_O = I_O\alpha;$

$0.5(196.2)(0.4) - 100 = -1.8\alpha$

$\alpha = 33.8 \text{ rad/s}^2$ *Ans.*

F17–16. Sphere $I_G = \frac{2}{5}(20)(0.15)^2 = 0.18 \text{ kg} \cdot \text{m}^2$

$\zeta + \Sigma M_{IC} = (\mathcal{M}_k)_{IC};$

$20(9.81)\sin 30°(0.15) = 0.18\alpha + (20a_G)(0.15)$

$0.18\alpha + 3a_G = 14.715$

$a_G = \alpha r = \alpha(0.15)$

$\alpha = 23.36 \text{ rad/s}^2 = 23.4 \text{ rad/s}^2$ *Ans.*

$a_G = 3.504 \text{ m/s}^2 = 3.50 \text{ m/s}^2$ *Ans.*

F17–17. $+\uparrow \Sigma F_y = m(a_G)_y;$

$N - 200(9.81) = 0 \quad N = 1962 \text{ N}$

$\xrightarrow{+} \Sigma F_x = m(a_G)_x;$

$T - 0.2(1962) = 200a_G$ (1)

$\zeta + \Sigma M_A = (\mathcal{M}_k)_A; \quad 450 - 0.2(1962)(1)$

$\qquad = 18\alpha + 200a_G(0.4)$ (2)

$(a_A)_t = 0 \quad a_A = (a_A)_n$

$\mathbf{a}_G = \mathbf{a}_A + \boldsymbol{\alpha} \times \mathbf{r}_{G/A} - \omega^2 \mathbf{r}_{G/A}$

$a_G\mathbf{i} = -a_A\mathbf{j} + \alpha\mathbf{k} \times (-0.4\mathbf{j}) - \omega^2(-0.4\mathbf{j})$

$a_G\mathbf{i} = 0.4\alpha\mathbf{i} + (0.4\omega^2 - a_A)\mathbf{j}$

$a_G = 0.4\alpha$ (3)

Solving Eqs. (1), (2), and (3),

$\alpha = 1.15 \text{ rad/s}^2 \quad a_G = 0.461 \text{ m/s}^2$

$T = 485 \text{ N}$ *Ans.*

F17–18. $\xrightarrow{+} \Sigma F_x = m(a_G)_x; \quad 0 = 12(a_G)_x \quad (a_G)_x = 0$

$\zeta + \Sigma M_A = (\mathcal{M}_k)_A$

$-12(9.81)(0.3) = 12(a_G)_y(0.3) - \frac{1}{12}(12)(0.6)^2\alpha$

$0.36\alpha - 3.6(a_G)_y = 35.316$ (1)

$\omega = 0$

$\mathbf{a}_G = \mathbf{a}_A + \boldsymbol{\alpha} \times \mathbf{r}_{G/A} - \omega^2 \mathbf{r}_{G/A}$

$(a_G)_y\,\mathbf{j} = a_A\mathbf{i} + (-\alpha\mathbf{k}) \times (0.3\mathbf{i}) - \mathbf{0}$

$(a_G)_y\,\mathbf{j} = (a_A)\mathbf{i} - 0.3\,\mathbf{j}$

$a_A = 0$ *Ans.*

$(a_G)_y = -0.3\alpha$ (2)

Solving Eqs. (1) and (2)

$\alpha = 24.5 \text{ rad/s}^2$

$(a_G)_y = -7.36 \text{ m/s}^2 = 7.36 \text{ m/s}^2 \downarrow$ *Ans.*

Chapter 18

F18–1. $I_O = mk_O^2 = 80(0.4^2) = 12.8 \text{ kg} \cdot \text{m}^2$

$T_1 = 0$

$T_2 = \frac{1}{2}I_O\omega^2 = \frac{1}{2}(12.8)\omega^2 = 6.4\omega^2$

$s = \theta r = 20(2\pi)(0.6) = 24\pi \text{ m}$

$T_1 + \Sigma U_{1-2} = T_2$

$0 + 50(24\pi) = 6.4\omega^2$

$\omega = 24.3 \text{ rad/s}$ *Ans.*

F18–2. $T_1 = 0$

$T_2 = \frac{1}{2}m(v_G)_2^2 + \frac{1}{2}I_G\omega_2^2$

$\quad = \frac{1}{2}\left(\frac{50}{32.2}\text{ slug}\right)(2.5\omega_2)^2$

$\qquad + \frac{1}{2}\left[\frac{1}{12}\left(\frac{50}{32.2}\text{ slug}\right)(5 \text{ ft})^2\right]\omega_2^2$

$T_2 = 6.4700\omega_2^2$

Or,

$I_O = \frac{1}{3}ml^2 = \frac{1}{3}\left(\frac{50}{32.2}\text{ slug}\right)(5 \text{ ft})^2$

$\quad = 12.9400 \text{ slug} \cdot \text{ft}^2$

So that

$T_2 = \frac{1}{2}I_O\omega_2^2 = \frac{1}{2}(12.9400 \text{ slug} \cdot \text{ft}^2)\omega_2^2$

$\quad = 6.4700\omega_2^2$

$T_1 + \Sigma U_{1-2} = T_2$

$T_1 + [-Wy_G + M\theta] = T_2$

$0 + \left[-(50 \text{ lb})(2.5 \text{ ft}) + (100 \text{ lb} \cdot \text{ft})(\frac{\pi}{2})\right]$

$\quad = 6.4700\omega_2^2$

$\omega_2 = 2.23 \text{ rad/s}$ *Ans.*

F18–3. $(v_G)_2 = \omega_2 r_{G/IC} = \omega_2(2.5)$

$I_G = \frac{1}{12}ml^2 = \frac{1}{12}(50)(5^2) = 104.17 \text{ kg} \cdot \text{m}^2$

$T_1 = 0$

$T_2 = \frac{1}{2}m(v_G)_2^2 + \frac{1}{2}I_G\omega_2^2$

$\quad = \frac{1}{2}(50)\left[\omega_2(2.5)\right]^2 + \frac{1}{2}(104.17)\omega_2^2 = 208.33\omega_2^2$

$U_P = Ps_P = 600(3) = 1800 \text{ J}$

$U_W = -Wh = -50(9.81)(2.5 - 2) = -245.25 \text{ J}$

$T_1 + \Sigma U_{1-2} = T_2$

$0 + 1800 + (-245.25) = 208.33\omega_2^2$

$\omega_2 = 2.732 \text{ rad/s} = 2.73 \text{ rad/s}$ *Ans.*

F18–4. $T = \frac{1}{2}mv_O^2 + \frac{1}{2}I_O\omega^2$

$\qquad = \frac{1}{2}(50 \text{ kg})(0.4\omega)^2 + \frac{1}{2}\left[50 \text{ kg}(0.3 \text{ m})^2\right]\omega^2$

$\qquad = 6.25\omega^2 \text{ J}$

Or,

$T = \frac{1}{2}I_{IC}\omega^2$

$\qquad = \frac{1}{2}\left[50 \text{ kg}(0.3 \text{ m})^2 + 50 \text{ kg}(0.4 \text{ m})^2\right]\omega^2$

$\qquad = 6.25\omega^2 \text{ J}$

$s_O = \theta r = 10(2\pi \text{ rad})(0.4 \text{ m}) = 8\pi \text{ m}$

$T_1 + \Sigma U_{1-2} = T_2$

$T_1 + P\cos 30° \, s_O = T_2$

$0 + (50 \text{ N})\cos 30°(8\pi \text{ m}) = 6.25\omega^2 \text{ J}$

$\omega = 13.2 \text{ rad/s}$ \hfill *Ans.*

F18–5. $I_G = \frac{1}{12}ml^2 = \frac{1}{12}(30)(3^2) = 22.5 \text{ kg} \cdot \text{m}^2$

$T_1 = 0$

$T_2 = \frac{1}{2}mv_G^2 + \frac{1}{2}I_G\omega^2$

$\qquad = \frac{1}{2}(30)[\omega(0.5)]^2 + \frac{1}{2}(22.5)\omega^2 = 15\omega^2$

Or,

$I_O = I_G + md^2 = \frac{1}{12}(30)(3^2) + 30(0.5^2)$

$\qquad = 30 \text{ kg} \cdot \text{m}^2$

$T_2 = \frac{1}{2}I_O\omega^2 = \frac{1}{2}(30)\omega^2 = 15\omega^2$

$s_1 = \theta r_1 = 8\pi(0.5) = 4\pi \text{ m}$

$s_2 = \theta r_2 = 8\pi(1.5) = 12\pi \text{ m}$

$U_{P_1} = P_1 s_1 = 30(4\pi) = 120\pi \text{ J}$

$U_{P_2} = P_2 s_2 = 20(12\pi) = 240\pi \text{ J}$

$U_M = M\theta = 20[4(2\pi)] = 160\pi \text{ J}$

$T_1 + \Sigma U_{1-2} = T_2$

$0 + 120\pi + 240\pi + 160\pi = 15\omega^2$

$\omega = 10.44 \text{ rad/s} = 10.4 \text{ rad/s}$ \hfill *Ans.*

F18–6. $v_O = \omega r = \omega(0.4)$

$I_O = mk_O^2 = 20(0.3^2) = 1.8 \text{ kg} \cdot \text{m}^2$

$T_1 = 0$

$T_2 = \frac{1}{2}mv_G^2 + \frac{1}{2}I_G\omega^2$

$\qquad = \frac{1}{2}(20)[\omega(0.4)]^2 + \frac{1}{2}(1.8)\omega^2$

$\qquad = 2.5\omega^2$

$U_M = M\theta = M\left(\frac{s_O}{r}\right) = 50\left(\frac{20}{0.4}\right) = 2500 \text{ J}$

$T_1 + \Sigma U_{1-2} = T_2$

$0 + 2500 = 2.5\omega^2$

$\omega = 31.62 \text{ rad/s} = 31.6 \text{ rad/s}$ \hfill *Ans.*

F18–7. $v_G = \omega r = \omega(0.3)$

$I_G = \frac{1}{2}mr^2 = \frac{1}{2}(30)(0.3^2) = 1.35 \text{ kg} \cdot \text{m}^2$

$T_1 = 0$

$T_2 = \frac{1}{2}m(v_G)_2^2 + \frac{1}{2}I_G\omega_2^2$

$\qquad = \frac{1}{2}(30)[\omega_2(0.3)]^2 + \frac{1}{2}(1.35)\omega_2^2 = 2.025\omega_2^2$

$(V_g)_1 = Wy_1 = 0$

$(V_g)_2 = -Wy_2 = -30(9.81)(0.3) = -88.29 \text{ J}$

$T_1 + V_1 = T_2 + V_2$

$0 + 0 = 2.025\omega_2^2 + (-88.29)$

$\omega_2 = 6.603 \text{ rad/s} = 6.60 \text{ rad/s}$ \hfill *Ans.*

F18–8. $v_O = \omega r_{O/IC} = \omega(0.2)$

$I_O = mk_O^2 = 50(0.3^2) = 4.5 \text{ kg} \cdot \text{m}^2$

$T_1 = 0$

$T_2 = \frac{1}{2}m(v_O)_2^2 + \frac{1}{2}I_O\omega_2^2$

$\qquad = \frac{1}{2}(50)\left[\omega_2(0.2)\right]^2 + \frac{1}{2}(4.5)\omega_2^2$

$\qquad = 3.25\omega_2^2$

$(V_g)_1 = Wy_1 = 0$

$(V_g)_2 = -Wy_2 = -50(9.81)(6\sin 30°)$

$\qquad\qquad = -1471.5\text{J}$

$T_1 + V_1 = T_2 + V_2$

$0 + 0 = 3.25\omega_2^2 + (-1471.5)$

$\omega_2 = 21.28 \text{ rad/s} = 21.3 \text{ rad/s}$ \hfill *Ans.*

F18–9. $v_G = \omega r_G = \omega(1.5)$

$I_G = \frac{1}{12}(60)(3^2) = 45 \text{ kg} \cdot \text{m}^2$

$T_1 = 0$

$T_2 = \frac{1}{2}m(v_G)_2^2 + \frac{1}{2}I_G\omega_2^2$

$\qquad = \frac{1}{2}(60)[\omega_2(1.5)]^2 + \frac{1}{2}(45)\omega_2^2$

$\qquad = 90\omega_2^2$

Or,

$T_2 = \frac{1}{2}I_O\omega_2^2 = \frac{1}{2}\left[45 + 60(1.5^2)\right]\omega_2^2 = 90\omega_2^2$

$(V_g)_1 = Wy_1 = 0$

$(V_g)_2 = -Wy_2 = -60(9.81)(1.5\sin 45°)$

$\qquad\qquad = -624.30 \text{ J}$

$(V_e)_1 = \frac{1}{2}ks_1^2 = 0$

$(V_e)_2 = \frac{1}{2}ks_2^2 = \frac{1}{2}(150)(3\sin 45°)^2 = 337.5 \text{ J}$

$T_1 + V_1 = T_2 + V_2$

$0 + 0 = 90\omega_2^2 + [-624.30 + 337.5]$

$\omega_2 = 1.785 \text{ rad/s} = 1.79 \text{ rad/s}$ \hfill *Ans.*

F18–10. $v_G = \omega r_G = \omega(0.75)$

$I_G = \frac{1}{12}(30)(1.5^2) = 5.625 \text{ kg} \cdot \text{m}^2$

$T_1 = 0$

$T_2 = \frac{1}{2}m(v_G)_2^2 + \frac{1}{2}I_G\omega_2^2$

$\quad = \frac{1}{2}(30)[\omega(0.75)]^2 + \frac{1}{2}(5.625)\omega_2^2 = 11.25\omega_2^2$

Or,

$T_2 = \frac{1}{2}I_O\omega_2^2 = \frac{1}{2}\left[5.625 + 30(0.75^2)\right]\omega_2^2$

$\quad = 11.25\omega_2^2$

$(V_g)_1 = Wy_1 = 0$

$(V_g)_2 = -Wy_2 = -30(9.81)(0.75)$

$\quad = -220.725 \text{ J}$

$(V_e)_1 = \frac{1}{2}ks_1^2 = 0$

$(V_e)_2 = \frac{1}{2}ks_2^2 = \frac{1}{2}(80)\left(\sqrt{2^2 + 1.5^2} - 0.5\right)^2 = 160 \text{ J}$

$T_1 + V_1 = T_2 + V_2$

$0 + 0 = 11.25\omega_2^2 + (-220.725 + 160)$

$\omega_2 = 2.323 \text{ rad/s} = 2.32 \text{ rad/s}$ *Ans.*

F18–11. $(v_G)_2 = \omega_2 r_{G/IC} = \omega_2(0.75)$

$I_G = \frac{1}{12}(30)(1.5^2) = 5.625 \text{ kg} \cdot \text{m}^2$

$T_1 = 0$

$T_2 = \frac{1}{2}m(v_G)_2^2 + \frac{1}{2}I_G\omega_2^2$

$\quad = \frac{1}{2}(30)[\omega_2(0.75)]^2 + \frac{1}{2}(5.625)\omega_2^2 = 11.25\omega_2^2$

$(V_g)_1 = Wy_1 = 30(9.81)(0.75 \sin 45°) = 156.08 \text{ J}$

$(V_g)_2 = -Wy_2 = 0$

$(V_e)_1 = \frac{1}{2}ks_1^2 = 0$

$(V_e)_2 = \frac{1}{2}ks_2^2 = \frac{1}{2}(300)(1.5 - 1.5 \cos 45°)^2$

$\quad = 28.95 \text{ J}$

$T_1 + V_1 = T_2 + V_2$

$0 + (156.08 + 0) = 11.25\omega_2^2 + (0 + 28.95)$

$\omega_2 = 3.362 \text{ rad/s} = 3.36 \text{ rad/s}$ *Ans.*

F18–12. $(V_g)_1 = -Wy_1 = -[20(9.81) \text{ N}](1 \text{ m}) = -196.2 \text{ J}$

$(V_g)_2 = 0$

$(V_e)_1 = \frac{1}{2}ks_1^2$

$\quad = \frac{1}{2}(100 \text{ N/m})\left(\sqrt{(3 \text{ m})^2 + (2 \text{ m})^2} - 0.5 \text{ m}\right)^2$

$\quad = 482.22 \text{ J}$

$(V_e)_2 = \frac{1}{2}ks_2^2 = \frac{1}{2}(100 \text{ N/m})(1 \text{ m} - 0.5 \text{ m})^2$

$\quad = 12.5 \text{ J}$

$T_1 = 0$

$T_2 = \frac{1}{2}I_A\omega^2 = \frac{1}{2}\left[\frac{1}{3}(20 \text{ kg})(2 \text{ m})^2\right]\omega^2$

$\quad = 13.3333\omega^2$

$T_1 + V_1 = T_2 + V_2$

$0 + [-196.2 \text{ J} + 482.22 \text{ J}]$

$\quad = 13.3333\omega_2^2 + [0 + 12.5 \text{ J}]$

$\omega_2 = 4.53 \text{ rad/s}$ *Ans.*

Chapter 19

F19–1. $\zeta + I_O\omega_1 + \Sigma \int_{t_1}^{t_2} M_O \, dt = I_O\omega_2$

$0 + \int_0^{4\text{ s}} 3t^2 \, dt = \left[60(0.3)^2\right]\omega_2$

$\omega_2 = 11.85 \text{ rad/s} = 11.9 \text{ rad/s}$ *Ans.*

F19–2. $\zeta + (H_A)_1 + \Sigma \int_{t_1}^{t_2} M_A dt = (H_A)_2$

$0 + 300(6) = 300(0.4^2)\omega_2 + 300[\omega(0.6)](0.6)$

$\omega_2 = 11.54 \text{ rad/s} = 11.5 \text{ rad/s}$ *Ans.*

$\xrightarrow{+} \quad m(v_1)_x + \Sigma \int_{t_1}^{t_2} F_x dt = m(v_2)_x$

$0 + F_f(6) = 300[11.54(0.6)]$

$F_f = 346 \text{ N}$ *Ans.*

F19–3. $v_A = \omega_A r_{A/IC} = \omega_A(0.15)$

$\zeta + \Sigma M_O = 0; \quad 9 - A_t(0.45) = 0 \quad A_t = 20 \text{ N}$

$\zeta + (H_C)_1 + \Sigma \int_{t_1}^{t_2} M_C \, dt = (H_C)_2$

$0 + [20(5)](0.15)$

$\quad = 10[\omega_A(0.15)](0.15)$

$\quad\quad + \left[10(0.1^2)\right]\omega_A$

$\omega_A = 46.2 \text{ rad/s}$ *Ans.*

F19–4. $I_A = mk_A^2 = 10(0.08^2) = 0.064 \text{ kg} \cdot \text{m}^2$

$I_B = mk_B^2 = 50(0.15^2) = 1.125 \text{ kg} \cdot \text{m}^2$

$\omega_A = \left(\dfrac{r_B}{r_A}\right)\omega_B = \left(\dfrac{0.2}{0.1}\right)\omega_B = 2\omega_B$

$\zeta + I_A(\omega_A)_1 + \Sigma \int_{t_1}^{t_2} M_A \, dt = I_A(\omega_A)_2$

$0 + 10(5) - \int_0^{5\text{ s}} F(0.1)dt = 0.064[2(\omega_B)_2]$

$\int_0^{5\text{ s}} F dt = 500 - 1.28(\omega_B)_2$ (1)

$\zeta + I_B(\omega_B)_1 + \Sigma \int_{t_1}^{t_2} M_B \, dt = I_B(\omega_B)_2$

$0 + \int_0^{5\text{ s}} F(0.2)dt = 1.125(\omega_B)_2$

$\int_0^{5\text{ s}} F dt = 5.625(\omega_B)_2$ (2)

Equating Eqs. (1) and (2),

$500 - 1.28(\omega_B)_2 = 5.625(\omega_B)_2$

$(\omega_B)_2 = 72.41 \text{ rad/s} = 72.4 \text{ rad/s}$ *Ans.*

F19–5. $(\underset{\rightarrow}{+})$ $m[(v_O)_x]_1 + \Sigma \int F_x\,dt = m[(v_O)_x]_2$

$$0 + (150\text{ N})(3\text{ s}) + F_A(3\text{ s})$$
$$= (50\text{ kg})(0.3\omega_2)$$

$\zeta + I_G\omega_1 + \Sigma \int M_G\,dt = I_G\omega_2$

$$0 + (150\text{ N})(0.2\text{ m})(3\text{ s}) - F_A(0.3\text{ m})(3\text{ s})$$
$$= [(50\text{ kg})(0.175\text{ m})^2]\,\omega_2$$

$$\omega_2 = 37.3\text{ rad/s} \qquad\qquad Ans.$$
$$F_A = 36.53\text{ N}$$

Also,

$$I_{IC}\omega_1 + \Sigma \int M_{IC}\,dt = I_{IC}\omega_2$$

$$0 + [(150\text{ N})(0.2 + 0.3)\text{ m}](3\text{ s})$$
$$= [(50\text{ kg})(0.175\text{ m})^2 + (50\text{ kg})(0.3\text{ m})^2]\omega_2$$
$$\omega_2 = 37.3\text{ rad/s} \qquad\qquad Ans.$$

F19–6. $(+\uparrow)$ $m[(v_G)_1]_y + \Sigma \int F_y\,dt = m[(v_G)_2]_y$

$$0 + N_A(3\text{ s}) - (150\text{ lb})(3\text{ s}) = 0$$
$$N_A = 150\text{ lb}$$

$\zeta + (H_{IC})_1 + \Sigma \int M_{IC}\,dt = (H_{IC})_2$

$$0 + (25\text{ lb}\cdot\text{ft})(3\text{ s}) - [0.15(150\text{ lb})(3\text{ s})](0.5\text{ ft})$$
$$= \left[\tfrac{150}{32.2}\text{ slug}(1.25\text{ ft})^2\right]\omega_2 + \left(\tfrac{150}{32.2}\text{ slug}\right)\left[\omega_2(1\text{ ft})\right](1\text{ ft})$$
$$\omega_2 = 3.46\text{ rad/s} \qquad\qquad Ans.$$

Answers to Selected Problems

Chapter 12

12–1. $v_2 = 59.5\ \text{ft/s}$
$t = 1.29\ \text{s}$

12–2. $v = 3.93\ \text{m/s}$
$s = 9.98\ \text{m}$

12–3. The car must be dropped from the 9th floor.

12–5. $t = 25\ \text{s}$
$s = 312.5\ \text{m}$

12–6. $\Delta s = 48.3\ \text{ft}$

12–7. $a_c = 1.74\ \text{m/s}^2$
$t = 4.80\ \text{s}$

12–9. $v = 29.4\ \text{m/s}$
$h = 44.1\ \text{m}$

12–10. $s|_{t=6\,s} = -27.0\ \text{ft}$
$s_{\text{tot}} = 69.0\ \text{ft}$

12–11. $s = 20\ \text{ft}$

12–13. Normal: $d = 517\ \text{ft}$
Drunk: $d = 616\ \text{ft}$

12–14. $t = 21.9\ \text{s}$

12–15. $s = 28.4\ \text{km}$

12–17. $v = 32\ \text{m/s}$
$s = 67\ \text{m}$
$d = 66\ \text{m}$

12–18. $s = 123\ \text{ft}$
$a = 2.99\ \text{ft/s}^2$

12–19. $s = 7.87\ \text{m}$

12–21. (a) When $t = 5\ \text{s}$,
$v = 45.5\ \text{m/s}$
(b) $v_{\max} = 100\ \text{m/s}$

12–22. At $t = 6\ \text{s}$, $s = -18\ \text{ft}$.
$s_T = 46\ \text{ft}$

12–23. $d_A = 41.0\ \text{ft}$
$d_B = 200\ \text{ft}$
$\Delta s_{AB} = 152\ \text{ft}$

12–25. $s = 54.0\ \text{m}$

12–26. $s_{BA} = \left| \dfrac{2v_A v_B - v_A^2}{2a_A} \right|$

12–27. $s|_{t=2.667\,s} = 3.56\ \text{m}$

12–29. $t = 2.47\ \text{s}$
$s|_{t=2\,s} = 18.7\ \text{m}$

12–30. $v = 14\ \text{m/s}$
$t = 250\ \text{s}$

12–31. (a) $s = -30.5\ \text{m}$
(b) $s_{Tot} = 56.0\ \text{m}$
(c) $v = 10\ \text{m/s}$

12–33. $t = 10.3\ \text{s}$
$h = 4.11\ \text{km}$

12–34. $v_1 = 3.68\ \text{m/s}\!\uparrow$
$t_2 = 1.98\ \text{s}$
$v_2 = 15.8\ \text{m/s}\!\downarrow$

12–35. $t = 0.549\left(\dfrac{v_f}{g}\right)$

12–37. $v = 11.2\ \text{km/s}$

12–38. $v = -R\sqrt{\dfrac{2g_0(y_0 - y)}{(R + y)(R + y_0)}}$

$v = 3.02\ \text{km/s}\!\downarrow$

12–39. $t' = 27.3\ \text{s}$
When $t = 27.3\ \text{s}$, $v = 13.7\ \text{ft/s}$.

12–41. $v_{\max} = 16.7\ \text{m/s}$
$v = v_{\max}$ for $2\ \text{min} < t < 4\ \text{min}$.

12–42. $a|_{t=0} = -4\ \text{m/s}^2$
$a|_{t=2\,s} = 0$
$a|_{t=4\,s} = 4\ \text{m/s}^2$
$v|_{t=0\,s} = 3\ \text{m/s}$
$v|_{t=2\,s} = -1\ \text{m/s}$
$v|_{t=4\,s} = 3\ \text{m/s}$

12–43. $s = 2\sin\left(\dfrac{\pi}{5}t\right) + 4$

$v = \dfrac{2\pi}{5}\cos\left(\dfrac{\pi}{5}t\right)$

$a = -\dfrac{2\pi^2}{25}\sin\left(\dfrac{\pi}{5}t\right)$

12–45. $t = 7.48\ \text{s}$
When $t = 2.14\ \text{s}$,
$v = v_{\max} = 10.7\ \text{ft/s}$, $h = 11.4\ \text{ft}$.

12–46. $s = 600\ \text{m}$
For $0 \le t < 40\ \text{s}$, $a = 0$.
For $40\ \text{s} < t \le 80\ \text{s}$, $a = -0.250\ \text{m/s}^2$.

12–47. At $s = 50\ \text{m}$, $a = 0.32\ \text{m/s}^2$.
At $s = 150\ \text{m}$, $a = -0.32\ \text{m/s}^2$.
At $s = 100\ \text{m}$, a changes from $a_{\max} = 0.64\ \text{m/s}^2$
to $a_{\min} = -0.64\ \text{m/s}^2$.

12–49. When $t = 0.1\ \text{s}$, $s = 0.5\ \text{m}$ and a changes from
$100\ \text{m/s}^2$ to $-100\ \text{m/s}^2$.
When $t = 0.2\ \text{s}$, $s = 1\ \text{m}$.

12–50. For $0 \le t < 5$ s, $a = 4$ m/s^2.
For 20 s $< t \le 30$ s, $a = -2$ m/s^2.
At $t = 5$ s, $s = 50$ m. At $t = 20$ s, $s = 350$ m.
At $t = 30$ s, $s = 450$ m.

12–51. $t' = 133$ s
When $t = 30$ s, $v = 45$ m/s and $s = 450$ m.
When $t = 75$ s, $v = v_{max} = 112.5$ m/s and
$s = 4500$ m.
When $t = 133$ s, $v = 0$ and $s = 8857$ m.

12–53. When $t = 15$ s:
$v = 270$ m/s
$s = 2.025$ km
When $t = 20$ s:
$v = 395$ m/s
$s = 3.69$ km

12–54. $v_{max} = 400$ ft/s
$t' = 33.3$ s
When $t = 5$ s, $v = v_{max} = 400$ ft/s and $s = 1000$ ft.
When $t = 33.3$ s, $s = 8542$ ft.

12–55. When $t = 6$ s, $v = 12.0$ m/s.
When $t = 10$ s, $v = 36.0$ m/s and $s = 114$ m.

12–57. $(v_{sp})_{Avg} = 37.8$ ft/s
When $t = 10$ s, $s = 133$ ft.
When $t = 30$ s, $s = s_T = 1.13(10^3)$ ft.

12–58. When $t = 25$ s, $a = a_{max} = 9$ m/sols^2 and $s = 469$ m.
When $t = 50$ s, $s = 1406$ m.

12–59. $t' = 33.3$ s
$s|_{t=5\,s} = 550$ ft
$s|_{t=15\,s} = 1500$ ft
$s|_{t=20\,s} = 1800$ ft
$s|_{t=33.3\,s} = 2067$ ft

12–61. When $t = 30$ s, $s = 90$ m.
When $t = 48$ s, $s = 144$ m.

12–62. When $t = 5$ s, $s = 83.3$ ft and $a = 20$ ft/s^2.
When $t = 10$ s, $s = 583$ ft.

12–63. $s_T = 980$ m

12–65. When $t = 5$ s, $s_B = 62.5$ m.
When $t = 10$ s, $v_A = (v_A)_{max} = 40$ m/s and
$s_A = 200$ m.
When $t = 15$ s, $s_A = 400$ m and $s_B = 312.5$ m.
$\Delta s = s_A - s_B = 87.5$ m

12–66. When $t = 30$ s, $v = 90$ m/s.
When $t = 60$ s, $v = 540$ m/s.

12–67. When $t = 30$ s, $s = 675$ m.
When $t = 60$ s, $s = 10{,}125$ m.

12–69. $a|_{s=0} = 1.00$ ft/s^2
$a|_{s=500\,ft} = 6.00$ ft/s^2
$t = 17.9$ s

12–70. When $s = 100$ m, $t = 10$ s.
When $s = 400$ m, $t = 16.9$ s.
$a|_{s=100\,m} = 4$ m/s^2
$a|_{s=400\,m} = 16$ m/s^2

12–71. At s = 100 s, a changes from $a_{max} = 1.5$ ft/s^2
to $a_{min} = -0.6$ ft/s^2.

12–73. $v = 9.68$ m/s
$a = 16.8$ m/s^2

12–74. $\Delta \mathbf{r} = \{6\mathbf{i} + 4\mathbf{j}\}$ m

12–75. (4 ft, 2 ft, 6 ft)

12–77. $\Delta r = 3.61$ km
$d = 5$ km

12–78. $s = 9$ km
$\Delta r = 3.61$ km
$v_{avg} = 2.61$ m/s
$(v_{sp})_{avg} = 6.52$ m/s

12–79. $\mathbf{a}_{AB} = \{0.404\mathbf{i} + 7.07\mathbf{j}\}$ m/s^2
$\mathbf{a}_{AC} = \{2.50\mathbf{i}\}$ m/s^2

12–81. $v = 8.55$ ft/s
$a = 5.82$ m/s^2

12–82. $v = 1003$ m/s
$a = 103$ m/s^2

12–83. $d = 4.00$ ft
$a = 37.8$ ft/s^2

12–85. $\mathbf{v}_{avg} = \{10\mathbf{i} - 10\mathbf{j}\}$ m/s

12–86. $v = 201$ m/s
$a = 405$ m/s^2

12–87. $v = 10.4$ m/s
$a = 38.5$ m/s^2

12–89. $\theta_A = 30.5°$
$v_A = 23.2$ m/s

12–90. $\theta = 58.3°$
$(v_0)_{min} = 9.76$ m/s

12–91. $v_A = 27.3$ ft/s

12–93. $\theta = 15.0°, t = 1.45$ s or
$\theta = 75.0°, t = 5.40$ s

12–94. $v_0 = 67.7$ ft/s
$\theta = 58.9°$

12–95. $d = \dfrac{v_0^2}{g \cos \theta}(\sin 2\phi - 2 \tan \theta \cos^2\phi)$

12–97. $h = 11.1$ ft

12–98. $\theta = 6.41°$

12–99. $v_A = 36.7$ ft/s
$h = 11.5$ ft

12–101. $v_A = 19.4$ m/s
$v_B = 40.4$ m/s

12–102. $d = 166$ ft

12–103. $x = 32.3$ ft
$y = 6.17$ ft
$v = 71.8$ ft/s

12–105. $v_A = 39.7$ ft/s
$s = 6.11$ ft

12–106. $x = 1.95$ ft
$y = -0.153$ ft

12–107. $v_A = 16.5$ ft/s
$v_B = 29.2$ ft/s

12–109. $\Delta t = \dfrac{2v_0 \sin (\theta_1 - \theta_2)}{g (\cos \theta_2 + \cos \theta_1)}$

12–110. $R_{\min} = 0.189$ m
$R_{\max} = 1.19$ m

12–111. $\theta_1 = 24.9° \searrow$
$\theta_2 = 85.2° \nearrow$

12–113. $\theta = 38.4°$
$h = 14.8$ ft

12–114. $a = 9.50$ m/s^2

12–115. $v = 38.7$ m/s

12–117. $a = 0.488$ m/s^2

12–118. $v = 4.58$ m/s
$a = 0.653$ m/s^2

12–119. $a = 42.6$ ft/s^2

12–121. $a = 8.43$ m/s^2
$\theta = 38.2° \searrow$

12–122. $a = 6.03$ m/s^2

12–123. $a = 1.05$ m/s^2

12–125. $a = 2.75$ m/s^2

12–126. 1.68 m/s^2

12–127. $a = 5.02$ m/s^2

12–129. $v = 51.1$ ft/s
$a = 10.3$ ft/s^2

12–130. $v_B = 62.9$ ft/s
$a = 0.0133$ ft/s^2

12–131. $a_t = 3.62$ m/s^2
$\rho = 29.6$ m

12–133. $v = 3.19$ m/s
$a = 4.22$ m/s^2

12–134. $a = 7.42$ ft/s^2

12–135. $a = 2.36$ m/s^2

12–137. $a = 3.05$ m/s^2

12–138. $a = 0.525$ m/s^2

12–139. $a = 0.763$ m/s^2

12–141. $y = -0.0766x^2$
$v = 8.37$ m/s
$a_n = 9.38$ m/s^2
$a_t = 2.88$ m/s^2

12–142. $a = 3.05$ m/s^2

12–143. $x = 0, y = -4$ m
$(a)_{\max} = 50$ m/s^2

12–145. $a = 7.48$ ft/s^2

12–146. $t = 1.21$ s

12–147. $t = 2.63$ s

12–149. $s_A = 1.40$ m
$s_B = 3$ m
$\mathbf{r}_A = \{1.38\mathbf{i} + 0.195\mathbf{j}\}$ m
$\mathbf{r}_B = \{-2.82\mathbf{i} + 0.873\mathbf{j}\}$ m
$\Delta r = 4.26$ m

12–150. $t = 14.3$ s
$a_B = 0.45$ m/s^2

12–151. $a = 32.2$ m/s^2

12–153. $a = 35.0$ m/s^2
$s = 67.1$ ft

12–154. $y = \{0.839x - 0.131x^2\}$ m
$a_t = -3.94$ m/s^2
$a_n = -8.98$ m/s^2

12–155. $t = 10.1$ s
$v = 47.6$ m/s
$a = 11.8$ m/s^2

12–157. $v_n = 0$
$v_t = 7.21$ m/s
$a_n = 0.555$ m/s^2
$a_t = 2.77$ m/s^2

12–158. $a_{\max} = \dfrac{a}{b^2}v^2$

12–159. $a = 14.3$ in./s^2

12–161. $v_r = -1.66$ m/s
$v_\theta = -2.07$ m/s
$a_r = 4.20$ m/s^2
$a_\theta = 2.97$ m/s^2

12–162. $v = 464$ ft/s
$a = 43.2(10^3)$ ft/s^2

12–163. $v = 120$ ft/s
$a = 76.8$ ft/s^2

12–165. $v_r = -2 \sin t$
$v_\theta = \cos t$
$a_r = -\dfrac{5}{2}\cos t$
$a_\theta = -2 \sin t$

12–166. $v_r = -2.33$ m/s
$v_\theta = 7.91$ m/s
$a_r = -158$ m/s^2
$a_\theta = -18.6$ m/s^2

12–167. $v_r = 0$
$v_\theta = 10$ ft/s
$a_r = -0.25$ ft/s^2
$a_\theta = -3.20$ ft/s^2

12–169. $\mathbf{a} = (\ddot{r} - 3r\dot{\theta}^2 - 3r\dot\theta\ddot\theta)\mathbf{u}_r$
$\quad + (3\ddot{r}\dot\theta + r\dddot\theta + 3\dot{r}\ddot\theta - r\dot\theta^3)\mathbf{u}_\theta + (\ddot{z})\mathbf{u}_z$

12–170. $a = 48.3$ in./s^2

12–171. $v_r = 1.20$ m/s
$v_\theta = 1.26$ m/s
$a_r = -3.77$ m/s^2
$a_\theta = 7.20$ m/s^2

12–173. $v_r = 1.20$ m/s
$v_\theta = 1.50$ m/s
$a_r = -4.50$ m/s^2
$a_\theta = 7.20$ m/s^2

12–174. $v_r = 16.0$ ft/s
$v_\theta = 1.94$ ft/s
$a_r = 7.76$ ft/s^2
$a_\theta = 1.94$ ft/s^2

12–175. $v_r = -\dfrac{20}{\sqrt{1 + \theta^2}}$

$v_\theta = \dfrac{20\theta}{\sqrt{1 + \theta^2}}$

$v_r = -14.1 \text{ ft/s}$

$v_\theta = 14.1 \text{ ft/s}$

12–177. $\dot\theta = 0.378 \text{ rad/s}$

12–178. $\mathbf{v} = \{-116\mathbf{u}_r - 163\mathbf{u}_z\} \text{ mm/s}$

$\mathbf{a} = \{-5.81\mathbf{u}_r - 8.14\mathbf{u}_z\} \text{ mm/s}^2$

12–179. $v = 12.6 \text{ m/s}$

$a = 83.2 \text{ m/s}^2$

12–181. $v = 16.8 \text{ ft/s}$

$a = 199 \text{ ft/s}^2$

12–182. $\dot\theta = 0.75 \text{ rad/s}$

12–183. $v_r = a\dot\theta$

$v_\theta = a\theta\dot\theta$

$a_r = -a\theta\dot\theta^2$

$a_\theta = 2a\dot\theta^2$

12–185. $v = 10.7 \text{ ft/s}$

$a = 24.6 \text{ ft/s}^2$

12–186. $v = 10.7 \text{ ft/s}$

$a = 40.6 \text{ ft/s}^2$

12–187. $v_r = 25.9 \text{ mm/s}$

$a_r = -195 \text{ mm/s}^2$

12–189. $a = 7.26 \text{ m/s}^2$

12–190. $v = 4.16 \text{ m/s}$

$a = 33.1 \text{ m/s}^2$

12–191. $v_r = -2.80 \text{ m/s}$

$v_\theta = 19.8 \text{ m/s}$

12–193. $v_r = 32.0 \text{ ft/s}$

$v_\theta = 50.3 \text{ ft/s}$

$a_r = -201 \text{ ft/s}^2$

$a_\theta = 256 \text{ ft/s}^2$

12–194. $v_r = 32.0 \text{ ft/s}$

$v_\theta = 50.3 \text{ ft/s}$

$a_r = -161 \text{ ft/s}^2$

$a_\theta = 319 \text{ ft/s}^2$

12–195. $v = 5.95 \text{ ft/s}$

$a = 3.44 \text{ ft/s}^2$

12–197. $v = 8.21 \text{ mm/s}$

$a = 665 \text{ mm/s}^2$

12–198. $v = 8.21 \text{ mm/s}$

$a = 659 \text{ mm/s}^2$

12–199. $v_B = 0.5 \text{ m/s}$

12–201. $v_P = 12 \text{ ft/s}$

12–202. $t = 160 \text{ s}$

12–203. $\Delta s_B = 1.33 \text{ ft} \rightarrow$

12–205. $v_A = 1.67 \text{ m/s}$

12–206. $v_A = 4 \text{ ft/s}$

12–207. $v_B = 1 \text{ ft/s}\uparrow$

12–209. $\Delta s_B = 2 \text{ ft}\uparrow$

12–210. $v_A = 24 \text{ ft/s}$

12–211. $v_A = 1.33 \text{ m/s}$

12–213. $v_B = 0.671 \text{ m/s}$

12–214. $v_B = 1.41 \text{ m/s}\uparrow$

12–215. $v_B = 2.40 \text{ ft/s} \rightarrow$

$a_B = 3.85 \text{ ft/s}^2 \rightarrow$

12–217. $v_C = 1.2 \text{ m/s}\uparrow$

$a_C = 0.512 \text{ m/s}^2\uparrow$

12–218. $v_b = 5.56 \text{ m/s}$

$\theta = 84.4°$

12–219. $\Delta s_C = 2 \text{ ft}$

12–221. $\dot s_B = 0.809 \text{ ft/s}$

12–222. $v_{B/A} = 875 \text{ km/h}$

$\theta = 41.5° \searrow$

12–223. $v_{B/A} = 28.5 \text{ mi/h}$

$\theta = 44.5° \nearrow$

$a_{B/A} = 3418 \text{ mi/h}^2$

$\theta = 80.6° \nearrow$

12–225. $v_{A/B} = 49.1 \text{ km/h}$

$\theta = 67.2° \searrow$

12–226. $v_w = 58.3 \text{ km/h}$

$\theta = 59.0° \searrow$

12–227. $v_{A/B} = 21.7 \text{ ft/s}$

$\theta = 18.0° \searrow$

$t = 36.9 \text{ s}$

12–229. $v_{A/B} = 98.4 \text{ ft/s}$

$\theta = 67.6° \searrow$

$a_{A/B} = 19.8 \text{ ft/s}^2$

$\theta = 57.4° \nearrow$

12–230. $v_{A/B} = v\sqrt{2(1 - \sin\theta)}$

12–231. $v_{B/A} = 36.6 \text{ mi/h}$

$\theta = 46.9° \searrow$

$a_{B/A} = 3737 \text{ mi/h}^2$

$\phi = 12.9° \searrow$

12–233. $v_r = 34.6 \text{ km/h}$

12–234. $v_m = 4.87 \text{ ft/s}$

$t = 10.3 \text{ s}$

12–235. $v_{w/s} = 19.9 \text{ m/s}$

$\theta = 74.0° \searrow$

12–237. $v_{B/C} = 18.6 \text{ m/s}$

$\theta_v = 66.2° \nearrow$

$a_{B/C} = 0.959 \text{ m/s}^2$

$\theta_a = 8.57° \searrow$

12–238. $v_B = 5.75 \text{ m/s}$

$v_{C/B} = 17.8 \text{ m/s}$

$\theta = 76.2° \searrow$

$a_{C/B} = 9.81 \text{ m/s}^2\downarrow$

12–239. $v_{A/B} = \sqrt{v_A^2 + v_B^2 - 2v_A v_B \cos\theta}$

$\theta = \tan^{-1}\left(\dfrac{v_A - v_B\cos\theta}{v_B\sin\theta}\right) \searrow$

Chapter 13

13–1. $s = 97.4$ ft

13–2. $v = 46.2$ ft/s
$s = 66.2$ ft

13–3. $v = 3.36$ m/s
$s = 5.04$ m

13–5. $v = 77.9$ ft/s

13–6. $v = 3.29$ m/s

13–7. $P = 392$ N

13–9. $a = 1.66$ m/s^2

13–10. $a = 1.75$ m/s^2

13–11. $a_B = 2.30$ ft/s$^2 \uparrow$

13–13. $v = \left(\dfrac{F_0 t_0}{\pi m} \right)\left(1 - \cos\left(\dfrac{\pi t}{t_0} \right) \right)$

$v = \dfrac{2 F_0 t_0}{\pi m}$

$s = \left(\dfrac{F_0 t_0}{\pi m} \right)\left(t - \dfrac{t_0}{\pi}\sin\left(\dfrac{\pi t}{t_0} \right) \right)$

13–14. $T = 11.25$ kN
$F = 33.75$ kN

13–15. $a_E = 0.75$ m/s$^2 \uparrow$
$T = 1.32$ kN

13–17. $a = 3.69$ ft/s^2

13–18. $R = 5.30$ ft
Total time $= 1.82$ s

13–19. $R = 5.08$ ft
Total time $= 1.48$ s

13–21. $s = 5.43$ m

13–22. $m_A = 13.7$ kg

13–23. $T = 4.92$ kN

13–25. $T = 49.2$ N

13–26. $v = 30$ m/s

13–27. $A_x = 35.6$ lb
$A_y = 236$ lb
$M_A = 678$ lb·ft

13–29. $v = 2.01$ ft/s

13–30. $v = 0.301$ m/s

13–31. $T = 1.63$ kN

13–33. $a_B = 0$
$a_C = 4.11$ m/s$^2 \rightarrow$
$a_D = 0.162$ m/s$^2 \rightarrow$

13–34. (a) $a_A = \dfrac{P}{m} - 3\mu g$

(b) $a_A = \dfrac{P}{2m} - 2\mu g$

13–35. $t = 2.04$ s

13–37. $F = \dfrac{m(a_B + g)\sqrt{4y^2 + d^2}}{4y}$

13–38. $v_B = 4.52$ m/s

13–39. $v = \dfrac{1}{m}\sqrt{1.09 F_0^2 t^2 + 2 F_0 t m v_0 + m^2 v_0^2}$

$x = \dfrac{y}{0.3} + v_0\left(\sqrt{\dfrac{2m}{0.3 F_0}} \right) y^{\frac{1}{2}}$

13–41. (a) $a_C = 6.94$ m/s^2
(b) $a_C = 6.94$ m/s^2
(c) $a_C = 7.08$ m/s^2
$\theta = 56.5°$ ⦧

13–42. $a_C = 7.49$ m/s^2
$\theta = 22.8°$ ⦧

13–43. $t = 5.66$ s

13–45. $v = 32.2$ ft/s

13–46. $P = 2mg \tan\theta$

13–47. $P = 2mg\left(\dfrac{\sin\theta + \mu_s \cos\theta}{\cos\theta - \mu_s \sin\theta} \right)$

13–49. $v_{B/AC} = a_0 (\sin\theta) t$

$s = \dfrac{1}{2} a_0 (\sin\theta) t^2$

13–50. $x = \dfrac{m v_0}{k} \cos\theta_0 (1 - e^{-(k/m)t})$

$y = -\dfrac{m g t}{k} + \dfrac{m}{k}\left(v_0 \sin\theta_0 + \dfrac{mg}{k} \right)(1 - e^{-(k/m)t})$

$x_{max} = \dfrac{m v_0 \cos\theta_0}{k}$

13–51. $d = \dfrac{(m_A + m_B)g}{k}$

13–53. $r = 1.36$ m

13–54. $v = 10.5$ m/s

13–55. $P = 17.3$ N
$N = 120$ N$\downarrow$

13–57. $T = 1.82$ N
$N_B = 0.844$ N

13–58. $v = 0.969$ m/s

13–59. $v = 1.48$ m/s

13–61. $v = 9.29$ ft/s
$T = 38.0$ lb

13–62. $N = 5.88$ lb
$a_t = 23.0$ ft/s^2

13–63. $N = 277$ lb
$F = 13.4$ lb

13–65. $v = 6.30$ m/s
$F_n = 283$ N
$F_t = 0$
$F_b = 490$ N

13–66. $t = 7.39$ s

13–67. $\theta = 26.7°$

13–69. $F_f = 1.11$ kN
$N = 6.73$ kN

13–70. $v_C = 19.9$ ft/s
$N_C = 7.91$ lb
$v_B = 21.0$ ft/s

13–71. $T = 414$ N
$\theta = 37.2°$

13–73. $v = \sqrt{gr}$
$N = 2mg$

13–74. $v = 80.2$ ft/s

13–75. $\theta = \cos^{-1}\left(\dfrac{m_B}{m_A}\right)$
$v_B = \sqrt{\dfrac{g(l-h)(m_A^2 - m_B^2)}{m_A m_B}}$

13–77. $N = 1.02$ kN

13–78. $\theta = 31.3°$
$l = 2.585$ ft

13–79. $N_P = 2.65$ kN
$\rho = 68.3$ m

13–81. $L = 50.8$ kN
$r = 3.60$ km

13–82. $N = 7.69$ kN

13–85. $F_A = 4.46$ lb

13–86. $F = 210$ N

13–87. $F = 1.60$ lb

13–89. $N = 2.77$ lb

13–90. $F_r = -29.4$ N
$F_\theta = 0$
$F_z = 392$ N

13–91. $F = 0.143$ lb

13–93. $F_{OA} = 12.0$ lb

13–94. $N_C = 54.4$ N
$F = 54.4$ N

13–95. $\dot\theta = 4.00$ rad/s
$T = 8$ N

13–97. $N = 6.37$ N
$F = 2.93$ N

13–98. $N = 24.8$ N
$F = 24.8$ N

13–99. $F_r = -20.0$ N
$F_\theta = 0$
$F_z = 2.45$ kN

13–101. $\theta = \tan^{-1}\left(\dfrac{4r_c\dot\theta_0^2}{g}\right)$

13–102. $N = 0.883$ N
$F = 3.92$ N

13–103. $T = 509$ lb

13–105. $F = -0.0155$ lb

13–106. $F = -0.0108$ lb

13–107. $F = 0.163$ lb

13–109. $F = 7.67$ N

13–110. $F = 7.82$ N

13–111. $r = 0.198$ m

13–113. $v_0 = 30.8$ km/s
$\dfrac{1}{r} = 0.502(10^{-12})\cos\theta + 6.11(10^{-12})$

13–114. $h = 35.9$ Mm
$v_s = 3.07$ km/s

13–115. $v_0 = 7.45$ km/s

13–118. $v_B = 7.71$ km/s
$v_A = 4.63$ km/s

13–119. $v_a = 2.59$ km/s
$T = 9.33$ hr

13–121. $v_A = 7.47$ km/s

13–122. $\Delta v_A = 466$ m/s
$\Delta v_B = 2.27$ km/s

13–123. $r_a = 10.8(10^9)$ km

13–125. (a) $r = 194(10^3)$ mi
(b) $r = 392(10^3)$ mi
(c) $194(10^3)$ mi $< r < 392(10^3)$ mi
(d) $r > 392(10^3)$ mi

13–126. $\theta = \pi - \cos^{-1}\left(\dfrac{k^2n - 1}{1 - k^2}\right)$

13–127. $\Delta v = 530$ m/s

13–129. $\Delta v = 2.57$ km/s

13–130. $v_A = 3.08$ km/s
$T = 15.1$ hr

13–131. $\Delta v = \sqrt{\dfrac{GM_e}{r_0}}\left(\sqrt{2} - \sqrt{1+e}\right)$

Chapter 14

14–1. $v = 10.7$ m/s

14–2. $x_{max} = 3.24$ ft

14–3. $s = 1.35$ m

14–5. $v_2 \approx 2.12$ km/s

14–6. $v_A = 10.5$ m/s

14–7. Observer A: $v_2 = 6.08$ m/s
Observer B: $v_2 = 4.08$ m/s

14–9. $P = 207$ N

14–10. $s = 178$ m

14–11. $\mu_k = 0.255$

14–13. $v_B = 31.5$ ft/s
$d = 22.6$ ft
$v_C = 54.1$ ft/s

14–14. $v_B = 3.34$ m/s

14–15. $s = 9.29$ ft

14–17. $v_B = 27.8$ ft/s

14–18. $v_A = 3.52$ ft/s

14–19. $N_B = 20.5$ lb

14–21. $k_B = 11.1$ kN/m

14–22. $s_{Tot} = 3.88$ ft

14–23. $s_1 = 0.628$ m
$s_2 = 0.377$ m

14–25. $v_B = 30.0$ m/s
$s = 130$ m

14–26. $v_A = 28.3$ m/s

14–27. $s = 2.48$ ft
$T_A = 36.0$ lb

14–29. $s = 0.730$ m

14–30. $s = 3.33$ ft

14–31. $R = 2.83$ m

$v_C = 7.67$ m/s

14–33. $v_D = 17.7$ m/s

$R = 33.0$ m

14–34. $s = 1.90$ ft

14–35. $v_C = 2.36$ m/s

14–37. $h_A = 22.5$ m

$h_C = 12.5$ m

14–38. $v_B = 26.1$ ft/s

$N_B = 135$ lb

14–39. $v_A = 1.98$ m/s$\downarrow$

14–41. $l_0 = 2.77$ ft

14–42. $\theta = 47.2°$

14–43. power input $= 4.20$ hp

14–45. $P = 8.32\,(10^3)$ hp

14–46. $t = 46.2$ min

14–47. $P = 12.6$ kW

14–49. $P_{in} = 113$ kW

$(P_{in})_{avg} = 56.5$ kW

14–50. $P = 1.12$ kW

14–51. power input $= 1.60$ kW

14–53. $P = 0.229$ hp

14–54. $P = 0.0364$ hp

14–55. $P_i = 22.2$ kW

14–57. $P_i = 622$ kW

14–58. $P = 58.1$ kW

14–59. $P = 5.33t$ MW

14–61. $P = [400(10^3)t]$ W

14–62. $P = (160\,t - 533t^2)$ kW

$U = 1.69$ kJ

14–63. $P_{max} = 10.7$ kW

14–65. $P = 0.231$ hp

14–66. $F = 227$ N

14–67. $k_B = 287$ N/m

14–69. $v_2 = 11.7$ ft/s

14–70. $d = 8.53$ m

$v_D = 10$ m/s

14–71. $v = 80.2$ ft/s

$N_G = 952$ lb

14–73. $v_A = 3.46$ m/s

14–74. $v = 32.3$ ft/s

14–75. $k = 8.57$ lb/ft

14–77. $y = 213$ mm

14–78. $v_C = 19.4$ ft/s

14–79. $v_B = 25.4$ ft/s

14–81. $h = 416$ mm

14–83. $F = GM_e m \left(\dfrac{1}{r_1} - \dfrac{1}{r_2}\right)$

14–85. $v_B = 34.8$ Mm/h

14–86. $l = 90$ ft

$h = 219$ ft

$a = 483$ ft/s^2

14–87. $v_B = 29.5$ ft/s

14–89. $\theta = 22.3°$

$s = 0.587$ m

14–90. $v_0 = \sqrt{g(5\rho - 2h)}$

14–91. $v_1 = 7.43$ m/s

$N = 86.7$ N

14–93. $(v_s)_2 = 1.68$ m/s

14–94. $s_A = 1.29$ ft

14–95. $x = 17.7$ ft

$y = 8.83$ ft

14–97. $d = 1.34$ m

Chapter 15

15–1. $t = 0.280$ s

$v_B = 15.6$ ft/s

15–2. $v = 29.4$ ft/s

15–3. $t = 0.439$ s

15–5. $\displaystyle\int F dt = 0.706$ N $\cdot$ s $\angle 40°$

15–6. $F = 19.4$ kN

$T = 12.5$ kN

15–7. $F_{AB} = 16.7$ lb

$v = 13.4$ ft/s

15–9. $(F_D)_{avg} = 16.6$ kN

15–10. $P = 205$ N

15–11. $I = 63.4$ N $\cdot$ s

15–13. $\mu_k = 0.340$

15–14. $v_2 = \dfrac{2Ct'}{\pi m}$

$s = \dfrac{Ct'^2}{\pi m}$

15–15. $v = 89.8$ ft/s

15–17. $\displaystyle\int F dt = 0.311$ lb $\cdot$ s

15–18. $v_2 = 12.0$ m/s $\rightarrow$

15–19. $v_A = 27.6$ ft/s$\uparrow$

$v_B = 55.2$ ft/s$\downarrow$

15–21. $T = 14.9$ kN

$F = 24.8$ kN

15–22. $v = 8.80$ ft/s

15–23. $\displaystyle\int F dt = 63.4$ N $\cdot$ s

15–25. $F_{avg} = 847$ lb

15–26. Observer A: $v = 7.40$ m/s

Observer B: $v = 5.40$ m/s

15–27. $v_2 = 21.8$ m/s

15–29. $v = 14.3 \text{ m/s}$
$F_D = 15.7 \text{ kN}$

15–30. $v_B = 2.10 \text{ m/s} \uparrow$
$v_A = 8.41 \text{ m/s} \downarrow$

15–31. $(v_A)_2 = 10.5 \text{ ft/s} \rightarrow$

15–33. $v_2 = 7.65 \text{ m/s}$

15–34. $v_2 = 1.92 \text{ m/s}$

15–35. $v = 3.5 \text{ ft/s} \rightarrow$

15–37. $v_2 = 5.21 \text{ m/s} \leftarrow$
$\Delta E = 32.6 \text{ kJ}$

15–38. $v_2 = 0.5 \text{ m/s}$
$\Delta T = 16.9 \text{ kJ}$

15–39. $v_2 = 0.600 \text{ ft/s} \leftarrow$

15–41. $v_b = 0.379 \text{ m/s} \rightarrow$

15–42. $v_{b/c} = 0.632 \text{ m/s} \rightarrow$

15–43. $v_{ABC} = 2.18 \text{ m/s} \leftarrow$

15–45. $v_2 = \sqrt{v_1^2 + 2gh}$

$\theta_2 = \sin^{-1}\left(\dfrac{v_1 \sin\theta}{\sqrt{v_1^2 + 2gh}}\right)$

15–46. $s_b = 18.2 \text{ ft}$

15–47. $s_{\max} = 481 \text{ mm}$

15–49. $(v_B)_2 = 3.29 \text{ ft/s}$

15–50. $W_B = 75 \text{ lb}$

15–51. $v = 4.29 \text{ m/s}$

15–53. $v_b = 2.73 \text{ ft/s} \rightarrow$
$v_g = 2.27 \text{ ft/s} \leftarrow$
$s_C = 1.09 \text{ ft}$

15–54. $v_c = 0.276 \text{ ft/s} \leftarrow$

15–55. $x = 0.221 \text{ m}$

15–57. $d = 6.87 \text{ mm}$

15–58. $e = 0.901$

15–59. $e = 0.75$
$\Delta E = 9.65 \text{ kJ}$

15–61. $(v_A)_2 = 0.400 \text{ m/s} \rightarrow$
$(v_B)_2 = 2.40 \text{ m/s} \rightarrow$
$d = 0.951 \text{ m}$

15–63. $v_B = ve \rightarrow$
$v_A = ve \leftarrow$
$v' = 0$

15–65. $t = 0.226 \text{ s}$

15–66. $d = 8.98 \text{ ft}$

15–67. $(v_A)_2 = 1.04 \text{ ft/s}$
$(v_B)_3 = 0.964 \text{ ft/s}$
$(v_C)_3 = 11.9 \text{ ft/s}$

15–69. $v'_B = 22.2 \text{ m/s}$
$\theta = 13.0°$

15–70. $(v_B)_2 = \dfrac{e(1 + e)}{2}v_0$

15–71. $v_A = 29.3 \text{ ft/s}$
$v_{B2} = 33.1 \text{ ft/s}$
$\theta = 27.7° \measuredangle$

15–73. $d = 13.8 \text{ ft}$

15–74. $e = 0.502$
$d = 7.23 \text{ ft}$

15–75. $d = 3.84 \text{ m}$

15–77. $(v_B)_3 = 3.24 \text{ m/s}$
$\theta = 43.9°$

15–78. $v'_B = 31.8 \text{ ft/s}$

15–79. $v_b = \left(\dfrac{1 + e}{7}\right)v \leftarrow$

15–81. (a) $(v_b)_1 = 8.81 \text{ m/s}$
$\theta = 10.5°$
(b) $(v_b)_2 = 4.62 \text{ m/s}$
$\phi = 20.3°$
(c) $s = 3.96 \text{ m}$

15–82. $s = 0.456 \text{ ft}$

15–83. $d = 1.15 \text{ ft}$
$h = 0.770 \text{ ft}$

15–85. $\mu = 0.25$

15–86. $(v_A)_2 = 4.06 \text{ ft/s}$
$(v_B)_2 = 6.24 \text{ ft/s}$

15–87. $v_A = 1.35 \text{ m/s} \rightarrow$
$v_B = 5.89 \text{ m/s}$
$\theta = 32.9° \measuredangle$

15–89. $v_B = 3.50 \text{ m/s}$
$v_A = 6.47 \text{ m/s}$

15–90. $(v_B)_2 = \dfrac{\sqrt{3}}{4}(1 + e)v$

$\theta_B = 30° \measuredangle$

$\Delta U_k = \dfrac{3mv^2}{16}\left(1 - e^2\right)$

15–91. $v'_A = 11.0 \text{ ft/s}$
$\theta_A = 18.8° \measuredangle$
$v'_A = 11.0 \text{ ft/s}$
$\theta_B = 58.3° \measuredangle$

15–93. $v'_A = 8.19 \text{ m/s}$
$v'_B = 9.38 \text{ m/s}$

15–94. $\mathbf{H}_O = \{42\mathbf{i} + 21\mathbf{k}\} \text{ kg} \cdot \text{m}^2/\text{s}$

15–95. $\mathbf{H}_O = \{-16.8\mathbf{i} + 14.9\mathbf{j} - 23.6\mathbf{k}\} \text{ slug} \cdot \text{ft}^2/\text{s}$

15–97. $\mathbf{H}_O = \{12.5\mathbf{k}\} \text{ kg} \cdot \text{m}^2/\text{s}$

15–98. $(H_A)_O = 72.0 \text{ kg} \cdot \text{m}^2/\text{s} \measuredangle$
$(H_B)_O = 59.5 \text{ kg} \cdot \text{m}^2/\text{s} \measuredangle$

15–99. $(H_A)_P = 66.0 \text{ kg} \cdot \text{m}^2/\text{s} \measuredangle$
$(H_B)_P = -73.9 \text{ kg} \cdot \text{m}^2/\text{s} \measuredangle$

15–101. $t = 3.41 \text{ s}$

15–102. $t = 0.910 \text{ s}$
$v = 17.8 \text{ ft/s}$

15–103. $v_2 = 9.22 \text{ ft/s}$

$\sum U_{1-2} = 3.04 \text{ ft} \cdot \text{lb}$

15–105. $v_2 = 19.3 \text{ ft/s}$

15–107. $v_2 = 13.8 \text{ ft/s}$

15–109. $v_2 = 45.1$ ft/s
$U_F = 2641$ ft·lb
15–110. $v_2 = 4.60$ ft/s
15–111. $v = 20.2$ ft/s
$h = 6.36$ ft
15–113. $v_B = 10.2$ km/s
$r_B = 13.8$ Mm
15–114. $T = 40.1$ kN
15–115. $F_x = 55.3$ lb
$F_y = 25.8$ lb
15–117. $C_x = 4.97$ kN
$D_x = 2.23$ kN
$D_y = 7.20$ kN
15–118. $P = 42.8$ hp
15–119. $F_x = 9.87$ lb
$F_y = 4.93$ lb
15–121. $F_x = 19.5$ lb
$F_y = 1.96$ lb
15–122. $F_x = 6.28$ kip
$F_y = 2.28$ kip
15–123. $F = 22.4$ lb
15–125. $T = 82.8$ N
$N = 396$ N
15–126. $A_x = 3.98$ kN
$A_y = 3.81$ kN
$M = 1.99$ kN·m
15–127. $m' = \dfrac{M}{s}$
15–129. $C_x = 4.26$ kN
$C_y = 2.12$ kN
$M_C = 5.16$ kN·m
15–130. $T = 1.12$ kN
15–131. $A_y = 4.18$ kN
$B_x = 65.0$ N →
$B_y = 3.72$ kN↑
15–133. $M_D = 10.7$ kip·ft
$D_y = 5.82$ kip
$D_x = 2.54$ kip
15–134. $v_{max} = 625$ m/s
15–135. $P = 452$ Pa
15–137. $v = \sqrt{\dfrac{P}{m_0} - \dfrac{2}{3}gy}$
15–138. $a_i = 133$ ft/s²
$a_f = 200$ ft/s²
15–139. $v_{max} = 580$ ft/s
15–141. $F = 3.55$ kN
15–142. $F_D = 11.5$ kN
15–143. $a = 37.5$ ft/s²
15–145. $F = m'v^2$
15–146. $v_{max} = 2068$ ft/s
15–147. $F = (7.85t + 0.320)$N

15–149. $v = \sqrt{\dfrac{2}{3}g\left(\dfrac{y^3 - h^3}{y^2}\right)}$

Review 1
R1–1 $v = 77.5$ ft/s
R1–2. (a) $a_A = a_B = 2.76$ ft/s²
(b) $a_A = 70.8$ ft/s²
$a_B = 3.86$ ft/s²
R1–3. $v_B = 47.8$ ft/s
R1–5. $(v_R)_2 = 0.840$ m/s
R1–6. $v_2 = 21.5$ ft/s
R1–7. $F = 85.7$ N
R1–9. $F_r = -68$ N
$N = 153$ N
R1–10. $t = 1.02$ s
R1–11. $v_2 = 31.7$ m/s
R1–13. $P = 946$ kW
R1–14. $s = 640$ m
R1–15.

s	t
0	0
2.01	0.25
3.83	0.50
5.49	0.75
7.03	1.00
8.48	1.25
9.87	1.50
11.2	1.75
12.5	2.00

R1–17. $T = 158$ N
R1–18. $v_{min} = 0.838$ m/s
$v_{max} = 1.76$ m/s
R1–19. $s = 0.933$ m
R1–21. $h = 0.390$ m
R1–22. $t = 23.8$ s
R1–23. $y = 0.0208x^2 + 0.333x$
R1–25. $v = 2.68$ ft/s
R1–26. $v_A = 26.8$ ft/s↓
R1–27. $(v_B)_2 = 1.34$ m/s ←
$(v_A)_2 = 1.30$ m/s
$(\theta_A)_2 = 8.47°$
R1–29. $v_B = 10.4$ m/s
R1–30. $a_A = 0.755$ m/s²
$a_B = 1.51$ m/s²
$T_A = 90.6$ N
$T_B = 45.3$ N
R1–31. $v = 0.686$ m/s
$T_B = 206$ N
$T_C = 103$ N
R1–33. $\theta = 14.6°$
$\phi = 14.0°$
R1–34. $a = 0.603$ ft/s²
R1–35. $v_2 = 13.8$ ft/s

R1–37. $v = 25.4$ ft/s

R1–38. $s = 1.84$ m

R1–39. $a = -24$ m/s^2
$\Delta s = -880$ m
$s_T = 912$ m

R1–41. $v_B = 0.5$ m/s

R1–42. $v = 5.38$ ft/s

R1–43. $v = 5.32$ ft/s

R1–45. $a|_{t=4} = 1.06$ m/s^2

R1–46. $v = 3t^2 - 6t + 2$
$a = 6t - 6$

R1–47. $P_i = 1.80$ hp

R1–49. $e = 0.901$

R1–50. $v_B = 3.33$ ft/s
$v_{B/C} = 13.3$ ft/s↑

Chapter 16

16–1. $v_A = 2.60$ m/s
$a_A = 9.35$ m/s^2

16–2. $a_t = 0.562$ ft/s^2
$a_n = 3600$ ft/s^2
$s = 3000$ ft

16–3. $v = 22$ ft/s
$a_t = 12.0$ ft/s^2
$a_n = 242$ ft/s^2

16–5. $v_C = 8.81$ in./s
$a_C = 32.6$ in./s^2

16–6. $\theta = 3.32$ rev
$t = 1.67$ s

16–7. $t = 6.98$ s
$\theta_D = 34.9$ rev

16–9. $v_A = 70.9$ ft/s
$v_B = 35.4$ ft/s
$(a)_A = 252$ ft/s^2
$(a)_B = 126$ ft/s^2

16–10. $v_A = 40.0$ ft/s
$v_B = 20.0$ ft/s
$(a)_A = 80.6$ ft/s^2
$(a)_B = 40.3$ ft/s^2

16–11. $v_A = v_B = 2.4$ ft/s
$a_A = 0.4$ ft/s^2
$a_B = 17.3$ ft/s^2

16–13. $\omega_B = 12$ rad/s

16–14. $\omega = 42.7$ rad/s
$\theta = 42.7$ rad

16–15. $a_t = 2.83$ m/s^2
$a_n = 35.6$ m/s^2

16–17. $v_B = 0.394$ m/s

16–18. $v = 1.34$ m/s

16–19. $\omega_B = 156$ rad/s

16–21. $v_B = 1.37$ m/s
$a_B = 0.472$ m/s^2

16–22. $\omega_D = 22.4$ rad/s

16–23. $\omega_D = 11.8$ rad/s

16–25. $v_P = 2.42$ ft/s
$a_P = 34.4$ ft/s^2

16–26. $\omega_C = 1.68$ rad/s
$\theta_B = 1.68$ rad

16–27. $\omega_B = 528$ rad/s
$\theta_B = 288$ rad

16–29. $(\omega_B)_2 = 10.6$ rad/s

16–30. $a = \dfrac{s}{2\pi}\omega^2$

16–31. $v_F = 0.318$ mm/s

16–33. $\omega_B = 312$ rad/s
$\alpha_B = 176$ rad/s^2

16–34. $a = \dfrac{\omega^2}{2\pi}\left(\dfrac{r_2 - r_1}{L}\right)d$

16–35. $\mathbf{v}_C = \{-4.8\mathbf{i} - 3.6\mathbf{j} - 1.2\mathbf{k}\}$ m/s
$\mathbf{a}_C = \{38.4\mathbf{i} - 64.8\mathbf{j} + 40.8\mathbf{k}\}$ m/s^2

16–37. $v_C = 2.50$ m/s
$a_C = 13.1$ m/s^2

16–38. $v_C = 21.2$ ft/s
$a_C = 106$ ft/s^2

16–39. $v_{AB} = \omega l \cos\theta$
$a_{AB} = -\omega^2 l \sin\theta$

16–41. $\omega = 8.70$ rad/s
$\alpha = -50.5$ rad/s^2

16–42. $\dot{y} = \dfrac{v\tan\theta}{3}$ ↓

$\ddot{y} = \dfrac{v^2}{9L\cos^3\theta}$ ↓

16–43. $v_P = 18.5$ ft/s ←

16–45. $v = -\left(\dfrac{r_1^2\omega\sin 2\theta}{2\sqrt{r_1^2\cos^2\theta + r_2^2 + 2r_1 r_2}} + r_1\omega\sin\theta\right)$

16–46. $\omega = 0.0808$ rad/s

16–47. $v_A = e\omega\sin\theta$ →
$a_A = e\omega^2\cos\theta$ ←

16–49. $v_C = L\omega$ ↑
$a_C = 0.577 L\omega^2$ ↑

16–50. $\omega = \dfrac{v_0}{a}\sin^2\theta$

$\alpha = \left(\dfrac{v_0}{a}\right)^2\sin 2\theta\sin^2\theta$

16–51. $\omega = 1.08$ rad/s
$v_B = 4.39$ ft/s

16–53. $\dot{\theta} = \dfrac{v\sin\phi}{L\cos(\phi - \theta)}$

16–54. $\dot{x} = \dfrac{lr(r + d\cos\theta)}{(d + r\cos\theta)^2}\omega$

        $\ddot{x} = \dfrac{lr\sin\theta\,(2r^2 - d^2 + rd\cos\theta)}{(d + r\cos\theta)^3}\omega^2$

16–55. $\omega_A = 2\left(\dfrac{\sqrt{2}\cos\theta - 1}{3 - 2\sqrt{2}\cos\theta}\right)$

16–57. $\phi = \theta$

        $\omega = \dfrac{2v}{h}\cos\theta$

16–58. $\omega_{AB} = 2.00$ rad/s

16–59. $v_B = 2.83$ ft/s

        $\omega_{BC} = 2.83$ rad/s

        $\omega_{AB} = 2.83$ rad/s

16–61. $v_C = 1.64$ m/s

16–62. $\omega_{BC} = 0.693$ rad/s

16–63. $\omega_{CB} = 2.45$ rad/s $\circlearrowright$

        $v_C = 2.20$ ft/s $\leftarrow$

16–65. $\omega = 20$ rad/s

        $v_A = 2$ ft/s $\rightarrow$

16–66. $\omega = 3.11$ rad/s

        $v_O = 0.667$ ft/s $\rightarrow$

16–67. $v_A = 5.16$ ft/s

        $\theta = 39.8°$ ⟋

16–69. $v_A = 2.45$ m/s ↑

16–70. $\omega_{BC} = 2.69$ rad/s

        $\omega_{AC} = 4.39$ rad/s

16–71. $v_C = 0.776$ m/s ↙

        $v_D = 1.06$ m/s ↖

16–73. $v_A = \left(\dfrac{2R}{R - r}\right)v \rightarrow$

16–74. $v_C = 1.80$ m/s

        $\omega_{BC} = 6.24$ rad/s

16–75. $v_B = 3.00$ m/s $\rightarrow$

16–77. $\omega_P = 5$ rad/s

        $\omega_A = 1.67$ rad/s

16–78. $\omega_C = 50$ rad/s

16–79. $v_C = \dfrac{\omega}{2}(R - r)$ ↑

16–82. $\omega_{AB} = 1.24$ rad/s

16–83. $v_A = 0$

        $v_B = 1.2$ m/s ↓

        $v_C = 0.849$ m/s ↙ 45°

16–85. $\omega_{AB} = 6$ rad/s $\circlearrowright$

        $v_E = 4.76$ m/s

        $\theta_E = 40.9°$ ⟍

16–86. $v_G = 6.00$ m/s $\leftarrow$

16–87. $v_O = 1.04$ m/s $\rightarrow$

16–89. $v_H = 18.0$ ft/s

16–90. $v_C = 2.50$ ft/s

        $v_D = 9.43$ ft/s

        $\theta = 55.8°$ ⟍

16–91. $v_C = 2.50$ ft/s

        $v_E = 7.91$ ft/s

        $\theta = 18.4°$ ⟍

16–93. $\omega_{AB} = 13.1$ rad/s

16–94. $\omega = 1.12$ rad/s

        $v_C = 3.38$ ft/s

16–95. $v_A = 60.0$ ft/s $\rightarrow$

        $v_C = 220$ ft/s $\leftarrow$

        $v_B = 161$ ft/s $^{60.3°}$ ⟍

16–97. $v_E = 2$ ft/s $\leftarrow$

16–98. $\omega = 5.33$ rad/s

        $v_O = 2$ ft/s $\leftarrow$

16–99. $\omega_C = 26.7$ rad/s $\circlearrowright$

        $\omega_B = 28.75$ rad/s $\circlearrowleft$

        $\omega_A = 14.0$ rad/s $\circlearrowright$

16–101. $\omega_{BC} = 1.98$ rad/s $\circlearrowleft$

16–102. $\omega_{CD} = 0.0510$ rad/s $\circlearrowright$

16–103. $\alpha = 0.332$ rad/s^2 $\circlearrowleft$

        $a_B = 7.88$ ft/s^2 $\leftarrow$

16–105. $\alpha = 0.0962$ rad/s^2 $\circlearrowleft$

        $a_A = 0.385$ ft/s^2 $\rightarrow$

16–106. $a_C = 3.82$ m/s^2↙

        $\alpha_{BC} = 9.60$ rad/s^2 $\circlearrowright$

16–107. $\alpha_{A/B} = 25.4$ rad/s^2 $\circlearrowright$

        $a_B = 5.21$ m/s^2↓

16–109. $\alpha_{AB} = 6.59$ rad/s^2

        $\alpha_{BC} = 5.16$ rad/s^2

16–110. $a_A = 12.5$ m/s^2 $\leftarrow$

16–111. $a_C = 10.3$ m/s^2 $\leftarrow$

16–113. $a_B = 2.25$ m/s^2

        $\theta = 32.6°$ ⟍

16–114. $a_D = 10.0$ m/s^2

        $\theta = 2.02°$ ⟋

16–115. $v_B = 4v \rightarrow$

        $v_A = 2\sqrt{2}v$ ⟋$_{45°}$

        $a_B = \dfrac{2v^2}{r}$↓

        $a_A = \dfrac{2v^2}{r} \rightarrow$

16–117. $a_B = 63.55$ ft/s^2

        $\theta = 87.7°$ ⟍

16–118. $\alpha_C = 10.7$ rad/s^2 $\circlearrowright$

        $a_D = 14.1$ in./s^2

16–119. $a_B = 1.58\,\alpha a - 1.77\omega^2 a$

16–121. $\omega_{AC} = 0$

        $\omega_F = 10.7$ rad/s $\circlearrowleft$

        $\alpha_{AC} = 28.7$ rad/s^2 $\circlearrowleft$

16–122. $\alpha_B = 1.43$ rad/s^2

16–123. $a_E = 0.0714$ m/s^2↑

16–125. $\omega = 4.73$ rad/s $\circlearrowright$

        $\alpha = 131$ rad/s^2 $\circlearrowleft$

16–126. $\omega = 0.25$ rad/s $\circlearrowleft$
$v_B = 5.00$ ft/s$\downarrow$
$\alpha = 0.875$ rad/s^2 $\circlearrowright$
$a_B = 1.51$ ft/s^2
$\theta_B = 85.2°$ $\nwarrow$

16–127. $\alpha_{AB} = 3.70$ rad/s^2 $\circlearrowleft$

16–129. $\mathbf{v}_C = \{-8.12\mathbf{i} - 8.12\mathbf{j}\}$ ft/s
$\mathbf{a}_C = \{61.9\mathbf{i} - 61.0\mathbf{j}\}$ ft/s^2

16–130. $v_B = 1.30$ ft/s
$a_B = 0.6204$ ft/s^2

16–131. $\mathbf{v}_m = \{7.5\mathbf{i} - 5\mathbf{j}\}$ ft/s
$\mathbf{a}_m = \{5\mathbf{i} + 3.75\mathbf{j}\}$ ft/s^2

16–133. $\mathbf{v}_C = \{-3\mathbf{i} + 3\mathbf{j}\}$ m/s
$\mathbf{a}_C = \{-36.75\mathbf{i} - 17.5\mathbf{j}\}$ m/s^2

16–134. $\mathbf{a}_A = \{-5.60\mathbf{i} - 16\mathbf{j}\}$ m/s^2

16–135. (a) $\mathbf{a}_B = \{-1\mathbf{i}\}$ m/s^2
(b) $\mathbf{a}_B = \{-1.69\mathbf{i}\}$ m/s^2

16–137. $\omega_{CD} = 9.00$ rad/s $\circlearrowleft$
$\alpha_{CD} = 249$ rad/s^2 $\circlearrowleft$

16–138. $v_B = 7.7$ m/s $\leftarrow$
$a_B = 201$ m/s^2

16–139. $\omega_{CD} = 0.750$ rad/s $\circlearrowleft$
$\alpha_{CD} = 1.95$ rad/s^2 $\circlearrowright$

16–141. $\omega_{CD} = 0.866$ rad/s $\circlearrowright$
$\alpha_{CD} = 3.23$ rad/s^2 $\circlearrowright$

16–142. $\mathbf{v}_C = \{-0.944\mathbf{i} + 2.02\mathbf{j}\}$ m/s
$\mathbf{a}_C = \{-11.2\mathbf{i} - 4.15\mathbf{j}\}$ m/s^2

16–143. $\omega_{AB} = 5$ rad/s
$\alpha_{AB} = 2.5$ rad/s^2

16–145. $\mathbf{v}_D = \{-31.5\mathbf{i} - 13.0\mathbf{j}\}$ ft/s
$\mathbf{a}_D = \{13.0\mathbf{i} - 79.5\mathbf{j}\}$ ft/s^2

16–146. $\omega_{CD} = 13.8$ rad/s

16–147. $(\mathbf{v}_{rel})_{xyz} = \{29\mathbf{j}\}$ m/s
$(\mathbf{a}_{rel})_{xyz} = \{4.3\mathbf{i} - 0.2\mathbf{j}\}$ m/s^2

16–149. $\omega_{AB} = 3$ rad/s $\circlearrowleft$
$\alpha_{AB} = 32.7$ rad/s^2 $\circlearrowleft$

16–150. $\omega_{DC} = 2.96$ rad/s $\circlearrowleft$

16–151. $\omega_{BC} = 0.720$ rad/s $\circlearrowright$
$\alpha_{BC} = 2.02$ rad/s^2 $\circlearrowright$

Chapter 17

17–1. $I_y = \dfrac{1}{3}ml^2$

17–2. $m = \pi h R^2\left(k + \dfrac{aR^2}{2}\right)$
$I_z = \dfrac{\pi h R^4}{2}\left[k + \dfrac{2aR^2}{3}\right]$

17–3. $I_z = mR^2$

17–5. $I_z = \dfrac{2}{5}mr^2$

17–6. $I_x = \dfrac{3}{10}mr_0{}^2$

17–7. $k_y = 1.56$ in.

17–9. $I_z = m\left(R^2 + \dfrac{3}{4}a^2\right)$

17–10. $I_O = 5.27$ kg·m^2

17–11. $I_A = 2.17$ slug·ft^2

17–13. $I_A = 7.67$ kg·m^2

17–14. $I_A = 222$ slug·ft^2

17–15. $I_O = 6.23$ kg·m^2

17–17. $L = 6.39$ m
$I_O = 53.2$ kg·m^2

17–18. $I_O = 6.99$ kg·m^2

17–19. $\bar{y} = 0.888$ m
$I_G = 5.61$ kg·m^2

17–21. $\bar{y} = 1.78$ m
$I_G = 4.45$ kg·m^2

17–22. $I_x = 3.25$ g·m^2

17–23. $I_{x'} = 7.19$ g·m^2

17–25. $F = 5.96$ lb
$N_B = 99.0$ lb
$N_A = 101$ lb

17–26. $M_A = 5.08(10^3)$ lb·ft

17–27. $a_G = 1.07$ ft/s^2
$N_B = 86.7$ lb
$N_A = 213$ lb

17–29. $a = 10.3$ ft/s^2

17–30. $T = 15.7$ kN
$C_x = 8.92$ kN
$C_y = 16.3$ kN

17–31. $a_G = 2.33$ m/s^2

17–33. $N_B = 237$ N
$N_A = 744$ N

17–34. $a = 4$ m/s^2 $\rightarrow$
$N_B = 1.14$ kN
$N_A = 327$ N

17–35. $a = 13.2$ ft/s^2
$N_A = 640$ lb
$N_B = 910$ lb

17–37. $N_A = 2.35$ kip
$N_B = 2.15$ kip
$a_{max} = 13.4$ ft/s^2

17–38. $N_A = 2.40$ kip
$N_B = 2.10$ kip
$a_{max} = 12.0$ ft/s^2

17–39. $\theta = \tan^{-1}\left(\dfrac{a}{g}\right)$

17–41. $a = 3.96$ m/s^2

17–42. $a = 2.01$ m/s^2
The crate slips.

17–43. $a = 2.68$ ft/s^2
$N_A = 26.9$ lb
$N_B = 123$ lb

17–45. $N_B = 10.7$ kN
$F_A = 2.25$ kN
$N_A = 8.89$ kN

17–46. $t = 3.94$ s

17–47. $h_{max} = 3.16$ ft
$F_A = 248$ lb
$N_A = 400$ lb

17–49. $F_{AB} = 112$ N
$C_x = 26.2$ N
$C_y = 49.8$ N

17–50. $P = 765$ N

17–51. $T = 1.52$ kN
$\theta = 18.6°$

17–53. $N_B = 402$ N
$N_A = 391$ N

17–54. $\alpha = 5.95$ rad/s^2

17–55. $\alpha = 8.72$ rad/s^2
$F_{AB} = 402$ lb

17–57. $\omega = 56.2$ rad/s
$A_x = 0$
$A_y = 98.1$ N

17–58. $A_x = 0$
$A_y = 262$ N

17–59. $F_A = \dfrac{3}{2}mg$

17–61. $t = 6.71$ s

17–62. $\omega = 10.9$ rad/s

17–63. $\omega = 9.45$ rad/s

17–65. $\alpha = 24.5$ rad/s^2
$A_x = 90$ N
$A_y = 29.4$ N

17–66. $m(a_G)_t r_{OG} + I_G \alpha = m(a_G)_t(r_{OG} + r_{GP})$

17–67. $r_P = 2.67$ ft
$A_x = 0$

17–69. $\omega = 2.71$ rad/s

17–70. $k = 27.2$ N $\cdot$ m/rad

17–71. $F_O = 299$ N

17–73. $N_A = \dfrac{mg}{4} \cos \theta$
$F_f = \dfrac{5mg}{2} \sin \theta$
$\theta = \tan^{-1}\left(\dfrac{\mu}{10}\right)$

17–74. $\alpha = 14.2$ rad/s^2

17–75. $A_x = 89.2$ N
$A_y = 66.9$ N
$t = 1.25$ s

17–77. $\omega_A = 38.3$ rad/s
$\omega_B = 57.5$ rad/s

17–78. $a = \dfrac{8}{25}(2 - \mu_k)g$

17–79. $a = 2.97$ m/s^2

17–81. $A_x = 0$
$A_y = 289$ N
$\alpha = 23.1$ rad/s^2

17–82. $N_A = 177$ kN
$V_A = 5.86$ kN
$M_A = 50.7$ kN $\cdot$ m

17–83. $M = 0.3 \, gml$

17–85. $N = wx\left[\dfrac{\omega^2}{g}\left(L - \dfrac{x}{2}\right) + \cos \theta\right]$
$S = wx \sin \theta$
$M = \dfrac{1}{2}wx^2 \sin \theta$

17–86. $\omega = 4.83$ rad/s
$\theta = 92.2°$

17–87. $v = 4.57$ m/s$\downarrow$

17–89. $\omega = 800$ rad/s

17–91. $\alpha = 5.62$ rad/s^2
$T = 196$ N

17–92. $\alpha = 7.86$ rad/s^2
$a_G = 34.1$ ft/s^2

17–93. $\alpha = 0.274$ rad/s^2
$a_B = 49.0$ ft/s^2
$\theta = 80.3°$ ⬎

17–94. $\alpha = 4.32$ rad/s^2

17–95. $\theta = 46.9°$

17–97. $\alpha = 0.500$ rad/s^2

17–98. $\alpha = 15.6$ rad/s^2

17–99. $\omega = 5.21$ rad/s

17–101. $a_A = 167$ ft/s^2

17–102. $T = 5.61$ N
$\alpha = 28.0$ rad/s^2

17–103. $\alpha = 0.109$ rad/s^2

17–105. $\alpha = 18.9$ rad/s^2
$P = 76.4$ lb

17–106. $\alpha = 1.25$ rad/s
$T = 2.32$ kN

17–107. $T = 3.13$ kN
$\alpha = 1.68$ rad/s
$a_C = 1.35$ m/s^2

17–109. $\alpha = 3$ rad/s^2

17–110. $s = 1.40$ in.

17–111. $\alpha_A = 43.6$ rad/s^2
$\alpha_B = 43.6$ rad/s^2
$T = 15.7$ N

17–113. $a_G = \mu_k g \leftarrow$

$$\alpha = \frac{2\mu_k g}{r}$$

17–114. $\omega = \frac{1}{3}\omega_0$

$$t = \frac{\omega_0 r}{3\mu_k g}$$

17–115. $s = 4.06$ ft

17–117. $T = 19.6$ N

17–118. $\alpha = 23.4$ rad/s^2
$B_y = 9.62$ lb

17–119. $\alpha = 6.67$ rad/s^2

Chapter 18

18–2. $\omega = 14.0$ rad/s

18–3. $\omega = 14.1$ rad/s

18–5. $s = 0.661$ m

18–6. $s = 0.859$ m

18–7. $T = 283$ ft · lb

18–9. $\omega = 7.57$ rad/s

18–10. $\omega = \frac{1}{k_G}\sqrt{\frac{\pi F d}{m}}$

18–11. $T = 0.0188$ ft · lb

18–13. $v_C = 16.9$ ft/s↑

18–14. $v_C = 11.8$ ft/s↑

18–15. $v_C = 1.97$ m/s

18–17. $\omega = \sqrt{\omega_0^2 + \frac{g}{r^2}s\sin\theta}$

18–18. $v_C = 6.11$ ft/s

18–19. $\omega = 4.90$ rad/s

18–21. $s = 0.304$ ft

18–22. $v_C = 19.6$ ft/s

18–23. $\omega_{AB} = \omega_{CD} = 2.24$ rad/s

18–25. $\omega = 3.48$ rad/s

18–26. $\omega_2 = 5.37$ rad/s

18–27. $v_B = 5.05$ ft/s

18–29. $v_A = 26.7$ ft/s

18–30. $\omega = 2.50$ rad/s

18–31. $\omega = \sqrt{\frac{3g}{l}}\sin\theta$

18–33. $\theta = 0.891$ rev, regardless of orientation

18–34. $\omega_2 = \frac{1}{r}\sqrt{v_G^2 - \frac{10}{7}gR}$

18–35. $v_G = 3\sqrt{\frac{3}{7}gR}$

18–37. $\theta = 25.4°$

18–38. $s = 0.301$ m
$T = 163$ N

18–39. $v_A = 1.29$ m/s

18–41. $v_C = 13.3$ ft/s

18–42. $v_C = 3.07$ ft/s

18–43. $v_C = 31.9$ m/s

18–45. $\omega = 1.73$ rad/s

18–46. $\omega = 2.57$ rad/s

18–47. $d = 3.38$ m

18–49. $\omega = 39.3$ rad/s

18–50. $v_A = 1.40$ m/s

18–51. $\omega_2 = 3.09$ rad/s

18–53. $\omega_{AB} = \omega_{BC} = \sqrt{\frac{3g}{L}}$

18–54. $v_b = 15.5$ ft/s ↓

18–55. $(\omega_{ABC})_2 = 7.24$ rad/s

18–57. $\omega = 12.8$ rad/s

18–58. $k = 42.8$ N/m

18–59. $k = 206$ N/m

18–61. $\theta = 48.2°$

18–62. $\omega = 1.80$ rad/s

18–63. $\omega = 85.1$ rad/s

18–65. $\omega = 1.82$ rad/s

18–66. $k = 100$ lb/ft

18–67. $k = 814$ N · m/rad

Chapter 19

19–5. $\int M\,dt = 0.833$ kg · m^2/s

19–6. $\omega_2 = 0.386$ rad/s

19–7. $\omega = 0.0178$ rad/s

19–9. $(v_O)_2 = 6$ m/s
$\omega_2 = 26.7$ rad/s

19–10. $t = 0.6125$ s

19–11. $\omega = 16$ rad/s
$v = 4$ m/s

19–13. $y = \frac{2}{3}l$

19–14. $\omega_0 = 2.5\left(\frac{v_0}{r}\right)$

19–15. $\omega_2 = 9.40$ rad/s

19–17. $\omega_B = 127$ rad/s

19–18. $\omega = 6.04$ rad/s

19–19. $v_A = 24.1$ m/s

19–21. $v_B = 34.0$ ft/s

19–22. $r_P = 1.39$ ft

19–23. $t = 9.74$ s

19–25. $v_A = 1.66$ m/s
$v_B = 1.30$ m/s

19–26. $\omega = 9$ rad/s

19–27. $y = \frac{\sqrt{2}}{3}a$

19–29. (a) $\omega_M = 0$

(b) $\omega_M = \dfrac{I}{I_z}\omega$

(c) $\omega_M = \dfrac{2I}{I_y}\omega$

19–30. $\omega = \dfrac{m_A k_A^2 \omega_A + m_B k_B^2 \omega_B}{m_A k_A^2 + m_B k_B^2}$

19–31. $\omega = 0.244$ rad/s

19–33. $\omega_2 = 2.55$ rev/s

19–34. $\omega_B = 10.9$ rad/s

19–35. $\omega = 0.0906$ rad/s

19–37. $\omega_2 = 6.94$ rad/s

19–38. $\omega = 3.70$ rad/s

19–39. $\omega_3 = \dfrac{3}{2}\sqrt{\dfrac{3g}{L}}$

19–41. $\omega_2 = \dfrac{1}{3}\omega_1$

19–42. $\omega_2 = 57$ rad/s
$U_F = 367$ J

19–43. $\omega_2 = 3.47$ rad/s

19–45. $v = 5.96$ ft/s

19–46. $\theta = 22.4°$

19–47. $\theta_2 = 39.8°$

19–49. $(v_D)_3 = 1.54$ m/s
$\omega_3 = 0.934$ rad/s

19–50. $\omega_2 = 7.73$ rad/s

19–51. $\theta = \tan^{-1}\left(\sqrt{\dfrac{7}{5}e}\right)$

19–53. $\omega_3 = 2.73$ rad/s

19–54. $\omega_1 = 1.02\sqrt{\dfrac{g}{r}}$

19–55. $(v_G)_{y2} = e(v_G)_{y1}\uparrow$

$(v_G)_{x2} = \dfrac{5}{7}\left((v_G)_{x1} - \dfrac{2}{5}\omega_1 r\right)\leftarrow$

Review 2

R2–1. $R = 33.3$ lb

R2–2. $N'_A = 383$ N
$N'_B = 620$ N

R2–3. $a = 16.1$ ft/s^2

R2–5. $\omega = 39.5$ rad/s

R2–6. $a_A = 282$ ft/s$^2 \rightarrow$

R2–7. $a_C = 50.6$ m/s^2
$\theta = 9.09°$ ⤢
$a_B = 8.00$ m/s$^2\uparrow$

R2–9. $\omega = 0.275$ rad/s
$\alpha = 0.0922$ rad/s^2

R2–10. $\alpha_B = \dfrac{\omega_A^2 T}{2\pi r_B^3}(r_A^2 + r_B^2)$

R2–11. $a_G = 3.22$ ft/s^2
$N_B = 14.8$ lb
$N_A = 65.3$ lb

R2–13. $a_G = 39.2$ m/s^2
It cannot be done, since
$$39.2 \text{ m/s}^2 > a_{max} = 5.89 \text{ m/s}^2.$$

R2–14. $F_{B'} = 4.50$ N
$N_{A'} = 1.78$ kN
$N_{B'} = 5.58$ kN

R2–15. $\omega_F = 784$ rev/min

R2–17. $v_B = 2.58$ m/s
$P = 141$ N

R2–18. $s = 9.75$ m

R2–19. $h = 0.2$ m: $a_G = 1.41$ m/s^2
$h = 0$: $a_G = 1.38$ m/s^2

R2–21. $a_B = 40.2$ ft/s^2
$\theta = 53.3°$ ⤢

R2–22. $v_B = 0.860$ m/s

R2–23. $T = 218$ N
$\alpha = 21.0$ rad/s^2

R2–25. $F_C = 12.4$ lb
$N_{B'} = 3.09$ kip
$N_{A'} = 1.31$ kip
$F_{A'} = 68.3$ lb

R2–26. $\alpha = 73.3$ rad/s^2
$t = 0.296$ s

R2–27. $\alpha = 12.1$ rad/s^2
$F = 30.0$ lb

R2–29. $\omega = 100$ rad/s

R2–30. $v = 4.78$ m/s

R2–31. $v_C = 9.38$ in./s ↙
$a_C = 54.7$ in./s^2 ↙

R2–33. (a) $a_G = 1.38$ m/s^2
(b) $a_G = 1.56$ m/s^2

R2–34. $(v_G)_2 = 9.42$ m/s

R2–35. $\alpha = \dfrac{1.30g}{l}$
$O_x = 0.325\,mg$
$O_y = 0.438\,mg$

R2–37. $\omega_A = 1.70$ rad/s
$\omega_B = 5.10$ rad/s

R2–38. $\omega = 28.6$ rad/s
$\theta = 24.1$ rad
At $y = 0.5$ m:
$v_P = 7.16$ m/s
$a_P = 205$ m/s^2

R2–39. $\dot{y} = 1.5\cot\theta$

R2–41. $v_2 = 3.46$ m/s

R2–42. $h = 1.80$ m

R2–43. $E_x = 9.87$ lb
$E_y = 4.86$ lb
$M_E = 7.29$ lb·ft

R2–45. $v_C = 1.34 \text{ m/s}$

R2–46. $\theta = \tan^{-1}\left[\dfrac{\mu(k_O^2 + r^2)}{k_O^2}\right]$

R2–47. $v = -r\omega \sin\theta$
$a = -r\omega^2 \cos\theta$

R2–49. $B_y = 180 \text{ N}$
$A_y = 252 \text{ N}$
$A_x = 139 \text{ N}$

R2–50. $B_y = 143 \text{ N}$
$A_y = 200 \text{ N}$
$A_x = 34.3 \text{ N}$

R2–51. $v_W = 14.2 \text{ ft/s}$
$a_W = 0.25 \text{ ft/s}$

Chapter 20

20–1. $\mathbf{v}_A = \{-7.61\mathbf{i} - 1.18\mathbf{j} + 2.54\mathbf{k}\} \text{ m/s}$
$\mathbf{a}_A = \{10.4\mathbf{i} - 51.6\mathbf{j} - 0.463\mathbf{k}\} \text{ m/s}^2$

20–2. $\boldsymbol{\omega} = \{-8.0\mathbf{j} + 4.0\mathbf{k}\} \text{ rad/s}$
$\boldsymbol{\alpha} = \{32\mathbf{i}\} \text{ rad/s}^2$

20–3. $\mathbf{v}_A = \{-5.20\mathbf{i} - 12\mathbf{j} + 20.8\mathbf{k}\} \text{ ft/s}$
$\mathbf{a}_A = \{-24.1\mathbf{i} - 13.3\mathbf{j} - 7.20\mathbf{k}\} \text{ ft/s}^2$

20–5. $\omega = 41.2 \text{ rad/s}$
$v_P = 4.00 \text{ m/s}$
$\alpha = 400 \text{ rad/s}^2$
$a_P = 100 \text{ m/s}^2$

20–6. $\mathbf{v}_P = \{-1.60\mathbf{i}\} \text{ m/s}$
$\mathbf{a}_P = \{-0.640\mathbf{i} - 12.0\mathbf{j} - 8.00\mathbf{k}\} \text{ m/s}^2$

20–7. $\mathbf{v}_P = \{-7.79\mathbf{i} - 2.25\mathbf{j} + 3.90\mathbf{k}\} \text{ ft/s}$
$\mathbf{a}_P = \{8.30\mathbf{i} - 35.2\mathbf{j} + 7.02\mathbf{k}\} \text{ ft/s}^2$

20–9. $\mathbf{v}_B = \{-7.06\mathbf{i} - 7.52\mathbf{k}\} \text{ ft/s}$
$\mathbf{a}_B = \{77.3\mathbf{i} - 28.3\mathbf{j} - 0.657\mathbf{k}\} \text{ ft/s}^2$

20–10. $\mathbf{v}_B = \{-90\mathbf{i} - 15\mathbf{j} + 6\mathbf{k}\} \text{ ft/s}$
$\mathbf{a}_B = \{243\mathbf{i} - 1353\mathbf{j} + 1.5\mathbf{k}\} \text{ ft/s}^2$

20–11. $\mathbf{v}_B = \{410\mathbf{i} - 15\mathbf{j} + 6\mathbf{k}\} \text{ ft/s}$
$\mathbf{a}_B = \{293\mathbf{i} - 1353\mathbf{j} + 1.5\mathbf{k}\} \text{ ft/s}^2$

20–13. $\mathbf{v}_A = \{5\mathbf{i}\} \text{ ft/s}$
$\mathbf{a}_A = \{-1.5\mathbf{i} - 6\mathbf{j} - 32\mathbf{k}\} \text{ ft/s}^2$

20–14. $\mathbf{v}_C = \{1.8\mathbf{j} - 1.5\mathbf{k}\} \text{ m/s}$
$\mathbf{a}_C = \{-36.6\mathbf{i} + 0.45\mathbf{j} - 0.9\mathbf{k}\} \text{ m/s}^2$

20–15. $\mathbf{v}_A = \{-8.66\mathbf{i} + 8.00\mathbf{j} - 13.9\mathbf{k}\} \text{ ft/s}$
$\mathbf{a}_A = \{-24.8\mathbf{i} + 8.29\mathbf{j} - 30.9\mathbf{k}\} \text{ ft/s}^2$

20–17. $\mathbf{v}_A = [100\mathbf{j} - 40\mathbf{k}] \text{ ft/s}$
$\mathbf{a}_A = \{-580\mathbf{i} + 60\mathbf{j} - 30\mathbf{k}\} \text{ ft/s}^2$

20–18. $\boldsymbol{\omega}_P = \{-40\mathbf{j}\} \text{ rad/s}$
$\boldsymbol{\alpha}_B = \{-6400\mathbf{i}\} \text{ rad/s}^2$

20–19. $\boldsymbol{\omega} = \{4.35\mathbf{i} + 12.7\mathbf{j}\} \text{ rad/s}$
$\boldsymbol{\alpha} = \{-26.1\mathbf{k}\} \text{ rad/s}^2$

20–21. $\boldsymbol{\omega} = \{30\mathbf{j} - 5\mathbf{k}\} \text{ rad/s}$
$\boldsymbol{\alpha} = \{150\mathbf{i}\} \text{ rad/s}^2$

20–22. $\mathbf{v}_A = \{10\mathbf{i} + 14.7\mathbf{j} - 19.6\mathbf{k}\} \text{ ft/s}$
$\mathbf{a}_A = \{-6.12\mathbf{i} + 3\mathbf{j} - 2\mathbf{k}\} \text{ ft/s}^2$

20–23. $\omega_A = 47.8 \text{ rad/s}$
$\omega_B = 7.78 \text{ rad/s}$

20–25. $\boldsymbol{\omega}_{BC} = \{0.204\mathbf{i} - 0.612\mathbf{j} + 1.36\mathbf{k}\} \text{ rad/s}$
$\mathbf{v}_B = \{-0.333\mathbf{j}\} \text{ m/s}$

20–26. $\boldsymbol{\omega} = \{0.980\mathbf{i} - 1.06\mathbf{j} - 1.47\mathbf{k}\} \text{ rad/s}$
$v_B = 12.0 \text{ ft/s}$

20–27. $\mathbf{a}_B = \{-96.5\mathbf{i}\} \text{ ft/s}^2$

20–28. $\mathbf{v}_B = [-1.33\mathbf{i} + 1\mathbf{j}] \text{ m/s}$

20–29. $\mathbf{a}_B = \{4.99\mathbf{i} - 3.74\mathbf{j}\} \text{ m/s}^2$

20–30. $\boldsymbol{\omega}_{BC} = \{-2\mathbf{k}\} \text{ rad/s}$

20–31. $\boldsymbol{\omega}_{BD} = \{-2.00\mathbf{i}\} \text{ rad/s}$

20–33. $v_B = 4.71 \text{ ft/s}$
$\boldsymbol{\omega}_{AB} = \{1.17\mathbf{i} + 1.27\mathbf{j} - 0.779\mathbf{k}\} \text{ rad/s}$

20–34. $a_B = 17.6 \text{ ft/s}^2$
$\boldsymbol{\alpha}_{AB} = \{-2.78\mathbf{i} - 0.628\mathbf{j} - 2.91\mathbf{k}\} \text{ rad/s}^2$

20–35. $\boldsymbol{\omega}_{BC} = \{0.769\mathbf{i} - 2.31\mathbf{j} + 0.513\mathbf{k}\} \text{ rad/s}$
$\mathbf{v}_B = \{-0.333\mathbf{j}\} \text{ m/s}$

20–37. $\mathbf{v}_C = \{-1.00\mathbf{i} + 5.00\mathbf{j} + 0.800\mathbf{k}\} \text{ m/s}$
$\mathbf{a}_C = \{-28.8\mathbf{i} - 5.45\mathbf{j} + 32.3\mathbf{k}\} \text{ m/s}^2$

20–38. $\mathbf{v}_C = \{-1\mathbf{i} + 5\mathbf{j} + 0.8\mathbf{k}\} \text{ m/s}$
$\mathbf{a}_C = \{-28.2\mathbf{i} - 5.45\mathbf{j} + 32.3\mathbf{k}\} \text{ m/s}^2$

20–39. $\mathbf{v}_B = \{-2.75\mathbf{i} - 2.50\mathbf{j} + 3.17\mathbf{k}\} \text{ m/s}$
$\mathbf{a}_B = \{2.50\mathbf{i} - 2.24\mathbf{j} - 0.00389\mathbf{k}\} \text{ ft/s}^2$

20–41. $\mathbf{v}_A = \{5.20\mathbf{i} - 5.20\mathbf{j} - 3.00\mathbf{k}\} \text{ ft/s}$
$\mathbf{a}_A = \{25\mathbf{i} - 26.8\mathbf{j} + 8.78\mathbf{k}\} \text{ ft/s}^2$

20–42. $\mathbf{v}_B = \{-17.8\mathbf{i} - 3\mathbf{j} + 5.20\mathbf{k}\} \text{ m/s}$
$\mathbf{a}_B = \{9\mathbf{i} - 29.4\mathbf{j} - 1.5\mathbf{k}\} \text{ m/s}^2$

20–43. $\mathbf{v}_B = \{-17.8\mathbf{i} - 3\mathbf{j} + 5.20\mathbf{k}\} \text{ m/s}$
$\mathbf{a}_B = \{3.05\mathbf{i} - 30.9\mathbf{j} + 1.10\mathbf{k}\} \text{ m/s}^2$

20–45. $\mathbf{v}_C = \{-6.75\mathbf{i} - 6.25\mathbf{j}\} \text{ m/s}$
$\mathbf{a}_C = \{28.75\mathbf{i} - 26.25\mathbf{j} - 4\mathbf{k}\} \text{ m/s}^2$

20–46. $\mathbf{v}_B = \{-17.3\mathbf{i} + 18.8\mathbf{j} + 10.0\mathbf{k}\} \text{ ft/s}$
$\mathbf{a}_B = \{-19.1\mathbf{i} + 24.0\mathbf{j} - 8.66\mathbf{k}\} \text{ ft/s}^2$

20–47. $\mathbf{v}_B = \{-17.3\mathbf{i} + 18.8\mathbf{j} + 10.0\mathbf{k}\} \text{ ft/s}$
$\mathbf{a}_B = \{-45.0\mathbf{i} + 24.0\mathbf{j} - 6.34\mathbf{k}\} \text{ ft/s}^2$

20–49. $\mathbf{v}_C = \{-2.7\mathbf{i} - 6\mathbf{k}\} \text{ m/s}$
$\mathbf{a}_C = \{-72\mathbf{i} - 13.5\mathbf{j} + 7.8\mathbf{k}\} \text{ m/s}^2$

20–50. $\mathbf{v}_P = \{-25.5\mathbf{i} - 13.4\mathbf{j} + 20.5\mathbf{k}\} \text{ ft/s}$
$\mathbf{a}_P = \{161\mathbf{i} - 249\mathbf{j} - 39.6\mathbf{k}\} \text{ ft/s}^2$

20–51. $\mathbf{v}_P = \{-25.5\mathbf{i} - 13.4\mathbf{j} + 20.5\mathbf{k}\} \text{ ft/s}$
$\mathbf{a}_P = \{161\mathbf{i} - 243\mathbf{j} - 33.9\mathbf{k}\} \text{ ft/s}^2$

20–53. $\mathbf{v}_B = \{-10.2\mathbf{i} - 28\mathbf{j} + 52.0\mathbf{k}\} \text{ m/s}$
$\mathbf{a}_B = \{-33.0\mathbf{i} - 159\mathbf{j} - 90\mathbf{k}\} \text{ m/s}^2$

20–54. $\mathbf{v}_C = \{1.8\mathbf{i} + 0.9\mathbf{j} + 2.25\mathbf{k}\} \text{ m/s}$
$\mathbf{a}_C = \{0.9\mathbf{i} - 9.075\mathbf{j} - 12.75\mathbf{k}\} \text{ m/s}^2$

Chapter 21

21–2 $I_{\bar{y}} = \dfrac{3m}{80}(h^2 + 4a^2)$

$I_{y'} = \dfrac{m}{20}(2h^2 + 3a^2)$

21–3. $I_y = \dfrac{1}{3}mr^2$

$I_x = \dfrac{m}{6}(r^2 + 3a^2)$

21–5. $I_{yz} = \dfrac{m}{6}ah$

21–6. $I_{xy} = \dfrac{m}{12}a^2$

21–7. $I_{xy} = 636\rho$

21–9. $\begin{pmatrix} \dfrac{2}{3}ma^2 & \dfrac{1}{4}ma^2 & \dfrac{1}{4}ma^2 \\[2mm] \dfrac{1}{4}ma^2 & \dfrac{2}{3}ma^2 & -\dfrac{1}{4}ma^2 \\[2mm] \dfrac{1}{4}ma^2 & -\dfrac{1}{4}ma^2 & \dfrac{2}{3}ma^2 \end{pmatrix}$

21–10. $I_{x'} = \dfrac{m}{12}(a^2 + h^2)$

21–11. $I_{aa} = \dfrac{m}{12}(3a^2 + 4h^2)$

21–13. $I_{yz} = 0$

21–14. $I_{xy} = 0.32 \text{ kg} \cdot \text{m}^2$
$I_{yz} = 0.08 \text{ kg} \cdot \text{m}^2$
$I_{xz} = 0$

21–15. $I_{xy} = 0.0966 \text{ slug} \cdot \text{ft}^2$
$I_{yz} = 0.0483 \text{ slug} \cdot \text{ft}^2$
$I_{xz} = 0.0372 \text{ slug} \cdot \text{ft}^2$

21–17. $\bar{y} = 0.5 \text{ ft}$
$\bar{x} = -0.667 \text{ ft}$
$I_{x'} = 0.0272 \text{ slug} \cdot \text{ft}^2$
$I_{y'} = 0.0155 \text{ slug} \cdot \text{ft}^2$
$I_{z'} = 0.0427 \text{ slug} \cdot \text{ft}^2$

21–18. $I_x = 4.50 \text{ kg} \cdot \text{m}^2$
$I_y = 4.38 \text{ kg} \cdot \text{m}^2$
$I_z = 0.125 \text{ kg} \cdot \text{m}^2$

21–19. $I_{aa} = 1.13 \text{ slug} \cdot \text{ft}^2$

21–21. $I_z = 0.429 \text{ kg} \cdot \text{m}^2$

21–22.
$I^3 - (I_{xx} + I_{yy} + I_{zz})I^2 + (I_{xx}I_{yy} + I_{yy}I_{zz} + I_{zz}I_{xx} - I_{xy}^2 - I_{yz}^2 - I_{zx}^2)I$
$\quad - (I_{xx}I_{yy}I_{zz} - 2I_{xy}I_{yz}I_{zx} - I_{xx}I_{yz}^2 - I_{yy}I_{zx}^2 - I_{zz}I_{xy}^2) = 0$

21–25. $T = \dfrac{9mh^2}{20}\left[1 + \dfrac{r^2}{6h^2}\right]\omega^2$

21–26. $\omega_2 = 61.7 \text{ rad/s}$

21–27. $\omega_2 = 87.2 \text{ rad/s}$

21–29. $\omega_{OB} = 15.1 \text{ rad/s}$

21–30. $h = 2.24 \text{ in.}$

21–31. $v_A = 18.2 \text{ ft/s}$

21–33. $\omega_p = 4.82 \text{ rad/s}$

21–34. $\mathbf{H}_A = \{-2000\mathbf{i} - 55\,000\mathbf{j} + 22\,500\mathbf{k}\} \text{ kg} \cdot \text{m}^2/\text{s}$

21–35. $T = 37.0 \text{ MJ}$

21–37. $\mathbf{v}_G = \{-10.2\mathbf{k}\} \text{ ft/s}$

21–38. $H = 0.625 \text{ kg} \cdot \text{m}^2/\text{s}$

21–39. $\omega_s = 28.7 \text{ rad/s}$

21–41.
$\Sigma M_x = (I_x\dot{\omega}_x - I_{xy}\dot{\omega}_y - I_{xz}\dot{\omega}_z) - \Omega_z(I_y\omega_y - I_{yz}\omega_z - I_{yx}\omega_x)$
$\quad\quad + \Omega_y(I_z\omega_z - I_{zx}\omega_x - I_{zy}\omega_y)$

21–43. $B_z = 4 \text{ lb}$
$A_x = -2.00 \text{ lb}$
$A_y = 0.627 \text{ lb}$
$B_x = 2.00 \text{ lb}$
$B_y = -1.37 \text{ lb}$

21–45. $A_Z = 1.46 \text{ lb}$
$A_X = B_X = 0$
$B_Z = 13.5 \text{ lb}$

21–46. $F_A = 277 \text{ N}$
$F_B = 166 \text{ N}$

21–47. $F_A = 213 \text{ N}$
$F_B = 128 \text{ N}$

21–49. $A_y = B_y = 0$
$A_z = B_z = 24.5 \text{ N}$

21–50. $D_y = -12.9 \text{ N}$
$\dot{\omega}_z = 200 \text{ rad/s}^2$
$D_x = -37.5 \text{ N}$
$C_x = -37.5 \text{ N}$
$C_y = -11.1 \text{ N}$
$C_z = 36.8 \text{ N}$

21–51. $\dot{\omega}_y = -102 \text{ rad/s}^2$
$A_x = B_x = 0$
$A_y = 0$
$A_z = 297 \text{ N}$
$B_z = -143 \text{ N}$

21–53. $\Sigma M_x = \dfrac{I_{yz}}{\rho^2}v^2 G$

$\Sigma M_y = -\dfrac{I_{zx}}{\rho^2}v^2 G$

$\Sigma M_z = 0$

21–54. $\dot{\omega}_y = 25.9 \text{ rad/s}^2$
$B_x = -0.0791 \text{ N}$
$B_z = 12.3 \text{ N}$
$A_x = -1.17 \text{ N}$
$A_z = 12.3 \text{ N}$

21–55. $A_Y = 0$
$A_Z = B_Z = 98.1 \text{ N}$
$A_X = -3.75 \text{ kN}$
$B_X = 3.75 \text{ kN}$

21–57. $M_x = -\dfrac{4}{3}ml^2\omega_s\omega_p \cos\theta$

$M_y = \dfrac{1}{3}ml^2\omega_p^2 \sin 2\theta$

$M_z = 0$

21–58. $A_z = -1.09$ kN
$B_z = 1.38$ kN

21–59. $\Sigma M_x = 0$
$\Sigma M_y = (-0.036 \sin \theta)$ N $\cdot$ m
$\Sigma M_z = (0.003 \sin 2\theta)$ N $\cdot$ m

21–61. $\alpha = 69.3°$
$\beta = 128°$
$\gamma = 45°$
No, the orientation will not be the same for any order. Finite rotations are not vectors.

21–62. (a) $A_y = 1.49$ kN
$B_y = 2.43$ kN
(b) $A_y = -1.24$ kN
$B_y = -5.17$ kN
(c) $A_y = 1.49$ kN
$B_y = 2.43$ kN

21–63. $M_x = -56.25$ N $\cdot$ m

21–65. $\omega_s = 105$ rad/s

21–66. $\Delta F = 53.4$ N

21–67. $\dot{\phi} = \left(\dfrac{2g \cot \theta}{a + r \cos \theta} \right)^{1/2}$

21–69. $\omega_p = 1.19$ rad/s

21–70. $\omega_s = 3.63(10^3)$ rad/s

21–71. $\theta = 68.1°$

21–73. $\dot{\phi} = 23.3$ rad/s

21–75. $\dot{\phi} = 12.8$ rad/s

21–77. $\dot{\phi} = 17.8$ rad/s
$\dot{\psi} = 22.3$ rad/s
$\beta = 19.8°$

21–78. $H_G = 0.352$ kg $\cdot$ m²/s
$\dot{\psi} = 35.1$ rad/s

21–79. $H_G = 17.2$ Mg $\cdot$ m²/s

Chapter 22

22–1. $x = -0.05 \cos (12.2t)$ m

22–2. $f = 4.98$ Hz
$\tau = 0.201$ s

22–3. $y = 0.107 \sin (7.00t) + 0.100 \cos (7.00t)$
$\phi = 43.0°$

22–5. $\omega_n = 49.5$ rad/s
$\tau = 0.127$ s

22–6. $x = -0.126 \sin (3.16t) - 0.09 \cos (3.16t)$ m
$C = 0.155$ m

22–7. $\omega_n = 19.7$ rad/s
$C = 1$ in.
$y = (0.0833 \cos 19.7t)$ ft

22–9. $x = -0.05 \cos (8.16t)$
$C = 50$ mm

22–10. $f = \dfrac{1}{\pi} \sqrt{\dfrac{k}{m}}$

22–11. $\tau = 1.18$ s

22–13. $\tau = 2\pi \sqrt{\dfrac{k_G^2 + d^2}{gd}}$

22–14. $d = 146$ mm
$k_G = 0.627$ m

22–15. $\tau = 2\pi \sqrt{\dfrac{2r}{g}}$

22–17. Let k_1 be the larger value.
$k_1 = 2067$ N/m
$k_2 = 302$ N/m

22–18. $\tau = 0.167$ s

22–19. $\omega_n = 9.47$ rad/s

22–21. $y = 1.10 \sin (5.48t)$ m

22–22. $\omega_n = \sqrt{\dfrac{3g (4R^2 - l^2)^{\frac{1}{2}}}{6R^2 - l^2}}$

22–23. $\tau = 2.67$ s

22–25. $\omega_n = \sqrt{\dfrac{k_1 + k_2}{m}}$

22–26. $\tau = 2\pi k_O \sqrt{\dfrac{m}{C}}$

22–27. $k_B = a \sqrt{\dfrac{m}{M} \left(\dfrac{\tau_2^2}{\tau_1^2 - \tau_2^2} \right)}$

$k = \dfrac{4\pi^2}{\tau_1^2 - \tau_2^2} m$

22–29. $k_z = \dfrac{\tau r}{2\pi} \sqrt{\dfrac{g}{L}}$

22–30. $\ddot{x} + 333x = 0$

22–31. $\tau = 1.52$ s

22–33. $\tau = 0.339$ s

22–34. $\tau = \pi \sqrt{\dfrac{m}{k}}$

22–35. $\tau = 0.487$ s

22–37. $\omega_n = 11.3$ rad/s

22–38. $\omega_n = 14.1$ rad/s

22–39. $\tau = 4.25 \sqrt{\dfrac{r}{g}}$

22–41. $y = A \sin \omega_n t + B \cos \omega_n t + \dfrac{F_0}{(k - m\omega^2)} \cos \omega t$

22–42. $c = 18.9$ N $\cdot$ s/m

22–43.
$y = (-0.0232 \sin 8.97t + 0.333 \cos 8.97t + 0.0520 \sin 4t)$ ft

22–45. $y = A \sin \omega_n t + B \cos \omega_n t + \dfrac{\delta_0}{1 - \left(\dfrac{\omega_0}{\omega_n} \right)^2} \sin \omega t$

22–46. $y = \big(361 \sin 7.75t + 100 \cos 7.75t$
$- 350 \sin 8t \big)$ mm

22–47. $\omega_n = 2.83$ rad/s

22–49. $(x_p)_{max} = 29.5$ mm

22–50. $(x_p)_{max} = 1.89$ in.

22–51. $C = \dfrac{3F_O}{\dfrac{3}{2}(mg + Lk) - mL\omega^2}$

22–53. $\omega = 14.0$ rad/s

22–54. $(x_p)_{max} = 14.6$ mm

22–55. $(x_p)_{max} = 35.5$ mm

22–57. $\omega = 19.7$ rad/s

22–58. $C = 0.490$ in.

22–59. $\omega = 19.0$ rad/s

22–61. $\omega = 6.55$ rad/s

22–62. $Y = \dfrac{mr\omega^2 L^3}{48EI - M\omega^2 L^3}$

22–63. $\tau_d = 0.666$ s

22–65. $\phi' = 9.89°$

22–66. MF $= 0.997$

22–67. $y = [-0.0702\, e^{-3.57t} \sin(8.540)]$ m

22–69. $\theta_{max} = 0.106$ rad

22–71. $\omega = 21.1$ rad/s

22–73. $x = [0.119 \cos(6t - 83.9°)]$ m

22–74. $c_c = \sqrt{8(m + M)k}$

$x_{max} = \left[\dfrac{m}{e}\sqrt{\dfrac{1}{2k(m + M)}}\right]v_0$

22–75. $x_{max} = \dfrac{2mv_0}{\sqrt{8k(m + M) - c^2}} e^{-\pi c/(2\sqrt{8k(m+M)-c^2})}$

22–77. $L\ddot{q} + R\dot{q} + \left(\dfrac{1}{C}\right)q = E_0 \cos \omega t$

22–78. $L\ddot{q} + R\dot{q} + \left(\dfrac{2}{C}\right)q = E_0 \cos \omega t$

22–79. $L\ddot{q} + R\dot{q} + \dfrac{1}{C}q = 0$

Index

Geometric Properties of Line and Area Elements

Centroid Location	Centroid Location	Area Moment of Inertia

Circular arc segment

Circular sector area

$$I_x = \tfrac{1}{4} r^4 (\theta - \tfrac{1}{2} \sin 2\theta)$$

$$I_y = \tfrac{1}{4} r^4 (\theta + \tfrac{1}{2} \sin 2\theta)$$

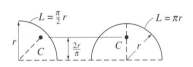

Quarter and semicircle arcs

Quarter circle area

$$I_x = \tfrac{1}{16} \pi r^4$$

$$I_y = \tfrac{1}{16} \pi r^4$$

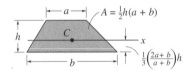

Trapezoidal area

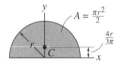

Semicircular area

$$I_x = \tfrac{1}{8} \pi r^4$$

$$I_y = \tfrac{1}{8} \pi r^4$$

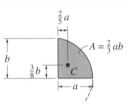

Semiparabolic area

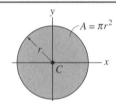

Circular area

$$I_x = \tfrac{1}{4} \pi r^4$$

$$I_y = \tfrac{1}{4} \pi r^4$$

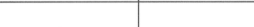

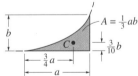

Exparabolic area

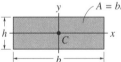

Rectangular area

$$I_x = \tfrac{1}{12} bh^3$$

$$I_y = \tfrac{1}{12} hb^3$$

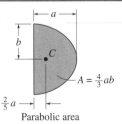

Parabolic area

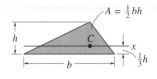

Triangular area

$$I_x = \tfrac{1}{36} bh^3$$